Handbook of Applied Mycology

Volume 3: Foods and Feeds

edited by

Dilip K. Arora
Banaras Hindu University
Varanasi, India

K. G. Mukerji
Department of Botany
University of Delhi
Delhi, India

Elmer H. Marth
University of Wisconsin—Madison
Madison, Wisconsin

MARCEL DEKKER, INC. New York • Basel • Hong Kong

ISBN 0-8247-8491-X

This book is printed on acid-free paper.

MARCEL DEKKER, INC.
270 Madison Avenue, New York, New York 10016

Current printing (last digit):
10 9 8 7 6 5 4 3 2 1

PRINTED IN THE UNITED STATES OF AMERICA

Handbook of Applied Mycology

Volume 3: Foods and Feeds

Handbook of Applied Mycology

Series Editor
Dilip K. Arora

Banaras Hindu University
Varanasi, India

Volume 1 Soil and Plants, edited by *Dilip K. Arora, Bharat Rai, K. G. Mukerji*, and *Guy R. Knudsen*

Volume 2 Humans, Animals, and Insects, edited by *Dilip K. Arora, Libero Ajello*, and *K. G. Mukerji*

Volume 3 Foods and Feeds, edited by *Dilip K. Arora, K. G. Mukerji*, and *Elmer H. Marth*

Volume 4 Fungal Biotechnology, edited by *Dilip K. Arora, Richard P. Elander, and K. G. Mukerji*

Volume 5 Mycotoxins in Ecological Systems, edited by *Deepak Bhatnagar, Eivind B. Lillehoj, and Dilip K. Arora*

HANDBOOK INTRODUCTION

Applied mycology is one of the most stimulating and rapidly evolving areas of the biological sciences. It encompasses many facets of agricultural, industrial, pharmacological, medical, and food sciences, all of which include organisms or processes that are subject to biotechnological manipulation. However, published information on the practical significance of fungi is fragmented in many specialized publications. There is an obvious need for a comprehensive treatment of the basic principles, methods, and applications of mycology as an applied, integrated, and multidisciplinary subject. The potential usefulness of such a handbook is suggested by the practical significance of mycological research, by the uniqueness and increasing coherence of the field's conceptual directions, and by the international scope of the subject. These five volumes of the *Handbook of Applied Mycology* attempt to fulfill this need. These volumes present and collate the major aspects of applied mycology and are designed as a modern standard reference for the work in this field.

In developing these volumes we decided that the treatment of the subject should be comprehensive but concise enough to enable completion within a set of five volumes published closely together rather than a continuous encyclopedic series taking several years to complete. However, this does not preclude the possibility of publishing supplementary volumes, as appropriate, to communicate new information. Further, we wanted to develop a treatise that would offer a detailed overview of applied aspects of mycology in terms of its theoretical, methodological, and empirical contributions. By emphasizing the interdisciplinary aspects of the subject, we hope that the handbook will highlight recent prospective linkages among diverse research paradigms.

Since the information presented in these volumes is expected to serve students as well as veterans in the field, both basic and advanced specialized materials are included. In addition to addressing researchers in mycology, the handbook will prove useful to a broad spectrum of professional groups, including biotechnologists, medical scientists, molecular biologists, food scientists, pharmacologists, soil microbiologists, microbial ecologists, and plant pathologists. Each chapter is intended to provide a balanced view of the current state of knowledge within an expanding field of interest. The selection of the chapters and overall organization of the book were guided by several insightful colleagues from around the world. These volumes were authored by a large team of international experts. Every chapter within the five volumes has been edited by a distinguished group of scientists. As a result, nine editors and over 200 authors representing 20 countries have contributed to this project.

Because of the multidisciplinary nature of the subject, judgment had to be made about the relative importance and significance of the hundreds of potential topics related to applied mycology. Also, a balance had to be struck between comprehensive coverage and intellectual appeal of the field. Readers should find this work useful for in-depth information and as a route to additional information contained in its critical discussions and wide range of references provided. Each of the five volumes is intended to be self-contained. Therefore, some degree of duplication of materials, especially basic principles, is inevitable.

No work of this magnitude can be accomplished without the support and contributions of many individuals. The planning and eventual production of this handbook were a collaborative effort. I am deeply indebted to the group of coeditors and many colleagues who have assisted me throughout all stages of the project. I also appreciate the dedication and hard work of the chapter authors, who cooperated persistently in the production of these volumes. I acknowledge with gratitude the support of M. V. Wiese, N. W. Schaad, S. Stotzsky, J. S. Singh, Bharat Rai, R. S. Upadhyay, V. S. Jaiswal, B. Thompson, Jon Olson, A. K. Pandey, Rejeev Gaur, and Shusma Gupta, whose interest and cooperation have been invaluable to me. I am grateful to the University of Idaho for providing assistance in the form of staff support and facilities, and the Banaras Hindu University, India, for granting leave during the period I edited these volumes. I appreciate the encouraging support provided by Ms. Sandra Beberman of Marcel Dekker, Inc., at all stages of production of this book. My gratitude is expressed to my teacher, J. L. Lockwood, and to my parents, whose example has always been my guiding principle. Finally, to Meenakshi, Sidhartha, and Gautam for their patience, encouragement, and willingness to forgo other interests so that the handbook might be completed, I am grateful.

Dilip K. Arora

PREFACE

Molds probably spoil more food than any other group of microorganisms. Some molds produce highly toxic mycotoxins, whereas other molds are used in fermentations to produce a variety of foods. Yeasts are also important as spoilage organisms and in fermentations, particularly for the production of alcohol.

In spite of the importance of these fungi, they have traditionally received only minimal attention in textbooks on food microbiology and hence by food microbiologists. This began to change when some molds were recognized as able to produce highly toxic secondary metabolites during growth on foods and feeds. Afterward, scientists having diverse backgrounds, including food microbiologists, mycologists, physiologists, toxicologists, and others, became interested in and studied molds. Thus, much new information about this group of microorganisms became available. Other investigators, for different reasons, studied yeasts and developed new information.

Such new information, when blended with older knowledge, is the substance of a book. Indeed, several books on food mycology appeared during the 1980s. However, the subject of fungi as related to foods is so broad that no single book on this subject can provide all the available information. Consequently, this volume supplements the other books on this subject by providing much information that is not readily available elsewhere. The book opens with a taxonomy of fungi in foods and feeds and then considers ecology, spoilage, and mycotoxin production by fungi in foods and feeds. This is followed by a series of discussions on xerophilic fungi and on fungi as they affect grain, vegetables, fruits, dairy products, nonproteinaceous and proteinaceous fermented foods, and food processing. Edible fungi including mushrooms and single-cell proteins are discussed as are products and uses of yeasts. Control of fungi through acceptable food additives is also considered.

We hope that this book will be useful to food and grain scientists, food microbiologists, mycologists with an interest in foods, persons in regulatory agencies who must deal with both the problems and benefits resulting from molds in foods and feeds, and students who wish to gain an appreciation for the importance of fungi in foods and feeds.

Our thanks go to the contributors for their efforts in providing thorough and up-to-date discussions of the various topics. We also thank Ms. Sandra Beberman and Mr. Andrew Berin of Marcel Dekker, Inc., for their efforts in turning manuscript pages into printed pages.

Dilip K. Arora
K. G. Mukerji
Elmer H. Marth

CONTENTS

CONTRIBUTORS

Dilip K. Arora Centre of Advanced Study in Botany, Banaras Hindu University, Varanasi, India

Zakia Bano Fruit and Vegetable Technology Department, Central Food Technological Research Institute, Mysore, India

Ramunas Bigelis Biotechnology Division, Amoco Research Center, Amoco Technology Company, Naperville, Illinois

Robert E. Brackett Department of Food Science and Technology, Georgia Agricultural Experimental Station, The University of Georgia, Griffin, Georgia

Marlene A. Bulgarelli* Department of Food Science and Technology, Georgia Agricultural Experimental Station, The University of Georgia, Griffin, Georgia

Shu-Ting Chang Department of Biology, The Chinese University of Hong Kong, Shatin, Hong Kong

Clyde M. Christensen Department of Plant Pathology, University of Minnesota, St. Paul, Minnesota

Jens C. Frisvad Department of Biotechnology, The Technical University of Denmark, Lyngby, Denmark

Alison Hamer Plant Pathology Department, Agricultural and Food Research Council Institute of Arable Crops Research, Rothamsted Experimental Station, Harpenden, Hertfordshire, England

---

**Present affiliation:* Center for Infectious Diseases, Division of Bacterial Diseases, Meningitis and Special Pathogens Branch, Centers for Disease Control, Atlanta, Georgia

Ailsa D. Hocking Food Research Laboratory, Division of Food Processing, Commonwealth Scientific and Industrial Research Organization, North Ryde, New South Wales, Australia

Surinder Singh Kahlon Department of Microbiology, Punjab Agricultural University, Ludhiana, Punjab, India

John Lacey Plant Pathology Department, Agricultural and Food Research Council Institute of Arable Crops Research, Rothamsted Experimental Station, Harpenden, Hertfordshire, England

Michael B. Liewen Microbiology and Regulatory Services, General Mills, Inc., Minneapolis, Minnesota

Naresh Magan Biotechnology Centre, Cranfield Institute of Technology, Cranfield, Bedford, England

Ian C. Marfleet* Biotechnology Centre, Cranfield Institute of Technology, Cranfield, Bedford, England

Elmer H. Marth Department of Food Science, University of Wisconsin—Madison, Madison, Wisconsin

Tilak W. Nagodawithana Technical Center, Universal Foods Corporation, Milwaukee, Wisconsin

S. Rajarathnam Fruit and Vegetable Technology Department, Central Food Technological Research Institute, Mysore, India

Nannapaneni Ramakrishna Plant Pathology Department, Agricultural and Food Research Council Institute of Arable Crops Research, Rothamsted Experimental Station, Harpenden, Hertfordshire, England

Robert A. Samson Centraalbureau voor Schimmelcultures, Baarn, The Netherlands

Don F. Splittstoesser Department of Food Science and Technology, New York State Agricultural Experimental Station, Cornell University, Geneva, New York

Tamotsu Yokotsuka Research Division, Kikkoman Corporation, Noda City, Chiba, Japan

Ahmed E. Yousef[†] Department of Food Science, University of Wisconsin—Madison, Madison, Wisconsin

---

Present affiliations:

*School of Applied Biology, Lancashire Polytechnic, Preston, England

[†]Department of Food Science, Ohio State University, Columbus, Ohio

Handbook of Applied Mycology
Volume 3: Foods and Feeds

1

TAXONOMY OF FILAMENTOUS FUNGI IN FOODS AND FEEDS

ROBERT A. SAMSON *Centraalbureau voor Schimmelcultures, Baarn, The Netherlands*

JENS C. FRISVAD *The Technical University of Denmark, Lyngby, Denmark*

DILIP K. ARORA *Banaras Hindu University, Varanasi, India*

I. INTRODUCTION

The importance of the filamentous fungi in food microbiology has gained general recognition. The importance of mutual understanding of food mycology and taxonomy as they relate to food processing and preservation has been stressed by Beuchat [1] for the quality control of foods and feed. The identification of the contaminating mycoflora is essential. The determination up to species level will provide important data about the biology and biochemistry of the organism, including the possible production of mycotoxins. Books and manuals for identification and isolation of foodborne fungi are available [2-4]. However, in the past the identification of foodborne fungi has often been neglected possibly due to lack of appropriate literature and techniques for carrying out a reliable identification, or the researcher, usually a microbiologist trained as a bacteriologist, lacked mycological training. Consequently, fungal isolates were not thoroughly identified and were incorrectly named. The literature on foodborne fungi has greatly suffered from the misidentified isolates. In some instances this has led to serious errors, e.g., isolates which were incorrectly identified as toxinogenic species. The taxonomic schemes of the most important foodborne genera such as *Penicillium*, *Aspergillus*, and *Fusarium* were varied and confusing. For many decades different species concepts were used. Although the research on the taxonomies of these genera is improving, mostly with the aid of modern techniques, the systematics of these genera are still in a state of flux. Recent international collaboration on the systematics of *Penicillium* and *Aspergillus* has improved the standardization of the species concepts, and in the important groups a consensus has been reached [5,6].

Presently the taxonomy of most foodborne genera is elucidated, and hence, identification up to species level is possible. Samson [7] has compiled the genera that occur in foods and feeds. There is a small difference between the two lists compiled by Samson et al. [8] and Pitt and Hocking [4] with respect to certain genera, and this may be explained by deviation in geographical distribution (Australia vs. Europe) and the different concept of food mycology.

In this chapter the identification of the filamentous fungi occurring in foods and feeds is discussed in brief. References to conventional methods of identification are given, while new approaches are shortly mentioned. A list of common and less common genera with the most relevant literature for identification is provided.

II. OCCURRENCE

Although there are many papers describing the mycoflora of specific products and commodities, only a few general lists of foodborne fungi and their specific substrates exist. Williams [9] lists spoiled foods from which species of *Penicillium* and *Aspergillus* have been isolated. Frisvad [10] has compiled a similar list, and the ecology of foodborne fungi is discussed in detail by Pitt and Hocking [4] and Frisvad and Samson (see Chapter 2).

During the last 15 years, *Penicillium* and *Aspergillus* have dominated the spoilage mycoflora of moldy food samples examined in our laboratory. Williams and Bialkowska [11] reported that from 294 samples with visible growth of one or more mold species, *Penicillium* accounted for 53.1% of isolates, *Aspergillus* and associated teleomorphs 15.1%, *Cladosporium* 8.9%, *Mucor* 8.5%, *Rhizopus* 4.8%, and other genera 9.6%. Of the Penicillia, subgenus *Penicillium* (90.1%) was dominant.

It is hoped that other laboratories will be encouraged to publish ecological data on foodborne fungi, so that further information may become available on the relationship between species and pattern of spoilage. This would not only supplement information already available about foods at risk from particular mold species, but would also highlight those ubiquitous species whose presence gives no information on likely contamination source.

III. IDENTIFICATION

A. Morphology

For the classification of filamentous fungi, morphology is still used as a primary criterion. This often facilitates a rapid identification, when the sporulating structures of the molds are present in or on the food product. Examination of the morphological structures can be easily carried out by making a simple microscope preparation and studying it with a well-equipped light microscope. However, traditional morphology-based taxonomy is now supplemented by other biochemical and physiological characteristics, in order to produce more sensitive taxonomies appropriate to the demands now being made by those who use many of the fungi concerned.

Until recently most mycologists were not interested in using data obtained by morphological, physiological, and biochemical studies of foodborne fungi for the purpose of numerical taxonomy. Numerical taxonomy can be a useful tool for grouping of fungi that have large number of characters in common. This technique encompasses a polythetic approach and can provide much information and indicate strain variations.

For identification, several general textbooks describing the most important fungal genera and species found in foods should be consulted [12-20].

Most foodborne Ascomycotina belong to the order Eurotiales (= Plectomycetes) and are characterized having an anamorph. These anamorphs belong to *Aspergillus*, *Penicillium*, *Basipetospora*, and other hyphomycetes. Benny and Kimbrough [21] and von Arx [12] have provided keys to the genera of the Eurotiales.

The Deuteromycotina includes the most important genera found in foods. The taxonomy of genera of these fungi is based on characters of conidiogenesis. Several reviews and classifications based on conidial development are available [22-24]. For the identification of the most common foodborne taxa, the reader is referred to Pitt and Hocking [4] or Samson and van Reenen-Hoekstra [3].

Coelomycete species, which are mostly accommodated in the Deuteromycotina, are usually not true saprobic contaminants, but often occur as weak plant pathogens [12,25]. In addition, many members of the Coelomycetes may not grow on selective laboratory media and will not sporulate adequately on agar media.

The classifications of foodborne yeasts will not be treated here. Published information about the given modern classification of yeasts can be found elsewhere [26-28].

B. Cultivation

For the morphological examination it is important to culture each species on an appropriate medium to achieve typical growth and sporulation. Malt-extract agar (MEA) and/or oatmeal agar (OA) are suitable media for most species. A number of species require special media such as potato-carrot agar (PCA), Czapek yeast autolysate agar (CYA), carnation leaf agar (CLA), etc. Recommendations and instructions for cultivation and media are given by Samson and van Reenen-Hoekstra [3] (Table 1). For other mycological media see King et al. [29], Gams et al. [30], and Stevens [31].

Most fungi can be incubated in light or in the dark at 25°C and identified after 5-10 days. For instance, *Fusarium*, *Trichoderma*, and *Epicoccum* show typical sporulation in diffuse daylight. Fungi such as *Phoma* should be cultivated in darkness followed by a period of alternating darkness/diffuse (day)light. To stimulate sporulation, irradiation with near UV ("black" light: greatest effect at 310 nm with a maximum emission about 360 nm) can be useful. The details of techniques for examining foodborne fungi are given by Samson and van Reenen-Hoekstra [3].

C. Secondary Metabolites

Patterns of secondary metabolites have now become an effective taxonomic tool, especially when used in conjunction with traditional taxonomy. Recently, Filtenborg, Frisvad, and coworkers have developed and described a very simple screening method for toxigenic fungi [32-37].

TABLE 1. Recommended Media for Cultivation and Identification of Filamentous Fungi

Fungi	Media
Acremonium	OA, CMA, MEA (2% MEA CBS)
Alternaria	MEA, PCA
Aspergillus	Cz, CYA, 2% MEA (CBS); for xerophilic species: Cz or MEA with additional sugar or other low water activity media
Aureobasidium	MEA
Botrytis	PCA, MEA, hay infusion agar
Byssochlamys	MEA, OA
Chrysonilia	OA
Cladosporium	MEA
Emericella	MEA, OA
Epicoccum	MEA, OA, PCA
Eurotium	MEA or Cz with sucrose 20 or 40% or other low water activity media
Fusarium	OA, PSA, PDA, SNA, CLA
Geotrichum	MEA
Monascus	MEA, OA
Moniliella	MEA, PCA
Mucorales, Absidia, Mucor, Rhizopus, Syncephalastrum	MEA (4%); for zygospore formation MYA (*Absidia*), CH (*Mucor* spp.)
Neosartorya	MEA, OA
Paecilomyces	MEA, OA
Penicillium	MEA, Cz, CYA, YES
Phialophora	MEA, OA
Phoma	OA, MEA (4% MEA CBS)
Scopulariopsis	MEA, OA
Stachybotrys	MEA, OA
Talaromyces	OA, MEA, CMA
Trichoderma	OA, MEA
Trichothecium	MEA
Ulocladium	MEA, PCA
Verticillium	MEA
Wallemia	MEA + additional sugar 20 or 40% or other low water activity media

TABLE 1. (Continued)

CH = Cherry decoction agar; CLA = carnation leaf agar; CMA = cornmeal agar; CYA = Czapek yeast extract agar; CY20S = Czapek yeast extract agar with 20% sucrose; Cz = Czapek agar; C20S = Czapek agar with 20% sucrose; DG 18 = dichloran glycerol agar base; MEA = malt extract agar; MEA + 20% sucrose = malt extract agar with 20% sucrose; MYA = malt yeast agar; OA = oatmeal agar; PCA = potato carrot agar; SNA = synthetischer Nährboden; YES = yeast extract sucrose agar.

Source: Ref. 3.

Frisvad and Filtenborg [38] and El-Banna et al. [39] have published detailed thin-layer chromatographic solvent systems and R_f values for a wide variety of *Penicillium* metabolites, so that metabolite profiles can now be used by the determinative taxonomist.

Identification of mold species is a time-consuming process, which usually takes at least 14 days, including isolation and subculturing. However, using TLC screening of secondary metabolite profiles, direct identification can be made on isolation media [40]. An identification procedure was proposed using standard isolation media, based on comparison of secondary metabolite profiles from a selected number of isolates. The procedure was tested on a number of food samples, and a high degree of agreement was found with results from normal identification procedures with pure cultures.

Frisvad and Thrane [41] used high-performance liquid chromatography (HPLC) with diode array spectra as a sophisticated method for comparing secondary metabolite profiles of isolates against standard isolates. This is a highly accurate and reproducible technique, though only accessible to well-equipped research laboratories.

Recently Kuraishi et al. [42] investigated the occurrence of 9 or 10 isoprene units of ubiquinones in *Aspergillus* and its teleomorphs and suggested a modification of the classification based on morphological criteria. This technique may be useful for fungal systematics.

D. Enzyme and Protein Patterns

Isozyme analysis is a valuable tool in the study of fungal taxonomy and genetics. The importance of isozyme analysis in fungal taxonomy has been discussed in detail in Volume 4 of the *Handbook of Applied Mycology* [44]. Isozyme analysis is not a new technique, but its use in fungal taxonomy is new. This technique will now become a useful tool for resolving several taxonomic disputers.

Cruickshank [45] and Cruickshank and Wade [46] developed effective methods for separation of species of *Sclerotinia, Botrytis*, and other genera by examining patterns of pectic enzymes after separation by gel electrophoresis. Small samples of culture fluid were subjected to electrophoresis at low temperatures, then the separated enzymes were allowed to act on methoxy pectin incorporated into the gel, and the sites of enzyme action visualized by ruthenium red staining.

To enable the differentiation of the many species in subgenus *Penicillium*, Cruickshank's technique was broadened by including the examination of amylase and ribonuclease isoenzymes. For amylase production, soluble starch was incorporated in the gels as a substrate, and for ribonucleases, ribosomal RNA. Fungi were cultured on wheat grains for the production of both these sets of enzymes [48]. Cruickshank and Pitt [49] and Pitt and Cruickshank [50] published a study of the isoenzyme patterns (zymograms) for all species accepted by Pitt [51] in subgenus *Penicillium*.

Recently Sugiyama and Yamatoya [52] and Yamatoya et al. [53] used polyacrylamide slab gel electrophoresis with specific staining for five enzymes as a chemotaxonomic criterion for comparing numerous isolates of *Aspergillus*.

Protein electrophoresis is a relatively simple, reproducible, and relatively inexpensive technique, which has many applications in food and feed mycology. This technique has been extensively used in fungal taxonomy for the last 15 years [54-57]. Like isozyme patterns, many reports have shown that protein patterns are useful for mycologists to resolve taxonomic disputes including identification of fungi below the species or virulent strains. Protein patterns have been used to examine the relationship of teleomorphs with their asexual counterparts. Immuno-blotting of separated proteins or cross-line immunoelectrophoresis studies can also be done to facilitate the distinction among species at various taxonomic levels. This method is faster than DNA:DNA hybridizations. The results obtained by protein electrophoresis can be compared and grouped by computer-assisted techniques.

E. Immunological and Molecular Methods

Serological tests have been applied to study bacterial pathogens since the 1920s. However, only in the last two decades have immunological techniques become powerful tools in many mycological laboratories for the identification of foodborne fungi. Serology has not only been applied for the identification and relationship among fungi and host-pathogen interactions, but also to the studies of in vitro translation products of nucleic acid and to intimate structural changes within the cells as a result of fungal invasion. Two important discoveries have contributed in the taxonomic application of serological methods: the development of very sensitive enzyme immunoassay systems and production of monoclonal antibodies which identify unique epitopes of the antigens. Recent literature includes the use of several techniques of great sensitivity, such as enzyme-linked immunosorbentancy (ELISA), immunofluorescent and latex agglutination test, radioimmunoassay (RIA), immunoelectron microscopy (IEM), and immunoprecipitation. The use of these procedures is great, and some of these may be used for taxonomy and identification purposes. Several excellent review articles and books are available describing the application and use of these techniques in fungal and bacterial systematics [58-62].

Recently Kaufman and Standard [58] summarized the work on specific and rapid identification of medically important fungi by exoantigen detection. This analysis can be useful for determining antigenic relationships among various isolates of a particular species as well as between different strains of fungi. The current use of exoantigen tests necessitate the use of control antigens and monospecific antibodies [58]. This exoantigen analysis may be useful in resolving taxonomic problems related to the foodborne fungi. However, Notermans and collaborators [63-69] have reported practical serological methods for the detection of heat-stable extracellular polysaccharides produced by various molds and they developed an ELISA and a latex agglutination method for *Penicillium* and *Aspergillus*. Also Stynen et al. [70] described a latex agglutination test for the detection of *Aspergillus* in the serum of patients with invasive aspergillosis, which has now been modified for the detection of foodborne Aspergilli and Penicillia.

In the past, several stable biochemical techniques have been used to differentiate between fungal groups [71-74]. Recent developments in molecular detection biology are of great help for fungal taxonomists, especially those evaluating taxa or groups of fungi that have been considered difficult to classify due to problems in defining discontinuities in morphological criterion and lack of phylogenetic relationships and complementary mating types. In recent years a number of review articles have been published on molecular fungal taxonomy [75-82] and genetic nomenclature [83].

Presently four approaches are used in the molecular identification of filamentous fungi: (1) determination of G+C molar percentage, (2) DNA-DNA complementarity, (3) rRNA sequence comparisons, and (4) restriction fragment length polymorphism (RFLP). The use of these methods for the study of taxonomy and other research applications will be described in detail in Volume 4 of the *Handbook of Applied Mycology* (see Chapter 2) and elsewhere [84].

In this chapter we will describe in brief the use of RFLP in fungal taxonomy. The use of RFLPs and RFLPHs has been increasing in fungal taxonomy and phylogenetic studies. The results obtained by these techniques may be analyzed by computers. RFLP analysis may also reveal several regions of DNA that may be a good source for the development of diagnostic probes. Generally, RFLPs of mtDNA have been used as a taxonomic tool for many fungi. This is because of the smaller size of mtDNA (18.9-121 kb), then genomic DNA, which gives 10-20 fragments and distinct bands on agarose gels. By using only classified methods, taxonomic studies of the genus *Fusarium* is a most difficult task. This is because of lack of morphological stability of species within the group. The RFLPs are useful for the taxonomic differentiation of species and isolates of *Fusarium* and its formae specialis races [85,86].

Despite the importance of genus *Aspergillus* in food biotechnology, the classification of some groups is still unclear. Kozlowski and Stepien [86] used RFLPH to test the phylogenetic relationship between several species of the genus *Aspergillus*. By using plasmids containing cloned fragments of *A. nidulans* mtDNA as probes, they concluded that restriction enzyme analysis of mtDNA is a useful method in resolving the taxonomic disputes of the genus *Aspergillus*. The relationship, postulated on the basis of results obtained, differed greatly in two cases from the classification of the genus presented by Raper and Fennell [87]. Klich and Mullaney [88,89] used RFLP in differentiating two very closely related

Aspergillus species, e.g., *A. oryzae* and *A. flavus*. Gomi et al. [92] described a rapid procedure for differentiating *Aspergillus flavus* and related taxa. By agarose gel electrophoresis of genomic DNA digested with *SMA1* restriction enzyme, bands showed patterns typical of *A. oryzae*, *A. sojae*, *A. flavus*, and *A. parasiticus*.

Several recent genetic studies in *Penicillium* and *Aspergillus* and their teleomorphs determining ribosomal DNA or RNA sequences, mitochondrial DNAs, etc. were carried out by various researchers [93-98], and the taxonomic value is apparent but the results need careful evaluation, particularly since type cultures were not included in these studies.

Variation among several species in *Neurospora* mtDNA has been studied by using RFLPH analysis with cloned fragments of mtDNA from a wild strain of *N. crassa* as a probe [96]. The RFLP techniques were also used for taxonomic studies of several species of *Sclerotinia* [97], isolates of pathogenic filamentous ascomycetes *Cochliobolus heterostrophus* [98], and many species of the genus *Agaricus* [99].

The molecular fungal systematics is certainly becoming a new tool to attempt resolving taxonomic problems. New molecular techniques are bound to develop. Many laboratories are not equipped to deal with radioactive materials; therefore, development of new nonradioactive probes may provide a wider acceptance of these techniques. Certain techniques such as rapid sequencing of rRNA and methods for polymerase chain reaction amplification of DNA and RNA may be integrated into fungal systematics.

F. Computer-Assisted Keying

During the last few decades, practical identification schemes for bacteria and yeasts have abandoned the use of dichotomous keys in favor of the recognition of profiles of reactions typical of the organisms sought. Examples of such profile schemes include the A.P.I. bacterial and C.O.M.P.A.S.S. yeast identification systems. Identification by profile has several advantages over dichotomous keys, including the use of a constant range of criteria, the capacity to incorporate variable results, and ready adaptation for speedy use on a microcomputer. Additionally, the principles of such identification schemes are readily understood by microbiologists who only occasionally need to identify molds.

Now that more data on physiological characteristics of molds are available to supplement existing morphological information, it is possible to construct identification keys that take into account various aspects of the whole organism. Several workers [100-103] have discussed and provided computer-assisted keying for some groups of *Penicillium* and *Aspergillus*, and it is hoped that this approach could apply to all foodborne taxa.

G. Nomenclature

The changing nomenclature of fungi of microbiological, industrial, and medical importance is often difficult to understand for workers in food microbiology. Inadequate taxonomy or uncertain nomenclature can lead to tremendous confusion. However, nomenclatural changes arise either through reclassification or through correction of errors in identification of type and other relevant isolates.

Hawksworth [104] reviewed the main reasons for instability in names in *Aspergillus* and *Penicillium*, emphasizing the differences between those due to nomenclatural and taxonomic changes. He drew attention to the options now available under the Code for the conservation or rejection of names in the rank of species, including the conservation of types. Recent international initiatives to improve the stability of names are reviewed by Cannon [105-107].

A greater nomenclatural threat is the discovery of earlier names for accepted species. For example, many names in *Aspergillus* and *Penicillium* are of uncertain application: Raper and Fennell [87] listed 70 in *Aspergillus* and Pitt [51] 175 in *Penicillium*. However, recently a possibility was opened up for the first time for specific names of "species of major economic importance," and while "case law" needs to develop to define "major economic importance," species such as *Aspergillus fumigatus* Fres., *A. niger* van Tieghem, *Penicillium chrysogenum* Thom, and *P. roqueforti* Thom would clearly be expected to fulfill this requirement. Frisvad et al. [108] have discussed and proposed the conservation of three economically important species, namely, *A. niger*, *A. nidulans*, and *P. chrysogenum*.

IV. LIST OF COMMON FUNGAL GENERA OCCURRING IN FOODS AND FEEDS

In the following list, the common and less common genera are compiled. In this list the genera that occur more commonly are set in boldface. However, please note that for specific foods and feeds, a mold genus or species can be dominant depending on the conditions of growth or available nutrients. In some cases the mold flora can deviate considerably from the generally known foodborne taxa.

A. Zygomycotina

1. *Absidia* van Tieghem

The most common species is *A. corymbifera* [109].

Amylomyces Calmette

A. rouxii is often associated with Asian fermented food products [110]. Ellis [119] in his DNA renaturation experiments concluded that *Amylomyces rouxii* could be accommodated in *Rhizopus arrhizus* as *R. arrhizus* var. *rouxii*.

Cunninghamella Thaxter

Most species of this genus are not foodborne, although *C. echinulata* and *C. elegans* can sometimes be found on nuts or other tropical food products [111,112].

2. *Mucor* Mich

Common species on food are *M. plumbeus*, *M. hiemalis*, *M. racemosus*, and *M. circinelloides*. For detailed keys, see Schipper [113-116].

3. *Rhizomucor* Schipper

This genus includes the thermophilic species, *R. pusillus* and *R. mieheii*, and is closely related to *Mucor* [116].

4. *Rhizopus* Ehrenb

Common species in food are *R. oryzae*, *R. stolonifer*, and *R. microsporus*. The latter taxon has been separated into four varieties [118] including var. *oligosporus* and var. *rhizopodiformis*, but in most recent taxonomies the taxa are recognized at the species level. For keys, see Refs. 3, 117, 118, and 120.

5. *Syncephalastrum* Schroeter

The only common species is *S. racemosum* Cohn [3,121].

Thamnidium Link

T. elegans sometimes occurs in food products [122].

B. Ascomycotina

1. *Byssochlamys* Westling
Anamorph: *Paecilomyces* Bain

Four species are known, of which *B. nivea* and *B. fulva* are commonly encountered. A key is given by Stolk and Samson [123] and Samson [124].

Chaetomium Kunze
Anamorph: *Acremonium* and others

Most species of *Chaetomium* are found in soil and on plant debris. *C. globosum* and several other members may cause biodeterioration of food (corn, rice) and feedstuffs [125-127]. Keys to *Chaetomium* are found in Ames [128] and von Arx et al. [129,130].

2. *Emericella* Berk. and Br.
Anamorph: *Aspergillus*

About 20 species are known, of which several are regularly found on food. Synoptic keys are given by Horie [131] and Christensen and States [132]. The latter authors treated this genus as an anamorphic group, the *A. nidulans* group, in the sense of Raper and Fennell [87].

3. *Eupenicillium* Ludwig
Anamorph: *Penicillium* Link

Several taxa are known (see Refs. 51, 133). *E. brefeldianum*, *E. ochrosalmoneum*, *E. euglaucum* (= *E. hirayamae*), and related sclerotial species are found in food and may occur as a heat-resistant contaminant.

4. *Eurotium* Link
Anamorph: *Aspergillus*

About 20 species are known, but common in food are *E. chevalieri*, *E. amstelodami*, and *E. herbariorum* (= *E. repens*). Note that the teleomorph is optimally produced on low water activity media (MEA or Czapek agar with additional sucrose or DG18 agar). Keys to the species are provided by Blaser [134], Pitt [135], and Kozakiewicz [136].

5. *Monascus* van Tieghem
Anamorph: *Basipetospora*

Three species are recognized, which Hawksworth and Pitt [132] have keyed using cultural and microscopic characters.

6. *Neosartorya* Malloch and Cain
Anamorph: *Aspergillus*

Most species of *Neosartorya* are soilborne, but they are also encountered in heat-treated products (e.g., pasteurized fruit juices). A key is given by Raper and Fennell [87] (as the *Aspergillus fischeri*-group); consult also Malloch and Cain [138] for the correct nomenclatural treatment of *Neosartorya*. Recently Kozakiewicz [136] and Samson et al. [137] reexamined the genus and reclassified some taxa.

7. *Neurospora* Shear and Dodge
Anamorph: *Chrysonilia* von Arx

About 10 species are known. *Neurospora sitophilia, N. crassa,* and *N. intermedia* are infrequently found in food [140]. Often the *Chrysonilia* anamorph is abundantly produced on the food product, and the teleomorph is produced in older cultures or can only be obtained after crossing of heterothallic strains.

8. *Talaromyces* C. R. Benjamin
Anamorph: *Penicillium* Link or *Paecilomyces* Bain

About 20 species are known and some taxa are infrequently found in insufficiently heat-treated product (e.g., pasteurized fruit juices). A common foodborne species is the large-spored variety of *T. flavus*, which was recently recognized as a separate species *T. macrosporus* [141]. Keys are given by Stolk and Samson [142] and Pitt [51].

9. *Xeromyces* Fraser

X. bisporus (= *Monascus bisporus* (Fraser) v. Arx) is obligately xerophilic and is often overlooked because of its failure to grow on commonly used media [143]. The fungus grows extremely slowly and is found on dry products, e.g., tobacco, sweets, etc.

C. Deuteromycotina

1. *Acremonium* Link
Teleomorphs: *Emericellopsis* v. Beyma and others

Most species are saprophytic and isolated from dead plant material and soil, but some species are pathogenic to plants and humans. The genus is worldwide in distribution. For detailed descriptions and keys, see Gams [144] and Domsch et al. [15].

TABLE 2. Ascomycetous Genera with an *Aspergillus* Anamorph

Section *Aspergillus*:	*Eurotium* Link:Fr., *Dichlaena* Mont. and Durieu, *Edyuillia* Subram.
Section *Fumigati*:	*Neosartorya* Malloch and Cain
Subgenus *Ornati*:	*Warcupiella* Subram., *Sclerocleista* Subram., (?) *Hemicarpenteles* Sarbhoy and Elphick
Section *Nidulantes*:	*Emericella* Berk. and Br.
Section *Flavipedes*:	*Fennellia* Wiley and Simmons
Section *Circumdati*:	*Petromyces* Malloch and Cain
Section *Cremei*:	*Chaetosartorya* Subram.

2. *Alternaria* Ness
Teleomorph: *Pleospora*

Many species are known; common foodborne species are *A. alternata* and *A. tenuissima* [23,24,145].

Arthrinium Kunze

About 20 species are known; *A. apiospermum* is sometimes found in food [23,24].

Ascochyta Lib
Teleomorph: *Didymella*

About 40 species are known [146].

3. *Aspergillus* Mich
Teleomorphs: *Eurotium, Emericella, Neosartorya,* and others (see Table 2).

For a general key, see Raper and Fennell [87]. For a compilation of Aspergilli described since 1965, see Samson [147]. Several groups have been reinvestigated since then: *A. ochraceus* group [149]; *A. nidulans* group [132], *A. flavus* group [150,152,153], *A. fumigatus* [154], *A. restrictus* [155]. Gams et al. [156] provided new names for the different sections in *Aspergillus* (see Table 3). See also Samson and Pitt [5,6] for a more recent account on advances in *Aspergillus* systematics. Keys to common foodborne Aspergilli are provided by Pitt and Hocking [4], Samson and van Reenen-Hoekstra [3], and Klich and Pitt [151].

On the basis of studies of DNA homology among *A. flavus* and related species, Kurtzman et al. [157] reduced several well-known species to subspecies or varietal status. Specifically, *A. parasiticus* became *A. flavus* subspecies *parasiticus* Kurtzman et al.; *A. oryzae* became *A. flavus* subspecies *flavus* variety *oryzae* Kurtzman et al.; and *A. sojae* became *A. flavus* subspecies *parasiticus* variety *sojae* Kurtzman et al.

Klich and Mullaney [88] disagreed with this conclusion, arguing that DNA restriction enzyme fragment polymorphism showed differences between the DNA of *A. flavus* and *A. oryzae*. Klich and Pitt [152,153] also did not accept the species concept of Kurtzman et al. [157], because morphological differences existed between these two species, and on the more

TABLE 3. Nomenclature of Infrageneric Taxa of the Genus *Aspergillus* According to Gams et al. [156]

Subgenus *Aspergillus*
Section *Aspergillus* (*A. glaucus*-group)[a]
Section *Restricti* (*A. restrictus*-group)
Subgenus *Fumigati*
Section *Fumigati* (*A. fumigatus*-group)
Section *Cervini* (*A. cervinus*-group)
Subgenus *Ornati* (*A. ornatus*-group)
Subgenus *Clavati*
Section *Clavati* (*A. clavatus*-group)
Subgenus *Nidulantes*
Section *Nidulantes* (*A. nidulans*-group)
Section *Versicolor* (*A. versicolor*-group)
Section *Usti* (*A. ustus*-group)
Section *Terrei* (*A. terreus*-group)
Section *Flavipedes* (*A. flavipes*-group)
Subgenus *Circumdati*
Section *Wentii* (*A. wentii*-group)
Section *Flavi* (*A. flavus*-group)
Section *Nigri* (*A. niger*-group)
Section *Circumdati* (*A. ochraceus*-group)
Section *Candidi* (*A. candidus*-group)
Section *Cremei* (*A. cremeus*-group)
Section *Sparsi* (*A. sparsus*-group)

[a]Group names per Raper and Fennell [87] in parentheses.

practical grounds that it was important that species used for food fermentations possess different species names from those that are known to be mycotoxigenic. In their study of Cruickshank and Pitt [49] on isoenzyme patterns of *A. flavus* and related species, no differences between the taxa could be found.

Kozakiewicz [135] in her publication of *Aspergillus* species from food products investigated many species and based her classification on scanning electron micrographs of air-dried conidia. Many of her taxonomic conclusions are debatable, and her proposed nomenclature will be discussed elsewhere.

4. *Aureobasidium* Viala and Boyer

Hermanides-Nijhof [158] keyed and described 15 species in this genus. *A. pullulans* is very common.

Bipolaris Shoemaker
Teleomorph: *Cochliobolus*

A genus similar to *Drechslera* [23].

5. *Botrytis* Mich.
Teleomorph: *Botrytinia* (= *Sclerotinia* Fuckel pro parte).

The genus includes important plant pathogens with a worldwide distribution. For more detailed descriptions, see Ellis [23], Jarvis [159], and Domsch et al. [15].

6. *Chrysonilia* von Arx
Teleomorph: *Neurospora*

Two or three species are known, representing the anamorphs of *Neurospora* [160]. In the western world, the species are considered as spoilage organisms, while in the Orient the fungus is sometimes used in food fermentation (often referred to in the literature as *Monilia*).

7. *Chrysosporium* Corda

Most species are isolated from soil or human and animal tissue. However, some species, e.g., *C. xerophilum*, *C. sulfureum*, and *C. farinicola*, are found on dried food [161,162].

8. *Cladosporium* Link
Teleomorph: *Mycosphaerella* Johanson

A genus with a worldwide distribution. Several species are plant pathogens or saprobic and more or less host-specific on old or dead plant material. Common species are *C. cladosporioides*, *C. herbarum*, *C. macrocarpum*, and *C. sphaerospermum*. For detailed descriptions and keys, see Refs. 15, 23, and 163.

Colletotrichum Corda

C. gloeosporioides (= *Glomerella cingulata*) is common. For a key to species, see von Arx [164].

Curvularia Boedijn
Teleomorph: *Pseudocochliobolus*

About 30 species [23,165].

Drechslera Ito
Teleomorph: *Pyrenophora*

About 20 species exist, mostly found on grasses [23,24,166,167].

9. *Epicoccum* Link

Two species are known, of which *E. nigrum* (= *E. purpurascens*) is very common [3,15].

10. *Fusarium* Link
Teleomorphs: *Nectria* Fr., *Plectosphaerella* Kleb. and others

Most *Fusarium* spp. are soil fungi and have a worldwide distribution. Some are plant parasites causing root and stem rot, vascular wilt, and fruit rot. Several species are known to be pathogenic to human and animals; others cause storage rot and are toxin producers. For modern detailed descriptions and keys, the reader is referred to Gerlach and Nirenberg [163], Burgess and Liddell [169], and Nelson et al. [170]. For identification, cultivation both on PDA and especially carnation leaf agar (CLA) is strongly recommended by modern *Fusarium* workers.

11. *Geotrichum* Link
Teleomorph: *Dipodascus*

Several species are known, of which *G. candidum* is very common [171, 172].

Gliocladium Corda
Teleomorphs: *Nectria, Hypocrea,* and others

Common species are *G. roseum* and *G. viride* [15].

Lasiodiploida Ellis and Everhart
Teleomorph: *Botryosphaeria*

L. theobromae (= *Botryodiplodia theobromae*) is often found as the anamorph of *Botryosphaeria rhodina* from (sub)tropical material [25].

Memnoniella Höhnel

A similar genus to *Stachybotrys*, but conidia occur in chains. Three species are known, of which *M. echinata* sometimes occurs in food [173].

Moniliella Stolk and Dakin

For more detailed descriptions and a key, see De Hoog [174,175].

Myrothecium Tode

Tulloch [176] accepted and keyed out eight species. Nguyen et al. [177] described the seedborne taxa.

12. *Paecilomyces* Bain
Teleomorphs: *Byssochlamys, Thermoascus, Talaromyces*

Common foodborne species are *P. variotii* and the anamorphs of *Byssochlamys* spp. [124].

13. *Penicillium* Link
Teleomorphs: *Eupenicillium, Talaromyces,* and others

Many species of *Penicillium* are common contaminants on various substrates and are known as potential mycotoxin producers. Correct identification is therefore important when studying possible *Penicillium* contamination.

New classifications [51,178] were published replacing the outdated monograph by Raper and Thom [179]. However, for a very recent account on advances in *Penicillium* taxonomy, the reader is referred to Samson and

TABLE 4. Recent Nomenclatural Changes for Common *Penicillium* Names

Old names	Correct name
P. candidum	*P. camemberti* Thom
P. caseicolum	*P. camemberti* Thom
P. claviforme	*P. vulpinum* (Cooke and Massee) Seifert and Samson
P. corymbiferum	*P. hirsutum* Dierckx
P. cyclopium	*P. aurantiogriseum* Dierckx
P. frequentans	*P. glabrum* (Wehmer) Westling
P. glaucum	Doubtful name
P. granulatum	*P. glandicola* (Oud.) Seifert and Samson
P. janthinellum	*P. simplicissimum* (Oud.) Thom, *pro parte*
P. martensii	*P. aurantiogriseum* Dierckx
P. nigricans	*P. janczewskii* Zaleski
P. notatum	*P. chrysogenum* Westling
P. palitans	*P. commune* Thom
P. patulum	*P. griseofulvum* Dierckx
P. puberulum	*P. aurantiogriseum* Dierckx
P. urticae	*P. griseofulvum* Dierckx
P. verrucosum var. *cyclopium*	*P. aurantiogriseum* Dierckx

Pitt [5,6]. In the new taxonomic schemes the nomenclature of many "old" and well-known species has been corrected (Table 4). In Samson and Pitt [6] new taxonomies on subgenus *Penicillium* [180-182], *Furcatum* [183], and the *P. funiculosum* complex [184] are presented. Common foodborne taxa are treated and keys to the most common species are given by Samson and van Reenen-Hoekstra [3], Pitt [185], and Pitt and Hocking [4].

14. *Phialophora* Medlar
Teleomorphs: *Pyrenopeziza*, *Coniochaeta*, and others

Phialophora species have been isolated from decaying wood, foodstuffs (e.g., butter, margarine, apples), soil-diseased human and animal tissue, and also occur as parasites or saprobes in plant material. For detailed descriptions and keys, see Refs. 15, 186.

15. *Phoma* Sacc
Teleomorphs: *Pleospora*, *Leptosphaeria*, and others

Common species like *P. exigua* and *P. herbarum* are described by Boerema et al. [187-189], Domsch et al. [15], Dorenbosch [190], and Samson and van Reenen-Hoekstra [3].

Pithomyces Berk. and Br.

About 15 species exist; a well-known species that produces toxins is *P. chartarum* [23,24].

16. *Scopulariopsis* Bain
Teleomorph: *Microascus*

For a key and descriptions, see Morton and Smith [191] and Domsch et al. [15]. Common species are *S. brevicaulis*, *S. fusca*, and *S. candida*.

Sporendonema Desmazieres

Two species are accepted by Sigler and Carmichael [192]. *S. casei*, found mostly on cheese, is psychrophilic and grows and sporulates well at 8°C.

Stachybotrys Corda

About 15 species are known; *S. chartarum*, in particular, can be common in food [173].

Stemphylium Wallr.
Teleomorph: *Pleospora*

About six species [23,24].

17. *Trichoderma* Pers.
Teleomorph: *Hypocrea* (in most species, not produced in culture)

A very common genus, especially in soil and decaying wood. *Gliocladium* (with strongly convergent phialides) and *Verticillium* (with straight and moderately divergent phialides) are closely related. For more detailed descriptions, see Rifai [193] and Domsch et al. [15].

18. *Trichothecium* Link

The only species in this genus, *T. roseum*, may sometimes be foodborne [3,15].

19. *Ulocladium* Preuss

This genus includes about 10 species; *U. atrum* and *U. consortiale* are common [23,24].

20. *Verticillium* Nees
Teleomorph: *Nectria* and related genera

About 35 species are known, mostly found on plant material. Some common species are described by Domsch et al. [15].

21. *Wallemia* Johan-Olsen

W. sebi, the only species, is a very common xerophile [3,4,158]. This fungus is also known by the names *Sporendonema epizoum*, *Hemispora stellata*, and *Sporendonema sebi*.

ACKNOWLEDGMENTS

A major part of this manuscript arose from the international collaborative NATO project (0216/86) between R.A.S. and J.C.F. These authors thank the NATO Scientific Affairs Division (Brussels, Belgium) for this research grant.

REFERENCES

1. Beuchat, L. R. Bridging the gap: Taxonomists and food mycologist. In *Advances in Penicillium and Aspergillus Systematics* (R. A. Samson and J. I. Pitt, eds.), Plenum Publishers, New York, 1985, pp. 113-118.
2. Beuchat, L. R. *Food and Beverage Mycology*, 2nd ed. Avi Publishing, Westport, CT, 1987.
3. Samson, R. A., and van Reenen-Hoekstra, E. S. *Introduction to Food-borne Fungi*, 3rd ed. Centraalbureau voor Schimmelcultures, Baarn, 1988.
4. Pitt, J. I., and Hocking, A. D. *Fungi and Food Spoilage*. Academic Press, Sydney, 1985.
5. Samson, R. A., and Pitt, J. I. (eds.). *Advances in Penicillium and Aspergillus Systematics*. Plenum Press, New York, 1985.
6. Samson, R. A., and Pitt, J. I. (eds.). *Modern Concepts in Pencillium and Aspergillus Systematics*. Plenum Press, New York, 1990.
7. Samson, R. A. Filamentous fungi in foods and feeds. *J. Applied Bacteriology* (Symp. Suppl. 1989):27S-35S, 1989.
8. Samson, R. A., Frisvad, J. C., and Filtenborg, O. Taxonomic schemes for foodborne fungi. In *Methods for the Mycological Examination of Food* (A. D. King, J. I. Pitt, L. R. Beuchat, and J. L. Corry, eds.), Plenum Press, New York, 1988, pp. 247-270.
9. Williams, A. P. *Penicillium* and *Aspergillus* in the food microbiology laboratory. In *Modern Concepts in Penicillium and Aspergillus Systematics* (R. A. Samson and J. I. Pitt, eds.), Plenum Press, New York, 1990, pp. 67-71.
10. Frisvad, J. I. Fungal species and their production of mycotoxins. In *Introduction to Food-borne Fungi* (R. A. Samson and E. S. van Reenen-Hoekstra, eds.), Centraalbureau voor Schimmelcultures, Baarn, 1988, pp. 239-249.
11. Williams, A. P., and Bialkowska, A. Moulds in mould spoiled foods and foodproducts. Leatherhead Food R.A. Research Report No. 527, 1985.
12. von Arx, J. A. *The Genera of Fungi Sporulating in Pure Culture*, 3rd ed. J. Cramer Verlag, Vaduza, 1981.
13. Ainsworth, G. C., Sparrow, F. K., and Sussman, A. S. *The Fungi. An Advanced Treatise*, Vol. 4A. A Taxonomic Review with Keys: Ascomycetes and Fungi Imperfecto. Academic Press, New York and London, 1973.
14. Ainsworth, G. C., Sparrow, F. K., and Sussman, A. S. *The Fungi. An Advanced Treatise*, Vol. 4B. A Taxonomic Review with Keys: Basidiomycetes and Lower Fungi. Academic Press, New York and London, 1973.

15. Domsch, K. H., Gams, W., and Anderson, T. H. *Compendium of Soil Fungi*, Vols. I and II. Academic Press, London, 1980.
16. Carmichael, J. W., Kendrick, W. B., Conners, I. L., and Sigler, L. *Genera of Hyphomycetes*. University of Alberta Press, Edmonton, 1980.
17. Zycha, H., Stepmann, R., and Linnemann, G. *Mucorales. Eine Beschreibung aller Gattungen und Arten dieser Pilzgruppe*. J. Cramer-Verlag, Lehre, 1964.
18. Hanlin, R. T. *Keys to Families, Genera and Species of the Mucorales*. J. Cramer Verlag, Vaduz, 1973.
19. Hesseltine, C. W., and Ellis, J. J. Mucorales. In *The Fungi*, Vol. IVb (G. C. Ainsworth, F. K. Sparrow, and A. S. Sussman, eds.), Academic Press, New York, 1973, pp. 187-217.
20. O'Donnell, K. L. Zygomycetes in culture, Palfrey contributions in Botany 2. University of Georgia Press, Athens, GA, 1979.
21. Benny, G. L., and Kimbrough, J. W. A synopsis of the orders and families of Plectomycetes with keys and genera. *Mycotaxon*, *12*:1-91, 1980.
22. Cole, G. T., and Samson, R. A. *Patterns of Development in Conidial Fungi*. Pitman Press, London, 1979.
23. Ellis, M. B. *Dematiaceous Hyphomycetes*. Commonwealth Mycological Institute Press, Kew, Surrey, 1971.
24. Ellis, M. B. *More Dematiaceous Hyphomycetes*. Commonwealth Mycological Institute Press, Kew, Surrey, 1976.
25. Sutton, B. C. *The Coelomycetes, Fungi Imperfecti with Pycnidia, Acervuli and Stromata*. Commonwealth Mycological Institute Press, Kew, Surrey, 1980.
26. Barnett, J. A., Pane, R. W., and Yarrow, D. *Yeasts, Characteristics and Identification*. Cambridge University Press, 1983.
27. Kreger-van Rij, N. J. W. (ed.). *The Yeasts. A Taxonomic Study*. Elsevier, Amsterdam, 1984.
28. Smith, M. T., and Yarrow, D. Yeasts. In *Introduction to Food-borne Fungi* (R. A. Samson and E. S. van Reenen-Hoekstra, eds.), Centraalbureau voor Schimmelcultures, Baarn, 1988, pp. 210-218.
29. King, A. D., Pitt, J. I., Beuchat, L. R., and Corry, J. L. (eds.). *Methods for the Mycological Examination of Food*. Plenum Press, New York, 1986.
30. Gams, W., Plaats-Niterink, H. A. van der, Samson, R. A., and Stalpers, J. A. *CBS Course in Mycology*, 3rd ed. Centraalbureau voor Schimmelcultures, Baarn, 1987.
31. Stevens, R. B. (ed.). *Mycology Guidebook*. University of Washington Press, 1974.
32. Filtenborg, O., and Frisvad, J. C. A simple screening method for toxigenic moulds in pure culture. *Lebensmit. Wissensch. Technol.*, *13*:128-130, 1980.
33. Filtenborg, O., Frisvad, J. C., and Svendsen, J. Simple screening procedure for molds producing intracellular mycotoxins in pure culture. *Appl. Environm. Microbiol.*, *45*:581-585, 1983.
34. Frisvad, J. C. Profiles of primary and secondary metabolites of value in classification of *Penicillium viridicatum* and related species.

In *Advances in Penicillium and Aspergillus Systematics* (R. A. Samson and J. I. Pitt, eds.), Plenum Press, New York, 1985, pp. 311-325.
35. Frisvad, J. C. Taxonomic approaches to mycotoxin identification. In *Modern Methods in the Analysis and Structural Elucidation of Mycotoxins* (R. J. Cole, ed.), Academic Press, Orlando, FL, 1986, pp. 415-457.
36. Thrane, U. Detection of toxigenic *Fusarium* isolates by thin layer chromatography. *Lett. Appl. Microbiol.*, *3*:93-96,
37. Frisvad, J. C. Secondary metabolites as an aid to *Emericella* classification. In *Advances in Penicillium and Aspergillus Systematics* (R. A. Samson and J. I. Pitt, eds.), Plenum Press, New York, 1985, pp. 437-444.
38. Frisvad, J. C., and Filtenborg, O. Classification of terverticillate Penicillia based on profiles of mycotoxins and other secondary metabolites. *Appl. Environm. Microbiol.*, *46*:1301-1310, 1983.
39. El-Banna, A. A., Pitt, J. I., and Leistner, L. Production of mycotoxins by *Penicillium* species. *Syst. Appl. Microbiol.*, *10*:42-46, 1987.
40. Filtenborg, O., and Frisvad, J. C. Identification of *Penicillium Aspergillus* species in mixed cultures without subculturing. In *Modern Concepts in Penicillium and Aspergillus Systematics* (R. A. Samson and J. I. Pitt, eds.), Plenum Press, New York, 1990, pp. 27-36.
41. Frisvad, J. C., and Thrane, U. Standardized high-performance liquid chromatography of 182 mycotoxins and other fungal metabolites based on alkylphenone retention indices and UV-VIS spectra (diode array detection). *J. Chromatography*, *404*:195-214, 1987.
42. Kuraishi, H., Itoh, M., Tsuzaki, N., Katayama, Y., Yokoyama, T., and Sugiyama, J. Ubiquinone system as a taxonomic tool in *Aspergillus* and its teleomorphs. In *Modern Concepts in Penicillium and Aspergillus Systematics* (R. A. Samson and J. I. Pitt, eds.), Plenum Press, New York, 1990, pp. 407-420.
43. Micales, J. A., Bonde, M. R., and Peterson, G. L. The use of isozyme analysis in fungal taxonomy and genetics. *Mycotaxon*, *27*: 405-449, 1986.
44. Micales, J. A., Bonde, M. R., and Peterson, G. L. Isozyme analysis in fungal taxonomy and molecular genetics. In *Handbook of Applied Mycology*, Vol. 4, *Fungal Biotechnology* (D. K. Arora, R. P. Elander, and K. G. Mukerji, eds.), Marcel Dekker, New York (in press).
45. Cruickshank, R. H. Distinction between *Sclerotinia* species by their pectic zymograms. *Trans. Br. Mycol. Soc.*, *80*:117-119, 1963.
46. Cruickshank, R. H., and Wade, G. C. Detection of pectic enzymes in pectin-acrylamide gels. *Anal. Biochem.*, *107*:177-181, 1980.
47. Cruickshank, R. H., and Pitt, J. I. The zymogram technique: Isoenzyme patterns as an aid in *Penicillium* classification. *Microbiol. Sci.*, *4*:14-17, 1987.
48. Cruickshank, R. H., and Pitt, J. I. Identification of species in *Penicillium* subgenus *Penicillium* by enzyme electrophoresis. *Mycologia*, *79*:614-620, 1987.
49. Cruickshank, R. H., and Pitt, J. I. Isoenzyme patterns in *Aspergillus flavus* and closely related taxa. In *Modern Concepts in Penicillium and Aspergillus Systematics* (R. A. Samson and J. I. Pitt, eds.), Plenum Press, New York, 1990, pp. 259-264.

50. Pitt, J. I., and Cruickshank, R. H. Speciation and synonymy in *Penicillium* subgenus *Penicillium*—towards a definitive taxonomy. In *Modern Concepts in Penicillium and Aspergillus Systematics* (R. A. Samson and J. I. Pitt, eds.), Plenum Press, New York, 1990, pp. 103-119.
51. Pitt, J. I. *The Genus Penicillium and Its Teleomorphic States. Eupenicillium and Talaromyces.* Academic Press, London, 1979.
52. Sugiyama, J., and Yamatoya, K. Electrophoretic comparison of enzymes as a chemotaxonomic tool among *Aspergillus* taxa: (1) *Aspergillus* sects. *Ornati* and *Cremei.* In *Modern Concepts in Penicillium and Aspergillus Systematics* (R. A. Samson and J. I. Pitt, eds.), Plenum Press, New York, 1990, pp. 385-393.
53. Yamatoya, K., Sugiyama, J., and Kuraishi, H. Electrophoretic comparison of enzymes as a chemotaxonomic tool among *Aspergillus* taxa: (2) *Aspergillus* sect. *Flavi.* In *Modern Concepts in Penicillium and Aspergillus Systematics* (R. A. Samson and J. I. Pitt, eds.), Plenum Press, New York, 1990, pp. 395-405.
54. Franke, R. G. Electrophoresis and the taxonomy of saprophytic fungi. *Bull. Torrey Bot. Club, 100*:287-296, 1973.
55. Shecter, Y. Symposium on the use of electrophoresis in the taxonomy of algae and fungi. *Bull. Torrey Bot. Club, 100*:253-259, 1973.
56. Snider, R. D. Symposium on the use of electrophoresis in the taxonomy of algae and fungi. III. Electrophoresis and the taxonomy of phytopathogenic fungi. *Bull. Torrey Bot. Club, 100*:272-276, 1973.
57. Hearn, V. M., Moutaouakil, M., and Latge, J. Analysis of components of *Aspergillus* and *Neosartorya* mycelial preparations by gel electrophoresis and Western blotting techniques. In *Modern Concepts in Penicillium and Aspergillus Systematics* (R. A. Samson and J. I. Pitt, eds.), Plenum Press, New York, 1990, pp. 235-245.
58. Kaufman, L., and Standard, P. G. Specific and rapid identification of medically important fungi by exoantigen detection. *Ann. Rev. Microbiol., 41*:209-225, 1987.
59. Miller, S. A., and Martin, R. R. Molecular diagnosis of plant diseases. *Ann. Rev. Phytopathology, 26*:409-432, 1988.
60. Schaad, N. W. (ed.). *Laboratorium Guide for Identification of Plant Pathogenic Bacteria.* APS Press, St. Paul, MN, 1989.
61. Saettler, A. W., Schaad, N. W., and Roth, D. A. (eds.). *Detection of Bacteria in Seeds and Other Plant Materials.* APS Press, St. Paul, MN, 1989.
62. De Hoog, G. S., Gueho, E., and Boekhout, T. Experimental fungal taxonomy. In *Handbook of Applied Mycology,* Vol. 2, *Humans, Animals, and Insects* (D. K. Arora, L. Ajello, and K. G. Mukerji, eds.), Marcel Dekker, New York, 1991, pp. 369-394.
63. Notermans, S., and Heuvelman, C. J. Immunological detection of moulds in food by using the enzyme-linked immunosorbent assay (ELISA); preparation of antigens. *Int. J. Food Microbiol., 2*:247-258, 1985.
64. Notermans, S., Heuvelman, C. J., Beumer, R. R., and Maas, R. Immunological detection of moulds in food: Relation between antigen in production and growth. *Int. J. Food Microbiol., 3*:253-261, 1986.

65. Notermans, S., Heuvelman, C. J., van Egmond, H. P., Paulsch, W. E., and Besling, J. R. Detection of mould in food by the enzyme-linked immunosorbent assay. *J. Food Protect.*, *49*:786-791, 1986.
66. Cousin, M. A., Notermans, S., Hoogerhout, P., and van Boom, J. H. Detection of beta-galactofuranosidase production by *Penicillium* and *Aspergillus* species using 4-nitrophenyl beta-D-galactofuranoside. *J. Appl. Bact.*, *66*:311-317, 1989.
67. Kamphuis, H. J., Notermans, S., Veeneman, G. H., van Boom, J. H., and Rombouts, F. M. A rapid and reliable method for the detection of moulds in foods: Using the latex agglutination assay. *J. Food Prot.*, *52*:244-247, 1989.
68. Notermans, S., Veeneman, G. H., van Zuylen, C. W. E. M., Hoogerhout, P., and van Boom, J. H. (1-5)-Linked beta-D-galactofuranosides are immunodominant in extracellular polysaccharides of *Penicillium* and *Aspergillus* species. *Mol. Immunolog.*, *25*:975-979, 1988.
69. Kamphuis, H., and Notermans, S. Basic principles of immunological detection of molds. In *Modern Methods in Food Mycology* (R. A. Samson, A. D. Hocking, J. I. Pitt, and A. D. King, eds.), Elsevier, Amsterdam, 1991 (in press).
70. Stynen, D., Meulemans, L., Goris, A., Braendlin, N., and Symons, N. Characteristics of a latex agglutination test, based on monoclonal antibodies, for the detection of mould antigen in foods. In *Modern Methods in Food Mycology* (R. A. Samson, A. D. Hocking, J. I. Pitt, and A. D. King, eds.), Elsevier, Amsterdam, 1991 (in press).
71. Hall, R. Molecular approaches to taxonomy of fungi. *Bot. Rev.*, *35*:285-304, 1969.
72. Tyrrell, D. Biochemical systematics and fungi. *Bot. Rev.*, *35*:305-316, 1969.
73. Storck, R. Molecular mycology. In *Molecular Microbiology* (J. B. G. Kwapinski, ed.), John Wiley and Sons, New York-London, 1974, pp. 423-479.
74. Szecsi, A., and Dobrovolsky, K. Phylogenetic relationships among *Fusarium* species by DNA reassociation. *Mycopathologia*, *89*:89-94, 1985.
75. Bennett, J. W. Prospects of a molecular mycology. In *Gene Manipulation in Fungi* (J. W. Bennett and L. L. Lasure, eds.), Academic Press, Orlando, FL, 1985, pp. 35-63.
76. Croft, J. H. Genetic variation and evolution in *Aspergillus*. In *Evolutionary Biology of the Fungi* (A. D. Rayner, C. M. Brasier, and D. Moore, eds.), Cambridge University Press, New York-London, 1986, pp. 311-323.
77. Croft, J. H., Bhattacherjee, V., and Chapman, K. E. RFLP analysis of nuclear and mitochondrial DNA and its use in *Aspergillus* and *Penicillium* systematics. In *Modern Concepts in Penicillium and Aspergillus Systematics* (R. A. Samson and J. I. Pitt, eds.), Plenum Press, New York, 1990, pp. 309-320.
78. Kurtzman, C. P. Molecular taxonomy of the fungi. In *Gene Manipulation in Fungi* (J. W. Bennett and L. L. Lasure, eds.), Academic Press, Orlando, FL, 1985, pp. 35-63.

79. Kurtzman, C. P. Impact of nucleic acid comparisons on the classification of fungi. *Proc. Indian Acad. Sci. (Plant Sci.)*, *97*:185-201, 1987.
80. Blanz, P. A., and Unseld, M. Ribosomal RNA as a taxonomic tool in mycology. *Stud. Mycology, 30*:247-258, 1987.
81. Klich, M. A., and Mullaney, E. J. Molecular methods for identification and taxonomy of filamentous fungi. In *Handbook of Applied Mycology*, Vol. 4, *Fungal Biotechnology* (D. K. Arora, R. P. Elander, and K. G. Mukerji, eds.), Marcel Dekker, New York (in press).
82. Yoder, O. C., Valent, B., and Chumley, F. Genetic nomenclature and practice for plant pathogenic fungi. *Phytopathology, 76*:383-385.
83. Keller, G., and Manak, P. *DNA Probes*. Stockton Press, Hant, U.K., 1989.
84. Coddington, A., Mathews, P. M., Cullis, C., and Smith, K. H. Restriction digest pattern of total DNA from different races of *Fusarium oxysporum* f. sp. *pisi*-an improved method for race classification. *J. Phytopathology, 118*:9-20, 1987.
85. Manicom, B. Q., Bat-Joseph, M., Rosner, A., Vigodsky-Hass, H., and Kotze, J. M. Potential applications of random DNA probes and restriction fragment polymorphisms in the taxonomy of Fusaria. *Phytopathology, 77*:669-672, 1987.
86. Kozlowski, M., and Stepien, P. R. Restriction enzyme analysis of mitochondrial DNA of members of the genus *Aspergillus* as an aid in taxonomy. *J. Gen. Microbiol., 128*:471-476, 1982.
87. Raper, K. B., and Fennell, D. I. *The Genus Aspergillus*. Williams and Wilkins Co., 1965.
88. Klich, M. A., and Mullaney, E. J. DNA restriction enzyme fragment polymorphism as a tool for rapid differentiation of *Aspergillus flavus* from *Aspergillus oryzae*. *Exp. Mycol., 11*:170-175, 1987.
89. Mullaney, E. J., and Klich, M. A. A review of molecular biological techniques for systematic studies of *Aspergillus* and *Penicillium*. In *Modern Concepts in Penicillium and Aspergillus Systematics* (R. A. Samson and J. I. Pitt, eds.), Plenum Press, New York, 1990, pp. 301-307.
90. Kusters-van Someren, M., Kester, H. C. M., Samson, R. A., and Visser, J. Variation in pectinolytic enzymes of the black Aspergilli: A biochemical and genetic approach. In *Modern Concepts in Penicillium and Aspergillus Systematics* (R. A. Samson and J. I. Pitt, eds.), Plenum Press, New York, 1990, pp. 321-334.
91. Kusters-van Someren, M., Samson, R. A., and Visser, J. The use of RFLP analysis in classification of the black Aspergilli: Reinterpretation of the *Aspergillus niger* aggregate. *Curr. Genetics* (in press).
92. Gomi, K., Tanaka, A., Iimura, Y., and Takahashi, K. Rapid differentiation of four related species of koji molds by agarose gel electrophoresis of genomic DNA digested units SMAI restriction enzyme. *J. Gen. Appl. Microbiol., 35*:225-232, 1989.
93. Logrieco, A., Peterson, S. W., and Wicklow, D. T. Ribosomal RNA comparisons among taxa of the terverticillate Penicillia. In *Modern Concepts in Penicillium and Aspergillus Systematics* (R. A. Samson and J. I. Pitt, eds.), Plenum Press, New York, 1990, pp. 343-354.

94. Sekiguchi, J., Oksaki, T., Yamamoto, M., Koichi, K., and Smiba, T. Characterization and comparison of mitochondrial DNAs and rRNAs from *Penicillium urticae* and *P. chrysogenum*. *J. Gen. Microbiol.*, *136*:535-543, 1990.
95. Taylor, J. W., Pitt, J. I., and Hocking, A. D. Ribosomal DNA restriction studies of *Talaromyces* species with *Paecilomyces* and *Penicillium* anamorphs. In *Modern Concepts in Penicillium and Aspergillus Systematics* (R. A. Samson and J. I. Pitt, eds.), Plenum Press, New York, 1990, pp. 357-369.
96. Taylor, J. W., Smolich, B. D., and May, G. An evolutionary comparison of homologous mitochondrial plasmid DNAs from three *Neurospora* species.
97. Kohn, L. M., Petsche, D. M., Bailey, S. R., Novak, L. A., and Anderson, J. B. Restriction fragment length polymorphisms in nuclear and mitochondrial DNA of *Sclerotinia* species. *Phytopathology*, *78*:1047-1051,
98. Garber, R. C., and Yoder, O. C. Mitochondrial DNA of the filamentous ascomycetes *Cochliobolus heterostrophus*. *Curr. Genet.*, *8*:621-628, 1984.
99. Castle, A. J., Horgen, P. A., and Anderson, J. B. Restriction fragment length polymorphism in the mushrooms *Agaricus brunnescens* and *Agaricus bitorquis*. *Appl. Environ. Microbiol.*, *53*:816-822, 1987.
100. Williams, A. P. Identification of *Penicillium* and *Aspergillus*. In *Modern Concepts in Penicillium and Aspergillus Systematics* (R. A. Samson and J. I. Pitt, eds.), Plenum Press, New York, 1990, pp. 289-294.
101. Bridge, P. D. Identification of terverticillate Penicillia from a matrix of percent positive test results. In *Modern Concepts in Penicillium and Aspergillus Systematics* (R. A. Samson and J. I. Pitt, eds.), Plenum Press, New York, 1990, pp. 283-287.
102. Klich, M. A. Computer applications in *Penicillium* and *Aspergillus* systematics and identification. In *Modern Concepts in Penicillium and Aspergillus Systematics* (R. A. Samson and J. I. Pitt, eds.), Plenum Press, New York, 1990, pp. 269-278.
103. Pitt, J. I. PENNAME, a new computer key to common *Penicillium* species. In *Modern Concepts in Penicillium and Aspergillus Systematics* (R. A. Samson and J. I. Pitt, eds.), Plenum Press, New York, 1990, pp. 279-281.
104. Hawksworth, D. L. Problems and prospects for improving the stability of names in *Aspergillus* and *Penicillium*. In *Modern Concepts in Penicillium and Aspergillus Systematics* (R. A. Samson and J. I. Pitt, eds.), Plenum Press, New York, 1990, pp. 75-82.
105. Cannon, P. F. International Commission on the Taxonomy of Fungi. Name changes in fungi of microbiological, industrial and medical importance. Parts 1-2. *Microbiol. Sci.*, *3*:168-171, 285-287, 1986.
106. Cannon, P. F. International Commission on the Taxonomy of Fungi. Name changes in fungi of microbiological, industrial and medical importance. Part 3. *Microbiol. Sci.*, *5*:23-26, 1988.
107. Cannon, P. F. International Commission on the Taxonomy of Fungi. Name changes in fungi of microbiological, industrial and medical importance. Part 4. *Mycopathologia*, *111*:75-83, 1990.

108. Frisvad, J. C., Hawksworth, D. L., Kozakiewicz, Z., Pitt, J. I., Samson, R. A., and Stolk, A. C. Proposal to conserve important species names in *Aspergillus* and *Penicillium.* In *Modern Concepts in Penicillium and Aspergillus Systematics* (R. A. Samson and J. I. Pitt, eds.), Plenum Press, New York, 1990, pp. 83-89.
109. Hesseltine, C. W., and Ellis, J. J. Species of *Absidia* with ovoid sporangia. *Mycologia, 58*:761-785, 1966.
110. Ellis, J. J., Rhodes, I. H., and Hesseltine, C. W. The genus *Amylomyces. Mycologia, 68*:131-143, 1976.
111. Samson, R. A. Revision of the genus *Cunninghamella* (fungi, Mucorales). *Proc. Konink. Nederlandse Akad. Wetenschappen. ser. C, 72*:322-335, 1969.
112. Lunn, J. A., and Shipton, W. A. Re-evaluation of taxonomic criteria in *Cunninghamella. Trans. Br. Mycol. Soc., 81*:303-312, 1983.
113. Schipper, M. A. A. A study on variability in *Mucor hiemalis* and related species. *Stud. Mycol., Baarn, 4*:1-40, 1973.
114. Schipper, M. A. A. On *Mucor mucedo, Mucor flavus* and related species. *Stud. Mycol., Baarn, 10*:1-33, 1975.
115. Schipper, M. A. A. On *Mucor circinelloides, Mucor racemosus* and related species. *Stud. Mycol., Baarn, 12*:1-40, 1976.
116. Schipper, M. A. A. 1. On certain species of *Mucor* with a key to all accepted species. 2. On the genera *Rhizomucor* and *Parasitella. Stud. Mycol., Baarn, 17*:1-70, 1978.
117. Schipper, M. A. A. A revision of the genus *Rhizopus.* I. The *Rh. stolonifer*-group and *Rh. oryzae. Stud. Mycol., Baarn, 25*:1-19, 1984.
118. Schipper, M. A. A., and Stalpers, J. A. A revision of the genus *Rhizopus.* II. The *Rhizopus microsporus*-group. *Stud. Mycol., Baarn, 25*:19-34, 1984.
119. Ellis, J. J. Species and varieties in the *Rhizopus arrhizus-Rhizopus oryzae* group as indicated by their DNA complementarity. *Mycologia, 77*:243-247, 1985.
120. Liou, G. Y., Chen, C. C., and Hsu, W. H. Atlas of the genus *Rhizopus* and its allies. *Mycol. Mem., 3*:1-32, 1990.
121. Benjamin, R. K. The merosporangiferous Mucorales. *Aliso, 4*:321-433, 1959.
122. Hesseltine, C. W., and Anderson, P. The genus *Thamnidium* and a study of the information of its zygospores. *Amer. J. Bot., 43*:696-702, 1951.
123. Stolk, A. C., and Samson, R. A. Studies in *Talaromyces* and related genera. I. *Hamigera,* gen nov. and *Byssochlamys. Persoonia, 6*:341-357, 1971.
124. Samson, R. A. *Paecilomyces* and some allied Hyphomycetes. *Stud. Mycol., Baarn, 6*:1-119, 1974.
125. Udagawa, S., Muroi, T., Kurata, H., Sekita, S., Yoshihira, K., Natori, S., and Umeda, M. The production of chaetoglobosins, sterigmatocystin, o-methylsterigmatocystin and chaetocin by *Chaetomium* spp. and related fungi. *Can. J. Microbiol., 25*:170-177.

126. Udagawa, S. Taxonomy of mycotoxin-producing *Chaetomium*. In *Toxigenic Fungi. Their Toxins and Health Hazard* (H. Kurata and Y. Ueno, eds.), Elsevier Press, Amsterdam, 1984, pp. 139-167.
127. Sekita, S., Yoshihira, K., Natori, S., Udagawa, S., Muroi, T., Sugiyama, Y., Kurata, H., and Umeda, M. Mycotoxin production by *Chaetomium* spp. and related fungi. *Can. J. Microbiol.*, *27*:766-772, 1979.
128. Ames, C. M. *A Monograph of the Chaetomiaceae*. U.S. Army Res. Div., Ser. 2, 1963, 125 pp.
129. von Arx, J. A., Dreyfuss, M., and Müller, E. A. A revaluation of *Chaetomium* and the *Chaetomiaceae*. *Persoonia*, *12*:169-179, 1984.
130. von Arx, J. A., Guarro, J., and Figueras, M. J. The Ascomycete genus *Chaetomium*. *Beihefte Nova Hedwigea*, *84*:162, 1988.
131. Horie, Y. Ascospore ornamentation and its application to the taxonomic re-evaluation in *Emericella*. *Trans. Mycol. Soc. Japan*, *21*: 483-493, 1980.
132. Christensen, M., and States, J. S. *Aspergillus nidulans* group: *Aspergillus navahoensis*, and a revised synoptic key. *Mycologia*, *74*:226-235, 1982.
133. Stolk, A. C., and Samson, R. A. The ascomycetes genus *Eupenicillium* and related *Penicillium* anamorphs. *Stud. Mycol., Baarn*, *23*:1-149, 1983.
134. Blaser, P. Taxonomische und physiologische Untersuchungen über die Gattung *Eurotium* Link. ex Fr. *Sydowia*, *28*:1-49, 1975.
135. Pitt, J. I. Nomenclatorial and taxonomic problems in the genus *Eurotium*. In *Advances in Penicillium and Aspergillus Systematics* (R. A. Samson and J. I. Pitt, eds.), Plenum Press, New York, 1985, pp. 383-395.
136. Kozakiewicz, Z. *Aspergillus* species on stored products. *Mycol. Papers*, *161*:1-188, 1989.
137. Hawksworth, D. L., and Pitt, J. I. A new taxonomy for *Monascus* species based on cultural and microscopical characters. *Australian J. Bot.*, *31*:51-61, 1983.
138. Malloch, D., and Cain, R. F. The Trichocomataceae: Ascomycetes with *Aspergillus*, *Paecilomyces* and *Penicillium* imperfect states. *Can. J. Bot.*, *50*:2613-2628, 1972.
139. Samson, R. A., Nielsen, P. V., and Frisvad, J. C. The genus *Neosartorya*: Differentiation by scanning electron microscopy and mycotoxin profiles. In *Modern Concepts in Penicillium and Aspergillus Systematics* (R. A. Samson and J. I. Pitt, eds.), Plenum Press, New York, 1990, pp. 455-467.
140. Frederick, L., Uecker, F. A., and Benjamin, C. R. A new species of *Neurospora* from soil of West Pakistan. *Mycologia*, *61*:1077-1084, 1970.
141. Frisvad, J. C., Filtenborg, O., Samson, R. A., and Stolk, A. C. Chemotaxonomy of the genus *Talaromyces*. *Antonie van Leeuwenhoek*, *57*:179-189, 1990.
142. Stolk, A. C., and Samson, R. A. The genus *Talaromyces*. Studies on *Talaromyces* and related genera. II. *Stud. Mycol., Baarn*, *2*:1-65, 1972.

143. Pitt, J. I., and Hocking, A. D. Food spoilage fungi. I. *Xeromyces bisporus* Fraser. *CSIRO Food Research Quarterly, 42*:1-6, 1982.
144. Gams, W. *Cephalosporium-artige Schimmelpilze (Hyphomycetes)*. G. Fischer, Stuttgart, 1971.
145. Simmons, E. G. Typification of *Alternaria, Stemphylium* and *Ulocladium*. *Mycologia, 59*:67-92, 1967.
146. Punithalingam, E. Graminicolous *Ascochyta* species. *Mycol. Pap., 142*:1-214, 1979.
147. Samson, R. A. A compilation of the Aspergilli described since 1965. *Stud. Mycol., Baarn, 18*:1-40, 1979.
148. Al-Musallam, A. Revision of the black *Aspergillus* species. Dissertation. University of Utrecht, Utrecht, Netherlands, 1980.
149. Christensen, M. The *Aspergillus ochraceus* group: Two new species from western soils and synoptic key. *Mycologia, 74*:210-225, 1982.
150. Christensen, M. A synoptic key and evaluation of species in the *Aspergillus flavus* group. *Mycologia, 73*:1056-1084, 1981.
151. Klich, M. A., and Pitt, J. I. A laboratory guide to common *Aspergillus* species and their teleomorphs. CSIRO Division of Food Processing, North Ryde, Australia, 1988.
152. Klich, M. A., and Pitt, J. I. The theory and practice of distinguishing species of the *Aspergillus flavus* group. In *Advances in Penicillium and Aspergillus Systematics* (R. A. Samson and J. I. Pitt, eds.), Plenum Press, New York, 1985, pp. 211-220.
153. Klich, M. A., and Pitt, J. I. Differentiation of *Aspergillus flavus* from *A. parasiticus* and other closely related species. *Trans. Br. Mycol. Soc., 91*:99-108, 1988.
154. Frisvad, J. C., and Samson, R. A. Chemotaxonomy and morphology of *Aspergillus fumigatus* and related taxa. In *Modern Concepts in Penicillium and Aspergillus Systematics* (R. A. Samson and J. I. Pitt, eds.), Plenum Press, New York, 1990, pp. 201-208.
155. Pitt, J. I., and Samson, R. A. Taxonomy of *Aspergillus* section *Restricti*. In *Modern Concepts in Penicillium and Aspergillus Systematics* (R. A. Samson and J. I. Pitt, eds.), Plenum Press, New York, 1990, pp. 249-257.
156. Gams, W., Christensen, M., Onions, A. H., Pitt, J. I., and Samson, R. A. Infrageneric taxa of *Aspergillus*. In *Advances in Penicillium and Aspergillus Systematics* (R. A. Samson and J. I. Pitt, eds.), Plenum Press, New York, 1985, pp. 55-62.
157. Kurtzman, C. P., Smiley, M. J., Robnett, C. J., and Wicklow, D. T. DNA relatedness among wild and domesticated species in the *Aspergillus flavus* group. *Mycologia, 78*:955-959, 1986.
158. Hermanides-Nijhof, E. J. *Aureobasidium* and allied genera. *Stud. Mycol., Baarn, 15*:141-177, 1977.
159. Jarvis, W. R. *Botryotinia* and *Botrytis* species. Taxonomy, physiology and pathogenicity. *Can. Dept. Agric. Harrow, Monograph 15*: 1-195, 1977.
160. von Arx, J. A. On *Monilia sitophila* and some families of Ascomycetes. *Sydowia, 34*:13-29, 1981.

161. Pitt, J. I. Two new species of *Chrysosporium*. *Trans. Br. Mycol. Soc.*, *49*:467-470, 1966.
162. Van Oorschot, C. A. N. A revision of *Chrysosporium* and allied genera. *Stud. Mycol., Baarn*, *20*:1-89, 1980.
163. de Vries, G. A. Contribution to the knowledge of the genus *Cladosporium*. Diss. Univ. Utrecht, Reprint J. Cramer Lehre, 1967.
164. von Arx, J. A. Die Arten der Gattung *Colletotrichum* Cda. *Phytopathol. Zeitschr.*, *29*:413-468, 1957.
165. Benoit, M. A., and Mathur, S. B. Identification of species of *Curvularia* on rice seed. *Proc. Int. Seed Testing Assoc.*, *35*:99-119, 1970.
166. Chidambaram, P., Mathur, S. B., and Neergaard, P. Identification of seed-borne *Drechslera* species. *Friesia*, *10*:165-207, 1973.
167. Alcorn, J. L. Generic concepts in *Drechslera bipolaris* and *Exserophilum*. *Mycotaxon*, *17*:1-86, 1981.
168. Gerlach, W., and Nirenberg, G. H. *The Genus Fusarium, A Pictorial Atlas*. Mitt. Biol. Bundesanst. Land. u. Forstwissensch., Berlin-Dahlem, 1988.
169. Burgess, L. W., and Liddell, G. M. *Laboratory Manual for Fusarium Research*. University of Sydney Press, Sydney, 1983.
170. Nelson, P. E., Tousson, T. A., and Marasa, W. F. O. *Fusarium Species. An Illustrated Manual for Identification*. Pennsylvania State University Press, University Park, PA, 1983.
171. von Arx, J. A. Notes on *Dipodascus*, *Endomyces* and *Geotrichum* with the description of two new species. *Antonie van Leeuwenhoek*, *43*:333-340, 1977.
172. De Hoog, G. S., Smith, M. T., and Gueho, E. A revision of the genus *Geotrichum* and its teleomorphs. *Stud. Mycol., Baarn*, *29*: 1-131, 1986.
173. Jong, S. C., and Davis, E. E. Contributions to the knowledge of *Stachybotrys* and *Memnoniella* in culture. *Mycotaxon*, *3*:409-485, 1976.
174. De Hoog, G. S. The black yeasts. II. *Moniella* and allied genera. *Stud. Mycol., Baarn*, *19*:1-34, 1979.
175. De Hoog, G. S., and Hermanides-Nijhof, E. J. Survey of the black yeasts and allied fungi. *Stud. Mycol., Baarn*, *15*:170-221, 1977.
176. Tulloch, M. The genus *Myrothecium*. *Mycol. Pap.*, *130*:1-42, 1972.
177. Nguyen, T. H., Neergaard, P., and Mathur, S. B. Seed-borne species of *Myrothecium* and their pathogenic potential. *Trans. Br. Mycol. Soc.*, *61*:347-354, 1973.
178. Ramirez, C. *Manual and Atlas of the Penicillia*. Elsevier Biomedical Press, Amsterdam, 1982.
179. Raper, K. B., and Thom, C. *A Manual of the Penicillia*. Williams and Wilkins Co., Baltimore, 1949.
180. Pitt, J. I., and Cruickshank, R. H. Speciation and synonymy in *Penicillium* subgenus *Penicillium*—towards a definitive taxonomy. In *Modern Concepts in Penicillium and Aspergillus Systematics* (R. A. Samson and J. I. Pitt, eds.), Plenum Press, New York, 1990, pp. 103-119.

181. Stolk, A. C., Samson, R. A., Frisvad, J. C., and Filtenborg, O. The systematics of the terverticillate Penicillia. In *Modern Concepts in Penicillium and Aspergillus Systematics* (R. A. Samson and J. I. Pitt, eds.), Plenum Press, New York, 1990, pp. 121-136.
182. Frisvad, J. C., and Filtenborg, O. Terverticillate Penicillia: Chemotaxonomy of and mycotoxin production. *Mycologia, 81*:837-861, 1989.
183. Frisvad, J. C., and Filtenborg, O. Revision of *Penicillium Furcatum* based on secondary metabolites and conventional characters. In *Modern Concepts in Penicillium and Aspergillus Systematics* (R. A. Samson and J. I. Pitt, eds.), Plenum Press, New York, 1990, pp. 157-170.
184. van Reenen-Hoekstra, E. S., Frisvad, J. C., Samson, R. A., and Stolk, A. C. The *Penicillium funiculosum*-complex: Well defined species and problematic taxa. In *Modern Concepts in Penicillium and Aspergillus Systematics* (R. A. Samson and J. I. Pitt, eds.), Plenum Press, New York, 1990, pp. 173-191.
185. Pitt, J. I. *A Laboratory Guide to Common Penicillium Species*, 2nd ed. CSIRO Division of Food Processing, North Ryde, Australia, 1988.
186. Schol-Schwartz, M. B. Revision of the genus *Phialophora* (Moniliades). *Persoonia, 6*:59-94, 1970.
187. Boerema, G. H., Dorenbosch, M. M., and van Kesteren, H. A. Remarks on species of *Phoma* referred to *Peyronella* I. *Persoonia, 4*:47-68, 1965.
188. Boerema, G. H., Dorenbosch, M. M., and van Kesteren, H. A. Remarks on species of *Phoma* referred to *Peyronella* III. *Persoonia, 6*:171-177, 1971.
189. Boerema, G. H., and Dorenbosch, M. M. J. The *Phoma* and *Ascocyta* species described by Wollenweber and Hochapfel in their study on fruit-rotting. *Stud. Mycol., Baarn, 3*:1-50, 1973.
190. Dorenbosch, M. M. J. Key to nine ubiquitous soil-borne *Phoma*-like fungi. *Persoonia, 6*:1-14, 1970.
191. Morton, F. J., and Smith, G. The genera *Scopulariopsis* Bainier, *Microascus* Zukal, and *Doratomyces* Corda. *Mycol. Pap., 86*:1-96, 1963.
192. Sigler, L., and Carmichael, J. W. Taxonomy of *Malbranchea* and some other Hyphomycetes with arthroconidia. *Mycotaxon, 4*:349-488, 1976.
193. Rifai, M. A. A revision of the genus *Trichoderma*. *Mycol. Pap., 116*:1-56, 1969.

2

FILAMENTOUS FUNGI IN FOODS AND FEEDS: ECOLOGY, SPOILAGE, AND MYCOTOXIN PRODUCTION

JENS C. FRISVAD *The Technical University of Denmark, Lyngby, Denmark*

ROBERT A. SAMSON *Centraalbureau voor Schimmelcultures, Baarn, The Netherlands*

I. INTRODUCTION

The presence of propagules of filamentous fungi in foods and feeds can easily be shown by direct microscopic observation or by dilution or direct plating on selective media [1]. This is only of concern if the fungi can actually grow on the particular food item under the given ecological conditions. Growth of filamentous fungi in foods or feeds (spoilage) causes concern, however, because this spoilage may be followed by problems such as unpleasant appearance and odor of the product, inability of seeds to germinate, allergy, human or animal infection, production of toxic fungal volatile compounds and mycotoxins [2].

Both fungal growth and mycotoxin production are dependent on environmental factors, with the limits for mycotoxin production usually being more narrow than those for growth [3-14]. It is not recommended to base prophylactic measures on these differences in environmental limits because other problems caused by fungal growth, rather than mycotoxin production, will remain. Fungal spoilage of foods and feeds should therefore be avoided wherever possible. Once fungal growth in particular foods or feeds is discovered, they should be monitored for important mycotoxins. Numerous mycotoxins are produced by different types of fungi growing on foods and feeds under various environmental conditions [15-18]. To reduce the number of mycotoxin analyses, detailed data on the associated microflora of different foods and feeds and mycotoxin production are needed [19]. This review deals with ecology of filamentous fungi on foods and mycotoxin production. The ecology and mycotoxin production by *Fusarium* species have been reviewed [20-26]; therefore, this chapter only emphasizes the influence of environmental factors on storage fungi, especially *Penicillium* and *Aspergillus* species.

II. ECOLOGY (INTRINSIC AND EXTRINSIC FACTORS)

A. Substrate Composition

The substrate composition (type of food or feed) has a major influence on the mycoflora in different foods, even when effects of important environmental factors like water activity and temperature are not considered. Usually phytopathogenic fungi have been considered specific to certain plants while most of the penicillia and aspergilli were regarded as common saprophytes. Notable exceptions among the penicillia are *P. italicum* and *P. digitatum*, spoiling citrus fruits; *P. expansum*, *P. crustosum*, and *P. solitum*, which often occur on and rot pomaceous fruits; *P. gladioli*, mostly found on gladiolus bulbs; *P. sclerotigenum*, which causes a rot in yams; *P. hirsutum* var. *allii*, which causes rot in garlic; and *P. hirsutum* var. *hirsutum*, which causes rot in many flower bulbs. Many of these penicillia also are important elements of the rhizosphere soil around vegetable root fruits together with *P. commune*, *P. hirsutum* var. *hordei*, *P. hirsutum* var. *albocoremium*, *P. aurantiogriseum*, and *P. glabrum* in temperate regions (J. C. Frisvad, unpublished).

The mycoflora of the same plant products in storage will be different under different environmental conditions. For example, several species of *Penicillium* such as *P. oxalicum*, *P. islandicum*, *P. funiculosum*, *P. pinophilum*, and *P. aethiopicum* are associated with cereals in the tropical and subtropical regions, whereas *P. verrucosum* and *P. solitum* are found only in temperate regions [19]. Other species of *Penicillium* such as *P. aurantiogriseum* var. *aurantiogriseum* and *P. aurantiogriseum* var. *viridicatum*, *P. commune* and *P. crustosum* are commonly distributed worldwide.

Regardless of geographic region, *P. aurantiogriseum* and its varieties are common on stored cereals, while *P. crustosum*, *P. commune*, and *P. echinulatum* are common only on nuts and other lipid- and protein-rich foods like cheese and meat [19]. With the change in the water content of food (lower water activity), the colonizing mycoflora become remarkably reduced with particular reference to *P. brevicompactum*, *P. chrysogenum*, *Aspergillus versicolor*, and *Eurotium* spp., which are more important in general. There is much information available on the mycoflora of stored food [27] but of little applied value since (a) the mycoflora may have been determined in non-surface disinfected grains, nuts, or seeds and a superficial flora will dominate on selective isolation media, while the internal mycoflora may be masked [28,29]; and (b) in many instances, several species of *Penicillium* and *Fusarium* are either not identified or misidentified [21,30].

B. Temperature

Temperature has a great impact on fungal growth and mycotoxin production. At low temperatures (-7 to 5°C) a number of fungi grow well including *Fusarium* spp., *Cladosporium* spp., *Geomyces pannorum*, *Thamnidium elegans*, and *Penicillium* spp. [27]. Toxigenic foodborne penicillia that grow well below 5°C include *P. aurantiogriseum*, *P. crustosum*, *P. commune*, *P. camemberti*, *P. solitum*, *P. roqueforti*, *P. brevicompactum*, *P. chrysogenum*, *P. echinulatum*, *P. verrucosum*, *P. expansum*, *P. hirsutum*, *P. griseofulvum*, *P. coprophilum*, *P. coprobium*, *P. glandicola*, *P. glabrum*, *P. thomii*, and

P. spinulosum [27]. *Talaromyces thermophilus, Aspergillus fumigatus, Humicola lanuginosa, Paecilomyces variotii, Aspergillus niger, Byssochlamys nivea,* and *Emericella nidulans* [31,32] are fungi inhabiting food stored at high temperatures. Even though many of these species are known mycotoxin producers under optimal conditions, very little is known about production of mycotoxins at elevated temperatures.

Most toxigenic foodborne penicillia grow at low temperatures and can produce mycotoxins [33]. At 7.2°C patulin production by *P. expansum, P. griseofulvum,* and *P. vulpinum* was substantial or even greater than at 12.8°C [33]. Most aspergilli do not grow at all or grow very poorly at 0-12°C. Toxin production by *Aspergillus ochraceus* is temperature dependent; at, e.g., 15 or 22°C under a low moisture condition production of penicillic acid was favored, but at 30°C and a high moisture condition ochratoxin A production was favored [34]. *P. verrucosum* produced ochratoxin A at 4°C, but the lower limit for ochratoxin production by *A. ochraceus* was 12°C [35]. Different species responded differently to different combinations of temperature, and water activity in production of mycotoxins was apparent [35]. It should be mentioned that all six strains of *Penicillium cyclopium* and *P. viridicatum* used by Northolt et al. [35] for investigations on the influence of environmental conditions on ochratoxin A production are correctly identified as *P. verrucosum* [17,30].

Verrucologen production by *Neosartorya fischeri* was optimal at 21 and 25°C, but the amount of the biosynthetic presuccessor fumitremorgin A was maximal at 25, 30, and 37°C. Growth was always followed by tremorgen production [36].

Temperature cycling only had a minor effect on aflatoxin production by *Aspergillus parasiticus*, but aflatoxins were only produced above 270 degree-hours per day [37]; Park and Bullerman [38,39], however, showed that 12-h alternations between 5 and 25°C promoted higher yields of aflatoxin than at constant temperature (15°C); lower yields at a constant 18 or 25°C resulted with *Aspergillus flavus*, while *A. parasiticus* produced more aflatoxin at cycling temperatures than at constant 15, 18, or 25°C.

C. Water Activity

The influence of water activity on germination, growth, and mycotoxin production of filamentous fungi has been reviewed by Corry [40] and Beuchat [41]. Important experimental data on germination and growth of fungi at different water activities have been given by several workers [24,42-50]. Detailed data on the influence of water activity on fungal growth and production of several mycotoxins have been provided by other workers [35, 51-58]. They have emphasized that temperature and water activity interact strongly on mycotoxin production. Data on the influence of water activity on growth and mycotoxin production by foodborne fungi have been compiled in Table 1, and misidentified cultures have been eliminated or corrected according to Frisvad [30].

D. pH

Most food spoilage fungi can grow from pH 2.5 to 9.5, and some fungi like *P. funiculosum* can grow even at pH 1.8 [59]. Usually the most important

TABLE 1. Minimal Water Activities for Germination and Toxin Production by Different Filamentous Fungi

	Minimal a_w for		
Fungus	Growth	Toxin production	Toxin
Alternaria alternata	0.85-0.88	0.90	Alternariol Altenuene Alternariol monomethyl ether
A. candidus	0.75-0.78 0.85 (NaCl)		
A. clavatus	0.85	0.99	Patulin
A. flavus	0.78-0.80	0.83-0.87	Aflatoxin
A. fumigatus	0.85-0.94		
A. ochraceus	0.76-0.83	0.83-0.87 0.81-0.88	Ochratoxin Penicillic acid
A. parasiticus	0.78-0.82	0.87	Aflatoxin
A. penicillioides	0.73[a]-9,66 0.75 (NaCl)		
A. restrictus	0.71-0.75		
A. sydowii	0.78 0.81 (NaCl)		
A. tamarii	0.78		
A. terreus	0.78		
A. versicolor	0.78		
A. wentii	0.73-0.75 0.79 (NaCl)		
Basipetospora (Oospora) halophila	0.77-0.78 0.75 (NaCl)		
Botrytis cinerea	0.93-0.95		
Byssochlamys nivea	0.84-0.92		
Chrysosporium xerophilum	0.71		
Chrysosporium fastidium	0.61		
Cladosporium cladosporioides	0.86-0.88		
Cladosporium herbarum	0.85-0.88		
Epicoccum nigrum	0.86-0.90		
Eurotium amstelodami	0.71-0.76 0.75 (NaCl)		
Eurotium chevalieri	0.71-0.73		

TABLE 1. (Continued)

Eurotium echinulatum	0.64		
Eurotium repens	(0.69[a])	0.72-0.74	
Eurotium rubrum	0.70-0.71		
Eurotium sp. (FRR 2471)	0.68-0.72 0.75 (NaCl)		
Exophiala werneckii	0.77-0.78 0.75 (NaCl)		
Fusarium avenaceum	0.87-0.91		
F. culmorum	0.87-0.91		
F. graminearum	0.89		
F. oxysporum	0.87-0.89		
F. poae	0.89		
F. solani	0.87-0.90		
F. sporotrichioides	0.86-0.88		
F. tricinctum	0.89		
F. verticilloides	0.87		
Geomyces pannorum	0.92		
Geomyces	0.73-0.75[a] 0.75[a] (NaCl)		
Mucor circinelloides	0.90		
M. racemosus	0.94		
M. spinosus	0.93		
Neosartorya fischeri	0.915	0.955 0.925 0.925	Fumitremorgin A Fumitremorgin C Verrucologen
Paecilomyces variotii	0.79-0.84 0.91 (NaCl)		
P. aurantiogriseum	0.79-0.85	0.97-0.99	Penicillic acid
P. brevicompactum	0.78-0.82		
P. charlesii	0.78-0.80		
P. chrysogenum	0.78-0.81		
P. citrinum	0.80-0.82		
P. commune	0.83		
P. digitatum	0.90		
P. expansum	0.82-0.85	0.99	Patulin
P. griseofulvum	0.81-0.85	0.95	Patulin
P. islandicum	0.83-0.86		
P. oxalicum	0.88		

TABLE 1. (Continued)

Fungus	Minimal a_W for Growth	Minimal a_W for Toxin production	Toxin
P. roqueforti	0.83		
P. rugulosum	0.85		
P. verrucosum	0.81-0.83	0.83-0.90	Ochratoxin
Phytophthora infestans	0.85		
Polypaecilum pisce	0.75-0.77[a] 0.83[a] (NaCl)		
Pythium splendens	0.90		
Rhizoctonia solani	0.96		
Rhizopus stolonifer	0.93		
Stachybotrys atra	0.94	0.94	Stachybotrin
Thamnidium elegans	0.94		
Trichothecium roseum	0.90		
Verticillium lecanii	0.90		
Wallemia sebi	0.69-0.75		
Xeromyces bisporus	0.61		

[a]Germination occurred, but not growth.

Source: Refs. 124, 140, 150, 157. Corrected according to recent taxonomic changes in aspergilli and penicillia.

effect of pH is on the dissociation of preservatives such as benzoic acid, sorbic acid, etc., but other effects have also been observed. Nielsen et al. [60] found that *Neosartorya fischeri* grew poorly at pH 2.5 with low mycotoxin production, but when citric, malic, or tartaric acid was added to the medium, growth and fumitremorgin production increased markedly. In some instances even small changes in pH will cause a drastic change in mycotoxin production. For example, at an initial pH value of 6.0 *Fusarium graminearum* yielded around 32 times more fusarin C than at pH 7 [61]. According to the "hurdle effect" [62], good growth at marginal environmental conditions requires optimal conditions for other environmental factors. Altering the pH from 6.5 to 4 in wheat extract agar increased the minimum water activity for germination by 0.02 unit at optimum temperature and 0.05 unit at marginal temperatures for field and storage fungi [63].

E. Atmosphere and Redox Potential

The amounts of oxygen and carbon dioxide in the atmosphere and especially of dissolved oxygen in the substrate [64] strongly influence growth and mycotoxin production by different filamentous fungi [65-70]. An excellent review on responses of fungi to modified atmospheres has been written by Hockling [71]. Carbon dioxide usually inhibits *Aspergillus* and *Penicillium* species at concentrations over 15%, while it may be stimulatory at lower concentrations [72,73]. *Penicillium roqueforti* [73], *Xeromyces bisporus* [74], and *Byssochlamys nivea* [75] grow quite well in atmospheres containing 80-90% carbon dioxide. Low oxygen values in the atmosphere (0.5-2%) had nearly no effect on growth and mycotoxin production by *Byssochlamys nivea, P. griseofulvum, P. expansum, and Aspergillus versicolor.* However, the combination of high carbon dioxide and low oxygen content inhibited growth of the same molds and prevented mycotoxin production by nearly all strains [76,77]. Production of aflatoxin by *A. parasiticus* [78-81], of penicillic acid by *P. aurantiogriseum* [82], and of ochratoxin by *Aspergillus ochraceus* [83] was also inhibited by combinations of low oxygen and high carbon dioxide concentrations. At least 1.5% oxygen was necessary for aflatoxin production when the carbon dioxide concentration was more than 3% [84,85].

Several fungi—*Penicillium glabrum, P. corylophilum, Fusarium oxysporum, F. equiseti, Mucor plumbeus, Absidia corymbifera,* and the xerophilic *Eurotium repens* and *Xeromyces bisporus*—can grow in pure nitrogen atmosphere [71]. In atmospheres of 97-99% carbon dioxide and trace amounts of oxygen most fungi grow poorly, but *Fusarium oxysporum* and *Mucor plumbeus* grow at rates from 0.5-4% of the growth in air [71]. We have isolated *P. glabrum* on several occasions from carbonated beverages without preservatives (Frisvad and Filtenborg, unpublished) but it is unknown whether *P. glabrum* can produce mycotoxins in these products. The presence of tremorgenic mycotoxins in a sample of canned beer [86] shows that a high concentration of carbon dioxide in a beverage may not be sufficient to prevent mycotoxin production in all instances. *Neosartorya fischeri* produced the tremorgenic mycotoxins verrucologen and fumitremorgin A and B (on a laboratory substrate) at oxygen concentrations as low as 0.1%, but the mycotoxin production was greatly reduced at 0.1% and 1% compared to 3% and 20.9% oxygen [87]. Farber and Sanders [61] found that shake cultures of some *Fusarium* spp. yielded significantly greater amounts of fusarin C than the corresponding nonshaken cultures, so high concentrations of oxygen usually stimulate mycotoxin production. Knowledge of growth and mycotoxin production of many other molds at low redox potentials or facultative anaerobic conditions is still quite meager.

F. Microbial Competition

A number of experiments have been done to test the influence of other microorganisms on production of aflatoxins by *Aspergillus flavus* [88-100] and on zearalenone production by *Fusarium graminearum* [92]. In most experiments only partial inhibition of aflatoxin production was obtained [99] or aflatoxin B_1 (or G_1) was degraded to other less toxic compounds [98, 101,102]. In some instances *A. flavus* inhibited production of patulin and griseofulvin by *P. griseofulvum,* penicillic acid production by *P. aurantiogriseum,* and citrinin production by *P. citrinum* [89]. The inhibition of aflatoxin biosynthesis by the same three species of *Penicillium* did not appear to be caused by poor competitiveness of *A. flavus,* metabolization of aflatoxin by the fungi, or known mycotoxin produced by the penicillia.

It appeared, however, that a water-soluble and heat-stable metabolite from the penicillia markedly inhibited production of aflatoxin. *Clostridium botulinum* could grow and produce toxins in the presence of fungi in tomatoes. By changing the pH of the tomatoes the fungi created good conditions for toxin production by either *Clostridium* or *Alternaria* [103]. *Lactobacillus casei* and *Streptococcus lactis* inhibited growth, but only to a certain extent of *A. flavus*, *A. parasiticus*, *A. ochraceus*, *P. commune*, *P. roqueforti*, *P. expansum*, and *P. griseofulvum* [104]. In combination with preservatives these bacteria may inhibit fungal growth and mycotoxin production more efficiently.

G. Arthropods

Insects and mites play an important role as contamination sources of fungi, and they usually cause an increase in metabolic activity, water activity, and temperature [105,106]. The interaction between insects, microorganisms, and environmental factors and mycotoxin production is often very complex [107-109]. Thus fungi may produce volatile compounds that either attract or repel different species of insects [106], and different mycotoxins may play a role in the interaction between arthropods and filamentous fungi on foods and feeds [99].

III. SPOILAGE AND PROCESSING FACTORS

A. General Problems

Physical damage of grains, nuts, etc. will usually favor *Penicillium* species rather than *Aspergillus* species [110,111] and speed up the deterioration process considerably. This is also true for infection of apples with *P. expansum* and citrus fruits with *P. italicum* or *P. digitatum* [112].

It is noteworthy that the mycoflora present at a particular combination of environmental conditions (water activity, pH, temperature, atmosphere) is constant. This "spoilage association" already was emphasized by several workers [24,113,114]. Recently it has been shown that fresh or stored foods and feeds contain a typical associated mycoflora [18,19]. This has been shown for the common terverticillate penicillia in important groups of foods such as cereals (*P. aurantiogriseum*, *P. verrucosum*, *P. hirsutum* var. *hordei*), nuts (*P. crustosum*, *P. commune*, *P. solitum*, *P. expansum*, *P. echinulatum*), onions (*P. hirsutum*), pomaceous fruits (*P. expansum*, *P. crustosum*, *P. solitum*), citrus fruits (*P. italicum*, *P. digitatum*), fermented or processed meat (*P. chrysogenum*, *P. nalgiovense*, *P. verrucosum*), etc. [18,19]. This spoilage association was less clear a few years ago, because species such as *P. aurantiogriseum*, *P. verrucosum*, *P. commune*, *P. solitum*, and *P. crustosum* were often all identified as *P. cyclopium* (or *P. verrucosum* var, *cyclopium*).

Under more extreme environmental conditions the mycoflora of different products can be more constant, i.e., at low water activity species such as *Eurotium repens*, *E. rubrum*, *E. amstelodami*, *E. chevalieri*, *P. brevicompactum*, and *P. chrysogenum* will dominate the mycoflora of many different kinds of food (cereals, nuts, meat, dried fruits) [19,114], and deviation of the mycoflora of these foods may be insignificant. For example, while the normal mycoflora of stored Canadian and North-European barley is clearly dominated by *P. aurantiogriseum*, *P. verrucosum*, and *P. hirsutum* var. *hordei* [19], species like *P. chrysogenum*, *P. brevicompactum*, *Eurotium* spp., *Aspergillus flavus*, or *A. versicolor* may be dominant when drier than usual conditions prevail. Factors that govern the special affinity of *P. verrucosum* and *P. aurantiogriseum* (and not, for example, *P. crustosum* or *P. commune*) for barley may be (a) high production of extracellular amylases or (b) resistance to phytoalexins and tannins of barley. Even though *P. crustosum* will grow on barley grains sterilized by radiation, we have never found this species actively growing in any barley sample (Frisvad and Filtenborg, unpublished). The reason for this difference between a more constant mycoflora at low water activities and a more specific associated mycoflora at higher water activities may be caused by inhibition of production of extracellular enzymes by the fungi and phytoalexins by the plants at low water activities. As just mentioned, the reason for good growth of *P. aurantiogriseum* and *P. verrucosum* on cereals and *P. crustosum* and *P. commune* on nuts [18,115] may be their good amylase and/or lipase production.

B. Heat-Treated Products

Some ascomycetes are heat resistant [116-125]. These include *Byssochlamys nivea* and *B. fulva*, *Neosartorya fischeri*, *N. fischeri* var. *glabra* [126], *N. fischeri* var. *spinosa*, *N. quadricincta* [127], *Talaromyces macrosporus* (reported as *T. flavus*, which is less heat resistant), *T. bacillisporus*, *T. trachyspermus*, *E. brefeldianum*, and *E. lapidosum* [125]. Z-values (the increase in temperature necessary to decrease or increase the decimal reduction time by a factor of 10) (see Ref. 117) of 5 to 12°C have been reported for the most heat-resistant isolates of *Byssochlamys*, *Neosartorya*, and *Talaromyces*, so the figures of 7.1 (grape juice adjusted to 65 Brix with sucrose) and 9.1°C (5 Brix) for an *Eurotium* isolate [124] are quite surprising, because *Eurotium* spp. are usually considered to be appreciably less heat resistant than *Byssochlamys*, *Neosartorya*, and *Talaromyces* spp. [128]. Recent studies have shown that the most common heat-resistant spoilage organisms in *Neosartorya* are *N. fischeri* var. *glabra* and var. *spinosa* [129].

Heating during drying of cereals is unlikely to influence the dormant conidia and spores, because they can tolerate 110°C for 3 min [130]. Subsequent germination, growth, and mycotoxin production on cereals depend on the interaction between the intrinsic and extrinsic environmental factors described above.

C. Low Water Activity Products

Filamentous fungi that actually spoil foods containing large amounts of either salt or carbohydrates include *Xeromyces bisporus*, *Wallemia sebi*, *Aspergillus penicillioides*, *Eurotium amstelodami*, *E. chevalieri*, *E. repens*, *E. rubrum*, *E. herbariorum*, *E. echinulatum*, *Chrysosporium inops*, *C. farinicola*, *C. fastidium*, and *C. xerophilum* (= *Sporotrichum pruinosum*; Ref. 131) and *Polypaecilum pisce*, and *Basipetospora* (*Oospora*) *halophila* on salted products [27]. The most common aspergilli and penicillia on dried foods include *A. clavatus*, *A. niger*, *A. flavus*, *A. versicolor*, *A. sydowii*, *A. wentii*,

A. ochraceus, *A. tamarii*, *A. candidus*, *P. chrysogenum*, *P. brevicompactum*, *P. corylophilum*, *P. thomii*, *P. smithii*, *P. citrinum*, *P. charlesii*, and *P. aurantiogriseum* and many other less xerotolerant penicillia [4,19,27,132]. Several of the latter species of *Penicillium* and *Aspergillus* produce known mycotoxins [30,133], but it is not known whether these toxins are produced at the low water activities of these foods.

D. Low Oxygen Products

Filamentous fungi found in carbonated beverages, beer, or wine include *Fusarium oxysporum*, *F. solani*, *Cladosporium* spp., *Penicillium glabrum*, *P. roqueforti*, and *P. crustosum* [59,86,134]. An apparent natural intoxication resulting from consumption by beer contaminated by *P. crustosum* was reported by Cole et al. [86]. Roquefortine C and isofumigaclavine A and B were isolated from *P. crustosum* when grown on laboratory substrates, but tremor symptoms in the person consuming the beer were indicative of presence of penitrem A in the original beer sample. We examined several contaminated beer samples, and in all instances a mixture of *P. roqueforti* and *P. crustosum* was recovered (Frisvad, unpublished). All isolates of *P. crustosum* examined by us produced penitrem A [30]. Natural occurrence of penitrem A resulting in dog mycotoxicosis has been reported for moldy cream cheese and moldy walnuts [135,136], but penitrem A has not yet been isolated from beverages or food.

The mycoflora of cereals under "airtight" storage [71,137-140,143a] is also very specific and resembles that of silage. The most common species present are *P. roqueforti*, *Aspergillus candidus*, *A. terreus*, *Paecilomyces variotii*, and *Scopulariopsis brevicaulis* [29,141]. Magan and Lacey [65] reported that *Fusarium culmorum*, *P. roqueforti*, and *A. candidus* were the most tolerant to low oxygen concentrations, which is consistent with the fungi actually found on cereals under airtight storage.

E. Acid-Treated Feeds

Silage has a very specific flora of filamentous fungi because of the low pH and the build-up of carbon dioxide, alcohols, and lower fatty acids. The most commonly found filamentous fungi are listed in Table 2. Of particular importance are *Aspergillus fumigatus* and its possible production of fumigaclavines [142], *Byssochlamys nivea* and its possible production of patulin and byssochlamic acid [143], and *Penicillium roqueforti* and its possible production of patulin and penicillic acid [18]. However, none of these toxins has yet been found naturally occurring in silage, even though silage has been implicated in mycotoxicosis [142]. It is remarkable that a great number of patulin producers have been isolated from silage (*P. glandicola*, *P. roqueforti* chemotype II, *Paecilomyces variotii*, and *Byssochlamys nivea*). Of particular importance is the selection for *Aspergillus flavus/parasiticus* and the stimulation of aflatoxin production in formic acid- and propionic acid-treated cereals mentioned below [144].

Acidic phosphate formulations were effective in protecting high moisture corn from molding and mycotoxin production, and addition of 2% tricalcium phosphate also inhibits insects [145].

TABLE 2. Filamentous Fungi Reported from Silage

Zygomycotina:
Absidia corymbifera
Mucor griseo-cyanus
Mucor hiemalis
Mucor racemosus
Ascomycotina and Deuteromycotina:
Aspergillus fumigatus
Byssochlamys nivea
Fusarium verticilloides
Geotrichum candidum
Monascus purpureus
Paecilomyces variotii
Penicillium roqueforti
Scopulariopsis brevicaulis
Trichoderma viride

Source: Adapted from Refs. 143a, b, c.

F. Preserved Products

Preservatives are used in some foods and feeds to prevent fungal growth. Some important preservatives are formic acid, acetic acid, propionic acid, sorbic acid, benzoic acid, and sulfur dioxide. The parabens and natamycin (pimaricin) are effective at a broad range of pH values, whereas the other preservatives mentioned above are most effective as undissociated acids, i.e., at pH values from 3 to 5. The undissociated acids have a much lower solubility than the potassium or sodium salts, however [146]. Results from Rusul and Marth [147] indicate that potassium sorbate is a more effective inhibitor of growth and aflatoxin production by *Aspergillus parasiticus* than potassium benzoate at pH 5.5 and 4.5. At pH 3.5 the maximum concentration of potassium sorbate or benzoate inhibiting growth of *A. parasiticus* is 0.025%, whereas it is 0.10% and 0.05%, respectively, for potassium benzoate and sorbate at pH 5.5. Most filamentous fungi are inhibited by concentrations of potassium sorbate or benzoate (0.01-0.3%) at pH 3.5-4.5 [148,149]. A number of penicillia, some of them isolated from sorbate-treated cheeses, were examined by Marth et al. [150] and Finol et al. [151]. Table 3 shows that only *P. roqueforti* and *P. commune* [152-154] are able to grow at sorbate concentrations above 0.3%. The isolates from sorbate-treated cheeses called *P. cyclopium, P. puberulum, P. viridicatum, P. crustosum,* and *P. lanosoviride* [151] have been reexamined by Frisvad [154,155]. They were all reidentified to the modern concept of *P. commune,* and they all produced cyclopiazonic acid. It is possible that the major

TABLE 3. Maximum Concentration of Potassium Sorbate (ppm) Which Permits Growth of Penicillia Found on Cheese at 25°C

Original identification	Reidentified as [30,154,155]	Maximum amount of sorbate permitting growth (ppm)
P. roqueforti	*P. roqueforti*	5400
P. roqueforti	*P. roqueforti*	7100
P. roqueforti	*P. roqueforti*	10000
P. roqueforti 22	*P. roqueforti*	9000
P. roqueforti K1	*P. roqueforti*	9000
P. roqueforti NRRL 849	*P. roqueforti*	2000 (a)
P. cyclopium	*P. commune*	3000
P. cyclopium 8	*P. commune*	3000
P. cyclopium (variant)	*P. commune*	12000
P. cyclopium (atypical) 25	*P. commune*	9000
P. cyclopium (atypical) 40	*P. commune*	9000
P. viridicatum	*P. commune*	6000
P. viridicatum 34	*P. commune*	9000
P. puberulum	*P. commune*	12000
P. puberulum 33	*P. commune*	9000
P. puberulum 37	*P. commune*	9000
P. lanoso-viride	*P. commune*	3000
P. lanosoviride 44	*P. commune*	3000
P. crustosum	*P. commune*	7000
P. crustosum 42	*P. commune*	6000
P. notatum	*P. chrysogenum*	2300
P. notatum	*P. chrysogenum*	500
P. cyaneo-fulvum	*P. chrysogenum*	1800
P. chrysogenum	*P. chrysogenum*	500
P. frequentans	*P. glabrum*	2800
P. frequentans	*P. glabrum*	500
P. expansum	*P. expansum*	1000
P. camemberti NRRL 877	*P. camemberti*	500
P. digitatum	*P. digitatum*	1000
P. patulum M59	*P. griseofulvum*	1000
A. flavus (10 strains)	*A. flavus*	500
A. flavus 93070	*A. flavus*	1000
A. parasiticus (4 strains)	*A. parasiticus*	500

TABLE 3. (Continued)

A. niger	*A. niger*	500
A. nidulans	*Emericella nidulans*	500
A. ochraceus	*A. ochraceus*	1000

Source: Adapted from Ref. 149.

fraction of the species isolated by Wei-Yun et al. [156] identified as *P. cyclopium* and *P. viridicatum* also were *P. commune* as only two of them produced penicillic acid, a mycotoxin associated with *P. aurantiogriseum*. The two strains of *P. cyclopium* and *P. viridicatum* producing ochratoxin A were probably *P. verrucosum* chemotype I, another species found occasionally on cheese [18,19].

The influence of sorbic acid, benzoic acid, and propionic acid on growth of different species of *Penicillium* has not been examined systematically, but results listed in Table 3 (see also Ref. 157) indicate that creatine-positive [158] penicillia such as *P. roqueforti*, *P. commune*, *P. crustosum*, *P. solitum*, and *P. echinulatum* are among the most preservative-resistant species. Not only can *Penicillium* species be resistant to preservatives, but a growing number of taxa belonging to *Eurotium*, *Acremonium*, etc. have been regularly found on preservative-treated products in the Netherlands [125].

Some reports indicate that organic acids such as acetic acid, propionic acid, and sorbic acid stimulate production of mycotoxins or other secondary metabolites when present in small concentrations [144,159]. Formic acid, acetic acid, and propionic acid stimulate penicillin production by *Penicillium chrysogenum* [159,160]. It is of greater significance for human and animal health that formic acid and propionic acid not only select for *A. parasiticus* and *A. flavus* in cereal grains but also stimulate aflatoxin biosynthesis [144, 161,162]. Stimulation of aflatoxin production by *A. flavus* and T-2 toxin production by *Fusarium acuminatum* by sorbate has also been reported [163-166]. Propionic acid was reported as stimulating aflatoxin biosynthesis [167]. Small concentrations of sorbate and benzoate (25 ppm) at pH 3.5 also stimulated verrucologen production by *Neosartorya fischeri* [60]. Subinhibitory concentrations of pimaricin also supported strong formation of aflatoxins [168]. Gourama and Bullerman [169] found that penicillic acid production was inhibited by potassium sorbate and natamycin at all concentrations. Thus, it is very important that preserved foods should contain sufficient undissociated acids to inhibit fungal growth.

Other preservatives such as sulfite seem to be quite effective in controlling growth and toxin production [60,170] because of toxin degradation by sulfur dioxide.

Farag et al. [171] found that the antioxidants butylated hydroxy anisole (BHA) and butylated hydroxy quinone (TBHQ) inhibited growth and aflatoxin production by *Aspergillus parasiticus*, and butylated hydroxy toulene (BHT) and dodecyl gallate (DDG) stimulated growth and aflatoxin production. Because of the latter observation the authors recommend that use of these antioxidants should be discontinued.

G. Irradiated Products

Gamma and X-ray irradiation of fungal spores may induce mutations in filamentous fungi even though they are well protected by melanin and certain secondary metabolites [172-174]. An increase in radiation resistance after several irradiation cycles has not yet been observed with *Aspergillus flavus* or other fungi [175,176]. Several studies [177-183] indicate that irradiated toxigenic cultures grown on heat-sterilized laboratory media resulted in an increase in mycotoxin production. This increase may have been caused by lower inoculation levels (it has been shown that low inoculation levels cause an increase in mycotoxin production compared to high inoculum levels) [184-186]. Jemmali and Guilbot [187] found that nontoxigenic strains of *Aspergillus flavus* could regain the ability to produce aflatoxin after gamma radiation. However, experiments with irradiation have always resulted in decreased mycotoxin formation [175,188-191]. White and blue light inhibit mycotoxin production, as has been shown for *Alternaria alternata* [192], *Aspergillus flavus* [193], and other fungi [194]. Nielsen et al. [36] found, however, that alternating light and darkness enhanced verrucologen and fumitremorgin production as compared to total darkness. Thus light may in some instances induce or stimulate mycotoxin production. G. W. van Eijk (personal communication) observed that anthraquinoine production by an atypical strain of *Talaromyces stipitaus* was only initiated by light, and toxin production was very poor in total darkness.

Certain species are resistant to irradiation, especially dark pigmented species such as *Aureobasidium pullulans* and *Aspergillus niger*. The prevalence of *A. niger* on sun-dried grapes and spices may be caused by this resistance [27]. However, the decrease in growth and aflatoxin production by *Aspergillus flavus* in raisins may be caused by the high temperature (45-50°C) during solar drying [195].

IV. MYCOTOXIN PRODUCTION

Mycotoxins have been defined by Bennett [196] as natural products produced by fungi (yeasts and mushrooms excluded) that evoke a toxic response when introduced in low concentrations to higher vertebrates and other animals by a natural route. A great number of foodborne fungi are able to produce one or more mycotoxins [3,17,30] and other secondary metabolites with toxic effects on other microorganisms, insects, or plants [197].

These secondary metabolites are of great chemical diversity, and they may be efficiently identified by chromatographic techniques, such as thin-layer chromatography (TLC) and high performance liquid chromatography (HPLC), preferably with specific detection techniques and comparison with known standards [198,199]. Confirmations of identity could be chemical treatments, directly on TLC plates or pre- or post-column derivatizations in HPLC, but specific detectors or scanners are more efficient in confirming identity of mycotoxins. These detectors include mass selective detectors (MSD), Fourier transform infrared detectors (FT-IRD) and diode array ultraviolet detectors (DAD) [200]. Fast atom bombardment mass spectrometry [201] or UV scanning [202] can be used directly on TLC plates. These methods are important to ensure correct identification of mycotoxins in food and feed samples.

A. Which Fungi Produce Which Toxins?

Most mycotoxins have been reported as being produced by several species, often from the same genus. The aflatoxins, for example, have only been reported as being produced by three closely related species in the genus *Aspergillus*: *A. flavus*, *A. parasiticus*, and *A. nomius* [203], even though several unverified reports of aflatoxin production by other species have been presented (i.e., *Penicillium puberulum*, *Aspergillus wentii*, *A. fumigatus*, *A. tamarii*, *Emericella rugulosa*, see Refs. 30, 204, 205). Other mycotoxins, like patulin, have been reported to be produced by a large number of species in different genera [30]. In some of these reports either the isolate or patulin was misidentified, but still *Penicillium*, *Aspergillus*, and *Paecilomyces* species are able to produce this mycotoxin. It appears that secondary metabolites with complicated biosyntheses are more restricted in distribution than those with a short biosynthetic route [206]. A large number of *Penicillium* and *Fusarium* strains have been "misidentified," and therefore the connection between species in these genera and mycotoxins have been somewhat obscured [21,25,30]. However, when these misidentifications are omitted in relation with our own findings, a rather clear picture of the mycotoxin-producing ability of different foodborne filamentous fungi emerges (Table 4). Furthermore, different isolates belonging to the same species produce the profiles of mycotoxins very consistently in pure culture, provided optimal production media are employed [154,207]. However, secondary metabolite production is very dependent on growth medium composition [208], and thus mycotoxins produced on laboratory media may not necessarily be produced in foods or feeds. The influence of environmental factors on mycotoxin production is usually examined on autoclaved substrates, but Cuero et al. [58,91,92,209] noted that irradiated seeds may retain viability and thus behave more like the natural substrate. A whole range of media and incubation conditions may be necessary to ensure that the full potential of each species of the filamentous fungi has been determined in the laboratory. It is still possible, however, that a fungus may produce a mycotoxin in cereals during storage while production of the same toxin is not expressed on common mycotoxin screening media like yeast extract sucrose (YES) agar.

B. Food Commodities and Their Specific Profile of Mycotoxin-Producing Species

Much information has accumulated concerning the specific non-*Penicillium* mycoflora of foods and feedstuffs and its potential for mycotoxin production [15,210-212]. However, because of the confusion concerning the taxonomy and mycotoxin production of *Penicillium* species and many misidentifications among the penicillia [30], the connection between foods and feedstuffs and their *Penicillium* mycoflora is less satisfactory. The data in Table 5 concerning the dominant penicillia in foods and feedstuffs are based on our own experience and confirmations of identity of culture collection *Penicillium* strains or strains requested from different scientists working on mycoflora

TABLE 4. Mycotoxin Production by Important Foodborne and Feedborne Filamentous Fungi

Species	Mycotoxins
Zycomycotina:	
Rhizopus oryzae	Isofumigaclavine A
Rhizopus stolonifer	Rhizonin
Ascomycotina:	
Byssochlamys fulva	Patulin, Byssotoxin, Byssochlamic acid
Byssochlamys nivea	Patulin, Byssochlamic acid, Malformins
Chaetomium globosum	Chaetoglobosins, Chaetomin, Cochiodinol, Chaetocin
Claviceps paspali	Ergot alkaloids, Paspalicin, Paspaline, Paspalinine, Paspalitrem A and B
Claviceps purpurea	Ergot alkaloids, Secalonic acid B and C
Emericella nidulans	Sterigmatocystin, Nidulotoxin
Eupenicillium baarnense	Penicillic acid
Eupenicillium egyptiacum	Xanthocillin X
Eupenicillium ehrlichii	Brefeldin A, Palitantin, Penicillic acid
E. lapidosum	Patulin
E. ochrosalmoneum	Citreoviridin
Eurotium amstelodami	Physicon, Sterimatocystin (?)
E. chevalieri	Emodin, Physicon, Gliotoxin (?), Xanthocillin X (?)
E. herbariorum	Sterigmatocystin (?)
Neosartorya fischeri	Fumitremorgin A, B, and C
	Verruculogen
Neosartorya fischeri var.	
glabra chemotype I	Canescin (?)
chemotype II	Mevinolins
chemotype III	Trypacidin, Tryptoquivalins
Neosartorya quadricincta	Cyclopaldic acid (?)
N. fischeri var. *spinosa*	Terrein, Tryptoquivalins
Talaromyces bacillisporus	Bacillosporins, Pinselin
T. flavus	Dehydroaltenusin, Mitorubrins, Vermicelline, Vermiculine, Vermistatine, Wortmannin, Wortmannolone
T. macrosporus	Duclauxins, Mitorubrins
T. stipitatus	Botryodiploidin, Duclauxin, Stipitatic acids, Talaromycin A and B
T. trachyspermus	Glauconic acid, Spiculisporic acid

TABLE 4. (Continued)

Talaromyces wortmannii	Chrysophanol, Flavomannin, Mitorubrins, Rugulosin, Skyrin, Vermiculine, Wortmin
Deuteromycotina:	
Alternaria alternata	Alternariols, Altertoxins, Tenuazonic acid
A. tenuissima	Alternariols, Tenuazonic acid
Aspergillus aculeatus	Aculeasins, Neoxaline, Secalonic acid D
A. caespitosus	Fumitremorgins, Verrucologen
A. candidus	Candidulin, Terphenyllin, Xanthoascin
A. carneus	Citrinin
A. clavatus	Ascladiol, Clavatol, Cytochalasin E, Kojic acid (?), Patulin, Tryptoquivalins, Xanthocillins (?)
A. flavus	Aflatoxin B_1 and B_2, Aflatrem, Aflavinins, Aspergillic acids, Cyclopiazonic acid, Kojic acid, 3-Nitropropionic acid, Paspalinins
A. fumigatus	Fumigaclavins, Fumagillin, Fumigatin, Fumitoxins, Fumitremorgin A and C, Gliotoxin, Spinulosin, Tryptoquivalins, Verrucologen
A. niger	Malformins, "Napthoquinones," Nigragillin
A. nomius	Aflatoxin B_1, B_2. G_1, G_2, Aspergillic acids
A. ochraceus	Emodin, Kojic acid (?), Neoaspergillic acids, Ochratoxins, Penicillic acid, Secalonic acid A, Viomellein, Xanthomegnin
A. oryzae	Cyclopiazonic acid, Kojic acid, 3-Nitropropionic acid
A. parasiticus	Aflatoxin B_1, B_2, G_1, G_2, Aflavinine, Aspergillic acids, Kojic acid
A. sydowii	Nidulotoxin, Sterigmatocystin (?), Griseofulvin (?)
A. terreus	Citreoviridin, Citrinin, Gliotoxin (?), Mevinolin, Patulin, Quadrone, Terrein, Terreic acid, Terretonin, Territrems, Terredionol, Terramide A
A. ustus	Austamid, Austdiol, Austins, Austocystine, Kojic acid (?), Xanthocillin X (?)
A. versicolor	Nidulotoxin, Sterigmatocystin
A. wentii	Emodin, Kojic acid, 3-Nitropropionic acid, Wentilacton, Physicon

TABLE 4. (Continued)

Species	Mycotoxins
Cladosporium herbarum	Epi- and fagi-cladosporic acid
Epicoccum nigrum	Flavipin
Fusarium acuminatum	"Butenolide," Enniatins, Moniliformin, Trichothecenes type A[a]
F. anthophilum	Moniliformin
F. avenaceum	Lateropyrone (= antibiotic Y), Enniatins, Fusarin C, Moniliformin
F. chlamydosporus	Moniliformin
F. cerealis (= *F. crookwellense*)	Butenolide, Culmorin, Fusarin C, Sambucoin, Trichothecenes type B,[b] Zearalenones[c]
F. culmorum	2-Acetylquinazolin-4(3H)-one, Aurofusarins,[d] Butenolide, Culmorin, Cyclonerotriol, Fusarin C, Sambucinol, Sambucoin, Trichothecenes type B, Zearalenones
F. equiseti	Equisetin, Fusarochromanone, Trichothecenes type A, Zearalenones
F. graminearum	4-Acetamido-2-butenoic acid, Aurofusarins, Butenolide, Culmorin, Fusarin C, Sambucoin, Trichothecenes type B, Zearalenones
F. larvarum	Fusarentins, Melleins, Monocerin
F. oxysporum	Bikaverins, Enniatins, Fusaric acid, Lycomarasmin, Moniliformin, Oxysporone
F. pallidoroseum (= *F. semitectum*)	Moniliformin, Zearalenones
F. poae	"Butenolide," Fusarin C, Trichothecenes type A
F. proliferatum	Moniliformin
F. sacchari	Moniliformin
F. sambucinum	"Butenolide," Enniatins, Fusarin, Sambucinol, Sambucoin, Trichothecenes type A
F. solani	Fusaric acid, Naphthoquinones[e]
F. sporotrichioides	"Butenolide," Fusarin C, Trichothecenes type A, Zearalenones
F. tricinctum	"Butenolide," Fusarin C
F. verticilloides (= *F. moniliforme*)	Bikaverins, Cyclonerodiol, Fumonisins, Fusarin C, Fusariocins, Gibberellins, Moniliformin, N-Jasmonoyl-isoleucins, Naphthoquinones

TABLE 4. (Continued)

Geotrichum candidum	Agroclavine, Elymochlavine, Ergosine, Lysergic acid
Paecilomyces variotii	Patulin, Variotin
Penicillium aculeatum	Mitorubrins, Penitricins, Vermicellin
P. aethiopicum (= *P. lanosocoeruleum*)	Griseofulvin, Tryptoquivalins, Viridicatumtoxin
P. allahabadense	Mitorubrins
P. atramentosum	Oxaline, Roquefortine C, Rugulovasine A
P. aurantiogriseum var. *aurantiogriseum*	Auranthine, Aurantiamine, Penicillic acid, Puberulic acid, Terrestric acid (few isolates), Verrucofortin, Verrucosidin, Viomellein, Vioxanthin, Viridicatins,[f] Xanthomegnin, Nephrotoxin glycopeptide
P. aurantiogriseum var. *melanoconidium*	Oxaline, Penicillic acid, Penitrem A, Verrucosidin, Viridicatins
P. aurantiogriseum polonicum	Penicillic acid, Verrucofortine, Verrucosidin, Viridicatins
P. aurantiogriseum viridicatum	Brevianamide A, Penicillic acid, Rubrosulphin, Verrucofortine, Viomellen, Viopurpurin, Vioxanthin, Viridamin, Viridicatins, Viridic acid, Xanthomegnin, Xanthoviridicatin D and G
P. bilaii	Gliotoxin
P. brasilianum	Fumitremorgins, Paraherquamide, Penicillic acid, Verrucologen, Viridicatumtoxin
P. brevicompactum	Brevianamide A, Botryodiploidin, Mycophenolic acid, Pebrolides
P. camemberti	Cyclopiazonic acid
P. charlesii	Carolic acids
P. chrysogenum	Emodic acid, Meleagrin, Penicillin, Roquefortine C
P. citreonigrum	Citreoviridin
P. citrinum	Citrinin
P. commune	Cyclopaldic acid, Cyclopiazonic acid, Isofumigaclavine A, Palitantin, Rugulovasine A
P. coprobium	Patulin, Roquefortine C
P. coprophilum	Griseofulvin, Meleagrin, Roquefortine C
P. corylophilum	—
P. cremeogriseum	Brefeldin A, Palitantin

TABLE 4. (Continued)

Species	Mycotoxins
Penicillium crustosum	Penitrem A, Roquefortine C, Terrestric acid, Viridicatins
P. digitatum	Tryptoquivalins
P. echinulatum chemotype I	Viridicatins
P. echinulatum chemotype II	Chaetoglobosin C, Palitantin, Viridicatins
P. expansum	Chaetoglobosin C, Citrinin, Patulin, Roquefortine C
P. funiculosum	Secalonic acid D (traces)
P. glabrum	Asterric acid, Citromycetin
P. glandicola	Meleagrin, Oxaline, Patulin, Penitrem A, Roquefortine C
P. graminicola	Verrucologen
P. griseofulvum	Cyclopiazonic acid, Griseofulvin, Patulin, Roquefortine C
P. hirsutum var. *hirsutum*	Compactin, Roquefortine C, Terrestric acid
P. hirsutum var. *albocoremium*	Citrinin, Meleagrin, Penicillic acid, Roquefortine C, Terrestric acid, Viridicatins
P. hirsutum var. *allii*	Meleagrin, Palitantin, Roquefortine C, Viridicatins
P. hirsutum var. *hordei*	Carolic acid, Roquefortine C, Terrestric acid
P. islandicum	Cyclochlorotine, Emodin, Erythroskyrin, Islanditoxin, Luteoskyrin, Rugulosin, Simatoxin
P. italicum	Deoxybrevianamide E, 5,6-Dihydro-4-methoxy-2H-pyran-2-one
P. janczewskii	Amauromine, Griseofulvin, Penicillic acid, Penitrem A
P. lanosum	Griseofulvin, Kojic acid
P. maniginii	Citreoviridin
P. miczynskii	Citreoviridin
P. minioluteum	Minioluteic acid, Mitorubrins, Secalonic acid D, Spiculisporic acid
P. nalgiovense	Penicillin
P. novae-zeelandiae	Patulin
P. oxalicum	Oxaline, Roquefortine C, Secalonic acid D

TABLE 4. (Continued)

P. paxilli	Paxillin
P. piceum	Emodin, Mitorubrins, Rugulosin
P. pinophilum	Mitorubrins, Vermicellin, Vermiculine
P. piscarium	Brefeldin A, Janthitrems
P. pulvillorum	Penicillic acid
P. purpurogenum	Glauconic acid, Mitorubrins
P. raistrickii	Griseofulvin, Penicillic acid
P. roqueforti chemotype I	Isofumigaclavine A and B, Marcfortines, Mycophenolic acid, PR-toxin, Roquefortine C
P. roqueforti chemotype II	Botryodiploidin, Mycophenolic acid, Patulin, Penicillic acid, Roquefortine C
P. rubrum	Rubratoxins, Rugulovasins
P. rugulosum	Mitorubrins, Rugulosin, Skyrin, Spiculisporic acid
P. simplicissimum (= *P. janthinellum*)	Viomellein, Xanthomegnin
P. smithii	Citreoviridin
P. solitum	Compactin, Viridicatins
P. soppii	Terrein
P. variabile	Rugulosin
P. verrucosum	Citrinin, Ochratoxin A
P. vulpinum	Oxaline, Patulin, Roquefortine C, Viridicatins
P. westlingii	Citrinin
Phomopsis lepstromiformis	Phomopsin
Stachybotrys chartarum	Satratoxins
Trichoderma viride	Trichodermin
Wallemia sebi	Wallemia A and B, Walleminol

[a]Type A trichothecenes: Diacetoxyscirpenol, HT-2 toxin, Neosolaniol, T-2 toxin.

[b]Type B trichothecenes: Deoxynivalenol, 3-Acetyldeoxynivalenol, Fusarenone X, Nivalenol.

[c]Zearalenones: Zearalenone, Zearalenol.

[d]Aurofusarins: Aurofusarin, Fuscofusarin, Rubrofusarin.

[e]Naphthoquinones: Fusarubin, Javanicin, Marticin, Solaniol.

[f]Viridicatins: Cyclopenol, Cyclopenin, Viridicatin, Viridicatol, Cyclopeptin, Dehydrocyclopeptin, 3-O-Methylviridicatin.

Source: Refs. 17, 18, 30, 154, 219.

TABLE 5. *Penicillium* Species of Common Occurrence on Different Types of Foods and Feedstuffs Under "Normal" Storage Conditions

Cereals:

P. aurantiogriseum var. *aurantiogriseum*

P. aurantiogriseum var. *melanoconidium*

P. aurantiogriseum var. *polonicum*

P. aurantiogriseum var. *viridicatum*

P. verrucosum chemotype II

P. hirsutum var. *hordei*

(*P. aethiopicum*, *P. citrinum*, *P. islandicum*, and *P. citreonigrum* "replace" *P. verrucosum* and *P. hirsutum* var. *hordei* under tropical conditions)

Possible mycotoxins: Penicillic acid, verrucosidin, verrucofortine, xanthomegnin, viomellein, ochratoxin A, citrinin, roquefortine C, penitrem A, nephrotoxic glycopeptides in colder climates; penicillic acid, verrucosidin, verrucofortine, xanthomegnin, viomellein, citrinin, citreoviridin, cyclochlorotine, islanditoxin, luteoskyrin, rugulosin, griseofulvin, viridicatumtoxin, nephrotoxic glycopeptides [213] in warmer regions.

Nuts and high lipid grains and fruits:

P. crustosum

P. commune

P. solitum

P. echinulatum chemotype II

P. funiculosum

P. oxalicum

P. citrinum

Possible mycotoxins: Penitrem A, terrestric acid, roquefortine C, cyclopiazonic acid, cyclopaldic acid, compactin, chaetoglobosin C, oxaline, secalonic acid D, citrinin.

Fruits and vegetables:

P. expansum

P. crustosum

P. solitum

P. hirsutum var. *hirsutum*

P. hirsutum var. *allii*

P. hirsutum var. *albocoremium*

P. italicum

P. digitatum

P. glabrum

P. thomii

Possible mycotoxins: Patulin, terrestric acid, penitrem A, compactins, oxaline, roquefortine C, citrinin, citromycetin.

Meat and eggs:

P. commune

P. crustosum

P. aurantiogriseum var. *polonicum*

P. verrucosum

P. chrysogenum

P. glabrum

P. variabile

P. corylophilum

P. roqueforti

P. expansum

TABLE 5. (Continued)

Possible mycotoxins: Cycloplazonic acid, penitrem A, terrestric acid, roquefortine C, rugulovasine, ochratoxin A, cyclopaldic acid, citromycetin, rugulosin, botrodiploidin, patulin, citrinin.

Fish (temperate regions):

P. verrucosum chemotype I

P. solitum

Possible mycotoxins: Ochratoxin A, compactin.

Fats, oils, margarines:

P. echinulatum chemotype I	*P. solitum*
P. commune	*P. spinulosum*

Possible mycotoxins: Cyclopiazonic acid, cyclopaldic acid, compactin.

Cheese:

P. commune (strongly dominating)	*P. roqueforti*
P. solitum	*P. verrucosum* chemotype I

Possible mycotoxins: Cyclopiazonic acid, cyclopaldic acid, compactin, botryodiploidin, ochratoxin A, patulin.

Source: Refs. 18, 19, 154.

and mycotoxin production. Thus, all fungal strains received as *P. cyclopium* from nuts proved to be *P. crustosum*, while all isolates called *P. cyclopium* reported from cereals are *P. aurantiogriseum* or *P. verrucosum* [17,30]. Correct identification of fungi in foods is very important. Even though several mycotoxins are known to be produced by *P. aurantiogriseum*, it has recently been shown that isolates of this species produce a new class of nephrotoxic mycotoxins [213]. Because standards of these nephrotoxins are not yet available, the species name is the only indication that certain foods could contain these mycotoxins.

V. CONCLUSIONS

The occurrence of molds in foods and feeds is strongly dependent on environmental factors such as temperature, water activity, pH, and redox potential, but more specific knowledge of the growth limits of individual widespread species of filamentous fungi and their interaction is still quite meager. Knowledge of the resistance of molds to processing conditions such as pasteurization and preservation is much more detailed, including data on potential mycotoxin production. Following recent advances in fungal taxonomy and mycotoxin identification, it has been shown that even though species of *Penicillium* and *Aspergillus* often can be regarded as saprophytic organisms, the mycoflora of different types of foods (cereals, nuts, meat, cheeses, spices, etc.) contain specific associated molds provided the food is within normal temperature and water activity limits. Since the mycotoxin profiles of fungal species are specific, it follows that each food or feed item has specific mycotoxin profiles under certain environmental conditions.

Much information concerning the influence of environmental factors on single species of important filamentous fungi has been accumulated, but data for many other species are lacking. It is important that species known to be specific for different types of foods and feedstuffs be tested together and individually on their substrates under the influence of several environmental factors, both concerning germination, growth, and mycotoxin production, because these environmental factors may strongly interact [214-218]. Thus, we do not know exactly why *P. aurantiogriseum* and *P. verrucosum* are so dominant on cereals and how they compete and produce their toxins under realistic conditions, or which factors make *P. commune* dominant on cheese or *P. crustosum* dominant on nuts. Research in these important areas will help greatly to improve the quality of the foods containing these organisms and might considerably reduce the number of necessary mycotoxin analyses in different products.

ACKNOWLEDGMENT

The authors thank NATO Scientific Affairs Division for a research grant for international collaboration.

REFERENCES

1. King, A. D., Pitt, J. I., Beuchat, L. R., and Corry, J. E. L. (eds.). *Methods for the Mycological Examination of Foods*, Plenum Press, New York, 1986.
2. Sauer, D. B. Effects of fungal deterioration on grain: Nutritional value, toxicity, germination. *Intl. J. Food Microbiol.*, *7*:267-275, 19 .
3. Betina, V. (ed.). *Mycotoxins, Production, Isolation, Separation and Purification*, Elsevier, New York, 1984.
4. Northolt, M. D., and Soentoro, P. S. S. Fungal growth on foodstuffs related to mycotoxin contamination. In *Introduction to Food-borne Fungi*, 3rd ed. (Samson, R. A., and van Reenen-Hoekstra, E. S., eds.), Centraalbureau voor Schimmelcultures, Baarn, 1988, pp. 231-238.
5. Beuchat, L. R. (ed.). *Food and Beverage Mycology*, 2nd ed., AVI, New York, 1987.
6. Northolt, M. D., and Bullerman, L. B. Prevention of mold growth and toxin production through control of environmental conditions. *J. Food Prot.*, *45*:519-526, 1982.
7. Bullerman, L. B., Schroeder, L. L., and Park, K.-Y. Formation and control of mycotoxins in food. *J. Food Prot.*, *47*:637-646, 1984.
8. Ray, L. L., and Bullerman, L. B. Preventing growth of potentially toxic molds using antifungal agents. *J. Food Prot.*, *45*:953-963, 1982.

9. Lillehoj, E. B., and Elling, F. Environmental conditions that facilitate ochratoxin contamination of agricultural commodities. *Acta Agric. Scand.*, *33*:113-128, 1983.
10. Moss, M. O. Conditions and factors influencing mycotoxin formation in the field and during storage of food. *Chem. Ind.*, 533-536, 1984.
11. Jarvis, B. Factors affecting the production of mycotoxins. *J. Appl. Bacteriol.*, *34*:199-213, 1971.
12. Panasenko, V. T. Ecology of microfungi. *Bot. Rev.*, *33*:189-215.
13. Magan, N., and Lacey, J. Ecological determinants of mould growth in stored grain. *Intl. J. Food Microbiol.*, *7*:245-256, 1988.
14. Paster, N., and Bullerman, L. B. Mould spoilage and mycotoxin formation in grains as controlled by physical means. *Intl. J. Food Microbiol.*, *7*:257-266, 1988.
15. Betina, V. *Mycotoxins. Chemical, Biological and Environmental Aspects*. Elsevier, New York, 1989.
16. Cole, R. J., and Cox, R. H. *Handbook of Toxic Fungal Metabolites*. Academic Press, New York, 1981.
17. Frisvad, J. C. Taxonomic approaches to mycotoxin identification. In *Modern Methods in the Analysis and Structural Elucidation of Mycotoxins* (Cole, R. J., ed.), Academic Press, New York, 1986, pp. 415-457.
18. Frisvad, J. C. Fungal species and their specific production of mycotoxins. In *Introduction to Food-borne Fungi*, 3rd ed. (Samson, R. A., and van Reenen-Hoekstra, E. S., eds.), Centraalbureau voor Schimmelcultures, Baarn, 1988, pp. 239-249.
19. Frisvad, J. C., and Filtenborg, O. Specific mycotoxin producing *Penicillium* and *Aspergillus* mycoflora of different foods. Proc. Jpn. Assoc. Mycotoxicol. Suppl. 1:163-166, 1988.
20. Burgess, L. General ecology of the Fusaria. In *Fusarium: Diseases, Biology and Taxonomy* (Nelson, P. E., Toussoun, T. A., and Cook, R. J., eds.), Pennsylvania State University Press, University Park, 1981, pp. 225-235.
21. Marasas, W. F. O., Nelson, P. E., and Toussoun, T. A. *Toxigenic Fusarium Species. Identity and Mycotoxicology*, Pennsylvania State University Press, University Park, 1984.
22. Shepherd, M. J., and Gilbert, J. *Fusarium* mycotoxins in cereals and other stored products. In *Spoilage and Mycotoxins of Cereals and Other Stored Products* (Flannigan, B., ed.), International Biodeterioration 22 (supplement), CAB International, London, 1986, pp. 61-69.
23. Mills, J. T. Ecology of toxigenic *Fusarium* species. *Proc. Jap. Assoc. Mycotoxicol. Suppl.*, *1*:167-170, 1988.
24. Lacey, J. Pre- and post-harvest ecology of fungi. *J. Appl. Bact. Symp. Suppl.*, *18*: 11S-25S, 1989.
25. Thrane, U. *Fusarium* species and their specific profiles of secondary metabolites. In *Fusarium—Mycotoxins, Taxonomy, Pathogenicity* (Chelkowski, J., ed.), Elsevier, New York, 1989, pp. 199-225.
26. Bottalico, A., Logrieco, A., and Visconti, A. *Fusarium* species and their mycotoxins in infected cereals in the field and in stored grains. In *Fusarium—Mycotoxins, Taxonomy and Pathogenicity* (Chelkowski, J., ed.), Elsevier, New York, 1989, pp. 85-119.

27. Pitt, J. I., and Hocking, A. D. *Fungi and Food Spoilage*. Academic Press, New York, 1985.
28. Andrews, S. Dilution plating versus direct plating of various cereal samples. In *Methods for the Mycological Examination of Food* (King, A. D., Pitt, J. I., Beuchat, L. R., and Corry, J. E. L., eds.), Plenum Press, New York, 1986, pp. 40-45.
29. Frisvad, J. C., Kristensen, A. B., and Filtenborg, O. Comparison of methods used for surface disinfection of food and feed commodities before mycological analysis. In *Methods for the Mycological Examination of Food* (King, A. D., Pitt, J. I., Beuchat, L. R., and Correy, J. E. L., eds.), Plenum Press, London, 1986, pp. 32-40.
30. Frisvad, J. C. The connection between the penicillia and aspergilli and mycotoxins with special emphasis on misidentified isolates. *Arch. Environ. Contam. Toxicol.*, *18*:452-467, 1989.
31. Apinis, A. E. Thermophilic fungi in certain grasslands. *Mycopath. Mycol. Appl.*, *48*:63-74, 1972.
32. Lacey, J. Colonization of damp organic substrates and spontaneous heating. In *Microbial Growth and Survival in Extreme Environment* (Gould, G. W., and Corry, J. E. L., eds.), Academic Press, London, 1980, pp. 53-70.
33. Lowett, J., and Thompson, R. G. Patulin production by species of *Aspergillus* and *Penicillium* at 1.7, 7.2, and 12.8°C. *J. Food Prot.*, *41*:195-197, 1978.
34. Bacon, C. W., Sweeney, J. G., Robbins, J. D., and Burdick, D. Production of penicillic acid and ochratoxin A on poultry feed by *Aspergillus ochraceus*: Temperature and moisture requirements. *Appl. Microbiol.*, *26*:155-160, 1973.
35. Northolt, M. D., van Egmond, H. P., and Paulsch, W. E. Ochratoxin A production by some fungal species in relation to water activity and temperature. *J. Food Prot.*, *42*:485-490, 1979.
36. Nielsen, P. V., Beuchat, L. R., and Frisvad, J. C. Growth and fumitremorgin production by *Neosartorya fischeri* as affected by temperature, light and water activity. *Appl. Environ. Microbiol.*, *54*: 1504-1510.
37. Stutz, H. K., and Krumperman, P. H. Effect of temperature cycling on the production of aflatoxin by *Aspergillus parasiticus*. *Appl. Environ. Microbiol.*, *32*:327-332, 1976.
38. Park, K. Y., and Bullerman, L. B. Increased aflatoxin production by *Aspergillus parasiticus* under conditions of cycling temperatures. *J. Food Sci.*, *46*:1147-1151, 1981.
39. Park, K. Y., and Bullerman, L. B. Effect of cycling temperatures on aflatoxin production by *Aspergillus parasiticus* and *Aspergillus flavus* in rice and cheddar cheese. *J. Food Sci.*, *48*:889-896, 1983.
40. Corry, J. E. L. Relationships of water activity to fungal growth. In *Food and Beverage Mycology*, 2nd ed. (Beuchat, L. R., ed.), AVI, New York, 1987, pp. 51-99.
41. Beuchat, L. R. Influence of water activity on sporulation, germination, outgrowth, and toxin production. In *Water Activity: Theory and Applications to Food* (Rockland, L. B., and Beuchat, L. R., eds.), Marcel Dekker, New York, 1987, pp. 137-151.

42. Pitt, J. I., and Christian, J. H. B. Water relations of xerophilic fungi isolated from prunes. *Appl. Microbiol.*, *16*:1853-1858, 1968.
43. Ayerst, G. The effect of moisture and temperature on growth and spore germination in some fungi. *J. Stored Prod. Res.*, *5*:127-141, 1969.
44. Mislivec, P. B., and Tuite, J. Temperature and relative humidity requirements of species of *Penicillium* isolated from yellow dent corn kernels. *Mycologia*, *62*:75-88, 1970.
45. Pitt, J. I., and Hocking, A. D. Influence of solute and hydrogen ion concentration on the water relations of some xerophilic fungi. *J. Gen. Microbiol.*, *101*:35-40, 1977.
46. Hocking, A. D., and Pitt, J. I. Water relations of some *Penicillium* species at 25°C. *Trans. Brit. Mycol. Soc.*, *73*:141-145, 1979.
47. Magan, N., and Lacey, J. The effect of water activity, temperature and substrate on interactions between field and storage fungi. *Trans. Brit. Mycol. Soc.*, *82*:83-93, 1984.
48. Magan, N., and Lynch, J. M. Water potential, growth and cellulolysis of fungi involved in decomposition of cereal residues. *J. Gen. Microbiol.*, *132*:1181-1187, 1986.
49. Andrews, S., and Pitt, J. I. Further studies on water relations of xerophilic fungi, including some halophiles. *J. Gen. Microbiol.*, *133*: 233-238, 1987.
50. Wheeler, K. A., and Hocking, A. D. Water relations of *Paecilomyces variotii*, *Eurotium amstelodami*, *Aspergillus candidus*, and *Aspergillus sydowii*, xerophilic fungi isolated from Indonesian dried fish. *Intl. J. Food Microbiol.*, *7*:73-78, 1988.
51. Northolt, M. D., Verhülsdonk, C. A. H., Soentoro, P. S. S., and Paulsch, W. E. Effect of water activity and temperature on aflatoxin production by *Aspergillus parasiticus*. *J. Milk Food Technol.*, *39*: 170-174, 1976.
52. Northolt, M. D., van Egmond, H. P., and Paulsch, W. E. Differences between *Aspergillus flavus* strains in growth and aflatoxin B_1 production in relation to water activity and temperature. *J. Food Prot.*, *40*:778-781, 1977.
53. Northolt, M. D., van Egmond, H. P., and Paulsch, W. E. Patulin production by some fungal species in relation to water activity and temperature. *J. Food Prot.*, *41*:885-890, 1978.
54. Northolt, M. D., van Egmond, H. P., and Paulsch, W. E. Penicillic acid production by some fungal species in relation to water activity and temperature. *J. Food Prot.*, *42*:476-484, 1979.
55. Reiss, J. Mycotoxins in foodstuffs. XII. The influence of the water activity of cakes on the growth of moulds and the formation of mycotoxins. *Z. Lebensm. Unters. Forsch.*, *167*:419-422, 1978.
56. Young, A. B., Davis, N. D., and Diener, U. L. The effect of temperature and moisture on tenuazonic acid production by *Alternaria tenuissima*. *Phytopathology*, *70*:607-610, 1980.
57. Magan, N., and Lacey, J. The effect of water activity and temperature on mycotoxin production by *Alternaria alternata* in culture and on wheat grain. In *Trichothecenes and Other Mycotoxins* (Lacey, J., ed.), John Wiley and Sons, New York, 1985, pp. 243-256.

58. Cuero, R. G., Smith, J. E., and Lacey, J. Interaction of water activity, temperature and substrates on mycotoxin production by *Aspergillus flavus, Penicillium viridicatum* and *Fusarium graminearum* in irradiated grain. *Trans. Brit. Mycol. Soc., 89*:222-226, 1987.
59. Pitt, J. I. *The Genus Penicillium and Its Teleomorphic States Eupenicillium and Talaromyces.* Academic Press, London, 1979.
60. Nielsen, P. V., Beuchat, L. R., and Frisvad, J. C. Growth and fumitremorgin production by *Neosartorya fischeria* as effected by food preservatives and organic acids. *J. Appl. Bacteriol., 66*:197-207, 1989.
61. Farber, J. M., and Sanders, G. W. Production of fusarin C by *Fusarium* spp. *J. Agric. Food Chem., 34*:963-966, 1986.
62. Leistner, L., and Rödel, W. Inhibition of microorganisms in foods by water activity. In *Inhibition and Inactivation of Vegetative Microbes* (Skinner, F. A., and Hugo, W. B., eds.), Academic Press, Longon, 1976, pp. 219-237.
63. Magan, N., and Lacey, J. Effect of temperature and pH on water relations of field and storage fungi. *Trans. Brit. Mycol. Soc., 82*: 71-81, 1984.
64. Miller, D. D., and Golding, N. S. The gas requirements of molds. V. The minimum oxygen requirements for normal growth and for germination of six mold cultures. *J. Dairy Sci., 32*:101-110, 1949.
65. Magan, N., and Lacey, J. Effects of gas composition and water activity on growth of field and storage fungi and their interactions. *Trans. Brit. Mycol. Soc., 82*:305-314, 1984.
66. Tabak, H. H., and Cooke, W. B. Growth and metabolism of fungi in an atmosphere of nitrogen. *Mycologia, 60*:115-146, 1968.
67. Tabak, H. H., and Cooke, W. B. The effects of gaseous environments on the growth and metabolism of fungi. *Bot. Rev., 34*:126-252, 1968.
68. Burges, A., and Fenton, E. The effect of carbon dioxide on the growth of certain soil fungi. *Trans. Brit. Mycol. Soc., 36*:104-108, 1953.
69. Calderon, O. H., and Staffeldt, E. E. Influence of high carbon dioxide concentration on selected fungi. *Dev. Ind. Microbiol., 14*: 218-228, 1973.
70. Yackel, W. C., Nelson, A. I., Wei, L. S., and Steinberg, M. P. Effect of controlled atmosphere on growth of mold on synthetic media and fruit. *Appl. Microbiol., 22*:513-516, 1971.
71. Hocking, A. D. Responses of fungi to modified atmospheres. *CSIRO Food Res. Q., 48*:56-65, 1988.
72. Golding, N. S. The gas requirements of molds. III. The effect of various concentrations of carbon dioxide on the growth of *Penicillium roqueforti* (three strains originally isolated from blue veined cheese) in air. *J. Dairy Sci., 23*:891-898, 1940.
73. Golding, N. S. The gas requirements of molds. IV. A preliminary interpretation of the growth rates of four common mould cultures on the basis of absorbed gases. *J. Dairy Sci., 28*:737-750, 1945.
74. Dallyn, H., and Everton, J. R. The xerophilic mould, *Xeromyces bisporus*, as a spoilage organism. *J. Food Technol., 4*:399-403, 1969.

75. Yates, A. R., Seaman, A., and Woodbine, M. Growth of *Byssochlamys nivea* in various carbon dioxide atmospheres. *Can. J. Microbiol.*, *13*: 1120-1123, 1967.
76. Orth, R. Wachstum und Toxinbildung von Patulin—und Sterigmatocystin-bildenden Schimmelpilzen unter kontrollierter Atmosphäre. *Z. Lebensm. Unters. Forsch.*, *160*:359-366, 1976.
77. Rice, S. L. Patulin production by *Byssochlamys* spp. in canned apple juice. *J. Food Sci.*, *45*:485-488, 495, 1980.
78. Shih, C. N., and Marth, E. H. Aflatoxin produced by *Aspergillus parasiticus* when incubated in the presence of different gases. *J. Milk Food Technol.*, *36*:421-425, 1973.
79. Fabbri, A. A., Fanelli, C., Serafini, M., and di Maggio, D. Aflatoxin production on wheat seed stored in air and nitrogen. *Trans. Brit. Mycol. Soc.*, *74*:197-199, 1980.
80. Wilson, D. M., and Jay, E. Influence of modified atmosphere storage on aflatoxin production in high moisture corn. *Appl. Microbiol.*, *29*: 224-228, 1975.
81. Landers, K. E., Davis, N. D., and Diener, U. S. Influence of atmospheric gases on aflatoxin production by *Aspergillus flavus* in peanuts. *Phytopathology*, *57*:1086-1090, 1967.
82. Lillehoj, E. B., Milburn, M. S., and Ciegler, A. Control of *Penicillium martensii* development and penicillic acid production by atmospheric gases and temperature. *Appl. Microbiol.*, *24*:198-201, 1972.
83. Paster, N., Lisker, N., and Chet, I. Ochratoxin A production by *Aspergillus ochraceus* Wilhelm grown under controlled atmosphere. *Appl. Environ. Microbiol.*, *45*:1136-1139, 1983.
84. Buchanan, J., Sommer, N. F., and Fortlage, R. J. Aflatoxin suppression by modified atmospheres containing carbon monoxide. *J. Amer. Hort. Soc.*, *110*:638-641, 1985.
85. Wilson, D. M., and Jay, E. Effect of controlled atmosphere storage on aflatoxin production in high moisture peanuts (groundnuts). *J. Stored Prod. Res.*, *12*:97-100, 1976.
86. Cole, R. J., Dorner, J. W., Cox, R. H., and Raymond, L. W. Two classes of alkaloid mycotoxins produced by *Penicillium crustosum* Thom isolated from contaminated beer. *J. Agric. Food Chem.*, *31*:655-657, 1983.
87. Nielsen, P. V., Beuchat, L. R., and Frisvad, J. C. Influence of atmospheric oxygen content on growth and fumitremorgin production by a heat resistant mold, *Neosartorya fischeri*. *J. Food Sci.*, *54*: 679-682, 685, 1989.
88. Trucksess, M. W., Stoloff, L., and Mislivec, P. B. Effect of temperature, water activity and other toxigenic mold species on growth of *Aspergillus flavus* and aflatoxin production on corn, pinto beans and soybeans. *J. Food Prot.*, *51*:361-363, 1988.
89. Mislivec, P. B., Trucksess, M. W., and Stoloff, L. Effect of other toxigenic mold species on aflatoxin production by *Aspergillus flavus* in sterile broth shaken cultures. *J. Food Prot.*, *51*:449-451, 1988.
90. Cuero, R. G., Lillehoj, E. B., Kwolek, W. F., and Zuber, M. S. Mycoflora and aflatoxin in preharvest maize. In *Trichothecenes and Other Mycotoxins* (Lacey, J., ed.), John Wiley and Sons, Chichester, 1985, pp. 109-117.

91. Cuero, R. G., Smith, J. E., and Lacey, J. Stimulation by *Hyphopichia burtonii* and *Bacillus amyloliquefaciens* of aflatoxin production by *Aspergillus flavus* in irradiated maize and rice grains. *Appl. Environ. Microbiol.*, *53*:1142-1146, 1987.
92. Cuero, R., Smith, J. E., and Lacey, J. Mycotoxin formation by *Aspergillus flavus* and *Fusarium graminearum* in irradiated maize grains in the presence of other fungi. *J. Food Prot.*, *51*:452-456, 1988.
93. Alderman, G. G., Emeh, C. O., and Marth, R. H. Aflatoxin and rubratoxin produced by *Aspergillus parasiticus* and *Penicillium rubrum* when grown independently, associatively or with *P. italicum* or *Lactobacillus plantarum*. *Z. Lebensm. Unters. Forsch.*, *15*:305-311, 1973.
94. Boller, R. A., and Schroeder, H. W. Influence of *Aspergillus chevalieri* on production of aflatoxin in rice by *A. parasiticus*. *Phytopathology*, *63*:1507-1510, 1973.
95. Boller, R. A., and Schroeder, H. W. Influence of *Aspergillus candidus* on production of aflatoxin in rice by *A. parasiticus*. *Phytopathology*, *64*:121-123, 1973.
96. Ashworth, L. J., Schroeder, H. W., and Langley, B. C. Aflatoxins: Environmental factors governing occurrence in Spanish peanuts. *Science*, *148*:1228-1229, 1965.
97. Lillehoj, E. B., Ciegler, A., and Hall, H. H. Aflatoxin B uptake by *Flavobacterium aurantiacum* and resulting toxic effects. *J. Bacteriol.*, *93*:464-468, 1967.
98. Lafont, P., and Lafont, J. Métabolisme de l'aflatoxine B_1 per *Aspergillus candidus* Link. *Ann. Microbiol (Inst. Past.)*, *125B*:451-457, 1974.
99. Wicklow, D. T., Horn, B. W., and Shotwell, O. L. Aflatoxin formation in preharvest maize ears coinoculated with *Aspergillus flavus* and *Aspergillus niger*. *Mycologia*, *79*:679-682, 1987.
100. Wicklow, D. T., Hesseltine, C. W., Shotwell, O. L., and Adams, G. L. Interference competition and aflatoxin levels in corn. *Phytopathology*, *70*:761-764, 1980.
101. Lillehoj, E. B., Ciegler, A., and Hall, H. H. Aflatoxin G uptake by cultures of *Flavobacterium aurantiacum*. *Can. J. Microbiol.*, *13*:624-627, 1967.
102. Ciegler, A., Lillehoj, E. B., Peterson, R. E., and Hall, H. H. Microbial detoxification of aflatoxin. *Appl. Microbiol.*, *14*:934-939, 1966.
103. Draughon, F. A., Chen, S., and Mundt, J. O. Metabiotic association of *Fusarium*, *Alternaria* and *Rhizoctonia* with *Clostridium botulinum* in fresh tomatoes. *J. Food Sci.*, *53*:120-123, 1988.
104. El-Gendy, S. M., and Marth, E. H. Growth of toxigenic and nontoxigenic aspergilli and penicilia at different temperatures and in the presence of lactic acid bacteria. *Arch. Lebensmittelhyg.*, *31*:192-195, 1980.
105. Mills, J. T. Insect-fungus associations influencing seed deterioration. *Phytopathology*, *73*:330-335, 1983.

106. Dunkel, F. The relationship of insects to the deterioration of stored grain by fungi. *Intl. J. Food Microbiol.*, *7*:227-244, 1988.
107. De Tempe, J., and Limonard, T. Seed-fungal-bacterial interactions. *Seed Sci. Technol.*, *1*:203-216, 1973.
108. Dowd, P. F. Synergism of aflatoxin B_1 toxicity with the co-occurring fungal metabolite kojic acid to two caterpillars. *Entom. Exper. Applic.*, *47*:69-71, 1988.
109. Sinha, R. N., Abramson, D., and Mills, J. T. Interrelations among ecological variables in stored cereals and associations with mycotoxin production in the climatic zones of western Canada. *J. Food Prot.*, *49*:608-614, 1986.
110. Tuite, J., Koh-Knox, C., Cantone, F. A., and Bauman, L. F. Effect of physical damage to corn kernels on the development of *Penicillium* species and *Aspergillus glaucus* in storage. *Phytopathology*, *75*:1137-1140, 1985.
111. Perez, R. A., Tuite, J., and Baker, K. Effect of moisture, temperature and storage time on subsequent storability of shelled corn. *Cereal Chem.*, *59*:205-209, 1982.
112. Raper, K. B., and Thom, C. *Manual of the Penicillia*, Williams and Wilkins, Baltimore, 1949.
113. Westerdijk, J. The concept "association" in mycology. *Antonie van Leeuwenhoek*, *15*:187-189, 1949.
114. Lacey, J. Factors affecting fungal colonization of grain. In *Spoilage and Mycotoxins of Cereals and Other Stored Products* (Flannigan, B., ed.), International Biodeterioration 22 (supplement), CAB International, London, 1986, pp. 29-34.
115. Kinderlerer, J. L. Volatile metabolites of filamentous fungi. *J. Appl. Bacteriol. Symp. Suppl.*, *18*:133S-144S, 1989.
116. Hocking, A. D., and Pitt, J. I. Food spoilage fungi. II. Heat-resistant fungi. *CSIRO Food Res. Q.*, *44*:73-82, 1984.
117. Baggerman, W. I., and Samson, R. A. Heat resistance of fungal spores. In *Introduction to Food-borne Fungi*, 3rd ed. (Samson, R. A., and van Reenen-Hoekstra, E. S., eds.), Centraalbureau voor Schimmelcultures, Baarn, 1988, pp. 262-267.
118. Beuchat, L. R., and Rice, S. L. *Byssochlamys* spp. and their importance in processed fruits. *Adv. Food Res.*, *25*:237-288, 1979.
119. Kavanagh, J., Larchet, N., and Stuart, M. Occurrence of a heat resistant species of *Aspergillus* in canned strawberries. *Nature* (London), *198*:1322, 1963.
120. Scott, V. N., and Bernard, D. T. Heat resistance of *Talaromyces flavus* and *Neosartorya fischeri* isolated from commercial fruit juices. *J. Food Prot.*, *50*:18-20, 1987.
121. Van der Spuy, J. E., Matthee, F. N., and Crafford, D. J. A. The heat resistance of moulds *Penicillium vermiculatum* Dangeard and *Penicillium brefeldianum* Dodge in apple juice. *Phytophylactica*, *7*: 105-108, 1975.
122. Beuchat, L. R. Extraordinary heat resistance of *Talaromyces flavus* and *Neosartorya fischeri* ascospores in fruit products. *J. Food Sci.*, *51*:1506-1510, 1986.

123. King, A. D., and Halbrook, W. U. Ascopore heat resistance and control measures for *Talaromyces flavus* isolated from fruit juice concentrate. *J. Food Sci.*, *52*:1252-1254, 1266, 1987.
124. Splittstoesser, D. F., Lammers, J. M., Downing, D. L., and Churey, J. J. Heat resistance of *Eurotium herbariorum*, a xerophilic mould. *J. Food Sci.*, *54*:683-685, 1989.
125. Samson, R. A. Food borne filamentous fungi. *J. Appl. Bacteriol. Symp. Suppl.*, *18*:27S-35S, 1989.
126. McEvoy, I. J., and Stuart, M. R. Temperature tolerance of *Aspergillus fischeri* var. *glaber* in canned strawberries. *Irish J. Agric. Res.*, *9*:59-67, 1970.
127. Okagbue, R. N. Heat-resistant fungi in soil samples from northern Nigeria. *J. Food Prot.*, *52*:59-61, 1989.
128. Pitt, J. I., and Christian, J. H. B. Heat resistance of xerophilic fungi based on microscopical assessment of spore survival. *Appl. Microbiol.*, *20*:682-686, 1970.
129. Samson, R. A., Nielsen, P. V., and Frisvad, J. C. The genus *Neosartorya*: Differentiation by scanning electron microscopy and mycotoxin profiles. In *Modern Concepts in Penicillium and Aspergillus Systematics* (Samson, R. A., and Pitt, J. I., eds.), Plenum Press, New York, 1990, pp. 449-461.
130. Kingsland, G. C. Relationships between temperature and survival of *Aspergillus flavus* Link ex Fries on naturally contaminated maize grain. *J. Stored Prod. Res.*, *22*:29-32, 1986.
131. Boekhout, T., van Oorschot, C. A. N., Stalpers, J. A., Batenburg-van der Vegte, W. H., and Weihman, A. C. M. The taxonomic position of *Chrysosporium xerophilum* and septal morphology in *Chrysosporium, Sporotrichum* and *Disporotrichum*. *Stud. Mycol.* (Baarn), *31*:29-39, 1989.
132. Wheeler, K. A., Hocking, A. D., Pitt, J. I., and Anggawati, A. M. Fungi associated with Indonesian dried fish. *Food Microbiol.*, *3*:351-357, 1986.
133. Mislivec, P. B., Dieter, C. T., and Bruce, V. R. Effect of temperature and relative humidity on spore germination of mycotoxic species of *Aspergillus* and *Penicillium*. *Mycologia*, *67*:1187-1189, 1975.
134. Moreau, M. La mycoflore des bouchons de liege. *Rev. Mycol.*, *42*: 155-189, 1978.
135. Richard, J. L., and Arp, L. H. Natural occurrence of the mycotoxins penitrem A in mouldy cream cheese. *Mycopathologia*, *67*:107-109, 1979.
136. Richard, J. L., Bacchetti, P., and Arp, L. H. Moldy walnut toxicosis in a dog, caused by the mycotoxin, penitrem A. *Mycopathologia*, *76*:55-58, 1981.
137. Hyde, M. B. Airtight storage. In *Storage of Cereal Grains and Their Products* (Christensen, C. M., ed.), American Association of Cereal Chemists, St. Paul, 1974, pp. 383-419.
138. Brecht, P. E. Use of controlled atmospheres to retard deterioration of produce. *Food Technol.*, 45-50, 1980.
139. Sherbal, J. *Controlled Atmosphere Storage of Grains*, Elsevier, Amsterdam, 1980.

140. Clarke, J. H., and Hill, S. T. Mycofloras of moist barley during sealed storage in farm and laboratory silos. *Trans. Brit. Mycol. Soc.*, *77*:557-565, 1981.
141. Ceynowa, J. Mykologische Untersuchungen an luftdicht gelagertem Getreide. Dissertation, Agrarwissenschaftlichen Fakultät der Christian-Albrechts-Universität Kiel, 1986.
142. Cole, R. J., Kirksey, J. W., Dorner, J. W., Wilson, D. M., Johnson, J. C., Neill Johnson, A., Bedell, D. M., Springer, J. P., Chexal, K. K., Clardy, J. C., and Cox, R. H. Mycotoxins produced by *Aspergillus fumigatus* species isolated from molded silage. *J. Agric. Food Chem.*, *25*:826-830, 1977.
143. Escoula, L. Moisissures toxinogène des fourrages ensilés. *Ann. Rech. Vétér.*, *6*:303-310, 315-324, 1975.
143a. Woolford, M. K. A review. The detrimental effects of air on silage. *J. Appl. Bacteriol.*, *68*:101-116, 1990.
143b. Gedek, B., Bauer, J., and Schreiber, H. Zur Mykotoxinbildung Silage-verderbender Schimmelpilze. *Wien. tierärzl. Mschr.*, *68*:299-301, 1981.
143c. Pelhate, J. Mycoflore des mais-fourrages ensilés. *Rev. Mycol.*, *24*:65-95, 1974/1975.
144. Pettersson, H., Göransson, B., Kiessling, K.-H., Tideman, K., and Johansson, T. Aflatoxin in a Swedish grain sample. *Nord. Vet. Med.*, *30*:482-485, 1978.
145. Lebron, C. I., Molins, R. A., Walker, H. W., Kraft, A. A., and Stahr, H. M. Inhibition on mold growth and mycotoxin production in high-moisture corn treated with phosphates. *J. Food Prot.*, *52*:329-336, 1989.
146. De Boer, E. Food preservatives. In *Introduction to Food-borne Fungi*, 3rd ed. (Samson, R. A., and van Reenen-Hoekstra, E. S., eds.), Centraalbureau voor Schimmelcultures, Baarn, 1988, pp. 268-273.
147. Rusul, G., and Marth, E. H. Growth and aflatoxin production by *Aspergillus parasiticus* NRRL 2999 in the presence of potassium benzoate or potassium sorbate and at different initial pH values. *J. Food Prot.*, *50*:820-825, 1987.
148. Chipley, J. R. Sodium benzoate and benzoic acid. In *Antimicrobials in Foods* (Branen, A. L., and Davidson, P. M., eds.), Marcel Dekker, New York, 1983, pp. 11-35.
149. Liewen, M. B., and Marth, E. H. Growth and inhibition of microorganisms in the presence of sorbic acid: A review. *J. Food Prot.*, *48*:364-375, 1985.
150. Marth, E. H., Capp, C. M., Hasenzahl, L., Jackson, H. W., and Hussong, R. V. Degradation of potassium sorbate by *Penicillium* species. *J. Dairy Sci.*, *49*:1197-1205, 1966.
151. Finol, M. L., Marth, E. H., and Lindsay, R. C. Depletion of sorbate from different media during growth of *Penicillium* species. *J. Food Prot.*, *45*:398-404, 1982.
152. Polonelli, L., Morace, G., Rosa, R., Castagnola, M., and Frisvad, J. C. Antigenic characterization of *Penicillium camemberti* and related common cheese contaminants. *Appl. Environ. Microbiol.*, *53*:872-878, 1987.

153. Pitt, J. I., Cruickshank, R. H., and Leistner, L. *Penicillium commune, P. camembertii*, the origin of white cheese moulds, and the production of cyclopiazonic acid. *Food Microbiol.*, *3*:363-371, 1986.
154. Frisvad, J. C., and Filtenborg, O. Terverticillate penicillia: Chemotaxonomy and mycotoxin production. *Mycologia, 81*:837-861, 1989.
155. Hermansen, K., Frisvad, J. C., Emborg, C., and Hansen, J. Cyclopiazonic acid production by submerged cultures of *Penicillium* and *Aspergillus* strains. *FEMS Microbiol. Lett.*, *21*:253-261, 1984.
156. Wei-Yun, T., Liewen, M. B., and Bullerman, L. B. Toxicity and sorbate sensitivity of molds isolated from surplus commodity cheeses. *J. Food Prot.*, *51*:457-462, 1988.
157. Frisvad, J. C. Physiological criteria and mycotoxin production as aids in identification of common asymmetric penicillia. *Appl. Environ. Microbiol.*, *41*:568-579, 1981.
158. Frisvad, J. C. Creatine-sucrose agar, a differential medium for mycotoxin producing terverticillate *Penicillium* species. *Lett. Appl. Microbiol.*, *1*:109-113, 1985.
159. Norregaard, P., Holm, H., and Emborg, C. Organic acids and penicillin production. *Appl. Microbiol. Biotechnol.*, *20*:221-224, 1984.
160. Jensen, E. B., Nielsen, R., and Emborg, C. The influence of acetic acid on penicillin production. *Eur. J. Appl. Microbiol. Biotechnol.*, *13*:29-33, 1981.
161. Kiessling, K.-H., Pettersson, H., Tideman, K., and Andersson, I.-L. A survey of aflatoxin and *Aspergillus flavus/parasiticus* in acid treated Swedish grain. *Swedish J. Agric. Res.*, *12*:63-67, 1982.
162. Holmberg, T., Pettersson, H., Nilsson, N. G., Göransson, B., and Grossman, R. A case of aflatoxicosis in fattening calves caused by aflatoxin formation in inadequate formic acid treated grain. *Zbl. Vet. Med. A*, *30*:656-663, 1983.
163. Gareis, M., Bauer, J., von Montgelas, A., and Gedek, B. Stimulation of aflatoxin B_1 and T-2 toxin production by sorbic acid. *Appl. Environ. Microbiol.*, *47*:416-418, 1984.
164. Shih, C. N., and Marth, E. H. Aflatoxin formation, lipid synthesis, and glucose metabolism by *Aspergillus parasiticus* during incubation with and without agitation. *Biochim. Biophys. Acta*, *338*:286-296, 1974.
165. Uraih, N., and Chipley, J. R. Effects of various acids and salts on growth and aflatoxin production by *Aspergillus flavus* NRRL 3145. *Microbios*, *17*:51-59, 1976.
166. Yousef, A. E., and Marth, E. H. Growth and synthesis of aflatoxin by *Aspergillus parasiticus* in the presence of sorbic acid. *J. Food Prot.*, *44*:736-741, 1981.
167. Al-Hilli, A. L., and Smith, J. E. Influence of propionic acid on growth and aflatoxin production by *Aspergillus flavus*. *FEMS Microbiol. Lett.*, *6*:367-370, 1979.
168. Rusul, G., and Marth, E. H. Growth and aflatoxin production by *Aspergillus parasiticus* in a medium at different pH values and with or without pimaricin. *Z. Lebensm. Unters. Forsch.*, *187*:436-439, 1988.

169. Gourama, H., and Bullerman, L. B. Effects of potassium sorbate and natamycin on growth and penicillic acid production by *Aspergillus ochraceus*. *J. Food Prot.*, *51*:139-144, 1988.
170. Doyle, M. P., and Marth, E. H. Bisulphite degrades aflatoxin: Effect of temperature and concentration of bisulphite. *J. Food Prot.*, *41*:774-780, 1978.
171. Farag, R. S., Daw, Z. Y., Higazy, A., and Rashad, F. M. Effects of some antioxidants on the growth of different fungi in a synthetic medium. *Chem. Mikrobiol. Technol. Lebensm.*, *12*:81-85, 1989.
172. Ashwood-Smith, M. J., and Horne, B. Response of *Aspergillus* and *Penicillium* spores to ultraviolet irradiation at low temperatures. *Photochem. Photobiol.*, *16*:89-92, 1972.
173. Wheeler, M. H. Comparisons of fungal melanin biosynthesis in ascomycetous, imperfect, and basidiomycetous fungi. *Trans. Brit. Mycol. Soc.*, *81*:29-36, 1983.
174. Killebrew, R., Bonner, F. L., Bronce, D., and Grodner, R. M. Irradiation with ^{60}Co gamma rays of hard wheat flour to destroy fungal spores and toxins. *Trans. Am. Nuclear Soc.*, *11*:85-86, 1968.
175. Faizur-Rahman, A. T. M., and Idziak, E. S. Gamma irradiation-recycling of *Aspergillus flavus* and its effect on radiation resistance and toxin production. *Can. Inst. Food Sci. Technol. J.*, *10*:5-8, 1977.
176. Farkas, J. Microbiological safety of irradiated foods. *Intl. J. Food Microbiol.*, *9*:1-15, 1989.
177. Applegate, K. L., and Chipley, J. R. Increased aflatoxin G_1 production by *Aspergillus flavus* via gamma irradiation. *Mycologia*, *65*: 1266-1273, 1973a.
178. Applegate, K. L., and Chipley, J. R. Increased aflatoxin G_1 production by *Aspergillus flavus* via cobalt irradiation. *Poultry Sci.*, *52*:1492-1496, 1973.
179. Applegate, K. L., and Chipley, J. R. Effects of ^{60}Co gamma irradiation on aflatoxin B_1 and B_2 production by *Aspergillus flavus*. *Mycologia*, *66*:435-445, 1974.
180. Applegate, K. L., and Chipley, J. R. Production of ochratoxin A by *Aspergillus ochraceus* NRRL 3174 before and after exposures to ^{60}Co irradiation. *Appl. Environ. Microbiol.*, *31*:349-353, 1976.
181. Priyadarshini, E., and Tulpule, P. G. Aflatoxin production in irradiated foods. *Food Cosmet. Toxicol.*, *14*:293-295, 1976.
182. Priyadarshini, E., and Tulpule, P. G. Effect of graded doses of irradiation on aflatoxin production by *Aspergillus parasiticus* in wheat. *Food Cosmet. Toxicol.*, *17*:505-507, 1979.
183. Schindler, A. F., Abadie, A. N., and Simpson, R. E. Enhanced aflatoxin production by *Aspergillus flavus* and *Aspergillus parasiticus* after gamma irradiation of the spore inoculum. *J. Food Prot.*, *43*:7-9, 1980.
184. Bullerman, L. B., Narnhart, H. M., and Hartung, T. E. Use of gamma-irradiation to prevent aflatoxin production in bread. *J. Food Sci.*, *38*:1238-1240, 1973.
185. Tantaoui-Eleraki, A. Influence de la densité de contamination en spores d'*Aspergillus flavus* sur la production d'aflatoxine B_1 dans les aliments. *Compt. Rend. Seances Acad. Agric. France*, *64*:69, 1978.

186. Sharma, A., Behere, A. G., Padwal-Desai, S. R., and Nadkarni, G. B. Influence of inoculum size of *Aspergillus parasiticus* spores on aflatoxin production. *Appl. Environ. Microbiol.*, *40*:989-993, 1980.
187. Jemmali, M., and Guilbot, A. Influence de l'irradiation gamma des spores d'*A. flavus* sur la production d'aflatoxine B_1. *C.R. Acad. Sc. Paris*, sér. D., 269:2271-2273.
188. Behere, A. G., Sharma, A., Padwaldesai, S. R., and Nadkarni, G. B. Production of aflatoxins during storage of gamma irradiated wheat. *J. Food Sci.*, *13*:1102-1103, 1978.
189. Ogbadu, G. Effect of gamma irradiation on aflatoxin B_1 production by *Aspergillus flavus* growing on some Nigerian foodstuffs. *Microbios*, *27*:19-26, 1980.
190. Bullerman, L. B., and Hartung, T. E. Effect of low level gamma irradiation on growth and patulin production by *Penicillium patulum*. *J. Food Sci.*, *40*:195-196, 1975.
191. Tsai, W. J., Shao, K. P., and Bullerman, L. B. Effects of sorbate and propionate on growth and aflatoxin production on sublethally injured *Aspergillus parasiticus*. *J. Food Sci.*, *49*:86-90, 1984.
192. Häggblom, P., and Unestam, T. Blue light inhibits mycotoxin production and increases total lipids and pigmentation in *Alternaria alternata*. *Appl. Environ. Microbiol.*, *38*:1074-1077, 1979.
193. Joffe, A. Z., and Lisker, N. Effects of light, temperature, and pH values on aflatoxin production in vitro. *Appl. Microbiol.*, *18*:517-518, 1969.
194. Reiss, J. Mycotoxin in foodstuffs. V. The influence of temperature, acidity and light on the formation of aflatoxins and patulin in bread. *Eur. J. Appl. Microbiol.*, *2*:183-190, 1975.
195. Hussein, A. M., Sommer, N. F., and Fortlage, R. J. Suppression of *Aspergillus flavus* in raisins by solar heating during sun drying. *Phytopathology*, *76*:335-338.
196. Bennett, J. W. Mycotoxins, mycotoxicoses, mycotoxicology and mycopathologia. *Mycopathologia*, *100*:3-5, 1987.
197. Wicklow, D. T. Metabolites in the coevolution of fungal chemical defence systems. In *Coevolution of Fungi with Plants and Animals* (Pirozynskii, K. A., and Hawksworth, D. L., eds.), Academic Press, London, 1987, pp. 173-201.
198. Frisvad, J. C., and Thrane, U. Standardized high-performance liquid chromatography of 182 mycotoxins and other fungal metabolites based on alkylphenone retention indices and UV-VIS spectra (diode array detection). *J. Chromatogr.*, *404*:195-214, 1987.
199. Frisvad, J. C., Filtenborg, O., and Thrane, U. Analysis and screening for mycotoxins and other secondary metabolites in fungal cultures by thin-layer chromatography and high-performance liquid chromatography. *Arch. Environ. Contam. Toxicol.*, *18*:331-335, 1989.
200. Cole, R. J. (ed.). *Modern Methods in the Analysis and Structural Elucidation of Mycotoxins*, Academic Press, Orlando, 1986.
201. Chang, T. T., Lay, J. O., and Francel, R. J. Direct analysis of thin-layer chromatography spots by fast atom bombardment mass spectrometry. *Anal. Chem.*, *56*:109-111, 1984.

202. Engel, G. Untersuchungen zur Bildung von Mykotoxinen und deren quantitative Analyse. VI. Citreoviridin. *J. Chromatogr.*, *130*:293-297, 1977.
203. Kurtzman, C. P., Horn, B. W., and Hesseltine, C. W. *Aspergillus nomius*, a new aflatoxin-producing species related to *Aspergillus flavus* and *Aspergillus tamarii*. *Antonie van Leeuwenhoek*, *53*:147-158, 1987.
204. Rehm, H.-J. Mycotoxine in Lebensmitteln. IV. Mitteilung. Aflatoxine verschiedener Pilzarten. *Z. Lebensm. Unters. Forsch.*, *150*: 146-151, 1972.
205. Frank, H. K. Zweifel uber das Vorkommen von Aflatoxin bei der Gettung *Penicillium*. *Z. Lebensm. Unters. Forsch.*, *150*:151-153, 1972.
206. Frisvad, J. C. The use of high-performance liquid chromatography and diode array detection in fungal chemotaxonomy based on profiles of secondary metabolites. *Bot. J. Linn. Soc.*, *99*:81-95, 1989.
207. Frisvad, J. C., and Filtenborg, O. Secondary metabolites as consistent criteria in *Penicillium* taxonomy and a synoptic key to *Penicillium* subgenus *Penicillium*. In *Modern Concepts in Penicillium and Aspergillus Classification* (Samson, R. A., and Pitt, J. I., eds.), Plenum Press, New York, 1990, pp. 373-384.
208. Filtenborg, O., Frisvad, J. C., and Thrane, U. The significance of yeast extract composition on metabolite production in *Penicillium*. *Modern Concepts in Penicillium and Aspergillus Classification* (Samson, R. A., and Pitt, J. I., eds.), Plenum Press, New York, 1990, pp. 433-440.
209. Cuero, R. G., Smith, J. E., and Lacey, J. The influence of gamma irradiation and sodium hypochlorite sterilization on maize microflora and germination. *Food Microbiol.*, *3*:107-113, 1986.
210. Natori, S., Hashimoto, K., and Ueno, Y. *Mycotoxins and Phycotoxins 1988*, Elsevier, Amsterdam, 1989.
211. Marasas, W. F. O., and Nelson, P. E. *Mycotoxicology*, The Pennsylvania State University Press, University Park, 1986.
212. Steyn, P. S., and Vleggaar, R. (eds.). *Mycotoxins and Phycotoxins*, Elsevier, Amsterdam, 1986.
213. Yeulet, S. E., Mantle, P. G., Rudge, M. S., and Greig, J. B. Nephrotoxicity of *Penicillium aurantiogriseum*, a possible factor in the etiology of Balkan Endemic Nephropathy. *Mycopathologia*, *102*: 21-30, 1988.
214. Horn, B. W., and Wicklow, D. T. Factors influencing the inhibition of aflatoxin production in corn by *Aspergillus niger*. *Can. J. Microbiol.*, *29*:1087-1091, 1983.
215. Magan, N., and Lacey, J. Interactions between field and storage fungi on wheat grains. *Trans. Brit. Mycol. Soc.*, *85*:29-37, 1985.
216. Weckbach, L. S., and Marth, E. H. Aflatoxin production by *Aspergillus parasiticus* in a competitive environment. *Mycopathologia*, *62*: 39-45, 1977.
217. Wilson, D. M., Huang, L. H., and Jay, E. Survival of *Aspergillus flavus* and *Fusarium moniliforme* in high moisture corn stored under modified atmosphere. *Appl. Microbiol.*, *30*:592-595, 1975.

218. Frank, H. K., Münzer, R., and Diehl, J. F. Response of toxigenic and non-toxigenic strains of *Aspergillus flavus* to irradiation. *Sabouraudia, 9*:21-26, 1971.

219. Frisvad, J. C., Filtenborg, O., Samson, R. A., and Stolk, A. C. Chemotaxonomy of the genus *Talaromyces*. *Antonie van Leeuwenhoek, 57*:179-189, 1990.

3

XEROPHILIC FUNGI IN INTERMEDIATE AND LOW MOISTURE FOODS

AILSA D. HOCKING *Division of Food Processing, Commonwealth Scientific and Industrial Research Organization, North Ryde, New South Wales, Australia*

I. INTRODUCTION

Food provides a rich habitat for microorganisms, as most foods contain an abundance of nutrients, such as carbohydrate, proteins, and lipids, and growth factors, such as vitamins and minerals. If food is to be stored, microbial growth must be inhibited or prevented, and man has devised a number of methods of food preservation. Preservation by reducing the water activity (a_w) of food is one of the earliest examples of food technology. It may be achieved directly by dehydration, such as sun or oven drying, or by the addition of solutes like sugar and/or salt. Often a combination of these methods is used. In many foods, a_w is the dominant factor governing food stability and the types of microorganisms able to grow in and spoil food. At high a_w (>0.95), bacteria are the dominant flora of most foods, and foods with a_w values below 0.85 are rarely spoiled by bacteria. The exceptions are brines and salted foods, which may be spoiled by moderately and extremely halophilic bacteria. However, bacteria do not appear to have adapted to sugar-rich environments. At a_w values below about 0.90, yeasts and molds take over as the major spoilage organisms, particularly in sugar-rich foods. The molds are by far the most numerous and diverse group of microorganisms found in low a_w habitats.

The interactions between a_w and pH are also extremely important in selecting the dominant microflora of a particular food (Fig. 1). In foods with neutral pH and high a_w, such as meat, bacteria dominate. In high a_w foods with a pH below about 4.0 (for example, fruit juices and yogurt), yeasts are likely to be dominant, although some lactic acid bacteria are equally successful in these conditions. Some filamentous fungi also compete quite well in high a_w, low pH environments.

In the a_w range of intermediate and low moisture foods, yeasts and molds are the dominant spoilage flora. Different types of intermediate moisture foods (IMF) have characteristic mycofloras associated with them. The major factors determining which types of fungi dominate on specific

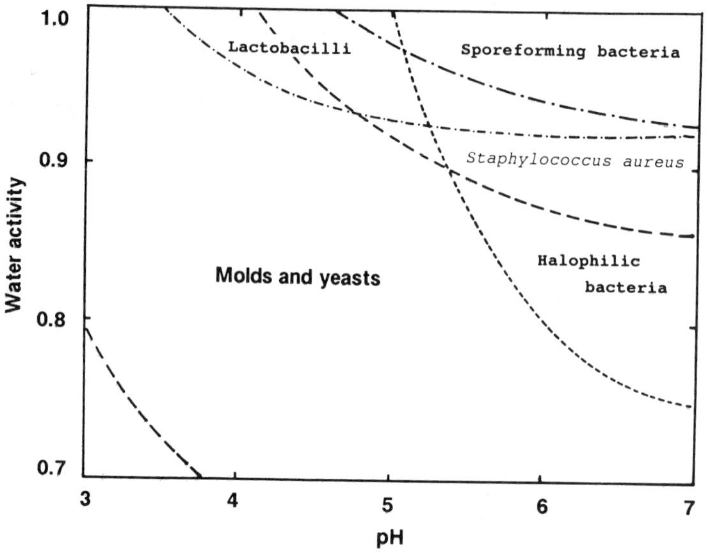

FIGURE 1. A schematic diagram of the combined influence of pH and a_w on microbial growth.

foods are a_w, the kind of solute (sugar or salt), and storage conditions, especially temperature. Other factors such as food additives (spices, flavors, preservatives) and the carbon:nitrogen balance of the foods (i.e., high in carbohydrate or high in protein) influence the composition of the mycoflora.

For the purpose of examining the characteristic mycofloras, reduced a_w foods have been divided into four broad groups, which are dealt with below: cereals, nuts, and spices (Section II), dried fruits, confectionery, and similar high sugar foods (Section III), dried meats (Section IV), and dried seafoods (Section V).

II. MYCOFLORA OF CEREALS, NUTS, AND SPICES

The mycoflora of cereals, nuts, spices, and similar commodities changes during harvest and storage. Immediately after harvest, the dominant fungi on grains are field fungi, such as *Alternaria, Fusarium, Cladosporium, Penicillium*, yeasts, and smuts [1]. However, these fungi cannot grow or cause spoilage unless the commodities are improperly dried before storage or become wet during storage. During storage, the field fungi gradually die out, and storage fungi take their place. The most important genera are *Eurotium* (also known as the *Aspergillus glaucus* series), other *Aspergillus* species, particularly members of the *A. restrictus* series, some of the more xerophilic *Penicillium* species, *Wallemia*, and occasionally *Chrysosporium* [1-4].

A. Wheat, Rice, Barley, and Maize

The mycoflora that develops on stored wheat is influenced particularly by storage temperature and moisture content [5]. In cooler climates, xerophilic *Penicillium* species may be most common [6], but under warmer conditions *Eurotium* and *Aspergillus* species predominate [4,7]. After *Eurotium* species, the most frequently reported Aspergilli from stored wheat are *A. restrictus*, *A. candidus*, *A. versicolor*, *A. ochraceus*, *A. niger*, and *A. sydowii* [6-8]. The Penicillia most frequently encountered have been *P. chrysogenum* [7] and *P. aurantiogriseum* (as *P. cyclopium*) [9]. *Penicillium citrinum* is common in Australian wheat flour [10] and has been reported in wheat flour from the United States [11] and Japan [12], but is not frequently reported from wheat.

The mycoflora of rice reflects the fact that it is grown in warmer climates than most other grain crops. *Eurotium repens*, *E. rubrum*, *E. amstelodami*, and *E. chevalieri* are the most common *Eurotium* species, and *A. restrictus*, *A. candidus*, *A. versicolor*, and *A. ochraceus* the most common Aspergilli. Of the *Penicillium* species identified, *P. rugulosum*, *P. citreonigrum*, *P. canescens*, *P. aurantiogriseum* (as *P. cyclopium*), and *P. citrinum* have been recorded most commonly [13,14].

The mycoflora of stored barley is similar to that of stored wheat. The most common storage species reported in an Egyptian study were *Aspergillus fumigatus*, *A. niger*, *A. flavus*, *A. sydowii*, *Penicillium citrinum*, and *P. funiculosum* [15]. From the cooler climate of Scotland, Flannigan [16] reported that *Eurotium* species were most common, followed by *Penicillium* species and a few species of *Aspergillus*.

Because maize requires high humidity for growth, persistent high moisture conditions during harvest and drying may allow some field fungi to become well established. Preharvest invasion by toxigenic species such as *Aspergillus flavus*, *Fusarium graminearum*, and *F. moniliforme* may lead to production of aflatoxins or trichothecene and other *Fusarium* toxins. However, these fungi are not normally the dominant flora of stored maize. In studies of the mycoflora of U.S. maize, Lichwardt et al. [17] and Barron and Lichwardt [18] showed that *Eurotium* species, especially *E. rubrum*, *E. amstelodami*, and *E. chevalieri*, were the most significant spoilage fungi, together with *Aspergillus restrictus* and *Penicillium* species, *P. aurantiogriseum* and *P. viridicatum* being the most common.

Commodities that are made from grains, like flour, semolina, bran, and related high carbohydrate foods such as sago and tapioca, usually have a similar mycoflora to the grains or substances from which they were derived.

B. Nuts and Spices

Dried nuts are quite susceptible to spoilage. Because nuts have relatively low soluble carbohydrate contents and high oil contents, small increases in moisture content cause significant increase in a_w. Such increases can easily be caused by moisture migration due to uneven temperatures in shipping containers or direct sunlight on one side of a poorly insulated silo.

Many different species of fungi have been isolated from stored nuts, but most are field fungi, and few of them have the capacity to cause spoilage unless storage conditions are poor. In stored peanuts, *Aspergillus flavus* is a recognized problem, but *A. niger* has also been reported as a significant spoilage species [19]. *Penicillium funiculosum* and *P. purpurogenum* are also commonly isolated from stored peanuts [19] but are unlikely to cause spoilage unless the a_w is quite high. In an extensive study on the mycoflora of pecan nuts [20], 119 species from 44 genera were isolated from 37 samples. The most common species, in order of dominance, were *A. niger, A. flavus, E. repens, A. parasiticus, A. ficuum, Rhizopus oryzae*, and *P. expansum*. A study of hazelnuts [21] found that *R. stolonifer* and *P. aurantiogriseum* were the most common fungi.

In Australia, the most common species causing spoilage of nuts and grains during storage and/or shipment are *Eurotium* species, *Aspergillus restrictus*, and *A. penicilloides*. If commodities become extremely wet, then other *Aspergillus* species or Penicillia may cause spoilage, and mycotoxins may be formed if the a_w rises above about 0.85. Xerophilic fungi cause rancidity and other off-flavors in nuts, oilseeds, and grains, as well as decreasing their germinability, but there is little evidence that they form significant mycotoxins.

Spices are frequently heavily contaminated with xerophilic fungi, with figures of up to 10^9 CFU per gram being recorded [22]. The fungi most commonly isolated from spices are *Eurotium* species, *Aspergillus restrictus, A. penicilloides*, other *Aspergillus* species, and *Wallemia sebi* [3]. Some of the more fastidious xerophiles, e.g., *Xeromyces bisporus* and xerophilic *Chrysoporium* species, have been isolated from spices in the author's laboratory, but there are no reports in the literature of spice contamination by these fungi. *Eurotium halophilicum*, a rare species, has recently been reported from cardamom seeds [23]. Growth of xerophilic fungi in spices can lead to loss of flavor and volatile components, production of off-flavors, and clumping in ground spices. Spices carrying high numbers of fungal spores may contaminate the products in which they are used, possibly shortening the shelf life of those products.

III. MYCOFLORA OF FOODS HIGH IN SUGARS

Foods that contain high concentrations of sugars, e.g., dried fruits, confectionery, jams and conserves, jellies, fruit cakes, and fruit concentrates, are susceptible to spoilage by the same range of xerophilic fungi as the previous group of foods. However, they also provide an ideal habitat for some less common, more fastidious xerophilic fungi: *Xeromyces bisporus*, xerophilic *Chrysosporium* species, *Eremascus* species, and the xerophilic yeast *Zygosaccharomyces rouxii*.

A. Dried Fruits, Jams, and Conserves

Many dried fruits (apricots, pears, peaches, bananas) contain high levels of SO_2 to prevent browning, and this usually effectively controls mold growth. However, high moisture apricots (a_w above 0.72) may be spoiled by the xerophilic yeasts *Z. rouxii* and *Z. bailii*, both of which can tolerate relatively high levels of SO_2.

Some kinds of dried fruits, particularly vine fruits (currants, sultanas, raisins) and prunes are produced without SO_2. Pitt and Christian [24] isolated a wide variety of xerophilic molds from Australian dried and high moisture prunes. *Eurotium* species, especially *E. herbariorum*, were most prevalent; *X. bisporus* and xerophilic *Chrysoporium* species were also frequently isolated. Prunes are the only substrate from which the rare fungi *Eremascus albus* and *E. fertilis* have been reported in recent years [24].

In a survey of the mycoflora of dried prunes, Tanaka and Miller [25] found 13 species of yeasts, the most common being *Z. rouxii*, followed by *Saccharomyces mellis*, *Torulopsis magnoliae*, and *T. stellata*. Of 124 mold isolates, the most common were *Eurotium* species (45%), *A. niger* (14.5%), and various *Penicillium* species (33%). *Xeromyces bisporus* was not reported, but may have been missed because of its requirement for low a_w media. However, *X. bisporus* appears to be quite common on dried fruits, having been found in prunes in the United Kingdom [26], in the author's laboratory in Australian muscatel raisins, a_w 0.66, dates from the Middle East, currants, a_w 0.66-0.67 [26], and Chinese dates, a_w 0.72 [27]. *Wallemia sebi* is rarely isolated from dried fruit, but has been reported spoiling dried paw-paw (papaya) [28].

Jams and conserves are usually formulated to about 0.78 to 0.75 a_w. They are hot-filled at temperatures between 65 and 80°C, and the jars are vacuum-sealed. Spoilage rarely occurs in properly sealed, unopened jars. However, spoilage may occur once jars are opened, and the fungi most often responsible are *Eurotium* species and *Z. rouxii*. In jams of higher a_w, xerophilic Penicillia like *P. corylophilum* may cause spoilage [3].

B. Confectionery, Fruit Concentrates, Syrups, Honey, and Brines

There is little in the literature on spoilage of confectionery by fungi. Most confectionery is manufactured with an a_w sufficiently low to prevent growth of common xerophiles like *Eurotium* species. However, soft-filled chocolate may be spoiled by *Z. rouxii*, which produces gas in the fillings, splitting the casing and allowing leakage [3]. Xerophilic *Chrysoporium* species (*C. inops* and *C. xerophilum*) have been isolated from table jellies, coconut [29], and from jelly confections [3]. Coconut can be spoiled by a number of xerophilic molds. *Eurotium* species and *P. citrinum* cause ketonic rancidity, while growth of *Chrysosporium* species causes cheesy butyric off-flavors [30]. *Xeromyces bisporus*, the most xerophilic of all fungi, has been isolated from licorice [3,27], table jelly [26], chocolate sauce, a_w 0.77 [26], and chocolate-coated marshmallow (A. D. Hocking, unpublished).

Low a_w liquid products (between 0.60 and 0.85 a_w) are susceptible to spoilage by xerophilic yeasts. In high sugar environments, the most common species are *Zygosaccharomyces rouxii*, *Z. bisporus*, *Z. bailii*, and some *Torulopsis* species [3,31]. In brines, *Debaryomyces hansenii* (*Torulopsis famata*) is probably the most significant spoilage yeast [3].

IV. DRIED AND MANUFACTURED MEAT PRODUCTS

Many manufactured meat products, such as salamis, hams, and other cured meats, have a relatively high a_w—0.80-0.95. These products rely on a combination of factors for their microbiological stability: presence of nitrite, reduced pH, addition of salt, reduction of Eh by vacuum packaging, and sometimes a heat treatment during manufacture [32]. Dried meats, for example, biltong and Chinese dried meat products (sougan), rely primarily on reduced a_w for their stability. Consequently, the mycofloras of these two classes of meat products are somewhat different.

A. Salamis, Wursts, and Cured Meats

These products are most likely to be spoiled by *Penicillium* species, though *Eurotium*, *Aspergillus*, and other molds also occur [33-35]. The most common *Penicillium* species on salamis imported into Japan were *P. aurantiogriseum* (as *P. cyclopium*), *P. miczynskii*, and *P. viridicatum*, with *Aspergillus*, *Eurotium*, *Cephalosporium*, and *Mucor* species also common [35].

Two extensive studies of the mycoflora of fermented sausages and cured hams in Europe [33] and the United States [34] reported that *Penicillium*, *Aspergillus*, *Eurotium*, and *Cladosporium* were the most prevalent mold genera on both these products. The most common *Penicillium* species were *P. expansum*, *P. janthinellum*, and *P. chrysogenum* in the U.S. study [34] and *P. aurantiogriseum*, *P. palitans* (*P. solitum*), and *P. chrysogenum* on European sausages [33]. Both studies reported *E. rubrum* and *E. repens* as the most common *Eurotium* species, and *A. versicolor* as the most common *Aspergillus* species. Other common Aspergilli were *A. fumigatus* and *A. tamarii* [33], and *A. restrictus*, *A. niger*, and *A. wentii* [34]. Hadlok [36] noted that spices play a major role in the contamination of meat products with *Penicillium* and *Aspergillus*.

B. Dried Meats

Dried meats have received less attention in the microbiological literature than manufactured meats, no doubt because they are intrinsically more stable.

African biltong is a dried meat product which is brined or dry-salted. Pickling mixtures may also include brown sugar, vinegar, pepper, coriander, and other spices. The meat is air-dried for 1-2 weeks after curing [37]. Leistner [32] reported that biltong was stable at <0.77 a_w and pH <5.5, but van der Riet [37] suggests that biltong should be 0.68 a_w (24% moisture) or less to prevent mold growth.

In a study of the mycoflora of 20 commercial biltong samples [38], *Eurotium*, *Aspergillus*, and *Penicillium* were the most common genera, with the significant spoilage species being *E. amstelodami*, *E. chevalieri*, *E. repens*, *E. rubrum*, *A. versicolor*, and *A. sydowii*. The most frequently isolated yeasts were *Debaryomyces hansenii*, *Candida zeylanoides*, and *Trichosporon cutaneum* [38]. Chinese intermediate moisture meat products may be spoiled by *Eurotium* species if the a_w is greater than 0.69 [32].

V. DRIED SEAFOOD PRODUCTS

Fish and other seafoods constitute a major source of protein in the diet of many people in tropical countries. Salting and drying is the most common and cheapest method of fish preservation, but in the humid conditions of the tropics, the a_w during drying and of the finished products is often not low enough to prevent mold spoilage. As with dried meat products, the fungi most commonly reported from dried seafoods are *Eurotium* species, followed by *Aspergillus* and *Penicillium* [8,39-45]. *Scopulariopsis, Cladosporium, Wallemia, Mucor,* and *Acremonium* species have also been reported, but in lower numbers.

The most common *Eurotium* species are *E. rubrum, E. repens, E. amstelodami,* and *E. chevalieri,* while *A. restrictus, A. niger, A. versicolor* series, and *A. flavus* are the most common Aspergilli encountered. In one study, *Aspergillus wentii* was reported from nearly 15% of samples of Indonesian dried fish [44], but has rarely been reported elsewhere. *Penicillium* species appear to play only a minor role in spoilage of dried seafoods in tropical areas, although they are frequently isolated. Penicillia are more important in temperate climates. In a study of the mycoflora of dried shrimps in the United States [45] almost half the fungi isolated were *Penicillium* species, and Penicillia were reported as comprising over 20% of the mycoflora of dried sardines in a Tokyo market [39].

In a recent study on the mycoflora of Indonesian dried fish and seafoods [44], the principal fungus isolated was a previously undescribed species, *Polypaecilum pisce* [46]. It was isolated from nearly half of the samples examined, in some cases covering the fish in a powdery white growth. In view of its prevalence on these samples, it is remarkable that this species has not been described before. The halophilic fungus *Basipetospora halophila* (*Scopulariopsis halophilica*) [46] was also isolated during the survey of molds on Indonesian dried seafoods, though it was uncommon [44]. *B. halophila* is a rare fungus, and all reported isolates have come from dried fish or seaweed [3].

Wallemia sebi was once regarded as the principal fungus spoiling dried and salt fish [47] on which it is known as "dun" mold. However, it is rare on tropical fish, and most reports of *W. sebi* on dried fish products have come from more temperate regions [39,47].

VI. FACTORS AFFECTING THE MYCOFLORA OF LOW a_w FOODS

Water activity is no doubt the principal factor that selects for a specific fungal flora on a particular foodstuff. Many other factors also exert a selective pressure, especially solute type (i.e., sugar or salt), and storage temperature. Numerous studies on the effects of a_w, solute, and temperature on the growth rates of spoilage fungi [48-56] indicated clearly that some fungi compete better in salty environments, while others are better adapted to high sugar environments.

Figures 2 and 3, derived from data from Wheeler et al. [53-55], show growth rates for five fungi on salt-based media at 20°C (Fig. 2) and 30°C (Fig. 3). Two of these fungi, *Polypaecilum pisce* and *Basipetospora halophila* are halphilic [48,55]. *Wallemia sebi* was once considered to be halophilic because it was isolated from salt fish [47], but it actually grows faster in sugary environments [52,53].

The competitiveness of a particular species in a particular environment relies very much on its growth rate and its ability to reproduce itself. For example, at 20°C in a salty environment at high a_w, *Aspergillus flavus*

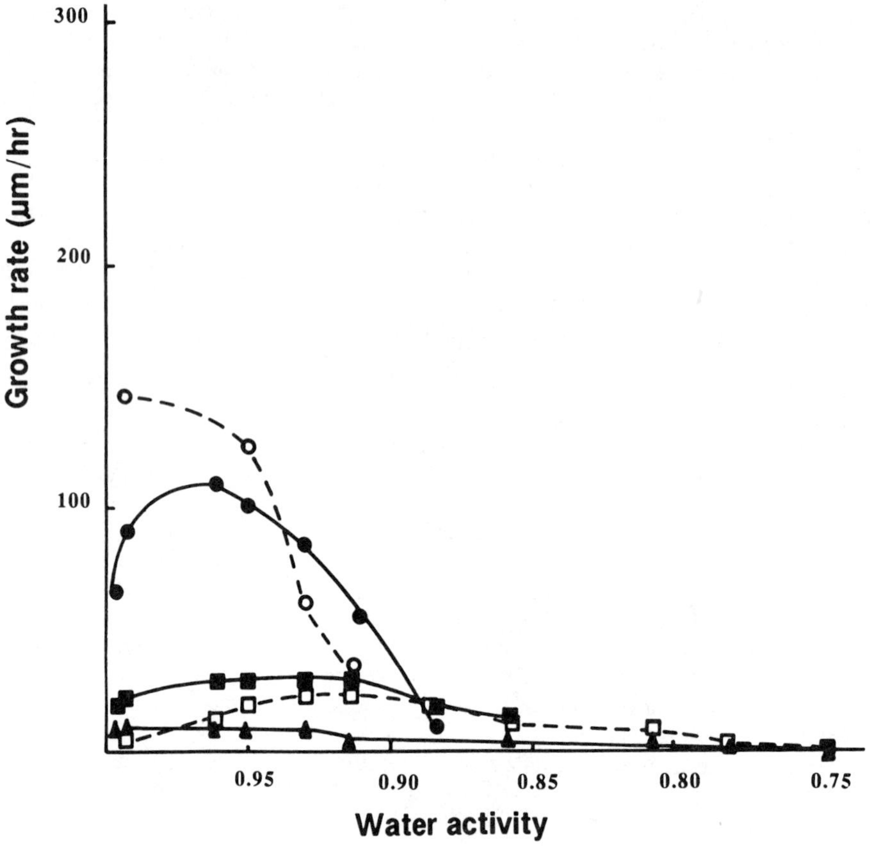

FIGURE 2. Comparative growth rates of five fungi on NaCl-based media at 20°C. (•) *Eurotium rubrum;* (○) *Aspergillus flavus;* (▪) *Polypaecilum pisce;* (□) *Basipetospora halophila;* (▲) *Wallemia sebi.*

will outgrow the other four fungi illustrated (Fig. 2) down to about 0.925 a_w. Between 0.925 and 0.89 a_w, *E. rubrum* has the fastest growth rate. Below 0.89 a_w, the two halophilic fungi and *W. sebi* grow more rapidly than the *Aspergillus* or *Eurotium* species. Despite the fact that *W. sebi* always grows slowly, it competes well because it produces hugh numbers of very small conidia, and it produces them very quickly, sometimes within 48 hours of germination [57]. *Polypaecilum pisce* competes relatively poorly at 20°C, only growing down to 0.86 a_w, while *Basipetospora halophila* and *W. sebi* both grow down to 0.75 a_w, i.e., saturated NaCl.

At 30°C (Fig. 3), *Aspergillus flavus* again dominates at the higher a_w values, down to 0.92–0.91 a_w, with *Eurotium rubrum* competing well around 0.93–0.90 a_w. *Polypaecilum pisce* is the most competitive species from about 0.90 down to 0.75 a_w, with *B. halophila* also growing well below 0.80 a_w. The growth data illustrated in Figures 2 and 3 provide evidence supporting the observation that *P. pisce* is often the dominant fungus on dried salted

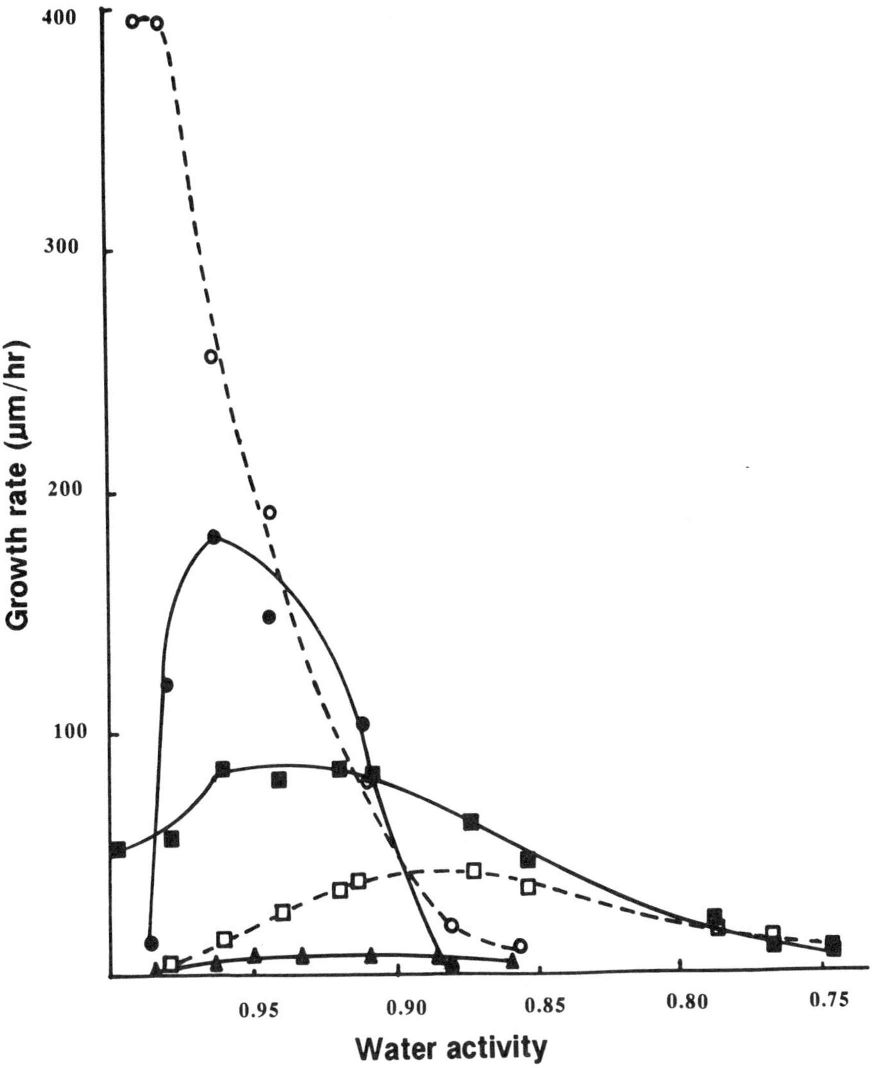

FIGURE 3. Comparative growth rates of five gungi on NaCl-based media at 30°C. (•) *Eurotium rubrum;* (○) *Aspergillus flavus;* (■) *Polypaecilum pisce;* (□) *Basipetospora halophila;* (▲) *Wallemia sebi.*

fish from the tropics, while *W. sebi* is rarely isolated from such samples [44]. Dried fish sampled in the Indonesian study were between 0.79 and 0.65 a_W [44], but during drying the a_W would be between 0.90 and 0.75 for some time: ideal conditions for growth of fungi like *P. pisce* and *B. halophila*. *Polypaecilum pisce* can grow down to 0.705 a_W on glucose-fructose-based media at 30°C [55], and the growth curve in Figure 3 implies that this species is capable of growth below 0.75 a_W in salty conditions as well.

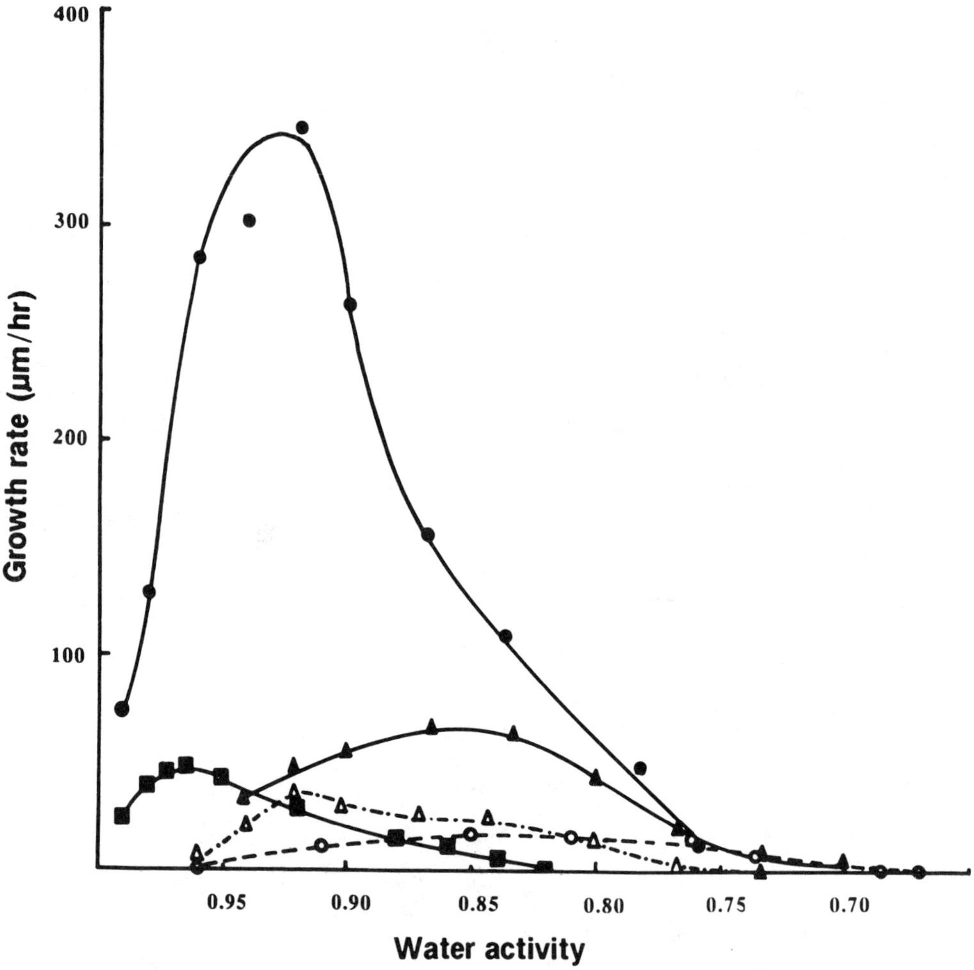

FIGURE 4. Comparative growth rates of five fungi on glucose-fructose-based media at 25°C. (●) *Eurotium rubrum;* (○) *E. halophilicum;* (■) *Polypaecilum pisce;* (▲) *Xeromyces bisporus;* (△) *Chrysosporium fastidium.*

Substrates with a high sugar content have a different mycoflora from salty foods. Figure 4 shows the comparative growth rates for five fungi on glucose-fructose-based media at 25°C: *Eurotium rubrum, E. halophilicum, P. pisce, Xeromyces bisporus,* and *Chrysosporium fastidium* (data derived from Refs. 48, 52, 55). *Eurotium rubrum* is a representative *Eurotium* species, growing extremely rapidly on sugary substrates, with an optimum between 0.98 and 0.85 a_w. Below about 0.77 a_w, growth rates become slower than those of the fastidious extreme xerophiles. *Polypaecilum pisce* competes very poorly with *Eurotium* in sugary environments at 25°C. The three obligate xerophiles illustrated, *X. bisporus, E. halophilicum,* and *C. fastidium,* compete poorly at the higher a_w values, but below about 0.77

a_w, these fungi will outgrow less extreme xerophiles, such as the common *Eurotium* species.

Eurotium halophilicum is exceptionally and obligately xerophilic: Its water relations were studied by Andrews and Pitt [48] (as *Eurotium* species, FRR 2471, subsequently identified as *E. halophilicum* [23]). *Eurotium halophilicum* is unable to grow above 0.94 a_w and has a very slow growth rate even under optimum conditions. It is also misnamed: it is xerophilic rather than halophilic, growing faster and at a lower a_w on glucose-fructose- than on salt-based media [48].

Xeromyces bisporus, the most xerophilic of all known fungi, has been reported as germinating at 0.605 a_w, and as able to produce ascospores at 0.67 and aleurioconidia at 0.66 a_w [24]. It is probably the most common of the extremely xerophilic fungi.

No doubt other factors are important in determining which fungi are most likely to colonize a particular habitat or foodstuff. The carbon-nitrogen ratio, the pH, presence of preservatives, type of packaging and gaseous atmosphere will all influence the mycoflora. However, few if any studies on the importance of these factors, or the interactions between them, have been carried out.

VII. PREVENTION OF SPOILAGE OF LOW a_w FOODS

If the a_w of a food cannot be reduced sufficiently to prevent fungal growth, then additional inhibitory hurdles can be used [32]. Normal heat processing such as pasteurization will kill most fungal spores, so heating at some stage during production will significantly lower the load of fungal spores in a food. However, ascospores are more resistant to heat than conidia. For example, most *Aspergillus* conidia are killed by heat treatment of 60°C for 10 minutes, but some ascospores, particularly those of *E. chevalieri*, can survive 80°C for 10 minutes [58]. The ascospores of *X. bisporus* are also quite heat resistant. Pitt and Christian [48] reported that a small proportion (0.1%) survived 10 minutes at 80°C, while Dallyn and Everton [26] observed that heat treatment of 2 minutes at 90°C, 4 minutes at 85°C, or 9 minutes at 80°C was required to kill 2000 *X. bisporus* ascospores in a medium of 0.9 a_w and pH 5.4. *Xeromyces bisporus* has caused spoilage of Australian fruit cakes, a_w 0.75-0.76, because the ascospores survived the baking process [27].

Preservatives such as potassium sorbate or benzoate can prevent or delay mold spoilage of reduced a_w foods and are added to some types of intermediate moisture foods to extend shelf life. Examples are high moisture prunes, some intermediate moisture meat products, and some baked goods [32,59,60]. A heat process in the presence of preservatives is more effective in killing fungal conidia than heat alone [61], so the combination is an even more effective way of prolonging shelf life of reduced a_w foods. New trends in vacuum packaging, modified atmosphere storage, and oxygen scavenging are also helping to extend the shelf life of many intermediate moisture foods by delaying staling, rancidity, and fungal spoilage [62-64].

VIII. ISOLATING AND ENUMERATING FUNGI FROM LOW a_w FOODS

Because of the wide range of a_w values over which various food spoilage fungi will grow, the choice of a plating medium is very important, as it will determine the types of fungi isolated or enumerated. The characteristics of the foodstuff being examined should be critically assessed, and a medium chosen which reflects those characteristics. High a_w media are suitable for high a_w foods, such as meat, seafood, fruits, vegetables, salads, etc., but for dried and intermediate moisture foods like nuts, grains, spices, and confectionery, high a_w media will not enumerate the significant microflora. It is important to realize that many xerophilic fungi cannot be recovered from foods using high a_w plating media.

If a medium of reduced a_w is used, then type of solute is the next consideration. For foods high in sugar, a glucose- or glycerol-based medium is most suitable, while for salty foods, the use of a reduced a_w medium containing some salt (but not necessarily based entirely on salt) is more appropriate.

A. Plating Techniques

Xerophilic fungi can be isolated from reduced a_w foods using common mycological techniques with modified plating media.

1. Dilution Plating

Many fungi from dried or intermediate moisture foods can be enumerated by dilution plating. The food sample may need to be rehydrated in diluent (such as 0.1% peptone water) for 30 minutes to 1 hour before stomaching or blending, then spread-plated onto a suitable growth medium. Gradual rehydration can aid resuscitation of yeast cells and fungal spores, shortening germination times and increasing subsequent growth rates [3].

2. Direct Plating

For particulate foods like nuts or grains, direct plating can often provide more meaningful results than dilution plating [65]. Direct plating provides an estimate of the extent of infection in a commodity, usually expressed as a percentage. Results are not directly comparable with those obtained by dilution plating. Direct plating is often the best way to study the degree of contamination of a commodity with a specific fungus, such as *Aspergillus flavus*, particularly if a selective medium such as *Aspergillus flavus* and *parasiticus* agar (AFPA, 66) is used. Direct plating is the only satisfactory way of isolating fastidious xerophiles like *Xeromyces bisporus*, *Eremascus* species, xerophilic *Chrysosporium* species, and a few others.

Grains, nuts, and similar commodities are likely to be carrying a high load of surface contaminants and should be surface-disinfected before plating. Soaking in a solution of sodium hypochlorite (0.35-0.4%) for 2 minutes, followed by a rinse of sterile water, effectively removes or kills most contaminant spores adhering to the surface of the sample [67]. This permits detection of hyphae that have penetrated and grown in the commodity.

For isolation of extreme xerophiles from low a_w foods, surface disinfection is unnecessary. Samples such as dried fruits, salted fish, and confectionery can be cut into small pieces and plated directly onto a suitable medium. Alternatively, the surface of the commodity can be sampled by pressing a piece onto the agar, then removing it, leaving an impression. Any spores or mycelium transferred will form colonies in a few days.

B. Moderately Xerophilic Fungi

"Moderately xerophilic" fungi are capable of growth below 0.85 a_w but are not fastidious in their nutrient or a_w requirements. Many members of the common food spoilage genera *Penicillium* and *Aspergillus* (especially the *Aspergillus restrictus* series), *Eurotium* species, *Wallemia sebi*, and perhaps a few others can be considered as "moderately xerophilic" (Fig. 5). These fungi are important in spoilage of a wide range of low moisture foods, especially stored grains, nuts, and spices.

Media of reduced a_w should be used to enumerate moderate xerophiles, as many of them grow slowly and compete poorly at high a_w. However, if the a_w of the medium is reduced sufficiently to allow good growth of *Wallemia, A. penicilloides*, and *A. restrictus*, *Eurotium* species rapidly overgrow these slowly developing species. This problem has been largely overcome with the development of Dichloran 18% Glycerol agar (DG18) [68]. DG18 agar contains 2 ppm dichloran (2,6-dichloro-4-nitroaniline) to inhibit the spreading growth of *Eurotium* colonies and 18% glycerol to reduce the a_w to 0.955. Media based on sugars are sufficiently rich to allow *Eurotium* to overcome the inhibitory effects of dichloran, but glycerol lowers the a_w without reducing the effectiveness of dichloran. DG18 allows slowly growing species like *A. penicilloides* and *W. sebi* to be enumerated in the presence of significant numbers of *Eurotium* colonies. Most species of *Penicillium* and *Aspergillus* also grow well on DG18. DG18 has the added advantage that it encourages the development of the characteristic colored hyphae and yellow cleistothecia of *Eurotium* colonies, allowing some species to be differentiated directly on the enumeration plates.

If a high a_w medium is used to enumerate fungi from a low a_w commodity, the results can be quite misleading. Table 1 shows counts obtained from a range of low a_w samples which were plated onto DRBC agar (0.997 a_w) [69] and DG18 (0.955 a_w). The differences in some instances were several orders of magnitude, and the fungi which grew on the high a_w medium were often of little significance as spoilage fungi [22].

C. Fastidious Extreme Xerophiles

Extreme xerophiles *require* reduced a_w for growth and, in addition, grow poorly on media based on solutes other than sugars. There are relatively few fastidious extreme xerophiles, but they are important causes of spoilage in foods high in sugar. Dried fruits, confectionery, fruit cakes and puddings, chocolate, and spices are susceptible if their a_w is between about 0.75 and 0.65. The species responsible are *Xeromyces bisporus*, xerophilic *Chrysosporium* species (*C. fastidium*, *C. inops*, *C. farinicola*, and *C. xerophilum*), and the two *Eremascus* species, *E. albus* and *E. fertilis*. Of these, the most common are *X. bisporus* and *C. inops*. *Eremascus* species

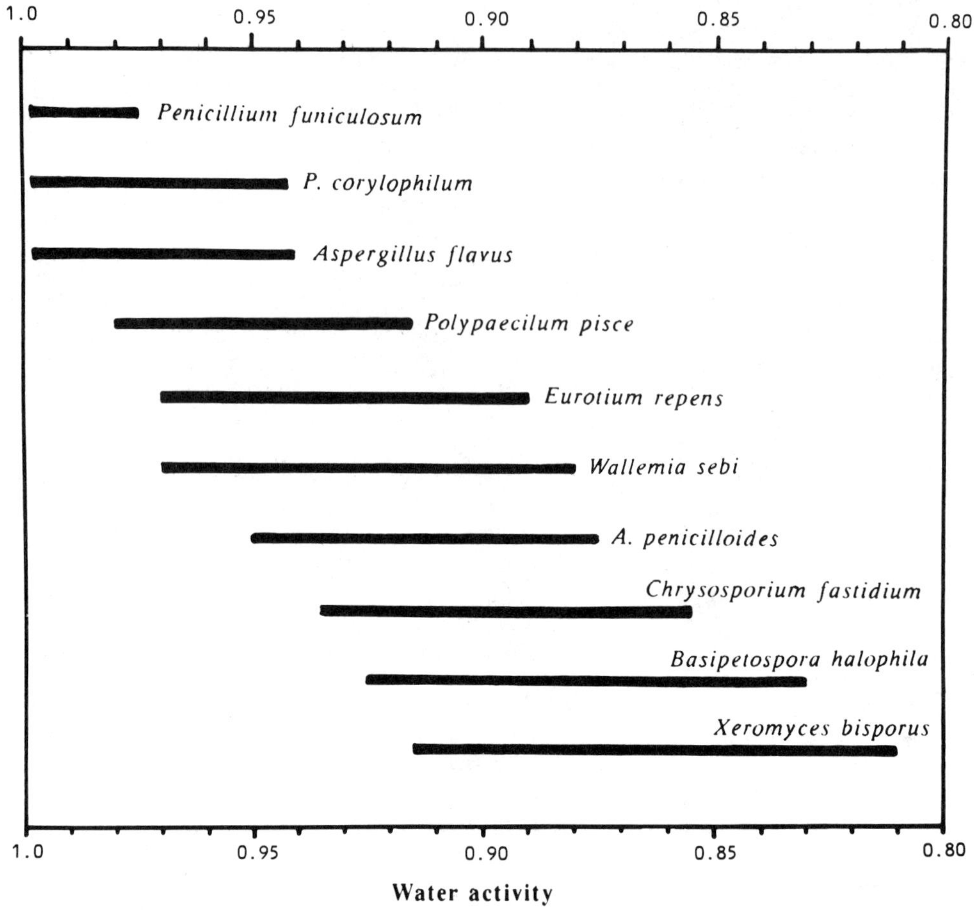

FIGURE 5. Water activity permitting 80% or more of the maximum growth rate of some food spoilage fungi.

are extremely uncommon. Foods with a_w values below 0.65 are generally stable. Although *X. bisporus* is capable of growth to near 0.61 a_w [24], growth is so slow that it would take many months for visible colonies to appear.

Fastidious extreme xerophiles grow extremely poorly, if at all, on conventional high a_w media. Some of the xerophilic *Chrysosporium* species are able to grow weakly on high a_w media, but *X. bisporus* will not grow on any medium above 0.97 a_w [52].

If a low a_w commodity like confectionery or dried fruit shows signs of white mold growth, it is very likely that the fungus responsible is an extreme xerophile. Direct plating of small pieces of the commodity onto a rich, low a_w medium such as Malt Extract Yeast Extract 50% Glucose agar (MY50G, 0.89 a_w) [3] is the best method for isolating these fungi. Alternatively, the food can be examined under a stereomicroscope, and mycelium transferred to MY50G with an inoculating needle. Colonies should develop after 1-3 weeks incubation at 25°C.

TABLE 1. Comparison of Counts Obtained on DRBC and DG18 for Low a_w Commodities

	Log total mold count per gram	
Commodity	DRBC	DG18
Semolina	5.4	6.6
Dried chilies 1	4.2	7.0
Dried chilies 2	4.9	8.2
Dried fish 1	0	4.9
Dried fish 2	5.9	6.4
Dried fish 3	4.1	7.7
Flour	5.3	6.4

Source: Ref. 22.

If *Eurotium* species are also present, isolation of fastidious xerophiles can be difficult, as they grow more slowly than *Eurotium* on MY50G. Under these circumstances it may be necessary to use Malt Extract Yeast Extract 70% Glucose Fructose agar (MY70GF) [3] which is about 0.76 a_w. In MY70GF, equal parts of glucose and fructose are used to prevent crystallization of the medium. Growth of even extreme xerophiles is slow on MY70GF, so plates should be incubated for at least 4 weeks at 25°C in closed containers to prevent drying out. Once growth is apparent, small portions of the colonies should be transferred to MY50G to allow more rapid growth and sporulation.

D. Halophilic Fungi

Two media based on mixtures of salt and glucose have been developed for isolation of the halophilic fungi *Polypaecilum pisce* and *Basipetospora halophila* from salt fish [3]. For *P. pisce*, Malt Extract Yeast Extract 5% Salt 12% Glucose agar (MY5-12) is recommended. *Basipetospora halophila* grows more rapidly on MY10-12, which contains 10% rather than 5% salt. *Wallemia sebi* also grows well on these media. Care should be taken in the preparation of these media, as they are heat sensitive, and gel poorly if overcooked.

E. Yeasts in Low a_w Commodities

Enumerating low numbers of yeasts in highly viscous syrups and concentrates can be difficult, but visible fermentation usually indicates sufficient numbers of yeasts for detection by dilution plating. Diluents should contain 20% sucrose or glucose to minimize osmotic shock.

Various media, all containing high levels of sugars, have been developed for enumeration of yeasts in syrups and fruit concentrates. Tilbury [31] recommends Scarr's Osmophilic Agar [70] as being simple to prepare and easy to use, but points out that the relatively high a_w (0.95) allows growth of some nonxerophilic yeasts, and occasionally bacteria as well. MY50G [3] is probably a better plating medium for xerophilic yeasts.

Restaino et al. [71] showed that the addition of 60% sucrose to Potato Dextrose Agar dramatically improved the recovery of the xerophilic yeast *Zygosaccharomyces rouxii*, from chocolate syrup, particularly during the log phase of growth. Only 1-10% of the *Z. rouxii* cells recovered from PDA/60% sucrose plating medium (a_w 0.92) were enumerated on standard PDA during log phase growth. The use of a 60% sucrose-phosphate buffer diluent in conjunction with the 60% sucrose PDA medium maximized recovery of *Z. rouxii* [68].

If yeasts are suspected to be present in very low numbers and the product is too viscous to be passed through a membrane filter, then a practical method of detection is enrichment in the product itself. This can be done by diluting the product 1:1 with distilled water, and incubating for up to 4 weeks. The dilution will not be sufficient to cause osmotic shock, and the increase in a_w allows any xerophilic yeasts present to grow more rapidly and thus be detected earlier.

A simple presence-absence test has been developed for detecting small numbers of xerophilic yeasts in high sugar products [72]. Samples were homogenized in yeast extract 50% glucose broth, incubated at 30° 2-30 days, and sampled at daily intervals from 2 days. Samples were examined microscopically and 0.03 ml plated onto YEG50 agar, with incubation 5-7 days at 30°C. Using this method, 28 strains of xerophilic yeasts (nearly all *Z. rouxii*) were isolated from 27 spoiled samples. The detection time was 4 days (microscopically) and 7 days (streak culture).

The direct epifluorescence filter (DEFT) technique has also been used to detect low numbers of yeasts in high sugar commodities [73]. Creme fondant samples were preincubated at 30°C for 24 hours, then analyzed by DEFT and plate counts. Pretreatment with trypsin and Triton-X 100 were necessary before filtration through 2 μm Nuclepore membrane filters. Numbers as low as 1 per gram were detected in 25 hours, and 1 per 10 grams in 49 hours.

IX. IDENTIFICATION OF XEROPHILIC FUNGI

It is outside the scope of a chapter such as this to fully describe all xerophilic fungi and methods for their identification, particularly for the most common ones such as *Penicillium* and *Aspergillus* species. For full descriptions and methods see Pitt and Hocking [3], Pitt [74], Raper and Fennell [75], and Klich and Pitt [76]. Set out below is a simplified key for identification of the most common *Eurotium* species, and some of the less common xerophiles. The key is based on those of Pitt and Hocking [3] and assumes that isolates have been grown on low a_w media such as MY50G. The key does not include any *Penicillium* species or most of the Aspergilli, even though many of them grow well on MY50G.

A. Key to Xerophilic Fungi

1. Bright yellow, barely macroscopic spherical bodies (cleistothecia) visible in the aerial mycelium of colonies. *Eurotium* species grow on CY20S agar for identification	2
Yellow cleistothecia not present	5

2.	Ascospores not exceeding 5 µm, with conspicuous ridges	3
	Ascospores often 5.5 µm or more, without conspicuous ridges or flanges	4
3.	Ascospores with rough walls and two prominent irregular longitudinal ridges. Colonies on CY20S bright yellow (from cleistothecia) and green (from conidial heads)	*E. amstelodami*
	Ascospores like pulley wheels, with smooth walls and two prominent logitudinal flanges. Colonies with conspicuous yellow to orange sterile hyphae	*E. chevalieri*
4.	Colonies with yellow or orange sterile hyphae, ascospores smooth walled with barely a trace of a longitudinal furrow	*E. repens*
	Colonies with orange to reddish-brown hyphae, becoming red-brown in age, ascospores with a distinct longitudinal furrow	*E. rubrum* *E. herbariorum*
5.	Colonies showing green conidial colors	6
	Colonies white, brown or black	7
6.	Conidia nearly cylindrical, borne in long chains on *Aspergillus* heads. Heads columnar in age	*A. restrictus*
	Conidia ellipsoidal, not adhering in chains. *Aspergillus* heads radiate in age	*A. penicilloides*
7.	Colonies chocolate brown, very small	*Wallemia sebi*
	Colonies white, pale brown or black	8
8.	Colonies white or pale brown	9
	Colonies black, or with black areas	*Bettsia* (see *Chrysosporium*)
9.	Colonies white, lemon-shaped conidia produced on polyphialides	*Polypaecilum*
	Polyphialides not produced	10
10.	On MY50G, solitary asci produced, containing or releasing mature ascospores within 14 days	*Eremascus*
	Mature asci not evident in culture on MY50G in 14 days	11
11.	Spherical or cylindroidal aleurioconidia or similar spores produced on MY50G in 14 days	12
	No aleurioconidia on MY50G in 14 days; developing cleistothecia containing fat globules present	*Xeromyces*

12. Aleurioconidia on short conidiophores; no intercalary arthroconidia present. Fungus isolated from salty substrate — *Basipetospora*
Intercalary arthroconidia and chlamydoconidia present; aleurioconidia on tiny pedicels. Fungus not isolated from salty substrate — *Chrysosporium*

B. Brief Descriptions of Some Xerophilic Fungi

1. *Eurotium* sp.

All *Eurotium* species are xerophilic and grow poorly on high a_w media. They are normally grown on Czapek Yeast Extract agar with 20% sucrose (CY20S) for identification purposes [3,75,76]. This medium (a_w 0.98) encourages the development of characteristic mycelial colors, while also allowing good development of cleistothecia and ascospores in most species.

The four most common *Eurotium* species, *E. amstelodami*, *E. chevalieri*, *E. repens*, and *E. rubrum*, can be distinguished from each other by the shape and size of their ascospores and the colors of their mycelium. They all produce *Aspergillus* heads (Fig. 6a), bearing spiny or rough conidia (Fig. 6b) on phialides only.

The colonies of *E. amstelodami* are colored bright yellow from the cleistothecia, with an overlay of dark to dull green *Aspergillus* heads. The ascospores are 4.5-5.0 μm long, rough walled, with wide, irregular longitudinal flanges (Fig. 6c).

Eurotium chevalieri colonies are more orange than those of *E. amstelodami* because of the presence of orange hyphae among the yellow cleistothecia, with an overlay of grey-green *Aspergillus* heads. The ascospores are very characteristic, shaped like pulley wheels, with two prominent, parallel longitudinal flanges and smooth walls (Fig. 6d).

Eurotium repens colonies are similar to those of *E. chevalieri* but the ascospores are quite different. *Eurotium repens* produces smooth-walled ascospores, 5.0-5.5 μm long, which are almost egg-shaped. They have no ridges or flanges, and there is usually no more than a trace of a longitudinal furrow (Fig. 6c).

Eurotium rubrum colonies are much redder than any of the other common *Eurotium* species, especially in age. The ascospores are similar to those of *E. repens*, except for a shallow, but distinct, longitudinal furrow (Fig. 6f).

2. *Aspergillus restrictus* Series

The two common species in the *A. restrictus* series, *A. restrictus* and *A. penicilloides*, are both xerophilic, though *A. penicilloides* is more xerophilic than *A. restrictus*. Both species produce phialides only. They can be distinguished from each other by the shapes of their *Aspergillus* heads and their conidia. *Aspergillus restrictus* has columnar heads, producing conidia in long, adherent chains from phialides on the upper half of the vesicle (Fig. 6g). The conidia are usually cylindroidal. The heads of *A. penicilloides* are radiate because the phialides cover three quarters or more of the vesicle. Conidia are ellipsoidal, are produced in shorter chains, and do not adhere to each other (Fig. 6h).

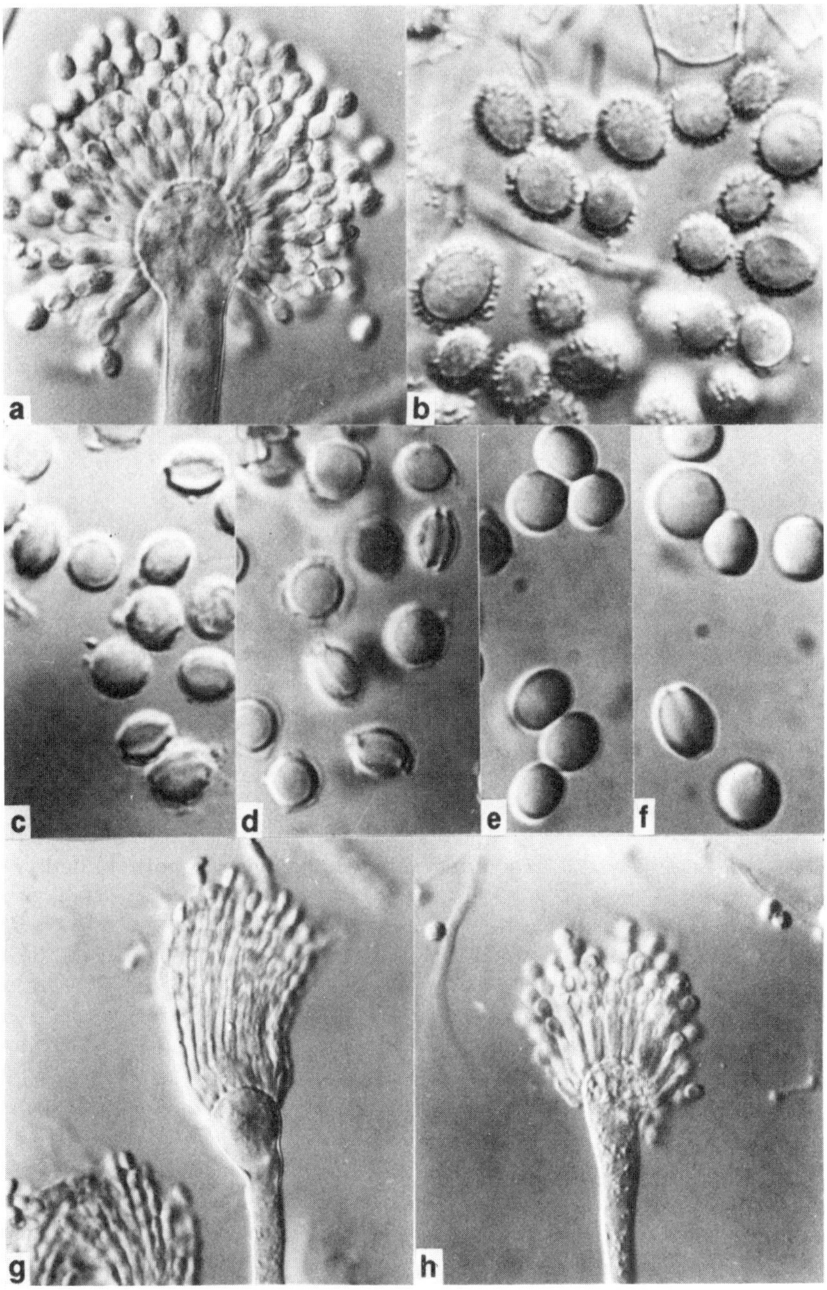

FIGURE 6. *Eurotium* and *Aspergillus restrictus* series: (a) *Aspergillus* head of *Eurotium rubrum* × 600; (b) conidia of *E. rubrum* × 1500; (c)-(f) ascospores of *Eurotium* species × 1500; (c) *E. amstelodami*; (d) *E. chevalieri*; (e) *E. repens*; (f) *E. rubrum*; (g) *Aspergillus restrictus* head, showing long conidial chains × 600; (h) *A. penicilloides* × 600.

3. Xerophilic *Chrysosporium* sp.

The xerophilic *Chrysosporium* species grow as white or pale yellow-brown colonies on MY50G agar. They produce solitary, hyaline, smooth-walled aleurioconidia either directly (sessile) or on small pedicels on the sides of vegetative hyphae. In some species, the vegetative hyphae may differentiate partially or wholly into conidia: intercalary chlamydoconidia, which are thick-walled and nearly spherical, and unswollen, heavy-walled hyphal segments (arthroconidia). One species, *C. farinicola*, sometimes produces a teleomorph classified in the genus *Bettsia*, with black cleistothecia and dark-walled ascospores [3].

Chrysosporium species are differentiated by the size, shape, and proportion of the various types of conidia they produce. *Chrysosporium fastidium* forms pale yellow-brown colonies which produce mainly aleurioconidia (Fig. 7a) with few intercalary arthroconidia or chlamydoconidia. *Chrysosporium inops* produces white colonies, and the conidia are predominantly arthroconidia and chlamydoconidia (Fig. 7b). *Chrysosporium xerophilum* is similar to *C. inops* except that it grows faster (25-30 mm on MY50G after 14 days, compared with 12-20 mm for *C. inops*), produces more aleurioconidia, and, in age, the hyphae differentiate almost entirely into arthroconidia and chlamydoconidia (Fig. 7c).

4. *Eremascus* sp.

The two *Eremascus* species, *E. albus* and *E. fertilis*, are both xerophilic, and both are rare [3]. Colonies are white and floccose; asci are borne singly without any surrounding wall or hyphae (Fig. 7d), and each ascus contains eight smooth-walled subglobose ascospores (Fig. 7e). The species are differentiated by slight differences in their ascus initials.

5. *Xeromyces bisporus*

Xeromyces bisporus is the only member of this genus and is the most xerophilic of all the fungi. On MY50G, *X. bisporus* forms low, sparse, translucent colonies which are either colorless or pale pinkish brown. After 4-6 weeks, "D"-shaped ascospores are formed in pairs in thin-walled, colorless cleistothecia (Fig. 8a-c). Thick-walled aleurioconidia may also be formed (Fig. 8d).

6. *Wallemia sebi*, *Basipetospora halophila*, and *Polypaecilum pisce*

These three fungi are grouped together here because they are associated with salty foods, though *W. sebi* is often found in other low moisture environments as well. *Wallemia sebi*, the only species in the genus, is easily recognized because it forms characteristic small, chocolate brown colonies, 2-5 mm in diameter after 1 week, on most media. It forms small, brown, rough-walled conidia from the ends of short, fertile hyphae (Fig. 9a). Although it is a xerophile, it can also grow on some high a_w media.

Basipetospora halophila and *P. pisce* both form white colonies and grow better on media which contain some salt [46,48,55]. *Basipetospora halophila* forms aleurioconidia in short chains from simple conidiophores which become shorter as successive conidia form (Fig. 9b-d). Chlamydoconidia and arthroconidia are never produced.

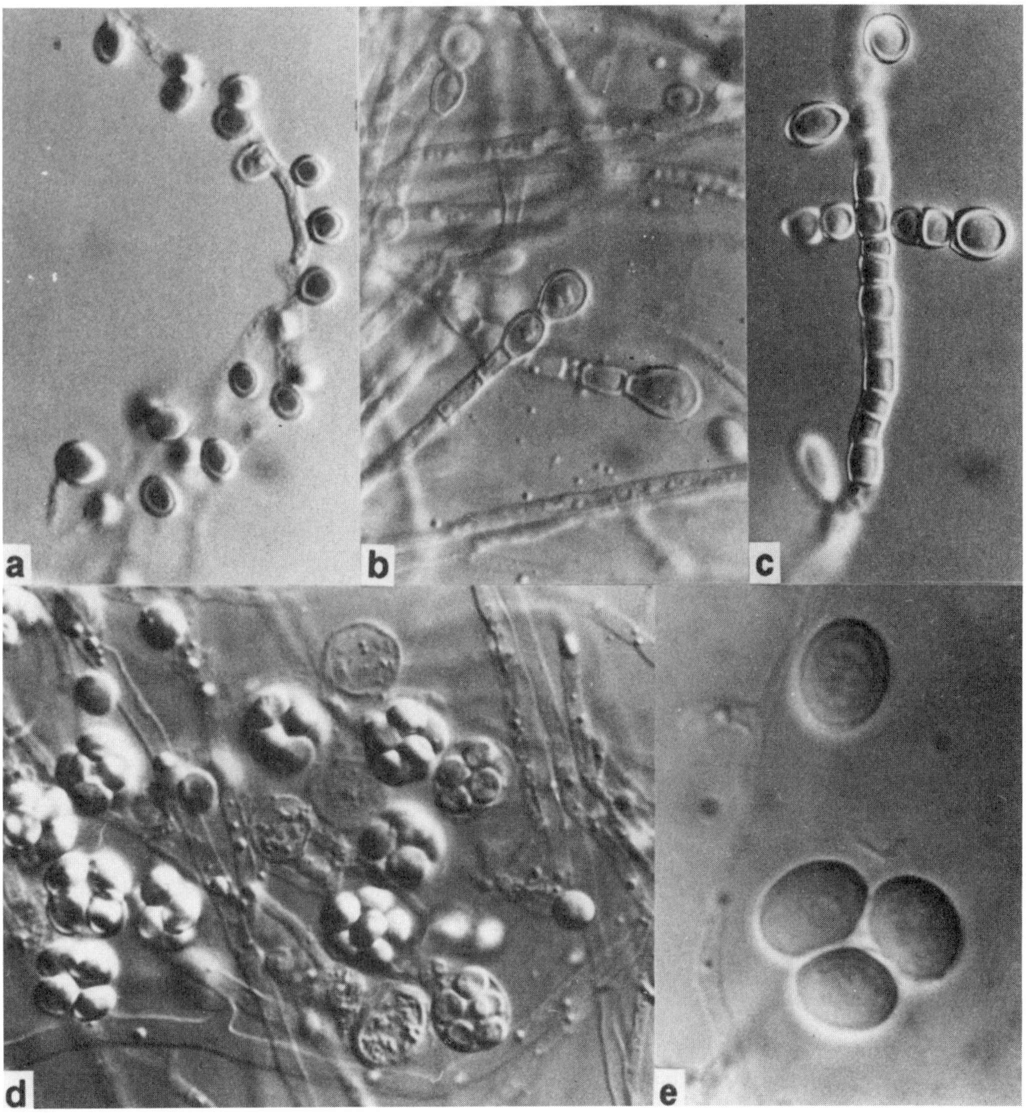

FIGURE 7. *Chrysosporium* and *Eremascus* species: (a) *C. fastidium* aleurioconidia × 750; (b) *C. inops* terminal chlamydoconidia × 750; (c) *C. xerophilum* showing complete differentiation of hyphae into arthroconidia and chlamydoconidia × 750; (d) *Eremascus albus* asci containing immature and mature ascospores × 750; (e) *E. albus* ascospores × 1875.

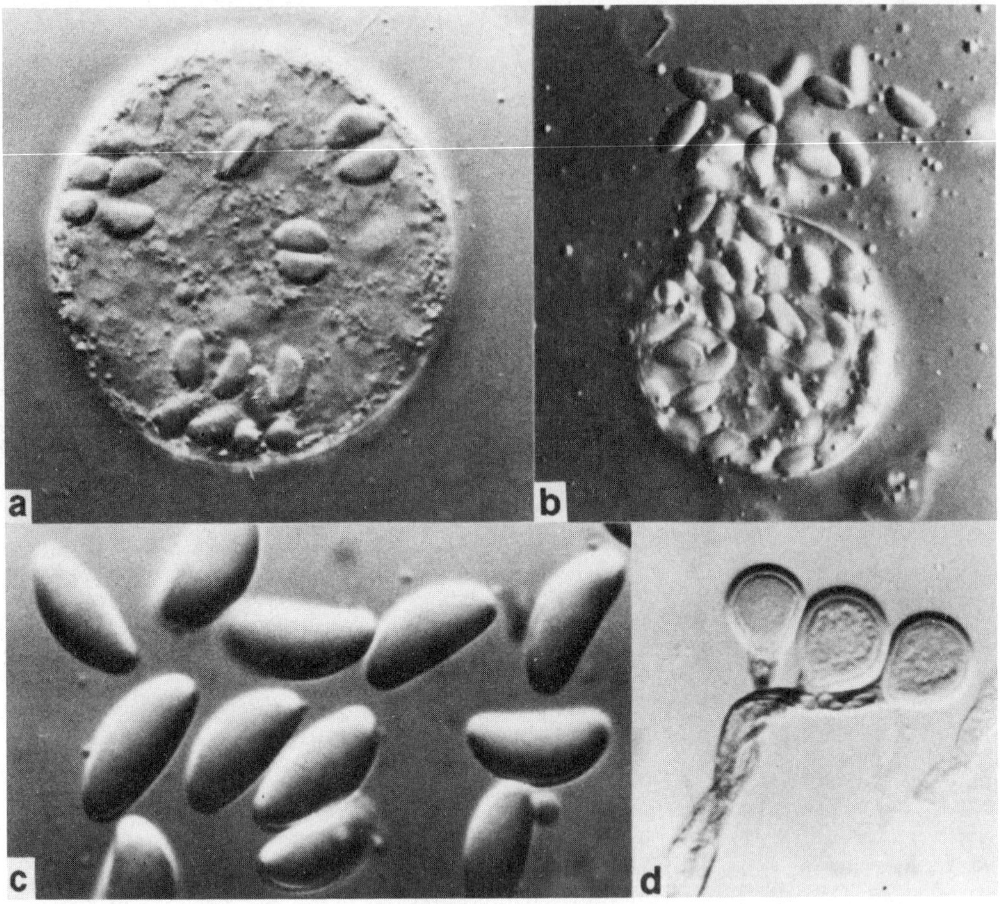

FIGURE 8. *Xeromyces bisporus*: (a) developing cleistothecium showing ascospores forming in pairs × 750; (b) mature cleistothecium liberating mature ascospores × 750; (c) mature ascospores × 1875; (d) aleurioconidia × 750.

Polypaecilum pisce is common on dried salted seafoods, particularly from tropical regions [46]. It forms polyphialides, phialides with several necks, each of which produces a delicate chain of lemon-shaped conidia (Fig. 9c,f). The conidiophores may be quite complex, with a number of branches (Fig. 9g).

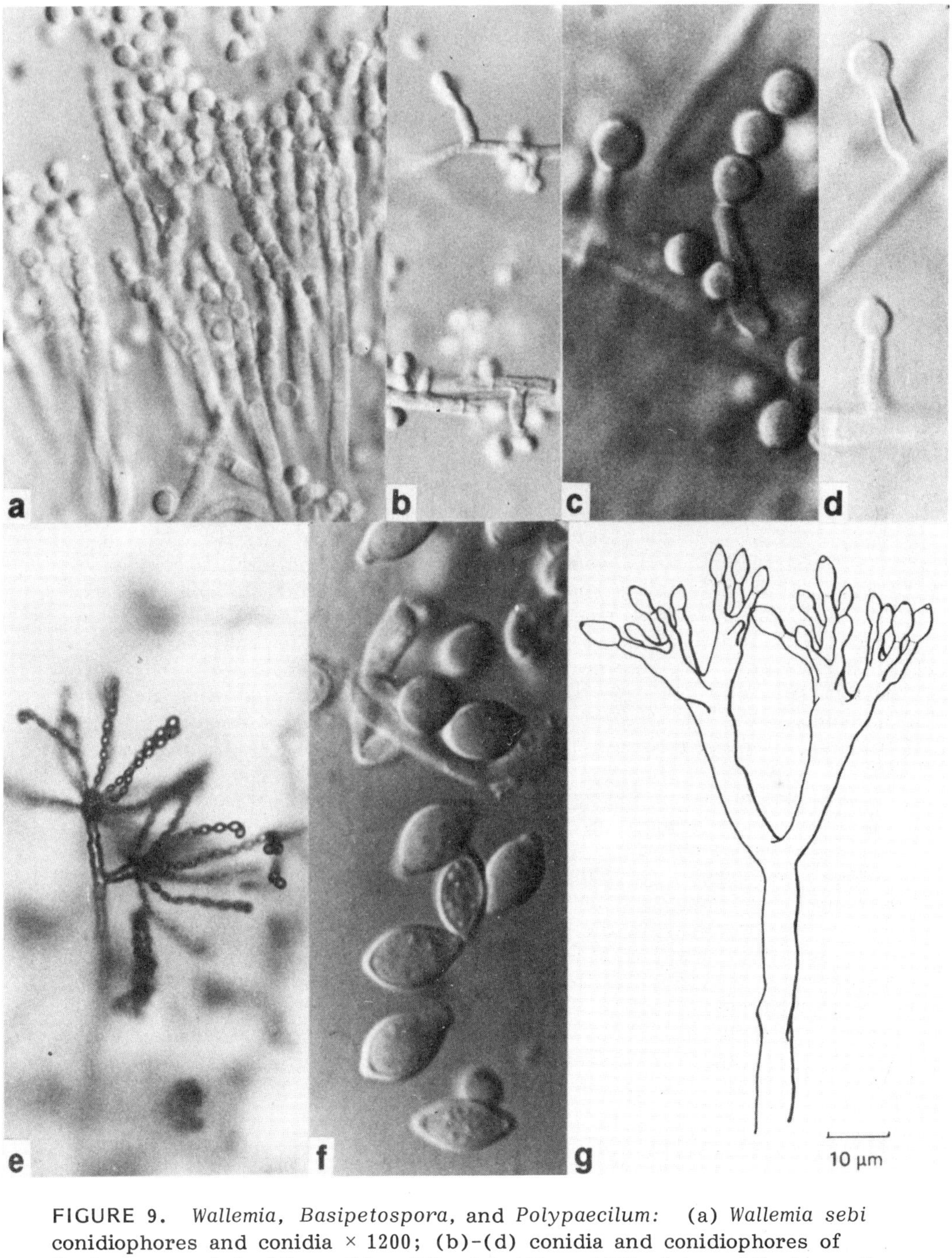

FIGURE 9. *Wallemia*, *Basipetospora*, and *Polypaecilum*: (a) *Wallemia sebi* conidiophores and conidia × 1200; (b)-(d) conidia and conidiophores of *Basipetospora halophila*: (b) × 750; (c),(d) × 1875; (e) conidiophore of *Polypaecilum pisce* showing delicate conidial chains × 300; (f) lemon-shaped conidia × 1875; (g) camera lucida sketch of branched conidiophores.

X. CONCLUSION

The mycoflora of intermediate and low moisture foods is dominated by relatively few fungal genera. By far the most widespread and probably the most important in biodeterioration are species of *Eurotium*, *Aspergillus*, and *Penicillium*. These fungi can be found in most foods with an a_w below 0.90.

From the point of view of food spoilage and loss, *Eurotium* species are probably the most destructive of all. They are usually the first colonizers of improperly dried stored commodities, and as they grow, they raise the a_w, allowing other species like potentially toxigenic Aspergilli and Penicillia to take over. Although *Eurotium* species do not represent a real mycotoxin problem, they do produce a variety of secondary metabolites, and they cause oxidative rancidity problems in grains and nuts.

Foods that contain very high concentrations of solutes like sugars or salts often present a more rigorous environment, limiting the number of fungal species that are able to spoil them. High-sugar foods like confectionery and dried fruits are an ideal habitat for some of the more specialized xerophilic fungi, particularly *Xeromyces bisporus* and the xerophilic *Chrysosporium* species. Foods with a high salt content such as salted, dried seafoods are the only known habitat for the specialized halophilic xerophiles *Polypaecilum pisce* and *Basipetospora halophila*.

It is important to remember that detection and enumeration of such specialized fungi requires specially adapted media and techniques. Many xerophilic fungi will not grow on normal, high a_w enumeration media. In choosing an isolation medium, the characteristics of the food being analyzed should be taken into consideration. For high sugar foods, a high sugar medium such as MY50G [3] may be most appropriate for isolating the spoilage fungi. For salted, dried fish, MY5-12 [3] would better reflect the composition of the food and thus provide more suitable growth conditions for halophilic xerophiles.

For routine use in enumerating the flora of stored commodities (grains, nuts, spices), a reduced a_w enumeration medium such as DG18 [3,68] is recommended. This medium is good for detecting common xerophiles like *Eurotium* species, *Aspergillus restrictus*, *A. penicilloides*, *Wallemia sebi*, and most *Aspergillus* and *Penicillium* species. However, it will not detect *X. bisporus* and some of the other more fastidious sugar-loving xerophiles.

Intermediate and low moisture foods provide some extremely specialized ecological niches, inhabited by highly evolved species of fungi. The physiological adaptations that enable microorganisms to grow at reduced a_w are complex and varied. The most successful group of microorganisms colonizing these low a_w niches, both in number of species and ability to grow at low a_w, has been the xerophilic fungi [77].

REFERENCES

1. Christensen, C. M., and Kaufmann, H. H. Deterioration of stored grain by fungi. *Annu. Rev. Phytopathol.*, *3*:69-84, 1965.
2. Christensen, C. M. Storage fungi. In *Food and Beverage Mycology* (L. R. Beuchat, ed.), AVI Publishing Company, Newport, CT, 1978, pp. 173-190.

3. Pitt, J. I., and Hocking, A. D. *Fungi and Food Spoilage*. Academic Press, Sydney, 1985.
4. Sauer, D. B., Storey, C. L., and Walker, D. E. Fungal populations in farm-stored grains and their relationship to moisture, storage, time, regions and insect infestation. *Phytopathology, 74*:1050-1053, 1984.
5. Magan, N., and Lacey, J. Interactions between field and storage fungi on grain. *Trans. Br. Mycol. Soc., 85*:29-37, 1985.
6. Wallace, H. A. H., Sinha, R. N., and Mills, J. T. Fungi associated with small wheat bulks during prolonged storage in Manitoba. *Can. J. Bot., 54*:1332-1343, 1976.
7. Moubasher, A. H., Elnaghy, M. A., and Abdel-Hafez, S. I. Studies on the fungus flora of three grains in Egypt. *Mycopathol. Mycol. Appl., 47*:261-274, 1972.
8. Ogasawara, K., Sekijo, I., Sunagawa, H., and Umemura, M. Studies on the toxigenic mold contamination of foodstuffs. 4. Mycoflora and frequency of toxigenic mold in seasoned dried marine products. *Rep. Hokkaido Inst. Public Health, 28*:26-31, 1978.
9. Pelhate, J. Inventaire de la mycoflore des bles de conservation. *Bull. Trimest. Soc. Mycol. Fr., 84*:127-143, 1968.
10. Eyles, M. J., Moss, R., and Hocking, A. D. The microflora of Australian wheat flour and the effect of flour milling procedures on microbial contamination. *Food Aust., 41*:704-708, 1989.
11. Graves, R. R., and Hesseltine, C. W. Fungi in flour and refrigerated dough products. *Mycopathol. Mycol. Appl., 29*:277-290, 1966.
12. Kurata, H., and Ichinoe, M. Studies on the population of toxigenic fungi in foodstuffs. I. Fungal flora of flour-type foodstuffs. *J. Food Hyg. Soc. Jpn., 8*:237-246, 1967.
13. Kurata, H., Udagawa, S., Ichinoe, M., Kawasaki, Y., Takada, M., Tazawa, M., Koizumi, A., and Tanabe, H. Studies on the population of toxigenic fungi in foodstuffs. III. Mycoflora of milled rice harvested in 1965. *J. Food Hyg. Soc. Jpn., 9*:23-28, 1968.
14. Mallick, A. K., and Nandi, B. Research: Rice. *Rice J., 84*:8-13, 1981.
15. Abdel-Kader, M. I. A., Moubasher, A. H., and Abdel-Hafez, S. I. I. Survey of the mycoflora of barley grains in Egypt. *Mycopathologia, 68*:143-147, 1979.
16. Flannigan, B. Microflora of dried barley grains. *Trans. Br. Mycol. Soc., 53*:371-379, 1969.
17. Lichwardt, R. W., Barron, G. L., and Tiffany, L. H. Mold flora associated with shelled corn in Iowa. *Iowa State Coll. J. Sci., 33*:1-11, 1958.
18. Barron, G. L., and Lichwardt, R. W. Quantitative estimations of the fungi associated with deterioration of stored corn in Iowa. *Iowa State J. Sci., 34*:147-155, 1959.
19. Joffe, A. Z. The mycoflora of fresh and stored groundnut kernels in Israel. *Mycopathol. Mycol. Appl., 39*:255-256, 1969.
20. Huang, L. H., and Hanlin, R. T. Fungi occurring in freshly harvested and in-market pecans. *Mycologia, 67*:689-700, 1975.
21. Senser, F. Untersuchungen zum Aflatoxingehalt in Haselnussen. *Gordian, 79*:117-123, 1979.

22. Hocking, A. D. Improved media for enumeration of fungi from foods. *CSIRO Food Res. Q., 44*:73-82, 1981.
23. Hocking, A. D., and Pitt, J. I. Two new xerophilic fungi and a further record of *Eurotium halophilicum. Mycologia, 80*:82-88, 1988.
24. Pitt, J. I., and Christian, J. H. B. Water relations of xerophilic fungi isolated from dried prunes. *Appl. Microbiol., 16*:1853-1858, 1968.
25. Tanaka, H., and Miller, M. W. Microbial spoilage of dried prunes. 1. Yeasts and molds associated with spoiled dried prunes. *Hilgardia, 34*:167-170, 1963.
26. Dallyn, H., and Everton, J. R. The xerophilic mould, *Xeromyces bisporus*, as a spoilage organism. *J. Food Technol., 4*:399-403, 1969.
27. Pitt, J. I., and Hocking, A. D. Food spoilage fungi. 1. *Xeromyces bisporus* Fraser. *CSIRO Food Res. Q., 42*:1-6, 1982.
28. Dallyn, H., and Fox, A. Spoilage of materials of reduced water activity by xerophilic fungi. In *Microbial Growth and Survival in Extremes of Environment* (G. W. Gould and J. E. L. Corry, eds.), Academic Press, London, 1980, pp. 129-139.
29. Kinderlerer, J. L. Fungi in dessicated coconut. *Food Microbiol., 1*:205-207, 1984.
30. Kinderlerer, J. L. Spoilage in dessicated coconut resulting from growth of xerophilic fungi. *Food Microbiol., 1*:23-28, 1984.
31. Tilbury, R. H. The microbial stability of intermediate moisture foods with respect to yeasts. In *Intermediate Moisture Foods* (R. Davis, C. G. Birch, and K. J. Parker, eds.), Applied Science Publishers, London, 1976, pp. 138-165.
32. Leistner, L. Hurdle technology applied to meat products of the shelf stable product and intermediate moisture food types. In *Properties of Water in Foods* (D. Simantos and J. L. Multon, eds.), Martinus Nijhoff Publishers, Dordrecht, The Netherlands, 1985, pp. 309-329.
33. Hadlok, R., Samson, R. A., Stolk, A. C., and Schipper, M. A. A. Mould contamination in meat products. *Fleischwirtschaft, 56*:374-376, 1976.
34. Leistner, L., and Ayres, J. C. Molds and meats. *Fleischwirtschaft, 48*:62-65, 1968.
35. Takatori, K., Takahashi, K., Suzuki, T., Udagawa, S., and Kurata, H. Mycological examination of sausages in retail markets and the potential production of penicillic acid of their isolates. *J. Food Hyg. Soc. Jpn., 16*:307-312, 1975.
36. Hadlok, R. Schimmelpilzkontamination von Fleischerzeugnissen durch naturbelassene Gewurze. *Fleischwirtschaft, 49*:1601-1609, 1969.
37. Van der Riet, W. B. Biltong, a South African dried meat product. *Fleischwirtschaft, 62*:1000-1001, 1982.
38. Van der Riet, W. B. Studies on the mycoflora of biltong. *S. Afr. Food Rev., 3*:105-111, 1976.
39. Hitokoto, H., Morozumi, S., Wauke, T., Sakai, S., Zen-Yoji, H., and Benoki, M. Studies on fungal contamination on foodstuffs in Japan: Fungal flora on dried small sardines marketing in Tokyo. *Ann. Rpt. Tokyo Metropol. Res. Lab. Publ. Health, 27*:36-40, 1976.

40. Ichinoe, M., Suzuki, M., and Kurata, H. Microflora of commercial sliced dried fish including bonito. *Bull. Nat. Inst. Hyg. Sci.*, *95*: 96-99, 1977.
41. Okafor, N. Fungi associated with mouldy dried fish. *Nigerian J. Sci.*, *2*:41-44, 1968.
42. Phillips, S., and Wallbridge, A. The mycoflora associated with dried salted tropical fish. *Proceedings of the Conference on the Handling, Processing and Marketing of Tropical Fish.* Tropical Products Institute, London, 1977, pp. 353-356.
43. Townsend, J. F., Cox, J. K. B., Sprouse, R. F., and Lucas, F. V. Fungal flora of South Vietnamese fish and rice. *J. Trop. Med. Hyg.*, *74*:98-100, 1971.
44. Wheeler, K. A., Hocking, A. D., Pitt, J. I., and Anggawati, A. Fungi associated with Indonesian dried fish. *Food Microbiol.*, *3*:351-357, 1986.
45. Wu, M. T., and Salunkhe, D. K. Mycotoxin producing potential of fungi associated with dried shrimps. *J. Appl. Bacteriol.*, *45*:231-238, 1978.
46. Pitt, J. I., and Hocking, A. D. New species of fungi from Indonesian dried fish. *Mycotaxon*, *22*:197-208, 1985.
47. Frank, M., and Hess, E. Studies on salt fish. V. Studies on *Sporendonema epizoum* from "dun" salt fish. *J. Fisheries Res. Bd Canada*, *5*:276-286, 1941.
48. Andrews, S., and Pitt, J. I. Further studies on the water relations of xerophilic fungi, including some halophiles. *J. Gen. Microbiol.*, *133*:233-238, 1987.
49. Avari, G. P., and Allsopp, D. The combined effect of pH, solute and water activity (a_w) on the growth of some xerophilic *Aspergillus* species. *Biodeterioration*, *5*:548-556, 1983.
50. Ayerst, G. The effects of moisture and temperature on growth and spore germination in some fungi. *J. Stored Products Res.*, *5*:669-687, 1969.
51. Hocking, A. D., and Pitt, J. I. Water relations of some *Penicillium* species at 25°C. *Trans. Br. Mycol. Soc.*, *73*:141-145, 1979.
52. Pitt, J. I., and Hocking, A. D. Influence of solute and hydrogen ion concentration on the water relations of some xerophilic fungi. *J. Gen. Microbiol.*, *101*:35-40, 1977.
53. Wheeler, K. A., Hocking, A. D., and Pitt, J. I. Effects of temperature and water activity on germination and growth of *Wallemia sebi*. *Trans. Br. Mycol. Soc.*, *90*:365-368, 1988.
54. Wheeler, K. A., Hocking, A. D., and Pitt, J. I. Water relations of some *Aspergillus* species isolated from dried fish. *Trans. Br. Mycol. Soc.*, *31*:631-637, 1988.
55. Wheeler, K. A., Hocking, A. D., and Pitt, J. I. Influence of temperature on the water relations of *Polypaecilum pisce* and *Basipetospora halophila*, two halophilic fungi. *J. Gen. Microbiol.*, *134*:2255-2260, 1988.
56. Wheeler, K. A., and Hocking, A. D. Water relations of *Paecilomyces variotii*, *Eurotium amstelodami*, *Aspergillus candidus* and *A. sydowii*. *Int. J. Food Microbiol.*, *7*:73-78, 1988.

57. Hocking, A. D. Effects of water activity and culture age on the glycerol accumulation patterns of five fungi. *J. Gen. Microbiol., 132*: 269-275, 1985.
58. Pitt, J. I., and Christian, J. H. B. Heat resistance of xerophilic fungi based on microscopical assessment of spore survival. *Appl. Microbiol., 20*:682-686, 1970.
59. Bolin, H. R., and Boyle, F. P. Use of potassium sorbate and heat for the preservation of prunes at high moisture levels. *J. Sci. Food Agric., 18*:289-291, 1967.
60. Schade, J. E., Stafford, A. E., and King, A. D. Preservation of high-moisture dried prunes with sodium benzoate instead of potassium sorbate. *J. Sci. Food Agric., 24*:905-911, 1973.
61. Beuchat, L. R. Combined effects of solutes and food preservatives on rates of inactivation and colony formation by heated spores and vegetative cells of molds. *Appl. Environ. Microbiol., 41*:472-477, 1981.
62. Yuda, S., Takahashi, J., and Yanagisawa, H. Studies on the quality preservation of foods by using ethanol vapor. *Jpn. Packag. Res., 1*:29-30, 1984.
63. Mizutani, K. Historical background and present status of oxygen absorber. *New Food Ind., 29*:12-17, 1987.
64. Pafumi, J., and Durham, R. Cake shelf life extension. *Food Technol. Aust., 39*:286-287, 1987.
65. Mislivec, P. B., and Bruce, V. R. Direct plating versus dilution plating in qualitatively determining the mold flora of dried beans and soybeans. *J. Assoc. Off. Anal. Chem., 60*:741-743, 1977.
66. Pitt, J. I., Hocking, A. D., and Glenn, D. R. An improved medium for the detection of *Aspergillus flavus* and *A. parasiticus*. *J. Appl. Bacteriol., 54*:109-114, 1983.
67. Andrews, S. Optimization of conditions for the surface disinfection of sorghum and sultanas using sodium hypochlorite solutions. In *Methods for the Mycological Examination of Food* (A. D. King, J. I. Pitt, L. R. Beuchat, and J. E. L. Corry, eds.), Plenum Press, New York, 1986, pp. 28-32.
68. Hocking, A. D., and Pitt, J. I. Dichloran-glycerol medium for enumeration of xerophilic fungi from low moisture foods. *Appl. Environ. Microbiol., 39*:488-492, 1980.
69. King, A. D., Hocking, A. D., and Pitt, J. I. Dichloran-rose bengal medium for enumeration and isolation of molds from foods. *Appl. Environ. Microbiol., 37*:959-964, 1979.
70. Scarr, M. P. Selective media used in the microbiological examination of sugar products. *J. Sci. Food Agric., 10*:678-681, 1959.
71. Restaino, L., Bills, S., and Lenovich, L. M. Growth response of an osmotolerant, sorbate-resistant *Saccharomyces rouxii* strain: Evaluation of plating media. *J. Food Prot., 48*:207-209, 1985.
72. Jermini, M. F. G., Geiges, O., and Schmidt-Lorenz, W. Detection, isolation and identification of osmotolerant yeasts from high-sugar products. *J. Food Prot., 50*:468-472, 1987.
73. Pettifer, G. L. Detection of low numbers of osmophilic yeasts in creme fondant within 25 h using a pre-incubated DEFT count. *Lett. Appl. Microbiol., 4*:95-98, 1987.

74. Pitt, J. I. *A Laboratory Guide to Common Penicillium Species*, 2nd ed. CSIRO Division of Food Processing, Sydney, 1988.
75. Raper, K. B., and Fennell, D. I. *The Genus Aspergillus*, Williams and Wilkins, Baltimore, 1965.
76. Klich, M. A., and Pitt, J. I. *A Laboratory Guide to Common Aspergillus Species and Their Teleomorphs*, CSIRO Division of Food Processing, Sydney, 1988.
77. Hocking, A. D. Strategies for microbial growth at reduced water activities. *Microbiol. Sci.*, 5:280-284, 1988.

4

FUNGI AND SEED QUALITY

CLYDE M. CHRISTENSEN *University of Minnesota, St. Paul, Minnesota*

I. INTRODUCTION

Seeds of cultivated plants constitute a major source of food for humans, feed for domestic animals, and the raw material for many industrial products. They constitute one of our basic renewable resources. From the time of their first formation on growing plants until their final consumption or use, they are subject to damage and destruction by a variety of biological agents, a major one of which is fungi. Fungi in seeds and their effects on the quality of the seeds for various uses have been studied for approximately 100 years. The present review aims to evaluate some of the techniques used in these studies and to present some of the more important problems posed by fungi in the seeds of our major crop plants.

II. TECHNIQUES FOR STUDYING FUNGI IN SEEDS

It may be appropriate to begin with a brief description and evaluation of the techniques used to determine the number and kinds of fungi on and in seeds. This is important because the results obtained and the conclusions drawn from them may differ according to the techniques used, and the uncritical use of these techniques, without regard to their limitations, has at times led to erroneous conclusions.

A. Examination of Seeds with the Unaided Eye and with the Microscope

1. Examination with the Unaided Eye

The presence of some fungi in seeds can be detected with the unaided eye, as *Fusarium* head blight or scab or wheat and barley, *Fusarium* ear rots of corn, and kernel smudge and staining of wheat and barley seeds caused by *Alternaria*, *Helminthosporium*, *Cladosporium*, and other common fungi. The Official United States Standards for Grains [1], under specifications for malting barley, states that malting barley is permitted to have 4% damaged kernels, defined as follows: "(m) Mold-damaged kernels (major): Kernels and pieces of kernels of barley which are weathered and contain considerable evidence of molds," and "(n) Mold-damaged kernels (minor): Kernels and pieces of kernels of barley containing slight evidence of mold." The difference might lie to some extent in the eye of the beholder—what to the seller might appear to be minor discoloration might, to the buyer, appear to be major discoloration. The buyers decide; bright barley with no evidence of molding is preferred.

Some investigators of problems relating to damage caused by storage fungi in grains, even in 1987, speak of molds "showing themselves," which appears to put the major burden of mold detection on the molds themselves—they have to "show themselves." Results of such tests can hardly carry much weight.

2. Microscopic Examination of Seeds or Seed Parts

Warnack and Preece [2] made longitudinal sections of barley seeds with the aid of a freezing microtome and examined microscopically 50 sections of each of 10 kernels of two samples of barley. They found mycelium in the parenchymal layer of the lemma and palea and in the pericarp of the caryopsis itself. The identify of the mycelium was not determined. Only a limited number of seeds can be examined in this way, and the information gained also is limited, but the technique does give some information on the location of the mycelium. Had sections also been plated out, additional interesting information might have been gleaned from the fungi that grew out, so that the mycelium could have been identified.

Christensen [3] removed pieces of the outer pericarps of wheat kernels, examined them microscopically, and found mycelium to be common and sometimes abundant on the inner surface of the outer pericarp layers. He also "scrubbed" the surface of wheat kernels with cotton swabs dipped in sterile water, removed small strips of the pericarps so scrubbed and put them on thin drops of agar on cover slips on van Tieghem cells, with the inner side of the strips in contact with the agar, so that it was directly observable with low and high power of the microscope. This permitted examination of the mycelium present on the inner side and within the cells of the pericarp layer, and of the mycelium that grew out. Fungi that grew out could be transferred to agar in petri dishes where they sporulated and could be identified. He found mycelium of *Alternaria* to be common and sometimes abundant on the inner side of the outer pericarp layers of wheat kernels from many sources, including dry-land farming areas of Montana and Wyoming in the western United States and from irrigated areas. *Alternaria* appeared to be universally present at harvest in all of the seeds of all of the wheat samples he studied. In seeds that had been stored for some time and were beginning to undergo invasion by storage fungi, mycelium and sometimes microsclerotia of species of *Aspergillus* were common in the inner layer of cells of the pericarp.

In this work it was found that even a brief wash or rinse of the seeds with a 2% solution of sodium hypochlorite (the usual "surface disinfectant" used in work with fungi in seeds) killed this mycelium within the cells of the outer pericarp layers and also that on the inner side of the outer pericarp. Even a single wipe of the seed surface with 40% ethyl alcohol, followed immediately by a wipe with sterile water, resulted in death of this mycelium. Unquestionably, the usual treatment of shaking seeds for 1-2 minutes in 1-2% sodium hypochlorite before plating them on an agar medium to determine fungi present within them, also kills this mycelium. This treatment also, as will be seen below, does not eliminate contaminating spores on the outside of the seeds.

Using the same technique we found mycelium and also yeast cells within the inner cells of the outer pericarp layers of sound, unstained barley seeds. Yeasts, in fact, sometimes are abundant on the outside and just within the outer layers of seeds of barley, which should remove any mystery that might remain concerning the discovery of fermentation in earlier times. Using the same examination procedure, mycelium has been found in the testa or thin covering of the endosperm or meat of coconuts [4] and in the same tissues of Brazil nuts. Presumably this mycelium comes from floral infection, in which the fungus or fungi grow down through the pollen tubes. As described below, floral infection by *Aspergillus flavus*, followed by invasion of the developing seeds, is common in some of the high-aflatoxin-risk crops, and presumably many other fungi enter by the same route.

Examination with the stereoscopic microscope, using magnifications of 10X to 100X or more, of seeds undergoing invasion by storage fungi often is useful in the inspection of sectioned seeds for presence of storage fungi—mycelium and sometimes sporophores can be detected in germ cavities and on the surface of the germ beneath the covering pericarps. Such microscopic examination also is useful in identifying fungi that grow out from seeds plated on agar. Often two or more species of fungi grow from the same seed, and examination with the unaided eye will miss some of these. If *Aspergillus candidus* is present in the seeds along with *A. glaucus*, as it often is, it will not appear until the seeds have been incubated for 7-10 days, by which time the seeds and the agar may be overgrown by *A. glaucus*, and *A. candidus* appears only as a few white heads among the jungle of *A. glaucus* sporophores. Examination with the unaided eye will miss the *A. candidus*. If one is evaluating the condition and storability of a given lot of seeds, this can be important, because the presence of *A. glaucus* may indicate only the early stages of deterioration, whereas the presence of *A. candidus* in even a small percentage of the seeds is cause for alarm, because it indicates that advanced deterioration is under way in that portion of the bin or bulk from which the sample came.

B. Plating Seeds on Agar

This technique, under various names, has been in common use for close to a 100 years in plant pathology laboratories and still enjoys wide use. In many studies on the microflora of seeds it is the only method used—it has become standard and routine; like other routine methods and procedures it may sometimes give only routine results.

Until the late 1930s or early 1940s the common procedure was to shake the seeds to be plated in a 1:1000 solution of $HgCl_2$, rinse them in a solution of 2-4% sodium hypochlorite or in sterile water, and plate them on agar—usually potato dextrose agar (PDA) or acidified potato dextrose agar

(APDA). The plates were incubated until fungi grew out and could be identified, and the fungi that grew out were presumed to be present within the kernels.

Moore and Olien [5] and Olien and Moore [6] showed that the mercury adsorbed to outer tissues of seeds so "disinfected" kept some fungi present in the seeds from growing, and if the mercury was precipitated in insoluble form by a 5-minute dip in a M/5 solution of sodium thiosulfate, other fungi present in the seeds but sensitive to mercury could grow. *Alternaria* and *Helminthosporium* were among the fungi that tolerated the mercury present in the tissues of the seeds after treatment with 1:1000 $HgCl_2$, and so were the ones most likely to grow. Some of the results of extensive surveys of the microflora of wheat and barley seeds, where this $HgCl_2$ treatment was used and almost only *Alternaria* and *Helminthosporium* were detected, may have to be reevaluated in the light of this. That *Alternaria* and *Helminthosporium* were the major fungi present may be true, but other fungi may also have been present but went undetected because of their sensitivity to mercury. Also, any agar medium is to some extent selective, including PDA and APDA; they will not reveal, for example, some of the common storage fungi such as *Aspergillus restrictus* or *A. glaucus*, which require or grow better on a medium containing 5-10% NaCl.

Since the late 1930s and early 1940s the standard surface disinfectant or disinfestant used when determining the fungi present in seeds and other plant parts has been a 1 or 2% solution of NaOCl, sometimes with ethyl alcohol added. Seeds are shaken for 1-2 minutes in the NaOCl, then either rinsed in sterile water or not rinsed but plated directly on the agar medium and incubated until fungi can grow out and be identified. The procedure has two limitations: (a) NaOCl may kill mycelium within and directly under the outer pericarps of such seeds as those of the cereal grains, and (b) it will not kill all of the contaminating spores or other inoculum on the surface of the seeds.

Lack of appreciation of the latter limitation has led to the "proving" of some things that were not so. Boller and Schroeder [7], for example, contaminated seeds of rough rice with spores of an aflatoxin-producing strain of *Aspergillus parasiticus*, then stored the seeds at relative humidities of 70-100% and, after intervals of 7, 14, and 28 days, removed portions of the seeds at each relative humidity, shook them in 1% NaOCl for one minute, rinsed them in sterile water, plated them on agar, and incubated the plates to determine the percentage of kernels from which *A. parasiticus* grew. After storage for 7 days, the fungus was recovered from 61.5, 64.0, and 73.0 percent of the kernels stored, respectively, at 75, 80, and 85% RH, and after 28 days it was recovered from 69.0, 75.0, and 68.5% of the kernels stored at 75, 80, and 85% RH. They also determined moisture contents of the seeds, and after 7 days samples at 75 and 80% RH had moisture contents of approximately 13.5%, and those at 85% RH had moisture contents of approximately 14.5%. They concluded that "*A. parasiticus* infects kernels of inoculated rice stored at RH of 75 percent or higher."

A. parasiticus is a member of the *A. flavus* group [8]. There is considerable evidence that members of this group require a moisture content in the range of 18.0-18.5% in starchy cereal grains, including rice, to grow [9-12]; Ayerst [13] gives 78% relative humidity as the absolute lower limit that permits germination of spores of *A. flavus*, and at that RH 95 days were required for the spores to germinate.

Boller and Schroeder cited none of these papers, and they neglected one important step in their work—they did not test the inoculated or contaminated rice seeds at zero time. Christensen and Mirocha [14] repeated this work, inoculating samples of rough rice with spores of *A. parasiticus* and storing them at 75, 80, 85% RH and at 25°C for 32 and 72 days. Using the same "surface disinfection" procedure as Boller and Schroeder—1% NaOCl for 1 minute, followed by a rinse in sterile water—they recovered *A. parasiticus* from 57% of the seeds at zero time, immediately after the seeds had been contaminated with spores, and from a decreasing number of seeds at all three relative humidities after 32 and 72 days.

This is discussed in some detail because it is a very important point; if *A. parasiticus* or *A. flavus* can grow and produce aflatoxin in seeds stored at moisture contents in equilibrium with 75% RH, there would be an increased hazard of aflatoxin production in stored grains—large quantities of corn are stored with moisture contents slightly above those in equilibrium with 75% RH. Boller and Schroeder found some aflatoxin in the rice seeds they inoculated, and so did Christensen and Mirocha. In both tests, rice seeds were initially free of aflatoxin. Seeds contaminated with *A. parasiticus* by Christensen and Mirocha were tested immediately after contamination, and were found to contain approximately 30 ppb of aflatoxin B_1, 25 ppb of G_1, and lesser amounts of aflatoxins B_2 and G_2. The source of the aflatoxin was the spores used to contaminate the rice. Boller and Schroeder were in error. Tests similar to those of Boller and Schroeder were made by Lillehoj et al. [15] with cracked corn; they obtained similar results and made similar and equally erroneous conclusions. Shaking seeds in 1 or 2% NaOCl for 1 or 2 minutes will not eliminate all surface contaminants. Recognizing this we discontinued many years ago using "surface disinfection" or "surface disinfestation" when describing our methods in research papers, but simply state that "seeds were shaken for 1 minute in a 2 percent solution of NaOCl to rid them of most surface contaminants."

C. Dilution Cultures

A weighed amount or counted number of seeds are comminuted for 1 minute in a suspension medium of sterile 0.12% agar in water (other suspension media work equally well), diluted further if necessary, in the same suspension medium, 1-ml aliquots placed in each of several petri dishes, agar cooled to about 35-40°C added, the dishes swirled to distribute the suspension evenly, then incubated until the colonies appear and can be counted and identified. The colonies that appear represent the living propagules in the comminuted sample—spores or viable hyphal fragments. We have determined the origin of literally thousands of colonies of fungi in these dilution cultures by microscopic examination of dishes on the stage of the microscope, just when the germlings are forming, and have never seen a colony arise from anything but a spore. The mycelium of fungi does not endure the chopping up it is subjected to in a blender. This procedure has been used to determine the relative amount of invasion (or the relative amount of sporulation) of different fungi in a given lot of seeds from storage, and thus evaluate the storability of the grain from which the sample was taken, but for that purpose, in our experience, it gives little information that cannot be gotten from the much simpler and faster plating out of whole seeds or kernels.

These different methods are by no means mutually exclusive—each gives some information that the others do not. Depending on the information desired, all of the methods, including some not described here, may be used to determine the number and kinds of fungi in a given lot or sample of seeds.

1. Agar Media

For decades PDA or APDA have been more or less standard media to detect fungi in agricultural seeds. The acid in the APDA is for the purpose of inhibiting the growth of bacteria so commonly present either on or within seeds—probably mostly on rather than within—but not eliminated by the surface disinfectant before the seeds are plated. These are excellent media to detect a great variety of fungi. One qualification attached to their use is that the lush growth that they encourage of some of the more common seed-inhabiting fungi such as *Alternaria*, *Helminthosporium*, and *Fusarium* may inhibit or conceal the growth of other fungi present. A variety of special media have been developed for the detection of certain fungi or groups of fungi, or for the identification of certain fungi. Examples of these are the media for the isolation and identification of species of *Fusarium* [16] for the isolation and identification of species of *Aspergillus*, and for the isolation and identification of species of storage fungi [8,17]. The need for special media is illustrated by *Aspergillus restrictus*, a group species of xerophytic fungi that grow in stored seeds and seed products (and in many other organic materials whose moisture contents are in equilibrium with relative humidities of about 68-72%). Media of moderately high osmotic pressure are required to detect members of this group. They are common in some lots of stored grains, especially in wheat that has been in storage for 6 months or more at moisture contents just below 15.0%. In some moderately extensive studies of fungi in stored wheat and corn, no members of this group have been detected, and a reasonable presumption is not that *A. restrictus* was not present, but that the media used were not suitable for its detection. Christensen and Meronuck [17] give the composition of most of the media used to detect storage fungi in seeds.

III. FIELD AND STORAGE FUNGI

Seedborne or seed-inhabiting fungi have been divided into two groups—"field fungi" and "storage fungi"—based on their habits, when they invade the seeds, and when they do their damage. The distinction is ecological, not taxonomic.

A. Field Fungi

These invade seeds of plants in the field, while the plants are still growing or when they are mature and still standing, or swathed, before threshing or combining. Once the seeds have been removed from the plants and have been naturally or artificially dried, the field fungi cease to grow; by harvest

TABLE 1. Equilibrium Moisture Contents[a] of Common Grains, Seeds, and Feed Ingredients at Relative Humidities of 65-90% and Fungi Likely to be Encountered

Relative humidity (%)	Starchy cereal seeds,[b] defatted soybean, and cottonseed meal, alfalfa pellets, most feeds	Soybeans	Sunflower, safflower seeds, peanuts, copra	Fungi
65-70	13.0-14.0	12.0-13.0	5.0-6.0	*Aspergillus halophilicus*
70-75	14.0-15.0	13.0-14.0	6.0-7.0	*A. restrictus*, *A. glaucus*, *Wallemia sebi*
75-80	14.5-16.0	14.0-15.0	7.0-8.0	*A. candidus*, *A. ochraceus*, plus the above
80-85	16.0-18.0	15.0-17.0	8.0-10.0	*A. flavus*, *Penicillium* plus the above
85-90	18.0-20.0	17.0-19.0	10.0-12.0	*Penicillium* plus the above

[a]Percent wet weight. The figures are approximations; in practice variations up to ±1.0% can be expected.

[b]Wheat, barley, oats, rye, rice, millet, maize, sorghum, and triticale.

they have done all the damage they can do; their injurious effects do not continue to increase in storage. Some of them are pathogenic and cause various sorts of damage in the seeds they invade and in the plants grown from these seeds; others are just among those present, as *Alternaria* that occurs almost universally under the outer pericarps of the seeds of wheat. All of the field fungi require, in the seeds they invade, a moisture content in equilibrium with relative humidities of 95-100%—an environment in which free, liquid water is available.

B. Storage Fungi

With only a few exceptions, storage fungi do not invade seeds before harvest, even when windrowed plants lie on the ground for some time in moist weather before being threshed. Field fungi may continue to grow in such seeds, but storage fungi invade them only after seeds have been threshed out and are in storage. Storage fungi are adapted to life without free water (Table 1), and some of them not only endure, but require, an environment without free water, and will not grow in materials whose moisture contents are in equilibrium with relative humidities about 85-90%. Some of these low moisture content fungi are found in strange places; *Aspergillus restrictus*, for example, invades the germs of wheat seeds stored with moisture contents of about 13.8-14.2% and grows in the interior of chocolate candies, in frosting concentrates, and in the coating of wallboards to which fungicides have been added. Fungicides that depend for their action on being dissolved in water have little effect on this fungus that grows without free water. Some of these storage fungi, such as *A. restrictus* and the even more xerophytic *A. halophilicus*, have not only sharply defined lower limits of moisture or relative humidity or water activity below which they cannot grow, but also sharply defined upper limits above which they cannot grow.

IV. FIELD FUNGI AND SEED QUALITY

A. Discoloration

Various field fungi, especially *Alternaria*, *Helminthosporium*, and *Cladosporium*, growing in the outer tissues of seeds during development of seeds or in seeds of windrowed plants before they are threshed can result in various degrees of discoloration, partly from pigments in their own mycelium, partly from melanins produced by the invaded host cells. This discoloration is likely to be especially pronounced toward the basal end of the seeds.

Such fungus-invaded and discolored seeds may result in reduced germinability and decreased vigor of seedlings that develop from them when they are planted or sprouted. Hence the seeds are discriminated against for planting or, if barley, for malting. Malters want sound, bright barley. Barley seeds stained by *Alternaria* and *Cladosporium* may have high germinability and good seedling vigor, and thus in themselves may not be harmful, but such seeds also may harbor *Fusarium* in amounts too small to be readily detectable but which, when the seeds are malted and brewed, produce "spouting" or "gushing" beer. Infection by *Fusarium* heavy enough to produce "scab" or "headblight" may render seeds not only useless for planting or for malting but also make them undesirable for feed, because some common species of *Fusarium* produce mycotoxins highly injurious to animals that consume them.

Pepper and Kiesling [18] list 148 species of fungi, in 68 genera, as having been reported on and within barley kernels. Since the seeds of barley and of other cereal plants except corn are exposed to all the winds that blow, and to all the microfloral inoculum borne by the winds, the list of externally borne contaminants probably could be greatly extended. Miles and Wilcoxson [19] have excellent summaries of the fungi and the problems associated with them in discoloration of barley seeds.

B. Pathogenicity

Some of the fungi that cause staining of seeds, such as *Helminthosporium* and *Fusarium*, also cause disease in the seedlings or growing plants that come from the infected seeds. At least a moderate number of plant pathogenic fungi are seedborne. The *Compendium of Wheat Diseases* [20], for example, lists the following as seedborne fungus diseases: *Fusarium*—root rot and head blight; *Alternaria triticina*—Alternaria leaf blight; *Dilophospora alopecuria*—leafblight; *Septoria nodorum*—leaf and glume blotch; *S. tritici*—leaf spot and blight; *Ustilago tritici*—loose smut; *Urocystis agropyri*—flag smut; *Neovossia indica*—Karnal bunt.

The *Compendium of Soybean Diseases* [21] lists the following seedborne pathogenic fungi in soybeans: *Peronospora manshurica*—downy mildew; *Cercospora sojina*—Cercospora leafspot; *Phyllosticta sojaecola*—leaf spot; *Corynespora cassiicola*—target spot; *Fusarium*—several species that may cause root rot; *Colletotrichum dematium*—anthracnose; *Alternaria tenuissima*—leaf spot; *Diaporthe phaseolorum*—pod and stem blight; *Sclerotium*—Sclerotium blight.

Fischer and Holton [22] list 623 species of smut fungi in 15 genera that occur on Gramineae, of which the cultivated cereals constitute the major food source of the world. Most of these smut fungi are seedborne, either by contaminating spores on the surface of the seeds or by mycelium within the tissues of the embryo. In some of these species infection may come mainly from spores in the soil, but even in those, seedborne spores may also contribute to infection. Some of these seedborne smut fungi affect the quality of seeds for marketing—China, for example, has essentially a total embargo against importation of wheat bearing spores of dwarf bunt (*Tilletia triticina* = *T. controversa*) which sometimes is common in wheat grown in Washington and Oregon. A single spore of this bunt detected by the sampling and examination procedures used at the port of entry will result in refusal of the entire cargo. Karnal smut of wheat recently was introduced into Mexico on wheat imported for food or feed, some of which, unfortunately, contaminated by this fungus, got into the hands of growers, was planted, and so established the disease there.

V. MYCOTOXINS

Some fungi when growing in seeds and other edible plant parts produce compounds that are toxic when consumed. These compounds are called mycotoxins, and the diseases they cause in the animals (including humans) that consume harmful amounts of them are called mycotoxicoses. That bread made from head-blighted (blighted by *Fusarium*) rye and barley was toxic when eaten was known in northern Europe and the U.S.S.R. more than a hundred years ago, and the toxicity of ergot was, of course, recognized long before that. But intensive study of mycotoxins and toxicoses dates from the early 1960s, when aflatoxin was first isolated, purified, characterized, and found to be responsible for widespread death of poultry in England. This stimulated the study of other mycotoxins, especially those produced by species of *Fusarium*, in many cases of the world. *Fusarium* toxins in feed grains are now recognized as being of worldwide importance in the health of domestic animals.

A. Aflatoxin

1. Production

Aflatoxin is produced by *Aspergillus flavus* and by *A. parasiticus* Speare, closely related species of the *A. flavus* group [8]. This group also contains the industrially important species of *A. oryzae*, long used in the

Orient to saccharify rice for fermentation into sake and for preparation of some fermented foods, and in Western countries for production of diastase for use in brewing and baking. Some of the strains used for diastase production may be *A. flavus*, not *A. oryzae*—intergrading forms among the species of the *A. flavus* group often make it difficult to place a given isolate in a given species. Most of those concerned with aflatoxin and aflatoxicoses simply refer to this fungus or these fungi as *A. flavus*—probably a sensible approach.

There appear to be more aflatoxin-producing strains in *A. parasiticus* than of *A. flavus*. *Aspergillus parasiticus* is more prevalent in tropical and subtropical areas, including the southern and southeastern portions of the United States where peanuts and cotton are grown, whereas *A. flavus* is more common in the cooler regions, as the northern portion of the Corn Belt in the United States, extending from Pennsylvania in the east to South Dakota and Nebraska in the west. As an illustration of this, aflatoxin in corn or maize in drought years can be an important problem in the southeastern United States, but is seldom a problem in Minnesota.

Of 283 isolates of *A. flavus* from rice in Texas, 268, or 84%, produced some aflatoxin when grown in pure culture in the laboratory, and 88, or 33%, of the isolates produced a lot of aflatoxin [23]. In contrast to this, none of the numerous isolates of *A. flavus* from feeds suspected of having been involved in illness in cattle in Minnesota produced any aflatoxin when grown in pure culture in the laboratory under conditions optimum for aflatoxin production; nor was any aflatoxin found in the suspect samples of feed from which these cultures were isolated [24]. Thus, the presence of the fungus in a suspect sample of feed is not evidence of the presence of the toxin. Rather special conditions are required for appreciable amounts of aflatoxin to be produced. In brief these are:

1. A strain of the fungus able to produce aflatoxin must be present.
2. The fungus must be present in practically pure culture, alone, not along with a mixture of other fungi such as almost inevitably occurs when grains and seeds in storage are undergoing microbiological spoilage. In corn, aflatoxin is a field problem, not a storage problem; the same probably is largely true of the other high risk aflatoxin crops—seeds in storage seldom are held at moisture contents high enough for *A. flavus* to grow.
3. The moisture content of the substrate must be at least as high as that in equilibrium with a relative humidity of 85%, or a water activity of 85. This is the lower limit of moisture that permits growth of *A. flavus*, and at that moisture content the fungus can grow only very slowly. For more rapid growth and appreciable toxin production, a moisture content in equilibrium with a relative humidity of 90% or more is required in corn and other starchy seeds, a moisture content of 20% or above.
4. A temperature between 12 and 42°C (54-108°F). *Aspergillus flavus* will grow well up to 55°C, and when growing in seeds and other plant materials, usually along with *A. candidus*, it will heat these up to 55°C and hold them at that temperature for some weeks, but it does not produce aflatoxin at a temperature above 40-42°C.

2. Durability

Once aflatoxin is formed, it is extremely durable under most conditions likely to be encountered in storage, handling, and processing of seeds or other plant parts. It endures boiling and the temperatures of ordinary cooking and it is not entirely destroyed by pressure cooking or even by autoclaving at 121°C for 15 minutes. The heat of pelletizing of pelleted feeds may kill the fungus but will not affect the toxin.

3. Effects on Animals

Aflatoxin consumed in the ration will cause various sorts of injury, depending on the amount consumed, the kind of animal, the age of the animal (young animals are more sensitive to injury than older ones), and the make-up of the ration (animals on a diet deficient in protein are more sensitive to aflatoxin injury than those on a well-balanced ration with adequate protein).

In all of the animals so far studied, the target organ of aflatoxin injury is the liver. Continuous consumption of even very small amounts of aflatoxin may result in cancerous tumors in the liver. Aflatoxin is said to be the most potent naturally occurring carcinogenic agent known. In one strain of rainbow trout, 0.5 parts per billion of aflatoxin in the ration will eventually result in lethal liver damage.

Aflatoxin consumed by domestic animals results in a variety of signs and lesions—decreased weight gain, decreased feed efficiency, stunting, increased susceptibility to bruising, hemorrhaging into the muscles and body cavities, general poor health, and increased susceptibility to infection by pathogenic organisms—a suppression of immunity somewhat like that caused by AIDS in humans.

4. Crops of High and Low Aflatoxin Risk

Cultivated crops differ greatly in their susceptibility to aflatoxin contamination. Peanuts or ground nuts, cottonseed, copra, Brazil and pistachio nuts and, in some areas in times of drought, maize or corn are of relatively high aflatoxin risk. Corn in the northern portion of its range, across the northern portion of the Corn Belt in the United States, and other major food and feed grains—wheat, oats, barley, rye, sorghum and millets—are of very low aflatoxin risk. Soybeans are an unsuitable substrate for aflatoxin production and are not known to have been significantly contaminated with aflatoxin anywhere where they are grown, stored, or processed.

In all of the high risk crops listed above in which aflatoxin is at times a problem, the major production of aflatoxin is in the field, from infection that occurs either through the flowers (*A. flavus* spores germinate on the pistils and grow down through the pollen tubes into the young embryo, and develop there as a pure culture) or through wounds made by insects, some of which carry spores of *A. flavus* along with them. There is no evidence that grains or seeds free of aflatoxin when stored will develop aflatoxin contamination in storage, even when they undergo spoilage caused by fungi; such deterioration almost always is caused by a mixture of fungi, seldom by a pure culture of *A. flavus* [25,26]. The aflatoxin problem in corn in the United States is summarized by Diener et al. [27] and in many tropical and subtropical countries by Zuber et al. [28].

5. Damage Caused by Aflatoxin

As an example of the damage that can be done by aflatoxin contamination under conditions favorable to it, in 1977 and again in 1980, drought prevailed during much of the growing season in the southern states of the United States, from Virginia south to Florida and west to Mississippi and Louisiana. Drought predisposes peanuts and developing ears of corn to infection by *A. flavus*, with consequent formation of aflatoxin. In those 2 years the aflatoxin problem in both corn and peanuts reached near-calamity proportions. Much corn was unfeedable and unsaleable because of its high aflatoxin content, and severe losses were suffered from injury to flocks and herds from consumption of aflatoxin-contaminated feed. Costs associated with the 1980 aflatoxin outbreak in that region were estimated at $100 million, and lawsuits amounting to $8 million were filed to recover damages resulting from injury to farm animals and commercial flocks and herds from aflatoxin-contaminated feed [27].

Control of damage by aflatoxin and other mycotoxins consists essentially of avoidance of grains and other food or feed plant materials contaminated with them. This requires establishing programs of sampling and testing to detect lots or parcels bearing toxins and diverting them to uses other than food or feed. Most countries now have strictly enforced regulations limiting the amounts of aflatoxin permitted in feed and food crops in commerce. High aflatoxin risk crops consumed or fed on the farms where they are grown and subject only to casual inspection must at times contribute aflatoxin along with nutriment to those who consume it.

B. Fusarium Toxins

Several of the more common species of *Fusarium* that cause disease in agricultural plants, especially head and seed blights and ear rots, are capable, under suitable growing conditions, of producing very potent toxins. These toxins are especially prevalent in maize, a major food and feed grain throughout much of the world. In the northern portion of the range in which corn is grown—the Corn Belt in the United States and in southern Canada—and in similar climates around the world, the damage done to domestic animals by *Fusarium* toxins probably greatly exceeds that done by aflatoxin. There are no legally established and enforced limits of contamination by *Fusarium* toxins, as there are for aflatoxin (there is a "suggested" limit of 2 ppm in the United States and Canada for deoxynivalenol in feed), and so they can be present in corn in intra- and interstate commerce without this being known. Also, corn grown for feed on the farm where it is used is likely to be subjected to only such inspection as the grower chooses to give it, although more and more growers are being alerted to the dangers of these toxins, and some of them are submitting samples of corn with obvious *Fusarium* ear rots to private testing laboratories or State Department of Agriculture laboratories for toxin analyses.

The principal known and widely occurring *Fusarium* toxins and toxicoses are summarized herewith.

1. Zearalenone

Zearalenone is produced by many isolates of *F. roseum* (= *F. graminearum*, = *Gibberella zeae*) [29]. Consumption by swine of corn or of mixed feed containing several hundred parts per billion (ppb) to 10-20 parts per million (ppm) can result in the estrogenic syndrome, in which the uterus of the prepuberal gilt becomes enlarged, edematous, and tortuous and the ovaries are shrunken. The vulva is enlarged, and this may proceed to vaginal or rectal prolapse. Litter size may be reduced. Young males undergo a feminizing effect, with enlarged nipples [30].

The estrogenic syndrome has been reported from many countries, and in some years in some areas it is very common; for many swine growers it constitutes a serious problem that, until the causal relationship of zearalenone was established, had no rational explanation and no means of control. Broiler chicks and laying hens tolerate relatively massive doses of zearalenone without detectable injury.

Fusarium roseum, or *F. graminearum*, the principal producer of zearalenone, is a common cause of ear rots in corn, sometimes occurring together with other toxin-producing species of *Fusarium*; it is not unusual to isolate two or more species of *Fusarium* from a single kernel of corn, or from a single small fragment of stem tissue. Little zearalenone is produced during the growing season in the infected ears, but it may increase rapidly in amount in ear corn stored in cribs. The combination of conditions that result in production of relatively large amounts of the toxin are:

1. Moderate prevalence of *F. roseum* ear rot in corn at harvest.
2. Storage of infected ears in cribs, at moisture contents of 22-25% or above, which favor continued growth of the fungus.
3. A period of low or fluctuating moderately low and somewhat higher temperatures, with the higher temperatures favoring slow continued growth of the fungus and the lower temperatures promoting toxin production.

These are the conditions that frequently prevail throughout much of the Corn Belt after harvest in the United States and in many other countries where the estrogenic syndrome is an important problem in swine.

2. Deoxynivalenol (DON, Vomitoxin) and Feed Refusal in Swine

DON is produced by *F. graminearum*, but evidently by different strains than those that produce zearalenone. The conditions that favor its production are the same as those of zearalenone—at least a moderate amount of ear infection of corn in the field before harvest, followed by crib storage at moisture contents high enough to permit continued slow growth of the fungus, and regular or intermittent low temperatures that stimulate production of the toxin.

Although DON also is known as vomitoxin, it seldom causes vomiting in swine, because they will not eat enough of the contaminated feed to cause vomiting—they simply refuse to eat the feed, and will starve rather than consume it [31]. Feed containing 1 ppm of DON may result in reduced feed intake and lower than normal weight gain. Other effects are reduced fertility, failure of sows to come into estrus, high mortality of nursing pigs, and enteritis and diarrhea in young pigs. Widespread outbreaks of DON are not unusual in the Corn Belt of the United States following years of abundant *Fusarium* ear rot. Cattle and poultry tolerate moderately large amounts of DON [32]. There is no evidence to suggest that DON, zearalenone or any other *Fusarium* toxins will continue to increase in amount in stored shelled corn, and they would not be expected to, since *Fusarium* requires a minimum moisture content of 22-25% to grow in corn and, at that moisture content, rapid invasion and decay by a variety of microflora would be expected. DON and zearalenone may be found in the same feed.

3. T-2 Toxin and Diacetoxyscirpenol

Episodes of toxicoses resulting from consumption of feeds contaminated by these toxins have been reported in the temperate zones around the world. These toxins are produced by species that Snyder and Hansen [33] lumped together as *F. tricinctum* but which other *Fusarium* taxonomists consider to be *F. poae* and *F. sporotrichioides* [16].

These toxins have been found in barley, wheat, millet, and safflower seed at harvest, as well as in field and sweet corn, and in many mixed feeds in which one or more of these grains was a major constituent. Evidently crib storage is not necessary for production of injurious amounts of these toxins, but crib storage of corn that comes from the field bearing these toxins may result in greatly increased production of the toxins.

All domestic animals are susceptible to injury by a few ppm of dietary T-2 and diacetoxyscirpenol. In cattle they cause decreased feed consumption, lowered milk production, sterility, gastrointestinal hemorrhaging, and death. In swine they cause infertility and lesions in various internal organs. In poultry, amounts of 2-10 ppb of T-2 resulted in decreased egg production, eggs with thin shells, and beak lesions. These toxins definitely are hazardous to the health of animals that consume them in even small amounts [10].

4. *Fusarium equiseti* and Tibial Dyschondroplasia in Poultry

Recently it has been found that a previously unknown toxin, produced by *F. equiseti*, added to the ration of broiler chicks caused tibial dyschondroplasia, a deformation of the legs that is important in commercial production of chickens and turkeys [34]. It also was lethal to chick embryos in fertile eggs.

C. Ergot

Ergot toxicity has been recognized for several hundred years, but it still is very much with us, and veterinary diagnostic laboratories continue to encounter cases of ergotism, especially in cattle. Several species of *Claviceps* may be involved, but in the cereal grains *C. purpurea* is the offender. Ergot sclerotia in whole grains can be readily detected and removed, but sometimes the ergot removed from wheat, rye, or triticale is ground along with other screenings and added to feeds, where special techniques are required to detect it and, if these are available, they are not in common commercial use, and ergotism in cattle doubtless will continue to be a problem.

VI. STORAGE FUNGI

A. The Problem

Seeds of our major cereal crops—wheat, rice, barley, rye, oats, corn, sorghum, and millets—are both extremely durable and highly perishable, depending on how they are kept. Held at low moisture contents and moderately low temperatures, sound seeds of these crops can be kept for years and still be of high quality for processing into food, feed, or other products, and even of high germinability. If stored at moisture contents and temperatures that permit invasion by storage fungi, they will lose germinability, develop the first stages of spoilage within days or weeks, and, in worst-case conditions, become decayed into a black and foul-smelling mess within a month or two. In the absence of insects and mites, and sometimes when these are present, storage molds are a major cause of spoilage and reduction in quality of grains stored or transported under conditions that permit molds to grow. No accurate figures are available on the losses caused by storage fungi, but judging from cases encountered by those familiar with them, losses are moderately sizable year after year, and for some individual farmers and grain merchants or elevator operators they can be catastrophic. These losses are needless. The principles and practices of good grain storage have been firmly established, and where these are known and followed, losses from storage fungi can be held to a minimum [17,25,26-35].

B. Genera and Species of Storage Fungi

The storage fungi comprise a relatively few species of *Aspergillus* (mainly *A. restrictus*, *A. glaucus*, *A. candidus*, *A. ochraceus*, and *A. flavus*—listed in order of increasing moisture content required for growth), plus, under certain circumstances, *Wallemia sebi*, *Chrysosporium fastidium*, and *Candida*, and, at low temperature and moisture contents in equilibrium with relative humidities of 85-90%, several species of *Penicillium*.

All of these storage fungi can grow at moisture contents in equilibrium with relative humidities below 90%, where no free water is available. Some of them not only prefer an environment without free water, they require it; for their detection in seeds in the laboratory special agar media have been developed. Some isolates of *A. halophilicus*, for example, which is not listed above because it is not found commonly and does not seem to be a primary cause of much damage in stored seeds, can be detected only when seeds are plated on agar media containing up to 10% NaCl, and will grow in agar saturated with NaCl. Each of the several common species of storage fungi has its own rather sharply delimited moisture content below which it cannot grow.

With the exception of *A. flavus*, which under some conditions can invade peanuts, cottonseed, and drought-stressed corn in hot weather in some areas, the storage fungi do not invade seeds to any significant extent before harvest, even in years of wet preharvest or harvest weather; the damage that they do develops only after the crops have been harvested and put into storage, often at moisture contents that those in charge of the frains consider safe for storage. Inoculum of these fungi is, however, present on seeds of all cereal plants by harvest. Seeds of legumes, borne in pods, are free of storage fungi at harvest, but become contaminated, presumably by airborne inoculum, as soon as they are threshed. Given moisture contents and temperatures favorable for their growth, these fungi will invade not only grains and grain products in storage, but an array of other organic materials. The moisture contents that permit invasion by these several species of common storage fungi are given in Table 1.

C. Types of Damage Caused by Storage Fungi

Invasion of seeds by storage fungi will result in: reduction in germinability; discoloration of germs or of entire seeds or kernels; mustiness and caking; weight loss; heating, sometimes to the point of ignition; and total decay (Figs. 1-4).

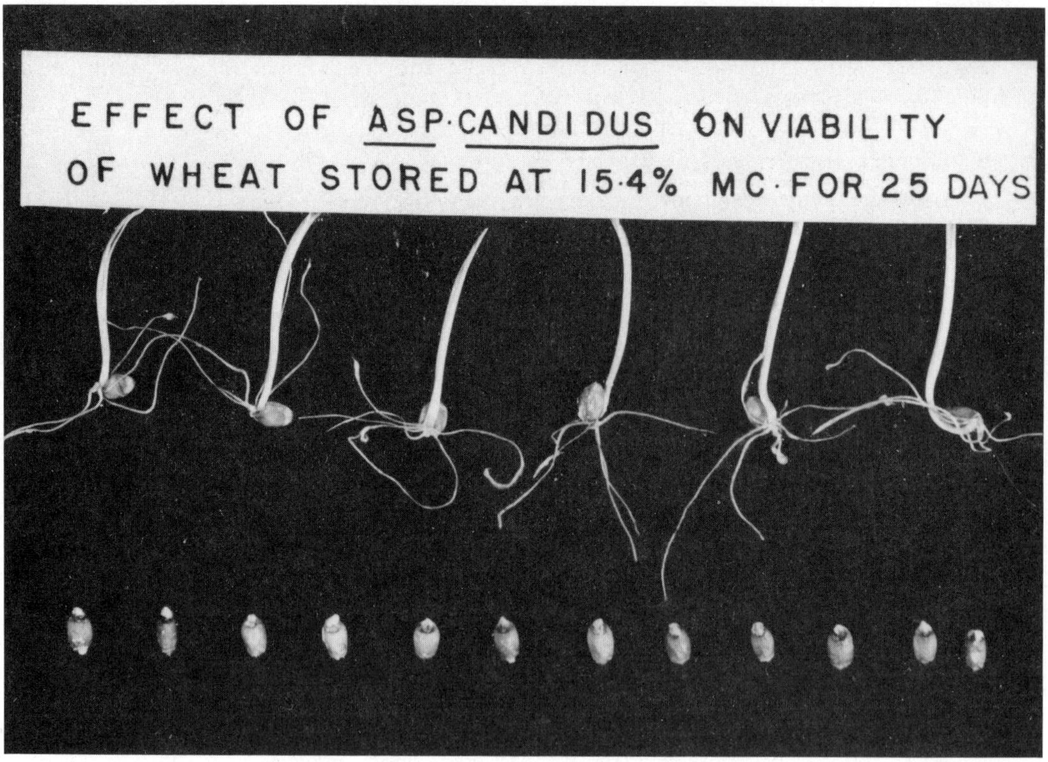

FIGURE 1. Reduction in germinability of wheat resulting from invasion of the germs by *Aspergillus candidus*. Top row, seeds free of fungi, germination 100%; bottom row, seeds inoculated with *Aspergillus candidus*, germination 0%.

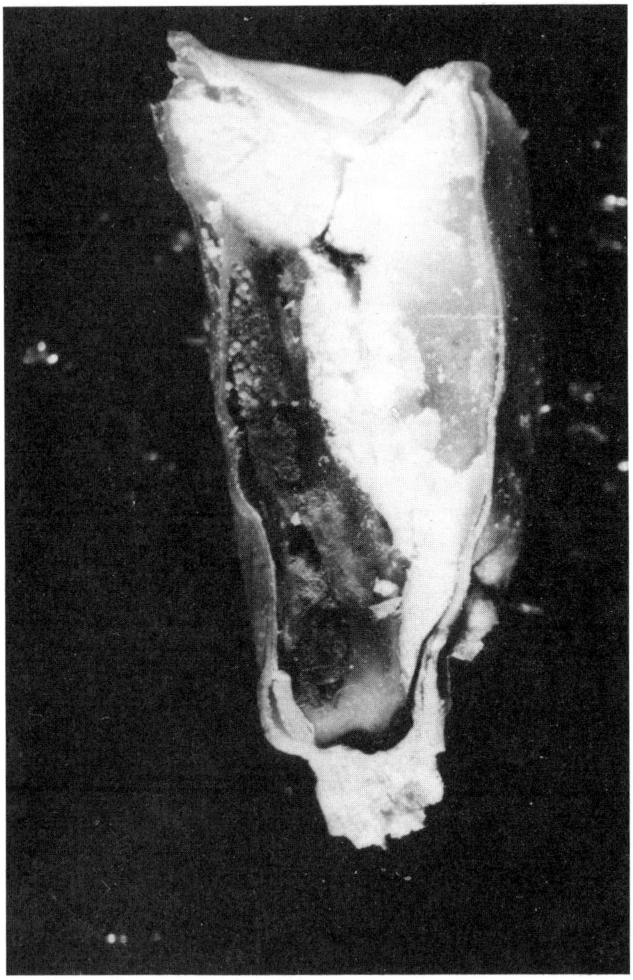

FIGURE 2. Corn kernel from commercial storage, split to show dark germ resulting from invasion by a storage fungus, probably *Aspergillus glaucus*.

1. Reduction in Germinability

Seeds of course can lose germinability from many causes—mechanical damage, processes inherent in their own makeup, age, and so on [36,37]. However, seeds held at moisture contents and temperatures in which their own activities are greatly reduced may be invaded by storage fungi and suffer reduction in germinability fairly rapidly, and in these circumstances the storage fungi can be the primary causes of loss in germinability [6,30].

Some of the species of *Aspergillus* involved in this, when growing in seeds at moisture contents near the lower limit that permit them to grow, invade and decay the germs of the seeds primarily or almost exclusively, and under such circumstances it is to be expected that germinability would be reduced.

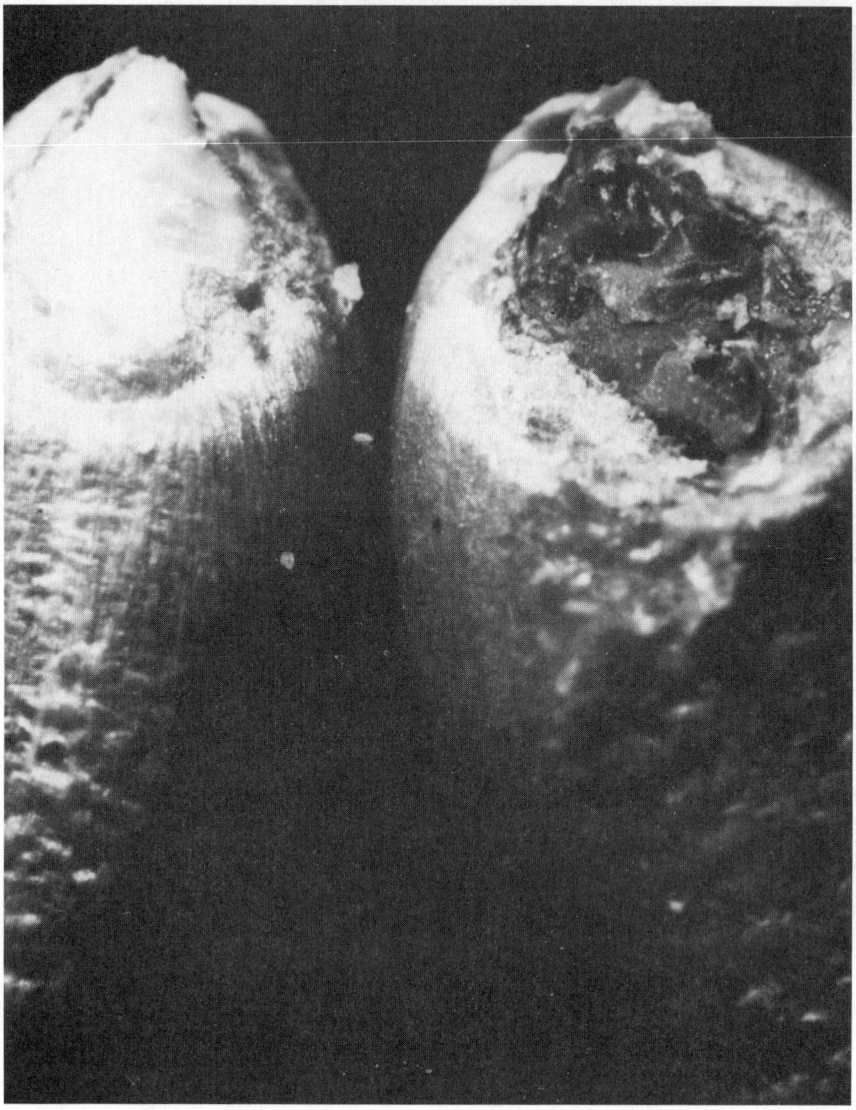

FIGURE 3. Wheat seed with the pericarps removed to expose the germs. Left, germ sound; right, germ blackened by a storage fungus.

2. Discoloration of Seeds, Mustiness, and Caking

Aspergillus restrictus and *A. candidus* especially, but also *A. glaucus* to a lesser extent, will cause the germs of seeds they are invading to turn jet black; in wheat this for a long time was known as "sick" wheat, and its cause was a mystery; it is a mystery no longer, but a product of invasion by known storage fungi. More extensive invasion may result in

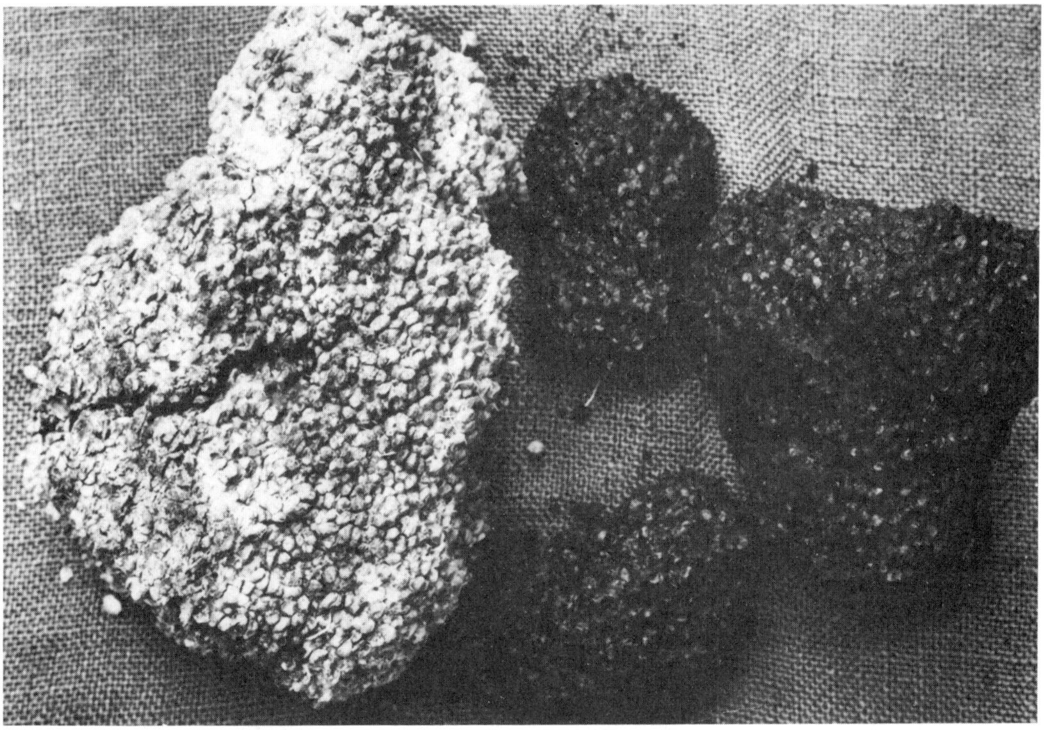

FIGURE 4. Soybeans caked (left) and blackened (right) by storage fungi (samples from commercial storage).

discoloration of the whole seeds or kernels. This discoloration may be preceded or accompanied by the development of caking, from abundant fungus mycelium, sometimes into solid masses, and the development of strong musty odors. By the time spoilage reaches these stages, where it can be detected with the unaided eye, or the unaided nose, it is fairly far advanced. Depending on the conditions that prevail, it may then proceed to the final stages, as described below.

3. Heating

Spontaneous heating, occasionally to the point of ignition, occurs in plant materials such as hay, manure piles, sugar cane bagasse, sugar beet pulp, baled cotton, wood chips, brewers' grains, pelleted feeds, and bulk-stored seeds and grains. Insects and mites may initiate this heating in stored grains (but not in the other materials listed in which heating also occurs), but such heating may be initiated by molds alone. *Aspergillus candidus* and *A. flavus* can carry the heating up to a temperature of 55°C and maintain it there for some weeks. The hotspots may then "burn themselves out" as the grain men say, or, under the right circumstances, thermophilic fungi and thermophilic bacteria may take over and raise the temperature up to 75°C. At that temperature short-chain hydrocarbons may be produced that, when exposed to air, oxidize rapidly and result in actual fire—authentic cases are known in which chunks of caked, heating corn and soybeans, when broken out from the solid heating mass, burst into flame. This heating to ignition occurs more commonly in soybeans than in the cereal grains, but does occur in corn also. This destroys not only the seeds or grains, but sometimes the storage structures as well. Detailed accounts of specific cases of heating in grains and seeds in commercial storage are given in Christensen and Kaufmann [25,26] and in Christensen and Meronuck [17].

VII. CONCLUSION

Fungi that invade seeds before harvest may cause discolorations, various blights and blemishes, provide inoculum to produce disease in the next planting, and affect the quality of the seed for planting, for food, feed, fermentation and for processing. Some seed-invading fungi produce potent toxins that may cause injury to the animals that consume them. Fungi that invade seeds in storage may cause reduction in germinability, discoloration, especially of the embryos, mustiness, caking, heating, and total spoilage.

REFERENCES

1. U.S. Department of Agriculture Federal Grain Inspection Service. *The Official United States Standards for Grains*. U.S. Government Printing Office, Washington, D.C., 1978.
2. Warnack, D. W., and Preece, T. F. Location and extent of fungal mycelium in grains of barley. *Trans. Br. Mycol. Soc.*, *56*:267-273, 1971.
3. Christensen, C. M. Fungi on and in wheat seed. *Cer. Chemistry*, *28*:408-415, 1951.
4. Pusposendjojo, N., and Christensen, C. M. Fungus flora of coconuts. *Turrialba*, *27*:255-258, 1977.
5. Moore, M. B., and Olien, C. R. Mercury bichloride as a surface disinfectant for cereal seeds. (Abstr.) *Phytopathology*, *42*:471, 1952.
6. Olien, C. R., and Moore, M. B. Certain mercurial seed treatments do not kill fungi on seed prior to planting. (Abstr.) *Phytopathology*, *44*:500, 1954.
7. Boller, R. A., and Schroeder, H. W. Influence of relative humidity on production of aflatoxin in rice by *Aspergillus parasiticus*. *Phytopathology*, *64*:17-21, 1974.
8. Raper, K. B., and Fennell, D. I. *The Genus Aspergillus*. Williams and Wilkens, Baltimore, Maryland, 1965.
9. Diener, U. L., and Davis, N. D. Limiting temperature and relative humidity of growth and production of aflatoxin and free fatty acids by *Aspergillus flavus* in sterile peanuts. *J. Am. Oil Chem. Soc.*, *44*:259-263, 1967.
10. Davis, N. D., and Diener, U. L. Environmental factors affecting the production of aflatoxin. In *Toxic Microorganism* (M. Herbzer, ed.), Published by the UJN-Joint Panels on Toxic Microorganisms and the U.S. Dept. of Interior, 1975, pp. 43-47.

11. Koehler, B. Fungus growth in shelled corn as affected by moisture. *J. Agric. Res.*, *56*:291-302, 1938.
12. Lopez, L. C., and Christensen, C. M. Effect of moisture content and temperature on invasion of stored corn by *Aspergillus flavus*. *Phytopathology*, *57*:588-590, 1963.
13. Ayerst, G. The effects of moisture and temperature on growth and spore germination of some fungi. *J. Stored Prod. Res.*, 5:127-141, 1969.
14. Christensen, C. M., and Mirocha, C. J. Relation of relative humidity to the invasion of rough rich by *Aspergillus parasiticus*. *Phytopathology*, *66*:204-205, 1976.
15. Lillehoj, E. B., Fennell, D. I., and Hesseltine, C. W. *Aspergillus flavus* infection and aflatoxin production in mixtures of high moisture and dry maize. *J. Stored Prod. Res.*, *12*:11-18, 1976.
16. Nelson, P. E., Toussoun, T. A., and Marasas, W. F. *Fusarium Species: An Illustrated Manual for Identification*. Pennsylvania State University Press, University Park, 1983.
17. Christensen, C. M., and Meronuck, R. A. *Quality Maintenance in Stored Grains and Seeds*. The University of Minnesota Press, Minneapolis, 1986.
18. Pepper, E. J., and Kiesling, R. L. A list of bacteria, fungi, yeasts, nematodes and viruses occurring on and within barley kernels. *Proc. Asso. Official Seed Analysts*, *53*:199-208, 1963.
19. Miles, M. R., and Wilcoxson, R. D. Kernel discoloration of barley. In *Vistas in Plant Pathology* (A. Varma and J. P. Varma, eds.), Malhotra Pub., New Delhi, 1986, pp. 139-155.
20. Wiese, M. V. (ed.). *Compendium of Wheat Diseases*, American Phytopathology Society, 1977.
21. Sinclair, J. B. (ed.). *Compendium of Soybean Diseases*, Amer. Phytopath. Soc. Press, St. Paul, 1982.
22. Fischer, G. W., and Holton, C. S. *Biology and Control of the Smut Fungi*, The Ronald Press, New York, 1957.
23. Snyder, W. C., and Hansen, H. N. The species concept in *Fusarium* with reference to discolor and other sections. *Amer. J. Bot.*, *32*:657-666, 1945.
24. Christensen, C. M., Mirocha, C. J., and Meronuck, R. A. Molds and mycotoxins in feeds. Univ. of Minnesota Miscellaneous Report 142, revised, 1987.
25. Christensen, C. M., and Kaufmann, H. H. Deterioration of stored grains by fungi. *Annu. Rev. Phytopathology*, *3*:69-84, 1965.
26. Christensen, C. M., and Kaufmann, H. H. *Grain Storage. The Role of Fungi in Quality Loss*. University of Minnesota Press, Minneapolis, 1969.
27. Diener, U. L., Asquith, R. L., and Dickens, J. W. (eds.). Aflatoxin and *Aspergillus flavus* in Corn. Southern Cooperative Series Bulletin 279, Auburn University, ALA, 1983.
28. Zuber, M. S., Lillehoj, E. B., and Renfro, B. L. (eds.). Aflatoxin in maize. A Proceedings of the Workshop, CIMMYT, Mexico, 1987.
29. Mirocha, C. J., Pathre, S. V., and Christensen, C. M. Zearalenone. In *Mycotoxins* (J. V. Rodericks, C. W. Hesseltine, and M. A. Mehlman, eds.), Pathotox Publishers, Park Forest South, IL, 1977, pp. 345-364.

30. Mirocha, C. J., Christensen, C. M., and Nelson, G. H. Estrogenic metabolite produced by *Fusarium graminearum* in stored corn. *Appl. Microbiol.*, *15*:497-503, 1967.
31. Cote, L. M., Reynolds, D. J., Vesonder, R. F., Buck, W. B., Swanson, S. P., Coffey, R. T., and Brown, D. C. Survey of vomitoxin-contaminated feed grains in midwestern United States and associated health problems in swine. *J. Am. Vet. Med. Assn.*, *184*:189-192, 1984.
32. Trenholm, H. L., Hamilton, R. M. G., Friend, D. W., Thompson, B. K., and Hartin, K. E. Feeding trials with vomitoxin (deoxynivalenol)-contaminated wheat: Effects on swine, poultry and dairy cattle. *J. Am. Vet. Med. Ass.*, *185*:527-531, 1984.
33. Snyder, W. C., and Hansen, H. N. The species concept in *Fusarium* with reference to discolor and other sections. *Amer. J. Bot.*, *32*:657-666, 1945.
34. Lee, Y. W., Mirocha, C. J., Schroeder, D. J., and Walser, M. M. TDP-1, a toxic component causing tibial dyschondroplasia in broiler chickens, and trichothecenes from *Fusarium roseum* 'Graminearum'. *Appl. Environ. Microbiol.*, *50*:102-107, 1985.
35. Christensen, C. M., and Sauer, D. B. Microflora. In *Storage of Cereal Grains and Their Products*, 3rd ed. (C. M. Christensen, ed.). American Association of Cereal Chemists, St. Paul, MN, 1982, pp. 219-240.
36. Boller, R. A., and Schroeder, H. W. Aflatoxin producing potential of *Aspergillus flavus-oryzae* isolated from rice. *Cer. Sci. Today*, *11*:342-344, 1966.
37. Roberts, E. H. (ed.). *Viability of Seeds*. Chapman and Hall, Ltd., London, 1972.

5

GRAIN FUNGI

JOHN LACEY, NANNAPANENI RAMAKRISHNA, and ALISON HAMER
Agricultural and Food Research Council Institute of Arable Crops Research, Rothamsted Experimental Station, Harpenden, Hertfordshire, England

NARESH MAGAN and IAN C. MARFLEET* *Biotechnology Centre, Cranfield Institute of Technology, Cranfield, Bedford, England*

I. INTRODUCTION

Cereal grains are rarely, if ever, entirely free of microbial contamination. The numbers and types of microorganisms present differ with stage of development and ripening of the grain before harvest and with storage conditions afterwards. They comprise bacteria, fungi, and actinomycetes, and they have differing effects on the quality of grain and on ways in which they contribute to losses.

Estimates of the losses of cereal grains in store from all causes differ widely. They may amount to about 10% worldwide [1] but may reach 50% in tropical countries [2]. However, Rohani et al. [3] found storage losses of paddy in West Malaysia of only 1%. By contrast, Vasan [4] estimated losses of high moisture paddy in southern India exceeding 15-25% in only 9 days. Further losses of soluble materials occurred in subsequent parboiling, through shattering during milling, and as gruel during cooking.

Fungi contribute greatly to these losses, either alone or together with insects. They cause undesirable discoloration and contribute to heating and to losses of dry matter through the utilization of carbohydrates as energy sources; they may degrade lipids and proteins or alter their digestibility; they may produce volatile metabolites giving off odors; they cause loss of germination and of quality for baking, malting and other food uses, for use as animal feed or as seed; and they may produce mycotoxins that are highly toxic or carcinogenic or that cause feed refusal and emesis (see Chapter 4). Their spores may present respiratory disease hazards to exposed workers [5,6].

Little is known of how individual fungal species contribute to these losses, although they differ widely in their responses to environmental conditions with greatly different maximum, minimum, and optimum temperatures and minimum and optimum water potentials for spore germination, growth, and sporulation, and in their tolerance of unfavorable intergranular carbon

Present affiliation: Lancashire Polytechnic, Preston, England dioxide and oxygen concentrations. Indeed, the species present can give such information on the way in which grain has been stored and on the possible health hazards consequent on its use.

In this chapter, we describe the colonization of grain pre- and postharvest, the environmental responses of grain fungi, the dynamics of their growth, and their effects on quality.

II. FUNGAL COLONIZATION OF CEREAL GRAINS

A. Preharvest Colonization of Grain

1. Environment

Microorganisms start to colonize grain soon after ear emergence and continue to develop throughout its maturation and ripening. They have to contend with extremes of environmental conditions since the ear is exposed to lower vapor pressures, greater wind speeds, and more radiation than other parts of the plant, especially by day, although temperatures may be less extreme than at the level of maximum leaf area. During grain development, the availability of water to superficial microorganisms is determined, at least in part, by the water potential of the underlying tissue, but it is increasingly controlled by rainfall and atmospheric humidity as the grain ripens and dries and is subject to wide fluctuations. Nutrient availability also changes with ripening. While the plant is still green, surface water may carry solutes derived from plant cells, microorganisms, and pollutants, which may be either nutritious or inhibitory. With ripening and senescence, chlorophyll and proteins are lost as degradative enzyme activity increases and nutrients are lost from cells, as membranes lose their integrity, becoming more readily available to superficial microorganisms. These changes are reflected in the development of the superficial microflora.

2. Temperate Small Grains

Propagules of a wide range of microorganisms become deposited on aerial plant parts, either from the air, through splashing from the soil or through contact with other plants. The first colonizers of wheat and barley grain (Table 1) after ear emergence are bacteria [7,8], but these are soon replaced by yeasts and then by filamentous fungi, especially after anthesis

TABLE 1. Principal Preharvest Fungi of Wheat and Barley Grain in the United Kingdom

Yeasts	*Cryptococcus albidus*, *C. laurentii*, *C. macerans*, *Rhodotorula glutinis*, *Sporobolomyces roseus*, *Torulopsis ingeniosa*, *Trichosporon cutaneum*
Yeastlike fungi	*Aureobasidium pullulans*, *Stephanoascus rugulosus*
Filamentous fungi	*Alternaria alternata*, *Cladosporium cladosporioides*, *C. herbarum*, *Epicoccum purpurascens*, *Fusarium culmorum*, *Verticillium lecanii*

Source: Refs. 7, 8, 10.

(growth stages (G.S.) 61-69 [9]). In studies of wheat and barley grain at Rothamsted over several seasons, populations of bacteria always exceeded those of fungi and usually peaked between G.S. 70 and 80. Occasionally, however, populations continued to increase up to harvest, especially when hot, dry weather restricted their development early in ear development and then cooler wetter weather occurred during ripening. The species of bacteria have seldom been identified but, by analogy with *Lolium perenne* [11, 12], might be expected to include *Pantoea agglomerans* (*Enterobacter agglomerans, Erwinia herbicola*), *Xanthomonas campestris, Flavobacterium* spp., and *Pseudomonas flavescens.*

Yeasts may be differentiated into two groups on the basis of color. Both white yeasts, including *Cryptococcus laurentii, C. albidus,* and *Rhodotorula* (*Torulopsis*) *ingeniosa,* and pink yeasts, including *Sporobolomyces roseus* and *Rhodotorula glutinis*, followed a similar pattern to bacteria, although populations were usually smaller and they fluctuated more widely [10]. However, *Cryptococcus macerans* and *Trichospororon beigelii* (*T. cutaneum*) showed distinctive seasonal patterns. *Cryptococcus macerans* occurred only early in grain development, from G.S. 60 to 85, with maximum populations at G.S. 70, while *T. beigelii* appeared only at G.S. 75, peaked at G.S. 85, and then disappeared again by G.S. 90.

The remaining fungi can also be divided into two groups: yeastlike fungi and filamentous fungi. The yeastlike fungi, comprising *Aureobasidium pullulans* and *Stephanoascus rugulosus* (previously referred to as *Hyalodendron* spp.), behaved similarly to yeasts [7,8]. *Stephanoascus rugulosus* has been isolated in such great abundance in studies of the superficial microflora of wheat and barley at Rothamsted [7,8] that it is surprising that they have not been reported elsewhere. Of the filamentous fungi, *Cladosporium* spp. and *Verticillium lecanii,* a mycoparasite, were usually first to appear, between G.S. 50 and 60, and they generally increased steadily to harvest. *Alternaria alternata* appeared on dilution plates at about G.S. 70 and *Epicoccum purpurascens* at about G.S. 75. However, *A. alternata* can be isolated from surface-sterilized direct-plated grain about 2 weeks before this, indicating infection and penetration of the pericarp some time before sporulation. The presence of subepidermal mycelium in cereal grains has been widely reported [13] and much of this can be attributed to *A. alternata.* By harvest, almost all wheat and barley grains may carry *Alternaria,* possibly resulting in decreased kernel size [7,8,14,15].

The incidence of *Fusarium* spp., especially *F. culmorum* in England and *F. graminearum* (*Gibberella zeae*) in Canada, is greatly affected by weather at anthesis (G.S. 61-69). For instance, at Rothamsted in 1982, populations of *F. culmorum* exceeded those of *A. alternata* in contrast with other years when only occasional isolations were made. *Fusarium graminearum* ear blight of wheat in Canada is favored by periods of 48-60 h at anthesis when surfaces are wet and temperatures are 20-30°C [16]. Pollen may stimulate infection, as with *Botrytis* [17], and stimulants, identified as choline and betaine, have been isolated from anthers [16].

Interrelationships of microorganisms with environmental factors and with one another have been examined using principal component analysis [18]. This enables large bodies of data from random samples to be analyzed to identify any underlying order and structure in the system.

Correlation matrices are used to identify the variables or principal components that account for portions of the total variance. The first principal component accounts for the greatest proportion of the total variance, and each subsequent principal component accounts for a successively smaller proportion. Studies of samples of barley seed from different parts of Canada suggest that both competition and antagonism occur. The first principal component, accounting for almost 25% of the total variability, included *Cochliobolus*, *Alternaria*, and *Acremonium* with positive loadings and *Cladosporium* and *Epicoccum* with negative loadings. Within the two groups, species seem to compete spatially for the same niche while antagonism occurs between the two groups. For instance, *Alternaria* and *Cochliobolus* both penetrate the pericarp, and *Acremonium* is a parasite of *Cochliobolus*. *Epicoccum* is a strong antagonist. The first group occurred most abundantly in Manitoba, which has a short growth period, and the second in Alberta, where the longer growing season favored the slower-growing *Cladosporium*. *Alternaria* and *Fusarium* are also antagonistic, resulting in a lower incidence of *Alternaria* in grain with a high incidence of *F. culmorum* [8,19,20].

Colonization of the ripening grain can be modified by prolonged wet weather and by lodging, which allows the grain to remain wet for long periods after rain. This favors growth of *Fusarium* spp., especially *F. culmorum* [19,21], and of *Penicillium* spp. usually considered characteristic of storage, such as *P. hordei*, *P. granulatum*, and *P. aurantiogriseum*. Otherwise, *Penicillium* spp. are only occasionally isolated before harvest with only *P. pedemontanum* and *Eupenicillium* spp. including *E. shearii* characteristic [22].

In Canada, early snowfall can result in swathed crops being left on the ground over winter. By spring, the grain is well colonized by preharvest fungi, with *Alternaria*, *Fusarium*, *Cochliobolus*, and *Epicoccum* recorded, respectively, on 87, 35, 12, and 2% of the grains [23]. The *Fusarium* spp. may include well-known mycotoxin producers, such as *F. tricinctum*, *F. poae*, *F. sporotrichioides*, *F. avenaceum*, *F. acuminatum*, and *F. sambucinum*. The microflora closely resembled that found in the U.S.S.R. in mild winters when millet was overwintered under a deep snow cover and then subjected to a slow thaw with frequent thaw-freeze cycles. Under these conditions the mycotoxicosis alimentary toxic aleukia was common in people consuming the grain [24].

In warmer climates, the most abundant components of the microflora are similar to those in temperate climates, with almost all grain carrying *Alternaria alternata* and *Cladosporium cladosporioides* and 13% carrying *Epicoccum purpurascens*. However, the principal *Fusarium* may be *F. semitectum* (56% of grains), and this may be accompanied by *Aspergillus* (43%), *Curvularia* (25%), *Drechslera* and *Penicillium* spp. (both 18%). The *Aspergillus* spp. included *A. fumigatus*, *A. flavus*, *A. nidulans*, *A. sydowii*, *A. terreus*, and *A. versicolor*, all species typical of storage, while the *Penicillium* included *P. chrysogenum*, *P. funiculosum*, and *P. oxalicum* [25]. *Alternaria alternata* may be an important cause of black point disease in India and elsewhere when rains are frequent during grain development [26]. In Nigeria, *Phoma sorghina* and *Fusarium semitectum* may occur on >90% of grains before harvest [27].

TABLE 2. Principal Preharvest Fungi of Maize Grain in the United States

Yeasts	*Cryptococcus* spp., *Sporobolomyces* spp.
Yeastlike fungi	*Aureobasidium pullulans*
Filamentous fungi	*Acremonium strictum, Alternaria alternata, Aspergillus flavus, A. niger, Cladosporium* spp., *Curvularia lunata, Fusarium subglutinans, Gibberella fujikuroi (F. moniliforme), G. zeae (F. graminearum), Mucor* spp., *Nigrospora oryzae, Penicillium funiculosum, P. oxalicum, Rhizopus* spp., *Trichoderma viride*

Source: Refs. 28-30.

3. Maize

The colonization of maize is restricted to a large extent by the husks that enclose the ears. Although the silks may carry *Penicillium* spp., mucoraceous fungi and yellow-pigmented bacteria at anthesis, the kernels remain sterile for up to 3 weeks. However, 8 weeks after anthesis, all ears carry some contaminated kernels [28]. The principal species are listed in Table 2. Although some of the species are the same as those on wheat and barley, there are also notable differences. One of the most marked differences is the frequent occurrence of *Fusarium moniliforme*. This first occurs about 2 weeks after mid-silk, and by 10 weeks after, up to 66% of kernels may carry the fungus endogenously [31]. *Acremonium strictum* is more variable in its occurrence and infection occurs later; first infections occur 3-4 weeks after mid-silk, and numbers increase to 10 weeks when about 45% of kernels are infected [31,32]. These species also differ in their distribution on the ear, with *F. moniliforme* most common at the tip end and *A. strictum* at the butt end. In South Africa, *F. moniliforme* is favored by the subtropical climate of northern Transvaal and *Fusarium subglutinans* by the cooler, wetter weather of eastern Transvaal. *Gibberella zeae* (*Fusarium graminearum*) is most common and infects up to 8% of the grains in the intermediate climate of western Transvaal [33]. Both Wicklow [34] and Hill et al. [35] found negative associations between *F. moniliforme* and other fungi, suggesting that prior invasion by the *Fusarium* could inhibit colonization by other species.

Gibberella zeae is also important in the cooler temperate climate of Canada where infection may be followed by mycotoxin formation. Maize is susceptible from soon after ear emergence, but infection in Ontario was correlated with rainfall in August. Infection was frequent when >70-80 mm rain fell within 6 to 9 days and infrequent when <60 mm fell over 10 to 16 days [16].

Penicillium spp. are usually considered as storage fungi, but *P. funiculosum* and *P. oxalicum* in the midwest United States, *P. funiculosum, P. purpurogenum,* and *P. citrinum* in southeastern United States, and *P. chrysogenum, P. steckii,* and *P. purpurogenum* in Spain seem characteristic of preharvest colonization [34-37]. *Penicillium oxalicum* can cause lesions on husks and kernels, affecting seed viability and causing ears to rot, while *P. funiculosum* causes streaking of the pericarp but does not affect seed viability or cause ear rotting [38].

Aspergillus spp. are also characteristic storage fungi, but *A. flavus* and, to a lesser extent, *A. parasiticus* are now known to frequently colonize maize and to produce aflatoxins before harvest, especially in the southeastern United States. Incidence of *A. flavus* and aflatoxin in ears before harvest is correlated with weather conditions, being favored by high temperatures and drought at the late milk to dough stage, when maize is most susceptible, and by insect damage. Consequently, there are differences in the time and amount of aflatoxin production between crops sown on different dates when the weather during kernel development differs. More aflatoxin can occur in the drier, more mature grain of an earlier crop through exposure to drought and/or temperatures above 30°C at the most susceptible stage or because of the longer period from then until harvest than with a later crop. This is despite the earlier kernel invasion after silking that occurs in later sown crops than in those sown earlier. Thus, a crop sown on March 29 showed kernel invasion commencing 30 days after full silk, while one sown on May 8 showed invasion commencing after only 15 days [35]. Damaged ears are more likely to be infected with *A. flavus* and to contain aflatoxin than undamaged, although in the southeastern United States only 10% of infection could be accounted for by insect damage. Elsewhere, aflatoxin was found where 21% of ears were damaged by ear worms but none where only 5% were damaged [39]. A positive association has been noted between *A. flavus* and *A. niger* and between *A. niger* and *P. funiculosum* on maize [34] which could affect aflatoxin production. Hill et al. [40] found that aflatoxin production in groundnuts was inhibited when the ratio of seeds yielding *A. flavus* to those yielding *A. niger* exceeded 19:1 but not when it was less than 9:1.

Infection of maize with *Helminthosporium maydis* ear rot can also increase the incidence of other fungi [41]. Thus, *A. flavus*, *Penicillium* spp., and *Fusarium* spp. were all more frequent in seeds from *H. maydis*-infected ears than in healthy ears.

4. Sorghum

Sorghum is an important crop in the drier tropics, usually flowering just as or after rains cease. However, earlier flowering cultivars may be exposed to rain during grain development, giving rise to extensive development of fungi, causing grain mold. Many fungi have been isolated from ears affected by grain mold. The most important are probably *Fusarium moniliforme* and *Curvularia lunata*, which both colonize young developing kernels from the base and penetrate the endosperm, while *Cladosporium* spp., *Alternaria* spp., *Phoma* spp., *Fusarium semitectum*, and, sometimes, *F. equiseti* colonize the exposed tip of the grain [26,42-45]. *Aspergillus flavus* may also occur and produce aflatoxins [46].

5. Rice

Despite its importance as a crop, there is little information on the preharvest colonization of rice by fungi. Those found by Kuthubutheen [47] are listed in Table 3. The number of species increased from ear emergence to harvest, but few occurred on more than 60% of grains.

TABLE 3. Preharvest Fungi of Malaysian Rice

Ear emergence	*Aspergillus niger*, *Aureobasidium pullulans*, *Choanephora cucurbitarum*, *Curvularia lunata*, *Fusarium* spp. (*F. culmorum*, *F. oxysporum**), *Nigrospora sphaerica*, *Penicillium* spp., *Pithomyces maydicus*
Milk stage	*Aspergillus* spp. (*A. fumigatus*, *A. niger*), *Aureobasidium pullulans*, *Choanephora cucurbitarum*, *Cladosporium* spp. (*C. cladosporioides*, *C. oxysporum*), *Cunninghamella echinulata*, *Curvularia* spp. (*C. lunata**, *C. pallescens*, *C. senegalensis*), *Fusarium* spp. (*F. culmorum*, *F. oxysporum**, *F. solani*), *Nigrospora sphaerica**, *Penicillium* spp., *Pestalotiopsis* spp. (*P. quepinii*, *P. versicolor*), *Pithomyces maydicus*, *Rhizopus stolonifer*, *Syncephalastrum racemosum*
Harvest	*Alternaria* spp. (*A. alternata**, *A. padwickii*), *Aspergillus* spp. (*A. flavus*, *A. fumigatus*, *A. niger*, *A. terreus*), *Cercospora oryzae*, *Chaetomium globosum*, *Cladosporium* spp. (*C. cladosporioides*, *C. oxysporum*), *Coprinus* sp., *Curvularia* spp. (*C. eragrostidis*, *C. lunata**, *C. lunata* var. *aeria**, *C. pallescens**, *C. senegalensis*), *Dichotomophthoropsis nymphearum*, *Drechslera* spp. (*D. halodes*, *D. oryzae*), *Fusarium* spp. (*F. culmorum**, *F. oxysporum**, *F. solani*), *Gibberella fujikuroi*, *Nigrospora* spp. (*N. oryzae*, *N. sphaerica**), *Penicillium* spp., *Pestalotiopsis quepinii*, *Phoma* spp., *Pithomyces* spp. (*P. chartarum*, *P. maydicus*), *Rhizopus stolonifer*, *Syncephalastrum racemosum*, *Trichoderma viride*

*Predominant species, on ≥ 60% of seeds.
Source: Ref. 47.

B. Harvest

Harvest marks a profound change in the grain ecosystem, from the environmental extremes of the ear to the comparatively equable environment of the grain store. However, it also provides an opportunity for the redistribution of fungal inoculum in the grain and for the introduction of further inoculum. In temperate regions, storage fungi are few before harvest, but during combine harvesting the inoculum of *Penicillium* spp. may be increased more than 250 times by contamination with residues from the harvester and with soil. By contrast, inoculum of *Eurotium* spp. was little changed [48].

C. Grain Drying

Drying is used to decrease the water content of grain so that fungi are unable to grow in store. However, without due care it can have the opposite effect. Grain driers can be classified into three groups [49].

1. Continuous Flow Driers

Thin layers of grain are exposed to high temperatures, which may differ according to the crop, its water content, and its end use [50], for relatively short periods of time. To ensure that viability is not affected, the upper limit for seed barley dried from 20% to 15% moisture is 75°C. Milling wheat can be dried with air temperatures up to 90°C before bread-making qualities are affected [51], while the maximum temperature for drying feed wheat, barley, and oats is 104°C. However, too rapid heating may cause stress cracks to form in maize grain, increasing susceptibility to fungal invasion, while overheating may alter the relationship between water availability and water content, perhaps leading to greater susceptibility to molding at given water contents [52]. This method can be applied to most crops with virtually any water content, but it is costly in its use of energy. However, it can be efficient when working at the maximum allowable temperatures and when drying the crop by only 4-5%. After continuous flow drying, grain is still warm and must be allowed to cool, perhaps aided by intermittent aeration (see Sec. II.C.3), to prevent moisture migration.

2. Batch Drying

Discrete batches of grain are dried to completion, using low temperatures, before placing in store. Radial ventilated bins are perhaps the most popular type of batch dryer, operating at 6-12°C above ambient air temperatures. Output rates have been slow until the recent development of fully automated systems, which are able to rotate and thoroughly mix the grain.

3. Low Temperature and Ambient Air In-Store Drying

The stored crop is ventilated with air at near-ambient temperatures in order to remove water and respiratory heat before grain deterioration occurs. This is the most popular type of drying system used in the United Kingdom, since it can be the least expensive under optimum operating conditions. Air usually enters the grain through ducts in the silo or barn floor and then passes through the grain bed. Heat is supplied to the system, when necessary, by powerful gas or electrical fan heaters adjacent to the air inlets. The extent of heating is determined by the temperature and relative humidity of the ambient air at the time of drying. With ambient air drying, no heat may be supplied. The relative humidity of the air moving through the grain equilibrates with the water content of the grain until it is saturated. Consequently, the grain dries from the bottom upwards, and that at the surface may remain damp for a long period after the store is filled, during which time it is highly susceptible to fungal spoilage. Optimization of drying strategies must retain air within the grain bed as long as possible to maximize its moisture carrying capacity while ensuring that the drying front reaches the surface before spoilage occurs [53].

D. The Development of Fungi in Stored Grain

1. Environment

Conditions in stored grain are more stable than those in the field. The insulating effect of the grain leads to temperatures within the bulk showing smaller diurnal fluctuations than ambient air, while the humidity of the intergranular air is largely controlled by the availability of water in the grain. Except in outdoor maize cribs, air flows only through convection or artificial ventilation. The gaseous content is controlled by the amount of respiratory activity by both grains and microflora, both of which are determined by the availability of water in the grain, but in moist grain stores it can contain >50% CO_2 and <1% O_2. However, respiration is reported to be inhibited when CO_2 concentrations exceed 10% [54].

The concept of water activity: Total water content, on either a fresh weight or dry weight basis, is the easiest and most commonly used method for assessment of the condition of stored grain. However, although it is easily determined, it gives little indication of the availability of water for microbial growth and allows few comparisons between grains and seeds of different types since each has a different relationship between water content and water availability.

The water contained in materials like cereal grains consists of bound water (water of constitution), which is chemically absorbed to the substrate surface initially in a monolayer of molecules, and free water, which is more weakly bound as more and more layers of molecules accumulate on surfaces. The free water is more readily available for microbial growth and metabolism. Water availability in hygroscopic materials can be more satisfactorily expressed in terms of water activity. Water activity (a_w) is the ratio between the vapor pressure of water in a substrate (P) and the vapor pressure of pure water (P_0) at the same temperature and pressure, as expressed by the formula:

$$a_w = \frac{P}{P_0}$$

Microorganisms react in response to the a_w of a substrate, not the water content, and information obtained on the a_w responses of fungi colonizing one product is applicable to any others. Thus, the range of a_w allowing fungal growth is 1.00 (pure water) down to about 0.6 a_w. A_w also determines the rate of microbial growth and respiration, and hence the output of heat that may lead to spontaneous heating (see Sec. II.D.1).

Other measures of water availability are equilibrium relative humidity (ERH) and water potential (ψ). ERH is the relative humidity of the intergranular air in equilibrium with water in the grain substrate and is numerically the same as a_w, except that it is expressed as a percentage, whereas a_w is a decimal fraction of one. Water potential, predominantly used in soil microbiology, is the sum of osmotic, matric, and turgor potentials and is measured in pascals (Pa). The relationships between these three measures of water availability are shown in Table 4.

Moisture sorption isotherms are used to portray the relationship between a_w and total water content in different materials. Each is a sigmoid curve, but the relationship differs with temperature and it also depends on whether water is being gained (absorbed) or lost (desorbed) because of a hysteresis phenomenon [55].

Spontaneous heating: Grain temperature changes not only in response to changes in the ambient air temperature but also in response to the

TABLE 4. A Comparison of Water Activity (a_w), Equilibrium Relative Humidity (ERH), and Water Potential Values at 25°C

a_w	ERH (%)	Water potential (MPa)
1.00	100	0
0.95	95	-7.1
0.90	90	-14.5
0.85	85	-22.4
0.80	80	-30.7
0.75	75	-39.6
0.70	70	-49.1
0.65	65	-59.2
0.60	60	-70.3

microbial and arthropod activity, a process known as spontaneous heating. Energy is released by respiration under aerobic conditions according to the following formula [56]:

$$C_6H_{12}O_6 + 6\ O_2 \rightarrow 6\ H_2O + 6\ CO_2 + 2835\ kJ$$

Heating occurs when this energy is released faster than it can escape from the substrate. By contrast, little energy is released by anaerobic respiration and little or no heating occurs in the absence of oxygen. The requirement for oxygen increases with temperature, to a maximum at 40°C but does not decrease greatly until the temperature exceeds 65°C. At this temperature, microbial growth is largely inhibited and heating results from exothermic chemical oxidation. Thus, the respiratory quotient may be 0.7 to 0.9 up to 66°C but less than 0.5 at higher temperatures [57].

Initially, some of the energy may be released by respiring grain but this is soon surpassed by microbial respiration and the grain is killed at about 40°C. The availability of water in the substrate determines the amount of activity that the substrate can support, the species that can grow, and, consequently, the rate of energy release and heating. The role of fungi in this process has been clearly demonstrated [58]. Heating commences at about 0.85-0.90 a_w (-22 to -15 MPa) and reaches a maximum of 65-70°C at 1.0 a_w (0 MPa). Heating usually occurs rapidly when uniformly damp grain is placed in store, reaching a maximum within a few days. Cooling then occurs gradually as microbial activity declines with drying of the grain and depletion of nutrient sources. Heating may also occur more locally, in "hot spots," if water is distributed unevenly in the bulk as a result of changes in water content during harvesting, leaks in the grain store allowing wetting by rainwater, or from moisture migration or insect infestation (see below). Low ambient temperatures may inhibit or delay heating, but in Canada heating to 20-45°C or higher can occur from initial temperatures of -5-0°C [59].

Spontaneous heating leads to convection within the grain bulk and moisture migration, which can lead to condensation, germination, molding, and spontaneous heating in other parts of the bulk. The development of molding in grain is determined by the interaction of environmental factors with microorganisms and arthropods as modified by chemical treatments and management strategies. The influence of these different aspects will be described below.

Moisture migration and hot spots: Diurnal and seasonal temperature changes in the external environment can cause the migration of water within grain bulks both in storage and during shipping. Water vapor may be carried from warm areas by convection or by moving down a temperature gradient to increase the relative humidity of air in cooler areas and, if dew point is reached, to condense and increase the water content of the surrounding grain. Thus, the amount of grain able to support fungal growth is increased [60]. Grain is a poor conductor of heat [61], so that ambient temperature changes principally affect grains adjacent to the walls of the storage container causing movement of water vapor toward the center or edge of the grain bulk, depending on the relative temperatures of grain and air. Such movement can be counteracted by aeration of the grain regularly as the seasons change, in order to buffer temperature effects.

The contents of a grain store are seldom uniform. Sound, whole grains are accompanied by green, shriveled, and broken grains, fragments of glumes and hairs, starch grains, soil, and other dust. These may accumulate in discrete zones within the silo, depending on the handling system used. Dust cones can form when silos are loaded by auger while broken grains accumulate below loading chutes in ships [62]. Such debris may differ in water availability from the sound grain, have a greater respiratory rate and provide a focus for insect invasion and fungal colonization that leads to the formation of "hot spots," localized areas of elevated temperature [63]. Moldy grain clumps together, inhibiting the air movement and allowing yet more heating.

Regular monitoring of relative humidity, temperature, and airflow at many points within the grain bulk following drying will enable the formation of hot spots to be detected [64]. Discovery of an area of high temperature or humidity then requires rapid action. Grain may be aerated with cool air, but this may lead to the temporary moistening of the cooler surface layers and, perhaps, sprouting and further mold development. Alternatively, the crop may be redried in batch or continuous flow dryers. Grain may be improved cosmetically in this way. Superficial fungal growth and spores may be removed but mycelium and mycotoxins may still remain undetected within the grains [65]. The airborne spores released may present serious health hazards to farm workers and livestock [66].

Similar effects may occur during the shipping of grain across different climatic zones. The problems of transporting grain by ship and the effects of changing wall temperatures when shipping crops from warm to cooler regions have been described [62,67]. Cargoes "sweat" as ships enter cooler waters because moisture migrates to the periphery and condenses on the walls of the hold. When warmer waters are encountered, the peripheral grain dries, causing moisture to condense on the cooler interior grain, again risking fungal spoilage.

These problems are almost inevitable with current trading practices which require grain to be sold on a wet weight basis. It is thus in the trader's interest to sell grain with as large a water content as possible. Trading grain on the basis of dry matter might prevent many spoilage problems.

2. Fungi Colonizing Grain Stored Conventionally in Aerobic Conditions

Many species of fungi have been found colonizing stored grain. These represent a range of different genera and include psychrotolerant, mesophilic, thermophilic, xerophilic, and hydrophilic species [68-72]. The most characteristic of these are species with *Aspergillus* and *Penicillium* anamorphs, although taxonomic revisions mean that many of the species recorded previously have been reduced to synonymy with others [73]. They include the extremely xerophilic *Eurotium* spp. (members of the *Aspergillus glaucus* group) and *A. restrictus*, the moderately xerophilic *A. candidus* and *A. flavus*, the only slightly xerophilic *A. fumigatus* with its broad temperature range for growth, the psychrotolerant *P. aurantiogriseum* and *P. verrucosum*, the mesophilic, slightly xerophilic *P. corylophilum* and *P. rugulosum*, and the thermophilic *Talaromyces thermophilus* (*P. dupontii*). Among the hydrophiles are *Fusarium* spp., Mucorales, and other thermophiles such as *Malbranchea cinnamomea*, *Rhizomucor pusillus*, *Thermoascus crustaceus*, and *Thermomyces lanuginosus*. Often species of *Penicillium* have not been identified because of difficulties in their recognition, but it is evident that they differ in their ecological requirements and in their tolerance of low a_w and high temperatures [23,37,74]. These species are widespread in their occurrence throughout the world and between them cover almost the whole range of environmental conditions likely to be found in stored grain.

Fungi can penetrate undamaged grain, although there is still some controversy as to whether the grain dies before or as a consequence of fungal invasion (E. Moreno, personal communication). However, Tuite and Christensen [75] concluded that fungal invasion either caused or greatly contributed to loss of viability. Perez et al. [76] found mold activity to be a more sensitive indicator of unfavorable storage conditions than germination. *Aspergillus* and *Penicillium* spp. may penetrate through the aleurone layer of wheat grains to the subaleurone and endosperm cells [77]. However, high moisture maize kernels remained mold-free for up to 9 days at 23°C provided there was no mechanical damage [78]. It is possible that latent colonization occurred since apparently uncolonized grain can mold rapidly when the grain is placed at high humidity [76]. Damage to the kernels not only increased susceptibility to fungal invasion, especially if this occurred near the embryo, but also increased the likelihood of visible molding [76,79]. *Penicillium* colonization of barley grain also increased with increasing damage but incidence of *Aspergillus* appeared to be independent of the degree of injury [80].

Varietal differences in susceptibility to molding have been reported in maize which remain discernible irrespective of the degree of damage [79, 81,82]. However, no differences related to endosperm properties could be detected in rice [83].

Differences in the microflora of different types of grain are compounded with those resulting from different geographic locations and climate. However, the general pattern of colonization is similar for all grains [84]. Many of the species found on maize, rice, sorghum, and other cereals in tropical climates are the same as those found in temperate regions except for a small number of additional species and, perhaps, a greater frequency of *Aspergillus* and fewer or different *Penicillium* spp. Thus, *Botryodiplodia theobromae* is reported on maize in Nigeria [85] and *Curvularia* spp., *Stachybotrys* spp., and *Fusarium dimerum* on rice in Egypt and Nigeria [86,87]. *Aspergillus parasiticus*, *A. flavus*, and *Penicillium islandicum* may be important in rice harvested in wet conditions in India when drying is delayed [88], and unusual species with *Aspergillus* anamorphs have been found in Iranian wheat, including *Emericella aurantiobrunneus* and *Aspergillus egyptiacus* [15]. *Aspergillus* spp., similar to *E. aurantiobrunneus*, producing only hulle cells, have also been found in Egyptian wheat and barley [86,89].

Water availability, temperature, and colonization: Temperature and a_w affect the growth of fungi and also their ability to compete with other microorganisms. As a result, characteristic microfloras develop in grain stored under different conditions, which can be good indicators of the conditions in which a sample has been stored, particularly its water activity and temperature when stored (Fig. 1; Table 5). Fungi that have been numerous in the field, such as *Alternaria*, persist as long as the grain is stored well with no molding or heating, but they disappear rapidly if these occur. The a_w/temperature relationships of colonization are best known for species with *Aspergillus* anamorphs, but there is a similar spectrum of reactions for species with *Penicillium* anamorphs, although the latter are mostly favored by lower temperatures than the former [22,72,90].

The most xerophilic fungi in stored grain are *Eurotium* spp. (members of the *Aspergillus glaucus* group), *Aspergillus restrictus*, and *Wallemia sebi*, which can grow down to 0.70 a_w (-49.1 MPa). As water becomes more readily available, more species are able to grow and spontaneous heating occurs. However, the conditions under which individual species are most numerous are not necessarily those at which they grow best in pure culture but often appear to be those at which they survive or compete better than other species. Thus, *Penicillium aurantiogriseum* and *P. viridicatum* are most abundant at 0°C and 1.0 a_w (0 MPa) although they grow best in pure culture at 25°C and 1.0 a_w. Similarly, *Eurotium* spp. are most abundant in grain at 30°C and 0.70 a_w (-49 MPa) but grow best at 30°C and 0.90 a_w (-15 MPa).

Chilling, either because of low ambient temperatures or through ventilation with cool air, increases the predominance of *Penicillium* spp., especially at high a_w. This is particularly important in Canada, where *Penicillium* spp. initiate heating despite the cold winter temperatures. Heating is initiated by *P. aurantiogriseum* and this is followed, successively, by *P. funiculosum*, *Aspergillus flavus*, *A. versicolor*, *Absidia* spp., and *Streptomyces* spp. as the temperature increases from -5 to 0°C to 45 to 65°C [59].

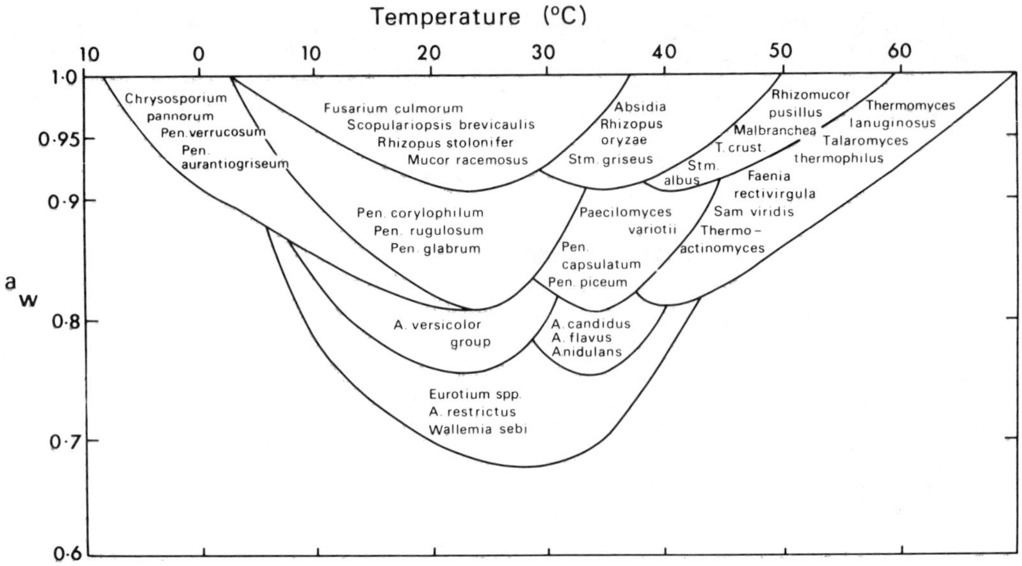

FIGURE 1. Prominant fungi and actinomycetes colonizing stored grain at different temperatures and water activities. A., *Aspergillus;* Pen., *Penicillium;* T. crust., *Thermoascus crustaceus;* Sam., *Saccharomonospora;* Stm., *Streptomyces.*

Colonization by different species overlapped, but the disappearance of individual species could be sudden, as with *A. flavus*, as the temperature approached 64°C. With the higher ambient temperatures following harvest in England (15-20°C), damp grain can heat to 60°C within 7-10 days and remain above ambient for 40-100 days [72].

The interrelationships of different environmental and biotic factors in the stored grain ecosystem have been studied in Canada by principal component analysis [18]. The importance of water availability in determining the behavior of a stored grain bulk was emphasized by its occurrence in the first four principal components, which together accounted for 40% of the total variability. Of these, temperature contributed to three and storage fungi for two. In the first principal component, storage period and water availability contributed to the increase in *Aspergillus* spp., including *A. versicolor, Penicillium* spp., *Rhizopus, Chaetomium,* and *Streptomyces* spp. and to a decrease in grain germinability and in the occurrence of the preharvest fungi *Alternaria, Cochliobolus,* and *Nigrospora.* The second principal component contained no fungal variables, but the third showed, surprisingly, an inverse relation of storage fungi with water and temperature, perhaps because of interactions with mite infestation. Two separate bulks were sampled in this study, and opposite relationships between *Cladosporium* and *A. flavus* with the "bin" variable perhaps suggest differences in inoculum or in another, unmeasured factor.

Where flooding occurred in Canada in spring, stratification of the wetted grain was characteristic [91]. Submerged grain became putrid but above

TABLE 5. Effects of Water Availability and Spontaneous Heating on Colonization of Cereal Grains by Fungi with *Aspergillus* and *Penicillium* Anamorphs and By Other Fungi

a_w	Max. temp. (°C)	Germ. (%)	Predominant fungi		
			Aspergillus	*Penicillium*	Other fungi
0.60	Ambient	90-100			
0.75	Ambient	75-90	*A. restrictus*		
0.85	Ambient	45-76	*Eu. amstelodami* *Eu. repens* *Eu. ruber*		
0.87	Ambient	N.D.		*P. brevicompactum*	
0.88	Ambient	N.D.		*P. verrucosum*	
0.89	Ambient	N.D.		*P. expansum* *P. granulatum* *P. griseofulvum* *P. rugulosum*	
0.90	25	15-45	*A. versicolor*	*P. aurantiogriseum* *P. citrinum* *P. funiculosum* *P. hordei* *P. janthinellum* *P. variabile*	
0.92	30	N.D.		*P. capsulatum*	
0.93	35	N.D.		*P. piceum*	
0.95	50	0-15	*A. candidus* *A. flavus* *A. niger* *A. ochraceus* *A. terreus* *Em. nidulans*	*P. purpurogenum* *P. rugulosum*	*Absidia corymbifera*
0.95	60	0	*A. fumigatus*	*T. emersonii* *T. thermophilus*	*Malbranchea cinnamomea* *Rhizomucor* spp. *Thermomyces* spp.

A., *Aspergillus*; *Em.*, *Emericella*; *Eu.*, *Eurotium*; *P.*, *Penicillium*; *T.*, *Talaromyces*; N.D., not determined.
Source: Refs. 71, 72, 74.

this was a layer of damp germinated grain colonized by *A. flavus*, *A. versicolor*, *Penicillium* spp., and *Trichothecium roseum*. *Penicillium aurantiogriseum* occurred throughout the profile but *Alternaria alternata*, *Wallemia sebi*, and *Eurotium* spp. only in the drier surface grain. Grain temperatures in outdoor stores were below ambient, reflecting the cold Canadian winter conditions, but in laboratory experiments, moisture was carried upwards through the bulk through spontaneous heating to 40°C.

Period of storage: Water availability and temperature interact to determine how long a bulk of grain can safely be stored. Thus, Snow et al. [92] recommended drying to 0.72 a_w (-45 MPa) for storage up to 3 months and 0.65 a_w (-59 MPa) for long-term storage up to 3 years. Grain, with an initial water content of 12.6% (0.60 a_w, -70 MPa), has been stored experimentally for up to 16 years with periodic assessment of viability and microflora [93-95]. The water content increased by up to 0.7% during this period. Field fungi survived better at 4-5°C than at ambient temperatures although species differed, with *Acremonium* and *Verticillium* dying out within 2 years and *Aureobasidium pullulans* surviving best. *Penicillium* spp. declined at ambient temperatures but not at 4.5°C, while *Aspergillus* spp. populations remained constant at both temperatures.

The storage period can be increased by decreasing a_w, by drying, or by decreasing temperature, by ambient or chilled air ventilation. Ambient air ventilation can be used to cool grain below 10°C to slow mold development and extend the safe storage period. Thus, grain containing 18% water (about 0.87 a_w, -19 MPa) cooled to 5-10°C in England can be kept for up to 8 months. In warmer climates, drying to 12-13% (about 0.52-0.60 a_w, -90 to -70 MPa) may be necessary before ventilation if fungal growth is to be prevented, and care must be taken to prevent addition of water when relative humidities are high. The data of Kreyger [96] suggest that germinative capacity can deteriorate even below 0.60 a_w (-70 MPa), below the limits for growth of all microorganisms. They also suggest that temperatures below 15-20°C would be necessary to store grain with 0.72 a_w (-45 MPa) for 3 months without visible mold, with oats and rye requiring the lower temperature.

The a_w necessary for safe storage are often below those normal in commercial trading and in grading standards. For instance, the specified water contents for grades 2 and 3 maize grains was 15.5 and 17.5% (about 0.80 and 0.88 a_w, -31 and -18 MPa) [90]. Water content standards can also be abused by the mixing of wet and dry grain to achieve the desired standard, rendering the whole more susceptible to molding, especially if the wetter grain was already invaded by fungi. Uniform drying at harvest is also essential since green, unripe grains are still metabolically active and more liable to heat and mold when mixed with adequately dried grain than suggested by the mean water content [72].

Field and storage fungi: Fungi colonizing grain have commonly been classified into *field* and *storage* fungi (see Chapter 4). Field fungi are those that colonize the grain before harvest but rarely develop further in store. Most common and widespread are *Cladosporium* spp., *Alternaria alternata* (= *A. tenuis*), *Epicoccum purpurascens*, and *Aureobasidium pullulans* with dark hyphae and spores containing melanin that may be protection against harmful UV radiation. Storage fungi are typically few before harvest and become abundant only in stored grain. Most characteristic in this group are *Aspergillus* and *Penicillium* spp.

Unfortunately, fungi are not so easily classified. The concept of field and storage fungi was derived in north temperate cereal-growing regions, and it becomes much less easy to maintain in humid, tropical areas. Here, *Aspergillus* spp. are much more frequent in the outdoor environment, and colonization of grain may be widespread before harvest. Thus, maize may be contaminated widely with *A. flavus* and aflatoxin during grain development and ripening, especially if the plants have become stressed by drought and/or heat [35]. Even in temperate regions, *P. aurantiogriseum*, regarded as a typical storage fungus, can grow on barley and wheat grain before harvest if the crop is lodged and the weather wet. Conversely, typical fungi such as *Fusarium* spp. can grow in store when water availability is high and temperatures low. Pelhate [97] proposed an intermediate group for fungi like *Fusarium* but did not include species more typical of storage that may colonize grain before harvest in this group.

Although such groupings as field and storage fungi are convenient, it is important not to be blinded by them. More information would be conveyed if organisms were classified by their ecological requirements, especially their water and temperature relationships and competitiveness under different environmental conditions. Most fungi colonizing grain before harvest require at least 0.85 a_w (-22 MPa) for germination, 0.90 a_w (-15 MPa) for sporulation, and grow at temperatures from -5 to +3°C to about 30-34°C. Species common during storage germinate with 0.72-0.80 a_w (-45 to -31 MPa) and grow over temperature ranges of -4 to +15°C to 30-55°C. Thermophilic fungi tend to have high water requirements but also grow only at temperatures higher than those normal out of doors at least in temperate regions. Fungi that occur both before and after harvest tend to have requirements close to the borderlines between the two groups. Thus, *A. flavus* requires at least 0.80 a_w (-31 MPa) for germination and 0.85 a_w (-22 MPa) for sporulation and has a temperature range for growth from 6-8 to 42-45°C. *Fusarium culmorum*, on the other hand, has a minimum of 0.90 a_w (-15 MPa) for growth and a temperature range of 3-36°C. It is found mainly in cool, damp grain [98].

Fungal/arthropod interactions: Insects and mites are important in their own right as causes of deterioration of stored cereal grains, but they also interact with fungal colonization in many different ways. Thus, in the multivariate study of Canadian wheat stores referred to above [18], the second principal component, which accounted for 7.9% of the total variation, contained only arthropod and environmental variables while the third (6.8% of total variation), sixth (4.3%), ninth (3.2%), tenth (3.2%), and eleventh (3.0%) components all contained both fungal and arthropod variables. These indicated that temperature affected insect development more than water content, that mites are favored by increasing temperature and, to a lesser extent, by increasing water content but that fungi, in the presence of mites showed an unexpected inverse relationship with both water and temperature; and that the beetles *Cryptolestes*, *Cryptophagus*, and *Ahasverus* affected fungi differently depending on species and whether they occurred alone or together.

Insects require at least 17°C to infest grain, but their occurrence is independent of water availability, within the range normally found in grain. Mite infestations occur between 3 and 30°C and with water contents exceeding 12% (about 0.52 a_w, -90 MPa). Like fungi, their respiratory activity can cause heating and the release of water. However, in contrast, insects can cause heating in dry grain. For instance, heating in grain containing 12-14% water (0.52-0.65 a_w, -90 to -59 MPa) can be initiated by *Oryzaephilus surinamensis*. This is replaced by *Cryptolestes ferrugineus* at

25-30°C, but this is killed as the temperature nears 40°C. During this process, the water content of the grain may be increased by up to 3%, favoring the growth of *Penicillium* and *Streptomyces* [99]. Similarly, *Sitophilus oryzae*, in wheat grain initially containing 15% (about 0.75 a_w, -40 MPa) water, may cause temperature rises up to 20°C [100]. Insect respiration also leads to the release of water and to moisture migration, both favoring fungal colonization, while the damage that they cause and their fecal material may provide new substrates for colonization. Fungi provide food for insects and mites, may affect their development and reproduction with their metabolites, and some may even be pathogenic.

The reactions detected by principal component analysis have been interpreted in various ways. The inverse relation between fungal occurrence and temperature and moisture in the presence of mites could have been caused by *Acarus siro* feeding and breeding on *Penicillium*, especially *P. aurantiogriseum*, and by *Acarus gracilis* feeding on both *Penicillium* and *Aspergillus versicolor*, as found by Armitage and George [101], or by the fungi occurring only as superficial contaminants that sporulated only on the isolation plates. An inhibitory secretion by mites has also been suggested [101], while quinones secreted by *Tribolium castaneum* can suppress fungal colonization [102]. In component six, the two genera of beetle, *Cryptophagus* and *Ahasverus*, are fungivorous, while the mites, *Lepidoglyphus*, *Tydeus*, and *Tarsonemus*, scavenge decomposed microorganisms and grain debris. In the remaining components, *Cryptophagus* and *Ahasverus*, both deterred *Rhizopus* by feeding on it, while *Cryptolestes* provided substrates by damaging the grain embryo. However, if all three beetles occurred together, their roles were in some way reversed, while *Streptomyces* was favored, perhaps because the insect excreta formed a good substrate for their growth.

In other studies, fungi were favored by the presence of mites on which they were pathogenic—*Aspergillus restrictus* by *Lepidoglyphus destructor* and *Wallemia sebi* by *A. siro* [101]. A wide range of fungi can provide substrates for the growth and reproduction of insects and mites. Field fungi, such as *Alternaria alternata* and *Cladosporium cladosporioides*, are particularly favorable, while *Aspergillus fumigatus*, *A. niger*, and *Streptomyces* spp. supported few species. Among the mites, *A. siro* could grow on 17 of the 24 fungal species tested and *L. destructor* on 12, while, of the beetles, *C. ferrugineus* and *O. surinamensis* could grow on 10, *Tribolium castaneum* on 8, and *T. confusum* on 7 [103]. Insects also disperse fungi through the grain mass by carrying spores in their gut or externally on hairs, ovipositors, and mouthparts [104,105]. For instance, *Sitophilus* spp. may inoculate grain during oviposition, either when the pericarp is first damaged by the female's mouthparts or subsequently with the ovipositor.

Conversely, insects may also be inhibited by fungi, with *Aspergillus candidus*, *A. flavus*, *A. niger*, and *A. ochraceus* inhibiting egg laying by *Ahasverus advenus* [102]. *Penicillium aurantiogriseum*, *P. citrinum*, and *P. purpurogenum* can repel *T. confusum*, although *P. crustosum*, *P. urticae*, and *P. oxalicum* were neutral and *P. viridicatum*, *P. chrysogenum*, and other isolates of *P. aurantiogriseum* and *P. citrinum* were attractive. Most neutral species and even some that were attractive to *T. confusum* inhibited larval development, while *P. purpurogenum* caused high adult mortality [106,107]. Although only small amounts were detected, some of the observed effects could be caused by mycotoxins. Ochratoxin, citrinin, rubratoxin B, and patulin all inhibited one or more of three grain insects. T-2 toxin at 100 ppm can increase fecundity of *T. confusum* but decrease the hatch of eggs by 10-12%, retard larval development, and increase larval mortality. However, zearalenone, which has estrogenic activity, only increased the reproductive life of *T. confusum* [108].

Further information on fungal/arthropod interactions may be found in a review by Dunkel [105].

Preservatives: A wide range of chemicals have been evaluated for the prevention of molding in damp grain. Although some conventional fungicides have appeared promising in vitro, they have often been disappointing when applied to grain. Among the more successful have been volatile fatty acids, including acetic, formic, propionic, and iso-butyric acids, and sorbic acid and citral have also been suggested. However, acetic and formic acids have performed less well than propionic acid in laboratory tests, and only propionic acid has been used widely on farms. Failures have sometimes occurred in grain treated with propionic acid on farms leading to contamination with aflatoxin [109].

For treatment with propionic acid to be successful, application rates need to be adequate for the water content of the grain, probably about 1.3 g per 100 g of water in the grain, and treatment needs to be uniform, ensuring that there are no untreated or undertreated pockets that can allow invasion by propionate-tolerant organisms. Propionic acid is fungistatic rather than fungicidal and spore germination is more sensitive to inhibition than mycelial growth. Although most fungi are inhibited by only 0.2% of propionate in agar, *Paecilomyces variotii, Eurotium* spp., and *Aspergillus flavus* can tolerate up to 1%. They may also metabolize the propionate to allow invasion by more sensitive fungi [110-112]. Once growth is established in undertreated grain, it can then spread to that which had been otherwise adequately treated. Small concentrations of propionate can also stimulate aflatoxin production by *A. flavus* [113].

Sorbic acid has been much less used, but it provides an attractive alternative to propionic acid because it is insecticidal as well as fungistatic [105]. The minimum concentrations of sorbic acid necessary to inhibit different fungi range from 0.01-0.3% and for insects from 0.05-0.5%. More evaluation of its use is required.

3. Moist Grain Storage

The respiration of damp grain with its associated microflora produces CO_2 and utilizes O_2, thus modifying the composition of the intergranular atmosphere within the grain bulk. This characteristic has been utilized to store damp maize and barley grain for animal feed in sealed, bolted or welded steel and in unsealed, concrete staved silos although fungal colonization of maize is more affected by decreasing a_w from 1.00 to 0.70 (0 to -49 MPa) than by decreasing O_2 from 21 to 1% [114-117]. Maize grain has been stored at water contents of 20-36% (about 0.92 to 1.0 a_w, -11 MPa) in sealed silos and barley grain with 19-22% (about 0.89-0.93 a_w, -16 to -10 MPa) in sealed and 20-40% (about 0.90-1.0 a_w, -15 to 0 MPa) in unsealed silos. Peak CO_2 concentrations 0f 60->90% and minimum O_2 concentrations < 10% can be attained within 3 weeks of placing grain in store. Control of fungal colonization then depends on maintaining high CO_2 and low O_2 levels throughout storage, but peak levels are seldom maintained. Air exchange between the silo and external atmospheres occurs as a result of imperfect sealing, through daily temperature and pressure fluctuations, and as a result of removing grain. Thus, a peak concentration of 50% CO_2, after 30 days' storage of barley grain (20-22% water content, about 0.90-0.92 a_w, -15 to -11 MPa) in a bolted steel silo, declined to only 15% within 200 days. When the silo is nearly empty, respiration of the grain remaining may be insufficient to maintain an inhibitory atmosphere, and microbial development may occur with spontaneous heating. In unsealed silos, air exchange may occur at the exposed grain surface, allowing microbial activity and heating in the top 15-30 cm.

High moisture maize, stored in sealed silos in which CO_2 concentrations reached 99% after 3 weeks but declined to 25% after 120 days and O_2 concentrations were between 0.5 and 1.0%, showed a rapid increase in filamentous fungi in the first few days' storage, from 10^4 to 10^7 cfu/g. Numbers then declined to $<10^2$ cfu/g. Yeasts numbered 10^3 to 10^7 cfu/g throughout the storage period, but the species composition changed. They were more abundant in the upper levels of the silo than lower and most numerous around 24% water content (0.97 a_w, -4 MPa). There was a 90% decline in numbers from 23 to 21% water content (0.94-0.92 a_w, -9 to -11 MPa). *Candida parapsilosis* and *C. intermedia* were common in the grain before harvest but were replaced during the first few days' storage by *Hansenula anomala* (68% of identified isolates) and *Issatchenkia orientalis* (*Candida krusei*) (26%). *I. orientalis* tended to be more common in the warmer grain. Other yeasts identified included *Pichia membranaefaciens*, *Candida* (*Torulopsis*) *glabrata*, *C. tropicalis*, *Hyphopichia burtonii*, and *Metschnikowia* (*Candida*) *pulcherrima*, with, occasionally, *Hanseniaspora uvarum* (*Kloeckera apiculata*), *Rhodotorula* spp., and *Torulaspora delbrueckii* (*Saccharomyces rosei*) [114, 118-120].

Moist barley grain stored in unsealed silos is initially covered with a layer of straw, grass, and/or a polyethylene sheet. Despite this, up to 30 cm of the uppermost grain becomes moldy, least under a thick layer of grass that formed good silage. Under straw, the grain was colonized by *Absidia corymbifera*, *Aspergillus flavus*, *A. fumigatus*, *Penicillium* spp., *Rhizomucor pusillus*, and, at water contents about 23% (about 0.93 a_w, -10 MPa), *Thermoascus crustaceus*. The actinomycetes *Faenia rectivirgula*, *Saccharopolyspora hordei*, *Streptomyces albus*, and *Thermoactinomyces* spp. also occurred, often in discrete layers. After the moldy layer was removed, the microflora depended on the rate at which grain was removed from the exposed surface. Removal of about 8 cm daily keeps pace with the depletion of CO_2 and penetration of O_2 into the surface layers of grain and minimizes fungal development. As rate of use declines, more fungal development occurs, accompanied by increasing spontaneous heating. A sequence of fungal colonization has been identified corresponding to different degrees of heating (Table 6) [116].

The predominant yeast species were similar to those found by Burmeister and Hartman [118] including *Candida tropicalis*, *Endomyces* (*Endomycopsis*)

TABLE 6. Sequence of Fungi Colonizing Moist Barley Stored in Unsealed Concrete-Staved Silos

Maximum temperature of surface grain (°C)	Predominant species
Ambient	*Hansenula anomala, Hyphopichia burtonii*
Ambient	Yeasts, *Penicillium roquefortii*
25-30	*Penicillium rugulosum, Eurotium* spp., *Aspergillus candidus*
30-35	*Absidia corymbifera, Rhizomucor pusillus*
35-50	*Aspergillus fumigatus*
50-60	*Thermomyces lanuginosus*

Source: Ref. 116.

fibuliger, Geotrichum candidum, Hansenula anomala, Hyphopichia burtonii, Issatchenkia orientalis (*Candida krusei*), and *Pichia* (*Candida*) *guilliermondii*. *Hansenula anomala* and *H. burtonii* were usually the most abundant yeasts, initially occurring alone but subsequently accompanied by *Penicillium roquefortii* and then *P. rugulosum*. With heating, yeasts declined and were replaced by thermotolerant and thermophilic filamentous fungi and actinomycetes. Fungal populations ranged from 4×10^5 to 1.5×10^8 cfu/g and actinomycetes from 2×10^5 to 1.4×10^6 cfu/g.

A similar sequence of fungi was found in sealed silos containing moist barley grain with increasing penetration of air as the grain warms in spring and the silos are emptied [117,121]. In Sweden, *Fusarium* survived sealed storage conditions better than other field fungi and low ambient temperatures prevented mold development before spring. This comprised yeasts and, later, *P. roquefortii*. Molding was decreased and storage periods prolonged by the provision of external breather bags, with capacities up to 25% of the silo volume, to decrease air exchange and by introducing CO_2, N_2, or NH_3 gases into the silos.

4. Crib Storage

Cribs have been used widely in the past to store maize. These are constructed of wire mesh or wooden strips and are preferably about 50 cm wide to allow free air circulation. Maize in the northern United States may still contain >30% water (about 1.0 a_w, 0 MPa) at harvest and even 40-50% (1.0 a_w, 0 MPa) in France. Shelled and stored in this condition, it will spoil rapidly but when stored on the ear in cribs, after dehusking, slow drying occurs and low winter temperatures prevent molding. If temperatures are slow to fall to freezing in mild spells or in warmer climates if drying is insufficiently rapid, *Fusarium* may colonize the grain and form mycotoxins [122].

5. Underground Storage

Grain has traditionally been stored underground in many parts of the world. Pits are still used in Africa and across Asia, from the Mediterranean to China

[15,123-128]. They were also used in prehistoric Europe [129]. The pits may be up to 3.5 m deep and beehive shaped, 2 m diameter at the base and 1 m at the top, or where the water table is high, only 0.5-1.5 m deep and square or round with sides or diameters of 2-5 m. Only occasionally has the grain stored in these pits been studied mycologically.

In Iran, pit-stored grain was generally of good quality with *Aspergillus restrictus* predominant especially close to the walls [128]. Occasionally grain was heavily colonized with *A. candidus*, *A. flavus*, *A. niger*, *A. ochraceus*, *Eurotium*, *Penicillium*, and *Rhizopus* spp., but the reasons for this failure are not clear. Sorghum stored in pits in Ethiopia showed some visible molding [124]. *Aspergillus candidus*, *Cylindrocarpon tonkinense*, and *Fusarium solani* occurred in most samples, and *Penicillium* spp. were also widespread.

In England, barley grain stored in Iron Age-type pits maintained its viability, despite colonization by *Penicillium* and *Eurotium* spp., unless rainfall was sufficient to increase grain water content markedly or leaks occurred in the clay seal [130,131]. Grain from the outsides of pits was always wetter than that close to the center, and CO_2 concentrations were usually 25%. Spontaneous heating only occurred in grain wetter than 24% (about 0.94 a_w, -9 MPa), increasing the CO_2 concentration to 33%. Higher CO_2 levels in moist grain prevented heating and allowed yeasts to proliferate.

6. Modified Atmosphere Storage

The intergranular atmosphere in stored grain can also be manipulated artificially by the introduction of gases, especially CO_2 or N_2 [132,133]. However, recommendations have usually been designed to control insect infestations and are often insufficient to control fungi. For instance, treatment of Australian wheat grain containing 12% water (0.52 a_w, -90 MPa) should exceed 70% CO_2 initially and remain above 35% for 10 days. With N_2, O_2 concentrations $<1\%$ are recommended for at least 6 weeks. However, welded steel bins lose 3-6% CO_2 daily and concrete bins 10-20%, and although grain was stored for up to 12 months, insect infestations were found at the end. Fungal contamination was not assessed but would not be expected with the low initial water content except as a consequence of insect infestation or moisture migration. In Italy, mold growth limited storage of grain with $>14.5\%$ water (about 0.75 a_w, -40 MPa) in nitrogen atmospheres containing only 0.2% O_2, with *Eurotium* spp., *Penicillium* spp., and *Aspergillus flavus* predominant. Mold growth in N_2 atmospheres was prevented only if these were totally free of O_2 [134].

Atmospheres containing 99.7% N_2 + 0.3% O_2, 61.7% CO_2 + 8.7% O_2 + 29.6% N_2, and 13.5% CO_2 + 0.5% O_2 + 84.8% N_2 all delayed deterioration of maize grain by *A. flavus* and *Fusarium moniliforme* but did not stop their growth. *Aspergillus flavus*-inoculated grain was visibly moldy after 4 weeks in the 61.7% CO_2 + 8.7% O_2 + 29.6% N_2 atmosphere. Aflatoxin production was limited to $<15\%$ aflatoxin B_1 and $<20\%$ total aflatoxins/kg, but removal from the modified atmospheres led to rapid deterioration and aflatoxin formation [135,136].

Sulphur dioxide (SO_2) and ammonia (NH_3) have been utilized in low temperature systems because of their ability to inhibit either spore germination or mycelial growth of contaminant grain fungi with small doses [137,

138]. Fungi, with the exception of *Penicillium* species, were controlled in maize grain by the addition of 0.1-0.3% SO_2.

7. Irradiation

Irradiation is widely used for food preservation and could also be used to sterilize grain before storage. However, large doses may be necessary. Although *Aspergillus* and *Pencillium* spp. are killed by 1.2 kGy, *Mucoraceae* require a dose of 6 kGy and *Fusarium*, yeasts, and *Bacillus* up to 12 kGy [139]. With low doses of radiation, there is a risk that mycotoxin production may be enhanced although reports differ, perhaps due to differences in irradiation conditions or water content. Applegate and Chipley [140] found increased aflatoxin production by *A. flavus* in a wheat medium after irradiation with up to 2 kGy, while Chang and Markakis [141] found decreased production in barley grain. Nontoxigenic isolates of *A. flavus* may or may not be stimulated to produce aflatoxin [142,143].

III. WATER, TEMPERATURE, AND GAS RELATIONSHIPS OF GRAIN FUNGI

A. Effect of Water Activity and Temperature on Grain Fungi

1. Germination

At harvest, spores of a range of different species are present on the surface of grain, and water availability and temperature are critical in determining their rates of germination and growth and their ability to colonize grain. Studies of spore germination in grain fungi have included at least one of the following measurements:

1. The minimum a_w allowing germination
2. The minimum and maximum temperatures allowing germination
3. The lag time for germination
4. The rate of germ tube extension

In general, *Aspergillus* and *Penicillium* species can germinate at lower a_w than field fungi. For example, *Alternaria alternata* and *Cladosporium herbarum* require at least 0.85-0.86 a_w (-22 to -21 MPa) and *Acremonium strictum* at least 0.97 a_w (-4 MPa), while *Eurotium rubrum* can germinate with 0.70 a_w (-49 MPa) and *Aspergillus restrictus* with 0.72 a_w (-45 MPa) [98,144,145]. Spore germination has most commonly been studied at only 25°C, although some workers have used a wider range, between 20 and 30°C, to demonstrate effects of temperature on water relationships [98,146]. Thus, *Penicillium* spp. isolated from maize germinated with 0.81-0.83 a_w (-29 to -26 MPa) at 16°C, 0.81 a_w (-29 MPa) at 23°C, and 0.83-0.86 a_w (-26 to -21 MPa) at 30°C [36]. The minimum a_w allowing germination generally occurs at the optimum temperature for growth, and it is often assumed that fungi can grow over very limited a_w ranges at low temperatures. However, many field fungi and *Penicillium* spp. are still able to grow between 0.995 and 0.90 a_w at 5°C. At low a_w, germination may sometimes occur without any subsequent growth [98,146], e.g., conidia of *A. flavus* germinated at 0.75 a_w (-40 MPa) did not grow further and subsequently died [147]. By contrast, Schneider [148] found that *Fusarium* spp. grew over a wider a_w range than that for germination but this appears to be the only example of this in the literature. Close to the a_w limits for germination, germ tube growth is often abnormal, especially in *A. alternata* and some *Fusarium* and *Penicillium* spp. [36,98,149].

The lag time for germination often increases with decreasing a_w [98, 150,151]. Spores of grain fungi usually germinate rapidly at high a_w (>0.98 a_w, -3 MPa), taking from a few hours to 1-2 days. However, at low a_w, this period may extend to weeks, months, or indeed years. For example, conidia of *Aspergillus echinulatus* only germinated after 730 days at 0.62 a_w (-66 MPa), although the germ tubes produced were abnormal and failed to produce mycelium [151]. Because lag times may be so long, the absolute minimum a_w for germination may not be accurately determined when experiments extend only for very short periods of time. This probably explains why Galloway [152] found minimim a_w for germination larger than those reported by Snow [150]. Both spore age and nutrition can also affect the lag times for germination and spores.

Germ tube elongation is rapid at high a_w, and usually linear if temperature is constant [150,153]. As a_w decreases, rate of elongation decreases and it further slows with time. For *A. alternata* this occurs at 0.93 a_w (-10 MPa) and for *Eurotium repens* at <0.80 a_w (-31 MPa). At constant a_w, germ tube elongation is maximum at the optimum temperature [153].

2. Mycelial Growth

Like germination, minimum a_w levels for growth of grain fungi have usually been determined at only one temperature [144,150,154], although information on a wide range of a_w and temperatures is much more relevant to spoilage of stored grains. This is particularly important in trying to understand the roles of different species and their interrelationships with one another. Most preharvest and some storage fungi grow optimally at close to 1.00 a_w (0 MPa), but some groups are exceptional in that they grow much better at lower a_w. For instance, *E. repens*, *E. amstelodami*, and *A. versicolor* are all able to grow fastest between 0.90 and 0.95 a_w (-15 to -7 MPa) at optimum temperature. However, close to limiting temperatures, growth only occurs near to 0.995 a_w (-0.7 MPa). Several *Penicillium* spp. also grow better at 0.98 a_w (-3 MPa) than 1.00 a_w (0 MPa) [144]. Although a range of solutes have been used to alter a_w in vitro and a range of nutrient levels, the growth rates of individual species on a specified medium are thought to be relatively consistent [144,155]. As a consequence, Pitt [73] has used growth rates at different temperatures and a_w as important criteria in the identification of *Penicillium* species.

Few workers have studied the combined effects of a_w and temperature on mycelial growth of grain fungi [98,146]. Detailed measurements of growth rates at different a_w and temperature have been plotted and data points showing equal growth rates joined to give diagrams of the water/temperature relationships of each species (Fig. 2). This provides a useful guide to at least two dimensions of the fundamental niche of each fungus. These diagrams indicate the optimum a_w and temperature for growth of each species and its temperature and a_w limits. This has shown that most field fungi, including *Epicoccum nigrum* and *Fusarium culmorum*, require at least 0.90 a_w (-15 MPa) for growth, although *A. alternata* and *C.*

herbarum can grow with 0.88 a_w (-18 MPa) on wheat extract agar. Many *Aspergillus* and *Penicillium* spp. are tolerant of much lower a_w, with *P. brevicompactum* growing at 0.82 a_w (-27 MPa) and *E. amstelodami* able to grow well at 0.72 a_w (-45 MPa). In general, germination of *Aspergillus* and *Penicillium* spp. occurred at lower a_w than mycelial growth [98]. While these field fungi, *Aspergillus* and *Penicillium* spp., differ significantly in their minimum a_w for growth, there is some overlap, particularly at marginal temperatures for growth.

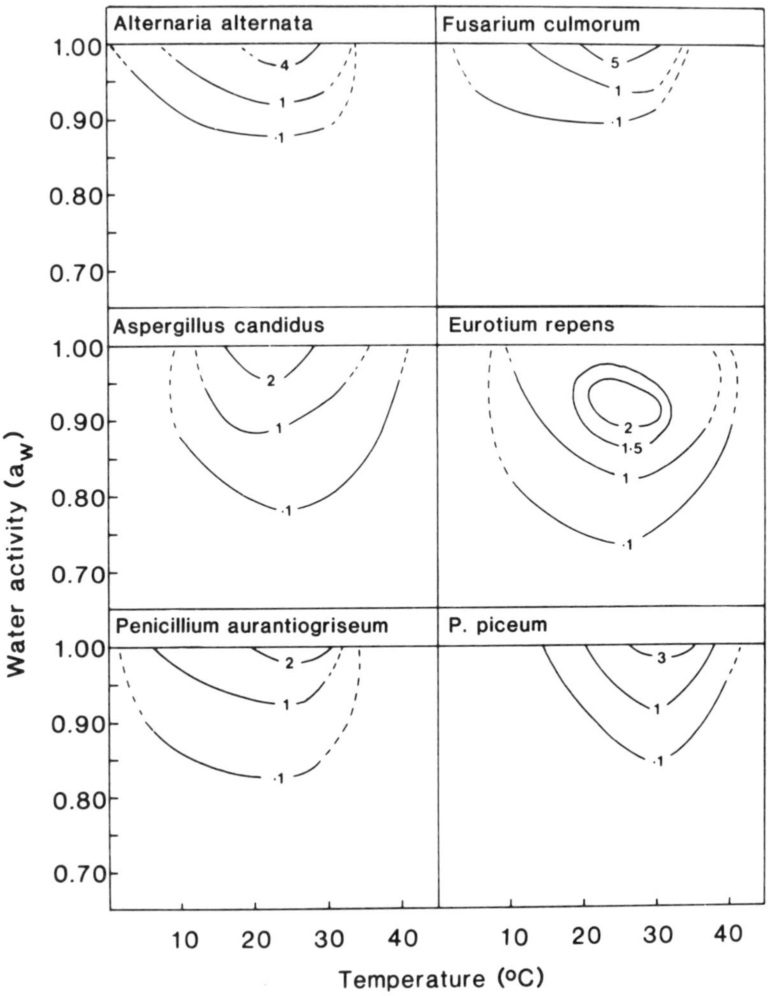

FIGURE 2. Growth rates (mm/day) of grain fungi at different temperatures and water activities. P., *Penicillium*.

Germination times and linear growth rates of *A. restrictus* and *A. versicolor* show significant inverse correlations over a range of a_w and temperature conditions [145,146]. Such relationships may exist for other grain fungi, but this area has not been investigated.

Grain fungi have been differentiated on the basis of their water relations. Xerophilic fungi have been distinguished on the basis of a_w limits for either germination or growth, but there is disagreement over the criteria to use. Both Heintzeller [156] and Pelhate [97] defined xerophilic fungi as those with spores able to germinate with less than 0.80 a_w (-31 MPa) and with optimum growth below 0.95 a_w (-7 MPa). Indeed, Pelhate [97] devised three categories—xerophilic, mesophilic, and hygrophilic—based on ability to grow at 0.95 a_w (-7 MPa), 0.95-1.00 a_w (-7 to 0 MPa), and 1.00 a_w (0 MPa), respectively. The use of the term mesophilic in relation to a_w tolerance is confusing since this term is already used in relation to temperature requirements. Subsequently, Pitt [144] defined a xerophile as a fungus "capable of growth under at least one set of environmental conditions at 0.85 a_w (-22 MPa) or less." This definition has now been widely accepted.

3. Sporulation

Sporulation is a vital phase in the life cycle of fungi as it provides a mechanism for surviving both nutrient and environmental stress. However, there is little detailed information on how a_w and other environmental factors affect anamorph and teleomorph formation in grain fungi. Table 7 compares the minimum a_w for germination, mycelial growth, and sporulation for a range of grain fungi. Field fungi, e.g., *A. alternata, Cladosporium* spp., and *Fusarium culmorum*, have minima for sporulation between 0.90 and 0.92 a_w (-15 to -11 MPa). Some *Penicillium* species, e.g., *P. brevicompactum* and *P. aurantiogriseum*, produce conidia down to 0.85 a_w (-22 MPa), while others, e.g., *P. piceum*, require at least 0.86 a_w (-21 MPa). *Aspergillus candidus* and *A. nidulans* produce conidia down to 0.83 and 0.80 a_w (-26 to -31 at 25°C [66,98]. Conidal production of *E. chevalieri* and *E. intermedius* was enhanced when water activity was decreased to about 0.90 a_w (-15 MPa), but not that of *E. amstelodami, E. repens*, and *E. ruber* [157]. In general, the time taken for sporulation increases with decreasing water activity while the minimum a_w at which it will occur changes with temperature [74,98] and both may be affected by the nutrient status of the substrate and spore age [150].

Teleomorph formation often occurs over a narrower range of a_w than conidial production since it is usually less tolerant of low a_w. For example, *Emericella (Aspergillus) nidulans* produces conidia down to 0.85 a_w (-22 MPa) but ascospores only above 0.995 a_w (-0.7 MPa) [150], while *Eurotium* spp. form conidia down to 0.73 a_w (-43 MPa) and only immature cleistothecia at 0.77 a_w (-36 MPa) that fail to mature ascospores, even in 120 days [144,157]. More quantitative and qualitative information on the ability of grain fungi to sporulate under different environmental conditions is still required, especially since visible molding, largely sporulation, is often used as a criterion of spoilage.

TABLE 7. Minimum Water Activity Levels Permitting Germination, Growth, and Sporulation of Fungi from Grain

Species	Germination	Growth (linear)	Sporulation (asexual)
Alternaria alternata	0.85	0.88	0.90
Cladosporium cladosporioides	0.86	0.88	0.90
Fusarium culmorum	0.87	0.90	0.92
Aspergillus candidus	0.78	0.80	0.83
A. versicolor	0.76	0.78	0.80
Emericella nidulans	0.83	0.80	0.80
Eurotium repens	0.72	0.75	0.78
Penicillium aurantiogriseum	0.80	0.83	0.85
P. brevicompactum	0.80	0.82	0.85
P. piceum	0.79	0.85	0.89
P. roquefortii	0.83	0.83	0.83

Source: Ref. 98.

B. Water Activity, pH, and Growth of Grain Fungi

Hydrogen ion concentration may be changed by the application of pesticides or preservatives to grain and thus affect spores and mycelium present on the grain. Hydrogen ion concentration can affect metabolic processes, particularly those involved with spore germination, sporulation, and morphogenesis [158]. Therefore, it may also have a selective effect on the minimum a_w at which individual fungi germinate and grow. Grain fungi grow over a wide pH range, e.g., *A. niger*, 1.5 to 9.8; *A. candidus*, 2.1 to 7.7; and *E. repens*, 1.8 to 8.5 on media with 0.995 a_w (-0.7 MPa) [159]. Pitt and Hocking [151] found little difference in germination time between pH 4 and 6.5 for a number of *Aspergillus* spp., but in other studies when pH was changed from 6 to 4.5 the lag times for germination of "field" fungi, *Penicillium* and *Aspergillus* spp. increased by 1-2 days at 0.995 a_w (-0.7 MPa) and by 6-7 days at 0.75-0.85 a_w (-40 to -22 MPa) [98]. The same change in pH also increased the minimum a_w for germination by about 0.02 a_w at optimum temperatures and by up to 0.05 a_w at marginal temperatures.

C. Interactions of Water Activity, Temperature, and Gas Composition

Although fungi are usually considered obligate aerobes, the ability of many species to tolerate high CO_2 and low O_2 concentrations is often underestimated and some species may even be microaerophilic. *Aspergillus* and *Penicillium* spp. appear to be particularly tolerant of very low O_2 concentrations while *Penicillium roquefortii* can tolerate up to 75% CO_2 [160]. However, CO_2 and O_2 levels may interact so that atmospheres with high CO_2 concentrations are more inhibitory when O_2 concentrations are small than

TABLE 8. Concentrations of O_2 (%) Required to Halve Growth of Some Field and Storage Fungi on Wheat Extract Agar at 14 and 23°C

	23°C			14°C		
Water activity (a_w)	0.95	0.90	0.85	0.95	0.90	0.85
Alternaria alternata	<0.14	0.14	N.G.	3.80	5.00	N.G.
Fusarium culmorum	2.60	<0.14	N.G.	12.50	5.00	N.G.
Aspergillus candidus	1.00	0.45	5.00	<0.17	9.40	N.G.
A. fumigatus	5.40	6.20	N.G.	N.T.	N.T.	N.T.
Eurotium repens	3.00	5.00	10.20	0.90	4.00	N.G.
Penicillium aurantiogriseum	5.30	2.40	13.00	<0.17	10.20	N.G.
P. brevicompactum	0.60	0.40	1.00	<0.17	<0.17	N.G.
P. hordei	<0.14	1.30	12.50	0.80	1.60	N.G.

N.G., no growth; N.T., not tested.

Source: Ref. 161.

when they are close to ambient levels. Other interactions, as between gas composition, water availability, and temperature, have generally been neglected [161].

Usually, delays or lag times prior to germination of groups of field fungi, *Aspergillus*, and *Penicillium* spp. were negatively correlated with both a_w and O_2 concentration [161]. At 0.98 a_w (-3 MPa), decreasing O_2 concentration only slightly increased the lag phase for growth initiation but with 0.80-0.85 a_w (-31 to -22 MPa), decreasing the O_2 concentration to <1.0% increased the lag phase markedly. This effect was more pronounced with *Aspergillus* and *Penicillium* spp., which gave lag times of only a few days at 0.98-0.95 a_w (-3 to -7 MPa) in normal air but 10-20 days at 0.90-0.85 a_w (-15 to -22 MPa) and 1% O_2. Miller and Golding [162] found growth of *Aspergillus* spp. to be affected when atmospheres contained less than 5% O_2 at 0.95 a_w (-7 MPa). However, the concentration required to halve growth can change markedly with a_w and temperature (Table 8) [161]. Arbab [163], in comparisons of *Penicillium* spp., found that at 0.92 a_w (-11 MPa) growth of the preharvest species, *P. funiculosum* and *P. oxalicum*, was inhibited by different high CO_2/low O_2 mixtures from storage species such as *P. aurantiogriseum* and *P. viridicatum*. Sporulation was also decreased by the mixtures compared to air. *Fusarium moniliforme* was able to tolerate 60% CO_2 at 23°C but it was more sensitive at lower temperatures and when a_w was decreased from 0.99 to 0.94 (-1 to -8 MPa) [164].

Large concentrations of CO_2 are necessary to inhibit spore germination and growth of grain fungi. For instance, only with at least 60% CO_2 were spores of *P. aurantiogriseum* almost completely inhibited if O_2 was adequate [165]. The temperature range permitting germination decreased and the lag time before growth increased as CO_2 concentration was increased. A range of field fungi, *Aspergillus* and *Penicillium* spp. took up to 4 days to germinate at 0.90 a_w (-15 MPa) in air but 16-18 days with 15% CO_2 at the same a_w and 21% O_2 [161].

There was no effect on sporulation of field fungi, such as *A. alternata* and *Botrytis cinerea*, with up to 15% CO_2 at 0.995 a_w (-0.7 MPa), although growth was inhibited by 20% [166]. However, storage fungi contaminating spring wheat were able to grow with 50-79% CO_2 [167]. *Penicillium* spp. were found to be more tolerant of high CO_2 levels than *A. candidus* and *A. flavus*. By contrast, Pelhate [167] found *F. culmorum* and *P. roqueforti* could grow in 20% CO_2 at 20°C but not *Alternaria tenuissima*, *A. fumigatus*, or *P. aurantiogriseum*. *Aspergillus ochraceus* grew with 60% CO_2, but not with 80% CO_2. Sclerotium production was only observed in normal air [168]. Unfortunately, interaction with a_w was not considered in these studies.

Although the ability of fungi to sporulate in grain under different controlled atmospheres is important, it has received very little attention. In vitro work suggests that field fungi such as *A. alternata* and *Cladosporium* spp. produce conidia equally well in normal air and with less than 1% O_2, with ambient CO_2 levels. However, *Penicillium* spp. produced few spores and floccose mycelium with 1.00-0.14% O_2, while *E. repens* produced only conidia in 1% O_2 on media with 0.85 a_w (-22 MPa), cleistothecia only with 10% O_2 and 0.80 a_w (-31 MPa) and both conidia and cleistothecia in normal air and 0.98 a_w (-3 MPa) [161]. Sporulation was found to be unaffected by up to 15% CO_2, even at decreased a_w.

Investigations of the effects of sulphur dioxide (SO_2) on grain fungi are scarce. At atmospheric pollutant levels (100-200 ppb SO_2), germination of filamentous field fungi, such as *A. alternata*, *Cladosporium* spp., and *F. culmorum*, at 0.995 a_w (-0.7 MPa) was little affected [169] but germ tube extension was significantly decreased. By contrast, the phyllosphere yeasts that are common on grain prior to harvest were very sensitive to these concentrations. The responses of some fungi to SO_2 have perhaps been studied more in the food industry, where it is used as a preservative [170], and more information on the sensitivity to SO_2 of spore germination and mycelial growth of grain fungi is still required.

D. Water Activity, Temperature, and Interactions Between Grain Fungi

1. In Vitro Studies

During growth, harvesting, and storage, grain becomes contaminated with propagules of a wide range of different microorganisms. These may be deposited close to one another or be dispersed widely on the grain depending on their numbers. Under suitable environmental conditions, they develop and colonize the grain, but not all species survive because they interact with one another and compete for the same substrate. Such interactions may benefit one or both the competing fungi (commensalism or mutualism), be inhibitory to one or both (amensalism), or the fungi may coexist on the grain without any adverse effect on either and, perhaps, with mutual intermingling of the hyphae (neutralism). Five possible reaction types have been recognized between hyphae of two colonies growing toward one other on culture media [20,171]:

1. Hyphae of both colonies intermingle freely
2. Hyphae of both colonies are inhibited on contact

3. Hyphae of both colonies are inhibited at a distance
4. Hyphae of one colony are inhibited on contact while those of the other continue to grow normally
5. Hyphae of one colony are inhibited at a distance while those of the other continue to grow normally

Wicklow et al. [171] then calculated an index of antagonism (IA) for different fungi by giving numerical values to these reaction types (A = 0, B = 1, C = 2, D = 3, E = 4) and applying the formula:

$$IA = (n_B \times 1) + (n_C \times 2) + (n_D \times 3) + (n_E \times 4)$$

where n_B, n_C, n_D, and n_E are the number of pairings giving, respectively, B, C, D, and E reaction types. The index of antagonism was greater (>25) for *Penicillium funiculosum*, *P. variabile*, and *Trichoderma viride* than for *A. niger*, *Alternaria alternata*, *Curvularia lunata*, *Cladosporium cladosporoides*, and *Candida guilliermondi* (<5). *Penicillium funiculosum*, *P. variable*, *T. viride*, and *P. oxalicum* all inhibited *A. flavus* at a distance (reaction type C or E), while *A. alternata* and *A. niger* were inhibited on contact with *A. flavus* (reaction type B or D).

Magan and Lacey [20] modified the scoring system slightly to derive an *Index of Dominance* (I_D) in an attempt to develop an in vitro system to assay the ability of grain fungi to dominate in different systems. This index was used to measure the capability of individual species, competing in pairs, to dominate over a range of a_w, temperatures on two nutrient substrates. The Index of Dominance changed markedly with these factors but was not directly related to growth rate. However, Ayerst [154] suggested that the relative speed of germination and growth may be the main criteria for utilization of nutrient rich substrata like grain. Although *P. brevicompactum* and *P. hordei* are slow-growing fungi, they were the most competitive of the *Penicillium* spp. tested at 0.98 and 0.95 a_w (-3 to -7 MPa). *Aspergillus candidus* and *A. nidulans* were particularly competitive at 30°C, while *E. repens* and *A. versicolor* were uncompetitive at all temperatures, from 15 to 30°C, and all a_w tested, even though they are common on stored grain. Of the field fungi, *Epicoccum nigrum* and *F. culmorum* gave the highest I_D scores at 0.95 and 0.98 a_w (-7 and -3 MPa), while *A. alternata*, *Cladosporium* spp. were uncompetitive with hyphae intermingling freely with those of other species.

Hyphal characteristics of the interacting fungi may be observed by mounting a portion of agar from the zone of interaction onto a glass slide and examining it using a light microscope [20]. The interacting hyphae often appeared granulated or vacuolated with malformations. *Aspergillus candidus* caused swelling of hyphal tips of *A. fumigatus*, while *P. roquefortii* stimulated branching in *P. aurantiogriseum*. In the presence of *P. aurantiogriseum*, *P. viridicatum* hyphae were thickened and distorted, and *Fusarium culmorum* hyphae were partially lysed with cytoplasmic granulations. *Bacillus amyloliquefaciens* caused swelling, distortion, and the formation of spherical structures in hyphae of *Aspergillus flavus* (Fig. 3).

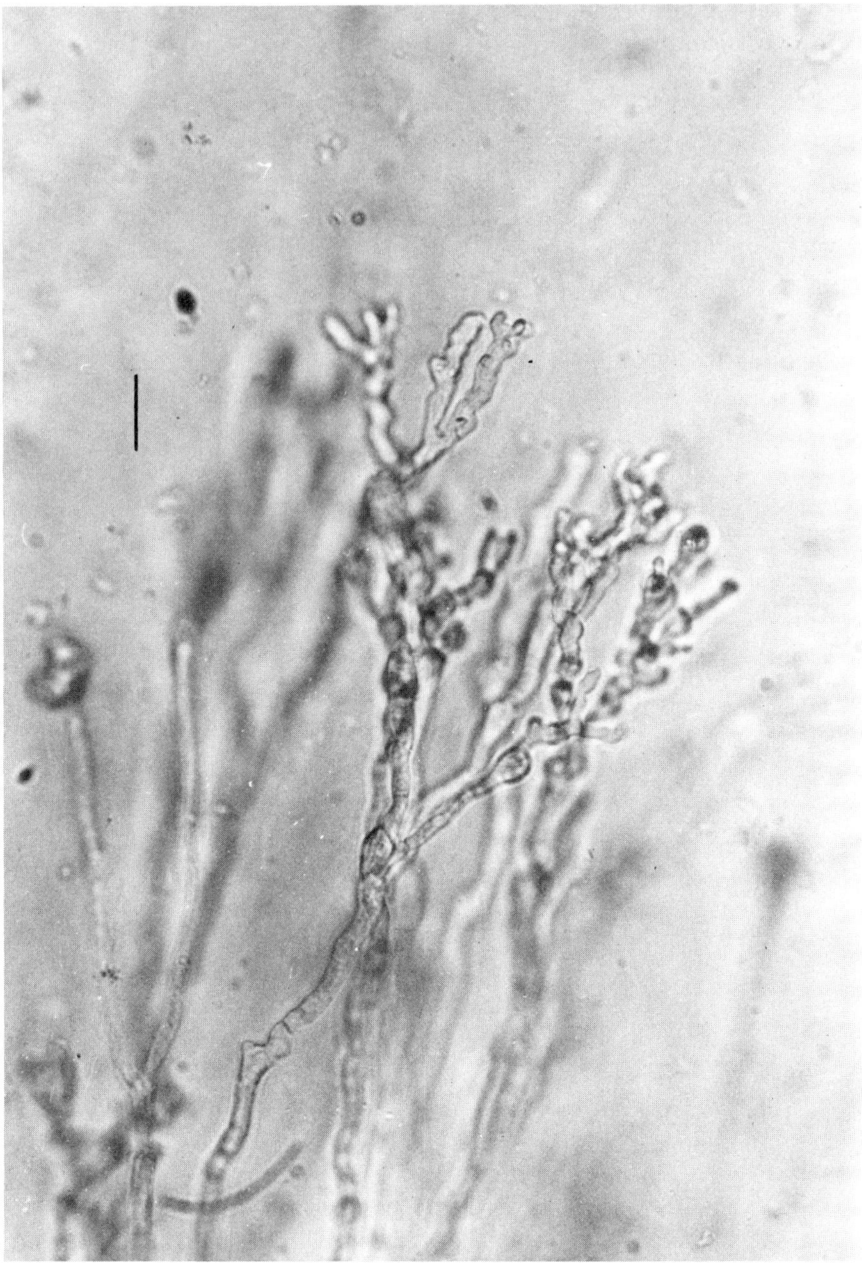

FIGURE 3. Swelling of *Aspergillus flavus* caused by *Bacillus amyloliquefaciens* (bar marker = 20 μm).

2. Studies on Developing Grain

Interactions between common fungi colonizing maize before harvest have been noted [34,172]. The fungi most commonly isolated in the southeastern United States are *Aspergillus flavus*, *A. niger*, *Fusarium moniliforme*, *Penicillium oxalicum*, *P. funiculosum*, and *Acremonium strictum* [34]. Interactions between these fungi were examined in kernels during development on maize grown in a controlled environment chamber following inoculation with a mixed conidial suspension containing all six fungi at silking, accompanied by wounding of the ears through the husk with a sterile toothpick or with inoculation of ears with a toothpick that had been dipped in the spore suspension [172].

After silk inoculation, *A. flavus*, *A. niger*, *F. moniliforme*, and *P. funiculosum* were isolated from 20-40% kernels but *P. oxalicum* and *A. strictum* were recovered from 2% only. After wound inoculation, *F. moniliforme* was most frequently isolated, then *A. niger* and *P. funiculosum*. *Aspergillus flavus* infection of kernels was decreased by the presence of competing fungi and aflatoxin levels were greatly decreased. Prior inoculation of the ears by *F. moniliforme* or *A. strictum* completely inhibited *A. flavus* infection. Under natural conditions, Lillehoj et al. [173] noted that with a high incidence of *F. moniliforme* in one maize ear, *A. flavus* was restricted to the tip.

When inoculated into wounds, *A. flavus* and *A. niger* occurred together in a large proportion of kernels, but aflatoxin production was unaffected [174]. However, similar inoculation of *A. flavus* with *P. oxalicum* resulted in total inhibition of aflatoxin production [175].

Evidence has also been presented of exchange of genetic information during interaction between competing molds [176]. A dsRNA virus was transferred from *P. chrysogenum* to *A. flavus* when the two fungi were cultured together and inhibited aflatoxin production by the latter.

Immunohistochemical techniques have also been used to examine fungal interactions on the host, using specific antisera to one or both fungi with either immunofluorescence or immunogold conjugates. Interactions between *Claviceps purpurea* and *Tilletia caries* on wheat were studied by Willingale and Mantle [177] using immunofluorescence. *Tilletia caries* was found to be displaced acropetally by *C. purpurea*.

3. Fungal Interactions in Stored Grain

Experiments on autoclaved wheat grain have often provided very different results from culture [177]. For instance, *Aspergillus versicolor*, although uncompetitive in culture, dominated the microflora in autoclaved wheat grain inoculated with a mixed spore suspension of storage fungi at 0.98-0.90 a_w (-3 to -15 MPa) and all temperatures tested. *Eurotium repens* and *A. candidus* were also present at 15°C and *A. nidulans* at 25-30°C and 0.90-0.95 a_w (-15 to -7 MPa). *Penicillium piceum* was seldom isolated at 15°C but was numerous at 30°C. *Penicillium aurantiogriseum* and *P. brevicompactus* competed well at 15-25°C and 0.95-0.90 a_w (-7 to -15 MPa). *Penicillium hordei* was the only storage fungus to behave similarly in grain and in culture, especially at 15 and 25°C. Of the field fungi, *F. culmorum* became dominant and contaminated most grains at all temperatures (15-30°C) and 0.99 and 0.95 a_w (-1 and -7 MPa), as might have been expected from its

Index of Dominance. However, *Alternaria alternata* had also colonized half the grain after 4 weeks' storage, but *Epicoccum nigrum* and *Cladosporium* spp. competed poorly. Fungal colonization was assessed in these studies by dilution plating and so was biased toward those fungi able to sporulate well. Further detailed ecological studies are still necessary to better understand fungal activity in stored grain.

4. Effects of Fungal Interactions on Mycotoxin Formation

Studies of *Aspergillus flavus* colonization of maize grain in the field (see above) have suggested that aflatoxin formation may be inhibited by interactions with some fungi. To determine which combinations might be most inhibitory of aflatoxin formation, mixed cultures have been prepared on autoclaved or irradiated cereal grains [172,179,180]. Areas of *A. flavus* sporulation were seen on the grain when this species was paired with *Cladosporium cladosporoides*, *Alternaria alternata*, *Curvularia lunata*, and *Nigrospora oryzae* on autoclaved maize but not when it was paired with *Trichoderma viride* [172]. *Aspergillus flavus* and *A. niger* occupied approximately equal portions of the kernel. Aflatoxin production was significantly decreased when *A. flavus* was paired with *T. viride* or *A. niger* in autoclaved maize but enhanced in the presence of *Candida guilliermondii*. The effect of *A. niger* has been attributed to a lowering of the substrate pH and, perhaps, to an inhibitory factor produced by the fungus [181]. Aflatoxin production was also inhibited when *A. parasiticus* was inoculated simultaneously with *Eurotium chevalieri* or *A. candidus* on rice and incubated at 20-35°C and 85-100% relative humidity, although both *A. parasiticus* and the competing fungi could be isolated from a large proportion of the grains [180,181].

Because of the large chemical and physical changes that occur in grain during autoclaving, Cuero et al. [182] used irradiated maize to study growth and mycotoxin production by *A. flavus* and *F. graminearum* in the presence of other fungi at different temperatures and water activities. When *A. flavus* was paired with *F. graminearum*, *P. viridicatum*, *A. niger*, or *A. oryzae* at 25°C, *A. flavus* formed considerably more visible growth at all a_w than the competing species. At 16°C, *A. flavus* again produced more visible growth in paired cultures than *F. graminearum* at all a_w tested and more than *A. niger* at 0.95 a_w (-7 MPa). By contrast, *P. viridicatum* grew better than *A. flavus* in paired culture at all a_w at 16°C. Compared to pure cultures grown under the same conditions, aflatoxin production in paired fungal cultures was decreased at high water activities (0.98 a_w, -3 MPa) but was enhanced when water activity was low (0.90 a_w, -15 MPa). Zearalenone production was markedly decreased at 16°C by the presence of *A. flavus* but was little affected at 25°C. When *A. flavus* was cultured with *Hyphopichia burtonii*, a yeast commonly isolated from stored maize, or with *Bacillus amyloliquefaciens*, aflatoxin production was enhanced at 16°C or 25°C and 0.95-0.98 a_w (-7 to -3 MPa) [183].

IV. THE KINETICS OF FUNGAL GROWTH IN STORED GRAIN

Measurement of the activity and biomass of fungi in stored grain has always presented problems. Traditionally, direct plating of whole grains onto agar and the plating of serial dilutions of washings or comminuted grains (dilution plating) have been used to quantify fungi (see Chapter 4). However, these methods favor isolation of heavily sporulating and fast-growing fungi, and may bear little relationship to the fungal biomass in the grain [184], although they enable the microflora of different samples to be compared [185]. Warnock [186,187] overcame some of these shortcomings by employing a fluorescent antibody technique that specifically detected mycelia of *Aspergillus, Penicillium,* or *Alternaria* spp. in barley grains. Where culture methods failed to isolate *Alternaria* and suggested that *Penicillium* and *Aspergillus* spp. were predominant, he found that *Alternaria* accounted for a greater length of mycelium than the other two taxa.

Other methods of quantifying fungal biomass have been described but most rely on either measurement of respiratory activity or analysis of biochemical markers characteristic of fungi. However, neither approach allows species of fungi to be differentiated.

A. Respiratory Measurement

During aerobic respiration by both storage microflora and grain, energy sources are metabolized through the catabolism, by oxidation, of complex molecules to their simpler component molecules, producing carbon dioxide and water [188,189]. The intensity of respiration is determined by water content, temperature, mechanical damage, and microbial contamination [190]. It is usually assessed by measuring changes in one or more components from the respiration equation (Sec. II.D.1) using methods described below.

1. Carbon Dioxide

Respired carbon dioxide can be absorbed in alkaline materials, e.g., soda lime [189], barium, sodium, or potassium hydroxides [191,192], or preweighed ascarite [193], and then measured by titration to neutrality with acids. Alternatively, gas chromatography [191] or infrared gas analyzers can be used to obtain direct measurements of CO_2 concentration [189]; Leach [195] exploited changes in conductivity in heated wires or alkaline solutions when exposed to differing gas concentrations.

2. Oxygen

Oxygen uptake can be assessed using Warburg-type respirometer systems or gas chromatography. A recently developed system [196] records oxygen uptake electronically and supplies controlled volumes of oxygen to the sealed units through an electrolytic process. In this system, carbon dioxide absorbed in sodium hydroxide solution can also be assessed after the experiment by titration, allowing respiratory quotients to be calculated. In some other systems, sealed respirometer units have been used in which the oxygen is not replaced and carbon dioxide is allowed to accumulate [83]. Respiration is inhibited when CO_2 concentrations exceed 10% [54], affecting the useful duration of experiments which depends on vessel and sample sizes. However, this situation could, perhaps, mimic more accurately the situation within a stored grain bulk, especially in sealed silos. Milner and Geddes [197] attempted an intermediate approach by designing a respirometer which aerated grain in a controlled manner.

3. Heat

Insulated Dewar flasks have been widely used to measure the heat output in respiring grain [198,199], using thermometers on thermocouples to measure temperature rise. Adiabatic incubation of fermentation apparatus or Dewar flasks has been used to monitor the thermal behavior of decomposing plant materials [198,200], while microcalorimetry has been used to measure small changes in heat output by the soil biomass [201] and may be applied to grain storage studies.

4. Respiratory Activity in Stored Grain and Dry Matter Loss

By utilizing the respiratory equation (Sec. II.D.1), CO_2 production can be translated into dry matter loss [193]. Typically, complete respiration of a carbohydrate gives a respiratory quotient, i.e., the ratio of oxygen absorbed to carbon dioxide evolved [56], of 1.0, and it is calculated that 14.7 g of carbon dioxide per kg of grain will be released for every 1% loss of grain dry matter [193,202]. During anaerobic fermentation, only about 0.493 g carbon dioxide is evolved from 1 kg grain for every 1% dry matter lost [194]. This process, which is considered negligible in cereal grain [193], can result in a respiratory quotient greater than 1.0. Alternatively, a respiratory quotient below 1.0 may result from lipid or protein metabolism [203]. For example, tripalmitin oxidation has a quotient at 0.7 [83].

The relative contributions of fungi and seeds to total respiration of grain has been argued for many years [61,204]. The respiratory processes of both are similar and dry matter losses in stored grain result from both. Seitz et al. [205] concluded that at high water content (22-27%, 0.93-0.99 a_w, -10 to -1 MPa), respiration by maize seed was sufficiently great for it to be the major component and for fungal respiration to account for only a small part. However, it is now well established that molds play a predominant role in grain deterioration, but the dry matter loss that they cause has still hardly been studied. Kreyger [96] pointed out, from work by Scholtz with wheat [206], that dry matter loss is unimportant, as long as there is no visible molding, despite his own work showing that barley, with 24% water content (about 0.94 a_w, -9 MPa) stored at 16°C for 10 weeks, lost 2% dry matter with visible molding. However, this loss was considered unimportant. More recently, Seitz et al. [65], working with maize, showed that fungal invasion and aflatoxin content could be unacceptable before the grain lost 0.5% dry matter and mold growth became visible. This work, while not directly comparable, suggests that Kreyger's assumptions may not be wholly accurate.

Grain seeds are living and, hence, respire in the same way as microorganisms. This means that seed dry matter is lost, this loss increasing as respiration rate increases. The rate at which a seed respires is largely governed by its moisture content and, to a lesser extent, by oxygen concentration, temperature, and damage to the seed. Mechanical damage increases carbon dioxide production and dry matter loss, especially if the embryo tissue is damaged [193,207].

With a respiratory quotient of one, CO_2 production and O_2 utilization can be related to dry matter loss by multiplying the weight (g) of CO_2 produced by 0.682, a figure obtained from the stoichiometry of the respiration equation in which 1 mol $C_6H_{12}O_6$ (180 g) yields 6 mol CO_2 (264 g) [56]. A formula was developed [192,202] to predict permissible storage times for aerated shelled maize

$$T = T_R \times M_T \times M_M \times M_D$$

where T = estimated allowable storage time before the consumption of 0.5% dry matter, T_R = time taken to lose 0.5% dry matter at 25% water content (0.98 a_w, -3 MPa), and M_T, M_M, M_D = multipliers relating to temperature, moisture content, and mechanical damage, obtained from graphs.

Fernandez et al. [192] evaluated one Saul and Steele formula [202] and then compared predicted results with those obtained in experiments in which CO_2 evolution, percentage kernel infection, and other factors were recorded. Predicted results were within 2% of experimental, except at low water contents where the error increased to 30%.

Under good storage conditions, with a water content of 12-14% (about 0.62-0.75 a_w, -66 to -40 MPa) and a temperature of 5-10°C, molding will not occur and dry matter losses will be due only to a slow, constant seed respiration. If the water activity exceeds a critical value, usually 0.75 a_w (-40 MPa), storage molds will start to proliferate [208] and lead, eventually, to a rapid increase in temperature and moisture content from respiration with considerable dry matter and quality loss.

White et al. [194] determined respiration rates in wheat grain stored in experimental silos under a range of environmental conditions. The data was fitted to several predictive equations in order to obtain allowable storage times based on the production of 1470 mg CO_2/kg grain, equivalent to a dry matter loss of 0.1% if the respiratory quotient is 1.0. However, when 55 days safe storage was predicted for grain containing 18.4% water (about 0.86 a_w, -21 MPa), visible molding was found after 23 days of storage. This suggests that 0.1% dry matter loss is far more than can be allowed, that determinations of respiration rates were inaccurate, or that respiratory quotients exceeded 1.0.

The observed respiratory quotient is the summation of several interacting factors. Carbohydrates are the predominant energy source in cereal grains and it is often assumed that anaerobic respiration and protein and lipid metabolism are negligible. However, CO_2 concentrations may be expected to increase within a grain bulk and may lead to anaerobic respiration. The respiratory quotients found by White et al. [194] were in the range 1.0 to 2.0, suggesting either anaerobic respiration or the conversion of carbohydrates to fats. Metabolism of proteins and fats, on the other hand, leads to respiratory quotients less than 1.0. Consequently, the predictions of 1470 mg CO_2 produced per g of grain per 1% dry matter lost and 0.682 g dry matter lost per g CO_2 produced must be corrected for the actual respiratory quotient found. Similar experiments with rapeseed [203], which contains more lipid than cereal grains, showed that storage fungi degraded lipids to free fatty acids.

Perhaps the most widely used predictions of safe storage times are based on the work of Kreyger [96], who used visible molding as a criterion of cereal grain spoilage to unacceptable levels, providing viability losses were unimportant. Although Kreyger considered dry matter losses of up to 2% to have little importance, many authorities have accepted his rationale (e.g., Refs. 49, 210). Seitz et al. [65] compared the measurement of fungal respiration in corn with ergosterol and aflatoxin levels and percentage kernel infection and found that unacceptable levels of spoilage could be attained before 0.5% dry matter was lost, advising the cautious use of previous recommendations and the inadequacy of visual inspection procedures. It seems clear that a redefinition of so-called "safe storage periods" is needed.

B. Biochemical Tests

Biochemical assays that directly measure fungal biomass in grain have been developed particularly over the last 15 years. Of these, assay of chitin and ergosterol have been most widely used, but assay of adenosine triphosphate (ATP) has also been suggested [189]. Chitin is an important component of fungal cell walls, while ergosterol is the major sterol in fungal membranes [210,211]. Ergosterol is absent from most higher plants and other microorganisms, but chitin is a major component of insect cuticles and bacterial cell walls. ATP is found in all metabolizing cells.

1. Chitin

The method involves the hydrolysis of chitin to glucosamine, followed by deamination to an aldehyde which is measured colorimetrically [212]. However, besides being a component of insects and bacteria, the glucosamine content of *Aspergillus oryzae* mycelium changes with age [213]. This makes it difficult to relate glucosamine assay to fungal biomass in grain and Jarvis et al. [184] comment that "where non-homogeneity of contamination occurs the variance of the results is such as to render the method (i.e., chitin assay) of little value."

2. Ergosterol

Ergosterol may be assayed by thin layer chromatography [213] or by high performance liquid chromatography [215], the latter assay being more expensive and complex than the chitin assay but quicker, and there is no risk of detecting nonfungal material. Both viable and nonviable mycelia are detected allowing total biomass to be quantified. Ergosterol assay has usually been preferred to chitin assay for wheat grain [189], rice [216], and maize [211]. However, there are indications that some substrates, e.g., oriental solid-state fermentations, affect ergosterol assays [217], and these need to be evaluated.

The effect of substrate composition on ergosterol levels is not yet fully understood, and so the test cannot be currently recommended as an index of biomass.

3. Adenosine Triphosphate Assay

Adenosine triphosphate (ATP) is assayed by measuring the luminescence produced by the oxidation of luciferin in the presence of luciferase and ATP. Only metabolically active material is detected, but a good extraction method is essential. This must release ATP quantitatively and inactivate all enzyme activity in the extract, and trichloroacetic acid has been found superior to most alternatives [189]. In artificial media, ATP has been used to measure biomass during colony growth. ATP activity increased during exponential growth but declined to zero once growth ceased and may be affected by environmental conditions and exposure to stress conditions such as low a_w. Living grain is also metabolically active, although quiescent under normal storage conditions with, perhaps, only low levels of ATP. An extraction method is needed that does not penetrate deeply into the grain, thus limiting its ability to detect total fungal biomass. However, Kaspersson [189] found correlations between ATP content and plate counts of microorganisms.

V. FUNGI AND GRAIN QUALITY

Cereal grains form rich nutrient sources for colonizing fungi. They are formed mostly of carbohydrates, proteins, and lipids, the relative amounts differing between varieties and especially between grain types [218,219]. Colonizing fungi form extracellular enzymes that utilize grain components leading to losses of dry matter and of quality. Quality loss occurs naturally during the aging of grain, but fungal activity accelerates this process [220].

Quality loss is a subjective term that can mean several things, depending on the grain type and its end use. For instance, with wheat for bread-making, gluten protein is the most important component, and tests such as the wheat meal fermentation test [221], sedimentation test [222], and extensibility, which rely on gluten quality for the final result, are used. Viability is also an important criterion for malting as well as for seed.

One consequence of fungal colonization of grain is that they produce enzymes which may affect the quality of the grain for some uses. Thus, a high α-amylase content in wheat is deleterious for bread-making, while in barley for malting it is required to hydrolyze the carbohydrate reserves. *Aspergillus* spp. differ in their production of α-amylase, with *A. flavus* producing most and *A. restrictus* least [223]. *Aspergillus fumigatus, Eurotium* sp., *Penicillium jensenii, A. sydowii*, and *A. niger* were intermediate. Other fungi may produce cellulase, polygalacturonase, protopectinase, pectin methyl esterase [224], 1-4-β-glucanase, β-glucosidase, and β-xylosidase [225] while Jain et al. (unpublished) found *N*-acetyl-β-D-glucosaminidase activity indicative of fungal invasion of wheat and barley grain and α-D-galactosidase indicative of growth of *Eurotium amstelodami*. Most of 75 fungi isolated from barley grain degraded arabinoxylan, but 30% were unable to degrade carboxymethyl cellulose [226,227].

A. Loss of Viability

Loss of viability is a sensitive indicator of deterioration, although there is some discussion as to whether seed death occurs before or as a consequence of fungal invasion (E. Moreno, personal communication). Some seeds can remain viable for 150 years or more, but cereal grains usually survive for less than 30 years [228]. The precise mechanism determining the survival of seeds is little understood. Living seeds are complex arrays of biochemical reactions, each affected by another and all under genetic control. Most research has been directed at the effects of aging on chromosomes since old seeds are known to produce more chromosomal and genetic abnormalities than freshly harvested seeds [229-231]. However, chronological age of the seed is not the only factor involved in producing chromosomal aberrations as increases in temperature [232-235], water content [208,235,236], and oxygen concentration [235] all increase chromosome damage and loss of viability. Concomitant with chromosome damage and loss of viability, protein production was also substantially decreased [237].

Under some circumstances, at least, fungi are primary causes of loss of viability in seeds [238]. The embryo of cereal grains is often preferentially invaded by fungi, mostly species of *Aspergillus* and *Penicillium* [238]. However, their role is often difficult to prove since mold-free seed is difficult to obtain and fungicides will not penetrate to the embryo [239]. This can lead to an inaccurate interpretation of the results of experiments. Thus, Malowan [240] concluded that, because fungicides did not prevent CO_2 evolution from seeds, the microflora did not cause the associated respiration. However, in much of the work by Christensen and his colleagues, it has been shown that where conditions are suitable for fungal proliferation, seed viability is decreased to near zero, while Hill and Lacey [72] found a linear relationship between initial water content and percentage germination. Wheat grain stored with 0.70 a_w (-49 MPa) or less maintained good viability for several months. However, increasing the water content of wheat without visible mold also resulted in loss of viability.

None of the biochemical processes involved in loss of viability caused by fungi are known. It is not known at what level of fungal invasion seed metabolism is sufficiently upset for chromosome damage to occur. Even a small contamination of the embryo could lead to some alteration in metabolism, possibly as a result of a defense mechanism being initiated [241]. Such an alteration would necessitate genetic involvement and could increase the rate of chromosome aberration and loss of viability.

B. Carbohydrate Degradation

About 75% of cereal grain is formed of carbohydrate [242] comprising both functional and storage saccharides, mostly as polysaccharides [243]. In the mature grain the major component is starch, composed of two types of polysaccharides: amylose, a single-chain polysaccharide, containing glucose molecules joined by α-1-4 glycosidic bonds; and amylopectin, also containing only glucose molecules, but branched regularly through α-1-6 glycosidic bonds [243].

During respiration, starch is hydrolyzed through the action of amylase on both amylose and amylopectin. This is specific for the α-1-4 glycosidic bond and produces the reducing sugar, maltose, from amylose but, because the α-1-6 glycosidic bonds prevent complete hydrolyzation, only large dextrin units are produced from amylopectin [244]. The amylase may come from both grain and fungal sources. Fungal invasion results in a corresponding increase in reducing sugars [208,242] but only Farag et al. [242] reported a proportional increase in amylase activity. The maltose is further hydrolyzed to glucose by maltase and dextrin to glucose by other enzymes [244], but there is no reference to the production of these enzymes in grain by fungi although it may be presumed that these or similar enzymes are produced. The resulting glucose is utilized as an energy source or to produce ribose sugars for nucleic acids.

C. Protein Degradation

Total protein accounts for about 14.5% of the dry matter in wheat [242,245], although varieties may differ somewhat [218] and quality of the protein is as important as amount. The amount and quality of protein are heritable characteristics that determine the physical and structural characteristics of baked wheat products and its suitability for different purposes. Thus, a high quality variety wheat will produce good bread over a fairly wide range of protein percentages, while a low quality variety will produce poor quality bread even when its protein content is high [246]. Within a variety, it has been shown that loaf volume is highly correlated with protein content [247, 248]. The protein fraction of grain and its quality are therefore important to human nutrition, although wheat lacks certain essential amino acids, e.g., lysine, threonine, and methionine [249].

Despite the importance of proteins in wheat, there have been virtually no studies of changes in proteins during molding. Such work as has been undertaken has usually involved total protein rather than the individual constituents [250]. Westermarck-Rosendahl [207] found no changes in total protein during spontaneous heating, while Cross [251], in similar experiments, found an increase in total protein. No explanations for these discrepancies are given. Infection of wheat grain with *Aspergillus flavus, A. niger, Fusarium solani,* and *Alternaria tenuis* increased their protein content [252]. The increase was attributed to fungi utilizing seed constituents for their growth. This may not be a full explanation. Microbial proteases facilitate invasion of seed tissue [253] which leads to destruction of the tissue. Autolysis by host enzymes then begins and seed proteins are hydrolyzed to peptides and free amino acids [254]. In peanuts, invasion results in the decomposition of proteins to low molecular weight components, the depletion of some enzymes, intensification of others, and also the production of new enzyme systems [46]. This increase in enzymatic activity during invasion may account for the increased total protein found by Farag in wheat while the absence of change may indicate cessation of the process.

D. Lipid Degradation

Cereal grain contains about 2.5% lipid [242], both storage and functional lipids. The storage lipids, mainly triglycerides, serve as energy reserves that are mobilized by specific enzymes whenever an energy source is required in addition to that provided by carbohydrates, as during germination or when the seed is damaged in any way [255]. Lipids can be degraded endogenously and through fungal activity, both by oxidation and by hydrolysis, with lipoxygenases and lipases, respectively [256]. The degradation of lipids, both endogenously and due to fungal activity, differs between whole grains and their products, being moisture dependent [69]. Fungal degradation of lipids was first reported by Milner and Geddes [257], but Dirks et al. [258] were the first to state conclusively that it was due to fungal lipases rather than to enzymatic activity in the grain. It is characterized by increased fat acidity [208], the amount depending on the form of damage and also on the types of invading fungi [259]. Zeleny [260] utilized fat acidity to develop a grain quality test, since modified by others [261], in which increasing damage, through molding and heating, but not damage before harvest, is indicated by increases in the fat acidity value [262]. This was achieved using specific enzyme inhibitors.

Farag et al. [242,263,264] have studied the effects of fungal lipases on different classes of lipids. Initially [242], it was thought that fungi could not utilize lipids as an energy source since the changes of crude lipid in culture media were so small. However, the component lipid classes were not analyzed, and further studies [263,264] showed that, although the relative amounts in each lipid class were little changed, the composition of lipids within each class altered greatly. The fungi hydrolyzed triglycerides to form a mixture of mono/diglycerides and free fatty acids. Some fatty acids increased, new fatty acids appeared, and some medium-length fatty acids disappeared.

E. Odors

Fungal colonization can lead to undesirable effects on the organoleptic quality of the grain through the production of volatile metabolites affecting the taste and smell. These odors have been variously described as musty, putrid, fungal, urinal, ammoniacal, acid, honey, floral, fruity, or rotten [265]. Such off-odors and off-flavors may result from the oxidation of triglycerides [266], but they could be prevented by sterols. However, fungi promote the conversion of sterols to hydrocarbons, thus causing rancidity [267]. More than 30 peaks have been produced on gas chromatograms, some of which have been identified by mass spectrometry as 1-octen-3-ol, 3-octanol, 1-octanol, 3-octanone, and 3-methyl-1-butanol [268]. More 1-octanol is produced in molding wheat grain than 3-octanone or 3-methyl-1-butanol [269].

VI. CONCLUSION

Fungi colonizing grain are thus important causes of losses of dry matter and quality. Their importance is, perhaps, often underestimated as they are less easy to see than insects and their visible growth may be removed when grain is handled and cleaned. Yet, much of the damage that they cause could be prevented if grain were dried adequately and stored in dry, weatherproof buildings. This may be readily achieved in some regions, where weather at harvest is normally dry, but is less easily achieved when harvest coincides with seasonal rains or when these arrive unexpectedly early. This may be a particular problem in tropical areas in monsoon seasons when drying grain adequately may be difficult. Even so, drying after harvest may be too late to prevent some fungal colonization and mycotoxin formation, e.g., with head molds of sorghum or preharvest aflatoxin formation in maize, or a delay before drying may allow aflatoxin formation between shelling maize and drying [270].

Control of molding may be achieved in different ways. However, it has been usual for only one factor to be altered, for instance, by drying, chemical treatment, or modification of the atmosphere. Tolerance of one unfavorable factor is usually greatest when all others are optimal and manipulation of only one factor does not allow any synergistic interactions. Manipulation of a range of environmental factors may allow the development of integrated management strategies that will control not only fungal spoilage of grain but also insect infestation.

Care should also be taken in shipment and marketing to ensure that grain is in an adequate condition to arrive in good condition. Without other protective measures, this requires that it is dry enough to minimize the risk of moisture migration during shipment and that mixture of good and slightly deteriorated batches is avoided. The norm in the United Kingdom is to dry wheat and barley grain to 14-15% moisture content (0.65-0.75 a_w, -59 to -40 MPa) [50], and in Canada to dry maize to 15-16% (0.76-0.81 a_w, -38 to -29 MPA) [64] to meet grade requirements. These levels may not be adequate for long-term safe storage, and only small changes in a_w may be necessary to allow large increases in fungal activity. Grade requirements should be modified to remove the risk of deterioration and perhaps grain should be marketed on a dry matter basis to remove the incentive to sell water. More attention is required to the maintenance of quality, and methods are needed to detect molding when the visible evidence has been removed during handling.

REFERENCES

1. Hall, D. W. Handling and storage of food grains in tropical and subtropical areas. *F.A.O. Agric. Dev. Pap. No. 90*, Food and Agriculture Organization, Rome, 1970.
2. Anon. *Post-Harvest Food Losses in Developing Countries.* National Academy of Science, Washington, D.C., 1979.
3. Rohani, M. Y., Shariffah Norin, S. A., and Samsudin, A. Post-harvest losses of paddy in the Krian/Sungei Manik areas. *MARDI Res. Bull.*, *13*:148-154, 1985.
4. Vasan, B. S. Handling of high moisture paddy during wet season—a practical approach to the existing problems. *Bull. Grain Technol.*, *18*:223-232, 1980.
5. Lacey, J. The microflora of grain dusts. In *Occupational Pulmonary Disease-Focus on Grain Dust and Health* (J. A. Dosman and D. J. Cotton, eds.), Academic Press, New York, 1980, pp. 417-440.
6. Lacey, J., and Crook, B. Fungal and actinomycete spores as pollutants of the workplace and occupational allergens. *Ann. Occup. Hyg.*, *32*:515-533, 1988.

7. Hill, R. A., and Lacey, J. The microflora of ripening barley grain and the effects of pre-harvest fungicide application. *Ann. Appl. Biol.*, *102*:455-465, 1983.
8. Magan, J., and Lacey, J. The phylloplane microflora of ripening wheat and effect of late fungicide applications. *Ann. Appl. Biol.*, *109*:117-128, 1986.
9. Zadoks, J. C., Chang, T. T., and Konzak, C. F. A decimal code for the growth states of cereals. *Weed Res.*, *14*:415-421, 1974.
10. Flannigan, B., and Campbell, I. Pre-harvest mould and yeast floras on the flag leaf, bracts and caryopsis of wheat. *Trans. Br. Mycol. Soc.*, *69*:485-494, 1977.
11. Dickinson, C. H., Austin, B., and Goodfellow, M. Quantitative and qualitative studies of phylloplane bacteria from *Lolium perenne*. *J. Gen. Microbiol.*, *91*:157-166, 1975.
12. Goodfellow, M., Austin, B., and Dickinson, C. H. Numerical taxonomy of some yellow-pigmented bacteria isolated from plants. *J. Gen. Microbiol.*, *91*:219-233, 1976.
13. Hyde, M. B., and Galleymore, H. B. The subepidermal fungi of cereal grain. II. The nature, identity and origin of the mycelium in wheat. *Ann. Appl. Biol.* *38*:348-356, 1951.
14. Jørgensen, J. Decline in incidence of some fungi of barley seed during storage under farm-like conditions in Denmark. *Acta Agric. Scand.*, *26*:59-64, 1976.
15. Lacey, J. The microbiology of cereal grains from areas of Iran with a high incidence of oesophageal cancer. *J. Stored Prod. Res.*, *24*:39-50, 1988.
16. Sutton, J. C. Epidemiology of wheat head blight and maize ear rot caused by *Fusarium graminearum*. *Can. J. Pl. Path.*, *4*:195-209, 1982.
17. Chu-Chou, M., and Preece, T. F. The effect of pollen grains on infections caused by *Botrytis cinerea* Fr. *Ann. Appl. Biol.*, *62*:11-22, 1968.
18. Wallace, H. A. H., and Sinha, R. N. Causal factors operative in distributional patterns and abundance of fungi: A multivariate study. In *The Fungal Community: Its Organization and Role in the Ecosystem* (D. T. Wicklow and G. C. Carrol, eds.), Marcel Dekker, New York, 1981, pp. 233-247.
19. Welling, B. Effect of lodging on germination capacity and fungus flora at harvest and after storage. *Tidsskr. PlAvl.*, *79*:243-253, 1975.
20. Magan, N., and Lacey, J. Effect of water activity, temperature and substrate on interactions between field and storage fungi. *Trans. Br. Mycol. Soc.*, *82*:83-93, 1984.
21. Lacey, J. Moulds of cereals. *Rep. Rothamsted Exp. Sta. 1974, Part 1*:215, 1975.
22. Hill, R. A., and Lacey, J. *Penicillium* species associated with barley grain in the U.K. *Trans. Br. Mycol. Soc.*, *82*:297-303, 1984.
23. Mills, J. T., and Frydman, C. Mycoflora and condition of grains from overwintered fields in Manitoba, 1977-1978. *Can. Pl. Dis. Survey*, *60*:1-7, 1980.

24. Joffe, A. Z. The mycoflora of overwintered cereals and its toxicity. *Bull. Res. Counc. Isr. Sect. D, 9*:101-126, 1960.
25. Mehrotra, B. S., and Dwivedi, P. K. Fungi associated with wheat in the field. *Int. Biodet. Bull., 16*:37-42, 1980.
26. Williams, R. J., and McDonald, D. Grain molds in the tropics: Problems and importance. *Ann. Rev. Phytopath., 21*:153-178, 1983.
27. Tyagi, P. D., and Olugbemi, L. B. Grain weathering in rainfed wheat in Nigeria as influenced by fungal pathogens and adverse weather conditions. *Samaru Misc. Paper, 9*:1-9, 1980.
28. Hesseltine, C. W., and Bothast, R. J. Mold development in ears of corn from tasseling to harvest. *Mycologia, 69*:328-340, 1977.
29. Tuite, J. F. Fungi isolated from unstored corn seed in Indiana in 1956-1958. *Pl. Dis. Reptr., 45*:212-215, 1961.
30. Wicklow, D. T. Patterns of fungal association within maize kernels harvested in North Carolina. *Pl. Dis., 72*:113-115, 1983.
31. King, S. B. Time of infection of maize kernels by *Fusarium moniliforme* and *Cephalosporium acremonium*. *Phytopathology, 71*:796-799, 1981.
32. Tuite, J. Fungi isolated from unstored corn seed in Indiana in 1956-1958. *Pl. Dis. Reptr., 45*:212-215, 1961.
33. Marasas, W. F. O., van Rensburg, S. J., and Mirocha, C. J. Incidence of *Fusarium* species and the mycotoxins, deoxynivalenol and zearalenone, in corn produced in oesophageal cancer areas in Transkei. *J. Agric. Food Chem., 27*:1108-1112, 1979.
34. Wicklow, D. T. Patterns of fungal association within maize kernels harvested in North Carolina. *Pl. Dis., 72*:113-115, 1988.
35. Hill, R. A., Wilson, D. M., McMillian, W. W., Cole, R. J., Sanders, T. H., and Blankenship, P. D. Ecology of the *Aspergillus flavus* group and aflatoxin formation in maize and groundnut. In *Trichothecenes and Other Mycotoxins* (J. Lacey, ed.), John Wiley, Chichester, 1985, pp. 79-95.
36. Mislivec, P. B., and Tuite, J. F. Species of *Penicillium* occurring in freshly-harvested and in stored dent corn kernels. *Mycologia, 62*:67-74, 1970.
37. Jiminez, M., Sanchis, V., Santamarina, P., and Hernandez, E. *Penicillium* in pre-harvest corn from Valencia (Spain). I. Influence of different factors on the contamination. *Mycopathologia, 92*:53-57, 1985.
38. Caldwell, R. C., Tuite, J., and Carlton, W. W. Pathogenicity of penicillia to corn ears. *Phytopathology, 71*:175-180, 1981.
39. Lillehoj, E., and Hesseltine, C. W. Aflatoxin control during plant growth and harvest of corn. In *Mycotoxins in Human and Animal Health* (J. Rodricks, C. W. Hesseltine, and M. A. Mehlman, eds.), Pathotox Publishers, Park Forest South, IL, 1977, pp. 107-119.
40. Hill, R. A., Blankenship, P. D., Cole, R. J., and Sanders, T. H. Effect of soil moisture and temperature on preharvest invasion of peanuts by *Aspergillus flavus* group and subsequent aflatoxin development. *Appl. Environ. Microbiol., 45*:628-633, 1983.
41. Doupnik, B. Maize seed predisposed to fungal invasion and aflatoxin contamination by *Helminthosporium maydis* ear rot. *Phytopathology, 62*:1367-1368, 1972.

42. Williams, R. J., and Rao, K. N. A review of sorghum grain moulds. *Trop. Pest.*, *27*:200-211, 1981.
43. Seitz, L. M., Mohr, H. E., Burroughs, R., and Glueck, J. A. Pre-harvest invasion of sorghum grain. *Cereal Chem.*, *60*:127-130, 1983.
44. Verma, V. S., and Khan, A. M. Fungi associated with sorghum seeds. *Mycopath. Mycol. Appl.*, *27*:314-320, 1965.
45. Siddiqui, M. R., and Khan, I. D. Fungi and factors associated with the development of sorghum ear-moulds. *Trans. Mycol. Soc. Japan*, *14*:289-293, 1973.
46. Mall, O. P., Pateria, H. M., and Chauhan, S. K. Mycoflora and aflatoxin in wet harvest sorghum. *Indian Phytopath.*, *39*:409-413, 1986.
47. Kuthubutheen, A. J. Effect of pesticides on the seed-borne fungi and fungal succession on rice in Malaysia. *J. Stored Prod. Res.*, *20*: 31-39, 1984.
48. Flannigan, B. Primary contamination of barley and wheat grain by storage fungi. *Trans. Br. Mycol. Soc.*, *71*:37-42, 1978.
49. Nash, M. J. *Crop Conservation and Storage in Cool Temperature Climates*, Pergamon Press, Oxford, U.K., 1978.
50. Anon. *Grain Drying and Storage No. 1. Grain Storage Methods*, MAFF booklet No. 2415, 1982.
51. Anon. *Grain Drying and Storage No. 3. High Temperature Grain Drying*, MAFF booklet No. 2417, 1982.
52. Tuite, J., and Foster, G. H. Effect of artificial drying on the hygroscopic properties of corn. *Cereal Chem.*, *40*:630-637, 1963.
53. Nellist, M. E. Near-ambient grain drying. *Agric. Engineer* (Autumn): 93-101, 1988.
54. Srour, S. Thermic properties of grains—production of heat and CO_2. In *Preservation and Storage of Grains, Seeds, and Their By-products* (J. Multon, ed.), Lavoisier Publishing, New York, 1988, pp. 189-202.
55. Pixton, S. W. Moisture content—its significance and measurement in stored products. *J. Stored Prod. Res.*, *2*:35-47, 1967.
56. Rees, D. V. H. A discussion of the sources of dry matter loss during the process of haymaking. *J. Agric. Eng. Res.*, *27*:469-479, 1982.
57. Currie, J. A., and Festenstein, G. N. Factors defining spontaneous heating and ignition of hay. *J. Sci. Fd. Agric.*, *22*:223-230, 1971.
58. Carter, E. P., and Young, G. Y. Role of fungi in the heating of moist wheat. Circular No. 138, United States Department of Agriculture, Washington, D.C., 1950.
59. Sinha, R. N., and Wallace, H. A. H. Ecology of a fungus induced hot spot in stored grain. *Can. J. Pl. Sci.*, *45*:48-59, 1965.
60. Burrell, N. J. Chilling. In *Storage of Cereal Grains and Their Products*, 2nd ed. (C. M. Christensen, ed.), American Association of Cereal Chemists, 1974, pp. 420-453.
61. Pomeranz, Y. Biochemical, functional and nutritive changes during storage. In *Storage of Cereal Grains and Their Products*, 2nd ed. (C. M. Christensen, ed.), American Association of Cereal Chemists, St. Paul, MN, 1974, pp. 56-114.
62. Milton, R. F., and Pawsey, R. K. Spoilage relating to the storage and transport of cereals and oilseeds. *Int. J. Food Microbiol.*, *7*: 211-217, 1988.

63. Bailey, C. H., and Gurjar, A. M. Respiration of stored wheat. *J. Agric. Res.*, *12*:685-713, 1918.
64. Bereza, K., Morris, D., and Clayton, R. Harvesting and storing quality grain corn. Factsheet, MAF, Ontario, 1981.
65. Seitz, L. M., Sauer, D. B., Mohr, H. E., and Aldis, D. F. Fungal growth and dry matter loss during bin storage of high moisture corn. *Cereal Chem.*, *59*:9-14, 1982.
66. Lacey, J. Potential hazards to animals and man from micro-organisms in fodders and grain. *Trans. Br. Mycol. Soc.*, *65*:161-184, 1975.
67. Pixton, S. W. The importance of moisture and equilibrium relative humidity in stored products. *Trop. Stored Prod. Inf.*, *43*:16-29, 1982.
68. Pepper, E. H., and Kiessling, R. L. A list of bacteria, fungi, yeasts, nematodes, and viruses occurring on and within barley kernels. *Proc. Ass. Off. Seed Analysts N. Am.*, *53*:199-208, 1963.
69. Semeniuk, G. Microflora. In *Storage of Cereal Grains and Their Products* (J. A. Anderson and A. W. Alcock, eds.), American Association of Cereal Chemists, St. Paul, MN, 1954.
70. Wallace, H. A. H. Fungi and other organisms associated with stored grain. In *Grain Storage, Part of a Storage* (R. N. Sinha and W. E. Muir, eds.), Avi Publishing Company, Westport, CT, 1973, pp. 71-97.
71. Lacey, J., Hill, S. T., and Edwards, M. A. Micro-organisms in stored grains: Their enumeration and significance. *Trop. Stored Prod. Inf.*, *39*:19-33, 1980.
72. Hill, R. A., and Lacey, J. Factors determining the microflora of stored barley grain. *Ann. Appl. Biol.*, *102*:467-483, 1983.
73. Pitt, J. I. *The Genus Penicillium and Its Teleomorphic States Eupenicillium and Talaromyces*, Academic Press, London, 1979.
74. Mislivec, P. B., and Tuite, J. F. Temperature and relative humidity requirements of species of *Penicillium* spp. isolated from yellow dent corn. *Mycologia, 62*:74-88, 1970.
75. Tuite, J. F., and Christensen, C. M. Grain storage studies. XVI. Influence of storage conditions upon the fungus flora of barley seed. *Cereal Chem.*, *32*:1-11, 1955.
76. Perez, R. A., Tuite, J., and Baker, K. Effect of moisture, temperature, and storage time on the subsequent storability of shelled corn. *Cereal Chem.*, *59*:205-209, 1982.
77. Chełkowski, J., and Cierniewski, A. Mycotoxins in cereal grain. Part X. Invasion of fungal mycelium into wheat and barley kernels. *Nahrung*, *6*:533-536, 1983.
78. Stoloff, L., Mislivec, P., and Kulik, M. M. Susceptibility of freshly picked ear corn to invasion by fungi. *Appl. Microbiol.*, *29*:123-124, 1975.
79. Tuite, J., Koh-Knox, C., Stroshine, R., Cantone, F. A., and Bauman, L. F. Effect of physical damage to corn kernels on the development of *Penicillium* species and *Aspergillus glaucus* in storage. *Phytopathology*, *75*:1137-1140, 1985.
80. Welling, B. The influence of thresh damage on microflora and germination of barley during storage. *Tidsskr. PlAvl.*, *72*:513-519, 1968.

81. Moreno-Martinez, E., and Christensen, C. M. Differences among lines and varieties of maize in susceptibility to damage by storage fungi. *Phytopathology*, *61*:1498-1500, 1971.
82. Cantone, F. A., Tuite, J., Bauman, L. F., and Stroshine, R. Genotype differences in reaction of stored corn kernels to attack by selected *Aspergillus* and *Penicillium* spp. *Phytopathology*, *73*:1250-1255, 1983.
83. Ilag, L. L., and Juliano, B. O. Colonisation and aflatoxin formation by *Aspergillus* spp. on brown rices differing in endosperm properties. *J. Sci. Fd. Agric.*, *33*:97-100, 1982.
84. Pelhate, J. Ecology of the microflora of grains and seeds. In *Preservation and Storage of Grains, Seeds and Their By-products* (J. Multon, ed.), Lavoisier Publishing, New York, 1988, pp. 244-262.
85. Oyeniran, J. O. Fungal deterioration of maize during storage in Nigeria. *Niger. J. Pl. Prot.*, *3*:102-105, 1977.
86. Abdel-Hafez, S. I. I., El-Kady, I. A., Mazen, M. B., and El-Maghraby, O. M. O. Mycoflora and trichothecene toxins of paddy grains from Egypt. *Mycopathologia*, *100*:103-112, 1987.
87. Ogundana, S. K. Fungi associated with the biodeterioration of stored rice in Nigeria. In *Biodeterioration* (T. A. Oxley, D. Allsopp, and G. Becker, eds.), Pitman, London, 1980, pp. 251-256.
88. Sahay, M. N., and Gangopadhyay, S. Effect of wet harvesting on biodeterioration of rice. *Cereal Chem.*, *62*:80-83, 1985.
89. Abdel-Kader, M. I. A., Moubasher, A. H., and Abdel-Hafez, S. I. I. Survey of the mycoflora of barley grains in Egypt. *Mycopathologia*, *68*:143-147, 1979.
90. Christensen, C. M., and Sauer, D. B. In *Storage of Cereal Grains and Their Products*, 3rd ed. (C. M. Christensen, ed.), American Association of Cereal Chemists, 1982, pp. 219-240.
91. Mills, J. T., and Abramson, D. Microflora and condition of flood-damaged grains in Manitoba, Canada. *Mycopathologia*, *73*:143-152, 1981.
92. Snow, D., Crichton, M. H. G., and Wright, N. C. Mould deterioration of feeding-stuffs in relation to humidity of storage. Part II. The water uptake of feeding-stuffs at different humidities. *Ann. Appl. Biol.*, *31*:111-116, 1944.
93. Pixton, S. W., Hyde, M. B., and Ayerst, G. Long-term storage of wheat. *J. Sci. Fd. Agric.*, *15*:152-161, 1967.
94. Pixton, S. W., and Hill, S. T. Long-term storage of wheat—II. *J. Sci. Fd. Agric.*, *18*:94-98, 1967.
95. Pixton, S. W., Warburton, S., and Hill, S. T. Long-term storage of wheat—III: Some changes in the quality of wheat observed during 16 years of storage. *J. Stored Prod. Res.*, *11*:177-185, 1975.
96. Kreyger, J. Drying and storing grains, seeds and pulses in temperate climates. IBVL publication 205, Wageningen, Holland, 1972.
97. Pelhate, J. A study of water requirements in some storage fungi. *Mycopath. Mycol. Appl.*, *36*:117-128, 1968.
98. Magan, N., and Lacey, J. The effect of temperature and pH on the water relations of field and storage fungi. *Trans. Br. Mycol. Soc.*, *82*:71-81, 1984.

99. Sinha, R. N., and Wallace, H. A. H. Ecology of an insect-induced hot spot in stored grain. *Res. Pop. Ecol.*, *8*:107-132, 1966.
100. Sinha, R. N. Effects of weevil infestation (Coleoptera: Curculionidae) infestation on abiotic and biotic quality of stored wheat. *J. Econ. Entomol.*, *77*:1483-1488, 1984.
101. Armitage, D. M., and George, C. L. The effect of three species of mites upon fungal growth on wheat. *Exp. Appl. Acarol.*, *2*:111-124, 1986.
102. Van Wyck, J. H., Hodson, A. C., and Christensen, C. M. Microflora associated with the confused four beetle, *Tribolium confusum*. *Ann. Entomol. Soc. Am.*, *52*:452-463, 1959.
103. Sinha, R. N. Adaptive significance of mycophagy in stored product arthropoda. *Evolution*, *22*:785-798, 1968.
104. Dix, D. E. Interactive bionomics of the maize weevil, *Sitophilus zeamais* Motschulsky, and *Aspergillus flavus* Link. Ph.D. thesis, University of Georgia, Athens, GA, 1984.
105. Dunkel, F. V. The relationship of insects to the deterioration of stored grain by fungi. *Int. J. Fd. Microbiol.*, *7*:227-244, 1988.
106. Wright, V. F., De las Casas, E., and Harein, P. K. The nutritional value and toxicity of *Penicillium* isolates for *Tribolium confusum*. *Env. Entomol.*, *9*:204-212, 1980.
107. Wright, V. F., De las Casas, E., and Harein, P. K. The preference of the confused flour beetle for certain *Penicillium* isolates. *Env. Entomol.*, *9*:213-216, 1980.
108. Wright, V. F., De las Casas, E., and Harein, P. K. The response of *Tribolium confusum* to the mycotoxins zearalenone (F-2) and T-2 toxin. *Env. Entomol.*, *5*:371-374, 1976.
109. Hacking, A., and Biggs, N. R. Aflatoxin B_1 in barley. *Nature*, *282*:128, 1979.
110. Burrell, N. J., Kozakiewicz, Z., Armitage, D. M., and Clarke, J. H. Some experiments on the treatment of damp maize with propionic acid. *Ann. Technol. Agric.*, *22*:595-603, 1973.
111. Lord, K. A., Lacey, J., and Cayley, G. R. Fatty acids as substrates and inhibitors of fungi from propionic acid treated hay. *Trans. Br. Mycol. Soc.*, *77*:41-45, 1981.
112. Lord, K. A., Lacey, J., and Cayley, G. R. Laboratory application of preservatives to hay and the effects of irregular distribution on mould development. *Anim. Fd. Sci. Technol.*, *6*:73-82, 1981.
113. Al-Hilli, A. L., and Smith, J. E. *FEMS Microbiol. Lett.*, *6*:367-370, 1979.
114. Burmeister, H. R., Hartman, P. A., and Saul, R. A. Microbiology of ensiled high moisture corn. *Appl. Microbiol.*, *14*:31-34, 1966.
115. Messer, H. J. M., Hill, J. M., Whittenbury, R., and Lacey, J. The use of concrete staved silos for storing high-moisture barley. *Rep. Agric. Res. Coun. Farm Build.* No. 4, 1965.
116. Lacey, J. The microbiology of moist barley stored in unsealed silos. *Ann. Appl. Biol.*, *69*:187-212, 1971.
117. Clarke, J. H., and Hill, S. T. Mycofloras of moist barley during sealed storage in farm and laboratory silos. *Trans. Br. Mycol. Soc.*, *77*:557-565, 1981.

118. Burmeister, H. R. A study of the yeasts in ensiled high moisture corn. Ph.D. thesis, Iowa State Unversity, Ames, IA, 1964.
119. Burmeister, H. R., and Hartman, P. A. Yeasts in ensiled high moisture corn. *Appl. Microbiol.*, *14*:35-38, 1966.
120. Pelhate, J. Maize silage: Incidence of moulds during conservation. *Folia Vet. Latina*, *7*:1-16, 1977.
121. Kaspersson, A., Lindgren, S., and Ekstrom, N. Microbial dynamics in barley grain stored under controlled atmosphere. *Anim. Fd. Sci. Technol.*, *19*:299-312, 1988.
122. Pelhate, J. Microbiology of moist grains. In *Preservation and Storage of Grains, Seeds and Their By-products* (J. Multon, ed.), Lavoisier Publishing, New York, 1988, pp. 328-346.
123. Gilman, G. A., and Boxall, R. A. The storage of food grain in traditional underground bins. *Trop. Stored Prod. Inf.*, *28*:19-38, 1974.
124. Niles, E. V. The mycoflora of sorghum stored in underground pits in Ethiopia.
125. Rännfelt, C. Controlled atmosphere storage in China. In *Controlled Atmosphere Storage of Grains* (J. Shejbal, ed.), Elsevier, Amsterdam, 1980, pp. 437-443.
126. Sigault, F. Significance of underground storage in traditional systems of grain production. In *Controlled Atmosphere Storage of Grains* (J. Shejbal, ed.), Elsevier, Amsterdam, 1980, pp. 3-13.
127. Kamel, A. H. Underground storage in some Arab countries. In *Controlled Atmosphere Storage of Grains* (J. Shejbal, ed.), Elsevier, Amsterdam, 1980, pp. 25-38.
128. Hyde, M. B., and Daubney, C. G. A study of grain storage in fossae in Malta. *Trop. Sci.*, *2*:115-129, 1960.
129. Bowen, H. C., and Wood, P. D. Experimental storage of corn underground and its implication for Iron Age settlements. *Univ. Lond. Inst. Archaeol. Bull.*, *7*, 1968.
130. Lacey, J. The microbiology of grain stored underground in Iron Age type pits. *J. Stored Prod. Res.*, *8*:151-154, 1972.
131. Hill, R. A., Lacey, J., and Reynolds, P. J. Storage of barley grain in Iron Age type underground pits. *J. Stored Prod. Res.*, *19*:163-171, 1983.
132. Shejbal, J. *Controlled Atmosphere Storage of Grains*. Elsevier, Amsterdam, 1980.
133. Hyde, M. B., and Burrell, N. J. Controlled atmosphere storage. In *Storage of Cereal Grains and Their Products*, 2nd ed. (C. M. Christensen, ed.), American Association of Cereal Chemists, St. Paul, MN, 1974, pp. 443-478.
134. Petersen, A., Schlegel, V., Hummel, B., Cuendet, L. S., Geddes, W. E., and Christensen, C. M. Grain storage studies. XXII. Influence of oxygen and carbon dioxide concentrations on mould growth deterioration. *Cereal Chem.*, *33*:53-66, 1956.
135. Wilson, D. M., and Jay, E. Influence of modified atmosphere storage on aflatoxin production in high moisture corn. *Appl. Microbiol.*, *29*:224-228, 1975.
136. Wilson, D. M., Huang, L. H., and Jay, E. Survival of *Aspergillus flavus* and *Fusarium moniliforme* in high-moisture corn stored under modified atmospheres. *Appl. Microbiol.*, *30*:592-595, 1975.

137. Eckhoff, S. R., van Cauwenberge, J. E., Bothast, R. J., Nofsinger, G. W., and Bagley, E. B. Sulphur dioxide-supplemented ambient air drying of high moisture corn. *Trans. A.S.A.E.*, *23*:1028, 1979.
138. Eckhoff, S. R., Tuite, J. F., Foster, G. H., Kirleis, A. W., and Okos, M. R. Microbial growth inhibition by SO_2 and SO_2 plus ammonia treatments during slow drying of corn. *Cereal Chem.*, *60*:185-188, 1983.
139. Cuero, R. G., Smith, J. E., and Lacey, J. The influence of gamma irradiation and sodium hypochlorite sterilization on maize seed microflora and germination.
140. Applegate, K. L., and Chipley, J. R. Effect of ^{60}Co gamma irradiation on aflatoxin B_1 and B_2 production by *Aspergillus flavus*. *Mycologia*, *66*:436-445, 1974.
141. Chang, H. G., and Markakis, P. Effect of gamma irradiation on aflatoxin production in barley. *J. Sci. Fd. Agric.*, *33*:559-564, 1982.
142. Applegate, K. L., and Chipley, J. R. Increased aflatoxin production by *Aspergillus flavus* via cobalt irradiation. *Poultry Sci.*, *52*:1492-1496, 1973.
143. Jemmali, M., and Guilbot, A. Influence de l'irradiation des spores d'*A. flavus* sur la production d'aflatoxins B_1. *C.R. Acad. Sci. Paris Ser. D.*, *269*:2271-2273, 1969.
144. Pitt, J. I. Xerophilic fungi and the spoilage of foods of plant origin. In *Water Relations of Foods* (R. B. Duckworth, ed.), Academic Press, London, U.K., pp. 273-307.
145. Smith, S. L., and Hill, S. T. Influence of temperature and water activity on germination and growth of *Aspergillus restrictus* *A. versicolor*. *Trans. Br. Mycol. Soc.*, *79*:558-560, 1982.
146. Ayerst, G. The effects of moisture and temperature on growth and spore germination in some fungi. *J. Stored Prod. Res.*, 5:127-141, 1969.
147. Teitell, L. Effects of relative humidity on viability of conidia of aspergilli. *Amer. J. Bot.*, *45*:748-753, 1958.
148. Schneider, R. Untersuchungen über Feuchtigkeitsansprüche parasitischer Pilze. *Phytopathol. Z.*, *21*:68-78, 1954.
149. Armolik, N., and Dickson, J. G. Minimum humidity requirements for germination of conidia associated with grain. *Phytopathology*, *46*:462-465, 1956.
150. Snow, D. The germination of mould spores at controlled humidities. *Ann. Appl. Biol.*, *36*:1-17, 1949.
151. Pitt, J. I., and Hocking, A. D. Influence of solutes and hydrogen ion concentration on the water relations of some xerophilic fungi. *J. Gen. Microbiol.*, *101*:35-40, 1977.
152. Galloway, L. D. The moisture requirements of mould fungi with special reference to mildew of textiles. *J. Text.*, *26*:123-129, 1961.
153. Magan, N., and Lacey, J. Ecological determinants of mould growth in stored grain. *Int. J. Fd. Microbiol.*, 7:245-256, 1988.
154. Griffin, D. M. Soil moisture and the ecology of fungi. *Biol. Rev. Camb. Philos. Soc.*, *38*:141-166, 1963.
155. Ayerst, G. Water and the ecology of fungi in stored products. In *Water, Fungi and Plants* (P. G. Ayres and L. Boddy, eds.), Cambridge University Press, 1986, pp. 359-373.

156. Heintzeller, I. The growth of moulds—dependence on hydration and various other limiting factors. *Arch. Mikrobiol.*, *10*:92-132, 1939.
157. Curran, P. M. T. Sporulation in some members of the *Aspergillus glaucus* group in response to osmotic pressure, illumination and temperature. *Trans. Br. Mycol. Soc.*, *57*:201-211, 1971.
158. Gottlieb, D. *The Germination of Fungus Spores*, Meadowfield Press, Durham, U.K., 1978.
159. Panasenko, V. T. Ecology of microfungi. *Bot. Rev.*, *33*:189-215, 1967.
160. Golding, N. S. The gas requirements of molds. III. The effects of various concentrations of carbon dioxide on the growth of *Penicillium roquefortii* in air. *J. Dairy Sci.*, *23*:891-898, 1940.
161. Magan, N., and Lacey, J. Effects of gas composition and water activity on growth of field and storage fungi and their interactions. *Trans. Br. Mycol. Soc.*, *82*:305-314, 1984.
162. Miller, D. D., and Golding, N. S. The gas requirements of moulds. V. The minimum oxygen requirements for normal growth and germination of six mould cultures. *J. Dairy Sci.*, *32*:191-210, 1949.
163. Arbab, A. K. Effect of modified atmospheres on the growth, sporulation and spore germination of selected storage and field penicillia of corn. M.Sc. thesis, Purdue University, Lafayette, IN, 1976.
164. Tuite, J., Haugh, C. G., Isaac, G. W., and Huxsoll, C. C. Growth and effect of moulds in stored high moisture corn. *Trans. Am. Soc. Agric. Eng.*, *10*:730-737, 1967.
165. Lillehoj, E. B., Milburn, M. S., and Ciegler, A. Control of *P. martensii* development by atmospheric gases and temperature. *Appl. Microbiol.*, *24*:198-201, 1972.
166. Wells, J. M., and Uota, M. Germination and growth of five fungi in low oxygen and high carbon dioxide atmospheres. *Phytopathology*, *60*:50-53, 1970.
167. Pelhate, J. Oxygen depletion as a method in grain storage. In *Controlled Atmosphere Storage of Grains* (J. Shejbal, ed.), Elsevier, Amsterdam, 1980, pp. 133-146.
168. Paster, N., Lisker, N., and Chet, I. Ochratoxin A production by *Aspergillus ochraceus* Wilhelm grown under controlled atmospheres. *Appl. Environ. Micro.*, *45*:1136-1139, 1983.
169. Magan, N., and McLeod, A. R. In vitro growth and germination of phylloplane fungi in atmospheric sulphur dioxide. *Trans. Br. Mycol. Soc.*, *90*:571-575, 1988.
170. Beuchat, L. R. *Food and Beverage Mycology*. Avi Publishing Co., Westport, CT, 1978.
171. Wicklow, D. T., Hesseltine, C. W., Shotwell, D. L., and Adamas, G. L. Interference competition and aflatoxin levels in corn. *Phytopathology*, *70*:761-764, 1980.
172. Wicklow, D. T., Horn, B. W., Shotwell, O. L., Hesseltine, C. W., and Caldwell, R. W. Fungal interference with *Aspergillus flavus* infection and aflatoxin contamination of maize grown in a controlled environment. *Phytopathology*, *78*:68-74, 1988.
173. Lillehoj, E. B., Fennell, D. I., and Kwolek, W. F. *Aspergillus flavus* and aflatoxin in Iowa corn before harvest. *Science*, *193*:495-496, 1966.

174. Wicklow, D. T., Horn, B. W., and Shotwell, O. L. Aflatoxin formation in preharvest maize ears coinoculated with *Aspergillus flavus* and *Aspergillus niger*. *Mycologia, 79*:679-682, 1987.
175. Ehrlich, K., Ciegler, M. K., and Lee, L. Fungal competition and mycotoxin production on corn. *Experientia, 41*:691-693, 1985.
176. Schmidt, F. R., and Esser, K. Aflatoxins: Medical, economic impact, and prospects for control. *Process Biochemistry, 20*:167-174, 1985.
177. Willingale, J., and Mantle, P. G. Interactions between *Claviceps purpurea* and *Tilletia caries* in wheat. *Trans. Brit. Mycol. Soc., 89*:145-153, 1987.
178. Magan, N., and Lacey, J. Interactions between field and storage fungi on wheat grain. *Trans. Br. Mycol. Soc., 85*:29-37, 19 .
179. Boller, R. A., and Schroeder, H. W. Influence of *Aspergillus chevalieri* on production of aflatoxin in rice by *Aspergillus parasiticus*. *Phytopathology, 63*:1507-1510, 1973.
180. Boller, R. A., and Schroeder, H. W. Influence of *Aspergillus candidus* on production of aflatoxin in rice by *Aspergillus parasiticus*. *Phytopathology, 64*:121, 1974.
181. Horn, B. W., and Wicklow, D. T. Factors influencing the inhibition of aflatoxin production in corn by *Aspergillus niger*. *Can. J. Microbiol., 29*:1087-1091, 1983.
182. Cuero, R., Smith, J. E., and Lacey, J. Mycotoxin formation by *Aspergillus flavus* and *Fusarium graminearum* in irradiated maize grains in the presence of other fungi. *J. Food Prot., 51*:542-556, 1988.
183. Cuero, R., Smith, J. E., and Lacey, J. Stimulation by *Hyphopichia burtonii* and *Bacillus amyloliquefaciens* of aflatoxin production by *Aspergillus flavus* in irradiated maize and rice grains. *Appl. Environ. Microbiol., 53*:1142-1146, 1987.
184. Jarvis, B., Seiler, D. A. L., Ould, A. J. L., and Williams, A. P. Observations on the enumeration of moulds in food and feeding stuffs. *J. Appl. Bact., 55*:325-336, 1983.
185. Lacey, J., Hill, S. T., and Edwards, M. A. Micro-organisms in stored grains: Their enumeration and significance. *Trop. Stored Prod. Inf., 39*:19-33, 1980.
186. Warnock, D. W. Assay of fungal mycelium in grains of barley, including the use of the fluorescent antibody technique for individual fungal species. *J. Gen. Microbiol., 67*:197-205, 1971.
187. Warnock, D. W. Use of immunofluorescence to detect mycelium of *Alternaria, Aspergillus* and *Penicillium* in barley grains. *Trans. Br. Mycol. Soc., 61*:547-552, 1973.
188. Greenhill, W. L. The respiration drift of harvested pasture plants during drying. *J. Sci. Food Agric., 10*:495-501, 1959.
189. Kaspersson, A. The role of fungi in deterioration of stored feeds, methodology and ecology. Report No. 31, Department of Microbiology, Swedish University of Agricultural Sciences, Uppsala, Sweden, 1986.
190. Bailey, C. H. Respiration of cereal grains and flaxseed. *Plant Physiol., 15*:257-274, 1940.
191. Larmour, R. K., Clayton, J. S., and Wrenshall, C. L. A study of

the respiration and heating of damp wheat. *Can. J. Res.*, *12*:627-645, 1935.

192. Fernandez, A., Stroshine, R., and Tuite, J. Mould growth and carbon dioxide production during storage of high moisture corn. *Cereal Chem.*, *62*:137-143, 1985.
193. Steele, J. L., Saul, R. A., and Hukill, W. V. Deterioration of shelled corn as measured by carbon dioxide production. *Trans. ASAE (Am. Soc. Agric. Eng.)*, *12*:685-689, 1969.
194. White, N. D. G., Sinha, R. N., and Muir, W. E. Intergranular carbon dioxide as an indicator of biological activity associated with the spoilage of stored wheat. *Can. Agric. Eng.*, *24*:35-42, 1982.
195. Leach, W. Further experimental methods in connection with the use of the Katharometer for the measurement of respiration. *Ann. Bot.*, *46*:583-596, 1932.
196. Tribe, H. T., and Maynard, P. A new automatic electrolytic respirometer. *The Mycologist*, *3*:24-27, 1989.
197. Milner, M., and Geddes, W. F. Grain storage studies. I. Influence of localised heating of soybeans on interseed air movements. *Cereal Chem.*, *22*:477-483, 1945.
198. Festenstein, G. N., Lacey, J., Skinner, F. A., Jenkins, P. A., and Pepys, J. Self-heating of hay and grain in Dewar flasks and the development of a farmer's lung hay antigens. *J. Gen. Microbiol.*, *41*:389-407, 1965.
199. Lacey, J. Colonisation of damp organic substrates and spontaneous heating. In *Microbial Growth in Extremes of Environment* (G. W. Gould and J. E. L. Corry, eds.), Society of Applied Bacteriology Technical Series No. 15, Academic Press, London, 1980, pp. 53-70.
200. Norman, A. G., Richards, L. A., and Carlyle, R. E. Microbial thermogenesis in the decomposition of plant materials, Part I. An adiabatic fermentation apparatus. *J. Bact.*, *41*:689-697, 1941.
201. Sparling, G. P. Microcalorimetry and other methods to assess biomass and activity in soil. *Soil Biol. Biochem.*, *13*:93-98, 1981.
202. Saul, R. A., and Steele, J. L. Why damaged shelled corn costs more to dry. *Agric. Engineering*, *47*:326-329, 1966.
203. White, N. D. G., Sinha, R. N., and Muir, W. E. Intergranular carbon dioxide as an indicator of deterioration in stored rapeseed. *Can. Agric. Eng.*, *24*:43-49, 1982.
204. Milner, M., and Geddes, W. F. Respiration and heating. In *Storage of Cereal Grains and Their Products* (J. A. Anderson and A. W. Alcock, eds.), American Association of Cereal Chemists, St. Paul, MN, 1954, pp. 152-193.
205. Seitz, L. M., Sauer, D. B., and Mohr, H. E. Storage of high moisture corn: Fungal growth and dry matter loss. *Cereal Chem.*, *59*:100-105, 1982.
206. Scholtz, B. Atmungsverluste bei Weizen in Abhängigkeit von Temperatur, Lagereit und Wassergehalt. *Landtech. Forschung.*, *212*:48-52, 1969.
207. Westermarck-Rosendahl, C., and Ylimaki, A. Spontaneous heating in newly harvested wheat and rye. 1. Thermogenesis and its effect on grain quality. *Acta Agric. Scand.*, *28*:151-158, 1978.

208. Hummel, B. C. W., Cuendet, L. S., Christensen, C. M., and Geddes, W. F. Grain storage studies. XIII. Comparative changes in respiration, viability, and chemical composition of mould-free and mould contaminated wheat upon storage. *Cereal Chem.*, *31*:143-149, 1954.
209. Anon. *Grain Drying and Storage.* The Electricity Council, London, 1987.
210. Deacon, J. W. *Introduction to Modern Mycology*, Vol. 7, Basic Microbiology Series, Blackwell, Oxford, U.K., 1984.
211. Seitz, L. M., Sauer, D. B., Burroughs, R., Mohr, H. E., and Hubbard, J. D. Ergosterol as a measure of fungal growth. *Phytopathology*, *69*:1202-1203, 1979.
212. Donald, W. W., and Mirocha, C. J. Chitin as a measure of fungal growth in stored corn and soybean seed. *Cereal Chem.*, *54*:466-474, 1977.
213. Sakurai, Y., Lee, T. H., and Shiota, H. On the convenient method for glucosamine estimation in Koji. *Agric. Biol. Chem.*, *41*:619-624, 1977.
214. Naewbanij, M., Seib, P. A., Burroughs, R., Seitz, L. M., and Chung, D. S. Determination of ergosterol using thin-layer chromatography and ultraviolet spectroscopy. *Cereal Chem.*, *61*:385-388, 1984.
215. Seitz, L. M., Mohr, H. E., Burroughs, R., and Sauer, D. B. Ergosterol as an indicator of fungal invasion in grains. *Cereal Chem.*, *54*:1207-1217, 1977.
216. Cahagnier, B., Richard-Molard, D., and Poisson, J. Evolution of the ergosterol content of cereal grains during storage—a possibility for a rapid test of fungal development in grains. *Sci. Aliments*, *3*: 219-244, 1983.
217. Nout, M. J. R., Bonants-van-Laarhoven, T. M. G., Jongh, P. de, and deKoster, P. G. Ergosterol content of *Rhizopus oligosporus* NRRL 5905 grown in liquid and solid substrates. *Appl. Microbiol. Biotechnol.*, *26*:456-461, 1987.
218. Shewry, P. R., Faulks, A. J., Pratt, H. M., and Miflin, B. J. The varietal identification of single seeds of wheat by sodium dodecyl-sulphate polacrylamide gel electrophoresis of gliadin. *J. Sci. Fd. Agric.*, *29*:847-849, 1978.
219. Shewry, P. R., Pratt, H. M., and Miflin, B. J. Varietal identification of single seeds of barley by analysis of hordein polypeptides. *J. Sci. Fd. Agric.*, *29*:587-596, 1978.
220. Quisenberry, K. S. Grain value to be safe-guarded during conditioning and storage. *Agric. Engng.*, *30*:586-588, 1949.
221. Pelshenke, P. A short method for the determination of gluten quality of wheat. *Cereal Chem.*, *10*:90, 1933.
222. Zeleny, L. A simple sedimentation test for estimating the bread-baking and gluten qualities of wheat flour. *Cereal Chem.*, *24*:465, 1947.
223. Ghosh, J., and Nandi, B. Deteriorative abilities of some common storage fungi of wheat. *Seed Sci. Technol.*, *14*:141-149, 1986.
224. Vidhyasekaran, P., Muthuswamy, G., and Subramanian, C. L. Role of seed borne microflora in paddy seed spoilage. I. Production of hydrolytic enzymes. *Indian Phytopath.*, *19*:333-341, 1966.

225. Sellars, P. N., McGill, C. E. C., and Flannigan, B. Degradation of barley by *Aspergillus fumigatus* Fres. In *Proceedings of the 3rd International Biodegradation Symposium* (J. M. Sharpley and A. M. Kaplan, eds.), Applied Science, London, U.K., 1976, pp. 633-643.
226. Flannigan, B. Degradation of arabinoxylan and carboxymethyl cellulose by fungi from barley kernels. *Trans. Br. Mycol. Soc.*, *55*:277-281, 1970.
227. Flannigan, B., and Sellars, P. N. Activities of thermophilous fungi from barley kernels against arabinoxylan and carboxymethyl cellulose. *Trans. Br. Mycol. Soc.*, *58*:338-341, 1972.
228. Turner, J. H. *Bull. Misc. Info. Royal Bot. Gdns. Kew.*, *6*:257-269, 1933.
229. Ashton, T. Genetical aspects of seed storage. In *The Storage of Seeds for Maintenance of Viability* (E. B. Owen, ed.), Commonwealth Agricultural Bureaux, Farnham Royal, U.K., 1956, pp. 34-38.
230. Barton, L. V. *Seed Preservation and Longevity.* Leonard Hill, London, 1961.
231. Gunthardt, H., Smith, L., Haferkamp, M. E., and Nilan, R. A. Studies on aged seeds. II. Relation of age of seeds to cytogenic effects. *Agron. J.*, *45*:438-441, 1953.
232. Peto, F. H. The effect of aging and heat on chromosome mutation rate in maize and barley. *Can. J. Res.*, *9*:261-264, 1933.
233. Navashin, M. S., and Shkvarnikov, P. Process of mutation in resting seed accelerated by increased temperature. *Nature*, *132*:482-483, 1933.
234. Thimmiah, C. M., and Gupta, P. C. Storage studies in triticale seed. In effect of location of seed production and storage environment on seed germinability loss in four triticale varieties. *Bull. Grain Technol.*, *17*:26-34, 1979.
235. Abdalla, F. H., and Roberts, E. H. Effects of temperature, moisture, and oxygen on the induction of chromosome damage in seeds of barley, broad beans, and peas during storage. *Ann. Bot. N.S.*, *32*:119-136, 1968.
236. Machacek, J. E., Robertson, E., Wallace, H. A. H., and Phillips, N. A. Effect of a high water content in stored wheat, oat, and barley seed on its germinability, susceptibility to invasion by moulds, and response to chemical treatment. *Can. J. Pl. Sci.*, *41*:288-303, 1961.
237. Osborne, D. J., Roberts, B. E., Payne, P. I., and Swati, S. Protein synthesis and viability in rye embryos. *Bull. R. Soc. N.Z.*, *12*:805-812, 1974.
238. Christensen, C. M. Loss of viability in storage: Microflora. *Seed Sci. Technol.*, *1*:547-562, 1973.
239. Christensen, C. M., Olfson, J. H., and Geddes, W. F. Grain storage studies. VIII. Relation of mould in moist stored cotton seed to increased production of carbon dioxide fatty acids and heat. *Cereal Chem.*, *26*:109-128, 1949.
240. Malowan, J. Some observations on the heating of cotton seed. *Cott. Oil Press.*, *4*:47-49, 1921.
241. Halloin, J. M. Deterioration resistance mechanisms in seeds. *Phytopathology*, *73*:335-339, 1983.

242. Farag, R. S., Mohsen, S. M., Khalil, F. A., and Basyony, A. E. Effect of fungal infection on the chemical composition of wheat, soybean and sesame seeds. *Grasas Aceites.*, *36*:357-361, 1985.
243. Aspinall, G. C., and Greenwood, C. T. Aspects of the chemistry of cereal polysaccharides. *J. Inst. Brew.*, *68*:167-178, 1962.
244. Vonk, H. J., and Western, J. R. H. *Comparative Biochemistry and Physiology of Enzymatic Digestion*, Academic Press, London, U.K., 1984.
245. Pomeranz, Y. Relation between chemical composition and bread-making potentialities of wheat flour. In *Advances in Food Research*, Academic Press, New York, 1968, pp. 340-376.
246. Binger, H. P. Current studies on flour composition of baking quality. *Bakers Dig.*, *39*:24-27, 1965.
247. Fifield, C. C., Weaver, R., and Hayes, J. F. Bread load volume and protein content of hard red wheats. *Cereal Sci. Today*, *4*:179-183, 1959.
248. Bayles, B. B. Developments and problems in wheat quality research. *Trans. A.S.C.C.*, *12*:97-102, 1954.
249. Zeleny, L. Criteria of wheat quality. In *Wheat: Chemistry and Technology* (Y. Pomeranz, ed.), American Association of Cereal Chemists, St. Paul, MN, 1971, pp. 19-49.
250. Ghetie, V. Degradation of reserve proteins in the seed during germination. *Studi Cercetari Biochim.*, *9*:91-105, 1966.
251. Cross, D. E., and Thompson, T. L. Thermogenic properties and nutrient retention of corn at different moisture levels during storage. *Trans. ASAE (Am. Soc. Agric. Eng.)*, *14*:665-668, 1971.
252. Nasuno, S. Polyacrylamide gel disc electrophoresis of alkaline proteinases from *Aspergillus* species. *Agric. Biol. Chem.*, *35*:1147-1150, 1971.
253. Pernollet, J. C. Protein bodies of seeds: Ultrastructure, biochemistry, biosynthesis and degradation. *Phytochem.*, *17*:1473-1480, 1978.
254. Cherry, J. P. Protein degradation during seed deterioration. *Phytopathology*, *73*:317-321, 1983.
255. St. Angelo, A. J., and Ory, R. L. Lipid degradation during seed deterioration. *Phytopathology*, *73*:315-317, 1983.
256. St. Angelo, A. J., Kuck, J. C., and Ory, R. L. Role of lipoxygenase and lipid oxidation in quality of oil seeds. *J. Agric. Food Chem.*, *27*:229-234, 1979.
257. Milner, M., and Geddes, W. F. Grain storage studies. III. The relation between moisture content, mould growth, and respiration of soybeans. *Cereal Chem.*, *23*:225-247, 1946.
258. Dirks, B. M., Boyer, P. D., and Geddes, W. F. Some properties of fungal lipases and their significance in stored grain. *Cereal Chem.*, *32*:356-373, 1955.
259. Goodman, J. J., and Christensen, C. M. Grain storage studies. XI. Lipolytic activity of fungi isolated from stored corn. *Cereal Chem.*, *29*:299-308, 1952.
260. Zeleny, L., and Coleman, D. A. Acidity in cereals and cereal products, its determination and significance. *Cereal Chem.*, *15*:580-595, 1938.

261. Baker, D., Neustadt, M. H., and Zeleny, L. Application of the fat acidity test as an index of grain deterioration. *Cereal Chem.*, *34*: 226-233, 1957.
262. Baker, D., Neustadt, M. H., and Zeleny, L. Relationships between fat acidity values and types of damage in grain. *Cereal Chem.*, *36*: 308-311, 1959.
263. Farag, R. S., Yousef, A. M., Sabet, K. A., Fahim, M. M., and Khalil, F. A. Chemical studies on corn embryos infected by various fungi. *J. Am. Oil Chem. Soc.*, *58*:722-728, 1981.
264. Farag, R. S., Khalil, F. A., Mohsen, S. M., and Bosyony, A. E. Effect of certain fungi on the lipids of wheat kernels, sesame and soybean seeds. *Grasas Aceites.*, *36*:362-367, 1985.
265. Stawicki, S., Kaminski, E., Niewiarowicz, A., Trojan, M., and Wasowicz, E. The effect of microflora on the formation of odors in grain during storage. *Ann. Technol. Agric.*, *22*:449-476, 1973.
266. Allen, J. C., Farag, R. S., and Crook, E. M. The metal catalysed oxidation of aqueous emulsions of linoleic and trilinolein. *J. App. Biochem.*, *1*: 15, 1979.
267. Farag, R. S., Rahim, E. A., Ibrahim, N. A., and Basyony, A. E. Biochemical studies on the unsaponifiables of wheat kernels, soybeans and sesame seed infected by some fungi. *Grasas Aceites.*, *36*:368-372, 1985.
268. Kaminski, E., Stawicki, S., and Wasowicz, E. Volatile flavor substances produced by molds on wheat grain. *Acta Aliment. Polon.*, *1*:153, 1975.
269. Abramson, D., Sinha, R. N., and Mills, J. T. Mycotoxin and odor formation in moist cereal grain during granary storage. *Cereal Chem.*, *57*:346-351, 1980.
270. Mora, M. Diagnosis on contamination levels of white maize with aflatoxins in Costa Rica (Abstract). Report. Grain and Seed Research Center, University of Costa Rica, 1988.
271. Christensen, C. M. Influence of small differences in moisture content upon the invasion of hard red winter wheat by *Aspergillus restrictus* and *Aspergillus repens*. *Cereal Chem.*, *40*:387-390, 1963.

6

THE IMPORTANCE OF FUNGI IN VEGETABLES

MARLENE A. BULGARELLI* and ROBERT E. BRACKETT *Georgia Agricultural Experimental Station, The University of Georgia, Griffin, Georgia*

I. INTRODUCTION

Vegetables are not easily defined botanically or morphologically. Inasmuch as they encompass a large number of botanical families and are comprised of organs in various stages of vegetative or reproductive growth, several different classification schemes have been developed. Classification is often based upon cultural requirements (e.g., temperature and soil adaptation, seasonal growth), botanical relationships, plant organs utilized (e.g., root, leaf, flower), sensitivity to cold, storage life, or metabolic/respiratory activity [1]. For the purpose of this chapter, vegetables will be classified according to the plant organ used (Table 1). Within this classification, several botanical fruits are included as fruit vegetables.

Vegetables are perishable products with active metabolism during the postharvest period [1]. Therefore, proper handling and environmental conditions after harvesting are essential in maintaining product quality and availability. The quality of a vegetable seldom, if ever, improves during the postharvest period. Factors which deleteriously affect vegetable quality include: (a) respiratory and metabolic activities resulting in loss of nutritive value and sensory quality, (b) transpiration which causes loss of turgidity and withering, (c) mechanical damage and bruising, consequences of handling, packing, and transport, (d) parasitic attack and physiological disorders, and (e) growth phenomena such as sprouting, rooting, and color changes [2]. Postharvest diseases are responsible for significant losses in vegetable production [3]. Their economic costs are considerably higher than for losses incurred in the field due to the added costs of harvesting, transport, and storage [4].

A. Vegetables Affecting Microbial Growth

Fruit, due to their low pH, are spoiled primarily by fungi. In contrast, vegetables are susceptible to both fungi and bacteria. Several intrinsic

*Present affiliation: Centers for Disease Control, Atlanta, Georgia

TABLE 1. Classification of Vegetables According to the Plant Organ Used

Class	Commodities (examples)
Root vegetables	Carrot, celeriac, garlic, horseradish, Jerusalem artichoke, onion, parsnip, potato, radish, rutabaga, salsify, scorzonera, sweet potato, table beet, turnip
Leafy vegetables	Brussels sprouts, cabbage, celery, chard, chicory, Chinese cabbage, collard, cress, dandelion, endive, green onion, kale, leek, lettuce, spinach
Flower vegetables	Artichoke, broccoli, cauliflower
Immature fruit vegetables	Bean, cucumber, gherkin, okra, pea, pepper, squash, sweet corn
Mature fruit vegetables	Melon, tomato

Source: Ref. 1.

factors predispose fresh vegetables to microbial attack and subsequent spoilage [4]. The high water content, nutrient composition, and pH of most vegetables make them capable of supporting the growth of many bacteria, yeasts, and molds. The close proximity of the plant to the soil, as well as the presence of natural openings (e.g., lenticels, stomata) through which microorganisms can penetrate, allow for ease of contamination. Wounds resulting from harvesting and handling such as stem scars, scratches, abrasions, and bruises also provide ports of entry for invading microorganisms. Plant cell wall susceptibility to degradation varies and is dependent on the age and type of vegetable, the chemical structure of the plant cell walls, as well as growth and storage conditions. As a general rule, vegetables become more susceptible to infection by postharvest pathogens as they ripen. This phenomenon is believed to involve nutritional factors, enzymes, toxins, and energy metabolism [5].

Metabolically active foods such as vegetables possess potent defense mechanisms against microbial invasion [6]. One of the first lines of defense is the production of a physical barrier to entrance of spoilage organisms. For example, potatoes, yams, and sweet potatoes have a specialized defense mechanism (lignification) employed in wound healing [7]. This process often occurs when the plant is invaded by insects or has been otherwise damaged. In this process, periderm layers are formed adjacent to injuries and function as a barrier to penetration of parasitic fungi and bacteria [5]. Similar cell wall modifications have been observed in tomatoes with the deposition of callose [8] and in cucumbers and beans which accumulate hydroxyproline-rich glycoproteins [9].

The presence of preformed antimicrobial substances or enzymes in some vegetables enable them to defend against microbial attack. One such glycoalkaloid, tomatine, is present in high concentrations in the peel of green tomatoes. Postinfection formation of antimicrobial compounds known as stress metabolites or phytoalexins is another natural defense mechanism associated with plant tissues.

TABLE 2. Bacteria Commonly Associated with Postharvest Diseases of Vegetables

Food	Organism	Spoilage condition
Fresh vegetables		
Green beans	*Xanthomonas*	Blight
Carrots	*Erwinia*	Soft rot
Onions	*Pseudomonas*	Brown rot
Potatoes	*Corynebacterium*	Ring rot
Tomatoes	*Xanthomonas*	Bacterial spot
Canned vegetables		
Green beans, corn, peas	*Bacillus* *Desulfotomaculum* *Clostridium*	Flat-sour Sulfide stinker Putrid swell, hard swell
Tomatoes	*Bacillus* *Clostridium*	Flat-sour Butyric fermentation
Fermented vegetables		
Pickles	*Bacillus*	Soft rot, black rot
Vegetable juice	*Lactobacillus*, *Acetobacter*	Souring

Source: Ref. 63.

B. Microorganisms Associated with Vegetables

The microflora present on fresh vegetables are varied and arise from several external sources such as soil, air, fertilizer, animals, and humans. In the case of most vegetables, the microenvironment is such that there is ample moisture and a neutral pH. Such conditions most often favor the growth of bacteria, which grow rapidly under such conditions. As mentioned earlier, those microorganisms which eventually grow and predominate on foods are largely dictated by the microenvironment of that product. In general, few of the many kinds of bacteria present on vegetables cause spoilage or health hazards. However, certain species survive well in or on plant tissues. Appropriate conditions may allow these bacteria to cause blemishes, lesions, and decay [10]. Bacteria commonly associated with vegetables include *Bacillus*, *Clostridium*, *Erwinia*, *Lactobacillus*, *Pseudomonas*, and *Xanthomonas* (Table 2) [11-13].

Yeast populations on vegetables generally range from less than 10^3 to more than 10^6 organisms per g of tissue [14]. Senter et al. found 10^4 yeasts/g of tissue on southern peas [15] and collard leaves [16] immediately after harvest. Moreover, they also found that the season (spring vs. fall) of the year had no appreciable effect on the populations recovered from the collards. In contrast, Deák et al. [17] found slightly fewer yeasts on fresh sweet corn as purchased at a produce market. Although yeasts are often isolated from vegetables, they typically do not compete well with bacteria and molds. However, they can still become important spoilage agents under appropriate growth conditions [14,18].

TABLE 3. Yeasts Commonly Associated with Postharvest Diseases of Vegetables

Food	Organism	Spoilage condition
Tomatoes	*Candida*	Fermentation
Fermented vegetables (pickles, olives, sauerkraut)	*Candida, Cryptococcus, Hansenula, Pichia, Rhodotorula, Saccharomyces, Sporobolomyces, Torulopsis*	Film formation

Source: Ref. 63.

The predominant kinds of yeasts identified on fresh vegetables are usually the asporogenous pigmented varieties which belong to the genera *Cryptococcus, Torulopsis, Rhodotorula, Sporobolomyces*, and *Tilletiopsis* (Table 3) [19,20]. In a study of sweet corn, Deák et al. [17] isolated some 17 different species of yeasts. These species were primarily distributed into two groups, one consisting of nonfermenting basidiomycetous *Rhodotorula* and *Cryptococcus*, the other of fermentative *Candida* and *Kloeckera* species.

Molds are typically present in low concentrations (ca. 10^3 CFU/g of tissue). For example, Senter et al. found mold populations to be as low as 10^2 in fresh tomatoes [21] to as high as over 10^4 in southern peas [15]. Although molds are often found at very low populations, further contamination may occur after harvest, in transit or storage, and during processing and marketing [22,23]. The presence of molds on vegetables is not necessarily indicative of subsequent infection and spoilage. Often the vegetables simply serve as transport vectors. However, several species of molds are known to cause both pre- and postharvest diseases of vegetables.

Several different genera of molds are associated with vegetables at the time of harvest (Table 4). Molds most commonly isolated include *Aureobasidium, Fusarium, Alternaria, Epicoccum, Mucor, Chaetomium, Rhizopus*, and *Phoma* [22]. The specific molds isolated may differ depending on the vegetable in question and the point at which sampling is done. For example, Senter et al. [21] isolated *Cladosporium, Alternaria, Aspergillus, Fusarium, Myrothecium, Stemphyllium*, and *Rhizoctonia* on tomatoes immediately after harvest. In contrast, Deák et al. [17] isolated primarily penicillia and, to a lesser extent, aspergilli and *Cladosporium* from sweet corn. However, molds were only isolated after the corn had become spoiled.

The increasing demand for fresh vegetables by consumers has stimulated interest in the microbiology of these products. Several excellent books and reviews addressing the mycology of vegetables have been written in recent years [24-27]. It is the intent of this chapter to provide updated information regarding the effect of fungi on the quality of vegetable products. In addition, the impact of the technologies being developed for vegetable processing and marketing will also be discussed.

TABLE 4. Molds Commonly Associated with Postharvest Diseases of Vegetables

Vegetable	Organism	Spoilage condition
Asparagus	*Fusarium*	Fusarium rot
Green beans	*Botrytis*	Grey mold rot
	Colletotrichum	Anthracnose (smudge)
	Pythium	Wilt
	Rhizopus	Soft rot
Cabbage	*Alternaria*	Black spot
	Botrytis	Grey mold rot
Carrots	*Alternaria*	Black rot
	Botrytis	Grey mold rot
	Fusarium	Fungal rot
	Rhizopus	Soft rot
	Sclerotinia	Watery soft rot
Cauliflower	*Alternaria*	Black spot
Celery	*Botrytis*	Grey mold rot
	Mucor	Fungal rot
	Sclerotinia	Pink rot
Lettuce	*Botrytis*	Grey mold rot
	Sclerotinia	Watery soft rot
Onions	*Aspergillus*	Black mold rot
	Botrytis	Neck rot
	Colletotrichum	Anthracnose
	Penicillium	Blue mold rot
Peppers	*Alternaria*	Grey mold rot
	Botrytis	Black rot
Potatoes	*Fusarium*	Dry rot
	Phytophthora	Pink rot
	Pythium	Watery wound rot
Tomatoes	*Alternaria*	Black rot
	Aspergillus, Colletotrichum, Rhizopus	Fungal rot
	Botrytis	Grey mold rot

Source: Refs. 4, 63.

II. IMPORTANCE OF FUNGI IN VEGETABLES

Although bacteria are usually the favored competitors on most vegetables, the fungi are still very important. Many of the most important postharvest defects and diseases are directly caused by molds and yeasts. Moreover, molds can also compromise the wholesomeness of vegetables either by producing toxic substances in the vegetable or by stimulating the vegetable to do so. Thus, any discussion of the vegetable microbiology that does not include the role of fungi would be incomplete.

A. Spoilage

Spoilage of vegetables is often categorized according to when or where the spoilage becomes apparent. Spoilage occurring before vegetables are harvested is usually referred to as preharvest or field spoilage. Such spoilage has traditionally been considered symptomatic of plant diseases and has been studied by plant pathologists. In contrast, spoilage occurring during the processing, distribution, and sale of fresh vegetables is often referred to as postharvest spoilage or marketing disease. However, one should always be aware that such a system of categorizing spoilage is often very misleading. For example, sweet potatoes are sometimes afflicted with a serious disease caused by *Ceratocystis fimbriata*. This fungus normally invades the sweet potato tuber in the field but does not actually cause spoilage until after the vegetable is stored [28]. Thus, terming this type of spoilage as either pre- or postharvest disease inaccurately describes the problem.

Postharvest losses in fresh vegetables primarily occur as a result of mechanical injuries, nonparasitic disorders, and parasitic diseases [3]. Mechanical injuries include defects that are caused by physical insults to the vegetables. Common mechanical injuries include cuts, bruises, punctures, abrasions, insect scars, and cracking. The source of such injuries can be both avoidable (e.g., mishandling during processing) or unavoidable (e.g., hail damage). Mechanical damage has dual importance because it can not only reduce the value of the vegetable but can also increase the chance for later microbial spoilage.

Nonparasitic disorders include various physiological responses of vegetables to the postharvest environment. Examples of such disorders can include shriveling, sprouting, or production of undesirable colors or odors [3]. Like mechanical damage, nonparasitic disorders often make the vegetable more susceptible to microbial spoilage.

Parasitic diseases that result in decay are usually caused by microbial invasion. However, the term "parasitic disease" is similar to "pre-" or "postharvest disease" in that it incompletely describes all the aspects of the spoilage. Although many types of microbial spoilage result from invasion by parasitic microorganisms, much spoilage also results when normally saprophytic microorganisms behave as opportunistic spoilers. Examples of such opportunistic spoilage include tomatoes spoiled by members of *Fusarium* or *Geotricum* [23]. Often such spoilage follows mechanical damage or physiological disorders.

Between 12 and 20% of vegetables harvested for human consumption are lost through microbial spoilage by one or more of 250 market diseases [3, 29,30]. The primary causative agents of microbial spoilage are the bacteria, yeasts, and molds. However, plant viruses and insects also contribute to vegetable spoilage, although to a lesser extent [11]. Although bacteria are responsible for a large percentage of spoilage, fungi are an equally important group of organisms in vegetable spoilage. This spoilage results in loss of aesthetic and organoleptic qualities of the food. Moreover, foods spoiled by molds may have diminished nutritive value and in some instances may present a potential health threat as a consequence of mycotoxin production [31].

1. Common Types of Fungal Spoilage

There are many forms of vegetable spoilage attributed to fungi. However, only a relatively few kinds of fungi are responsible for the majority of spoilage conditions which occur.

Grey mold rot, caused by *Botrytis cinerea*, is one of the most widespread fungal diseases and is characterized by a gray mycelium. Often the presence of the mold is accompanied by dark, watery lesions in the vegetable tissues. It is responsible for the serious market decay of many vegetables and is most often observed in vegetables which have been wounded or bruised. Storage conditions consisting of warm temperatures and high humidity favor this type of spoilage [11,23]. This rot occurs on a wide variety of vegetables.

Black rot, also called *Alternaria* rot, is caused by *Alternaria* species. This rot causes leaf or nail head spot defects in cruciferous vegetables, melons, cucumbers, and tomatoes and is also commonly responsible for calyx rot in peppers [20]. This rot is usually quite dark and, unlike the gray mold rots, the tissue remains quite firm. However, *Alternaria* rot may be followed by bacterial soft rot infections in some instances [3,11,23].

Sour rot is a market disease of vegetables caused by *Geotrichum* [32]. A related type of spoilage is the soft rot caused by *Rhizopus* species. Neither of these fungi are able to enter through unbroken vegetable skin. Often these molds gain access to internal tissues of vegetables via the common fruit fly, *Drosophila melanogaster*. The fruit fly deposits spores and mycelial fragments of molds, along with its eggs, into growth cracks and wounds of vegetables. Such spoilage is characterized by a sour or fermentative odor and a cottony growth of mold (often with small black specks of sporangia) [23].

Brown or *Fusarium rot* often occurs in wet weather and tends to develop in vegetables which have first undergone bacterial rots. Aging or overmature vegetables seem to be particularly susceptible to this type of spoilage. Fusarium rots are another example of a spoilage problem where the fungus invades the vegetable in the field but may not manifest itself until sometime after harvest. This type of spoilage is accompanied by brownish spots which eventually develop into a pinkish-white mold [11,20].

In addition to these common forms of spoilage, many other important fungal spoilages exist but may only be found in specific circumstances or with certain vegetables. Discussion of these rather limited forms of spoilage is beyond the scope of this chapter. The reader is referred to one of a number of reference books which detail the problems of specific vegetable products [33-35].

2. Mechanism of Degradation

Plant cell walls have traditionally been divided into three functional-structural regions: the middle lamella, the primary wall, and the secondary wall. The coherence of plant tissues is primarily dependent on the pectic substances in the middle lamella, the region which binds the cells together. Development of postharvest diseases of vegetables depends to a great extent on the ability of the invading pathogen to secrete cell wall-degrading pectinolytic enzymes such as pectinesterase and polygalacturonase. In many cases, progressive degradation of pectic substances leads to a rapid loss of tissue coherence (soft rot) as a result of cell separation, increased liquefaction of the tissue, and ultimately cell death. This phenomenon, commonly referred to as maceration, may be further complicated by the presence of endogenous pectic enzymes. Increases in levels of these enzymes during maturation and ripening of vegetables have been associated with increased solubilization of pectic substances and subsequent tissue softening [36]. In addition to pectinolytic enzymes, other enzymes such as hemicellulases, cellulases, and proteinases can also be involved in tissue maceration, albeit to a lesser extent [37]. Therefore, it appears that the mechanism of plant tissue degradation is quite complex and often involves host tissue as well as microbial enzymic attack.

Some fungi such as *Botrytis*, *Alternaria*, *Colletotrichum*, *Fusarium*, *Monilinia*, and *Rhizopus* are strongly pectinolytic. The majority of these fungi are opportunistic parasites dependent on physical injury or excessive softening to gain entry to host tissues [38]. Damage incurred during harvesting provides the primary point of infection for opportunistic microorganisms.

The most commonly encountered storage rots are the soft rots. Lesions produced with soft rots grow rapidly to occupy the majority of available tissue. Such rots are characteristically waterlogged, have little or no tissue coherence and are accompanied by darkening of infected tissue as a result of polyphenol oxidase activity [38]. As the rot proceeds, the cavities formed within the tissues become filled with mycelium and the epidermis is punctured by conidia. Consequently, water is gradually lost and the infected vegetable shrivels and dries to a hard mass [38].

B. Safety

Organoleptic quality of vegetables spoiled by fungi often generates more interest and concern than does the safety aspect of the food. Usually, the normal spoilage microflora of vegetables have a competitive growth advantage and make it difficult for most pathogens to grow and accumulate. Some notable exceptions to this are the production of mycotoxins, the selective growth conditions in a metabiotic relationship, and the production of stress metabolites which are sometimes toxic to humans.

1. Mycotoxins

Relatively few of the many molds frequently associated with foods are capable of synthesizing mycotoxins. The predominant mycotoxigenic molds include *Aspergillus*, *Penicillium*, *Alternaria*, and *Fusarium*. Foods are susceptible to attack by molds in the field and during processing, transport, and storage [39]. Mycotoxins are secondary metabolites of some molds. These compounds are toxic to humans and animals. However, the severity of the toxicity depends on the amount and type of mycotoxin ingested. Some mycotoxins are known to be carcinogenic, mutagenic, and/or teratogenic. Heating of mycotoxins has little effect on reducing the toxicity. Information regarding toxicity, extent of contamination, stability in foods, and public health risk is limited for many of the mycotoxins. Therefore, their control in foods is not easily accomplished. However, the health risk associated with consumption of moldy vegetables is relatively low due to an aversion for selling and buying such vegetables. In the United States, the commodities most susceptible to mycotoxin contamination include peanuts, corn, cottonseed, and grains [40-42].

2. Metabiosis

Metabiosis is the process whereby the growth of one organism makes conditions more favorable for the growth of another organism. Such a metabiotic relationship has been demonstrated between fungi and *Clostridium botulinum* in tomatoes [43-45]. In this relationship, molds growing on high acid foods are capable of raising the pH of the food such that spores of *C. botulinum* can germinate, reproduce, and produce toxin. Powers [46] suggested that in some instances, the botulinum toxin may survive the thermal process given to canned tomato products. The likelihood of metabiosis occurring between fungi, *C. botulinum*, and tomatoes in nature is remote. However, the possibility of botulism food poisoning occurring in fresh or processed tomatoes as a result of fungal spoilage does exist.

3. Phytoalexins

Phytoalexins are metabolites produced by plants as a defense reaction when exposed to stress (e.g., injuries, cold, fungal infestation). These compounds accumulate in plant stems, roots, fruits, and seeds when infected with a variety of fungi. While the potential for phytoalexin production is present in all healthy plants, fungi may liberate bound forms through release of enzymes or may modify host compounds [20].

Phytoalexins are primarily toxic to fungi but can also be toxic to humans. For example, several furanocoumarin compounds, collectively called psoralens, are capable of causing blisters or lesions on consumers or handlers of celery contaminated by *Sclerotinia sclerotiorum*. Following exposure to the infected celery, it is necessary that the skin be exposed to sunlight or to ultraviolet radiation (320-380 nm) for lesions to be produced [47].

Release of antifungal compounds has also been documented in the sweet potato, Irish potato, carrot, turnip, snap bean, soybean, broad bean, English pea, pepper, barley, and rice. Accumulation of solanidine and α-solanine in Irish potatoes infected with *Phytophthora infestans* has been associated with gastrointestinal and neurological disorders in humans and farm animals [20].

III. POSTHARVEST HANDLING AND PROCESSING

Vegetables come to the consumer in many forms, including fresh, frozen, canned, and fermented. In addition, many different varieties of each of these forms can also exist. Each form and variety requires a slightly different processing procedure. Consequently, these processing procedures will affect fungi differently.

A. Fresh Vegetables

Several factors affect the postharvest storage life of fresh vegetables. These include cultivation techniques, physiological maturity at harvest, and handling and environmental conditions after harvesting. Measures which minimize mechanical and microbial deterioration will increase the shelf life

TABLE 5. Mean Log Count/g of Yeasts and Molds Found in Commercially Harvested and Processed Collards (*Brassica oleracea*)

	Yeast		Mold	
Month sampled	May	Nov.	May	Nov.
Sample point				
1. Field	4.5	4.2	3.5	3.6
2. After transportation	4.4	4.8	3.4	4.0
3. 1st spray wash	4.1	0.5	3.1	1.2
4. Hot water wash	3.7	1.4	3.2	2.0
5. Postblanch	0.5	2.4	0.5	2.5
6. Inspection	0.0	5.1	0.5	4.3
7. Chopper	0.9	2.3	0.2	1.9
8. Metering	1.7	2.2	1.5	2.6
9. Postblast-freeze	2.2	1.6	0.3	2.1

Source: Ref. 16.

and availability of this commodity in the marketplace. Between 1981 and 1985, per capita consumption of fresh vegetables in the United States increased from 72.4 lb to 81.4 lb [48]. As consumption of fresh vegetables continues to increase, the need to reduce losses from fungal spoilage is essential.

In order to understand how the microbiology of foods is impacted by processing, it is helpful to also have an understanding of the processing system involved. Fresh vegetables usually undergo a preliminary screening and grading inspection soon after harvest. From there, acceptable products are washed. Sometimes this washing procedure includes treatment with surfactants or sanitizers. After washing, the vegetables are processed into their final form. This processing is usually minimal for fresh vegetables and may include such treatments as peeling, cutting, or shredding. At that point, the processed vegetables are packaged, either for wholesale distribution or for retail sale. The effects of various handling and processing steps on yeast and mold populations in collards are presented in Table 5. From this table, it is readily apparent that populations of fungi may vary widely at different steps in processing.

The ultimate microbial quality of vegetables is often affected before harvest. It is at this stage that many soil fungi gain access to vegetables, especially those in close proximity to the ground. The organism responsible for the infamous Irish potato famine, *Phytophtora infestans*, can infest soils and attack potato tubers for years after the initial infestation. However, the damage caused by the infection may not show up until long after harvest [23].

Cleanliness of harvesting equipment, proper handling to minimize breakage, scratches and crushing, and culling of damaged or diseased vegetables are steps which can be taken to maintain high product quality. Field heat and the heat of respiration increase the susceptibility of the vegetables to microbial attack [20]. Therefore, field heat should be removed from the produce as soon as possible after harvest. Soil adhering to root and tuber vegetables harbors high populations of molds and their spores and should be removed when possible.

Contaminated equipment used in processing is also an important source of much of the microbial contamination in vegetables. One of the most common contaminants of processing equipment is the mold *Geotrichum candidum* (machinery mold) [49,50]. Surveying populations of this mold in foods has been used with increasing frequency as an aid in assessing adequate sanitation. Both in-plant chlorination of postblanching waters and improved equipment design have improved sanitation through more effective cleansing.

Vegetables are sometimes washed in chlorine solutions or other chemicals prior to shipping to reduce the microbial load and prolong freshness. However, the use of sanitizers may not always have significant beneficial effects with regard to fungal growth in vegetables. For example, Beuchat and Brackett [51] found that treating fresh lettuce with over 200 ppm residual chlorine had no effect on either the survival or growth of yeasts and molds on the treated lettuce. In contrast, Senter et al. [21] reported that mold populations on tomatoes decreased from about 10^2 mold colonies/g to undetectable levels after the tomatoes received a chlorine wash. Similar findings were reported by Golden et al. [52] for bell peppers treated with various fungicides and chlorine sanitizers. They found that these chemicals were sometimes helpful in reducing initial populations of fungi but that the antimycotic effect did not persist during storage of the peppers. Thus, treatment with sanitizers may help in some situations with some vegetables but not always. However, chlorinating wash water may help to reduce populations of viable yeasts and molds in the water, therefore minimizing cross-contamination of uncontaminated vegetables.

Another common practice in the produce industry is the use of refrigerated storage. Storage of vegetables at reduced temperatures in environmentally controlled chambers enables the vegetables to be held under optimal conditions until time for transport into the marketplace. The lowered temperature helps maintain the quality of fresh vegetables, mainly by slowing down the ripening process. However, it likewise also slows down the growth of any microorganisms present. The specific temperature used for a given vegetable is largely dependent on its tolerance to chill injury [53, 54]. Some vegetables, such as bell peppers, become damaged by temperatures below ~13°C.

Improved market quality of vegetables may also be achieved through the use of modified or controlled atmospheres. In this storage treatment, the atmosphere surrounding a vegetable is changed by adding or eliminating gases (e.g., N_2, O_2, CO_2, C_2H_4). Increasingly, polymeric films with various degrees of gas permeability have been used to create and maintain modified atmospheres within flexible film packages [55]. The addition of absorbers or adsorbers of O_2, CO_2, C_2H_4, and water to the package add additional control in maintaining a desired atmosphere.

The theory behind modified atmosphere storage is quite simple. Respiration in plants involves the enzymic oxidation of complex substrates such as starch, sugars, and organic acids to CO_2 and water, accompanied by a release of energy (heat). The rate of respiration is often a good index of the storage life of the vegetable. The lower the rate, the longer the storage life, and conversely, the higher the rate, the shorter the storage life. Modified atmospheres are especially effective in reducing the respiratory rate of vegetables when the optimum temperature for prevention of losses cannot be maintained during transport. Benefits derived from this type of storage include reduction in the rates of substrate depletion, O_2 consumption, CO_2 production, and release of heat. Several of the biochemical and physiological effects of modified atmosphere on fruits and vegetables have been reviewed by Kader [56].

Vegetables differ greatly in their tolerances to modified atmospheric storage. While fungal spoilage may be decreased, a particular vegetable's metabolism may not respond favorably to a specific modified atmosphere. For example, cauliflower develops off-flavors when placed in an atmosphere of increased CO_2 [27]. Other physiological disorders such as brown stain of lettuce and blackheart of potatoes have also been associated with inappropriately modified atmospheric storage [47]. The varying responses of vegetables to modified atmospheres dictate that the optimum combination of O_2, CO_2, and nitrogen be determined for each commodity and each cultivar.

Because microorganisms are less able to attack ripening vegetables as readily as overripened vegetables, the use of modified atmospheres also delays microbial spoilage. Consequently, the direct inhibition of microorganisms by gas concentrations seems to be secondary. Low O_2 concentrations ($<1\%$) or increased CO_2 levels ($>10\%$) are fungistatic toward many common spoilage fungi including *Penicillium, Mucor, Candida, Schlerotinia,* and *Botrytis* [58-60]. However, the inappropriate use of modified atmospheric storage has been associated with delayed wound healing, acceleration of senescence, and initiation of physiological breakdown (increased susceptibility to postharvest pathogens) [55]. Maintenance of proper temperature and relative humidity to prevent moisture condensation within the controlled environment is necessary to avoid growth of plant pathogens such as *Botrytis* and *Geotrichum.* As a precaution, modified atmosphere packaged vegetables are often given a fungicidal treatment prior to marketing [55].

The use of modified atmosphere packaging may act to either encourage or discourage the growth of fungi and resulting spoilage. Brackett [61] found that individual seal packaging, also known as shrink-wrapping, encouraged the growth of yeasts and molds on fresh tomatoes (Fig. 1). Similar observations were noted by Golden et al. [52] for shrink-wrapped bell peppers. In contrast, similar studies with broccoli [62] showed that the packaging had no significant effect on the growth of yeasts and molds (Fig. 2). In the cases of both vegetables, shelf-life was extended. Therefore, fungi's ability to bring about spoilage was not strongly correlated with their populations in the vegetables.

Many factors will interact to influence the growth of fungi in packaged vegetables. The gas transmission properties of the film, the composition of the gas, and the physiological and biochemical properties of the vegetable being packaged are some of the more important factors. These factors differ widely depending on the vegetable in question. Therefore, it is difficult to make accurate predictions for fungal growth in all vegetables.

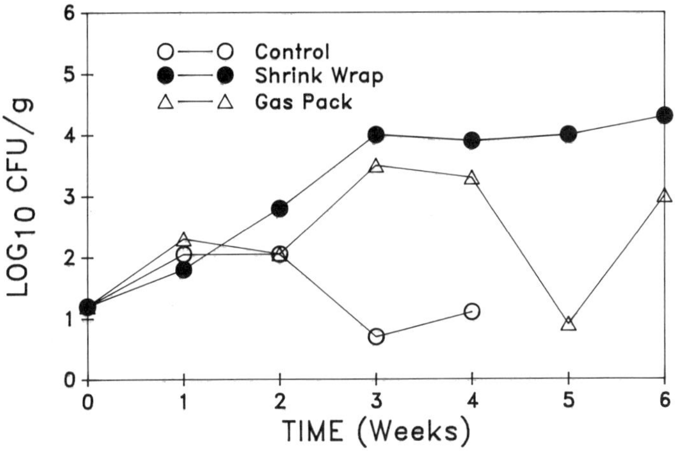

FIGURE 1. Yeast/mold populations on packaged and unpackaged fresh tomatoes stored at 13°C [61].

B. Heat Processed Vegetables

Heat processing, also known as canning, has been and continues to be an important food preservation technique. The specific processes for various foods are based upon the heat resistance of the organisms targeted for inactivation and pH. In general, vegetative bacteria are the most heat-sensitive and bacterial endospores are the most heat-resistant forms of microflora. Ascospores, conidia, and sclerotia of fungi are more heat resistant than vegetative bacteria but less so than bacterial endospores.

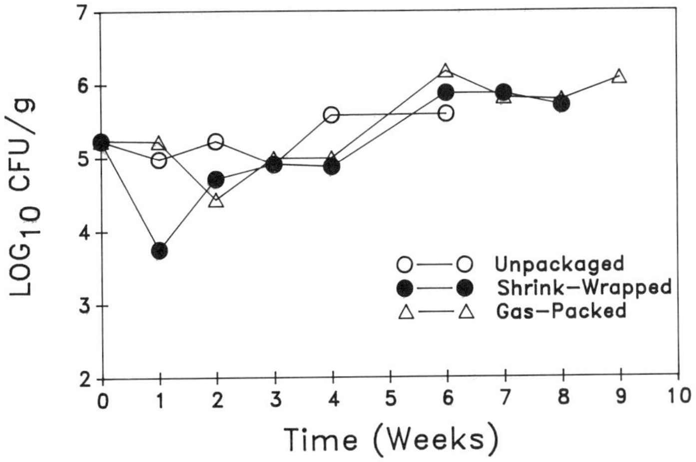

FIGURE 2. Yeast/mold populations on packaged and unpackaged fresh broccoli stored at 13°C [62].

There are two basic heat processes used for food. Low acid (pH ≥ 4.6) foods generally receive a relatively intense heat treatment sufficient to inactivate heat-resistant endospores of bacteria. Although it is desirable to eliminate all bacteria in these products, food processors are particularly concerned about eliminating the dangerous spore-former *Clostridium botulinum*. Heat processes for low acid foods are usually designed to accomplish the destruction of at least 12 $\log_{10}$ of *C. botulinum* endospores. Consequently, heat processing designed for low acid food is far in excess of what is needed to inactivate all known fungal ascospores and sclerotia.

Despite proper heat processing, canned goods still occasionally become spoiled by microorganisms. Often this spoilage results when containers leak, allowing air or cooling water to enter the container. It is theoretically possible for fungi to enter and grow within a leaky container. However, bacteria are normally able to more successfully compete with the fungi in low acid environments. Thus, such spoilages are almost always caused by bacteria.

Because *C. botulinum* is assumed not to grow in acidic environments, a much milder heat treatment is used for acidic vegetables, such as tomatoes. In these cases, the main purpose of the process is to inactivate vegetative bacteria and fungi rather than bacterial endospores. Moreover, the heating also serves to expel oxygen, further limiting the potential for growth of many fungi. However, any break in the hermetic seals of heat-processed acidic foods can result in spoilage by yeasts and molds. Such spoilage not only makes the food unpalatable, but can also result in subsequent growth of *C. botulinum* via metabiosis [45].

C. Frozen Vegetables

The degree to which processed vegetables are contaminated with microorganisms is a reflection of the condition of the raw product as well as the sanitary quality of the processing steps. With few exceptions, frozen vegetables are blanched in boiling water or in steam prior to freezing. The blanching inactivates many enzymes including catalase, peroxidase, and lipoxidase. The heat treatment also serves to destroy or damage many microbial cells. Blanching caused a dramatic decrease in the populations of both yeasts and molds present on collards (see Table 5) [16] and southern peas (Table 6) [15].

Cryoinjury of cells can occur either through thermal shock, concentration of extracellular solutes, dehydration, and/or internal ice crystal formation. The response of microorganisms to freezing is dependent upon many factors. Some of the more important factors include: (a) the cooling rate, (b) temperature of storage, (c) composition of product, (d) thawing rate, (e) type, species, and strain of the microorganism, and (f) age, population density, and nutritional status of the microbial cells [63]. In general, freezing will cause the death of 10-60% of the viable population and thus cannot be considered a method for sterilizing the food product [64]. However, yeast and mold spores are quite resistant to freeze injury. Moreau [65] indicated that spores of *Aspergillus niger*, *Rhizopus* sp., and *Mucor* sp. were able to withstand temperatures as low as -25°C. Such resistance to freezing by molds was likewise shown by Senter et al. [15,16] for several vegetables (see Tables 5 and 6).

TABLE 6. Log Count/g of Yeasts and Molds Found in Commercial Harvesting and Processing of Southern Peas

Sampling point	Yeast	Mold
Harvesting and transporting		
1. Freshly combined	4.5	4.4
2. 2 h Postharvest	5.1	4.2
3. 9 h Postharvest	4.3	4.4
Processing		
4. After 1st wash	3.2	2.9
5. After 2nd wash	2.7	2.6
6. Soak tank, time = 0 h	2.6	2.6
7. Soak tank, time = 10 h	3.2	3.0
8. After blanching	<1	<1
9. After inspection	1.4	0.9
10. Prefreezing	2.0	0.9
11. Postfreezing	2.3	1.7

Source: Ref. 15.

The storage temperatures normally used with frozen vegetables are not conducive to the growth of fungi. Consequently, the length of storage is often dependent on texture, flavor, color, and nutritional quality of the food upon thawing rather than the microbiology of the product. However, some molds are capable of growing at subfreezing temperatures [63]. Therefore, holding vegetables at improper subfreezing temperatures could allow the growth of some molds.

D. Fermented Vegetables

Vegetable fermentation has been practiced for millennia to both preserve foods and add variety to one's diet. The practice of fermentation begins with a brining stage whereby salt (sodium chloride) is either applied directly to the vegetables or the vegetables are soaked in a brine solution. The purpose of adding salt is twofold. First, osmotic pressure created by the added salt causes water to be drawn out from the vegetable tissues. This water is likewise accompanied by migration of sugars out of the tissue. Secondly, the addition of salt favors the growth of salt-tolerant lactic acid bacteria. It is these bacteria that utilize the sugars drawn out by the salt. The resultant drop in pH serves to inhibit spoilage and growth of pathogenic microorganisms.

Utilization of carbohydrates can also occur by fermentative yeasts in what is described as a secondary fermentation [66]. In this case, lactic acid bacteria have been inhibited by the low pH they have generated before all fermentable carbohydrates have been utilized. Fermentative yeasts are then able to metabolize the remaining fermentable carbohydrates.

Natural fermentation of cucumbers involves the lactic acid bacteria *Leuconostoc mesenteroides*, *Streptococcus faecalis*, *Pediococcus cerevisiae*, *Lactobacillus brevis*, and finally, *Lactobacillus plantarum*. Controlled fermentations of cucumbers rely on the addition of *P. cerevisiae* and *L. plantarum* inocula, or the latter alone [67]. Sauerkraut, the fermentation of shredded cabbage, is formed by the action of *L. mesenteroides*, *L. brevis*, and *L. plantarum*. Desirable fermenting bacteria of olives, the other major fermented vegetable in the United States, are primarily *L. mesenteroides*, *L. brevis*, and *L. plantarum*.

Although the lactic acid bacteria are primarily responsible for vegetable fermentations, yeasts also participate in the fermentation process. Several different types of fermentative yeasts have been isolated from fermentation brines including *Pichia subpelliculosa*, *P. anomala*, *Candida lactis/condensi*, *C. holmii*, *C. etchellsii*, *C. versatilis*, *Torulaspora delbrueckii*, *Zygosaccharomyces bailii*, and *Saccharomyces delbrueckii* [23].

Fungi can also be responsible for spoilage defects in fermented vegetables. Molds such as *Alternaria tenuis*, *Ascochyta cucumis*, *Cladosporium cladosporioides*, *Penicillium oxalicum*, *Fusarium oxysporum*, *Fusarium roseum*, and *Fusarium solani* have been found in the blossom, ovaries, or fruit of cucumber plants [68]. Often these flowers remain attached to the harvested fruit and thus allow entry of fungi into the fermentation process [69,70]. Pectinolytic and cellulolytic enzymes produced by these molds are extracted at the time of brining from remnants of flowers and become adsorbed to the surface of the cucumber [20]. Removal of the flowers has been found to significantly decrease the levels and deleterious effects of these enzymes in cucumber brines [70].

Another common spoilage problem, particularly with fermented cucumber "pickles," is excessive production of CO_2. This defect is primarily troublesome when fermentative yeasts produce CO_2 in internal tissues. This leads to inflated cucumbers known as "bloaters."

As with cucumber fermentations, yeasts can grow in or on the brine during fermentation of olives. Pink yeasts (*Rhodotorula glutinis*) and fermenting, pectinolytic yeasts (*Saccharomyces oleaginosus* and *Hansenula anomala*) can cause softening of olives [71,72].

Several techniques are used to prevent or control fungal spoilage and defects. Controlled fermentations are often used to completely utilize all fermentable carbohydrates so that none is left for fungi. In addition, the growth of oxidative yeasts can be minimized by restricting the supply of oxygen. This can be accomplished by increasing the depth of fermentation vats or purging vats with nitrogen [73].

IV. CONCLUSION

As vegetables continue to become more popular with consumers, the role of fungi in spoilage and safety of these products will likewise increase in importance. The increasing demand by consumers for minimally processed vegetables makes extending the shelf-life of these products of particular interest to researchers. Because yeasts and molds are so important in the spoilage of fresh vegetables, many of the new techniques for extending shelf life will necessarily focus on these microorganisms. Consequently, new research findings will likely give scientists even greater insight into behavior of fungi in vegetables.

REFERENCES

1. Weichmann, J. Introduction. In *Postharvest Physiology of Vegetables* (J. Weichmann, ed.), Marcel Dekker, New York, 1987, pp. 3-7.
2. Gorini, F. Market preparation methods and shelf life. In *Postharvest Physiology of Vegetables* (J. Weichmann, ed.), Marcel Dekker, New York, 1987, pp. 489-495.
3. Harvey, J. M. Reduction of losses in fresh market fruits and vegetables. *Annu. Rev. Phytopathol.*, *16*:321-341, 1978.
4. Dennis, C. Fungi. In *Postharvest Physiology of Vegetables* (J. Weichmann, ed.), Marcel Dekker, New York, 1987, pp. 377-411.
5. Eckert, J. W. Part I: General Principles. In *Postharvest Physiology, Handling and Utilization of Tropical and Subtropical Fruits and Vegetables* (E. B. Pantastica, ed.), AVI Publishing Co., Westport, CT, 1975, pp. 393-414.
6. Pitt, J. I., and Hocking, A. D. The ecology of fungal food spoilage. In *Fungi and Food Spoilage* (B. S. Schweigert and G. F. Stewart, eds.), Academic Press, New York, 1985, pp. 5-18.
7. Friend, J., Reynolds, S. B., and Aveyard, M. A. Phenylalanine ammonia lyase, chlorogenic acid and lignin in potato tuber tissue inoculated with *Phytophtora infestans*. *Physiol. Plant Pathol.*, *3*:495-507, 1973.
8. Beckman, C. H., Mueller, W. C., Tessier, B. J., and Harrison, N. A. Recognition and callose deposition in response to vascular infection in Fusarium wilt resistant or susceptible tomato plants. *Physiol. Plant Pathol.*, *20*:1-10, 1982.
9. Esquerré-Tugayé, M. T., Lafitte, C., Mazau, D., Toppan, A., and Touze, A. Evidence for the accumulation of hydroxyproline-rich glycoproteins in the cell wall of diseased plant as a defense mechanism. *Plant Physiol.*, *64*:320-326, 1979.
10. Meneley, J. C., and Stanghellini, M. E. Detection of enteric bacteria within locular tissue of healthy cucumbers. *J. Food Sci.*, *39*:1276-1268, 1974.
11. Friedman, B. A. Market diseases of fresh fruits and vegetables. *Econ. Bot.*, *14*:145-156, 1960.
12. Splittstoesser, D. F. Predominant microorganisms on raw plant foods. *J. Milk Food Technol.*, *33*:500-505, 1970.
13. Lund, B. M. Bacterial spoilage of vegetables and certain fruits. *J. Appl. Bacteriol.*, *34*:9-20, 1971.
14. Miller, M. W. Yeasts in food spoilage: An update. *Food Technol*, *33*:76-80, 1979.
15. Senter, S. D., Cox, N. A., Bailey, J. S., and Meredith, F. I. Effects of harvesting, transportation, and cryogenic processing on the microflora of southern peas. *J. Food Sci.*, *49*:1410-1411, 1437, 1984.

16. Senter, S. D., Bailey, J. S., and Cox, N. A. Aerobic microflora of commercially harvested, transported and cryogenically processed collards (*Brassica oleracea*). *J. Food Sci.*, *52*:1020-1021, 1987.
17. Deák, T., Heaton, E. K., Hung, Y. C., and Beuchat, L. R. Extending the shelf life of fresh sweet corn by shrink-wrapping, refrigeration, and irradiation. *J. Food Sci.*, *52*:1625-1631, 1987.
18. Walker, H. W. Spoilage of food by yeasts. *Food Technol.*, *31*:57-61, 65, 1977.
19. Last, F. T., and Warren, R. C. Non-parasitic microbes colonizing green leaves: Their form and functions. *Endeavor.*, *31*:143-150, 1972.
20. Mundt, J. O. Fungi in the spoilage of vegetables. In *Food and Beverage Mycology* (L. R. Beuchat, ed.), AVI Publishing Co., Westport, CT, 1978, pp. 110-128.
21. Senter, S. D., Cox, N. A., Bailey, J. S., and Forbus, W. R., Jr. Microbiological changes in fresh market tomatoes during packing operations. *J. Food Sci.*, *50*:254-255, 1985.
22. Webb, T. A., and Mundt, J. O. Molds on vegetables at the time of harvest. *Appl. Environ. Microbiol.*, *35*:655-658, 1978.
23. Brackett, R. E. Vegetables and related products. In *Food and Beverage Mycology*, 2nd ed. (L. R. Beuchat, ed.), Van Nostrand Reinhold, New York, 1987, pp. 129-154.
24. Pitt, J. L., and Hocking, A. D. (eds.). *Fungi and Food Spoilage*, Academic Press, New York, 1985.
25. Beuchat, L. R. (ed.). *Food and Beverage Mycology*, 2nd ed., Van Nostrand Reinhold, New York, 1987.
26. Brackett, R. E. Microbiological consequences of minimally processed fruits and vegetables. *J. Food Qual.*, *10*:195-206, 1987.
27. Deák, T. Microbial-ecological principles in controlled atmosphere storage of fruits and vegetables. In *Microbial Associations and Interactions in Foods* (I. Kiss, T. Deák, and K. Incze, eds.), Akademiai Kiado, Budapest, 1984, pp. 9-22.
28. Pitt, J. I., and Hocking, A. D. Spoilage of fresh and perishable foods. In *Fungi and Food Spoilage*, Academic Press, New York, 1985, pp. 365-381.
29. Beraha, L., Smith, M. A., and Wright, W. R. Control of decay of fruits and vegetables during marketing. *Dev. Indust. Microbiol.*, *2*:73-77, 1961.
30. LeClerg, E. L. Symposium on plant disease losses: Crop losses due to plant diseases in the United States. *Phytopathology*, *54*:1305-1319, 1964.
31. Jarvis, B., Seiler, D. A. L., Ould, A. J. L., and Williams, A. P. Observations on the enumeration of moulds in food and feedingstuffs. *J. Appl. Bact.*, *55*:325-336, 1983.
32. Moline, H. E. Comparative studies with two *Geotrichum* species inciting postharvest decays of tomato fruit. *Plant Dis.*, *68*:46-48, 1984.
33. Haard, N. F., and Salunkhe, D. K. In *Postharvest Biology and Handling of Fruits and Vegetables* (N. F. Haard and D. K. Salunkhe, eds.), AVI Publishing Co., Westport, CT, 1980.
34. Dennis, C. (ed.). *Post-harvest Pathology of Fruits and Vegetables*, Academic Press, London, 1983.

35. Weichmann, J. (ed.). *Postharvest Physiology of Vegetables*, Marcel Dekker, New York, 1987.
36. Gross, K. C., and Wallner, S. J. Degradation of cell wall polysaccharides during tomato fruit ripening. *Plant Physiol.*, *63*:117-120, 1979.
37. Codner, R. C. Pectinolytic and cellulolytic enzymes in the microbial modification of plant tissues. *J. Appl. Bact.*, *34*:147-160, 1971.
38. Chesson, A. Maceration in relation to the post-harvest handling and processing of plant material: A review. *J. Appl. Bact.*, *48*:1-45, 1980.
39. Gourama, J., and Bullerman, L. B. Mycotoxin production by molds isolated from 'Greek-style' black olives. *Int. J. Food Microbiol.*, *6*:81-90, 1988.
40. Hesseltine, C. W., Rogers, R. F., and Shotwell, O. L. Aflatoxin and mold flora in North Carolina in 1977 corn crop. *Mycologia*, *73*: 216-228, 1981.
41. Bullerman, L. B. Mycotoxins and food safety. *Food Technol.*, *40*: 59-66, 1986.
42. Harrison, M. A., Silas, J. C., and Carpenter, J. A. Incidence of aflatoxigenic isolates of *Aspergillus flavus/parasiticus* obtained from Georgia corn processing plants. *J. Food Qual.*, *10*:101-105, 1987.
43. Odlaug, T. E., and Pflug, I. J. *Clostridium botulinum* growth and toxin production in tomato juice containing *Aspergillus gracilis*. *Appl. Environ. Microbiol.*, *37*:496-504, 1979.
44. Mundt, J. O., and Norman, J. M. Metabiosis and pH of moldy fresh tomatoes. *J. Food Prot.*, *45*:829-832, 1982.
45. Draughon, F. A., Chen, S., and Mundt, J. O. Metabiotic association of *Fusarium*, *Alternaria*, and *Rhizoctonia* with *Clostridium botulinum* in fresh tomatoes. *J. Food Sci.*, *53*:120-123, 1988.
46. Powers, J. J. Effect of acidification of canned tomatoes on quality and shelf life. *Crit. Rev. Food Sci. Nutr.*, *7*:371-396, 1976.
47. Ayres, J. C. Antibiotic, inhibitory and toxic metabolites elaborated by microorganisms in foods. *Acta Aliment. Acad. Sci. Hung.*, *2*: 285-302, 1973.
48. United States Department of Agriculture. Agricultural Statistics. Consumption and Family Living, U.S. Government Printing Office, Washington, D.C., 1987, pp. 487-492.
49. Eisenberg, W. V., and Cichowicz, S. M. Machinery mold-indicator organism in food. *Food Technol.*, *31*:52-56, 1977.
50. Splittstoesser, D. F., Groll, M., Downing, D. L., and Kaminski, J. Viable counts versus the incidence of machinery mold (*Geotrichum*) on processed fruits and vegetables. *J. Food Prot.*, *40*:402-405, 1977.
51. Beuchat, L. R., and Brackett, R. E. Growth of *Listeria monocytogenes* on lettuce as influenced by shredding, chlorine treatment, modified atmosphere packaging, temperature and time. *J. Food Sci.*, *55*:755-758, 870, 1990.
52. Golden, D. A., Heaton, E. K., and Beuchat, L. R. Effect of chemical treatments on microbiological, sensory and physical qualities of individually shrink-wrapped produce. *J. Food Prot.*, *50*:673-680, 1987.

53. Lyons, J. M. Chilling Injury. In *Postharvest Physiology of Vegetables* (J. Weichmann, ed.), Marcel Dekker, New York, 1987, pp. 305-326.
54. Dennis, C. Salad Crops. In *Post-harvest Pathology of Fruits and Vegetables* (C. Dennis, ed.), Academic Press, London, 1983, pp. 158-167.
55. Zagory, D., and Kader, A. A. Modified atmosphere packaging of fresh produce. *Food Technol.*, *42*:70-77, 1988.
56. Kader, A. A. Biochemical and physiological basis for effects of controlled and modified atmospheres on fruits and vegetables. *Food Technol.*, *40*:99-104, 1986.
57. Kader, A. A., Zagory, D., Kerbel, E. L., and Wang, C. Y. Modified atmosphere packaging of fruits and vegetables. *CRC Crit. Rev. Food Sci. Nutr.*, *28*:1-30, 1989.
58. Eckert, J. W., and Sommer, N. F. Control of disease of fruits and vegetables by postharvest treatment. *Annu. Rev. Phytopathol*, *5*: 391-432, 1967.
59. Reyes, A. A., and Smith, R. B. Controlled atmosphere effects on the pathogenicity of fungi on celery and on the growth of *Botrytis cinerea*. *HortSci.*, *21*:1167-1169, 1986.
60. El-Goorani, M. A., and Sommer, N. F. Effects of modified atmospheres on postharvest pathogens of fruits and vegetables. *Hort. Rev.*, *3*:412-461, 1981.
61. Brackett, R. E. Changes in the microflora of packaged fresh tomatoes. *J. Food Qual.*, *11*:89-105, 1988.
62. Brackett, R. E. Changes in the microflora of packaged fresh broccoli. *J. Food Qual.*, *12*:169-181, 1989.
63. Banwart, G. J. Control of microorganisms by retarding growth. In *Basic Food Microbiology*, abridged ed., AVI Publishing Company, Westport, CT, 1981, pp. 546-641.
64. Ayres, J. C., Mundt, J. O., and Sandine, W. E. Prevention of food spoilage. In *Microbiology of Foods*, W. H. Freeman and Company, San Francisco, 1980, pp. 44-71.
65. Moreau, C., and Moss, M. Food-borne moulds. In *Moulds, Toxins and Food*, John Wiley & Sons, Chichester, 1979, pp. 3-26.
66. Fleming, H. P. Fermented vegetables. In *Economic Microbiology*, Vol. 7, *Fermented Foods* (A. H. Rose, ed.), Academic Press, New York, 1982, pp. 227-258.
67. Etchells, J. L., Fleming, H. P., and Bell, T. A. Factors influencing the growth of lactic acid bacteria during the fermentation of brined cucumbers. In *Lactic Acid Bacteria in Beverages and Food* (J. G. Carr, C. V. Cutting, and G. C. Whiting, eds.), Academic Press, New York, 1975, pp. 281-305.
68. Raymond, F. L., Etchells, J. L., Bell, T. A., and Masley, P. M. Filamentous fungi from blossoms, ovaries, and fruit of pickling cucumbers. *Mycologia*, *51*:492-511, 1960.
69. Etchells, J. L., Bell, T. A., and Jones, I. D. Studies on the origin of pectolytic and cellulolytic enzymes in commercial cucumber fermentations. *Food Technol.*, *9*:14-18, 1955.

70. Etchells, J. L., Bell, T. A., Monroe, R. J., Masley, P. M., and Demain, A. L. Populations and softening enzyme activity of filamentous fungi on flowers, ovaries and fruit of pickling cucumbers. *Appl. Microbiol.*, *6*:427-440, 1958.
71. Vaughn, R. H., Jakubczyk, T., MacMillan, J. D., Higgins, T. E., Dave, B. A., and Crampton, V. M. Some pink yeasts associated with softening of olives. *Appl. Microbiol.*, *18*:771-775, 1969.
72. Vaughn, R. H., Stevenson, K. E., Dave, B. A., and Park, H. C. Fermenting yeasts associated with softening and gas-pocket formation in olives. *Appl. Microbiol.*, *23*:316-320, 1972.
73. Fleming, H. P., Thompson, R. L., Bell, T. A., and Monroe, R. J. Effect of brine depth on physical properties of brine-stock cucumbers. *J. Food Sci.*, *42*:1464-1470, 1977.

7

FUNGI OF IMPORTANCE IN PROCESSED FRUITS

DON F. SPLITTSTOESSER *New York State Agricultural Experimental Station, Cornell University, Geneva, New York*

I. INTRODUCTION

The microorganisms of concern to the processor of different fruit-based foods are yeasts, molds, and a few groups of aciduric bacteria. Many of the fungi thrive in the low pH, high sugar environment afforded by fruits, and thus they often are the predominant microorganisms. Much of this chapter deals with methods for preventing the growth of fungi in different fruit products although a section is devoted to yeasts and their fermentation of fruit juices.

II. CONTAMINATION

A. Sources

The yeasts and molds found in fruit products may have been present on the raw fruit or may have been introduced at some stage during processing.

A wide range of fungal populations have been reported for different fruits as received at the processing plant. Yeast and mold counts on grapes have been found to range from 10^4 to 10^7 per gram with about half of the samples yielding figures of 10^6 and above [1]. Yeasts generally predominated over molds. Grapes with their thin skin and succulent flesh are especially susceptible to physical damage and subsequent growth of aciduric microorganisms. Other studies have shown that lower counts are generally to be observed on larger fruit which have less exposed surface area and on fruit that are protected by a firm skin or rind. Thus it has been reported that the mean count on sound apples was 10^4 per gram [2], the median population on tart cherries was 10^3 per gram [3], and the count on intact oranges was 10^5 per orange [4].

Processing lines are another source of the fungi that contaminate fruit products. The juices of most fruits provide an excellent growth medium for yeasts and molds and thus significant build-up may occur in areas where solubles collect. Generally yeasts grow more rapidly than molds and therefore predominate. Equipment which can serve as important sources of contamination include pitters, slicers, mills, presses, conveyors, pumps, and filling machines [4-6].

Certain fruit products may present special problems. Concord grape juice, for example, is stored in large tanks at -2°C for an extended time period to precipitate tartrates. Although the juice is pasteurized, chance contamination often occurs, and during storage psychrotrophic yeasts may multiply in the juice [7].

B. Detection

The methods for determining the incidence of fungi in fruit products include microscopic procedures, viable counts, chemical and serological tests, and analyses for metabolic products.

The Howard Mold Count is one of the oldest procedures for the microscopic examination of foods [8]. Although originally developed for preventing the use of moldy tomatoes in products such as catsup, it also is used to assess the quality of a variety of other processed fruits such as canned drupelet berries, citrus and pineapple juices, cranberry sauce, and various nectars, purees, and pastes [9]. The method consists of examining the food preparation on a special slide at a magnification × ~100. A prescribed number of fields are viewed under the microscope and the percentage containing mold mycelia are recorded. The U.S. Food and Drug Administration has defect action levels which define the percentage of positive fields that are acceptable for a given fruit product.

Another microscopic procedure involves the enumeration of hyphae of *Geotrichum candidum*. The mold grows on the surfaces of processing equipment, and therefore it has been used as an index of sanitation [10,11]. The procedure for "machinery mold" involves the careful rinsing of the canned or frozen fruits with water into sieves where the mold filaments are collected. After staining, the suspension is added to a rot fragment slide, and characteristic hyphae are counted using a magnification of × 30-45 [9]. Although fruit processed under unsanitary conditions show a higher incidence of *Geotrichum*, hyphal counts often do not correlate well with other indices of sanitation such as viable counts [3].

Various agar media are used for the enumeration of viable yeasts and molds in different fruit products [12]. Some of the more common are potato dextrose agar, malt extract agar, and plate count agar. These media can be made to be selective for fungi by acidification or by the inclusion of antibiotics. The addition of a prescribed amount of 10% tartaric acid to potato dextrose agar to reduce the pH to 3.5 is a common practice. The incorporation of chloramphenicol and/or chlortetracycline at concentrations not exceeding 100 μg/ml is also effective. When the spreading of mold colonies is a problem, it can be reduced by the addition of rose bengal dye or dichloran to the plating medium [13,14].

The fact that chitin is a component of fungal cell walls affords several methods for estimating the amount of mold in fruit products. Cousin, et al. [15] used alkaline hydrolysis to break down chitin and then analyzed for glucosamine colorimetrically. The species of mold and the age of the culture were found to influence the levels of glucosamine. Studies with foods to which a known amount of mold had been added showed that glucosamine levels correlated well with the quantity added but not well with Howard mold counts. More recently, Davies et al. [16] used near infrared spectroscopy to estimate the amount of chitin in freeze-dried tomato purees. Their results showed good agreement between spectroscopic readings and the amount of *Botrytis cinerea* and *Alternaria tenuissima* that had been added.

An enzyme-linked immunosorbent assay (ELISA) has been developed for the detection of *Penicillium* and *Aspergillus* species in different foods [17]. The water-extractable antigen was heat stable, which meant that molds could be detected even after food had undergone a heat treatment. Studies with three species of *Penicillium* showed that the antigen titer was related to the concentration of mold and that the ELISA procedure permitted the detection of as little as 38 ng of mycelium per g of food.

Another means for assessing a fruit product for fungal contamination is to analyze for various metabilites. Many yeasts, for example, produce ethanol as a major fermentation product. Sensitive analyses for alcohol, such as provided by gas chromatography, give early warning to the processor when yeasts are initiating growth in a food such as a fruit juice that is being stored in large tanks.

Glycerol, a product of mold metabolism, has been suggested as an indicator of quality for grapes that have been harvested mechanically [18]. Because the harvester breaks up the fruit, it is difficult to determine visually whether the grapes in the vineyard had been sound or heavily infested with mold. A difficulty in using glycerol as a criterion is that the amount produced has been found to be species related, and thus the extent of mold growth will not always correlate well with the level of metabolic product [19]. In this work it was found that *Aspergillus niger* and *Penicillium italican* produced little glycerol when propagated on grape juice, while *Rhizopus nigricans* sometimes generated over 2000 mg glycerol per g of dry mycelium. In this study it was also shown that *Botrytis cinerea*, one of the most common grape molds, was intermediate in that it produced about 100 mg glycerol per g of mycelium under most conditions.

Acetylmethylcarbinol and diacetyl are fermentation products of various yeasts and molds, and therefore the Voges-Proskauer test has been recommended as a means of assessing microbial quality of apple and citrus juices [20,21]. As with glycerol, the amounts of acetylmethylcarbinol and diacetyl do not always correlate well with levels of contamination. Some fungi produce both compounds while others generate only acetylmethylcarbinol and the amounts produced can vary significantly among species [22].

III. THERMALLY PROCESSED FRUITS

A variety of heat treatments are used for the preservation of different fruit products [23]. Canned fruits which may be in the form of whole fruit, halves, slices or diced product, are often processed in a boiling water bath or in a retort until a center temperature of 85-90°C is achieved. Apple sauce is filled hot into cans at a closing temperature of 93°C or above in order to achieve sufficient vacuum to prevent headspace detinning. Fruit juices such as apple, grape, and citrus are filled hot into glass or

TABLE 1. Heat Resistance of Different Yeasts

Species	Heating menstruum	D-value[a]
Debaromyces hansenii	Yeast-malt extract peptone glucose broth, pH 3.5	D_{48} = 10
Pichia membranaefaciens		D_{51} = 8.8
Candida krusel		D_{56} = 14
Kloeckera apiculata		D_{46} = 9.9
Rhodotorula rubra		D_{50} = 26
Zygosaccharomyces bailii	Citrate-PO_4 buffer pH 4.5	D_{60} = 0.10
Z. bailii ascospores		D_{60} = 14
Saccharomyces chevalieri		D_{60} = 0.10
S. chevalieri ascospores		D_{60} = 16
Saccharomyces uvarum		D_{55} = 0.07
S. uvarum ascospores		D_{55} = 8.0
Saccharomyces cerevisiae	Apple juice pH 3.6	D_{55} = 0.90
S. cerevisiae ascospores		D_{55} = 106
S. cerevisiae ascospores		D_{60} = 6.1

[a]D-value = minutes at a given temperature required to kill 90% of the cells, e.g., D_{48} = 10 signifies that at 48°C, 90% of the cells were destroyed in 10 minutes.

Source: Refs. 27, 87-89.

metal containers, with 88°C being a typical closing temperature. Aseptically processed juices are flash pasteurized to temperatures of 93-104°C, cooled, and then filled into previously sterilized containers under aseptic conditions. Fruit nectars are flash pasteurized for 30 seconds at 110°C and then filled into containers at temperatures of 88-90°C. Jams and jellies are generally filled into containers at temperatures of 80-85°C. Following filling, the jars containing the preserves may be held for a number of minutes before cooling to assure that contaminants on the cap are destroyed.

A. Heat Resistance of Yeasts

In general, yeasts possess little heat resistance and would not be able to survive the above processes. Temperatures as low as 46°C are lethal to some strains and, in general, asporogenous yeasts are more sensitive to heat than are ascomycetous species (Table 1). While the ascospores of yeasts are not highly resistant to heat, the D-values (minutes needed to kill 90% of the cells) at a given temperature exceed those for vegetative cells by a factor of 100 fold or more. The composition of the food influences their heat resistance. High concentrations of sugars afford protection, while ethanol, sorbate, and benzoate make yeasts more sensitive to heat [24-27].

When in a dry form, yeasts possess much higher degrees of heat resistance. For example, D-values as great as 5 minutes at the high temperature of 126.7°C have been observed for a strain of *Saccharomyces cereviseae* that had been equilibrated to 33-38% relative humidity [28]. Such resistance might be important in situations where dry heat was to be used for sterilizing packages that were to contain a moist fruit product. It should not present a problem for dry fruit, on the other hand, since the surviving yeasts would not be able to multiply in a food of such low water activity, that is, in a product equilibrated to 38% RH.

B. Heat Resistance of Molds

In general mold hyphae and conidia possess little heat resistance. The thermal death times for 10 different molds was found to be 5 minutes or less at 60°C [29], and D-values for conidia of *Aspergillus flavus* and *A. parasiticus* have been reported to range from 7.7 to 59 seconds at that temperature [30]. These molds would not survive the thermal processes that are given to fruit products, as described previously.

A few species do produce heat-resistant structures, and a number of spoilage outbreaks have been documented. Although some molds form heat-resistant sclerotia, structures which are hard compacted masses of mycelium, most spoilage has been due to the survival of the ascospores produced by a very few species. Spoilage may be manifest by visible mold growth, off-odors, the breakdown of fruit texture, or the solubilization of the starch or pectin in the suspending medium. It is the author's experience that sometimes the amount of mold growth is so limited that colonies or even mycelial fragments cannot be detected. In a spoilage outbreak involving growth of *Byssochlamys* in a canned lemon pie filling, the starch gel was solubilized, but no mycelium could be found when the filling was strained through the sieves used for the recovery of *Geotrichum*.

A classic outbreak due to sclerotia occurred in canned blueberries in 1938 [31]. The microorganism was identified as a species of *Penicillium*. The thermal death times at 85 and 90.5°C exceeded 270 and 30 minutes, respectively: thus, the sclerotia possessed a high degree of resistance. Surveys of blueberry fields revealed that sclerotia of this organizm were present in the soil.

The principal molds that produce heat-resistant ascospores are *Byssochlamys fulva* (anamorph: *Paecilomyces fulvus*), *Byssochlamys nivea* (anamorph: *Paecilomyces niveus*), *Neosartorya fischeri* (anamorph: *Aspergillus fischeri*), and *Talaromyces flavus* (anamorph: *Penicillium vermiculatum*, *Penicillum dangeardii*). The D-values cited for the different species (Table 2) are only approximations, because survivor curves generally do not show a logarithmic order of death [32]. Instead of obtaining a straight line when the logarithm of the survivor count is plotted against heating time, one usually sees a shoulder followed by an accelerated destruction rate [33].

Soil appears to be a common habitat for the different heat-resistant molds, and therefore low numbers of ascospores often can be recovered from fruits as delivered to the processor [34-38]. It appears that fruits such as pineapple that have direct soil contact may be especially susceptible to contamination. Ingredients of fruit products which have had soil contact also may contain significant levels of ascospores. Thus, it has

TABLE 2. Heat Resistance of Different Mold Ascospores

Species	Heating menstruum	D-value
Neosartorya fischeri	Cherry filling	$D_{85} = 52$
N. fischeri	Apple juice	$D_{87.8} = 1.4$
N. fischeri	Grape juice	$D_{85} = 60$
Talaromyces flavus	Cherry filling	$D_{85} = 26$
T. flavus	Apple juice	$D_{87.8} = 7.8$
T. flavus	Glucose-tartrate soln	$D_{85} = 12$
Byssochlamys fulva	Glucose-tartrate soln	$D_{90} = 17$
B. fulva	Grape juice	$D_{85} = 56$
Byssochlamys nivea	Grape juice	$D_{75} = 60$

Source: Refs. 90-96.

been the experience of the author that tapioca flour, which may be used as a thickener in fruit puddings and pie fillings and is derived from the root of the cassava plant, often contains the ascospores of *Byssochlamys*.

The ubiquitousness of heat-resistant molds coupled with the fact that most commercial pasteurization processes will not destroy their ascospores raises the question as to why spoilage is not more prevalent. One can speculate that the following may be involved: (a) the level of contamination on the raw fruit is usually low, generally only a few ascospores per 100 grams, (b) it usually takes a number of days of vegetative growth before ascospores are formed [39] and thus reasonable sanitation should prevent a significant buildup of spores on processing lines, and (c) different processing operations such as the washing and peeling of fruits and the filtration of fruit juices will remove most spores from the product [36].

Detailed descriptions of the different heat-resistant molds are provided by Samson et al. [40] and by Pitt and Hocking [41]. Some of their more important characteristics are as follows.

1. *Byssochlamys*

The asci are produced in open clusters rather than in sacs such as cleistothecia. The asci, which contain eight ascospores, usually remain intact unless subjected to mechanical treatment. The conidiophores resemble those of *Penicillium* species except that the tips of the phialides tend to bend away from the main axis. *Byssochlamys fulva* colonies range from tannish-yellow to olive brown in color, while those of *B. nivea* are white to cream colored. The colonies are never green. When ascospores are not observed, some strains have been incorrectly identified as species of *Paecilomyces*. The ability to produce ascospores can be permanently lost when cultures are grown at temperatures below 24°C [40].

2. *Neosartorya fischeri*

Asci are produced in ascocarps that give the colonies a granular appearance [41]. The asci may rupture spontaneously with the result that only free spores are observed. Ascospores, which may be smooth or slightly spiny, possess two equatorial crests. The conidiophores are characteristic of *Aspergillus* species with the phialides being borne directly on the vesicle. Conidial production is often sparse. The colonies are cream to white on malt extract agar.

3. *Talaromyces flavus*

The asci are produced in ascocarps that usually are abundant. The ascospores are elliptical in shape and possess spiny walls. The conidia are borne on penicillia which are produced in varying numbers, depending upon strain. The colonies are usually yellow due to the pigmentation of both the mycelium and the ascocarps. Some strains develop reddish hues.

IV. FROZEN FRUITS

Yeasts and molds have the ability to survive freezing and thus a certain level of viable fungi can be found in most frozen fruit products. In general, survival is highest in the fruits that were frozen most rapidly and held at the lowest storage temperatures [42]. The number of fungi present usually reflects processing methods and equipment sanitation. Some fruit products contain low fungal populations because they had been subjected to a lethal treatment; for example, the pasteurization of juices or the dipping of apple slices into strong sulfurous acid solutions.

Many yeasts [43] and molds [44] have the ability to grow at low temperatures and thus can spoil fruit products that are held at a defrost temperature for an extended period of time. Spoilage is generally not a problem for frozen fruits stored at recommended freezer temperatures (-18 to -23°C) because of the lack of available water.

V. LOW WATER ACTIVITY

A number of fruit products are preserved almost indefinitely or have extended shelf lives because of their low levels of available water (Table 3). Some of the more common products are citrus, apple, pear, and grape juice concentrates; dried figs, prunes, apricots, and raisins; and fruit preserves. In general, the products having a water activity of 0.70 or lower are shelf stable for one year or longer, whereas those with an activity of about 0.75 can be stored for at least 6 months without encountering spoilage [41].

Osmophilic yeasts and xerophilic molds have been defined as species that are capable of growing at a water activity of 0.85 or below. Although growth of many common spoilage fungi is inhibited by an a_w of 0.85 [45], there also are numerous species that do very well in such an environment (Table 4).

Two of the most osmo-tolerant fungi are *Zygosaccharomyces rouxii*, a yeast, and *Xeromyces bisporus*, a mold. Both are capable of growing in many of the commercial dried fruits and juice concentrates. Their growth

TABLE 3. Water Activity of Different Fruit Products

Food	Water activity
Raspberry juice concentrate, 65° Brix	0.79-0.80
Orange juice concentrate, 65° Brix	0.80-0.84
Raisins, 17% water	0.68
Prunes, 20% water	0.68
High moisture prunes, 35% water	0.94
Dates, under 24% water	0.60-0.65
Jams and jellies	0.75-0.94
Fruit cakes	0.75 and under

Source: Refs. 47, 97.

is slow, however, and they are readily destroyed by the heat treatments that often are given to these products. These factors may explain why spoilage outbreaks by the most osmoduric fungi are not more common.

Eurotium species (anamorph: *Aspergillus glaucus* group) produce ascospores and, as a result, are potential spoilers of heated fruit products such as jams and jellies. They rarely cause a problem, however, most likely because their ascospores generally do not possess sufficient resistance to heat.

TABLE 4. Minimum Water Activity for the Growth of Various Yeasts and Molds

Species	Water activity
Yeasts	
Saccharomyces cerevisiae	0.94
Zygosaccharomyces bailii	0.80
Hansenula anomala	0.75
Candida versatilis	0.70
Zygosaccharomyces bisporus	0.70
Zygosaccharomyces rouxii	0.65
Filamentous fungi	
Botrytis cinerea	0.93
Byssochlamys nivea	0.84
Penicillium expansum	0.82
Eurotium repens	0.72
Chrysosporium xerophilum	0.71
Xeromyces bisporus	0.61

Source: Refs. 41, 45, 47.

Studies on seven different species [46] indicated that a heat treatment of 10 minutes at 75°C, a relatively mild process, resulted in the destruction of most spores.

Many of the xerophilic molds require a medium of reduced water activity in order to produce vegetative growth [41]. Included in this group are the *Eurotium* species, *Xeromyces bisporus*, *Chrysosporium fastidium*, *Basipetospora halophila*, and *Xeromyces bisporus*. Media containing high concentrations (18-20%) of glycerol or sugars (glucose or sucrose) are often used for the propagation of these species [12]. Contrary to the molds, the osmophilic yeasts do not have a requirement for a low water activity environment for optimal growth [47].

VI. PRESERVATIVES

The principal preservatives that are added to fruit products are sulfur dioxide, sorbic acid, and benzoic acid. A common property of all three is that it is the undissociated acid that is the active, germicidal form. This means that the preservative becomes more effective as the pH is reduced (Table 5). Sorbic acid with a pK of 4.76 is considerably more active at higher pH values than is benzoic acid, which has a pK 4.19. For example, a fruit juice beverage of pH 3.8 treated with 300 mg/l preservative would contain 270 mg/l active sorbic acid compared with only 213 mg/l active benzoic acid.

TABLE 5. Percentage of Microbiologically Active Inhibitors at Different pH Values[a]

pH	Sulfur dioxide	Benzoic acid	Sorbic acid
3.0	9.1	94	98
3.1	7.4	92	98
3.2	5.9	91	97
3.3	4.8	88	97
3.4	3.8	86	96
3.5	3.1	83	95
3.6	2.5	80	94
3.7	2.0	76	92
3.8	1.6	71	90
3.9	1.2	66	88
4.0	1.0	61	85

[a]Henderson-Hasselbalch calculations based on pKa of the acids.

A. Sulfur Dioxide

Sulfur dioxide is used in the processing of a variety of fruit products. In addition to its ability to destroy microorganisms, it is used to prevent enzymatic and nonenzymatic browning. In general, it appears that molds are more sensitive than are yeasts to this preservative.

Dried or dehydrated fruits such as apricots, peaches, pears, apples, and grapes (raisins) contain concentrations of sulfur dioxide ranging from 800 to 2000 mg per kg. High levels also are present in the sulfite liquor that is used to preserve fruits that later will be remanufactured into jams, marmalades, candied and glace fruits. In the manufacture of maraschino cherries, concentrations as high as 15,000 mg per liter may be used because with this food another function of the sulfur dioxide is to bleach the fruit prior to coloring with a red dye.

A commercial process using sulfur dioxide exists for the preservation of grape juice that later will be used in wine making. Juice in large tanks is treated with liquefied sulfur dioxide gas to produce a final concentration of 1000-1200 mg/liter. When the juice is to be fermented, it is heated to 110°C to strip off the sulfur dioxide which is collected in calcium carbonate solution.

Grape juice can be preserved by much lower concentrations of sulfur dioxide when stored at a refrigerator temperature [1]. Juice samples treated with 200 mg per liter and held at 5°C failed to permit growth of yeasts and molds except when the initial population exceeded 10^7 per ml.

In winemaking, sulfur dioxide is added to the fruit prior to pressing to destroy wild yeasts and to reduce the amount of oxidation. Most wild yeasts are quite sensitive to sulfur dioxide, and it has been our observation that 100 mg per liter will reduce the number of viable yeasts by 4-5 $\log_{10}$ cycles during an exposure period of 24 hours.

At present, the total amount of sulfur dioxide permitted in wines in the United States is 350 mg per liter. Much of this would be combined with acetaldehyde, pyruvic acid, and other compounds and thus would not be in the active, free form. As a result, the maximal permitted concentrations will usually not protect a sweet wine from refermentation by yeasts.

There currently is a tendency to use less sulfur dioxide in fruit products or to find substitute compounds. This is because it has been recognized in recent years that certain asthmatics react violently to the compound and a number of deaths have been recorded [48]. To date a single compound that possesses both the antioxidative and fungicidal properties of sulfur dioxide remains to be discovered.

B. Benzoic and Sorbic Acids

These lipophilic acids appear to function in a similar manner. Thus fungi that are resistant to one of the compounds also will be resistant to the other [49]. The effectiveness of the two as preservatives is quite similar if one takes into consideration the molar concentrations of undissociated acid that are present in a given food system. In general, when a mixture of the two is used in a food, the effect is additive rather than synergistic.

A number of hypotheses have been advanced to explain the inhibitory activity of benzoic and sorbic acids [50,51]. It is believed that they accumulate on the cytoplasmic membrane and inhibit growth of fungi by blocking substrate transport and by uncoupling oxidative phosphorylation from the electron transport system [52]. It also has been shown that specific enzymes such as enolase [53], catalase [54], and coenzyme A [55] can be affected. It appears, therefore, that inhibition may be due to a number of mechanisms.

The acids are used to preserve a wide variety of fruit products including jams and jellies, soft drinks, high moisture prunes and raisins, refrigerated fruit salads, and wines [6,56-58]. The amounts used usually range from 200 to 1000 ppm, depending upon the food. Sweet cider may contain 1000 ppm, dried fruits 200-500 ppm, and wines 200 ppm or less. Low levels of sorbic acid are effective in wines because of the synergistic activity between the preservative and ethanol [59].

The fact that benzoic and sorbic acids impart off-flavors to fruit products has dictated, in part, the amounts that are used. Trained taste panels, for example, have been able to detect as little as 135 mg per liter of sorbic acid in table wines [56]. Sometimes the effects on flavor can be minimized, when permitted by law, by using a combination of two inhibitors.

Species and even strains of fungi may differ significantly in their resistance to these preservatives [49]. Some of the most tolerant are the yeasts *Zygosaccharomyces bailii* (formerly *Saccharomyces acidifaciens*) and *Zygosaccharomyces bisporus* [59,60]. Certain strains of *Z. bailii* have been found to be able to grow in pH 3 media containing 1000 ppm of sodium benzoate [61]. The mechanism for this resistance has been shown to be an inducible energy-requiring system which transports the preservatives out of the cell [62].

VII. ALCOHOLIC FERMENTATIONS

Most fruit juices present an excellent medium for yeasts due to their low pH and high content of fermentable sugars. Although alcoholic beverages can be made from many different fruits, those most popular in the Western world are wine from grapes and cider from apples. In this section some of the mycological considerations will be reviewed. Numerous publications describe wine and cider making in great detail [5,63].

A. Growth and Ethanol Production

Grape juice contains sufficient glucose, fructose, nitrogen compounds, growth factors, and minerals to permit the transformation of the juice into table wine containing 12% or more ethanol. Apple and many other fruit juices often contain less fermentable sugar and may be deficient in available nitrogen. It is not uncommon to supplement them with different yeast nutrients in order to carry out a complete fermentation.

During the course of the fermentation, yeast populations will reach figures of 10^8 per ml. Multiplication generally stops relatively early in the fermentation, often when only about one third of the sugar in grape juice has been converted to ethanol [64]. It is believed that the sterols required by yeasts for growth under anaerobic conditions become the limiting factor once oxygen has been depleted [65,66].

The theoretical alcohol yield from the hexose sugars is 51.1%. This is never achieved, however, because some of the carbon is converted into yeast cellular material as well as into trace amounts of other fermentation products. A common yield is 1% by volume of alcohol from 16 to 17 g per liter of sugar fermented [63].

B. Fermentation Products

By definition, wine yeasts are those that can tolerate the levels of ethanol that are commonly present in table wines. Most are strains of *Saccharomyces cerevisiae.*

A wide variety of yeast species can be recovered from grapes taken from the vineyard [67]. Often the wine yeasts are outnumbered by types that possess little ethanol tolerance. Most wineries inoculate their musts with established wine strains of *Saccharomyces cerevisiae,* although some winemakers believe that the most desirable flavors are produced when the fermentation is carried out by a natural mixed flora. In the latter situation, the inoculum usually is introduced from various contaminated surfaces within the winery. In these natural fermentations, *Kloekera apiculata, Torulopsis stellata,* and other species will be active along with *Saccharomyces cerevisiae* during the early stages [68]. Even when a pure culture inoculum is used, other species such as *Kloekera apiculata, Candida* sp., and *Hansenula anomala* may multiply and ferment until the concentration of ethanol becomes inhibitory [69].

The principal fermentation products are, of course, ethanol and carbon dioxide. Not all wine yeasts are highly ethanol tolerant; fermentation by some is arrested when the level of alcohol reaches as little as 7% by volume [70]. Glycerol is the next most important product resulting from the fermentation of hexose sugars. The amounts found in different wines range from 2 to 20 g per liter [71], or when related to ethanol, 7-8 g are produced per 100 g of alcohol [72]. The amount formed is influenced by yeast strain and fermentation conditions [73]. In addition to being slightly sweet, it is claimed that glycerol may affect the mouth feel of a wine.

Other by-products of fermentation are 2,3-butanediol and various higher alcohols [71]. The amounts produced are related to the yeast strain. The concentrations of 2,3-butanediol that have been found in wine range from 250 to over 1600 mg per liter. The main higher alcohols (fusel oils) are 1-propanol, isobutyl, active amyl, and isoamyl. Isoamyl is usually present in the highest concentrations with the values ranging from 80 to 400 mg per liter. Taste tests have shown that 300 ppm of isoamyl alcohol can be detected in wine, whereas the taste threshold values for isobutanol and 1-propanol exceed 500 ppm [70].

Yeasts produce varying amounts of acetic and succinic acids in wine and degrade varying amounts of malic acid. Succinic acid is the main carboxylic acid with as much as 2 g per liter being produced during fermentation. Its formation is influenced by the yeast strain and the amount of glutamate that is present [74]. Most wine yeasts produce less than 0.03% acetic acid during the alcoholic fermentation [71]. However, certain strains that have high aldehyde dehydrogenase activity may generate as much as 0.12%, a quantity which would result in the wine being spoiled [72].

Strains of *Saccharomyces* can decompose from 3 to 45% of the malic acid that is present in grape juice [70], while *Zygosaccharomyces bailii* and *Schizosaccharomyces pombe* metabolize it completely. Yeasts oxidatively decarboxylate malate to pyruvate, which in turn is decarboxylated to acetaldehyde, which then is reduced to ethanol. The end result is that one molecule of ethanol plus two molecules of carbon dioxide are produced from one molecule of malate [74].

Other important characteristics of wine yeasts are the amounts of hydrogen sulfide and sulfur dioxide that they produce. The concentration of hydrogen sulfide formed in a wine can range from undetectable levels to over 600 ppb [75]. It is commonly produced when the must contains elemental sulfur, a form introduced via vineyard sprays. Reducing substances formed by the yeasts cause a nonenzymatic reduction of the elemental sulfur [76,77]. Some yeasts also form hydrogen sulfide from sulfate and sulfite, a property that appears to be strain specific [78].

Although most yeasts produce only traces of sulfur dioxide in wines, a few strains generate as much as 80 ppm [79]. This has practical importance in that sulfur dioxide must be declared on the label if the wine contains more than 10 ppm.

C. Other Properties of Yeasts

In addition to fermentation products, yeasts possess a number of other variable products that are important to the winemaker.

Foaming and flocculation are two examples [80]. Foaming is important because the production of excessive amounts may result in the loss of juice unless larger fermentation tanks are used. Good flocculation, on the other hand, facilitates wine clarification through easier racking and filtration operations. Foaming appears to be related to the presence of hydrophobic proteins on the surface of the yeast. When these proteins are masked by mannan, the yeasts produce little or no foam [81]. It is believed that flocculation is related to anionic groups on the cell surface that react with divalent cations to form salt bridges. Both foaming and flocculation are genetically controlled [82,83].

In recent years there has been interest in killer yeasts, organisms that secrete a protein or glycoprotein that is lethal to closely related sensitive strains and species. Generation of the toxins is dependent upon the presence of double-stranded RNA. In addition to *Saccharomyces* species, killer toxins are generated by other genera such as *Candida, Cryptococcus,* and *Debaromyces.* Killer yeasts have been shown to be responsible for sluggish wine fermentations [84]. The killer factor has been bred into wine yeasts [85], but many wild yeasts along with certain strains of *Saccharomyces cerevisiae* are resistant [86]. Thus, the usefulness of killer yeasts in wine making remains to be established [72].

VIII. CONCLUSIONS

Most fruits and fruit products are an excellent growth medium for fungi due to their high sugar and acid content. Although certain aciduric bacteria can grow in these foods, fungi usually predominate when spoilage occurs or when intentional fermentations are carried out.

Yeasts and molds are widely distributed in nature. They can be readily found in orchards and vineyards, as well as on the surfaces of equipment used for processing fruits. It is important, therefore, that sound fruit be used and that processing equipment be maintained in a sanitary condition.

Most fungi possess little heat resistance and, as a result, pasteurization at temperatures of under 100°C usually will suffice for different canned fruit products. The ascospores of a few mold species will survive pasteurization, however, and at times have been responsible for spoilage outbreaks.

Certain species of yeasts and molds have the ability to grow at low water activities of 0.85 and below. These fungi have been responsible for the spoilage of high sugar fruit products such as preserves and juice concentrates.

The most widely used preservatives for food products are sulfur dioxide, benzoic acid, and sorbic acid. The effectiveness of all is increased as the pH is reduced because it is the undissociated acids that are the germicidal form. In general, the concentrations of acids that can be added to fruit products will extend shelf life but will not completely prevent spoilage of processed fruits.

The alcoholic fermentation of fruit juices by yeasts represents a beneficial role of fungi in fruits. The yeasts used in cider and winemaking vary in a number of important properties that markedly influence the quality of the finished beverage. The beverage maker usually tries to control the fermentation by using a heavy inoculum of a desired yeast strain.

REFERENCES

1. Splittstoesser, D. F., and Mattick, L. R. The storage life of refrigerated grape juice containing various levels of sulfur dioxide. *Am. J. Enol. Vitic.*, *32*:171-173, 1981.
2. Swanson, K. M. J., Leasor, S. B., and Downing, D. L. Aciduric and heat resistant microorganisms in apple juice and cider processing operations. *J. Food Sci.*, *50*:336-339, 1985.
3. Splittstoesser, D. F., Groll, M., Downing, D. L., and Kaminski, J. Viable counts versus the incidence of machinery mold (*Geotrichum*) on processed fruits and vegetables. *J. Food Protect.*, *40*:402-405, 1977.
4. Murdock, D. I., and Brokaw, C. H. Sanitary control in processing citrus concentrates. 1. Some specific sources of microbial contamination from fruit bins to extractors. *Food Technol.*, *12*:573-576, 1958.
5. Beech, F. W., and Carr, J. G. Cider and perry. In *Alcoholic Beverages* (A. H. Rose, ed.), Academic Press, London, 1977, pp. 139-313.
6. Berry, J. M. Yeast problems in the food and beverage industry. In *Food Mycology* (M. E. Rhodes, ed.), G. K. Hall, Boston, 1979, pp. 82-90.
7. Pederson, C. S., Albury, M. N., Wilson, D. C., and Lawrence, N. L. The growth of yeasts in grape juice stored at low temperatures. I. Control of yeast growth in commercial operations. *Appl. Microbiol.*, *7*:1-6, 1959.

8. Howard, B. J. Tomato ketchup under the microscope with practical suggestions to insure a cleanly product. USDA Circular No. 68, 1911.
9. Anon. *Official Methods of Analysis*, 14th ed. Arlington, Association of Official Analytical Chemists, 1984, pp. 928-935.
10. Cichowicz, S. M., and Eisenberg, W. V. Collaborative study of the determination of *Geotrichum* mold in selected canned fruits and vegetables. *J. Assoc. Offic. Anal. Chem.*, *57*:957-960, 1974.
11. Eisenberg, W. V., and Cichowicz, S. M. *Geotrichum*: Machinery mold. In *Food Mycology* (M. E. Rhodes, ed.), G. K. Hall, Boston, 1979, pp. 91-99.
12. King, A. D., Jr., Pitt, J. I., Beuchat, L. R., and Corry, J. E. L. *Methods for the Mycological Examination of Food*, Plenum Press, New York, 1986.
13. King, A. D., Jr., Hocking, A. D., and Pitt, J. I. Dichloran-rose bengal medium for enumeration and isolation of molds from foods. *Appl. Environ. Microbiol.*, *37*:959-964, 1979.
14. Splittstoesser, D. F., Einset, A., Wilkison, M., and Preziose, J. Effect of food ingredients on the heat resistance of *Byssochlamys fulva* ascospores. In *Proc. 4th Intl. Congress Food Sci. and Technol.*, CSIC, Valencia, 1974, pp. 79-85.
15. Cousin, M. A., Zeidler, C. S., and Nelson, P. E. Chemical detection of mold in processed foods. *J. Food Sci.*, *49*:439-445, 1984.
16. Davies, A. M. C., Dennis, C., Grant, A., Hall, M. N., and Robertson, A. Screening of tomato puree for excessive mould content by near infrared spectroscopy: A preliminary evaluation. *J. Sci. Food Agric.*, *39*:349-355, 1987.
17. Notermans, S., Heuvelman, C. J., Van Egmond, H. P., Paulsch, W. E., and Besling, J. R. Detection of mold in food by enzyme-linked immunosorbent assay. *J. Food Protect.*, *49*:786-791, 1986.
18. Kupina, S. A. Simultaneous quantitation of glycerol, acetic acid, and ethanol in grape juice by high performance liquid chromatography. *Am. J. Enol. Vitic.*, *35*:59-62, 1984.
19. Ravji, R. G., Rodriguez, S. B., and Thornton, R. J. Glycerol production by four common grape molds. *Am. J. Enol. Vitic.*, *39*:77-82, 1988.
20. Fields, M. L. Voges-Proskauer test as a chemical index to the microbial quality of apple juice. *Food Technol.*, *16*:98-100, 1962.
21. Hatcher, W. S., Hill, E. C., Splittstoesser, D. F., and Weihe, J. L. Fruit beverages. In *Compendium of Methods for the Microbiological Examination of Foods* (M. L. Speck, ed.), Amer. Pub. Health Assoc., Washington, D.C., 1984, pp. 644-650.
22. Fields, M. L. Acetylmethylcarbinol and diacetyl as chemical indexes of microbial quality of apple juice. *Food Technol.*, *18*:1224-1228, 1964.
23. Lopez, A. *A Complete Course in Canning. III. Processing Procedures for Canned Food Products*, 12th ed., Canning Trade, Baltimore, 1987, pp. 143-250.
24. Beuchat, L. R. Thermal inactivation of yeasts in fruit juices supplemented with food preservatives and sucrose. *J. Food Sci.*, *47*:1679-1682, 1982.

25. Corry, J. E. L. The effect of sugars and polyols on the heat resistance and morphology of osmophilic yeasts. *J. Appl. Bacteriol.*, *40*: 269-276, 1976.
26. Jermini, M. F. G., and Schmidt-Lorenz, W. Heat resistance of vegetative cells and asci of two *Zygosaccharomyces* yeasts in broths at different water activity values. *J. Food Protect.*, *50*:835-841, 1987.
27. Splittstoesser, D. F., Leasor, S. B., and Swanson, K. M. J. Effect of food composition on the heat resistance of yeast ascospores. *J. Food Sci.*, *51*:1265-1267, 1986.
28. Scott, V. N., and Bernard, D. T. Resistance of yeast to dry heat. *J. Food Sci.*, *50*:1754-1755, 1985.
29. Jensen, L. B. *Microbiology of Meats*, Garrard, Urbana, 1954.
30. Doyle, M. P., and Marth, E. H. Thermal inactivation of conidia from *Aspergillus flavus* and *Aspergillus parasiticus*. *J. Milk Food Technol.*, *38*:678-682, 1975.
31. Williams, C. C., Cameron, E. J., and Williams, O. B. A facultative anaerobic mold of unusual heat resistance. *Food Res.*, *6*:69-73, 1941.
32. King, A. D., Jr., Bayne, H. G., and Alderton, G. Nonlogarithmic death rate calculations for *Byssochlamys fulva* and other microorganisms. *Appl. Environ. Microbiol.*, *37*:596-600, 1979.
33. Splittstoesser, D. F. Fruits and fruit products. In *Food and Beverage Mycology* (L. R. Beuchat, ed.), Van Nostrand Reinhold, New York, 1987, pp. 101-128.
34. Fravel, D. R., and Adams, P. B. Estimation of U.S. and world distribution of *Talaromyces flavus*. *Mycologia*, *78*:684-686, 1986.
35. Hull, R. Study of *Byssochlamys* and control measures in processed fruits. *Anals. Appl. Biol.*, *26*:800-822, 1939.
36. King, A. D., jr., Michener, H. D., and Ito, K. A. Control of *Byssochlamys* and related heat-resistant fungi in grape products. *Appl. Microbiol.*, *18*:166-173, 1969.
37. Put, H. M. C. A selective method for cultivating heat resistant moulds, particularly those of the genus *Byssochlamys*, and their presence in Dutch soil. *J. Appl. Bacteriol.*, *27*:59-64, 1964.
38. Splittstoesser, D. F., Kuss, R. R., Harrison, W., and Prest, D. B. Incidence of heat resistant molds in eastern orchards and vineyards. *Appl. Microbiol.*, *21*:335-337, 1971.
39. Splittstoesser, D. F., Cadwell, M. C., and Martin, M. Ascospore production by *Byssochlamys fulva*. *J. Food Sci.*, *34*:248-250, 1969.
40. Samson, R. A., Hoekstra, E. S., and Van Oorshot, C. A. N. *Introduction to Food-borne Fungi*. Centraalbureau v. Schimmelcultures, Baarn, 1984.
41. Pitt, J. I., and Hocking, A. D. *Fungi and Food Spoilage*, Academic Press, New York, 1985.
42. Hsu, E. J., and Beuchat, L. R. Factors affecting microflora in processed fruits. In *Commercial Fruit Processing* (J. G. Woodruff and B. S. Luh, eds.), AVI, Westport, CT, 1986, pp. 129-161.
43. Davenport, R. R. Cold-tolerant yeasts and yeast-like organisms. In *Biology and Activities of Yeasts* (F. A. Skinner, S. M. Passmore, and R. R. Davenport, eds.), Academic Press, London, 1980, pp. 215-230.

44. Gunderson, M. F. Mold problem in frozen foods. In *Proc. Low Temperature Microbiology Symposium*, Campbell Soup, Camden, NJ, 1961, pp. 299-312.
45. Corry, J. E. L. Relationships of water activity to fungal growth. In *Food and Beverage Mycology* (L. R. Beuchat, ed.), Van Nostrand Reinhold, New York, 1987, pp. 51-99.
46. Pitt, J. I., and Christian, J. H. B. Heat resistance of xerophilic fungi based on microscopical assessment of spore survival. *Appl. Microbiol.*, *20*:682-686, 1970.
47. Tilbury, R. H. Xerotolerant (osmophilic) yeasts. In *Biology and Activities of Yeasts* (F. A. Skinner, S. M. Passmore, and R. R. Davenport, eds.), Academic Press, London, 1980, pp. 153-179.
48. Stevenson, D. D., and Simon, R. A. Sensitivity to ingested metabisulfites in asthmatic subjects. *J. Allergy Clin. Immunol.*, *68*:26-32, 1981.
49. Warth, A. D. Resistance of yeast species to benzoic and sorbic acids and to sulfur dioxide. *J. Food Protect.*, *48*:564-569, 1985.
50. Chipley, J. R. Sodium benzoate and benzoic acid. In *Antimicrobials in Foods* (A. L. Branen and P. M. Davidson, eds.), Marcel Dekker, New York, 1983, pp. 11-35.
51. Sofos, J. N., and Busta, F. F. Sorbates. In *Antimicrobials in Foods* (A. L. Branen and P. M. Davidson, eds.), Marcel Dekker, New York, 1983, pp. 141-175.
52. Freese, E., Sheu, C. W., and Galliers, E. Function of lipophilic acids as antimicrobial food additives. *Nature*, *241*:321-325, 1973.
53. Azukas, J. J., Costilow, R. N., and Sadoff, H. L. Inhibition of alcoholic fermentation by sorbic acid. *J. Bacteriol.*, *81*:189-194, 1961.
54. Troller, J. A. Catalase inhibition as a possible mechanism of the fungistatic action of sorbic acid. *Can. J. Microbiol.*, *11*:611-617, 1965.
55. Harada, K., Hizuchi, R., and Utsumi, I. Studies on sorbic acid. IV. Inhibition of the respiration in yeast. *Agr. Biol. Chem.*, *32*:940-946, 1968.
56. Ough, C. S., and Ingraham, J. L. Use of sorbic acid and sulfur dioxide in sweet table wines. *Am. J. Enol. Vitic.*, *11*:117-122, 1960.
57. Rushing, N. B., and Senn, V. J. Effect of preservatives and storage temperatures on shelf life of chilled citrus salads. *Food Technol.*, *16*:77-79, 1962.
58. Schade, J. E., Stafford, A. E., and King, A. D., Jr. Preservation of high-moisture dried prunes with sodium benzoate instead of potassium sorbate. *J. Sci. Food Agric.*, *24*:905-911, 1973.
59. Splittstoesser, D. F., Queale, D. T., and Mattick, L. R. Growth of *Saccharomyces bisporus* var. *bisporus*, a yeast resistant to sorbic acid. *Am. J. Enol. Vitic.*, *29*:272-276, 1978.
60. Cole, M. B., Franklin, J. G., and Keenan, M. H. J. Probability of growth of the spoilage yeast *Zygosaccharomyces bailii* in a model fruit juice system. *Food Microbiol.*, *4*:115-119, 1987.
61. Jermini, M. F. G., and Schmidt-Lorenz, W. Activity of Na-benzoate and ethyl paraben against osmotolerant yeasts at different water activity values. *J. Food Protect.*, *50*:920-927, 1987.

62. Warth, A. D. Mechanisms of resistance of *Saccharomyces bailii* to benzoic, sorbic and other weak acids used as food preservatives. *J. Appl. Bacteriol.*, *43*:215-230, 1977.
63. Amerine, M. A., Berg, H. W., Kunkee, R. E., Ough, C. S., Singleton, V. L., and Webb, A. D. *The Technology of Winemaking*, AVI Publ. Co., Westport, CT, 1980.
64. Amerine, M. A., and Joslyn, M. A. *Table Wines. The Technology of Their Production*, University of California Press, Berkeley, 1970.
65. LaRue, F., LaFon-LaFourcade, S., and Riberau-Gayon, P. Relationship between the sterol content of yeast cells and their fermentation activity in grape must. *Appl. Environ. Microbiol.*, *39*:808-811, 1980.
66. Ohta, K., and Hayashida, S. Role of tween 80 and monoolein in a lipid-sterol-protein complex which enhances ethanol tolerance of sake yeasts. *Appl. Environ. Microbiol.*, *46*:821-825, 1983.
67. Mrak, E. M., and McClung, L. S. Yeasts occurring on grapes and grape products in California. *J. Bacteriol.*, *40*:395-407, 1940.
68. Fleet, G. H., Lafon-LaFourcade, S., and Ribereau-Gayon, P. Evolution of yeasts and lactic acid bacteria during fermentation and storage of Bordeaux wines. *Appl. Environ. Microbiol.*, *48*:1034-1038, 1984.
69. Heard, G. M., and Fleet, G. H. Growth of natural yeast flora during the fermentation of inoculated wines. *Appl. Environ. Microbiol.*, *50*: 727-728, 1985.
70. Rankine, B. C. The importance of yeasts in determining the composition and quality of wines. *Vitis*, *7*:22-49, 1968.
71. Amerine, M. A., and Ough, C. S. *Methods for Analysis of Musts and Wines*. Wiley, New York, 1980.
72. Radler, F. Effect of yeast strains on fermentation. In *Proc. 2nd Intl. Symp. Cool Climate Viticulture and Oenology* (R. E. Smart, R. J. Thornton, S. B. Rodriguez, and J. E. Young, eds.), NZSOV, Auckland, 1988, pp. 298-303.
73. Radler, F., and Schutz, H. Glycerol production of various strains of *Saccharomyces*. *Am. J. Enol. Vitic.*, *33*:36-40, 1982.
74. Radler, F. Microbial biochemistry. *Experentia*, *42*:884-893, 1986.
75. Acree, T. E., Sonoff, E. P., and Splittstoesser, D. F. Effect of yeast strain and type of sulfur compound on hydrogen sulfide production. *Am. J. Enol. Vitic.*, *23*:6-9, 1972.
76. Schutz, M., and Kunkee, R. E. Formation of hydrogen sulfide from elemental sulfur during fermentation by wine yeast. *Am. J. Enol. Vitic.*, *28*:137-144, 1977.
77. Wainwright, T. Production of H_2S by yeasts: Role of nutrient. *J. Appl. Bacteriol.*, *34*:161-171, 1971.
78. de Mora, S. J., Eschenbruch, R., Knowles, S. J., and Spedding, D. J. The formation of dimethyl sulphide during fermentation using a wine yeast. *Food Microbiology*, *3*:27-32, 1986.
79. Suzzi, G., Romano, P., and Zambonelli, C. *Saccharomyces* strain selection in minimizing SO_2 requirement during vinification. *Am. J. Enol. Vitic.*, *36*:199-202, 1985.
80. Thornton, R. J. New yeast strains from old—the applications of genetics to wine yeasts. *Food Technol. Aust.*, *35*:46-50, 1983.
81. Suzzi, G., and Romano, P. Relation between foam and sulfur dioxide during the must fermentation. *Vini d'Italia*, *25*:157-162, 1983.

82. Thornton, R. J. Investigations on the genetics of foaming in wine yeasts. *Eur. J. Appl. Microbiol.*, *5*:103-107, 1978.
83. Thornton, R. J. The introduction of flocculation into a homothallic wine yeast. A practical example of the modification of winemaking properties by the use of genetic techniques. *Am. J. Enol. Vitic.*, *36*:47-49, 1985.
84. Vab Vuuren, H. J. J., and Wingfield, B. D. Killer yeasts—causes of stuck fermentations in a wine cellar. *S. Afr. J. Enol. Vitic.*, *7*: 113-118, 1986.
85. Hara, S., Iimura, Y., and Otsuka, K. Breeding of useful killer wine yeasts. *Am. J. Enol. Vitic.*, *31*:28-33, 1980.
86. Heard, G. M., and Fleet, G. H. Occurrence and growth of killer yeasts during wine fermentation. *Appl. Environ. Microbiol.*, *53*: 2171-2174, 1987.
87. Beuchat, L. R. Synergistic effects of potassium sorbate and sodium benzoate on thermal inactivation of yeasts. *J. Food Sci.*, *46*:771-777, 1981.
88. Put, H. M. C., and DeJong, J. The heat resistance of ascospores of four *Saccharomyces* species isolated from spoiled heat-processed soft drinks and fruit products. *J. Appl. Bacteriol.*, *52*:235-243, 1982.
89. Marco, F., Jermini, G., and Schmidt-Lorenz, W. Heat resistance of vegetative cells and asci of two *Zygosaccharomyces* yeasts in broths at different water activity values. *J. Food Sci.*, *51*:1506-1510, 1986.
90. Beuchat, L. R. Extraordinary heat resistance of *Talaromyces flavus* and *Neosartorya fischeri* isolated from commercial fruit juices. *J. Food Sci.*, *51*:1506-1510, 1986.
91. Scott, V. N., and Bernard, D. T. Heat resistance of *Talaromyces flavus* and *Neosartorya fischeri* isolated from commercial fruit juices. *J. Food Protect.*, *50*:18-20, 1987.
92. Splittstoesser, D. F., and Splittstoesser, C. M. Ascospores of *Byssochlamys fulva* compared with those of a heat resistant *Aspergillus*. *J. Food Sci.*, *42*:685-688, 1977.
93. King, A. D., Jr., and Halbrook, W. U. Ascospore heat resistance and control measures for *Talaromyces flavus* isolated from fruit juice concentrate. *J. Food Sci.*, *52*:1252-1254, 1266, 1987.
94. Bayne, H. G., and Michener, H. D. Heat resistance of *Byssochlamys* ascospores. *Appl. Environ. Microbiol.*, *37*:449-453, 1979.
95. Beuchat, L. R., and Toledo, R. T. Behavior of *Byssochlamys nivea* ascospores in fruit syrups. *Trans. Br. Mycol. Soc.*, *68*:65-71, 1977.
96. Conner, D. E., and Beuchat, L. R. Heat resistance of ascospores of *Neosartorya fischeri* as affected by sporulation and heating medium. *Intl. J. Food Microbiol.*, *4*:303-312, 1987.
97. Troller, J. A., and Christian, J. H. B. *Water Activity and Food*, Academic Press, New York, 1979.

8

CULTIVATED MUSHROOMS

SHU-TING CHANG *The Chinese University of Hong Kong, Shatin, Hong Kong*

I. INTRODUCTION

Mushrooms are generally considered as a special group of higher fungi, which are fleshy macrofungi with distinctive fruiting bodies bearing and discharging billions of spores. People are usually only interested in mushrooms that are edible and delicious; however, most mushrooms are unpalatable, and others even poisonous. Out of 250,000 species of fungi there are about 10,000 species of fleshy macrofungi and only a handful are lethal. There are no simple ways of distinguishing between the edible and the poisonous. You should eat a mushroom only if you know its name with considerable precision. About 2,000 species from more than 30 genera are regarded as prime edible mushrooms but only about 80 of them are experimentally grown, 40 economically cultivated, around 20 commercially cultivated, and only 5 to 6 have reached an industrial scale in many countries.

The word "mushroom" may mean different things in different countries. In this chapter, it refers to both wild and cultivated mushrooms that are edible. Those mushrooms that cannot be cultivated with present knowledge and technology, e.g., *Termitomycetes* spp. and *Tricholoma* spp., although they are edible and delicious, will be designated "wild mushrooms." Mushrooms that can be cultivated will be referred to as "cultivated mushrooms" or "edible mushrooms" and poisonous mushrooms called simply "poison mushrooms" or "toadstools."

Mushrooms were known and consumed as far back as prehistoric times, when most collectors had no knowledge of mycology or even of biology. The Greeks regarded mushrooms as providing strength for warriors in battle. The Pharoahs prized mushrooms as delicacies, and the Romans regarded mushrooms as the "food of the gods," served only on festive occasions. The annual world production of cultivated mushrooms was 1060 and 2182 thousand tons in 1978 and 1986 (Table 1), respectively. During this period of 8 years, production increased by 105%, with an average annual increase rate of about 13%.

TABLE 1. World Production of Edible Cultivated Mushrooms in 1986

Species	Common name	Metric tons[a] (× 1000)	%
Agaricus bisporus	Buttom mushroom	1227	56.2
bitorquis		314	14.4
Lentinus edodes	Shiitake or Oak mushroom	314	14.4
Volvariella volvacea	Straw mushroom	178	8.2
Pleurotus spp.	Oyster mushroom	169	7.7
Auricularia spp.	Wood Ear mushroom	119	5.5
Flammulina velutipes	Winter mushroom	100	4.6
Tremella fuciformis	White Jelly fungus/or "Silver Ear" mushroom	40	1.8
Pholiota nameko	"Nameko" or Viscid mushroom	25	1.1
Others including *Kuehneromyces mutabilis, Hericium erinaceus*, etc.		19	9,5
Total		2182	100.0

[a]Fresh equivalent weight.

Source: Ref. 1.

The diet of about one third of the world's population is deficient in protein. Edible mushrooms are considered to be a health food because their mineral content is higher than that of meat or fish and almost twice that of any vegetables. Furthermore, the protein content of fresh mushroom is about twice that of vegetables and four times that of oranges. Mushroom protein contains all nine amino acids essential to the diet of humans, and mushrooms are especially rich in lysine and leucine, which are lacking in most staple cereal foods (Table 2). Mushrooms also contain many vitamins (Table 3) and are devoid of starch and low in calories and carbohydrates. Mushrooms, therefore, are an ideal food for diabetics and for anyone counting calories. Mushroom protein can be produced with less land and with greater biological efficiency than proteins from animal sources. The overall nutritive value of mushrooms is somewhere between vegetables and meats (Table 4). With respect to essential amino acid indices and amino acid scores, the nutritive value of high grade mushrooms almost equals that of milk.

In view of current energy, food, and population problems, food and feed will have to be produced from waste materials on a massive scale within a few decades. Recently, many aspects of food protein production from wastes by yeasts, algae, and bacteria were considered by Birth et al. [6] and Tannenbaum and Wang [7]. Mushrooms and other macromycetes, the microbial foods used longest by man, can flourish successfully on a wide variety of inexpensive substrates/wastes, such as cereal straws, banana

TABLE 2. Comparison of Amino Acid Composition of Edible Mushrooms (g/100 g protein)

Amino acids	*A. bisporus*	*L. edodes*	*V. volvacea*	*Pleurotus*
Essential				
Isoleucine	4.3	4.4	4.2	4.0
Leucine	7.2	7.0	5.5	7.6
Lysine	10.0	3.5	9.8	5.0
Methionine	Trace	1.8	1.6	1.7
Phenylalanine	4.4	5.3	4.1	4.2
Threonine	4.9	5.2	4.7	5.1
Valine	5.3	5.2	6.5	5.9
Tyrosine	2.2	3.5	5.7	3.5
Tryptophan	ND	ND	1.8	1.4
Total	38.3	35.9	43.9	39.3
Nonessential				
Alanine	9.6	6.1	6.3	8.0
Arginine	5.5	7.0	5.3	6.0
Aspartic acid	10.7	7.9	8.5	10.5
Cystine	Trace	ND	ND	0.6
Glutamic acid	17.2	27.2	17.6	18.0
Glycine	5.1	4.4	4.5	5.2
Histidine	2.2	1.8	4.1	1.8
Proline	6.1	4.4	5.5	5.2
Serine	5.2	5.2	4.3	5.4
Total	61.6	64.0	56.1	60.7

Source: Ref. 2.

leaves, sawdust, and cotton wastes from textile factories. Cellulose, hemicellulose, and lignin are the main components of these wastes and are most resistant to biological degradation. However, mushrooms and a few other closely related organisms possess the enzyme complexes which enable them to attack and degrade these industrial and agricultural by-products. The great value in promoting cultivation of mushrooms also lies in their ability to grow on cheap carbohydrate materials and to transform various waste materials which are inedible by humans into a highly valued food protein for direct consumption. This is extremely important in rural areas where there is an enormous quantity of waste that has been found to be ideal as growing substrates for some edible mushrooms. Finally, the spent compost, which is the substrate left after mushroom harvesting, can be converted into stockfeed and plant fertilizer as a soil conditioner [8]. It is obvious

TABLE 3. Nutritional Values of Five Popular Edible Mushrooms

	Agaricus bisporus	*Lentinus edodes*	*Volvariella volvacea*	*Pleurotus ostreatus*	*Flammulina velutipe*
Protein, % (dry weight)	23.9-34.8	13.4-17.5	29.5-30.1	10.55-30.4	17.6-21
Calories, K Cal/ 100 g (dry weight)	328-368	387-392	338-374	345-367	378-39
Vitamins, mg/100 g (dry weight)					
Thiamin	1.0-8.9	7.8	1.2	4.8	6.1
Riboflavine	3.7-5.0	4.9	3.3	4.7	5.2
Niacin	42.5-57.0	54.9	91.9	108.7	106.5
Ascorbic acid	26.5-81.9	0	20.2	0	46.3
Ascorbic acid mg/100 g (fresh weight)	1.8	9.4	1.4	7.4 (*P. sajor-caju*)	
Fat, % (dry weight)	1.8-8.0	4.9-8.0	5.7-6.4	1.6-2.2	1.9

Source: Ref. 3.

TABLE 4. Comparison of Nutritive Value of Mushrooms with Various Foods

Essential amino acid indexes	Amino acid scores	Nutritional indexes
100 pork; chicken; beef	100 pork	59 chicken
99 milk	98 chicken; beef	43 beef
98 mushrooms (high)	91 milk	35 pork
96 *V. diplasia*	89 mushrooms (high)	31 pork
91 potatoes; kidney beans	71 *V. diplasia*	28 mushroom (high)
P. ostreatus	63 cabbage	27 *V. diplasia*
88 corn	59 potatoes	26 spinach
87 *A. bisporus*	*P. ostratus*	25 milk
86 cucumbers	53 peanuts	22 *A. bisporus*
79 peanuts	50 corn	21 kidney beans
76 spinach; soybeans	46 kidney beans	20 peanuts
74 *L. edodes*	42 cucumbers	17 cabbage
72 mushrooms	40 *L. edodes*	15 *P. osstreatus*
69 turnips	33 turnips	14 cucumbers

TABLE 4. (Continued)

53 carrots	32 mushrooms (low)	13
44 tomatoes	31 carrots	11 corn
	28 spinach	10 turnips
	23 soybeans	9 potatoes
	18 tomatoes	8 potatoes
		6 carrots
		5 mushrooms (low)

Source: Refs. 4, 5.

that mushroom cultivation opens the deadlock in the biological degradation of natural resources.

II. BIOLOGY OF CULTIVATED MUSHROOMS

Cultivated mushrooms are multicellular organisms. Most of them belong to the class Basidiomycetes. Species of the order Agaricales form the largest assemblage of edible mushrooms which are grown commercially for food.

Sporocaps (fruiting bodies) of members of Agaricales show great diversity in size, form, and consistency. In some species, such as *Volvariella volvacea* (Fig. 1) and *Dictyophora duplicata* (Fig. 2), the volva is very developed and remains distinct in the mature fruiting bodies. The volva is actually the ruptured universal veil of the young fruiting bodies. Other species like *Lentinus edodes* (Fig. 3) and *Pleurotus sajor-caju* (Fig. 4) lack the volva. However, all of them have a distinguishable pileus (cap) and stipe (stalk).

The vertical, radial plates on the lower surface of the pileus upon which the hymenium is extended are called lamellae or gills. The sexual spores, basidiospores, are produced on the gill surface. The spores can be white, pink, brown, and even black depending on the species and degree of maturity.

There are two main groups of sexual patterns in basidiomycetes. The common pattern of sexuality is heterothallism. Nearly 90% of all species investigated belong to this group [9], in which plasmogamy in cross-mating between different homokaryotic mycelia is necessary to establish the fertile dikaryotic mycelium. Homothallism occurs in a distinct minority of about 10% in which the mycelium forms a single germinated spore, which is self-fertile and has the competence to fruit in the absence of cross-mating. Although hyphal fusions occur both within and between homothallic mycelia, they are not essential to development of the fertile mycelium. Homothallism is considered as a sexual pattern because some sexual processes (karyogamy and meiosis) occur in the basidia of the fruiting bodies of homothallic species [10,11].

The patterns of sexuality are controlled by two mechanisms: (a) distribution of the four postmeiotic nuclei to the basidiospores, and (b) genetic factors of a mating system known as incompatibility factors. Under

FIGURE 1. Mature mushrooms of *Volvariella volvacea* cultivated on cotton waste compost showing the gills on the underside of pileus. The volva or universal veil is visible at the base of the mushroom in the left foreground.

control of these two mechanisms, the sexual patterns can be further divided into four types [12,13]. Primary homothallism involves a self-fertile mycelium that develops directly from a single spore with a single postmeiotic nucleus; presence of incompatibility factors has not yet been detected. Secondary homothallism has a distinctly different basis. It involves incompatible factors and is determined by the mechanism of nuclear distribution. The basidia of secondary homothallic species usually bear only two spores each, since two compatible postmeiotic nuclei migrate into each basidiospore. Therefore, when these spores germinate, the self-fertile mycelium is developed. In heterothallism, an incompatibility system always prevails, so when each spore receives a single postmeiotic nucleus there are two types of sexual patterns [9]: the unifactorial system in which sexuality is controlled by a single genetic factor, A, with multiple specificities (alleles) [6], and the bifactorial system in which sexuality is controlled by two unlinked genetic factors, A and B, each with multiple alleles. In Basidiomycetes, about 10% are homohallic, 25% are unifactorial heterothallic, and 65% are bifactorial heterothallic [9].

The characteristic rapid mycelial growth and high saprophytic colonization ability of most species of cultivated mushrooms make it possible to grow this group of fungi widely. In general, they have relatively simple nutritional requirements. This is because they have the metabolic ability to synthesize most of the complex compounds that living organisms require to carry out their growth activities. Therefore, their mycelia can grow on

FIGURE 2. Different stages of *Dictyophora duplicata*, the veiled lady mushroom, cultivated in a wooden box.

most of the common synthetic media as well as on corn cobs, wood shavings, cereal straw, and cotton wastes. Factors that influence mycelial growth are temperature, carbon dioxide, and oxygen concentration as well as pH of the substrates [3,14,15].

III. CONCEPT OF MUSHROOM CULTIVATION

There is tremendous appeal for a process which promises to produce a high nutritious food of excellent taste from materials without making extensive demands upon land or having requirements for expensive equipment. Mushroom cultivation is such a process at the conceptual level, and many people have been moved to undertake mushroom growing on a commercial basis because of this. Unfortunately, though simple in concept, mushroom growing is a complicated business, and when entered into by untrained individuals who are unaware of the various intricacies of the process, pitfalls are frequently encountered which commonly lead to failure of the venture. It is

FIGURE 3. *Lentinus edodes* grown on a compost of sawdust and used tea leaves.

FIGURE 4. Ready for harvesting. *Pleurotus sajor-caju* showing the one-sided growth of pileus which forms shell-shaped mushrooms.

especially distressing to hear of such misfortunes in developing countries where properly developed and managed mushroom farms can make important contributions to the nutrition and economic welfare of the people.

It has been pointed out previously by Miles and Chang [16] that mushroom cultivation can be a relatively primitive type of farming, or it can be a highly industrialized agricultural activity requiring a sizeable capital outlay for mechanized equipment. The straw mushroom, *Volvariella volvacea*, is commonly grown in southeast Asian countries on small, family-type farms. In contrast, cultivation of the *Agaricus* mushroom may be highly industrialized with a few farms producing a disproportionately large percentage of a country's yield as is true in many Western nations. In either event, the main objective of mushroom growers and researchers who work in this field is similar, namely, to increase the yield from a given surface area per period of time by use of a high yielding strain, by shortening the cropping period, or by increasing the number of high yielding flushes. To bring about this increase in yield requires an understanding of substrate materials and their preparation, selection of suitable media for spawn making, breeding of high quality and high yielding strains, and improved management of mushroom beds, including prevention of development of pests and mushroom diseases.

Thus, when one looks beyond the basic concept involved in growing mushrooms, one sees that there are a number of interesting operations, each of which can be rate limiting for total production. The successful grower requires the scientific knowledge that is provided through research and the practical experience obtained by personal participation involving observation of and training in the practices of mushroom cultivation. It is for this reason that I am outlining and describing the major steps of mushroom cultivation. People interested in establishing a mushroom farm/industry should be aware that it is a more complicated process than the layperson may realize, and those who are serious about establishing a mushroom farm should consult References 3, 14, and 15 about the various phases of mushroom technology and seek advice from experienced growers. The practices of mushroom cultivation, or mushroom technology, will be described as consisting of six major phases. These phases generally occur in the sequence shown in Table 5.

IV. MAJOR PHASES OF MUSHROOM CULTIVATION

Mushroom farming involves several different operations, each of which must be done properly if the enterprise is to be successful. Failure in any phase will result in a decreased harvest, at best, or a failure to achieve anything to harvest at worst. The different phases of mushroom technology that will be treated here are: (a) selection of an acceptable mushroom, (b) the requirement for and selection of a fruiting culture, (c) development of spawn, (d) preparation of compost, (e) mycelial (spawn) running, and (f) mushroom development. In addition, crop management and marketing are equally important operations.

TABLE 5. Major Phases of Mushroom Cultivation

Major phases	Main points to consider
Selection of an acceptable mushroom ↓	Location Climate Raw materials Acceptability
Selection of a fruiting culture ↓	Tissue culture Spore culture (a) without mating for homothallic sp. (b) mating with compatible isolates for heterothallic species Mixed culture Preservation
Development of spawn ↓ Spawning	Substrate Vigorous growth Free of contamination Avoid use of senescent and degenerate spawn Good survival of storage
Spawn running → Fructification (mushroom development) ↑ Composting	Establishment of mycelium Environmental requirements (a) temperature (b) light (c) aeration (O_2, CO_2) (d) pH (e) moisture Casing Watering and care
Preparation of compost	Concept of composting Microbial activity Softening of substrate for ease of colonization Physical characteristics Chemical components Aeration Water content

A. Selection of a Mushroom Species

Any food product has to be accepted by the indigenous people traditionally or through commercial promotion; otherwise, there is no market value. This also is true for mushroom production. If the mushroom is to be marketed fresh, it must be a species acceptable on the basis of its taste appeal to people in the area where it is cultivated. This can be determined for previously cultivated species by examination of import records, if available, or by testing for market acceptability with fresh mushrooms imported for that purpose. Once it has been determined that the species is acceptable to the local consumer, there remain other significant considerations before a decision to cultivate a particular species should be made. For example, are appropriate raw materials for substrate available? Is the environment such that it will not be excessively costly to maintain the necessary temperatures for mycelial running and for mushroom development? There is considerable variation among edible mushroom species in the temperatures suitable for vegetative growth (spawn running or mycelial running) and mushroom development. The temperature necessary for vegetative growth and for fruiting must be considered in selection of an acceptable mushroom (Tables 6 and 7), and it should be pointed out that strains or dikaryotic stocks of a species may differ in their required temperature ranges and optimal values so that even within a single species selections can be made.

B. Selection of a Fruiting Culture

After determination of a mushroom species that is acceptable as food to the indigenous population, for which suitable substrates are plentiful and the environmental requirements can be met without excessively costly systems of mechanical control, it is necessary to have a fruiting culture. A fruiting culture is one that can be used without further mating to make the spawn for mushroom production. The term "fruiting culture" is defined as a culture which has the genetic capacity to form fruiting bodies under suitable growing conditions. When grown under the proper conditions (conditions that will permit good vegetative growth and fruiting bodies without high costs for equipment and energy), this culture should produce fruiting bodies. With a heterothallic species the fruiting culture is a dikaryotic mycelium which was formed by a mating between two compatible single spore, monokaryotic isolates. With homothallic species, a single spore isolate is capable of forming fruiting bodies, and thus does not need to be mated with other isolates. Sometimes multispore cultures are used to obtain fruiting cultures of *Agaricus bisporus*, but this is not a suitable technique for heterothallic species. Tissue cultures derived from the stipe or pileus of the mushroom of either homothallic or heterothallic species can be used to establish fruiting cultures. Establishment of tissue cultures is the method used to isolate and propagate sporeless strains. Sporeless strains of species of *Pleurotus* are of great commercial interest, because these species shed spores early in the development of the fruiting body and continue to produce spores in abundance up to harvesting with the result that the spore density in the air in the mushroom houses becomes very heavy. Unfortunately, it is a common occurrence for mushroom workers under these conditions to suffer from respiratory tract problems and allergic reactions to these spores. Consequently, there is interest in developing sporeless mutants that will produce fruiting bodies equivalent to those of accepted commercial spore-forming stocks in yield, flavor, texture, fruiting time, etc. Obviously, a strain or stock that does not fruit, or fruits poorly, cannot be considered as a fruiting culture.

TABLE 6. Temperature Range, Substrate Type, Production-Cycle Time, and Approximate Yield of Edible Mushrooms from Nonaxenic Culture Methods

Species	Substrate	Temperature range		Production-cycle time	Yield[a]
		Mycelial growth	Fruiting		
Little or no pretreatment					
Lentinus edodes	Wood logs (outdoors, sometimes protected)	5-35 (24)[b]	6-25 (15) autumn (10) winter (20) spring	3-6 yr spring/autumn	40
Auricularia auricula	Wood logs (outdoors, sometimes protected)	15-34 (28)	15-28 (22-25)	2-5 yr spring/autumn	2-12
Auricularia polytricha	Wood logs (outdoors, sometimes protected)	10-36 (20-34)	15-28 (24-27)	1-2 yr	20-40
Tremella fuciformis	Wood logs	5-38 (25)	20-28 (20-24)	3-6 yr 7 months/yr	10-30
Some pretreatment					
Volvariella volvacea	(1) Rice straw (outdoor) (2) Cotton waste, rice straw (indoor)	15-45 (32-35) 15-45 (32-35)	22-38 (28-32) 22-38 (28-32	4-6 weeks 2-3 weeks	6-10 30-45
Pleurotus sajor-caju	(1) Pasteurized cereal straw (indoor) (2) Fermented cereal	14-32 (25-27) 14-32 (25-27)	10-26 (19-21) 10-26 (19-21)	4-10 weeks 4-10 weeks	80-100 or more 80-100 or more
Long composting process					
Agaricus bisporus	Composted cereal straw/ animal manure mixtures	3-32 (22-25)	9-22 (15-17)	14-16 weeks	65-80
Agaricus bitorquis	As above	3-35 (18-30)	18-25 (22-24)	14-16 weeks	40-65

[a]kg Fresh weight/kg d.m.

[b]Figures within parentheses are optimal values.

Source: Ref. 17.

TABLE 7. Temperature Range, Substrate Type, Production-Cycle Time, and Appropriate Yield of Edible Mushrooms from Axenic Culture Methods

Species	Substrate	Temperature range		Production-cycle time	Yield[a]
		Mycelial growth	Fruiting		
Flammulina velutipes	Sterilized sawdust, rice bran mixtures (polypropylene bottles)	3-34 (18-25)[b]	6-18 (8-12)	12-20 weeks	70-100
Lentinus edodes	As above (polythene bags)	5-35 (24)	6-25 (15) autumn (10) winter (20) spring	3-6 months	60-100
Auricularia auricula	As above (polythene bags)	15-34 (28)	15-28 (22-25)	8-10 weeks	20-25
Auricularia polytricha	As above (polythene bags)	10-36 (20-34)	15-28 (24-27)	6-8 weeks	70-85
Tremella fuciformis	As above (polythene bags)	5-38 (25)	20-28 (20-24)	6-8 weeks	80-100

[a]kg Fresh weight/100 kg d.m.

[b]Figures within parentheses are optimal values.

Source: Ref. 17.

C. Development of Spawn

A medium through which the mycelium of a fruiting culture has grown and which serves as the inoculum or "seed" for the substrate in mushroom cultivation is known as mushroom spawn. When the term "pure culture spawn" is used, it means that a strain or stock of known origin, free from contaminating organisms, is present. In spawn making, the entire process of preparation should be completed under aseptic conditions.

Failure to achieve a satisfactory harvest may frequently be traced to an unsatisfactory spawn. If the spawn has not been made from a genetically suitable fruiting culture, if a stock has degenerated, or if it is too old, the yield of mushrooms will be less than optimum. The successful spawn manufacturer must have a product which is a consistently good performer and thus has the confidence of the mushroom grower. For this a stable strain or stock possessing the genetic characteristics required by the growers is an absolute necessity. Ideal environmental conditions and management cannot overcome the limitations of a genetically inferior stock used to make spawn.

Although the potential of a spawn is set by the genetic constitution of the fruiting culture used in its manufacture, the substrate material is also very important. The substrates used in spawn manufacture may be different from the materials used in the cultivation of the mushroom, or they may be the same. Substrates may be used singly or in combination. Some of the substrates used in spawn making include various grains (rye, wheat, sorghum), rice straw cuttings, cotton waste, rice hulls, cotton seed hulls, etc. Spawns are frequently referred to as grain spawns or straw spawns, which tends to overemphasize the importance of the substrate, whereas, in reality, it is the strain or stock of the mushroom that is of prime importance in determining the merits of a spawn. The spawn substrate serves mainly as the vehicle that carries the vegetative mycelium of the mushroom that is used to inoculate the growing beds. This does not mean that the spawn substrate has no effect upon the success or failure of a spawn. The growth pattern of the mycelium may be influenced by the spawn substrate, as is seen by the more rapid growth and filling of the beds (spawn running) with some spawns than with others. The spawn manufacturer must also consider availability of the spawn substrate and, related to this, the cost.

In summary, in the manufacture of spawn, consideration must be given to the genetic capabilities of the fruiting culture: (a) for vegetative growth both in the spawn substrate and in the bed material following inoculation, and (b) for yield and quality of mushrooms produced. Consideration must also be given to the nature of the spawn material since this influences rapidity and thoroughness of mycelial growth in the spawn container as well as spawn running in the bed following inoculation. Considerations of availability and cost of substrate are also important. Some other obvious features of a good spawn include freedom from contamination, vigorous growth, and good survival in storage.

D. Preparation of Compost

In the *Agaricus* mushroom industry the process which renders horse manure suitable for growth of mushrooms is known as "composting." This is derived from the general term "compost," which refers to a mixture consisting largely of decayed organic matter that is used for fertilizing and conditioning land for horticultural purposes. It can be questioned whether the treatment of other substrates for growing should be regarded as "composting" since the starting materials and lengths of time for growth are not comparable, although the aims are similar, and we shall use the term "composting" in this general way. Consequently, we shall emphasize the purposes of composting and the general changes that are brought about by the process.

A substrate rich in available nutrients does not necessarily constitute a satisfactory medium for growing mushrooms. The reason for this is that the material first must be sterilized or else bacteria and molds will grow in abundance and curb growth of the mycelium and development of mushrooms. That is, when spawn is inoculated into a raw, unsterilized substrate, the naturally occurring microorganisms quickly gain dominance and retard or even prevent development of mushroom mycelium. The function of composting is to reverse that situation; i.e., to prepare a medium with characteristics that promote growth of mushroom mycelium to the practical exclusion of other organisms. To accomplish this, certain chemical properties and physical qualities must be built into the substrate. These features of the substrate are equally important and are interdependent.

When the best food materials that can serve the nutritional requirements of the mushroom have been accumulated, then the proper chemical state of the compost has been achieved. These nutrients must be in forms which are readily available to the mushrooms. For example, most mushroom species cannot use nitrates, and thus a substrate with nitrogen available only as nitrate would not be suitable. It is essential that there be nitrogen present in the compost materials in the form of protein. The compost also must be free of toxic substances which inhibit the growth of spawn. Such toxic substances may be produced by microorganisms present in nature which grow in the substrate during the composting process. Thus conditions which are unfavorable to the growth of these toxin-producing organisms must be employed during composting.

In regard to physical qualities of the substrate, such things as good aeration by the free admittance of air, ability to hold water without becoming waterlogged (which would curtail aeration), a proper pH, and good drainage are desirable.

Of course, the compost contains not only the substrate materials but also the microorganisms that bring about their breakdown and decay during the composting process. These microorganisms are essential for composting. The substrate during composting is never sterile but is literally teeming with millions of bacteria and fungi.

In practice, composting is accomplished by piling up the substrates for a period during which various changes take place so that the composted substrate is quite different from the starting material. A substrate consisting of agricultural and chemical materials other than horse manure, when composted, is called a "synthetic compost." Synthetic composts have been devised with numerous formulations of just about every type of agricultural waste product and residue. Synthetic composts comprise the general type of substrate used for growing most mushrooms and can also be used for *Agaricus*.

Straws and other plant wastes are mainly cellulose, hemicellulose, and lignin—the polysaccharides of plant cell walls. Some bacteria readily attack cellulose and hemicellulose and under suitable conditions easily bring about their decomposition. However, lignin is quite resistant to decomposition by bacteria. The easily decomposed carbohydrates serve as an excellent source of food for bacteria and fungi. Such carbohydrates diminish after composting as a consequence of the metabolic activities of the microorganisms in the compost, making the substrate less favorable for further growth of these potential competitors to the growth of mushrooms. The metabolic activities of the microorganisms also have some other effects: (a) conversion of simple nitrogenous materials such as nitrates and ammonia to complex proteins, thereby increasing the proteins required later for growth of the mushroom mycelium, and (b) a decrease in pH. The comprehensive results of research on mushroom composts have been extensively reviewed [18-23].

Principles of this complicated process of composting are known and guidelines are available, but in practice modifications are necessary to meet various situations that may be encountered. Such modifications might be necessitated by availability of raw materials, the microflora in the composting area, facilities in the growing area, and, especially important, the species of mushroom to be cultivated. The highly industrialized technology developed for cultivation of *Agaricus* cannot be followed unmodified for commercial production of the straw mushroom, *Volvariella volvacea*.

E. Spawn Running

Following composting of the substrate, the compost is put into beds where it is generally pasteurized by steam to kill the vegetative cells of microorganisms. After pasteurization, the bed temperatures must be allowed to fall below a certain level before the spawn is introduced into the bed as inoculum because high temperature damages the spawn. The amount of spawn used as inoculum per unit surface area is available, but, in general, larger amounts of spawn result in the bed becoming filled with mycelium more rapidly. Use of greater amounts of spawn increases production costs, however.

Brief mention will be made of the inoculation procedure (spawning) at this time. When the spawn is removed from the container, it is broken into small pieces by crushing and crumbling it with the fingers. Pieces of spawn may be broadcast over the bed surface and then pressed down firmly against the substrate to assure good contact, or they may be inserted 2-2.5 cm deep into the substrate.

Spawn running (mycelial running) is the phase during which mycelium grows from the spawn and permeates the substrate. Good mycelial growth is essential for mushroom production, and it is essential to maintain proper conditions of the beds and in the mushroom house during spawn running. The proper temperature and humidity for the species must be maintained. The bed surfaces must not be allowed to dry out, but they should be watered lightly with a fine water sprinkler. As the mycelium grows, it generates heat which contributes to water loss. Improper environmental control (of temperature, moisture, aeration) is a common cause of a poor spawn run. Mushroom house management is as important in this stage as it is in the following phase—mushroom development.

F. Mushroom Development

Under suitable environmental conditions, which are commonly different from those for spawn running, primordia formation occurs, followed by the production of fruiting bodies. Improper aeration, which leads to an increase in CO_2 in the vicinity of the mushroom beds where the mycelium is respiring, may inhibit primordia formation or later stages of mushroom development. Conditions of pH and temperature required for fruiting are also commonly different for fruiting than for mycelial running.

The appearance of mushrooms is commonly in rhythmic cycles called "flushes." Mushrooms are picked at different maturation stages depending upon the species and upon consumer preference and market value. The method of harvesting also varies with the species. Suitable temperature, humidity, and ventilation conditions must be maintained during the cropping period because these factors will affect the number of flushes and total yield that will be obtained.

V. CULTURE PRESERVATION

Cultures of cultivated mushrooms can be preserved either as spores or as vegetative mycelia. However, if single spore cultures are maintained, mating tests for heterothallic species would be required routinely as well as tests for fruiting ability; if spores are maintained for homothallic species, it would be necessary to check fertility of the culture by regular fruiting tests. Consequently, in practice vegetative mycelia of known origin are stored.

A. Short-Term Storage

The traditional method for maintaining mushroom cultures is by periodic transfer. Cultures maintained in this way are readily available upon subculturing, without any delay in recovering the culture as occurs in long-term storage in liquid nitrogen, for example.

Mycelial growth from the growing edge of the colony should be used in making subcultures. Deviations from the original cultures can be detected with mycelial transfers. The performance of the mycelium should be checked continuously, although not all degenerative symptoms can be detected in the mycelial stage. The degenerative symptoms commonly detected are sectors of slow growth, mycelium that is thin and of weak appearance, or mycelium that is matted or fluffy but has a normal growth rate. A slow-growing mycelium needs more time for colonization and tends to carry virus particles. A fluffy mycelium causes the grain to stick together and is harder to spread in the compost than normal grains. It tends to form "stroma," and it gives lower yields. Mycelia of these types should be discarded.

Laboratory media such as potato dextrose agar, malt agar, oatmeal agar, and Complete Agar Medium (0.5 g $MgSO_4 \cdot 7H_2O$; 0.46 g KH_2PO_4;

1 g K_2HPO_4; 2 g peptone; 20 g dextrose; 20 g agar; 2 g yeast extract; 0.5 mg thiamine HCl; 1 liter distilled water) are commonly used. As a general rule, a nutritionally weak medium is preferred because it lowers the metabolic rate of the organism and thus prolongs the period between transfers. However, it should be remembered that growth of a pure culture in laboratory medium on agar is not a natural condition, and to obtain the fullest development the conditions encountered in nature should be simulated as far as possible. Strains tend to become attenuated under artificial conditions of culture. It is known that the medium can act as a selective agent, and therefore, it can affect the frequency of appearance of some mutations, for example, the fluffy type of mycelium.

Tubes or culture bottles containing the subculture should have screw caps with liners and be sealed with paraffin film or aluminum foil (to avoid drying) and kept at a low temperature (ca. 5°C) to reduce the metabolic rate of the organism. Some mushrooms, however, are sensitive to chilling injury (e.g., *Volvariella volvacea*) and are best kept at 10-15°C [3].

B. Long-Term Storage

Long-term storage of mushroom cultures can be achieved by a single method or a combination of the following methods: starvation of nutrients, limitation of oxygen, lyophilization, and freezing. The rationale for these methods involves arrest or retardation of cellular metabolism. In practice, methods for long-term storage are usually adopted by research institutes and large spawn producers.

The small- and medium-size spawn makers or mushroom farmers only use the short-term methods to preserve their cultures. More detailed accounts of the storage of mushroom cultures, especially with the use of liquid nitrogen, can be found in the literature [24-26].

VI. CONCLUSIONS

I anticipate that future research will lead to an increase in the number of mushroom species that can be grown commercially, use of a wider variety of substrate materials, decreased cropping periods for a number of species, improved methods of protection against pests, and better techniques for preservation.

In spite of the many conceptual and technical problems in cultivation of mushrooms in tropical and subtropical regions, I foresee a more important role for them as a source of food protein to enrich human diet in those regions where the shortage of protein is marked. I also see no reason why both rural and urban areas cannot share in this new possibility and prospect. The introduction of new technology to maximize mushroom production per unit area at minimum cost so a cheap source of food protein derived from agricultural and industrial organic wastes in tropical and subtropical regions where these organic waste materials are abundant is a continuing challenge.

REFERENCES

1. Chang, S. T. World production of cultivated edible mushrooms in 1986. *Mush. J. Tropics*, *7*:117-120, 1987.
2. Chang, S. T. Mushrooms as human food. *BioScience*, *30*:399-401, 1980.
3. Chang, S. T., and Miles, P. G. *Edible Mushrooms and Their Cultivation*, CRC Press, Boca Raton, FL, 1989.
4. Crisan, E. V., and Sands, A. Nutritive value of edible mushrooms. In *The Biology and Cultivation of Edible Mushrooms* (S. T. Chang and W. A. Hayes, eds.), Academic Press, New York, 1978, pp. 137-165.
5. Li, G. S. F., and Chang, S. T. Nutritive value of *Volvariella volvacea*. In *Tropical Mushrooms—Biological Nature and Cultivation Methods* (S. T. Chang and T. H. Quimio, eds.), Chinese University Press, Hong Kong, 1982, pp. 199-219.
6. Birch, G. G., Paker, K. J., and Worgan, J. T. (eds.). *Food From Wastes*, Applied Science Publishers, London, 1976.
7. Tannenbaum, S. R., and Wang, D. I. C. (eds.). *Single Cell Protein II*, MIT Press, Cambridge, MA, 1975.
8. Chang, S. T., and Yau, C. K. Production of mushroom food and crop fertilizer from organic wastes. In *Proceedings of XIth International Conferences on Global Impacts of Applied Microbiology* (S. O. Emejuaiwe, O. Ogunbi, and S. O. Sanni, eds.), Academic Press, London, 1981, pp. 647-652.
9. Burnett, J. H. *Mycogenetics*, John Wiley and Sons, London, 1975.
10. Chang, S. T., and Ling, K. Y. Nuclear behaviour in the basidiomycete, *Volvariella volvacea*. *Am. J. Bot.*, *57*:165-171, 1970.
11. Chang, S. T., and Yau, C. K. *Volvariella volvacea* and its life history. *Am. J. Bot.*, *58*:552-561, 1971.
12. Raper, J. R. *Genetics of Sexuality in Higher Fungi*, Ronald Press, New York, 1966.
13. Elliott, T. J. Genetics and breeding of cultivated mushrooms. In *Tropical Mushrooms—Biological Nature and Cultivation Methods* (S. T. Chang and T. H. Quimio, eds.), The Chinese University Press, Hong Kong, 1982, pp. 11-26.
14. Chang, S. T., and Hayes, W. A. (eds.). *The Biology and Cultivation of Edible Mushrooms*, Academic Press, New York, 1978.
15. Chang, S. T., and Quimio, T. H. (eds.). *Tropical Mushrooms—Biological Nature and Cultivation Methods*, The Chinese University Press, Hong Kong, 1982.
16. Miles, P. G., and Chang, S. T. Application of biotechnology in strain selection and development of edible mushrooms. *Asian Food Journal*, *2*:3-10, 1986.
17. Smith, J. F., Fermor, T. R., and Zadrazil, P. Pretreatment of lignocellulosics for edible fungi. In *Treatment of Lignocellulosics with White Rot Fungi* (F. Zadrazil and P. Reiniger, eds.), Elsevier Applied Science, London, 1988, pp. 3-13.
18. Flegg, P. B. Mushroom composts and composting: A review of the literature. *Rep. Glasshouse Crops Res. Inst.*, 1960.

19. Gerrits, J. P. G. Compost treatment in bulk for mushroom growing. In *Treatment of Lignocellulosics With White Rot Fungi* (F. Zadrazil and P. Reiniger, eds.), Elsevier Applied Science, London, 1988, pp. 99-104.
20. Gray, K. R., Sherman, K., and Biddlestone, A. J. A review of composting—Part I. *Process Biochem.* (June);32, 1971.
21. Gray, K. R., Sherman, K., and Biddlestone, A. J. Review of composting—Part II: The practical process. *Process Biochem.* (October): 22, 1971.
22. Gray, K. R., Biddlestone, A. J., and Clark, R. Review of composting—Part III: Processes and products. *Process Biochem.* (October): 11, 1973.
23. Hayes, W. A. *Composting*, W. S. Maney and Sons, Ltd., Leeds, 1977.
24. Elliott, T. J., and Challen, M. P. The storage of mushroom strains in liquid nitrogen. *The Glasshouse Crops Res. Inst. Ann. Rep.* 194, 1979.
25. San Antonio, J. P., and Hwang, S. W. Liquid nitrogen preservation of the cultivated mushroom, *Agaricus bisporus* (Lange) Sing. *J. Am. Soc. Hortic. Sci., 95*:565-569, 1980.
26. Jodon, M. H., Royse, D. J., and Jong, S. C. Productivity of *Agaricus brunescens* stock cultures following 5, 7 and 10 year storage periods in liquid nitrogen. *Cryobiology, 19*:602-606, 1982.

9

BIOLOGICAL UTILIZATION OF EDIBLE FRUITING FUNGI

S. RAJARATHNAM and ZAKIA BANO *Central Food Technological Research Institute, Mysore, India*

I. INTRODUCTION

Some of the higher fungi, mostly belonging to the class Basidiomycetes, characteristically bear sexual spores in specialized fleshy structures of various shapes, sizes, colors, textures, and flavors. These structures are aptly designated as the "fruiting bodies," and such fungi are called the "fruiting fungi." In nature, as many as 2000 edible species have been recorded [1]. While the fruiting bodies serve the function of active dissemination of the spores for extended survival of the species, they also represent classic items of directly edible and highly palatable human food. On the surface of our planet, about 155.2 billion tons of organic matter is synthesized through photosynthesis every year [2]; while only a portion of this organic matter is directly edible in the form of cereal grains, fruits, vegetables, etc., much of it—taking varying forms such as cereal straws, dried leaves, twigs, sawdust, maize cobs, cotton waste, and so on—is inedible. The fruiting fungi are biochemically endowed with the ability to secrete a variety of hydrolyzing and oxidizing enzymes that aid in the degradation of these plant wastes (which are chemically lignocellulosics); they then use some of the degraded products for their growth and fructification and leave behind the rest in the form of spent substrate. Thus, these fungi are potent biological agents, given to us by nature to convert inedible organic wastes directly into palatable human food; moreover, they provide an alternative pathway for the production of food without having recourse to sunlight and independent of the photosynthetic route.

Exploitation of such fruiting fungi for the generation of edible biomass has several advantages:

1. They represent examples of the most efficient conversion of plant wastes into edible food [3,4].
2. Unlike many other single cell proteins, they are directly edible and are praised as food delicacies because of their characteristic texture and flavors.

TABLE 1. Production of Crude Protein and Gross Energy by Crops and Animals Compared with the Edible Fruiting Fungi

Species	Yield component	Yield[a] (kg)	DM (%)	Yield DM (kg/ha)	Crude protein (%)	Gross energy (MJ/kg)	Crude protein (kg/ha)	Gross energy (MJ/ha)
Grass (perennial)	Total harvested	60,000	10.0	12,000	17.5	18.5	2,100	222,000
Cabbage	Total harvested	54,545	11.0	6,000	13.6	17.5	316	105,000
Peas	Seed	2,511	86.0	2,159	26.1	18.9	566	40,805
Potatoes (main crop)	Tuber	27,621	21.0	5,800	9.0	17.6	52.2	102,080
Wheat (winter)	Seed	4,394	86.0	3,779	12.4	18.4	469	69,534
Maize	Seed	4,654	86.0	3,995	9.8	19.0	392	75,905
Rice	Seed	5,670	86.0	4,876	7.7	18.0	375	87,768
Cattle, beef (suckler)	Carcass			360	14.8	10.9	53	3,924
Cattle (dairy)	Milk			3,386	3.5	2.6	118.5	8,770
Sheep	Carcass			462	14.0	16.2	65	7,486
Pigs (baconer)	Carcass			875	12.0	16.5	105	14,438
Hen	Eggs (edible portion)			624	11.9	6.6	74	4,118
Pleurotus sajor-caju	Total harvested	248,888	10	24,888	22.5	12.5	5,599	312,344
Pleurotus flabellatus	Total harvested	133,333	10	13,333	21.6	11.3	2,880	151,196
Agaricus bisporus	Total harvested	189,000	10	18,900	26.3	13.7	4,970	259,308
Volvariella volvacea (on straw)	Total harvested	10,864	10	1,086	22.5	12.7	244	13,792
V. volvacea (on cotton waste)	Total harvested	22,222	10	2,222	22.5	12.7	500	28,219

[a]Fresh weight.

Source: Refs. 5-9.

3. Harvesting of fruiting bodies is the easiest possible method of separating edible biomass from the substrate in a solid state fermentation [3].
4. Their protein conversion efficiency per unit of land, and per unit of time, is far superior compared to animal sources of protein [5] (Table 1).
5. The spent substrate, in its degraded form, represents a spectrum of attractive ways of biological disposal [10,11].
6. They play a vital role in the ecology of the carbon cycle in nature, thus reducing the accumulation of organic plant material that accrues every year on the earth.

This review deals mainly with the various modes of using the fruiting fungi, based on their chemical and biochemical properties. Their ability to bioconvert a huge spectrum of inedible plant wastes into food is considered. Their food and bio-medical values, as consequences of their chemistry, are evaluated. Diversified applications and implications of the spent substrates are dealt with from the standpoint of their biochemical potentialities.

II. BIOMASS CONVERSION

After a period of incubation—sometime after inoculation of the substrate—mycelial cells of fruiting fungi begin to multiply, ultimately culminating in the production of the fruit bodies. The easiest way of separating edible biomass from the substrate would definitely be an added advantage when the fruiting fungi are compared with other microbes for biomass conversion. The potential substrates, various degrees of bioconversion, and factors influencing the bioconversion are dealt with in this chapter. The relative capacities of various fruiting fungi for bioconversion of lignocellulosic wastes into useful forms of food and feed have been recently reviewed by Zadrazil [3,12].

A. Potential Substrates

A number of agricultural wastes and industrial byproducts are useful to study biotransformation. Their worldwide availability will be considered here. These wastes are chemically composed mainly of cellulose, hemicellulose, and lignin.

Lignocellulose is a "recalcitrant" molecule. In nature, it is the building block of the stem which supports the plant until it attains its ultimate objective of reproduction in the form of seed. It is not an organic material that can be easily degraded in the laboratory; usually one has to employ a strong acid, accompanied by high temperature. Pretreatments to disrupt the lignin barrier—either with chemicals such as sodium hydroxide, or with solvents, or by physical treatments such as fine grinding—enable us to use a broader range of microorganisms. However, physical methods require a large amount of energy, while chemical methods lead to environmental pollution. A wide range of microorganisms produce enzymes which can degrade cellulose. Far fewer produce enzymes which can degrade the natural lignocellulosic wastes, because lignin limits access to cellulose. In fact, the only proven microbial means of converting unmodified lignocellulosics is found in the production of enzymes by various types of mushrooms [13]. Staniforth has discussed the potential of culturing *Pleurotus* for utilization of straw [14].

About two thirds of the earth's biomass production occurs on the land areas and about one third in the oceans [2]. Most terrestrial plant material occurs in the forests (about 65%), with a bit more than 15% generated in grasslands and cultivated lands. McHale [15] has estimated that about three quarters of the biomass generated on cultivated lands and grasslands is waste or residue. In many instances, these wastes, particularly cellulose, appear in forms and amounts suitable for consideration as raw materials for reclamation processes.

Type and availability of natural lignocellulosic wastes in any particular geographic region depend on factors such as the climate and environment, use and disuse, culture of people, and type and nature of the technology adopted or developed. Thus, while rice straw is more prevalent in the Far East and southern Asia, wheat straw and maize byproducts, like corn stover, are far more abundantly available in Europe and America. Wheat straw, soybean stover, and corn stover each are produced in excess of 100 million tons per year, while sorghum stover, oat straw, and corn cobs are produced in amounts from 15 to 56 million tons per year in the United States [16].

Annually about 3644 million tons of cereal straws are produced in the world [5]. Significant quantities of the cereal straws are disposed of by burning [17-19]. This lost energy could be conserved by subjecting straws to culturing by fruiting fungi. Calculating based on conversion efficiencies of different fruiting fungi and the probable availability of cereal straws in the various countries, possible biotransformation of straw into mushrooms and spent substrate is projected in Table 2.

Citronella (*Cymbopogon winterianus*) bagasse and lemon grass (*Cymbopogon citratus*) are the lignocellulosic residues of steam distillation of freshly harvested lemon grass and citronella leaves to recover their essential oils [22]. The essential oil content is low—0.5-1.3% by weight of fresh grass—and its recovery is not complete. After steam distillation, the bagasse is partially dried in the fields and a fraction is burned to generate steam for the stripping; the rest is left in fields where natural biodegradation takes place. Its use as a ruminant feed is limited due to animal rejection because of residual aroma and flavor. According to recent information on essential oil production [23], there is an estimated worldwide availability of about 200,000 tons of dry bagasse per year that could be used as a source of lignocellulosic wastes.

The fibrous residue remaining after extracting juice from sugar cane stalk is referred to as bagasse. In sugar mills, most bagasse is used as fuel; energy-efficient units can have 2 tons (dry weight) of surplus bagasse for every 100 tons of fresh sugar cane processed [24]. The lignocellulosic residue left after sucrose conversion into ethanol by the EX-FERM is referred to as EX-FERMented [25] sugarcane chips, which contain 44.8%, 37.7%, and 13.5%, respectively, of cellulose, hemicellulose, and lignin (bagasse samples contained 42.1%, 37.0%, and 14.6% of cellulose, hemicellulose, and lignin, respectively).

TABLE 2. Possible Production of *Pleurotus* Fruiting Bodies from Cereal Straws (Assuming 10% Dry:Dry Bioconversion) and Per Capita Availability Per Day (g)

Continent or country	Population (1000)	25% of Total cereal straw[a] used (1000 tons)	Mushrooms produced[b] (1000 tons, dry)	Availability (g, dry/person/day)	Fertilizer produced[c] (1000 tons)
World	4,836,960	794,009	79,400	44.9	210,412
Africa	555,116	39,296	3,929	19.3	10,413
Egypt	46,909	4,150	415	24.2	1,099
Kenya	20,600	1,757	175	23.3	465
North America	401,416	221,972	22,197	154.0	59,989
Canada	25,379	21,923	2,192	236.6	5,809
United States	238,840	187,231	28,723	214.7	49,616
Mexico	78,996	14,736	1,473	51.0	3,905
South America	267,749	37,979	3,797	38.9	10,064
Argentina	30,564	14,553	1,455	130.4	3,856
Brazil	135,564	17,628	1,762	35.6	4,671
Asia	2,818,260	241,834	24,183	25.6	64,086
Afghanistan	16,519	2,044	204	33.8	542
China	1,059,521	125,966	12,596	32.6	33,380
India	758,927	56,261	5,626	20.3	14,909
Iran	44,632	3,725	372	22.8	987
Pakistan	100,380	7,252	725	19.78	1,921
Europe	492,251	133,588	13,358	74.3	35,400
Australia	15,691	10,649	1,064	185.8	2,822

[a]Values for straw were obtained from the grain yield from *FAO Production Year Book 1985* [20] using as conversion factors 1.8 for wheat, 1.0 for rice, 2.4 for maize, and 1.7 for other cereals.

[b]Values can be arrived at for *Agaricus* and *Volvariella* using the yield conversion factors of 7 [21] and 1.5 [9], respectively, instead of 10 as used for *Pleurotus* in the table.

[c]Calculated based on the fact that starting with 1 ton rice straw, 265 kg spent straw (as fertilizer) is obtained after the harvest of fruiting bodies.

Source: Ref. 5.

In Israel, 1.5 tons of straw are produced per acre of cotton cultivated [26]. Its high lignin content (~25%) has limited the value of cotton straw as a direct ruminant feed [27]. Shive is a bulky, woody byproduct of flax (*Linum usitalissimum* L.) that is left after scutching and has little value [28]. For every ton of fiber produced, 2.5 tons of shive will be left after scutching. The high lignin content prevents ruminants from utilizing the cellulose [29].

Orange peel and distillery grape stalks are some of the wastes produced in large quantities in Italy [30], and they cause disposal and environmental problems. In 1985, the orange juice industries have processed 600,000 tons of citrus fruits, with a residual waste that constitutes 60% of the weight of the treated fruits. This waste contains a considerable amount of residual sucrose and macromolecular carbohydrates, but a low protein content; therefore, it has a good digestibility level but is of low nutritional value. Distillery grape stalks have a low carbohydrate content as well as a low protein content but have a high level of lignocellulosic materials. These characteristics limit their use as animal feed. A compost of the slime sludge has been found useful for mushroom production; this involves treatment of waste water from mandarin orange canneries, and thus its useful disposal [31].

Coffee pulp represents a major byproduct of the coffee industry, representing about 28.7% of the coffee fruit (on dry weight basis) during the wet coffee-processing method [32]. Millions of tons of coffee pulp are produced worldwide every year; in Mexico alone, about one million tons of pulp are produced annually. Difficult and improper handling of the pulp causes many water pollution problems in the rivers and insalubrious conditions in the land areas affecting agricultural activities in the fields of many countries. The relatively high content of lignin, potassium, caffeine, tannins, and phenols have limited use of coffee pulp as feed for cattle, hogs, and chickens [33,34]. Through research on use of *Pleurotus ostreatus* for bioconversion of this waste, done at the National Institute of Biotic Resources by Martinez and coworkers, it has been possible to set up a mushroom-producing plant on a semi-industrial scale in Mexico, designed to work on one ton of coffee pulp every day, to effect a total daily production of 110-130 kg of fresh mushrooms [35].

Zadrazil [36] has studied the possibilities of using sulfite waste liquors (from paper manufacture) to produce—after supplementation—a product called "Bycobact" suitable for edible fungal bioconversion.

B. Biomass Yield

Conversion of lignocellulosic wastes into edible fruiting bodies is the most vital aspect of any mushroom cultivation; it is usually expressed as biological efficiency (BE), which is defined as the dry weight of fruiting bodies produced per dry substrate. This varies from species to species, within the same species on different substrates, and further on using the same species and same substrate under different cultural conditions. Species of *Pleurotus* are known to colonize and yield on a wide range of unfermented natural lignocellulosic wastes [37]. Zadrazil [38] has reported a BE of 10% for *Pleurotus* sp. "*Florida*" on wheat straw. A BE of 11.10% is observed for *P. sajor-caju* when cultured on wet, chopped, unfermented rice straw in perforated polyethylene containers on a commercial scale [39]. The authors could draw a significant correlation between the loss of dry weight of the substrate and production of biomass during different seasons of the year.

Biological efficiency of 11.34% for *P. ostreatus* when cultured on fresh coffee pulp has been reported [40]. During a 40-day incubation period, *P. sajor-caju* on rice straw under nonsterile conditions effected a BE of about 10% [41]. Over a 45-day period, *P. ostreatus* colonized a peat-moss-based substrate and effected 10% bioconversion into fruiting bodies [42]. The capacity of *P. sajor-caju* to grow, degrade, and fructify on *Saccharum munja*, a profusely growing weed, has been reported recently. Although the biological efficiency was comparatively low on *S. munja* [43], the ready availability of this weed over a larger area and cost-free availability, particularly in the rural sectors, make it favorable for use in mushroom culturing. One kg of dry cotton straw yielded 600-700 g of fresh mushrooms [44] up to the first flush; there was a loss of 50% of the dry matter content of cotton straw due to fungal growth. During this period, there was a rapid increase in fungal respiration, measured by O_2-electrode, and this eventually reflects on the active metabolism of the fungus effecting fructification involving synthesis of biomass at a faster rate compared to the spawn run.

Coprinus comatus on soya-enriched waste cellulosic residues (clarified sludge) from the bleached Kraft pulp mill yielded about 24% of the wet substrate weight; the average weight of the fruiting bodies was two to three times higher on the soya-Kraft sludge substrate than on conventional manure compost [45].

A substrate of pure oak sawdust, supplemented with 10% wheat bran and filled in polypropylene bags yielded 150 g of fresh *Lentinum edodes* per kg [46]. Bech has observed a yield of 700 kg per ton of dry substrate with *A. bisporus* while preparing a productive commercial compost [7].

The biological efficiency of *Volvariella* varies from 25 to 65% [47]; the average efficiency is about 15% for straw substrate under natural conditions and 33% for cotton-waste compost under controlled conditions. The term "biological efficiency" here refers to the yield of mushroom in proportion to the dry weight of compost at spawning, as described by Tschierpe and Hartman [48].

Quimio and Abilay [49] have observed formation of abundant fruiting bodies of *Collybia reinakeana* on rice straw-corn meal substrate only when cased with garden soil; there is no mention of the quantitative bioconversion of the substrate. A substrate of sawdust (of suitable broad-leaved trees) with 20-25% rice bran loaded into a plastic bag measuring 50 × 10 cm could yield 350 g of dried fruiting bodies of *Tremella fuciformis*; neither biological efficiency nor weight of substrate per bag was mentioned [50]. On a substrate of sawdust, bagasse, and bamboo trash, fresh yield of *Dictyophora duplicata* was found to be as high as 148% [51]. A yield of 20% for *P. ostreatus* using *Cassia* substrate has been reported [52].

Of 10 fungi tested, *P. ostreatus* best used potato peels for protein production by solid state fermentation, increasing true protein content of the substrate from 7.5 to 11.6%; however, the biomass yield was not mentioned [53]. *Pleurotus ostreatus* and *Coprinus aratus* have been employed to promote protein production on bagasse and straw [54,55]. Yields of 3.5, 2.7, and 1.6 (based on culture bed weight) were produced by *Flammulina velutipes, Ganoderma lucidum,* and *Polyporus* spp., respectively, on a substrate of fermented sugar cane bagasse [56]. Chahal et al. [57] have observed the ability of *P. sajor-caju* to bioconvert hemicelluloses (left behind after use of cellulose from lignocellulosics for ethanol production). Utilization of industrial wastes such as corrugated paper, distillery slop, peat, saccharification waste, and grass choppings for bioconversion by *P. sajor-caju* has been reported; the first three substrates could yield fruiting bodies only when supplemented with rice bran [58]. Chopped rice straw with addition of rice bran (10-15%) effected better biomass yield by *P. ostreatus* than a substrate of pine sawdust or rice hulls [59].

C. Factors Affecting Bioconversion

This subject has been reviewed by several authors [60-63]. Major factors related to substrate and ambient atmosphere greatly influencing the degree of fungal bioconversion are enumerated below.

1. Temperature and Relative Humidity

Most species of *Agaricus* require a temperature not exceeding 20°C, Volvariella fruits well above 32°C, whereas species of *Pleurotus* thrive from 20 to 30°C. This specificity for the requirement of temperature might be ultimately related to various enzymatic systems involved in fructification influenced by temperature. Relative humidity (RH) is of little importance as long as the mycelium is contained in polyethylene bags; once the beds are exposed for fructification,, a minimum of 80% RH is essential to prevent surface desiccation. Temperature and RH requirements have been detailed by many authors [61,63,64].

2. Substrate Particle Size and Water Content

These are two interrelated factors affecting mycelial growth. Whole straw beds, while culturing *Pleurotus*, did not yield well because of the high interair spaces which led to active desiccation [37]. Rice straw pieces (2-3 cm) with 75% water content have been found ideal for cultivation. Moisture above 80% paves the way for bacterial growth accompanied by fungal contaminations like *Coprinus, Pluteus*, etc.

3. Oxygen and Carbon Dioxide in the Atmosphere

Many species of *Pleurotus* grow faster when atmospheric CO_2 is high—even as high as 22% [65]. This is a remarkable feature of *Pleurotus*, and it stands in sharp contrast to *Agaricus*, which is oxygen sensitive [66]. Ginterova also has confirmed that the O_2 requirement of *Pleurotus* is lower than many other fungi [67]. Oxygen assumes the greatest importance during fructification when the rate of O_2 consumption is rapid.

4. Chemical Nature of Substrate

Agaricus requires a preformed substrate wherein the natural lignocellulosic wastes are allowed to ferment through active microbial growth to narrow down the C:N ratio; the requirement for the substrate N content is also quite high compared to that for *Pleurotus* [68]. Even though *Pleurotus* degrades and metabolizes lignin, it colonizes quickly over the substrates having a low lignin content such as waste paper, maize cobs, cotton waste, etc. [37]. Fermentation, and particularly pasteurization of coffee pulp, reduces the level of phenolics, with consequent active growth of the mushroom mycelium [40].

5. Extra Nutrient Supplementation

Supplementation of mushroom beds with extra nutrients has invariably proved to effect better yields [69,70]; however, it is essential to know the time of supplementation, as well as the nature and concentration of the supplement. In general, supplementing the growth substrate during spawning with nitrogen sources leads to substrate contamination, whereas supplementation after spawn run increases the yield tremendously; increase in protein content, change in flavor, and change in texture also have been noted [71]. Organic nitrogen supplements such as yeast cake, cotton seed meal and soyabean meal, effected increased yields; inorganic nitrogen sources and urea, in particular, during *Pleurotus* culturing adversely affected mycelial growth, possibly due to ammonification [71,72]. The effect of delayed-release nutrient on the yield and size of *Pleurotus* fruit bodies has been recently reported [73]. Tokimoto and Kawai have examined the nutritional factors influencing fructification in *Lentinus edodes* [74].

6. Contaminations and Diseases

Trichoderma, Plicaria, Papulaspora, Conidiobolus, Paecilomyces, and *Peziza* sometimes are fungal contaminants in compost beds of *Agaricus* [75-77]. Likewise *Trichoderma, Monilia, Fusarium, Penicillium, Trichothecium, Mucor, Sclerotium, Coprinus, Pluteus*, and *Papulaspora* occur in straw beds of *Pleurotus* [78,79]. *S. rolfsii* is a serious substrate-borne contaminant that reduces mushroom yield by 80-96%; treating straw with carboxin [80], methyl bromide [81], ethyl formate [82], or liquid ammonia [83] effectively controlled the contaminant. *Aspergillus, Coprinus*, and *Sclerotium* occur in the straw substrate during culturing of *Volvariella volvacea* [84]. Parasitic diseases of the fruiting bodies [85,86] play a major role in affecting the yield. Relatively, species of *Pleurotus* were free from parasitic diseases until recently when bacterial diseases due to *Pseudomonas* were reported [62]. Ecological factors play a vital role in influencing the occurrence of weed molds in the substrate while culturing fruiting fungi [37,87].

III. USES OF FRUITING BODIES

A. Chemical Composition and Nutritional Value

Data in food composition tables are often presented in different units of measurement based on the highly variable fresh or dry weight, making comparison of such data difficult. To ensure universal comparison of the values among different species, it is advisable to express all of the analytical values on a dry weight basis. Significant compositional changes are associated with strain, age or stage of development, lapse of time after harvest, and different portions of a single fruiting body [88]. The chemical nature of the substrate has a bearing on the chemical composition of fruiting bodies [71,72]. Protein and amino acid contents in cultivated mushrooms are approximately two times higher than those in the wild ones

[89]. Fruiting bodies, in general, on a dry weight basis contain about 55% carbohydrates, 32% proteins, 2% fats, and the rest is the ash constituting the minerals. Crisan and Sands [88], Chang [90], and Bano and Rajarathnam [5,91] have reviewed the chemistry and food value of fruiting bodies. Mannitol and trehalose are the main free sugars (10-12%) and the remaining carbohydrate is in the polymeric form, namely, crude fiber; starch is absent, while glycogen is present (5-8%), and thus the fruiting bodies would be recommended for diabetics. Mushroom protein, unlike animal protein, is incomplete, lacking some of the essential sulfur-containing and aromatic amino acids. The fruiting bodies are good sources of vitamins [92]—particularly B-complex and folic acid, which counteracts pernicious anemia [93]. Of the six edible mushrooms investigated, *V. volvacea* had the highest content of ergosterol provitamin D_2 (0.4% on dry weight), followed by *L. edodes* (0.27%) and *A. bisporus* (0.23%) [94]. Ergosterol content of 44 widely found species of edible mushrooms ranged from 19 to 780 mg/100 g of dry matter [95]. Biological utilization of available iron from the fruit bodies of *P. flabellatus* and *P. sajor-caju* has been investigated [96]. Assessing their nutritional value, fruiting bodies, in general, stand in between high class vegetables and low grade proteins.

Flavor is the most important inducement for past and present widespread consumption of wild and commercially grown edible mushrooms. More importance has been placed in recent years on volatile fractions of mushroom flavor [97]. As many as 150 volatile compounds have been identified in various mushroom species [98]. A series of eight carbon (C_8) compounds are believed to be the most important volatile flavor compounds. It has been demonstrated that some of the C_8 and C_{10} compounds can be enzymatically formed from linoleic and linolenic acids, both of which are usually predominant in fruiting bodies [99]. Preponderence of 1-octen-3-ol in the fresh fruiting bodies of *Cantharellus cibarius*, *Coprinus atramentarius*, and *Leucocoprinus elaeidis* has been investigated. 1-Octen-3-one has the highest aroma value in fresh fruiting bodies of *Psalliota bispora*.

B. Biological Properties

Edible mushrooms display several interesting biological activities. *Pleurotus ostreatus* and other related species can destroy nematodes through the release of toxin [100]. Pleurotin, a polycyclic compound isolated from *P. griseus*, is reported to possess antibiotic properties [101]. Dihydrofolate synthetase activity, involved in the synthesis of folic acid, has been reported present in species of *Pleurotus*, *Lentinus*, *Flammulina*, and *Tricholoma* [102]. Species of *Pleurotus*, because of their high content of dietary fiber, could find a place in diet therapies for hyperlipemia, diabetes, etc. [103]. Various endogenous cytokinins isolated from the fruiting bodies of *P. sajor-caju* have been thought to contribute to the long storage life of the fruiting bodies [104].

Certain interesting hot water-soluble polysaccharides isolated from fruiting bodies have been known to possess antitumor activity since they can inhibit and regress the growth of ascites of sarcoma-180 injected into subcutaneously the right groin of mice [105-108]. Mori et al. [109] and Mouri [110] have reviewed information on the carcinostatic activity of the fruiting bodies of mushrooms. $(1 \rightarrow 3)$ β-D glucans were found to be more effective as antitumor agents (active at doses of 10-0.2 mg/kg) than α-glucans, which are active above 20 mg/kg [108]. 6-Deoxyilludin M, a new antitumor antibiotic, is obtained from the fermentation cultures of *P. japonicus*; the compound is weakly active against *Bacillus subtilis* and markedly active against *Murine leukemia* P_{388} [111]. The biological activities of polysaccharides from *Auricularia auricula* and *Tremella fuciformis* on biosynthesis of lymphocytic DNA/RNA, as well as their antiulcer effect, have been reported [112,113]. Species of *Boletus, Clavaria, Lactarius, Russula, Pholiota,* and *Flammulina* have been demonstrated to possess lectin activities, causing hemagglutination of erythrocyctes [114]. Two crystalline forms of lectins from *Flammulins velutipes* have been reported [115]. *Lentinus edodes* and *Grifola frondosa* decreased hypertension and plasma cholesterol in rats [116]. Hypocholesterolemic property of mushrooms is reported by others [117-119]. An antileukosis substance has been isolated from *Coprinus radiatus* [120].

Extracts of dried fruiting bodies, blended with a sugar solution and then fermented with yeast, resulted in an alcoholic beverage rich in proteins, organic acids, and minerals; it also contained vitamin D_2 and had serum cholesterol-lowering activity. The product has a novel flavor [121].

IV. BIODEGRADATION OF THE GROWTH SUBSTRATES

A. Chemical Changes

In this chapter relatively more emphasis is put on *Pleurotus* spp. because of their ability to colonize a wide range of unfermented natural lignocellulosic wastes and thrive over a range of cultural conditions. Fruiting fungi vary greatly in their relative abilities to colonize a lignocellulosic substrate, the type of substrate they can colonize, their rate of growth and degradation of the substrate, and ultimately the capacities to fructify and bioconvert the inedible waste into edible biomass (Table 3).

1. In Vitro Degradation

Several reports are available on ability of *Pleurotus* to degrade various types of natural lignocellulosic wastes, such as cereal straws [122-126], woods—alder, beech, oak, poplar—[127-130], and sawdust [131,132]. *Agaricus, Volvariella,* and *Lentinus* degrade, respectively, compost, straw holocellulose, and wood [68,133-135]. Experiments dealing with the growth of fruiting fungi on lignocellulosic substrates under sterile conditions (obviously with restricted oxygen supply) are discussed here. Such studies help to ascertain that the changes observed under the defined conditions are brought out by the fungal monoculture without inclusion of any foreign organisms; but such organisms invariably will be present when cultured on nonsterile or pasteurized substrates. A substrate of sterilized particulate lignocellulose can be degraded by the mycelium through inoculation with culture discs [136,137] or culture suspension [138].

Growth of a species produces, in time (i.e., after the incubation period), various changes (both qualitative and quantitative) in different constituents of lignocellulosic wastes. An estimation of the degree of substrate degradation is possible by a study of various parameters.

TABLE 3. Edible Fruiting Fungi Cultured on Cereal Straw Substrate, Growth Characteristics, and Digestibility of Substrate

Species	Growth on sterile straw substrate	Saprophytic colonization ability	Rate of decomposition	Fructification and yield			Straw lignin degradation	Increase of straw digestibility
Agaricus arvensis	1	4	7	10	12–13	14b	17	20
Agaricus bisporus	1	4	7	10	12	16a	18	20
Agaricus bitorquis	1	4	7	10	13	16a	18	20
Agrocybe aegerita	3	5	8	10	13	15b	17	20
Auricularia judea	2	5	8	10	12–13	14b	17	20
Coprinus comatus	2	5	7–8	10	12	14a	18	20
Coprinus fimentarus	2	5	7–8	10		14a		
Flammulina velutipes	3	5	7–8	10	12	14a	17	20
Kuchneromyces mutabilis	3	5	7–8	10	12	14b	18	20
Lentinus edodes	2	4	7–8	10	13	14–15b	18	21
Lepista nuda	1	5	7	10	12	14b		
Macrolepiota procera	1	5	7	10	12	14b	18	
Macrolepiota rhacodes	1	5	7	10	12	14b		

Pleurotus sp. *abalone*	3	6	9	11	13	15-16a	19	21
Pleurotus cornucopieae	3	6	9	11	13	16a	19	21
Pleurotus eryngii	3	5	8	11	13	16a	19	21
Pleurotus sp. "*Florida*"	3	6	9	11	13	16a	19	21
Pleurotus flabellatus	3	6	9	11	13	15a	19	21
Pleurotus ostreatus	3	6	9	11	12	16a	19	21
Pleurotus sapidus	3	6	9	11	13	16a	19	21
Pleurotus sajor-caju	3	6	9	10	13	15-16a	19	21
Pholiota nameko	2	5	8	10	13	14-15b	17	20
Stropharia rugosoannulata	3	6	8	11	12	14-15a	19	21
Volvaria volvacea	3	6	8	10	13	14-15a	17	20

1, slow; 2, good; 3, very good; 4, low; 5, good; 6, very good; 7, low; 8, medium, 9, high; 10, for good yield, extra supplementation of the substrate necessary; 11, for good yield, extra supplementation of the substrate not necessary; 12, fructation below 18°C; 13, fructification above 18°C; 14, yield of fruiting bodies low; 15, yield of fruiting bodies good; 16, yield of fruiting bodies very good; 17, no or low; 18, good; 19, very good; 20, no or low; 21, good or very good.

[a]Commercially cultivated.

[b]At experimental stage.

Source: Ref. 4.

Loss of dry matter (loss of organic matter, LOM): This is the simplest criterion adopted to evaluate, in a crude manner, the extent of degradation of a substrate. It takes into consideration the fact that, during growth of the fungus and consequent decomposition of the substrate, some CO_2 and H_2O are lost during the metabolic activity of the cultured fungus. Studying degradation of various substrates by *Pleurotus* sp. "*Florida*" and *P. cornucopiae*, Zadrazil [38] has found that rape and sunflower substrates showed higher LOM, while relatively low decomposition rates were found for beech sawdust and rice husks. Similar results were recorded with *Agrocybe aegerita*. The difference in degree of decomposition of the substrate by different species has been related to differences in fructification under conditions described.

Substrate solubility: This involves measurement of release of water-soluble substances or release of sugars [38]. *Pleurotus cornucopiae* and *P.* sp. "*Florida*" effected a decrease in water-soluble substances and reducing sugars of wheat straw substrate during the initial period of incubation; after 20 days, there was a progressive increase in these water-soluble substances and after 120 days, there was an eight and a five-fold increase in reducing sugars (at 25°C), respectively, for the just mentioned two species of *Pleurotus* [38]. These fungi start growing immediately after inoculation, utilizing, first, the available free sugars from the substrate, and then they shift to use of carbon from polysaccharides, as demonstrated by secretion of cellulolytic enzymes some time after inoculation [139]. *Agrocybe aegerita* had the least ability to degrade the substrates and to liberate water-soluble substances [38]. The amount of water-soluble substances released from cotton straw increased from 19.5 to 57% during the first 21 days of growth of *P. florida* [44].

Biomass yield: The inoculated substrate under sterile conditions is colonized by the mushroom mycelium, and the formation of fruiting bodies is not very common. Wood [140] employed measurement of laccase activity to estimate biomass of *Agaricus* mycelium in compost. It was shown that laccase activity is directly proportional to mycelial growth in axenic compost cultures.

Invariably *Pleurotus* with culturing either does not form fruiting bodies or they are formed immediately, possibly because of inadequate availability of oxygen. Zadrazil [38] and Nicolini et al. [30] observed the formation of fruiting bodies on a number of substrates. Our observation is that species like *P. flabellatus* invariably formed fruiting bodies in the head space of Erlenmeyer flasks above the straw substrate, but could grow completely only when the cotton plugs were removed, exposing them to the atmosphere. Biomass yield assumes greater importance during in vivo degradation of lignocellulosic wastes.

Hexosamine: This compound is measured as glucosamine in the degraded substrate; it serves as a measure of the synthesis of fungal biomass in the course of solid state degradation, because fungal cell walls are constituted mostly of chitin, which is composed of glucosamine. Increase in hexosamine content of straw degraded by *P. flabellatus* over a 65-day period of incubation has been observed [139]. On a substrate of wheat straw and orange peel plus grape stalks, *P. ostreatus* caused a decrease of hexosamine content after 70 days of incubation; this involved formation of fruiting bodies [30]. The biomass of *P. ostreatus "Florida"* during its growth on cotton straw reached a maximum at the end of 6 days of incubation [141].

Heat of combustion: This parameter decreased only slightly during fermentation, perhaps because of differences in rates of degradation of cellulose and lignin. It should be noted that lignin has a higher heat of combustion (ca. 5200 cal/g) than cellulose (ca. 4030 cal/g). Loss of energy caused by fungal metabolism correlated with the extent of decomposition of straw. While studying the decomposition of beech wood by *P. ostreatus*, it was observed that the heat of combustion of decayed samples was not affected by the amount of decomposition (i.e., weight loss) [142]. During degradation of beech and spruce by *P. ostreatus* and *Lentinus*, combustion heat was practically unchanged even at a high degree of fungal decomposition; for both fungi it fluctuated around the values measured in nonrotten wood [143].

pH: In general, the pH of the substrate decreases during its degradation. Thus the starting pH of 6.5-7.0 of wheat/rice straw drops to a final pH of 5.1 [38] or even 4.2 [139]. Secretion of oxalic acid by *Pleurotus Agaricus* [144] might be one reason for the decrease in pH. Daugulis and Bone [145] did not observe any drop in pH during the culturing of *P. sapidus* on maple and cedar barks, which is indicative of lack of growth.

Phenols and amino acids: The content of phenolics in the 70% alcohol-soluble fraction of degraded rice straw substrate during culturing with *P. flabellatus* progressively decreased as the incubation period increased; it is well known that *Pleurotus* mycelium secretes oxidizing enzymes that oxidize/degrade the phenols [139]. Likewise, amino acids increased during fungal growth, possibly due to protease activity [139,146,147].

Nitrogen: During solid state fermentation, the proportion of N in the substrate increased, due to CO_2 loss [136]. The greatest differences in the proportional accumulation of N were detected after 120 days with *P. cornucopiae* and *S. rugosoannulata*. The increase in N, however, did not correlate with in vitro digestibility. An increase in the nitrogen content of wheat straw, orange peel, and grape stalks was also observed during the growth of *P. ostreatus*; a tendency for reduction of N content was observed under the conditions of fructification (e.g., on wheat straw and orange peel plus grape stalks) [30].

Cellulose, hemicellulose, and lignin: Decreases of these components in wheat straw, rice straw, orange peel, grape stalks, and sugarcane bagasse have been reported [25,30,148]. In general, the ligninolytic property was more pronounced in *Pleurotus* spp., and they have a tendency to degrade more hemicellulose than cellulose. The relative merit of *Pleurotus* to affect lignin has been described [149].

In vitro digestibility: This is a simple criterion to assess the extent of substances/nutrients available in a substrate for microbial growth. Rumen microflora act on these substrates, thereby making them digestible by ruminants. This parameter is called in vitro dry matter enzymatic digestibility (IVDMED).

A detailed insight into this subject was gained by the observation of Zadrazil [137] that some white-rot fungi (including species of *Pleurotus*) can colonize wheat straw, liberate water-soluble substances from the substrate polymer complex during solid state fermentation [148], and use lignin [142]; during this process the digestibility of the degraded substrates increases for ruminants [150-154].

The in vitro digestibility of wheat straw at pH 9.0 with a moisture content of 65% in solid state fermentation by *Coprinus* sp. 386 at 37°C for 21 days increased from 40% (unfermented lot) to 65.5%. However, this period did not conform to the time required for maximum delignification. The author concluded that total delignification is not a prerequisite for maximizing digestibility [155]. Most of the fruiting fungi significantly improved digestibility of biodegraded residues of citronella bagasse and lemon grass [22].

Degradation of oat straw by *P. ostreatus* and *P. cornucopiae* did not increase digestibility, possibly due to slow growth rates of the fungi. Digestion of straw degraded by *P. ostreatus* and *Stropharia* was not affected [123]. However, this is surprising in spite of increased solubility and decreased lignin content of the substrate. Sommer et al. [156] made a similar observation—a decrease in digestibility of lignin and cellulose after culturing with and harvesting of *P. ostreatus*.

2. In Vivo Degradation

Changes in various constituents of growth substrates during growth and fructification are discussed here. In fact, a study of these changes is of great importance to know exactly what is going on in the substrate during culturing on a commercial scale for economical conversion of the growth substrates. Such a study should help to identify limiting constituents in the substrate so that supplementation with this constituent, in the right form and at the right stage of growth, might eventually aid in enhancing production of fruiting bodies.

Loss of dry matter (loss of organic matter, LOM): Concomitant with growth and fructification on lignocellulosic substrates, there is a decrease in organic matter. This is due to loss of CO_2 and H_2O during metabolism of the fungus and also removal of materials from the substrate for the construction of fruiting bodies. This loss, in general, is greater during fructification than during the spawn run [157,158], evidently indicating the high rate of fungal catabolic activity (during fructification) associated with anabolic activity of building up of fruiting bodies.

Water solubility of the substrate: There was a constant increase in the water solubility of the wheat straw substrate extracted in 0.1 N H_2SO_4 or 0.1 N NaOH, as the extraction temperature rose from 20 to 120°C. A similar trend was observed in the glucose content of the extracts, except for an initial decrease in the first 14 days of growth [38]. The spent straw after the harvest of *P. flabellatus* contained nearly four times the soluble sugars as the undegraded straw substrate [146]. Hong [159] reported changes in mannitol, trehalose, and free reducing sugars in rice straw and sawdust media during the growth and fructification of *P. ostreatus* and found a constant increase of these components. An increase in the amino nitrogen content of both rice straw and sawdust media after harvesting of the fruiting bodies of *Pleurotus* also was noted [146,159].

Hexosamine: An increase in hexosamine content from spawning until the end of the spawn run and a decrease during fructification, thus indicating its possible translocation into the growing, fruiting primordia, to form mature fruit bodies has been observed [139]. Matcham et al. also measured biomass yield in malt extract liquid culture and solid rye grain culture [160].

Cellulose, hemicellulose, and lignin: The decrease in cellulose content of the growth substrate varies, depending on species and yield. During growth and fructification of *P. ostreatus* on rice straw and poplar sawdust [159] and that of *P. flabellatus* on rice straw [157], it was observed that the decrease in cellulose content was greatest during fructification. This is understandable in view of the high rate of metabolic activity of the mushroom during fructification which is involved in construction of fruit bodies. A similar trend was observed while culturing *Agaricus* [134,158]. Giovannozi-Sermanni [161] also noted a progressive decrease in cellulose content from *P. ostreatus*, from spawning to the end of cropping. Zadrazil [38] found that absolute quantities of α-cellulose, hemicellulose, and lignin complexes (dry weight basis) remained unchanged during growth and fructification of *Pleurotus* on wheat straw. While culturing *P. sajor-caju* on *Saccharum munja, S. munja* plus rice straw (1:1), and rice straw, the amount of cellulose decomposed was more in the former and less in rice straw [43].

Hong [159] observed a progressive decrease in lignin content of both rice straw and poplar sawdust during growth and fructification of *P. ostreatus*. Decreases have been reported in lignin content of flax shieve [28] during growth of several strains of Pleurotus or of wheat straw [126, 162] during growth of several species of *Pleurotus*. All these reports converge toward one conclusion, namely, that lignin decrease is greatest during the spawn run. Possibly, the fungus forms a lignin-humus-protein complex that is utilized rapidly by a growing mycelium like that of *A. bisporus* [134].

Ash and minerals: Total ash was found to show a relative increase from the time of spawning until the end of cropping because of the constant use of organic matter. However, the total amount of ash could be assumed to remain constant although a fraction of it also entered into the developing fruit bodies. Hence, to determine the rate of disappearance of a constituent from a unit weight of substrate, it is advisable to calculate the degradation analytical values on a constant ash basis and to express all values as percentage of the dry matter during spawning, as described by Gerrits [134] after culturing *A. bisporus*.

Acid and alkali extractives: Zadrazil [163] has employed spectroscopic analysis to trace the changes in the wheat straw substrate degraded by *P.* sp. "*Florida*" and *P. eryngii* separately at various growth stages. The individual fractions were separated, and their solutions were read spectroscopically. With progressive growth, both species manifested a clear leveling off of the extinction curve at 280 nm, evidently indicating that water-soluble phenolics were attacked by both species. A similar observation was made when the wheat straw substrate was dissolved in 0.1 N NaOH and fulvic acid. However, the humic acid fraction peak remained very distinct at 280 nm until the end of cropping, in contrast to the NaOH and fulvic acid fractions, which flattened with progressive mushroom growth. Spectroscopic characteristics of the extracts are influenced by extent of mycelial growth, change in substrate, and reduction in phenolics (Zadrazil, unpublished).

Dry matter digestibility: Degraded lignocellulosic wastes have greater ruminant digestibility due to changes (qualitative and quantitative) brought about by the fungus under study during its growth and fructification. To confirm this effect, ruminants are to be fed for longer periods (many months), which obviously requires huge amounts of spent substrate. Further, if the number of replicates maintained is large, the consumption of spent substrate would naturally be very high. To avoid this difficulty, and in order to predict the digestibility, use of small amounts of spent substrates in a laboratory, stepwise enzymatic digestibility tests have been advocated by various authors. In place of rumen fluid, several researchers have used fungal cellulases to predict in vitro digestibility [164-167].

It is only recently that Tsang et al. [168] reported the in vitro digestibility of the spent wheat straw remaining after growing and harvesting various strains of *Pleurotus*. Studying digestibility of the spent straw with commercial cellulase and β-glucosidase [167], the authors concluded that it does not appear practical to produce simultaneously mushrooms and a highly digestible delignified residue from straw. The dry weight of mushrooms produced was much lower than the weight loss of straw; there was no apparent correlation between mushroom yield and weight loss, or digestibility, of spent straw. Zadrazil [136] showed that digestibility of straw can be increased by longer fermentation and by forgoing the harvest of fruiting bodies.

3. Factors Affecting Degradation of Lignocellulosic Wastes in Solid State Fermentation

Different degrees of substrate degradation have been observed by different workers in the course of experiments with different or the same substrates. These differences in relative abilities of the species/strains of the fruiting fungi have to be attributed ultimately to the interaction between the genome (heritable genetic material) and the environmental factors—physical, chemical, and biological. A study of these factors is essential to define conditions required for optimum degradation of a particular substrate by any particular species/strain under study. Achievement of maximum degradation of the substrate should eventually increase the substrate solubility and thereby increase biomass conversion efficiencies. Effects of incubation time and temperature, substrate particle size and water content, chemical nature and preparation of substrate, oxygen availability and substrate nitrogen content and species/strain under study as well as coculturing on the substrate degradation have been investigated [169-171].

4. Nature of Degraded Products

The spectrum of sugars and oligosaccharides formed in the straw substrate as well as that of amino acids and peptides during the growth and fructification of *Pleurotus* has been reported [146]. There was a relative increase in mannitol and trehalose contents in the rice straw and poplar sawdust medium during fructification of *P. ostreatus* [159]. Solubility and spectral characteristics of alkali lignin isolated from rice straw during growth of *P. flabellatus* have been done [146]. The supposedly chromophoric band at 320 nm disappeared as mycelial growth progressed. There was a decrease in the degree of polymerization of the degraded holocellulose (compared to native holocellulose) during the growth and fructification [28]. Decomposition products of lignin such as vanillic, p-hydroxy benzoic and protocatechuic acids were found in the culture supernatant liquids [172].

B. Biochemical Changes

1. Degradative Enzymes

The capacity of mushroom mycelium to grow on a wide spectrum of lignocellulosic materials can be attributed to its ability to secrete a range of degradatory enzymes, both saccharifying [173-178] and oxidative [179-188]. While assaying the activity of these degradatory enzymes during growth on natural lignocellulosic wastes, it is essential to observe:

1. The method of enzyme preparation
2. The method adopted for enzyme assay
3. The unit of expression of enzyme activity

To ensure universal comparisons among values of enzyme activities reported for various species/strains, it is advantageous to follow one defined set of parameters for the functions cited above. One such method followed for extraction of cellulases and laccase/protease from rice straw substrate during culturing with *P. flabellatus* is described [139]. The efficiency of enzyme extractions in acetate buffer (50 mM) in the pH range of 4 to 6 indicated that the extractions were very efficient at pH 5.4 and all the enzymes (C_1, C_x, FPD, and *p*-glucosidase) studied were stable at this pH. The acetone powder prepared from the rice straw substrate during growth of *P. flabellatus* on extraction with 100 mM phosphate buffer (pH 5.5) at 5°C for 1 h yielded stable activities of laccase and protease [146].

Cellulases: P. flabellatus secretes various classes of cellulases, namely, C_1, C_x, and β-glucosidases (necessary for complete saccharification of crystalline native cellulose as proposed by Reese et al. [189]), while being cultured on wet chopped unfermented rice straw. Development of CMC-ase activity in waste paper substrate, related to the growth and fructification of *P. sajor-caju*, has been studied [190]. The cellulase activity was high during the onset of fruiting body formation; it was maintained for some time, and then decreased as the yield decreased. Forty-seven cultures of *Pleurotus* were screened for their ability to secrete cellulase [191]. The cellulose activity was measured by liquefaction of gel from a modified cellulose [192]. The best producers of cellulase were obtained from cultures of *P. cornucopiae* and *P. japonicus*.

The in vitro activities of cellulases, however, are low in comparison with *Trichoderma*, the international standard for cellulase production; under in vivo cultural conditions, the enzymatic activities vary, depending upon the rate of growth of the species in question, the substrate on which it is grown, and the stage of growth—being maximum at the end of spawn run, or, more correctly, during the fructification and also at the end of cropping [157]. *Pleurotus sajor-caju*, characterized by its fast growth rate, displayed higher cellulase activity than *P. flabellatus*, when cultured on rice straw (unpublished data). A positive correlation was found to exist between the yield of *Volvariella* species and production of cellulase(s) by the fungus [193].

When cultured on a substrate of filter paper and cotton, the endoglucanase activity was highest on the 10th, 6th, and 3rd day of mycelial growth for *Flammulina velutipes*, *Lentinus tigrinus*, and *P. ostreatus*, respectively. Highest exoglucanase activity was observed in culture filtrates of *L. tigrinus*, *P. cornucopiae*, *F. velutipes*, and *Hypholoma sublateritium* [194]. The activities of carboxymethyl cellulase, avicelase, xylanase, polygalacturonase, β-glucosidase, and pectin transliminase of *L. edodes* on a sawdust medium as a function of pH and temperature have been studied [159].

The carbohydrases of the mushroom spawn from *P. flabellatus* have been studied [157]. The 3-month-old spawn (used for obtaining optimum yields of the mushroom on rice straw) did not show CMC-ase or hemicellulase activity but contained fairly high amounts of β-glucosidase activity. It was also rich in soluble carbohydrates and reducing sugars. Absence of a polysaccharase system might possibly be due to the presence of adequate amounts of soluble carbohydrates and reducing sugars. Various enzymes noted at various stages of spawn growth of *P. sajor-caju* were in the order: polygalacturonase transeliminase (PGTE) > polygalacturonase (PG) > cellulase (C_X) > polymethylgalacturonase (PMG) > pectin-transeliminase (PTE) [195]. The order of enzymes for *Calocybe indica* during the spawn growth was PG > PMG > C_X > PGTE > PTE [196].

Hemicellulase (s): These include xylanase, arabinase, and mannase, all of which are involved in degradation of the corresponding polymers representing the hemicellulose fraction of substrates. On studying the development of hemicellulase(s) activity as measured on hemicellulose-B isolated from the rice straw substrate, it was found that *P. flabellatus* displayed higher activity(ies) at the end of spawn run, and reached its maximum activity after cropping [139]. This could be correlated with corresponding amounts of hemicellulose(s) depleted from the substrate. A similar trend in development of hemicellulase(s) was also observed when *P. sajor-caju* was cultured on rice straw [176]. The secretion of xylanase by *P. ostreatus* [173] and *P.* sp. "*Florida*" on flax shive [28] has also been observed. An enzyme preparation from *Agaricus bisporus* hydrolyzed rice straw hemicellulose to xylobiose, glucose, and arabinose but not to xylose [197].

Phenoloxidases and ligninolytic enzymes: Mushroom species can secrete various types of oxidizing enzymes like phenoloxidases, peroxidases, and catalases. The lignocellulose-degrading activities exhibited by the different species of *Pleurotus* have been described [198] and compared with those of other white-rot fungi. The phenoloxidase activity by *P. flabellatus* is of a laccase type and is unable to oxidize catechol/tyrosin [146]. The inability of laccase-deficient strains of *Sporotrichum pulverulentum* to degrade lignin [199] or polymeric dyes [200], which are used as substrates for the fungal lignin degradation system and the involvement of laccase in demethoxylation processes [201], suggests that laccase has a role in the degradation of lignin. Laccase activity in the rice straw substrate increased during the spawn run and reached a maximum at the beginning of fructification, as can be deduced from the decrease of corresponding amount of lignin (estimated chemically) in the straw substrate [146]. A similar increase in laccase activity during the spawn run of *A. bisporus* in compost has been observed [202]; the extent of laccase activity can be an indirect measure of the fungal biomass accumulated in the substrate (compost). Sermanni [203] reported the comparative laccase activities of *A. bisporus* and *P. ostreatus*; it was greater in the former than in the latter, probably due to higher lignin content of the substrate or for some other unknown reason. He also has described the properties of the laccases from these two types of mushrooms. Laccase was more prominent and peroxidase activity was not significant when *P.* sp. "*Florida*" was grown on cotton straw [141]. The activities reached a maximum only after 6-8 days of growth. Different strains of *P. ostreatus* showed various levels of laccase activity; this appears to be the reason for different degrees of lignin and cotton straw degradation.

Extracellular ligninases isolated from pure cultures of *P. ostreatus* displayed substrate specificity for guaiacol and hydroquinone and yielded a positive syringaldazine test [180]. A mixed culture of *Fusarium culmonarum* and *P. ostreatus* produced the highest amounts of cellulase and ligninolytic enzymes in a synthetic medium; the ability of the former was limited in pure culture [204]. The enzymic activity of the white-rot fruiting fungi in lignin removal was stimulated by the addition of ferulic acid and 3-indolyl acetic acid to the nutrient medium [127]. *Pleurotus ostreatus* grown on fir-wood meal produced the highest levels of polyphenoloxidases, peroxidases, and laccase [172]. The dependence of production of humus-like substance on extracellular phenol-oxidase activity is reported [205].

Proteases: The fruiting bodies, on a dry weight basis, contain about 20-35% protein (with a conversion factor of 4.38). This protein is built essentially by obtaining the nitrogenous substances from the substrate; while only a fraction of the substrate is in the form of free amino acids, the rest is evidently complexed in the protein form. Obviously, the species in question should degrade the available substrate proteins to build up the fruiting bodies. *Pleurotus flabellatus* was found to secrete proteases when grown in vitro and in vivo on rice straw [139,146]; however, the in vivo activity was greater, since this set of cultural conditions involved aerobic production of fruiting bodies (with an economic turnover of ~650 g fresh mushrooms/kg dry straw). Protease activity increased continuously during the spawn run, and the activity was enhanced during fructification. Protein decrease in the straw substrate correlated with the increase in the protease activity [146]. There was also a similar increase in protease activity during differentiation of other fungi [206-208]. Evidently the increase in protease activity enables the fungus to degrade more protein in the straw substrate and to divert the degradation products to formation of fruiting bodies.

C. Properties of Degradatory Enzymes

When studying secretion of degradatory enzymes by the fruiting fungi, it is useful to know their properties (Table 4)—factors affecting their production and activity. This, in turn, helps to fix the cultural conditions so as to effect optimum generation of the degradatory enzymes and hence faster degradation of the growth substrates.

When *P. ostreatus* was grown in a medium of cotton, cellobiose, or filter paper, endoglucanase and cellobiose had highest stability at pH 4 and a temperature 30-40°C. The phenol oxidizing ability of enzyme preparations obtained from the culture filtrates of four strains of *P. ostreatus* was studied by rate of O_2 uptake using the Clark O_2 electrode [185]. These enzyme preparations were highly stable and catalyzed oxidation of a range of phenolic substrates such as *p*-cresol, *p*-hydroxy benzoic acid, hydroquinone, guaiacol, pyrocatechol, and 3,4-dihydroxybenzaldehyde, manifesting properties of both *o*-diphenoloxidase and *p*-diphenoloxidase. Of 15 species of *Pleurotus* investigated, four produced laccase but not tyrosinase [187]. Syringic acid and vanillin induced greatest increases in activity of laccase and peroxidase of *P. ostreatus* [182].

The induction of new forms of laccase by phenolic substrates like ferulic acid has been observed [218-221]. Addition of several components of lignin into the culture medium effected considerable stimulation of intra- and extra-cellular laccase production by *P. ostreatus*. Production of protocatechuate, 3,4-dioxygenase (protocatechuate: oxygen 3,4-oxidoreductase) from *P. ostreatus* as well as its properties have been studied [222].

D. Role of Degradatory Enzymes During Spawn Run and Fructification

Discovering and expressing the changes in physiological and biochemical activities in the substrate during fruiting body formation and distinguishing the differences typical of the various stages of development from spawning until the end of cropping of basidiomycetous fruiting fungi have been undertaken by several authors [223-228]. Jablonski [224], while studying activities of laccase, peroxidase, and proteinase in various growth layers of a sterilized shredded corn cob substrate incubated with *P. ostreatus*, *P.* sp. "*Florida*," *Lentinus edodes*, and *Agrocybe aegerita*, has confirmed that the complexity of fructification consists of the species specificity of metabolic pathways in different strains of fungi. Thus, both *Pleurotus* species displayed high levels of laccase and protease activities, while the opposite trend was observed in the other two species. It has been shown that laccase activity appears in compost during colonization by the mycelium of *A. bisporus* and that it declines rapidly during fruiting. Further work showed that cellulase(s) behaved in the opposite fashion—a large increase in activity taking place at the time of fruiting [202]. Wood and Goodenough [229] provide further evidence about the source and time of change of activity of these enzymes during fruiting, to see if the enzymes were produced by *A. bisporus* and to determine whether the changes were associated with fruiting initiation or with later events during fruiting.

Pleurotus, like *Agaricus*, behaves during the spawn run like a white-rot fungus with ability to degrade lignin; during fructification it behaves

TABLE 4. Properties of Degradatory Enzymes of Fruiting Fungi

Species	Enzyme	Properties	Ref.
Agaricus bisporus	CMC-ase	Induced by cellulose and cellobiose; repressed by glucose and cellobiose in presence of cellulose; two forms were discerned—one adsorbed strongly to cellulose and the other nonadsorbable.	209
A. bisporus	Laccase	Produced constitutively in defined or complex media; inhibited by sodium azide and potassium cyanide; on electrophoresis displayed 4 bands; enzyme contained 15% carbohydrate and 2 atoms of Cu per every molecule.	210
Volvariella volvacea	CMC-ase β-glucosidase	Optimum at 57°C, pH 7; inactivated at 60°C (2 h). Optimum at 65°C, pH 5.8; inactivated at 70°C (3 h).	211
V. volvacea	CMC-ase β-blucosidase	Optimum at 45°C and pH 7; molecular weight 40,000. Optimum at 45°C and pH 5.8; molecular weight 58,000.	212
Pleurotus ostreatus	Xylanase	Optimum at 50°C; inactivated at 70°C (10 min); optimum pH 5.0; inhibitor—Mn; activators—K, Mg, Ca.	213
P. ostreatus	Laccase	Optimum at 50°C; optimum pH 5.5, stable pH range (5-6); inhibitors—chlorogenic acid, esculantine; activator—ferulic acid	214
P. ostreatus	Cellulase	Optimum at 30°C; activator—$CaCO_3$ (5%).	215
	Xylanase	Optimum at 25°C; activator—$CaCO_3$ (2%).	
	Protease	Optimum at 25°C; rice straw medium with pH 5-6 was optimum for secretion.	
P. sajor-caju	Avicelase	Optimum at 50°C; inactivated at 70°C (10 min); optimum pH 5.5; inhibitor—Mg^2; activator—Ca.	216
	CMC-ase β-glucosidase	Optimum at 40°C and pH 4.5; inhibitor Ag^+. Optimum at 50°C and pH 6.0	217
P. flabellatus	CMC-ase Hemicellulase Laccase	Optimum at 50°C and pH 4.8; inactivated at 90°C (10 min). Optimum at 50°C and pH 5.4; inactivated at 90°C (10 min). Optimum at 40°C and pH 7.0.	139

like a brown-rot fungus, with high levels of cellulase activity [139]. Changes in amounts and properties of phenoloxidases have been correlated with formation of fungal fruit bodies [230-232]. Similarly, variations in the activities of cellulases also have been shown to be associated with fungal fruiting body [233,234]. It still remains to be proved that the changes observed in enzyme activity or properties are directly correlated with the differentiation process or are merely secondary events of the developmental sequence [235].

Increase in protease activity during fructification of *P. flabellatus* has been noted [146]. Likewise, fruiting body development in *A. bisporus* also is accompanied by large increases in activity of intracellular proteases. It is likely that several enzymes are involved, since activity increased at each of the three pH values (3.6, 7.0, and 9.0) studied. Fungi can synthesize at least three classes of proteases—acidic, neutral, and alkaline [236]. Intracellular proteases assuming prominence during fructification are known in basidiomycetes and other microorganisms [235,237,238]. It is thought that these proteases are involved in increasing the rates of protein turnover associated with development. The protein turnover supplies amino acids to allow synthesis of proteins not previously present during development. Chang and Chang [239] have shown that new proteins are formed both in different tissues and at different times of development of fruiting bodies by *Volvariella volvacea*. Netzer [207] reported that proteases are present in all developmental stages of *P. ostreatus*. The activity increases significantly during primorida formation. This leads to the assumption that they may be involved in the differentiation process itself. Iwahara et al. [240] also discuss the extracellular enzymes during the life cycle of *P. ostreatus*.

E. White-Rot Reactions

Fruiting fungi like *Pleurotus* and *Agaricus*, during their growth on a lignocellulosic waste, degrade lignin preferentially so the degraded substrate becomes more bright (whiter) in color due to the exposed cellulose; such fungal degradation is called "white-rot" decay. This characteristic has been studied in three categories:

1. Ability of the fungal cultures to color phenolic media
2. Secretion of phenoloxidases
3. Capacity to degrade lignin by fungal monocultures

The first will be discussed here; the latter two have already been considered.

Pleurotus flabellatus produced intense color reactions with most phenolic compounds, but not with tyrosine [146]. These reactions are similar to those of *P. ostreatus* [241]. The association of phenoloxidase production with lignin degrading ability was emphasized by Bavendamm [242], who used a simple detection test (the coloration of phenolic media by the fungus under study) to separate white-rot from brown-rot fungi. The correlation between the capacity of a fungus to degrade lignin as manifested by formation of white-rot decay and the secretion of extracellular phenoloxidases is well documented [243-245].

Platt et al. [246] demonstrated the ability of *Pleurotus* to decolorize the polymeric dye poly-blue (polyvinylamine sulfonate—anthroquinone). *Pleurotus ostreatus* sp. "*Florida*" decolorized the dye both in solid and liquid media. Decolorizing ability developed in the absence of the dye, but only when the fungus had been previously cultivated on lignin-containing substrates. Thus, it appears that decolorization is a useful indication of lignin-degrading ability and can serve as a tool for screening the lignolytic ability of an organism. However, it may throw light only on one of the many steps in the long and complex process of lignin degradation, since according to present-day knowledge about lignin degradation, lignin does not induce the lignin-degrading system [247].

F. Aspects on Lignocellulose Degradation

The mechanism of enzymatic cellulose and hemicellulose degradation is a vast subject. The requirement for various kinds of cellulolytic and hemicellulolytic enzymes for degradation of corresponding polysaccharides, along with secretion of different classes of the polysaccharases by the fruiting fungi, support the capacity of these fungi to degrade holocellulose. However, the ability of these fungi to delignify the substrate is a unique feature. Many reviews are appearing from time to time on this subject. Only a few references dealing with the edible fruiting fungi are included.

While growing *Pleurotus* on wheat straw, Ginterova and Lazarova [248] observed a decrease of 20% dry weight and a fast decrease of hemicellulose and lignin during spawn run; during fructification of the substrate loss ranged up to 45% and cellulose consumption was higher than lignin. Compared to *Agaricus bisporus* (which degraded 7-12% lignin), the highest rate of lignin degradation (37-40%) was observed with *P. ostreatus* and *P. florida*, when they were grown in a synthetic medium containing wheat straw as a carbon source [249]. The ability to degrade lignin by *Pleurotus* is more rapid than that of *Sporotrichum* and *Lentinus* [124]. Two strains of oyster mushrooms (K-6 and 108) differed by the dynamics of lignin and cellulose utilization when grown on aspen, ash, alder, and oak trees. The xylosis index during growth of these strains on various woods ranged from 0.26 to 0.68 [130]. *Pleurotus ostreatus*, when cultured on aspen wood, degraded more cellulose than that of lignin—44 and 10.5%, respectively, after 30 months, and 75 and 33% after 48 months [129].

At a high N content (30 mM) in the medium, lignin degradation was not affected by *L. edodes* and *P. ostreatus*; cultures of *Phlebia brevispora* and *Pholiota mutabilis* showed weak to moderate lignin-degrading activity, whereas low N (2.6 mM) stimulated ligninolytic activity by *P. brevispora* [250]. Commonday and Macy [251] raised *P. ostreatus* on solid media containing either growth-limiting (1 mM) or excess (10 mM) NH_4Cl and found that $^{14}CO_2$ released from ^{14}C-labeled corn stover lignin in low N medium was three times that in the N-rich medium. Supplementation of the low N medium with glucose (0.3%) further enhanced ligninolytic activity [251]. At 20 mM nitrogen, lignin degradation was suppressed by 15% by *P. ostreatus* [252].

Lignosulfonic acid fractions isolated from sulfite waste liquor were degraded by liquid cultures of *P. ostreatus* using glucose as an additional carbon source; the rate of degradation was inversely proportional to the molecular weight of the fraction [253].

Lignin degradation by *Lentinus edodes* and *Grifola frondosa* was faster in an atmosphere of oxygen than in air. *Pleurotus* degraded lignin at equal rates in oxygen and in air. This difference reflects on the ecological considerations of these fungi [254]. In a study of biodegradation of wood components by *P. ostreatus*, lignin in poplar bed-logs was strongly decomposed with a progressive decrease in methoxyl and ester groups. The syringyl moiety of lignin was decomposed more rapidly than the guaicyl moiety [255]. Two different pathways of lignin degradation may exist within the genus *Pleurotus*: the route in which splitting is the first process and the path whereby demethylation is followed by ring splitting [256].

By growing a culture of *P. ostreatus* on ^{13}C-enriched DHP-lignin followed by ^{13}C-NMR spectra of the degraded lignin, it was evident that both the aliphatic side chain and the aromatic nucleus were degraded [257]. ^{14}C-Lignosulfonic acid released more $^{14}CO_2$ when incubated with *P. ostreatus* than did use of natural plant lignin for degradation. Preincubation of the earlier substrate with *Pseudomonas putida* followed by *P. ostreatus* effected a marked increase of $^{14}CO_2$ release [258]. *Pleurotus salignus* and *P. ostreatus* displayed the highest rate of lignin degradation as measured by formation of $^{14}CO_2$ (when cultured in synthetic medium containing ^{14}C-labeled lignin from wheat straw) when compared with *Agaricus bisporus* and *Phyllotopsis nidulans* [259].

V. APPLICATIONS AND IMPLICATIONS OF BIOCHEMICAL PROPERTIES OF FRUITING FUNGI

Several uses—both applied and implied—have been advocated by many researchers for spent substrate. These are mainly based on the spectrum of qualitative and quantitative changes, chemical as well as biochemical (enzymatic), brought about in the spent substrate by growth and fructification of the fruiting fungi. Depending upon the species/strain studied, cultural conditions provided, and the substrate used, the output of biomass in the form of fruiting bodies, loss of substrate, as well as the amount and quality of spent substrate also vary. In general, the spent substrate is more soluble, consisting of greater amounts of free sugars (3-5 times compared to the undegraded substrate) and free amino acids (2-3 times that of undegraded substrate) and cellulose with lesser degree of polymerizaztion. Concomitant with the decrease in cellulose, hemicellulose, and lignin contents, is an increase in the ash content. The wet, degraded substrate also possesses a variety of degradatory enzymes like cellulase(s), hemicellulase(s), oxidizing enzymes, and proteases. Several uses of such a spent substrate and the role of fruiting fungi in ecology of degradation of natural lignocellulosic wastes as related to the carbon cycle are represented in Figure 1.

A. Spent Substrate as an Upgraded Ruminant Feed

Suitability of spent straw as cattle feed after the growth of *P. florida* was suggested by Schanel et al. [153] and Kaneshiro [131]. There was an

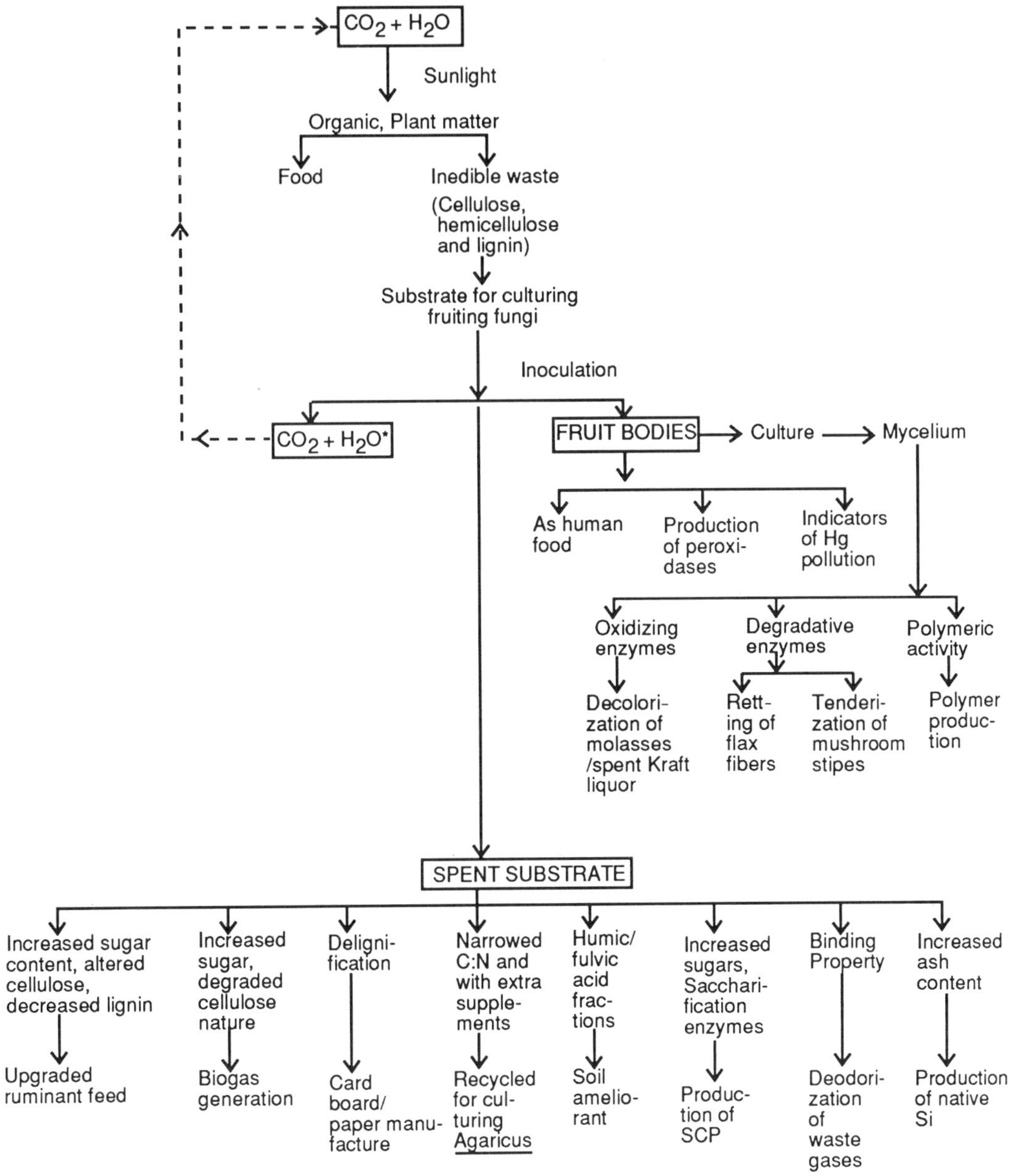

FIGURE 1. Role of edible fruiting fungi in the ecology of carbon cycle and applications and implications of their culturing. *As high as 70% of dry wheat straw substrate while culturing *Pleurotus florida* [38].

increase of 12% in units of in vitro digestibility of wheat straw incubated with *P. florida* [260]. *P. ostreatus* grew well on wheat straw, and after an incubation of 28 days improved the digestibility by only 6% [169]. Combined incubation of *P. ostreatus* with *Erwinia carotovora* increased the in vitro dry matter digestibility from 32.7 to 47.7%.

Zafar et al. [261] studied the ability of *P. sajor-caju* to degrade rice straw cellulose as well as the increase in digestibility in vivo of such a degraded straw. These authors observed that fungal growth effected an appreciable rise in the protein content of substrate from 1.7 to 10%; this may be considered an important development in the nutritive value of the straw. Furthermore, the authors did not mention the yield of fruiting bodies, thus the experiment appears to be an example of in vivo colonization of the hot water-treated rice straw (before the yield of fruiting bodies by the fungus) followed by in vivo digestibility. There was an increase in in vivo digestibility of the degraded straw by 34.1% as compared to the undegraded straw.

Before assessing the value of spent straw substrate as cattle feed, it is necessary to determine that the substrate is nontoxic. Although it is a well-established fact that *Pleurotus* spp., when cultivated, are edible, during commercial cultivation, when the substrate is neither pasteurized nor sterilized, rules out monoculturing because toxic substances might be produced by foreign organisms in the straw substrate. Bano et al. [262] used thin layer chromatography (TLC) of straw extracts and animal feeding trials to study the nontoxic nature of the spent straw after development of *P. sajor-caju* fruiting bodies. TLC studies did not show presence of any of the 12 mycotoxins, namely, aflatoxins B_1, B_2, G_1, G_2, ochratoxins A, B, and C, sterigmatocystin, zearalenone, patulin, penicillic acid, and citrinin. The spent straw, when fed to two sets of animals, namely, albino rats and mice, of both sexes did not produce any histological or histopathological abnormalties. No toxic effect or illness was observed in chicks or pigs fed with wheat straw that had been incubated with the oyster mushroom *P. ostreatus* [263].

Acid hydrolyzates of farming and food wastes containing 25-45% dry matter were neutralized with lime and inoculated with mycelia of *P. ostreatus*, *Phallus impudicus*, or *Coprinus comatus*. After 4-6 weeks at 25°C in perforated sacks, fermentation resulted in an increase of protein content of ~10% and decomposition of polysaccharides to easily digestible sugars. Such a fermented substrate has been suggested as a feed additive [264]. When rye straw was biologically hydrolyzed by *P. ostreatus* culture at 15-25°C for 6 weeks, there was an increase in dry matter digestibility and N-free extractives; the fodder value of such a substrate increased twofold. This is not inferior to treatment of straw with NaOH or NH_4OH to make the straw suitable as feed [265]. Biological treatment of wheat straw with *P. ostreatus* for 21 days at 30°C effected increase in crude protein content, followed by an increase in in vitro digestibility.

Production of a product with decreased lignin content, increased protein content, and increased digestibility by fermentation of feed lot waste with *P. ostreatus* and *Coprinus aratus* is reported [54]. Zadrazil [137] studied the quality of wheat straw after subjecting it to degradation by *P. ostreatus*, *P. eryngii*, *Flammulina velutipes*, and *Agrocybe aegerita*. The first effected maximum straw degradation. The fungi that release glucose from wheat straw, in order of increasing effect, are: *P. ostreatus, P. eryngii, A. aegerita,* and *F. velutipes.* Harvest of the fruiting bodies of the last two resulted in a decrease in the water-soluble fraction of the substrate, whereas with *Pleurotus* species the content increased. The significance of using such a degraded straw as an animal feed supplement as a function of duration of fungal growth is discussed.

B. Spent Substrate for Generation of Biogas

The possibility of utilizing straw degraded by fruiting fungi for production of biogas has been studied by Muller et al. [266,267]. In this study, 22 basidiomycetes were employed to degrade wheat straw in sterile, solid-state cultures. *Pleurotus ostreatus, P.* sp. "*Florida,*" and the mutants of *P.* sp. "*Florida*"—PFP 343 and PFP 551—isolated from the basidiospores of *P.* sp. "*Florida*" and described by Ericksson and Goodell [268] were used. Species like *P. ostreatus, P.* sp. "*Florida,*" and *Stropharia rugosoannulata* degraded more lignin as indicated by their high affinity of lignin. *Pleurotus*-treated straw released more of sugars by enzymatic hydrolysis. This straw was subjected to fermentation by methanogenic bacteria to produce biogas.

Japanese workers [269] reported an efficient means of biogas production during mushroom growth on cellulosic materials. While harvesting the mushrooms from time to time, the medium is replenished with cellulosic materials. When the sugar content of the medium increases and mushroom yield decreases, the spent medium is subject to methane fermentation. Biological pretreatment by *P. sajor-caju* enhanced the biogas yield from rice straw by 54% and by 52% for *Cyamopsis tetragonoloba.* Increased susceptibility due to degradation and increased N content (about three times that of undegraded substrate) as a consequence of fungal growth seem to be responsible for enhanced production of biogas [270].

C. Cardboard/Paper Manufacture

The active ligninolytic property of certain white-rot basidiomycete species could be exploited for production of pump from lignocellulosic wastes. The degraded substrate with its decreased lignin content (40-50% of the undegraded lot) would be suitable for production of cardboard or for paper manufacture. In fact, a patent [271] has been issued describing a method to delignify lignocellulosic materials with *P. ostreatus,* giving a product suitable for paper or board manufacture. The process is less expensive to operate than a chemical pulping process and avoids pollution problems. Its only drawback is that such biological delignification is quite slow; it is worth evolving species/strains of *Pleurotus* endowed with the ability to grow still faster and degrade lignin more actively.

Pilon et al. [272,273] reported on the biological pulping activity of *Pleurotus* along with that of other fungi. An examination of different fungal species has confirmed that it is among the white-rot fungi that potential candidates may be found for efficient biological treatment of pulp. The brown-rot fungi had a detrimental effect on paper strength, since cellulose and hemicellulose degradation can only lead to eventual deterioration of cellulose fibrils and, therefore, reduce fiber strength and ultimately paper strength. Since brown-rot fungi do not decompose lignin or lignin fragments significantly, and they degrade the carbohydrate moiety of wood, whereas white-rot fungi act on lignin, the latter are preferred examples for biological pulping. Biological treatment proceeds essentially the same way as chemical delignification, but more slowly and less drastically. Modification of lignin in the fibers occurs during the process as reflected by increased WRV (water retention value) and acid capacity of the pulp [272]. Fukuzumi et al. [274] also have screened the white-rot fungi, inclusive of *Pleurotus*, for biological pulping.

D. Recycling of Spent Substrate

The spent substrate consisting of used tea leaves and waste paper, after harvesting of *P. ostreatus*, provides the correct nutritional conditions for the growth of *A. bisporus* [275]. The function of gypsum to provide a more granular structure with increased water-holding capacity necessary for greatly improved mycelial growth with strand formation can be achieved by use of tea leaves together with waste paper and mycelium of *P. ostreatus*. Used tea leaves were also found to support the growth of many basidiomycetes [276]. The spent substrate appeared to supply the decomposition products of cellulose, hemicellulose, and lignin [277], which are provided in traditional mushroom cultivation by the compost and the end products of microbial degradation [278]. Several studies of Zadrazil [38,162, 163] have shown that during cultivation of *Pleurotus* species, cellulose-lignin complexes are decomposed by the fungus mycelium. In this process, free CO_2 is given off, the C:N ratio falls, and proportions of amino acids, proteins, vitamins, and minerals are altered. It is also known that proteins, organic N, a lowered C:N ratio Ca, Mg, K, and vitamins are necessary to support growth of *A. bisporus* [277]. Thus, utilization of used tea leaves and waste paper can result in production of two mushrooms in succession—first a crop of *P. ostreatus* followed by a crop of *A. bisporus*. Further, biodegradation of the waste materials has also been accomplished. The approximate economics of this process has been worked out.

Growth of *P. ostreatus* on a mixture of straw and cores of maize cobs left much organic matter unused, increased the protein content of the substrate, and formed humus, which leads to formation of N-rich lignin-humus complex, the material used by *A. bisporus* [279]. The possibility of using such a spent compost (after harvesting *P. ostreatus*) after suitable supplementation with horse or chicken manure and inorganic salts to obtain the correct C:N ratio and an optimum concentration of soluble salts for the growth of *A. bisporus* has been investigated [68].

E. Spent Substrate for Soil Amelioration

The spent straw substrate, particularly in its degraded form, can add to soil fertility, due to presence in it of constituents like humic/fulvic acid fractions. Grabbe [280] reported that the spent substrate can be used as a soil-improving agent by supplementing it with low-salt organic materials like peat and bark. A mixture of the spent substrate with these materials in a 1:1 ratio supplies humic substances suitable as a substrate for tree nurseries. In fact, the spent straw resulting from culturing *Volvariella* has proved to be effective in obtaining profitable yields of crops like tomato and radish [9]. Use of spent substrate after mushroom harvesting is reported to be more effective than fresh horse manure for composting and subsequent use in vegetable growing [281].

F. Deodorization of Waste Gases

An odorous waste gas can be deodorized by passing it through a layer of a compost-based medium containing rice hulls and spent mushroom culture medium obtained after culturing of *Pholiota nameko*. Removal of H_2S, NH_3, MeSH, Me_2S, and Me_2S_2 was ~ 99, 99, 99, 83, and 72%, respectively [282].

G. Decolorization of Spent Kraft Liquor and Molasses Pigment

Bleached Kraft mills usually discharge large volumes of brown effluents as a result of the different operations used in processing wood and pulp. To minimize pollution, these effluents are treated in most mills by biological oxidation in aerated lagoons and/or activated sludge systems before being released into streams. Such aerobic treatments have the ability to reduce BOD and COD of the effluents, but they fail to reduce color efficiently. The brown color of the effluents originates mainly from lignin degradation products, most of them chlorinated. It is well known that species of *Pleurotus*, as representative examples of white-rot fungi, are able to degrade and metabolize not only native lignin, but also Kraft lignins and lignosulfonates. *Pleurotus ostreatus* and *Polyporus versicolor* removed 40% of the color after 8 days [283]. *Phanerochaete chrysosporium* achieved over 60% decolorization after 4 days' cultivation, but thereafter the color increased slightly. It would be worthwhile to study the efficiency of *Pleurotus* for decolorizing and also to explore the possibilities of utilizing the enzyme systems in spent substrate for decolorizing activity.

Decolorization of the pigment of molasses is a major problem since molasses serves as one of the important raw materials for various fermentation industries due to its low cost and availability. In a fermentation factory using molasses, waste water contains much dark brown pigment (molasses pigment, MP), which is a kind of melanoidin and is not decolorized by usual biological treatments such as the activated sludge process. Decolorization of this MP remains a problem to be solved [284,285]. Aoshima et al. [285] found that *P. ostreatus* decolorized the MP, as determined by the giant colony method. Testing this property in the shake cultures, *P. ostreatus* displayed about 65% decolorization. *Pleurotus ostreatus* IFO 6519 has been selected among 24 white-rot basidiomycetes as suitable for decolorization of PVPP-treated molasses [286].

Fermentation wastes, especially those from alcoholic fermentation of molasses, are decolorized by incubation with *Pleurotus* or *Auricularia* [287]. Use of fruiting fungi for improving the biological treatment of phenol containing waste water is discussed [288-290]; this is based on the exploitation of the property of basidiomycetous fungal cultures to produce phenol-oxidizing enzymes.

H. Source of Saccharification Enzymes and Production of Single Cell Proteins

It is well known that several species of *Pleurotus* produce cellulolytic and hemicellulolytic enzymes in vitro. Further, after harvesting of fruiting bodies of *P. flabellatus* and *P. sajor-caju*, the wet rice straw substrate contained significant activities of the saccharifying enzymes [157,176]. Indeed, C_1, CX_1 FPD, and β-D glucosidase activities have been demonstrated in the wet substrate. A suitable extract of such a substrate, in simple buffer solutions (which are least expensive), and further incubation with simple cellulosic wastes like waste paper, cotton waste, etc. should aid in further saccharification of the waste. In addition, the enzymic extract by itself consists of considerable amounts of free sugars extracted from the spent straw. Saccharification of this type should help in the production of single-cell proteins (SCP). Production of several kinds of SCP on the straw hydrolyzate produced through the growth of *Pellicularia* [291] is worth trying with fruiting fungi.

Wheat straw degraded by *Pleurotus* species for about 14 days doubled in sugar content compared to undegraded straw [38]. Subjecting the straw to biodegradation after alkali treatment enhanced the effectiveness of saccharification; however, the fungus alone can serve the function of NaOH (without environmental pollution) but does so more slowly [293]. Exploitation of the spent straw substrate extract or enzymes for the production of SCP is worthy of study in view of the findings that certain yeast strains can ferment D-xylose to ethanol [294,295]. Biological modification of wheat straw with *P. ostreatus* followed by treatment with cellulase allowed a 72% conversion of cellulose to glucose, while chemical treatments with subsequent cellulase digestion effected 73-80% conversion; cellulase alone without any straw pretreatment saccharified only 10% of the cellulose [296]. Out of 39 strains of fungi selected for degradation of lignocellulosics, a mixed culture of *P. ostreatus* and *Fusarium culmonarum* produced the highest amounts of cellulolytic and ligninolytic enzymes, which can be exploited for saccharification of lignocellulosics [297].

I. Retting of Flax Fibers

Sharma [298] studied the use of polysaccharide-degrading enzymes for removal of noncellulosic materials (like hemicellulose, pectin, and lignin) present on flax fiber (dew retted at 45°C). An enzyme extract of flax shieve colonized by hybrids of *P. ostreatus* and *P. florida* also was used together with other enzyme sources. Xylanase, polygalacturonase, and laccase were found in high concentrations, along with minor amounts of pectinylase and cellulase, in extracts of *Pleurotus* (as compared to other enzyme sources). All the enzyme-treated roves produced high quality yarns compared with yarns spun from the untreated roves. Fluidity of all the yarns spun from enzyme-treated roves was low, suggesting that the enzymes did not affect the cellulose fibers. The importance of such an enzymatic treatment as an alternative to boiling with NaOH to save energy and to achieve the desired goal is discussed in this reference.

J. Peroxidase Manufacture

Coprinus macrorhizus has been employed to produce peroxidase in liquid cultures; a medium that contained iron salts enhanced enzyme yield [299].

K. Indicators of Mercury Pollution

Mushrooms collected in the vicinity of a chlor-alkali factory contained 10 times more mercury than those grown in other areas. Use of commonly occurring *Lycoperdon gemmatum*, *Mycena pura*, and members of collybiaceae as indicators in the study of mercury pollution has been advocated [300].

L. Tenderization of Mushroom Tough Stipes

A protein extract of *P. ostreatus* with chitinase, chitobiase, and glucanase (against alkali-insoluble glucan) was able to soften *P. ostreatus* stipes as well as those of *A. bisporus* and *L. edodes* [301].

M. Preparation of Native Silica from Lignocellulosic Substrates

The capacity of white-rot fungi, including *Pleurotus*, to degrade rice straw in order to obtain increased ash and silica (for use in manufacture of photovoltaic cells) has been recently reported [302]. The silica present in rice straw and husk is highly reactive, amorphous, and has a high surface area like silica gel. Use of heat/pyrolysis destroys the native amorphous structure and high surface area of the deposited silica. Incineration renders the native silica more coarse, crystalline, and dark in color so that it needs to be ground, using energy to be suitable for making cement. Light and scanning electron microscopy of degraded rice straw revealed an altered structure after *Pleurotus* degradation with a loss in dry weight of about 50% over 100 days of incubation. If the degradation efficiency of *Pleurotus* could be further enhanced over that of *Cyathus*, this might prove a promising way to obtain silica in its original deposited form without recourse to heating that is associated with the ashing process.

N. Polymer Production

Five strains of *Pleurotus* have been investigated for potential use as high viscosity polymer-producing organisms [303]. The high viscosity biopolymers find applications in foods, enhanced oil recovery, cosmetics and proprietary industrial preparations. Polymer production varied markedly from strain to strain; however, some of the strains surveyed were capable of producing several grams of polymer per liter under the test conditions. This approaches the concentrations obtained with industrial production strains of *S. rolfsii*, which produce a similar gum. *Pleurotus* strains required less N in the fermentation broth than did *S. rolfsii*.

VI. CONCLUSIONS

Most fruiting fungi, with their fast growth rate, can rapidly colonize lignocellulosic substrates. The search for new species, their identification, study of their growth rate and fructification involving biomass turnover and the quality of biomass produced will obviously be a continuous undertaking. Research on development of newer strains by genetic breeding or protoplast techniques is worth considering at this juncture. Such work would increase chances of selection for biological functions discussed thus far. Once the culture is purified and selected for desired traits, it is important to preserve its viability and vigor. The literature contains only a few articles dealing with this subject. Nevertheless, this subject should assume greater importance in the biotechnology of fruiting fungi. Their ability to degrade a wide range of substrates—fermented or unfermented, pasteurized or unpasteurized, sterilized or unsterilized—places them in the front rank of biological agents that yield biomass through conversion of plant wastes. Their requirement for less arable land, their independence of sunlight, their faster protein conversion efficiency compared to such animals as sheep, goat, pig, hen, etc. prompts their consideration as some of the most useful organisms on the globe for the production of human foods. In fact, FAO has recommended their use as food supplements by protein-deficient populations in developing and underdeveloped countries [304]. Being a labor-intensive project, this would obviously help to generate employment.

Their biochemical potential to decompose and delignify lignocellulosic substrates through secretion of a spectrum of degradative enzymes and to increase the substrate sugar content, as well as to modify the nature of cellulose, have led to diversified applications for the spent substrate. The magnitude of possible bioconversion of vast amounts of cereal straws available into biomass and fertilizer by employment of the fruiting fungi is quite remarkable (see Table 2). All these factors lead us to conclude that edible fruiting fungi, apart from yielding edible biomass, play a vital role in nature's carbon cycle. It is up to the ingenuity of mankind to further exploit their known potential and to identify biological abilities not yet observed.

REFERENCES

1. Chang, S. T., and Miles, P. G. Introduction to mushroom science. In *Tropical Mushrooms—Biological Nature and Cultivation Methods* (S. T. Chang and T. H. Quimio, eds.), Chinese Unversity Press, Hong Kong, 1982, pp. 3-10.
2. Bassham, J. A. General considerations. In *Cellulose as a Chemical and Energy Resources. Biotech. Bioeng. Symp.* No. 5 (C. R. Wilke, ed.), Interscience Publishers, New York, 1975.
3. Zadrazil, F., and Grabbe, K. Edible mushrooms. In *Biotechnology,* Vol. 3 (H. Dellweg, ed.), Verlag Chemie, Florida, 1983, pp. 145-165.
4. Zadrazil, F. White-rot fungi and mushrooms grown on cereal straw: Aim of the process, final products, scope for the future. In *Degradation of Lignocellulosics in Ruminants and Industrial Processes* (J. M. Van Der Meer, M. P. Rijkens, and B. A. Ferrante, eds.), Elsevier Applied Science, London, 1985, pp. 55-62.
5. Z. Bano and Rajarathnam, S. *Pleurotus* mushrooms. Part II. Chemical composition, nutritional value, post-harvest physiology, preservation and role as human food. *CRC Crit. Rev. Food Sci. Nutr., 27:* 87-158, 1988.

6. Philipps, R. W. The relative efficiencies of plants and animals. In *Biological Efficiency in Agriculture* (C. R. W. Spedding, J. W. Walsingham, and A. M. Hoxey, eds.), Academic Press, New York, 1981, pp. 355-383.
7. Bech, K. Preparing a productive commercial compost as a selective growing medium for *Agaricus bisporus*. *Mushroom Sci.*, *10*(2):77-83, 1979.
8. Hu, K. J., Song, S. F., and Liu, P. The comparison of composts made of different raw materials for *Volvariella volvacea*. *Mushroom Sci.*, *9*:687-690, 1974.
9. Chang, S. T. The prospects for mushroom protein in developing countries. In *Tropical Mushrooms: Biological Nature and Cultivation Methods* (S. T. Chang and T. H. Quimio, eds.), Chinese University, Hong Kong, 1982, pp. 463-474.
10. Rajarathnam, S., and Z. Bano. Biological significance of the natural cellulosic wastes degraded by *Pleurotus* mushrooms. *Indian Mushroom Science*, *2*:296-304, 1987.
11. Rajarathnam, S., and Z. Bano. *Pleurotus* mushrooms. Part III. Biotransformations of natural lignocellulosic wastes: Commercial applications and implications. *CRC Critic. Rev. Food Sci. Nutr.*, *28*(1): 31-113, 1989.
12. Zadrazil, F. Pretreatment of lignocellulosics for edible fungi. In *Treatment of Lignocellulosics with White-rot Fungi* (F. Zadrazil and Reiniger, eds.), Elsevier Applied Science, London, 1987, pp. 3-13.
13. Zetelaki-Hovrath, K. Protein enrichment of lignocellulosic agricultural wastes by mushrooms. *Biotech. Bioeng.*, *26*:389-396, 1984.
14. Staniforth, A. North American enterprises finds alternative to burning. *Farmers Weekly*, *99*:106-111, 1983.
15. McHale. *Ecological Context*, George Braziller, New York, 1970.
16. Dunlap, C. E. The evaluation of cellulose sources for unconventional uses. Final report on NSF Grant AER 76-17912 by the University of Missouri, Columbia, 1979.
17. Hayes, W. A., and Lim, W. C. Wheat and rice straw composts and mushroom production. In *Straw, Its Utilization and Disposal* (E. Grossbard, ed.), John Wiley and Sons, New York, 1979.
18. Penn, C. A. A brief outline of straw production, utilization and disposal in England and Wales. *Mushroom J.*, *37*:13-16, 1976.
19. Wood, D. A. Straw and the future. *Mushroom J.*, *164*:257-265, 1986.
20. *FAO Production Year Book*, Vol. 39, Food and Agriculture Organization, Rome, 1985.
21. Schisler, L. C., and Patton, G. The use of marine fishery products as mushroom compost additives. *Mushroom Sci.*, *9*:175-181, 1974.
22. Rolz, C., De Leon, R., De Arriola, M. C., and De Cabrera, S. Biodegradation of lemon grass and Citronella bagasse by white-rot fungi. *Appl. Environ. Microbiol.*, *52*:607-611, 1986.
23. Robbins, S. R. J. Selected markets for the essential oils of lemon grass, citronella and eucalyptus. *Trop. Prod. Inst. Rep.* 6171, 1981.
24. Paturau, J. M. *Byproducts of the Cane Sugar Industry*, Elsevier, Amsterdam, 1982.

25. Rolz, C., de Leon, R., de Arriola, M. C., and de Cabrera, S. White-rot fungal growth on sugarcane lignocellulosic residue. *Appl. Microbiol. Biotechnol.*, *25*: 535-541, 1987.
26. Platt, M. W., Chet, I., and Henis, Y. Lignocellulose degradation during growth of the mushroom *Pleurotus* sp. *'Florida'* on cotton straw. *Eur. J. Appl. Microbiol. Biotechnol.*, *13*:194-195, 1981.
27. Gohl, B. *Tropical Feeds*, FAO, Rome, 1981.
28. Sharma, H. S. S. Comparative study of the degradation of flax shive by strains of *Pleurotus*. *Appl. Microbiol. Biotechnol.*, *25*:542-546, 1987.
29. Jung, H. G., and Fahey, G. C. Nutritional implications of phenolic monomers and lignin: A review. *J. Anim. Sci.*, *57*:206-219, 1983.
30. Nicolini, L., Von Hunolstein, C., and Carilli, A. Solid state fermentation of orange peel and grape stalks by *Pleurotus ostreatus*, *Agrocybe aegerita* and *Armillaria mellea*. *Appl. Microbiol. Biotechnol.*, *28*: 95-98, 1987.
31. Mourin, T., Seki, H., Hasegawa, K., and Seki, Y. Studies on treatment of waste water from mandarin orange canneries. *Toyo Shokuhin Kenkyusho Kenkyu Hokokusho*, *14*:113-125, 1981.
32. Bressani, R. Subproducts del fruto del cafe. In *Pulpa de Cafe, Composicion, Technologia y Utilizacion* (J. E. Braham and R. Bressani, eds.), CIID, Bogota, 1979.
33. Adams, M. R., and Dougan, J. Biological management of coffee processing wastes. *Trop. Sci.*, *23*:178-196, 1981.
34. Braham, J. E., and Bressani, R. *Pulpa de Cafe Composicion, Technologia y Utilizacion*, CIID, Bogota, 1979.
35. Guzman, G., and Martinez, D. *Pleurotus* growing on coffee-pulp in a semi-industrial plant—a new promising mushroom cultivation technology in the subtropics of Mexico. *Mushroom News Lett. Trop.*, *6*:7-10, 1986.
36. Zadrazil, F. Possibilities for the use of sulfite liquors in mushroom culture and in fungus protein production. *Champignon*, *15*:23-26, 1975.
37. Rajarathnam, S., and Z. Bano. *Pleurotus* mushrooms. Part 1A. Morphology, life cycle, taxonomy, breeding and cultivation. *CRC Crit. Rev. Food Sci. Nut.*, *26*:157-223, 1987.
38. Zadrazil, F. The ecology and industrial production of *Pleurotus ostreatus*, *Pleurotus florida*, *Pleurotus cornucopiae*, and *Pleurotus eryngii*. *Mushroom Sci.*, *9*:621-652, 1976.
39. Z. Bano and Rajarathnam, S. Studies on the cultivation of *Pleurotus sajor-caju*. *Mushroom J.*, *115*:243-245, 1982.
40. Martinez, D., Guzman, G., and Soto, C. The effect of fermentation of coffee pulp in the cultivation of *Pleurotus ostreatus* in Mexico. *Mushroom News Lett. Trop.*, *6*:21-28, 1985.
41. Bisaria, R., and Madan, M. Lignin degradation by an edible mushroom *P. sajor-caju*. *Curr. Sci.*, *53*:322-324, 1984.
42. Manu-Tawaiah, W., and Martin, M. Cultivation of *Pleurotus ostreatus* mushroom in peat. *J. Sci. Food Agric.*, *37*:833-838, 1986.

43. Gujral, G. S., Bisaria, R., Madan, M., and Vasudevan, P. Solid state fermentation of *Saccharum munja* residue into food through *Pleurotus* cultivation. *J. Ferment. Technol.*, *65*:101-105, 1987.
44. Platt, M., Chet, I., and Henis, Y. Growth of *Pleurotus ostreatus* on cotton straw. *Mushroom J.*, *120*:425-427, 1982.
45. Mueller, J. C., Gawley, J. R., and Hayes, W. A. Cultivation of the shaggy mane mushroom (*Coprinus comatus*) on cellulosic residues from pulp mills. *Mushroom News Lett. Trop.*, *6*:15-20, 1985.
46. Campbell, A. C., and Slee, R. W. Commercial cultivation of shitake in Taiwan and Japan. *Mushroom News Lett. Trop.*, *7*:127-134, 1987.
47. Chang, S. T. *Volvariella volvacea*. In *The Biology and Cultivation of Edible Mushrooms* (S. T. Chang and W. A. Hayes, eds.), Academic Press, New York, 1978, pp. 573-603.
48. Tschierpe, H. J., and Hartman, K. A comparison of different growing methods. *Mushroom J.*, *60*:404-416, 1977.
49. Quimio, T. H., and Abilay, L. E. Notes on the cultivation of *Collybia reinakeana*, a seemingly potential edible mushroom. *Mushroom News Lett. Trop.*, *3*:6-7, 1983.
50. Huang, N. L. Cultivation of *Tremella fuciformis* in Fujian, China. *Mushroom News Lett. Trop.*, *2*:2-4, 1982.
51. Lin, J. N., Zheng, C. R., Zhang, Q. L., and Liu, Q. M. Studies on the artificial cultivation of *Dictyophora duplicata* (Bosc) Fisch. *Mushroom News Lett. Trop.*, *2*:14-15, 1982.
52. Muller, J. Cultivation of the oyster mushroom, *Pleurotus ostreatus* (Jacq. ex Fr.) Kummer, on *Cassia* substrate. *Mushroom J. Trop.*, *7*:89-96, 1987.
53. Kahlon, S. S., and Arora, M. Utilization of potato peels by fungi for protein production. *J. Food Sci. Technol.*, *23*:264-267, 1986.
54. Jauhari, K. S., and Sen, A. Production of protein by fungi from agricultural wastes. II. Effect of carbon/nitrogen ratio on the efficiency of substrate utilization and protein production by *Rhizoctonia melongina, Pleurotus ostreatus* and *Coprinus aratus, Zbl. Bakt.*, *133*: 597-603, 1978.
55. Jauhari, K. S., and Sen, A. Production of protein by fungi from agricultural wastes. V. Effect of various organic acids and growth promoters on the efficiency of substrate utilization and protein production by *Rhizoctonia melongina, Pleurotus ostreatus*, and *Coprinus aratus, Zbl. Bakt.*, *133*:614-618, 1978.
56. Nanci Togyo Col. Ltd. Cultivation of mushrooms. Jpn. Kokai Tokkyo Koho 8096, 090 (C1 C12 N1/14), 21 Jul 1980, Appl. 79/617, 10 Jan 1979.
57. Chahal, D. S., Ishaque, M., Bromillard, D., Chornet, E., Overend, R. P., Jauline, L., and Bonchard, J. Bioconversion of hemicelluloses into fungal protein. *J. Ind. Microbiol.*, *1*:355-361, 1987.
58. Terashita, T., and Kono, M. Utilization of the industrial wastes for the cultivation of *Pleurotus ostreatus*. *Kinki Daigaku Daigaku Nogakubu Kiyo*, *17*:113-119, 1984.
59. Takahashi, Z. Production of *Pleurotus ostreatus* on waste celluloses. *Kanzume Jiho*, *55*:153-157, 1976.

60. Chanter, D. O., and Cooke, D. Some factors affecting the growth rate of the sporophores of the cultivated mushroom. *Scientia Hortic.*, *8*:27-37, 1978.

61. Zadrazil, F., and Kurtzman, R. H. The biology of *Pleurotus* cultivation in the tropics. In *Tropical Mushrooms, Biological Nature and Cultivation Methods* (S. T. Chang and T. H. Quimio, eds.), The Chinese University, Hong Kong, 1982, pp. 277-303.

62. Rajarathnam, S., and Bano, Z. *Pleurotus* mushrooms. Part 1B. Pathology, in vitro and in vivo growth requirements and world status. *CRC Crit. Rev. Food Sci. Nat.*, *26*:243-311, 1988.

63. Kurtzman, R. H., and Chang-Ho, Y. Physiological considerations for cultivation of *Volvariella* mushrooms. In *Tropical Mushrooms, Biological Nature and Cultivation Methods* (S. T. Chang and T. H. Quimio, eds.), Chinese University, Hong Kong, 1982, pp. 139-168.

64. Wood, D. A., and Smith, J. F. The cultivation of mushrooms. Part III. *Mushroom J.*, *188*:665-674, 1988.

65. Zadrazil, F., and Schliemann, J. Okologische und biotechnologische Grundlagen der Domestikation von Speisepilzen. *Mushroom Sci.*, *9*: 199-217, 1974.

66. Tschierpe, H. J. Der Einfluss von Kohlendioxyd auf des Myzelwachstum des Kultur Champignons. *Mushroom Sci.*, *4*:211-218, 1959.

67. Ginterova, A. Dedikaryotization of higher fungi in submerged culture. *Folia Microbiol.* (Prague), *18*:277-280, 1973.

68. Sermanni, G. G., Basile, G., and Luna, M. Biochemical changes occurring in the compost during growth and reproduction of *Pleurotus ostreatus* and *Agaricus bisporus*. *Mushr. Sci.*, *10*:37-49, 1979.

69. Perry, F. G. The influence of supplementation on yield and composition of the mushroom. *Mushroom J.*, *171*:97-103, 1987.

70. Randle, P. E. Supplementation of mushroom composts—a review. *Crop Research* (Horticultural Research), *23*:51-59, 1983.

71. Rajarathnam, S., Bano, Z., and Patwardhan, M. V. Nutrition of the mushroom *Pleurotus flabellatus* during its growth on paddy straw substrate. *J. Hortic. Sci.*, *61*:223-232, 1986.

72. Zadrazil, F. Influence of ammonium nitrate, and organic supplements on the yield of *Pleurotus sajor-caju*. *Europ. J. Appl. Microbiol. Biotechnol.*, *9*:31-35, 1980.

73. Royse, D. H., and Schisler, L. C. Yield and size of *Pleurotus ostreatus* and *Pleurotus sajor-caju* as effected by delayed-release nutrient. *Appl. Microbiol. Biotechnol.*, *26*:191-194, 1987.

74. Tokimoto, K., and Kawai, A. Nutritional aspects on fruit body development in replacement culture of *Lentinus edodes*. *Rep. Tottori Mycol. Inst.*, *12*:25, 1975.

75. Fergus, C. L. The fungus flora of compost during mycelium colonisation by the cultivated mushroom, *Agaricus brunnescens*. *Mycologia*, *70*:636-650, 1978.

76. Fletcher, J. T. Mushroom, moulds and management. *Mushroom J.*, *56*:252-266, 1977.

77. Gandy, D. G. Weed moulds. *Mushroom J.*, *23*:428-429, 1974.

78. Rajarathnam, S., Bano, Z., and Patwardhan, M. V. Observations on the growth and in vitro degradation of rice straw by fungal

contaminants found in beds of the mushroom *Pleurotus flabellatus*; ecological considerations. *Ann. Appl. Biol.* (in press).
79. Saalbach, M. Untersuchungen über Schadpilze im Austernpilzanbau. *Mitt. Versuch. Pilz. Land. Rhein.*, *2*:74-79, 1978.
80. Rajarathnam, S., Singh, N. S., and Bano, Z. Efficacy of carboxin and heat treatment for controlling the growth of *Sclerotium rolfsii* during culture of the mushroom *Pleurotus flabellatus*. *Ann. Appl. Biol.*, *92*:323-328, 1979.
81. Rajarathnam, S., Bano, Z., and Muthu, M. Studies on the control of *Sclerotium rolfsii* contamination during the cultivation of *Pleurotus flabellatus*; efficacy of methylbromide. *Mushroom J.*, *57*:294-296, 1977.
82. Bano, Z., Rajarathnam, S., and Muthu, M. Use of ethyl formate in controlling the growth of *Sclerotium rolfsii* during the cultivation of *Pleurotus* species. *Mushroom Sci.*, *11*:541-547, 1981.
83. Rajarathnam, S., Bano, Z., and Singh, N. S. Efficacy of formaldehyde and liquor ammonia in controlling the growth of *Sclerotium rolfsii* during cultivation of the mushroom *Pleurotus flabellatus*. *Mushroom News Trop.*, *3*:3-11, 1983.
84. Chang-Ho, Y. Ecological studies of *Volvariella volvacea*. In *Tropical Mushrooms: Tropical Nature and Cultivation Methods* (S. T. Chang and T. H. Quimio, eds.), Chinese University, Hong Kong, 1982, pp. 187-198.
85. Gandy, D. G. Bacterial and fungal diseases. In *The Biology and Technology of the Cultivated Mushroom* (P. B. Flegg, D. M. Spencer, and D. A. Wood, eds.), John Wiley and Sons, New York, 1985, pp. 261-278.
86. Atkey, P. T. Viruses. In *The Biology and Technology of the Cultivated Mushroom* (P. B. Flegg, D. M. Spencer, and D. A. Wood, eds.), John Wiley and Sons, New York, 1985, pp. 241-260.
87. Sinden, J. W. Ecological control of weed molds in mushroom culture. *Ann. Rev. Phytopathol.*, *9*:411, 1971.
88. Crisan, E. V., and Sands, A. Nutritional value of edible mushrooms. In *The Biology and Cultivation of Edible Mushrooms* (S. T. Chang and W. A. Hayes, eds.), Academic Press, New York, 1978, pp. 137-181.
89. Gotoh, S., Aoyama, M., and Abe, H. Amino acid composition in the residues of ethanolic extraction of the spontaneous and cultivated edible mushrooms (40 spp.). *Nippon Eiyo Shakuryo Gakkaishi*, *38*:135-139, 1985 (Japan).
90. Chang, S. T. The production of straw mushroom on industrial and agricultural wastes as a method of food protein recovery in Southeast Asia, ASAIHL Lecture of the year 1978.
91. Bano, Z., and Rajarathnam, S. *Pleurotus* mushrooms as a nutritious food. In *Tropical Mushrooms: Biological Nature and Cultivation Methods* (S. T. Chang and T. H. Quimio, eds.), Chinese University Press, Hong Kong, 1982, pp. 363-382.
92. Yokokawa, H., and Sakairi, N. Analysis of vitamin compounds and sterol compositions in higher fungi. *Tachikawa Tandai Kiya*, *18*:45-52, 1985.
93. Bano, Z., and Rajarathnam, S. Vitamin values of *Pleurotus* mushrooms. *Qual. Plant—Plant Foods Humn. Nutr.*, *36*:11-15, 1986.

94. Huang, B., Yung, K. H., and Chang, S. T. The sterol composition of *Volvariella volvacea* and other edible mushrooms. *Mycologia, 77*: 959-963, 1985.
95. Koyama, N., Aoyagi, Y., and Sugahara, T. Fatty acid composition and ergosterol content of edible mushrooms. *Nippon Shokuhin Kogyo Gakkaishi, 31*:732-738, 1984.
96. Memuna, H., and Chakrabarti, C. H. A study on availability of iron in mushrooms. *Indian J. Diet., 19*:203-211, 1982.
97. Maga, J. A. Mushroom flavor. *J. Agric. Food Chem., 16*:517, 1976.
98. Pyysalo, H. On the formation of the aroma of some northern mushrooms. *Mushroom Sci., 10*(2):669-675, 1978.
99. Tressl, R., Bahri, D., and Engel, K. H. Formation of eight-carbon and ten-carbon components in mushrooms (*Agaricus campestris*). *J. Agric. Food Chem., 30*:89-93, 1982.
100. Thorn, R. G., and Barron, G. L. Carnivorous mushrooms. *Science, 224:*76-78, 1984.
101. Grandjean, J., and Huls, R. Structure of *Pleurotin*. Benzoquinon extracted from *Pleurotus griseus*. *Tetrahedron Lett., 22*:1893-1895, 1974.
102. Iwai, K., Ikeda, M., and Fujino, S. Studies on the biosynthesis of folic acid compounds. XI. Nutritional requirements for folate compounds and some enzyme activities involved in folate synthesis. *J. Nutr. Sci. Vitaminol., 23*:95-100, 1977.
103. Kurasawa, S., Sugihara, T., and Hayashi, J. Studies on dietary fiber of mushrooms and edible wild plants. *Nutr. Rep. Int., 26*:167-173, 1982.
104. Dua, I. S., and Jandaik, C. L. Cytokinins in two cultivated edible mushrooms. *Sci. Hortic.* (Amsterdam), *10*:301-306, 1979.
105. Chihara, G., Hamuro, J., and Fukuoka, F. *Carcinostatic polysaccharides.* Ajinomoto Co., Inc., Japan. 7250, 364 (C1, C12d, C08b, A 61 K) 18 December 1972, Appl. 69 21, 865, 22 March 1969.
106. Ikekawa, T., Uchara, N., Maeda, Y., Nakanishi, M., and Fukuoka, F. *Cancer Res., 29*:739, 1969.
107. Yoshioka, Y., Emori, M., Ikekawa, T., and Fukuoka, F. Antitumor activity of some fractions from Basidiomycetes. II. Isolation purification and structure of components from acidic polysaccharides of *Pleurotus ostreatus*. *Carbohydrate Res., 43*:305-320, 1975.
108. Yoshioka, Y., Tabeta, R., Saito, H., Uehara, N., and Fukuoka, F. Antitumor polysaccharides from *P. ostreatus* (Fr.) Qvel. isolation and structure of β-glucan. *Carbohydr. Res., 140*:93-100, 1985.
109. Mori, K., Toyomasu, T., Nanba, H., and Kuroda, H. Antitumor action of fruit bodies of edible mushrooms orally administered to mice. *Mushroom News Lett. Trop., 7*:121-126, 1987.
110. Mouri, T. Taste and pharmacological actions of mushrooms. *Kanzume Jiho, 55*:104-118, 1976.
111. Hara, M., Yoshida, M., Morimoto, M., and Nakano, H. 6-Deoxyllludin M, a new antitumor antibiotic: Fermentation, isolation and structural identification. *J. Antibiot., 40*:1643-1646, 1987.

112. Xia, E., Wang, S., and Chen, Q. Effect of polysaccharide from *Auricularia auricula, Tremella fuciformis,* and *T. fuciformis* spores on DNA and RNA biosynthesis by lymphocytes. *Zhongguo Yaoke Daxue Xuebao, 18*:141-143, 1987 (Chinese).
113. Xua, W., Wan, S., and Chen, Q. Antiulcer effect of the polysaccharides from *Tremella fuciformis,* spores of *T. fuciformis* and *Auricularia auricula. Zhongguo Yaoke Daxue Xuebao, 18*:45-47, 1987 (Chinese).
114. Jeune-Chung, K. H., Kim, M., and Chung, S. R. Studies on lectins from mushrooms (ii). Screening of bioactive substance, lectins from Korean wild mushrooms. *Yakhak Hoechi., 31*:213-218, 1987.
115. Hirano, S., Matsuura, Y., Kusunoki, M., Kitagawa, Y., and Katsube, Y. Two crystalline forms of a lectin from *Flammulina velutipes. J. Biochem., 102*:445-446, 1987.
116. Kabir, Y., Yamaguchi, M., and Kimura, S. Effect of Shitake (*Lentinus edodes*) and maitake (*Grifola frondosa*) mushrooms on blood pressure and plasma lipids of spontaneously hypertensive rats. *J. Nutri. Sci. Vitaminol., 33*:341-346, 1987.
117. Arakawa, N., Enomoto, K., Mukohyama, H., Nakajima, K., Tanabe, O., and Ingaki, C. Effect of basidiomycetes on plasma cholesterol in rats. *Eiyo To Shakuryo, 30*:29-33, 1977.
118. Huang, B. Antitumor and hypocholesterolemic constituents of edible mushrooms. *Yaoxue Tongbao, 17*:282-285, 1982 (Chinese).
119. Tam, S. C., Yip, K. P., Fung, K. P., and Chang, S. T. Hypotensive and renal effects of an extract of the edible mushrooms *Pleurotus sajor-caju, 38*:1155-1161, 1986.
120. Anisova, L. N., Bartoshevich, Yu, E., Efremenkova, O. V., Krasil' nikova, O. L., Kudinova, M. K., Murenets, N. V., Klyuev, N. A., Chernyshev, A. I., and Shorshnev, S. V. Isolation and identification of an antileukosis substance from *Coprinus radiatus. Antibiot. Med. Bioteknol., 32*:735-738, 1987 (Russian).
121. Terada, A., and Mizutani, N. Alcoholic beverage manufacture from mushrooms. Jpn. Kokai Tokyo Koho JP 6262,677 (8765,677) (C1.C12 G3/02), 24 March 1987, Appl. 85/191, 146, 30 Aug. 1985.
122. Chahal, D. S. Bioconversion of hemicelluloses into useful products in an integrated process for food/feed and fuel (ethanol) production from biomass. *Biotechnol. Bioeng. Symp., 14*:484, 1984.
123. Friedel, K., Seidel, D., Moegling, R., and Funk, I. Orientational studies on straw digestion by white-rot fungi. *Wiss Z. Wilhelm—Dieck-Univ. Rostock. Naturwiss., 32*:25-28, 1983.
124. Golovlev, E. L., Chermenskii, D. N., Okunev, O. N., Brustavetskaya, T. P., Golovlera, L. A., and Skryabin, G. K. Selection of fungal cultures for solid-phase fermentation of saw dust and straw. *Microbiol., 52*:78-82, 1983 (Russian).
125. Nizkovskaya, O. P., Pan'Kova, I. M., Kochetkova, G. I., and Manukovskii, N. S. Oxidation of wheat straw lignin by basidiomycetes. *Mikol. Fitopatol., 18*:133-135, 1984.
126. Zadrazil, F. Release of water soluble compounds in the breakdown of straw by basidiomycetes as a basis for the biological utilization of straw. *Z. Acka-Pflanzenbau., 142*:44-52, 1976.

127. Adamski, Z., and Zielinski, M. H. Mycological delignification of wood. *Chem. Technol. Chem.*, *7*:21-28, 1985 (Pol.).
128. Bis'ko, N. A., Fomina, V. I., and Bilai, V. T. Wood destruction caused by a fungus *Pleurotus ostreatus* (Fr.) Kumm. *Mikol. Fitopatol.*, *17*:199-202, 1983 (Russ.).
129. Bis'ko, N. A., Bilai, V. T., and Churikova, E. K. Decomposition of wood of various tree species by oyster fungus *Pleurotus ostreatus* (Fr.) Kumm. during its growth. *Mikol. Fitopatol.*, *18*:435-439, 1984 (Russ.).
130. Kawakami, H. Degradation of wood components by hiratake fungus, *Pleurotus ostreatus*. *Bokin Bobai.*, *7*:505, 1979.
131. Kaneshiro, T. Lignocellulosic agricultural wastes degraded by *Pleurotus ostreatus*. *Dev. Ind. Microbiol.*, *18*:591-597, 1977.
132. Kovacs-Ligetfalusi, I. Chemical test of poplar saw dust decayed by *Pleurotus ostreatus*. *Soil Biol. Conser. Biosphere Proc. 7th Meeting 1975*. 287, Szezi J. Akad kaido, ed., Budapest, Hungary, 1977.
133. Brodziak, L., and Wazny, J. Fruiting bodies of *Lentinus* and decomposed culture substrate as a source of protein and organic compounds. *Acta Mycol.*, *21*:13-21, 1985 (Pol.).
134. Gerrits, J. P. G. Organic compost constituents and water utilized by the cultivated mushroom during spawn run and cropping. *Mushroom Science*, *7*:111-126, 1969.
135. Chang-Ho, Y., and Yee, N. T. Comparative study of the physiology of *Volvariella volvacea* and *Coprinus cinereus*. *Trans. Brit. Mycol. Soc.*, *68*:167-172, 1977.
136. Zadrazil, F. The conversion of straw into feed by basidiomycetes. *Eur. J. Appl. Microbiol.*, *4*:273-281, 1977.
137. Zadrazil, F. Umwandlung von Pflanzenbefall in tierfutter durch nonere Pilze. *Mushroom Sci.*, *10*:231-239, 1979.
138. Zweck, S., Huttermann, A., and Chet, I. A convenient method for preparing incoula of homogenized inoculum. *Exper. Mycol.*, *2*:377, 1978.
139. Rajarathnam, S. Studies on the pathological and biochemical aspects during cultivation and storage of the mushroom *Pleurotus flabellatus*. Ph.D. thesis, University of Mysore, Mysore, India, 1981.
140. Wood, D. A. Enzyme assay for estimation of biomass of mycelium of *Agaricus bisporus* in a solid substrate, composted wheat straw. *Biotechnol. Lett.*, *1*:255-260, 1979.
141. Platt, M. W., Hadar, Y., and Chet, I. Fungal activities involved in lignocellulose degradation by *Pleurotus*. *Appl. Microbiol. Biotechnol.*, *20*:150-154, 1984.
142. Rypacek, V. *Biologie holzerstorenda Pilze*, VEBG Fischer Verlag, Jena, 1966.
143. Dobry, I., Dziurzynski, A., and Rypacek, V. Relation between combustion heat and chemical wood decomposition during white and brown rot. *Wood Sci. Technol.*, *20*:137-144, 1986.
144. Garibova, L. V., Kozlova, R. G., Losyakova, L. S., and Safnova, N. V. Oxalic acid production in the cultivation of seed mycelia of the edible fungi, *Agaricus bisporus* and *Pleurotus ostreatus* on granular nutritive substrates. *Biol. Nauki* (Moscow), *3*:79-84, 1982.

145. Daugulis, A. J., and Bone, D. H. Submerged cultivation of edible white-rot fungi or tree bark. *Eur. J. Appl. Microbiol.*, *4*:159-166, 1977.
146. Rajarathnam, S., Wankhede, D. B., and Bano, Z. Degradation of rice straw by *Pleurotus flabellatus*. *J. Chem. Technol. Biotechnol.*, *37*:203-214, 1987.
147. Wood, D. A. Biochemical changes during growth and development of *Agaricus bisporus*. *Mushroom Sci.*, *10*:401-417, 1979.
148. Zadrazil, F. Freisetzung wasserlöslicher Verbindungen während der Strohzersetzung durch Basidiomyceten als Grundlage für eine biologische Strohaufwestung. *Z. Aker. Pflanzenbau.*, *142*:44, 1976.
149. Riaz, M., Wilke, C. R., Yamanaka, Y., and Carroad, P. A. Decomposition of lignin and cellulose in relation to the enzymic hydrolysis of cellulose. *Proc. Symp. Res. Appl. Natl. Needs*, *2*:108-115, 1977.
150. Barrows, I., Seal, K. J., and Eggins, H. D. W. Biodegradation of barley straw by *Coprinus cinereus* for the production of ruminant feed. In *Straw Decay and Its Effect on Utilization and Disposal* (E. Grossbard, ed.), John Wiley and Sons, 1979, p. 147.
151. Pigden, W. J., and Heaney, D. P. Lignocellulose in ruminant nutrition. *Adv. Chem. Ser.*, *95*:245-261, 1969.
152. Jilek, R., Zezula, J., and Vodickova, M. Biological transformation of straw cellulose. *Mushroom Sci.*, *10*(1):303-309, 1979.
153. Schanel, L., Herzig, I., Dvorak, M., and Veznik. *Zlepsovaci Novrh C.*, *13*:1-14 (Patent CSSR), 1966.
154. Hartley, R. D., Jones, E. C., King, N. J., and Smith, G. A. *J. Agric. Sci.*, *25*:433, 1974.
155. Yadav, J. S. Solid state fermentation of wheat straw with alcaliphilic *Coprinus*. *Biotechnol. Bioeng.*, *31*:414-417, 1988.
156. Sommer, A., Skultetoyova, N., and Ginterova, A. Investigation into the nutritive value of the harvested substrate of *P. ostreatus*. *Polnohospodarstvo.*, *24*:152-158, 1978.
157. Rajarathnam, S., Wankhede, D. B., and Patwardhan, M. V. Some chemical and biochemical changes in straw constituents during growth of *Pleurotus flabellatus* (Berk & Br.) Sacc. *Eur. J. Appl. Microbiol. Biotechnol.*, *8*:125-134, 1979.
158. Wood, D. A., and Fermor, T. R. Nutrition of *Agaricus bisporus* in compost. *Mushroom Sci.*, *11*:63-71, 1981.
159. Hong, S. W., Shin, K., Yoon, Y., and Lee, W. Extracellular wood degradative enzymes from *Lentinus edodes*. *Hanguk Kyunhakhoechi.*, *14*:189-194, 1986.
160. Matcham, S. E., Jordon, B. R., and Wood, D. A. Estimation of fungal biomass in a solid substrate by three independent methods. *Appl. Microbiol. Biotechnol.*, *21*:108-112, 1985.
161. Sermanni, G. G., Basile, G., and Luna, M. Biochemical changes occurring in the compost during growth and reproduction of *Pleurotus ostreatus* and *Agaricus bisporus*. *Mushroom Sci.*, *10*:37-53, 1979.
162. Zadrazil, F. Straw decomposition by fungi (Basidiomycetes) with its subsequent use as feed supplement or compost. *Land. Wirtsch. Forsch. Sonderh.*, *32*:153, 1976.
163. Zadrazil, F. Cultivation of *Pleurotus*. In *The Biology and Cultivation of Edible Mushrooms* (S. T. Chang and W. A. Hayes, eds.), Academic Press, 1978, pp. 521-557.

164. Adegbola, A. A., and Paladines, O. Prediction of the digestibility of the dry matter of tropical forages from their solubility in fungal cellulase solutions. *J. Sci. Food Agric.*, *28*:775-785, 1977.
165. Goto, I., and Minson, D. J. Prediction of the dry matter digestibility of tropical grasses using a pepsin-cellulase bioassay. *Anim. Feed Sci. Technol.*, *2*:247, 1977.
166. McLeod, M. N., and Minson, D. J. A note on Onozuka 38 cellulase as a replacement for Onozuka SSP 1500 cellulase when estimating forage digestibility in vivo. *Anim. Feed Sci. Technol.*, 5:347, 1980.
167. Roughan, P. G., and Holland, R. Predicting in vitro digestibility of herbages by exhaustive enzymatic digestion of cell walls. *J. Sci. Food Agric.*, *28*:1057-1061, 1982.
168. Tsang, L., Reid, I. A., and Coxworth, E. Delignification of wheat straw by *Pleurotus* species under mushroom growing conditions. *Appl. Environ. Microbiol.*, *53*:1304-1306, 1987.
169. Streeter, C., Conway, K. E., Horn, G. W., and Mader, T. I. Nutritional evaluation of wheat straw incubated with the edible mushroom *Pleurotus ostreatus*. *J. Anim. Sci.*, *54*:183-188, 1982.
170. Zadrazil, F., and Brunnert, H. Influence of ammonium nitrate on the growth and straw decomposition of higher fungi. *Z. Pflanzenerhachr. Bodenkol.*, *142*:446, 1979.
171. Zadrazil, F., and Brunnert, H. Investigation of the physical parameters important for the solid state fermentation of straw by white-rot fungi. *Eur. J. Appl. Microbiol.*, *11*:183-191, 1981.
172. Feniksova, R. V., Ulezlo, I. V., and Pukit, N. Yu. Lignolytic activity of some wood destroying fungi. *Prikl. Biokhim. Microbiol.*, *8*:337-340, 1972 (Russ.).
173. Hong, J. S. Studies on the enzymes produced by *Pleurotus ostreatus*. IV. Properties of xylanase. *Misacngmul Hakhoe Chi.*, *14*:99, 1976.
174. Hong, J. S. Studies on the physico-chemical properties and the cultivation of the oyster mushroom. *Hanguk Nonghwa Hakhoe* Chi., *21*:150, 1978.
175. Jablonski, I. Changes in biochemical and physiological activities of substrates colonized by fungi *Pleurotus ostreatus, Lentinus edodes*, and *Agrocybe aergerita*. *Mushroom Sci.*, *11*(2):659-673, 1981.
176. Madan, M., and Bisaria, R. Cellulolytic enzymes from an edible mushroom, *Pleurotus sajor-caju*. *Biotechnol. Lett.*, 5:601-608, 1983.
177. Schmitz, H., and Eger, H. G. Degradation of fungal cell walls by enzymes of the basidiomycete *Pleurotus ostreatus* (Jacq. ex. fr.) Kummer. *Adv. Biotechnol.* (Proc. Int. Ferment. Symp.) (M. Moo-Young and C. W. Robinson, eds.), Pergamon Press, New York, 1981, pp. 505-509.
178. Trutneva, I. A. Study of the activity of extracellular encopolygalacturonases of agricultural fungi in culture. *Ukr. Bot. Zh.*, *35*:625, 1978.
179. Bollag, J. M., and Leonowicz, A. Comparative studies of extracellular fungal laccases. *Appl. Environ. Microbiol.*, *48*:849-854, 1984.
180. Hira, A., Barnett, S. M., Shiek, C. H., and Montecalvo, J., Jr. An extracellular lignase: A key to enhanced cellulose utilization. *AICHE Symp. Ser.*, *74*:17-20, 1978.

181. Jackuliak, O. Production of enzymes by some lignivorous fungi. *Drev. Vysk.*, *22*:227-235, 1978.
182. Kozlik, I. Effect of phenolic substances on the activity of oxidases of wood-destroying fungi. *Drev. Vysk.*, *25*:94-99, 1980.
183. Lamaison, J. L., Pourrat, A., and Pourrat, H. Catalasic activity of macromycetes. Preparation of a purified catalase. *Ann. Pharm. Fr.*, *33*:441-446, 1975.
184. Lobarzewski, J., Trojanwski, J., and Woytas-Wasilewska, M. The effect of fungal peroxidase on Na-lignosulfonates. *Holzforschung.*, *36*:173, 1982.
185. Semichaeviskii, V. D., Butovich, I. A., and Tsarevich, N. V. Phenol oxidizing ability of preparations of extracellular enzymes from *Pleurotus ostreatus*. *Dopor. Akad. Nank. Ukr. RSR.* Ser. B. *Geol. Khim. Biol. Nanki.*, *10*:75, 1984.
186. Semichaevskii, V. D., Dudchenko, L. G., and Melnichuk, G. G. Effect of pH on the formation of extracellular enzymes by *Pleurotus ostreatus*, a wood rotting basidiomycete. *Mikrobiol. Zh.*, *47*:72, 1985.
187. Tsuruta, T., and Kawai, M. Catechol oxidizing activities of basidiomycetous fungi. *Nippon-Kingakkai Kaiho.*, *24*:65, 1983.
188. Vetter, J. Study of the growth dynamics and of extracellular monophenol monooxygenase in some species of the genus, *Pleurotus*. *Ukr. Bot. Zh.*, *40*:74, 1983.
189. Reese, E. T., Sir, R. G., and Levinson, H. S. The biological degradation of soluble cellulose derivatives and its relationship to the mechanism of cellulose hydrolysis. *J. Bact.*, *59*:485-497, 1950.
190. Thayumanavan, B. Extracellular cellulase and laccase enzymes from *Pleurotus sajor-caju*. *Madras. Agric. J.*, *69*:132-134, 1982.
191. Ginterova, A., Janotkova, O., Zemek, J., Augustin, J., and Kunsik, L. Cellulase activity of higher fungi. *Folia Microbiol.* (Prague), *25*: 318-323, 1980.
192. Zemek, J., and Kunaik, L. The way of testing polysaccharide hydrolase activity of microorganisms. Czech. Pat. 186 059 (1978).
193. Rangaswamy, K., and Kandaswmay, T. K. Effect of certain amendments of cellulosics and yield of straw mushroom. *Indian J. Mushrooms*, *2*:8-11, 1976.
194. Melnichunk, G. G., and Danilyak, M. I. Dynamics of the activity of exocellular hydrolases of some species of Agaricales mushrooms. *Ukr. Bot. Zh.*, *38*:51, 1981.
195. Rawal, P. P., Singh, R. D., and Khandar, R. R. Pectinolytic and cellulolytic enzyme fluctuation at various stages of spawn growth of *Pleurotus sajor-caju*. *Indian J. Mushrooms*, *7*:14-17, 1981.
196. Doshi, A., Munot, J. F., and Chakravarti, B. P. Pectinolytic and cellulolytic enzyme production at various stages of spawn growth of *Calocybe indica*. *Mushroom J. Trop.*, *7*:83-85, 1987.
197. Hashimoto, K., and Takahashi, Z. Biochemical studies on the mushroom. VIII. Enzymic hydrolysis of xylan, *Toyo Shokuhin Kogyo Tanki Daigaku, Toyo Shokuhin Kenkyusho Kenkyu Hokokusho*, *60*: 320-326, 1970 (Japan).
198. Toyama, N., and Ogawa, K. Comparative studies on cellulolytic and oxidizing enzyme activities of edible and inedible wood rotters. *Mushroom Sci.*, *9*(1):745-760, 1974.

199. Ander, P., and Ericksson, K. E. Lignin degradation and utilization by microorganisms. In *Progress in Industrial Microbiology*, Vol. 14 (M. H. Bull, ed.), Elsevier, Amsterdam, 1978, p. 294.
200. Glenn, J. K., and Gold, M. H. Decolorization of several polymeric dyes by the lignin degrading basidiomycete *Sporotrichum pulverulantum: Appl. Environ. Microbiol.*, *45*:1741, 1983.
201. Ander, P., Ericksson, K. E., and Yu, H. S. Vanillic acid metabolism by *Sporotrichum pulverulentum*: Evidence for demethoxylation before ring cleavage. *Arch. Microbiol.*, *136*:1, 1983.
202. Turner, E. M., Wright, M., Ward, T., Osborne, D. M., and Self, R. Production of ethylene and other volatiles, and changes in cellulase and laccase activities during the life cycle of the cultivated mushroom *Agaricus bisporus*. *J. Gen. Microbiol.*, *91*:167-176, 1975.
203. Sermanni, G. G., and Luna, M. Laccase activity of *Agaricus bisporus* and *Pleurotus ostreatus*. *Mushroom Sci.*, *11*:485-496, 1981.
204. Afanas'eva, M. M., and Kadyrov, R. M. Selection of cellulose and lignin degrading fungi for use in an artificial closed ecological cycle. *Mikol. Fitopatol.*, *14*:410-416, 1980.
205. Haars, A., Majcherczyk, A., Trojanowski, J., and Huttermann, A. Bioconversion of organosolv lignin by different types of fungi. *Comm. Eur. Communities*:973-977, 1985.
206. Hoffmann, W., and Huttermann, A. Aminopeptidases of *Physarum polycephalum*: Activity, isoenzyme pattern and synthesis during fermentation. *J. Biol. Chem.*, *250*:7420, 1975.
207. Netzer, U. V. Investigations of primordia formation in the *Pleurotus ostreatus* dikaryon 868 x 381. *Mushroom Sci.*, *10*(1):703-711, 1979.
208. Schwalb, M. N. Developmental control enzymes of enzyme modification during fruiting of the basidiomycete *Schizophyllum commune*. *Biochem. Biophys. Res. Commun.*, *67*:478-482, 1975.
209. Manning, K., and Wood, D. A. Production and regulation of extracellular endocellulase by *Agaricus bisporus*. *J. Gen. Microbiol.*, *129*: 1839-1847, 1983.
210. Wood, D. A. Production, purification and properties of extracellular laccase of *Agaricus bisporus*. *G. Jen. Microbiol.*, *117*:327-338, 1980.
211. Wang, C. W. Cellulolytic enzymes of *Volvariella volvacea*. In *Tropical Mushrooms—Biological Nature and Cultivation Methods* (S. T. Chang and T. H. Quimio, eds.), Chinese University Press, Hong Kong, 1982, pp. 167-186.
212. Wang, C. W. Purification of cellulolytic enzymes of *Volvariella volvacea* by gel filtration. *Mushroom News Lett. Trop.*, *5*:4-8, 1985.
213. Hong, J. S. Studies on the enzymes produced by *Pleurotus ostreatus*. Properties of xylanase. *Korean J. Microbiol.*, *14*:99, 1976.
214. Lee, J. S., Lee, U. J., and Suh, D. S. Production, partial purification and physico-chemical characteristics of laccase from *Pleurotus ostreatus*. *Sanop Misaengmul Hakhoechi.*, *13*:65, 1985.
215. Danilyak, N. I. Kinetic characteristics of pH and temperature stability of exocellular cellulases. *Fermentn. Spirt. Prom.*, *4*:37, 1981.
216. Hong, J. S., and Kim, D. H. Studies on the enzymes produced by basidiomycetes. I. The production of crude enzymes. *Hanguk. Nonghuwa Hakhoe Chi.*, *24*:7, 1981.

217. Hong, J. S., Uhm, T. B., Jung, G. T., and Lee, K. B. Studies on the enzymes produced by *Pleurotus sajor-caju*. The production of cellulolytic enzymes. *Hanguk Kyunhakhoechi.*, *12*:59, 1984.
218. Leonowicz, A., and Trojanowski, J. Induction of new laccase form in the fungus *Pleurotus ostreatus*, by ferulic acid. *Microbios.*, *13*: 167, 1975.
219. Leonowicz, A., and Trojanowski, J. Induction of laccase by ferulic acid. *Acta Biochim.* Pol., *22*:291, 1975.
220. Leonowicz, A., and Trojanowski, J. Induction of laccase in basidiomycetes—the laccase colling messenger. *Acta Biochim.* Pol., *25*:147, 1978.
221. Molitoris, H. P. Wood degradation, phenoloxidases and chemotaxonomy of higher fungi. *Mushroom Sci.*, *10*:243-263, 1979.
222. Wojfas-Wasilewska, M., Trojanowski, J., and Luterek, J. Aromatic ring cleavage protocatechuic acid by the white-rot fungus *Pleurotus ostreatus*. *Acta Biochim.* Pol. , *30*:291, 1983.
223. Eger, G. Die Wirkung einiger N-Verbindungen auf Mycelwachstum und Primordienbildung des Basidiomyceten *Pleurotus* sp. *Florida*. *Arch. Microbiol.*, *74*:174, 1970.
224. Jablonsky, I., and Schanel, I. Biochemical and physiological activity of fungi *Lentinus edodes* and *P. ostreatus Florida*. *Third Symp. Physiol. Ecol. Cultivation Edible Fungi*, Prague, 1979.
225. Lyr, H. Die induktion der laccase bildung bei *Collybia velutipes*. *Arch. Mikrobiol.*, *28*:310-324, 1958.
226. Manchere, G. Research on the fruiting rhythm of a basidiomycete mushroom *Coprinus congregatus*. *J. Interdiscipl. Cycle Res.*, *2*:199, 1971.
227. Robert, J. C. Quelques aspects metaboliques de la fructification de basidiomycete *Coprinus congregatus*. *Mushroom Sci.*, *10*(1):683-702, 1979.
228. Thayumanavan, B. Biochemical changes brought about by the growth of *Pleurotus sajor-caju* on some natural substrates. *Farm Sci.*, *6*:33, 1979.
229. Wood, D. A., and Goodenough, P. Fruiting of *Agaricus bisporus*. Changes in extracellular enzyme activities during growth and fruiting. *Arch. Microbiol.*, *114*:161-165, 1977.
230. Leonard, T. J. Phenoloxidase activity and fruit body formation in *Schizophyllum commune*. *J. Bacteriol.*, *106*:162-167, 1971.
231. Hirsch, H. M. Environmental factors influencing the differentiation of protoperithecia and their relation to tyrosinase and melanin formation in *Neurospora crassa*. *Physiol. Plant.*, *7*:72, 1954.
232. Cutterbruck, A. Absence of laccase from yellow-spored mutants of *Aspergillus nidulans*. *J. Gen. Microbiol.*, *70*:423-435, 1972.
233. Thomas, D., and Mullins, J. T. Role of enzymatic wall softening in plant morphogenesis. Hormonal induction in *Achlya*. *Science*, *156*: 84-85, 1967.
234. Roseness, P. A. Cellulolytic enzymes during morphogenesis in *Dictyostelium discoideum*. *J. Bacteriol.*, *96*:639-645, 1968.
235. Mandelstam, J. Bacterial sporulation—A problem in the biochemistry and genetics of a primitive developmental system. *Proc. Royal Soc.* London, Ser. B., *193*:89-106, 1976.

236. Matsubara, H., and Feder, J. Other bacterial, mould and yeast proteases. In *The Enzymes*, Vol. 3 (P. D. Boyer, ed.), Academic Press, New York, 1971.
237. Iten, W., and Matile, P. Role of chitinase and other lysozymal enzymes of *Coprinus lagopus* in the autolysis of the fruiting bodies. *J. Gen. Microbiol.*, *61*:301-309, 1970.
238. Betz, H., and Weiser, U. Protein degradation and proteinases during yeast sporulation. *Eur. J. Biochem.*, *62*:65, 1976.
239. Chang, S. T., and Chan, K. Y. Quantitative and qualitative changes in proteins during morphogenesis of the basidiocarp of *Volvariella volvacea*. *Mycologia*, *65*:355-364, 1973.
240. Iwahara, H., Hoshimoto, T., and Fukuzumi, T. Changes in activities of extracellular enzymes during life cycle of the edible mushroom, *Pleurotus ostreatus*. *Mokuzai Gakkaishi*, *27*:331, 1981.
241. Kirk, T. K., and Kelman, A. Lignin degradation as related to the phenoloxidases of selected wood decaying basidiomycetes. *Phytopathol.*, *55*:739-745, 1965.
242. Bavendamm, W. The occurrence and detection of oxidases in wood destroying fungi. *Z. Pflanzen. Krankh. Pflanzenschutz.*, *38*:257, 1928.
243. Lyr, H. Detection of oxidases and phenoloxidases in higher fungi and the significance of these enzymes for the reaction of Bavendamm. *Planta*, *50*:359, 1958.
244. Rosch, R. Über die Fraktion des phenoloxidasen holzabbanendes Pilze. In *Holzu und Organismen* (G. Bercker, ed.), Duncker and Humbolt Press, Berlin, 1967, p. 173.
245. Kirk, T. K., Connors, W. J., and Zeikus, J. G. Requirement for a growth substrate during lignin decomposition by two wood rotting fungi. *Appl. Environ. Microbiol.*, *32*:192-194, 1976.
246. Platt, M. W., Hadar, Y., and Chet, I. The decolorization of the polymeric dye poly blue (polyvinalamine sulfonate anthraquinone) by lignin degrading fungi. *Appl. Microbiol. Biotechnol.*, *21*:394-396, 1985.
247. Kirk, T. K. Degradation and conversion of lignocelluloses. In *The Filamentous Fungi* (J. E. Smith, D. R. Sarry, and B. Kristainsen, eds.), Edward Arnold, London, 1983, p. 266.
248. Ginterova, A., and Lazarova, A. Degradation dynamics of lignocellulose materials by wood-rotting *Pleurotus* fungi. *Food Chem.*, *32*:434-437, 1987.
249. Friedel, K., Seidel, D., Moegling, R., and Funk, I. Orientational studies on straw digestion by white-rot fungi. *Wiss. Z. Wilhem.-Dieck-Univ. Rostock. Naturwiss.*, *32*:25-28, 1983.
250. Leatham, G. F., and Kirk, T. K. Regulation of ligninolytic activity by nutrient nitrogen in white-rot basidiomycetes. *FEMS Microbiol. Lett.*, *16*:65-67, 1983.
251. Commanday, F., and Macy, J. M. Effect of substrate nitrogen on lignin degradation by *Pleurotus ostreatus*. *Arch. Microbiol.*, *142*:61-65, 1985.

252. Freer, S. N., and Detroy, R. W. Biological delignification of carbon-14 labelled ligno-cellulose by basidiomycetes: Degradation and solubilization of the lignin and cellulose components. *Mycologia, 74*:943-951, 1982.
253. Wojtas-Wasilewska, M., Trojanowski, J., and Luterek, J. Biotransformation of sodium lignosulfonates of different molecular weights by the fungus *Pleurotus ostreatus. Acta Microbiol.* Pol., *29*:353-364, 1980.
254. Reid, I. D., and Seifert, K. A. Effect of an atmosphere of oxygen on growth, respiration and lignin degradation by white-rot fungi. *Can. J. Bot., 60*:252-260, 1982.
255. Haider, K., and Trojanowski, J. Decomposition of specifically carbon-14-labeled phenols and dehydropolymers of coniferyl alcohol as models for lignin degradation by soft and white-rot fungi. *Arch. Microbiol., 105*:33, 1975.
256. Platt, M., Trojanowski, J., Chet, I., and Huttermann, A. Differences in degradation of specifically carbon-14-labelled lignin model compounds within *Pleurotus genus. Microbiol. Lett., 23*:19-21, 1983.
257. Haider, K., Ellwardt, P. C., and Ernst, L. Lignin biodegradation studies by carbon-13 NMR spectroscopy of carbon-13-enriched DHP-lignin. *Int. Symp. Wood Pulping Chem.* SPCL. Stockholm, *3*:93-98, 1981.
258. Agosin, E., Daudin, J. J., and Odier, E. Screening of white-rot fungi on (^{14}C) lignin-labelled and (^{14}C) whole-labelled wheat straw. *Appl. Microbiol. Biotechnol., 22*:132-138, 1985.
259. Nizkovskaya, O. P., Pan'kov, I. M., Kochetova, G. A., and Manukovskii, N. S. Evaluation of lignolytic activity in basidiomycetes by a radioisotope method. *Mikol. Fitopatol., 15*:398-401, 1981 (Russ.).
260. Zadrazil, F. Conversion of different plant wastes into feed by basidiomycetes. *Eur. J. Appl. Microbiol. Biotechnol., 9*:243-248, 1980.
261. Zafer, S. I., Kausar, T., and Shah, F. H. Biodegradation of the cellulose component of rice straw by *Pleurotus sajor-caju. Folia Microbiol.* (Prague), *26*:394-397, 1981.
262. Bano, Z., Rajarathnam, S., and Narasimhamurthy, K. Studies on the fitness of "spent straw" obtained during cultivation of the mushroom *Pleurotus sajor-caju* as cattle feed. *Mushroom News Lett. Trop., 6*:11-16, 1986.
263. Herzig, I., Dvorak, M., and Veznik, J. Treatment of litter straw by application of the genus *Pleurotus. Biol. Chem. Vyz. Hospodarkskych. Zvirat., 3*:249, 1968.
264. Bartak, R., Ginterova, A., and Janotkova, O. Czech. 197,987 (C1 A23 K1/12) 15 February 1979, Appl. 76/480, 26 January 1976.
265. Czerpak, R., Obrusiewicz, T., and Szudyga, K. Studies on the improvement of the fodder value of straw by chemical hydrolysis and some biological methods. *Zesg. Probl. Postepow. Nauk Roln., 225*: 127-137, 1980 (Pol.).
266. Muller, H. W., and Trosch, W. Screening of white-rot fungi for biological pretreatment of wheat straw for biogas production. *Appl. Microbiol. Biotechnol., 24*:180-186, 1986.
267. Muller, H., Troesch, W., and Kulbe, K. D. Biological pretreatment of wheat straw for biogas production. *Eur. Congr. Biotechnol., 3*: 89-94, 1984.

268. Ericksson, K. E., and Goodell, E. W. Pleiotropic mutants of the wood rotting fungus *Polyporus adustus* lacking cellulase, mannase and xylanase. *Can. J. Microbiol.*, *20*:371-378, 1974.
269. Osaka Gas Co. Ltd., Japan. Production of mushrooms and methanne from cellulosic waste materials. Kokai Tokkyo Kho Jp. 59, 198, 987 (84,198,987) (C1 C12 P19/14) 10 November 1984, Appl. 83/73, 455, 25 April 1983.
270. Bisaria, R., Madan, H., and Mukhopadhyay, S. N. Production of biogas from residues from mushroom cultivation. *Biotechnol. Lett.*, *5*:811-812, 1983.
271. Eisenstein, A., Eisenstein, R., and Basler, A. Cellulose from wood or other lignocellulose plants by microbiological decomposition of lignocellulose. Eur. Pat. Appl. EP. 60,467 (C1 D21 C3/00) 22 September 1982, DE Appl. 3,110,117, 16 March 1981.
272. Pilon, L., Barbe, M. C., Desrochers, M., Jurasek, L., and Neumann, P. J. Fungal treatment of mechanical pumps—its effect on paper properties. *Biotechnol. Bioeng.*, *24*:2063-2076, 1982.
273. Pilon, L., Desrochers, M., Jurasek, L., and Neumann, P. J. Increasing water retention of mechanical pulp by biological treatments. *Tappi.*, *65*:93-96, 1982.
274. Fukuzumi, T., Yotukura, A., and Hayashi, Y. Screening of fungi for biological pulping. *Recent Adv. Lignin Biodegrad. Res.* (T. Higuchi, H. Chang, and T. K. Kirk, eds.), UNI Publ. Co. Ltd., Tokyo, 1983, pp. 246-256.
275. Harsh, N. S. K., and Biaht, N. S. Use of tea leaves as an aid to culture wood decaying fungi. *Int. Biodet. Bull.*, *17*:19, 1981.
276. Harsh, N. S. K., and Bisht, N. S. Spent substrate to serve as compost for the cultivation of *Agaricus bisporus*. *Int. Biodet. Bull.*, *20*:253, 1984.
277. Hayes, W. A. Nutritional factors in relation to mushroom production. *Mushroom Sci.*, *8*:663-674, 1972.
278. Hayes, W. A. Nutrition, substrates and principles of disease control. In *Biology and Cultivation of Edible Mushrooms* (S. T. Chang and W. A. Hayes, eds.), Academic Press, New York, 1978, p. 220.
279. Wood, D. A., and Fermor, T. R. Nutrition of *Agaricus bisporus* in compost. *Mushroom Sci.*, *11*:63-71, 1981.
280. Grabbe, K. Utilization of residues from mushroom cultivation. *Wiss. Umwelt*, *4*:221-226, 1983 (Ger.).
281. Gapinski, M. Utilization of substrates after mushroom cultivation. *Ogrodnictwo*, *19*:239-240, 1982 (Pol.).
282. Hitachi, K., and Kogyo, K. K. Deodorization of waste gases. Japan Kokkai Tokyo Koho Jp. 58,202,018, 25 November 1983. Appl. 82/87, 004, 21 May 1982.
283. Livernoche, D., Jurasek, L., Desrochers, M., Dorica, J., and Valiky, I. A. Removal of color from kraft mill waste-waters with cultures of white-rot fungi and with immobilized mycelium of *Coriolus versicolor*. *Biotechnol. Bioeng.*, *25*:2055-2065, 1983.
284. Amano, H., Mizunuma, K., and Kanai, Y. *Hakko Kyokaishi*, *23*: 177, 1965.

285. Aoshima, I., Tozawa, Y., Ohomomo, S., and Ueda, K. Production of decolorizing activity for molasses pigment by *Coriolus versicolor*. *Agric. Biol. Chem.*, *49*:2041-2045, 1985.
286. Tamaki, H., Kishihara, S., Fujii, S., Komoto, M., Arita, A. I., and Hiratuka, N. Decolorizing activity and sugar assimilation of basidiomycetes for polyvinylpolypyrrolidone treated molasses. Decolorization of cane sugar molasses by action of basidiomycetes. *Nippon Shokuhin Kogyo Gakkaishi*, *33*:270-273, 1986.
287. Hongo, M., Hayashida, S., and Tanaka, Y. Microbiological decolorization of fermentation waste fluid. Japan Kokai 74, 130,056 (Cl 91 C9) 12 Dec. 1974, Appl. 7344, 206, 20 April 1973.
288. Kolan Kaya, N., and Saglam, N. Biological decolorization of waste black kraft liquor and lignin biodegradation. *Doga Seri. A.*, *7*:513-517, 1983 (Turkish).
289. Timofeeva, S. S., and Vatyuzhina, G. S. Biochemical removal of phenolic compounds from waste water. USSR Su 1,130,540 (Cl CO2, F3/32), 23 December 1984, Appl. 3,504-972, 27 October 1982.
290. Timofeeva, S. S., and Ustyuzhina, G. S. Possible ways for improving the biological treatment of phenol containing waste-water by some fungi and yeasts. Deposited Doc. VINITI, 3283-83, 1983.
291. Taniguchi, M., Kometani, Y., Tanaka, M., Matsuno, R., and Kamikubu, T. *J. Appl. Microbiol. Biotech.*, *14*:74-80, 1982.
292. Hiroi, T. Enzymic hydrolysis of woods. VI. Susceptibility of rotted woods to cellulase from *Trichoderma viride*. *Mokuzai, Gakkaishi.*, *27*:684, 1981.
293. Hatakka, A. I. Pretreatment of wheat straw by white-rot fungi for enzymic saccharification of cellulose. *Eur. J. Appl. Microbiol. Biotechnol.*, *18*:350-357, 1983.
294. Schneider, H., Wang, P. Y., Chan, Y. K., and Maleszkar, R. *Biotech. Lett.*, *3*:89-92, 1981.
295. Slininger, P. J., Bothast, R. J., Vancanwenberge, P., and Kurtzman, C. P. *Biotech. Bioeng.*, *24*:371-384, 1982.
296. Detroy, R. W., Lindengelser, L. A., Julian, G., and Orton, W. L. Saccharification of wheat-straw cellulose by enzymatic hydrolysis following fermentative and chemical pretreatment. *Biotechnol. Bioeng. Symp.*, *10*:135-148, 1980.
297. Afans'eva, M. M., and Kadyrov, R. M. Selection of cellulose and lignin-degrading fungi for use in an artificial closed ecological cycle. *Mikol. Fitopatol.*, *14*:410-416, 1980.
298. Sharma, H. S. S. Enzymatic degradation of residual non-cellulosic polysaccharides present on dew-retted flax fibres. *Appl. Microbiol. Biotechnol.*, *26*:358-362, 1987.
299. Noda, S., and Matsui, S. Peroxidase manufacture by Japan Kokkai Tokyo Koho Jp. 6279,781 (87,79,781) (Cl C12 N9/08), 13 April 1987, Appl. 85/218, 610, 01 October 1985.
300. Rauter, W. Mushrooms as indicators of mercury emissions in the vicinity of a chlor-alkali plant. *Z. Lebensm.-unters. Forsch.*, *159*: 149-151, 1975.

301. Schmitz, H., and Eger, H. G. Degradation of fungal cell walls by enzymes of the basidiomycete *Pleurotus ostreatus* (Jacq. Fr.). *Adv. Biotechnol.* (Proc. Int. Ferment. Symp.) (M. Moo-Young and C. W. Robinson, eds.), Pergamon Press, 1981, pp. 505-509.
302. Rohtagi, K., Kuhad, R. C., and Johri, S. N. Enrichment of ash and silica in paddy straw by cyathus, *Pleurotus* and *Sporotrichum*. *J. Microbiol. Biotech.*, *1*:91, 1986.
303. Compere, A. L., Griffith, W. L., and Greeve, S. V. Polymer production of *Pleurotus*. *Dev. Ind. Microbiol.*, *21*:180, 1980.
304. Hayes, W. A. The protein in mushrooms. *Mushroom J.*, *30*:204-206, 1975.

10

NONPROTEINACEOUS FERMENTED FOODS AND BEVERAGES PRODUCED WITH KOJI MOLDS

TAMOTSU YOKOTSUKA *Kikkoman Corporation, Noda City, Chiba, Japan*

I. INTRODUCTION

Areas of the globe where seasonal winds such as tropical monsoons and storms bring high humidity and warm air from oceans to coastal regions of continents and islands have environmental conditions that favor development of numerous varieties of indigenous fermented foods. Typical tropical monsoon areas are south and southeast Asian countries; representative tropical stormy areas are the typhoon zone in southeast and east Asia; the cyclone zone of coastal areas of India, Pakistan, and East Africa; hurricane zones of the east coasts of New Guinea and Australia, and that of the southeast part of North America (Fig. 1).

Nomadic peoples of the Eurasian and African continents also have developed a great variety of fermented foods, such as cheese, butter, and fermented milk products from bovine milk, as well as milk of goats, sheep, and horses. Peoples everywhere have developed pickles and other fermented vegetables in diversity. It is principally lactobacilli that can grow at relatively lower temperatures used to produce both dairy products and pickles.

Traditional fermented foods and beverages around the world may be broadly divided into eight categories: (1) bread dough made with yeasts, (2) alcoholic foods and beverages made with yeasts, (3) vinegars made with *Acetobacter*, (4) fermented dairy products made with lactobacilli, (5) pickles made with lactobacilli in the presence of low salt concentrations, (6) fish or meat fermented with enzymes derived from inner organs of *Aspergillus* molds together with lactobacilli in the presence of high salt concentrations, (7) proteinaceous plant foods fermented with *Rhizopus* or *Aspergillus* molds with or without salt, and (8) proteinaceous plant foods fermented initially with *Aspergillus*, *Rhizopus*, or *Mucor* molds, followed with yeast and lactobacillus fermentations in the presence of high salt concentrations.

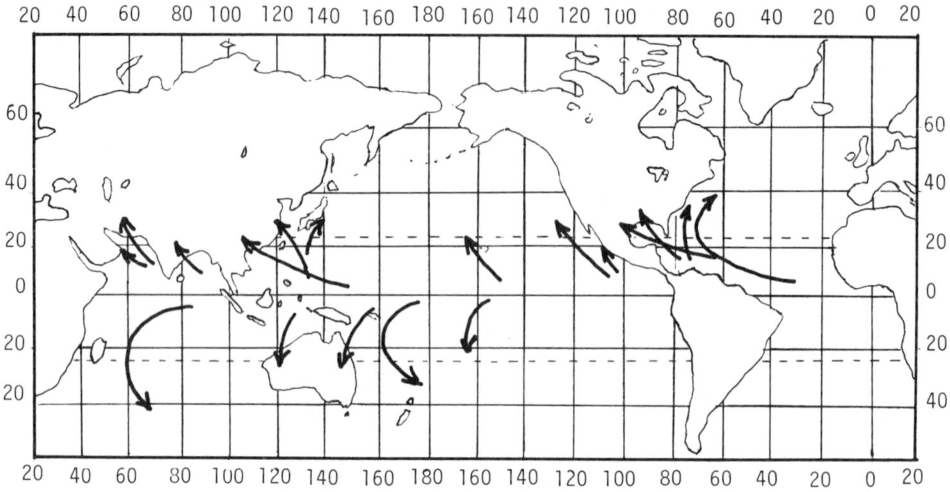

FIGURE 1. The courses of tropical storms. (From Ref. 134 with permission.)

The first encounter of humans with microbes in their dietary life is thought to date back some 10,000 years when they experienced spoilage of fish, poultry, or meat; the alcoholic fermentation of sweet fruits or starchy raw materials; the acetic fermentation of alcoholic beverages; or the lactic acid fermentation of milk and vegetables. Scientific reasons for and mechanisms of these changes were not known in those days.

With the dawn of civilizations in south Asia and north Africa, our ancestors had developed technology for commercial production of wine from grapes, beer from barley, and many other products made with alcoholic, acetic, or lactic fermentations. Three thousand years ago in China, mold enzymes were widely used for saccharification of starch and alcoholic beverage production and for fermenting proteinaceous foods with or without high salt concentrations. Almost all microbes associated with these traditional food fermentations are currently classified as fungi (Fig. 2) [1,2].

The microbes used for or found in food fermentations are also conveniently divided as follows: (a) filamentous fungi, (b) yeasts, and (c) bacteria. The filamentous fungi include the genera *Aspergillus*, *Penicillium*, *Monascus*, *Rhizopus*, and *Mucor*. Yeasts are in the genera *Saccharomyces*, *Zygosaccharomyces*, *Endomycopsis*, and *Torulopsis*, and the bacteria include the genera *Acetobacter*, *Lactobacillus*, *Pediococcus*, *Micrococcus*, *Clostridium*, and *Bacillus*.

Distribution of the numerous fermented foods in the world and their manufacturing processes are determined by the kinds of raw materials, principal microbes, environmental conditions, and food habits found in each region. Identification of the essential microorganisms and their sequences are important for research, but of equal importance is to realize how spoilage of substrates and undesirable microbial contamination can be avoided during food processing. This is of vital importance, especially in tropical areas, whenever one deals with proteinaceous raw materials subject to putrefaction. Development of undesirable microorganisms during

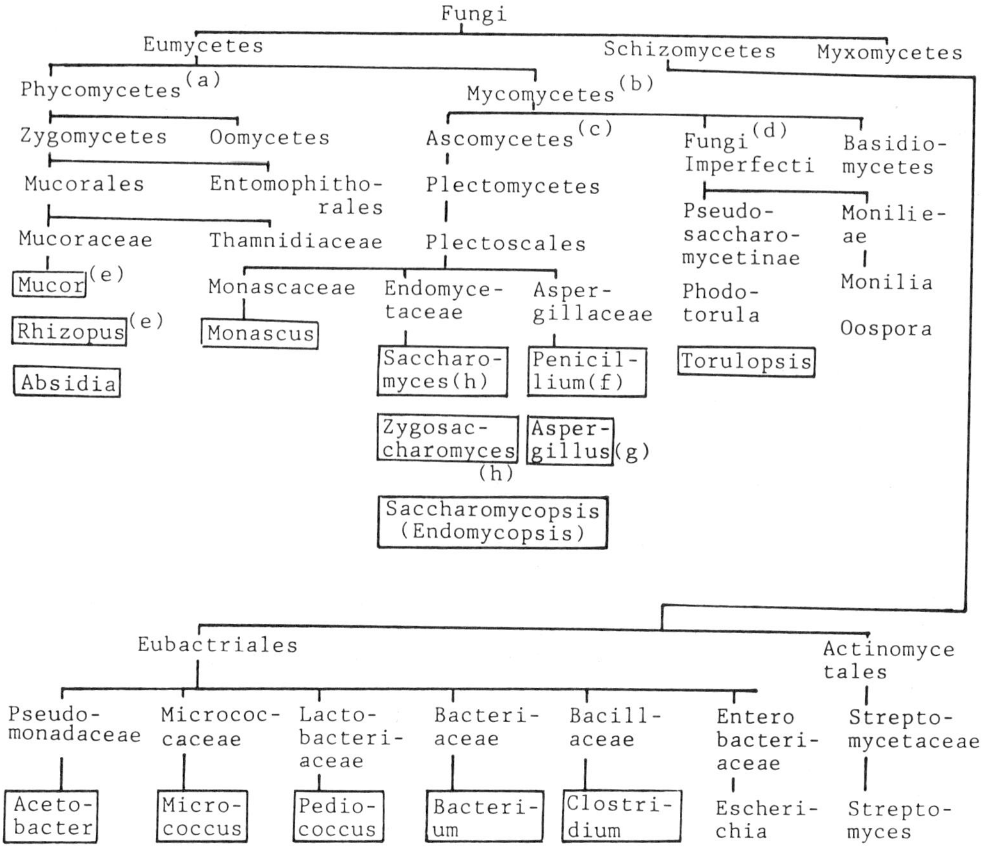

a: septate mycelia, sexual spore formation
b: nonseptate mycelia, asexual and sexual spore formation
c: sporeforming in special sacs or asci
d: spore nonforming
e: asexual spores are formed on the top of sporangiophore which is covered with a sporangium
f: no foot cells, no vesicle (penicillium means a brush)
g. asexual conidiophore are formed by splitting from sterigmata attached to the top of conidiophore (vesicle)
h: spore is formed by budding or conjugation of two cells

FIGURE 2. Classification of fungi.

fermentation is inhibited by the presence of high salt concentrations, by low pH values caused by lactic and or acetic fermentations, and by presence of alcohol produced by yeasts. Of crucial importance and the best way to control unwanted microorganisms is use of clean operations and facilities for manufacturing, including sterilization when appropriate.

This chapter covers as many traditional nonproteinaceous fermented foods and beverages as possible, but emphasis is on alcoholic beverages and vinegars prepared using edible filamentous molds or koji molds, notably in China and Japan.

The term *koji* is Japanese, written in Japanese-coined Chinese characters 麹 or 糀 , whereas Chinese say *chu*, *shui*, or *qu* for koji and use the characters 麯 or 曲 . Historically, the term meant naturally, spontaneously, or even artificially molded cereals, pulses, or beans, and other plant materials used as sources of hydrolytic enzymes for food processing. *Shu-jing* and *Chou-li*, written 3000 years ago in the Chou dynasty (1121-256 B.C.) in China [3-8], makes reference to chu and states that chu meant mold grown on millet or wheat, which was essential in making alcoholic beverages. Chu was also used to degrade proteinaceous foods in the presence of salt to make condiments named jiang. A book on agricultural technology entitled *Chi-min-yao-shu*, written in the sixth century (A.D. 530-550), gives detailed descriptions for preparation of several varieties of chu and some alcoholic and proteinaceous fermented foods made by using chu. Wheat, barley, millet or rice were the raw materials of chu. The book also records that chu was differentiated by the colors of the spores of molds, and called "yellow robe" or "five-color robe," for example, suggesting the former to be yellow *Aspergillus* (presumably *A. oryzae*), and the latter to be spontaneously mix-inoculated *Rhizopus*, *Mucor*, *Aspergillus*, *Monascus*, and or *Penicillium*. The yellow robe, such as steamed granular wheat kernels cultured with yellow *Aspergillus* molds, was mainly used for the preparation of fermented soybean foods. The five-color robe such as steamed crushed wheat dough spontaneously grown with various kinds of mold covered with multicolored spores was prepared while making alcoholic beverages or vinegars. There were two types of chu in those days: mold-cultured raw or cooked grains in granular form (*san-chu* in Chou dynasty) and mold-cultured balls or cakes made from grain flour (*ping-chu*, developed in the Han dynasty, 205 B.C.-A.D. 220). Ping-chu came to be used widely for alcoholic fermentation, because as we learned later, *Rhizopus* and yeast are easier to grow in ping-chu, which is less aerobic, whereas *Aspergillus oryzae* is used in sun-chu, which is more aerobic than its counterpart. Thus the so-called amylo-process was developed in the third century in China.

Interestingly, *Aspergillus* molds have been used and are still being used in Japan to prepare such foods as rice wine and fermented soybean condiments. While the term "koji molds" has come to be used worldwide in the professional literature, there are some discrepancies in the definition: Koji, when used by Japanese, refers to yellow, yellow-green, or black-colored spore-bearing *Aspergillus* molds such as *A. oryzae*, *A. sojae*, *A. tamarii*, *A. awamori*, *A. usami*, and others. In China and Southeast Asian countries, molds in the genera *Rhizopus* and *Mucor* are mainly used in wine making and those in the genera *Rhizopus* and *Neurospora* in tempe and oncom fermentations. Hence, in this chapter the term "koji molds" is used in a broader sense and refers to filamentous molds grown or cultured on solid plant materials in food processing as hydrolytic enzyme producers. This is the definition used in 1985 by Lotong [9] in her review article.

II. NONPROTEINACEOUS, NONALCOHOLIC, LACTIC ACID-FERMENTED FOODS

Pickles, which are prepared by fermenting vegetables predominantly with lactobacilli in the presence or absence of salt, are important examples of nonproteinaceous, nonalcoholic fermented foods produced worldwide. Mold enzymes, however, are rarely used in their production. It is only in Japan that rice-koji, steamed rice cultured with *Aspergillus* molds, is often added to pickles to provide a sweet taste which ameliorates saltines and simultaneously enhances flavor by increasing the content of free amino acids. The starch and protein in rice are degraded by amylolytic and proteolytic enzymes of koji during the storage of pickles just described.

Lactic-leavened bread and lactic-fermented gruels are made from cereals and legumes are popular in tropical areas, although they are not fermented foods made with koji-molds. Examples of lactic-leavened bread are idli in India, made from rice and black gram or mung-bean; puto in the Philippines, made from maize, sorghum, and/or barley; hopper in Sri Lanka, made from rice or wheat flour and coconut water; and kisra in Sudan, made from fermented sorghum flour. These all are sourdoughlike breads and are consumed as a staple of the diet. Yeasts and lactobacilli usually concurrently take part in fermentation of their doughs.

Ogi in Nigeria is a typical lactic-fermented cereal resembling yogurt. Maize, millet, sorghum, cassava, or sometimes soybean is washed with water, steeped, and then subjected to the lactic fermentation. It is then wet milled while the lactic fermentation continues, and is mashed. The resultant slurry is boiled to obtain ogi. Mahewu in South Africa, uji in Kenya, gari in Nigeria, and kenkey in Ghana are similar to ogi [28].

Tea is native to southeast Asia, but the practice of drinking tea made from fermented or nonfermented tea leaves is popular worldwide. Lactobacilli are usually isolated from fermented tea leaves. Miaeng (chewing tea) is a typical fermented tea food popular in northern Thailand. Steamed tea leaves are fermented for 6-12 months to prepare miaeng. The product contains about 4% protein and 20 mg% vitamin C. It is consumed with salt and/or peanuts [133].

III. ALCOHOLIC FERMENTED BEVERAGES

A. Alcoholic Beverages Made from Sugar-Containing Raw Materials

Alcoholic fermentation basically follows the Embden Meyerhof Parnas scheme for anaerobic dissimilation of sugars and is carried out by fermentation yeasts such as *Saccharomyces cerevisiae* as follows:

$$C_6H_{12}O_6 \xrightarrow[\textit{S. cerevisiae}]{\text{anaerobic}} 2C_2H_5OH + 2CO_2$$

Sugar-containing foods such as fruits, berries, honey, sugar cane, palm-sap, coconut juice, and others are the raw materials of alcoholic beverages and are easily fermented with yeasts. We have many alcoholic beverages of this type in the world, especially in tropical areas. Grape wine is believed to be the oldest fermented food but some think honey wine is the oldest. Currently, honey wine is manufactured in many civilized countries in Europe and in the United States. Tej is an Ethiopian honey wine, which is sometimes flavored with hops and spices [10]. Basi is a typical cane-sugar wine produced in the Philippines. Palm wines or toddies with various names are produced in large quantities in the Philippines, Sri Lanka, Nigeria, Ghana, Malaysia, and India [11-12]. Urwaga is an effervescent, slightly sour lactic and alcoholic beverage fermented from peeled bananas in Kenya. Ground roasted sorghum, millet, or maize is often added to the fermentation of urwaga juice [22]. Jack-fruit is a tree bearing large pungent fruits and is cultivated in moist tropical climates such as the eastern hilly area of India. Jack-fruit pulp is fermented to make a wine [23]. Pulque is the national drink of Mexico, which is made by fermenting agave juice with *S. cerevisiae* and *Zymomonas mobilis* [24-26]. Some lactobacilli also take part in the pulque fermentation and give it characteristic sourness and viscosity.

B. Saccharification Processes in Alcoholic Beverage Manufacture: Malt and Koji

Alcoholic beverages are also produced from many starchy raw materials. The starch must be saccharified into sugars, mainly glucose, to be fermented into ethanol by yeasts. The oldest method of converting starch into sugars is the salivation process using amylase from saliva. According to old records in Japan, professional virgin females chewed cooked rice until it was sweet so it could be fermented into the Japanese alcoholic beverage named sake. Astonishingly, chicha in South America is presently made from corn by the Andes Indians using the same method, although chicha can be produced by using the amylase contained in malt of corn or germinated corn kernels [27]. The alcohol content of chicha varies from 2 to 12% (v/v).

The method of hydrolyzing starch into sugars by malts has been popular since ancient times, especially in Western and some African countries. Typical examples are beer and whiskey made from barley in many industrialized countries. Kaffir beer produced in Johannesburg, Durban, and Pretoria [28] and other similar drinks in Africa such as pito in Nigeria [29], busaa in Kenya [30], mbege in Tanzania [31], maize beer in Zambia, bouza [32] in Egypt, and pozol and tesguino in Mexico [33] are of the same category. They are made principally by saccharifying the starch of sorghum, maize, millet, or wheat (only for bouza in Egypt) by the amylolytic enzymes in a germinated portion of these grains, and then fermenting it with yeasts, often accompanied by lactic fermentation. The fermented liquid is commonly neither pasteurized nor hopped. The product is opaque, effervescent, and sour with an alcohol content of 2-4%. Production steps for Kaffir beer involve sorghum malt prepared by soaking seeds in water for 1-2 days, draining them, and then allowing the seeds to germinate for a few days. This malt is mixed with water and a starchy substrate such as sorghum, maize, millet, or cassava, which is then subjected to concurrent saccharification and lactic fermentation. The mixture is then fermented with yeasts for a few days. Originally, the yeast for the fermentation was of natural origin, but now a portion from a previous batch is more commonly used as yeast starter. In making bouza, one part of wheat is made into malt, which is mixed with three parts of boiled wheat, followed by saccharification, lactic fermentation, and alcoholic fermentation.

The saccharification process, which originated in China 3000 years ago, is entirely different from that in the West and in African countries using malt. The amylolytic enzymes used in the Orient have been commonly prepared by culturing filamentous molds on raw materials of plant origin such as cereals, pulses, and/or beans. The malt enzyme was also used in ancient China, but it never became as popular as the mold enzymes. These mold-cultured materials have been called chu or qu in Chinese and koji in Japanese. *Aspergillus*, *Rhizopus*, and *Mucor* molds were selected as producers of amylolytic enzymes for making sugar-containing foods and alcoholic beverages. They are usually inoculated onto heat-treated cereals such as millets or wheat kernels in granular form, or onto ball- or cake-shaped dough made from pulverized and moistened raw materials such as raw rice or wheat.

C. Mixed-Culture Dough Inocula for Food and Beverage Fermentation in Southeast Asia

The mixed-culture dough inocula used in pulverized form for food and beverage fermentations are widely distributed in the East and in Southeast Asian countries and have different names such as chu in China, loog-pang in Thailand, ragi in Indonesia and Malaysia, budod in the Philippines, levain in Khasea and Sikkim, pab or phap in Tibet, bukhar and marcha in India, and nuruk and chuizu in Korea [9,134]. Frequently some spices such as garlic, ginger, chili, and/or pepper are ground, mixed with pulverized cereals such as rice or wheat and water, and made into a ball- or brick-shaped dough to prevent the starter cultivation from undesirable microbial contamination. The mixture is inoculated with powdered starter from the previous batch and a dough is made. This is made into the desired shape—flattened balls 2-4 cm in diameter, for example. After incubation for several days at room temperature, the dough is sun-dried and then powdered. The principal molds involved in this type of inoculum are varied, but commonly *Rhizopus* and *Mucor* are dominant. Such filamentous molds as *Aspergillus*, *Fusarium*, and *Amylomyces*, and yeasts in the genera *Candida*, *Saccharomyces*, *Hansenula*, *Endomycopsis*, and *Lactobacillus* also are likely to be present [20,34-53]. According to Ko [39], essential for a ragi is the presence of *Amylomyces rouxii*, which degrades starch into reducing sugars and ethanol. When combined with one or more yeast genera such as *Candida*, *Endomycopsis*, and *Saccharomyces*, which are present in ragi, ethanol formation is enhanced and a rich aroma and flavor are developed. *Endomycopsis* has both amylolytic and alcoholic enzyme activities. Therefore, this type of inoculum serves as the source of amylolytic, lipolytic, and/or proteolytic enzymes and of the microbial starters for alcoholic and lactic fermentations [39]. Kozaki [54,55] isolated many more yeasts than molds from budod, the solid microbial starter used in rice wine fermentation in the Philippines. The yeast was identified as *Saccharomycopsis* (same as *Endomycopsis*), which could convert starch into ethanol via glucose [54,55].

As for lactobacilli found in ball or cake inocula in Southeast Asia, only *Pediococcus pentosaceus* was found to be dominant, and *Lactobacillus* was not identified. Sumikawa et al. reported that both *Pediococcus* and *Lactobacillus* survived under dry conditions such as a_w 0.2 or less, but only *Pediococcus* could survive under wet conditions such as a_w 0.8 like in the Southeast Asian climate [56]. Twenty-two mix-cultured dough starters or chu for alcoholic beverages were purchased in Yunnan and Kweicho in China in 1984-1985. They were coarsely crushed wheat bran cultured with molds and were sold in bags containing 3-12 g each. Dominant molds were *Rhizopus* mixed with *Syncephalastrum* and were isolated from 20 starters at a level of 10^3-10^7/g. Four seed starters contained, per gram, 10^6-10^7 of the yeasts *Saccharomyces cerevisiae* and *S. fibligera*. Lactic streptococci were commonly isolated at an average level of 10^6/g; *Pediococcus* and *Lactobacillus* also were isolated [129,135].

Tape ketan in Indonesia [37] is a sweet-sour alcoholic pasty food fermented from glutinous rice. A dough microbial starter about 3 cm in diameter, made from rice flour and several spices and inoculated from the previous batch, is used in powder form. The alcohol content of tape ketan generally is less than 8% (v/v). There are products similar to tape ketan: tape ketela, made from cassava tuber, in Indonesia; tapai in Malaysia [46]; tapuy in the Philippines [48]; ruhi and madhu in India [57,58]; and some local products in Thailand [59]. These are generally produced from rice, glutinous or nonglutinous, in a somewhat similar manner. Chang is made from barley, wheat, glutinous rice, or corn by solid fermentation in Tibet and Nepal and belongs to the same group of products [60,61].

D. Alcoholic Beverages in China and Taiwan

Making chu for production of traditional alcoholic beverages in China, Korea, and Taiwan is characterized by the following: (a) the principal amylolytic enzyme producers are *Rhizopus* or *Mucor*, and (b) these are in most instances cultured on balls, cakes, or bricks (22 × 22 × 4.5 cm) and made from moistened raw rice, wheat, sorghum, or barley flour.

Chinese wines are classified on the basis of color into yellow wines (Huan jiu), red wines (anjiu), and others. They also are classified according to production sites; an example is Shaoxing jiu (Shaoxing is located south of Shanghai in Zheziang sheng). Shaoxing jiu is yellow wine having a 2000-year manufacturing history; it is produced in southeast China and also in Taiwan. Shaoxing jiu and Huan jiu in Shandong sheng, east of Beijing (Peking), constitute the two representative types of wine in China (Fig. 3).

According to Otani [62], in southeast China (Zheziang sheng and Fujian sheng), the main raw material of yellow wine is glutinous rice with or without nonglutinous long- or short-grained rice as submaterial and in northeast China (Shandong sheng, etc.) glutinous Boom Corn Millet with glutinous rice as submaterial. The raw materials of chu or koji vary depending on regions in China: wheat in the north, rice powder and wheat in the central part, and rice-powder in the south. The essential raw materials

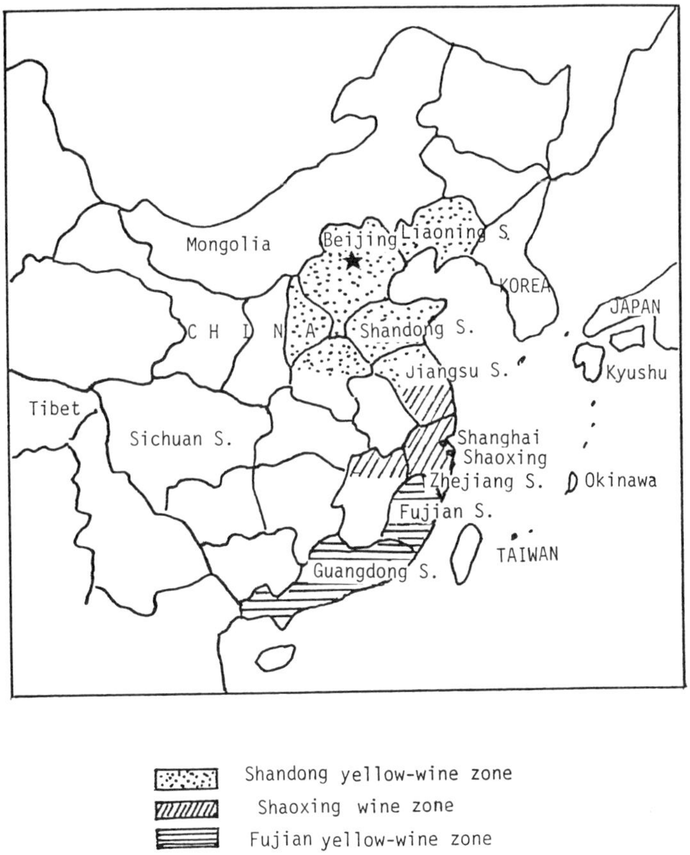

FIGURE 3. Famous Chinese wine-producing areas. (From Ref. 62.)

of Shaoxing jiu are glutinous rice and wheat-chu [60]. In making a typical Shaoxing jiu, a mixture of coarsely crushed rice, wheat bran, some spices, and water is formed into small cubes or balls about 2-5 cm in diameter, which are inoculated naturally or from the previous batch with *Rhizopus*, *Mucor*, *Aspergillus*, and *Absidia* molds, and yeasts taking 3 weeks to make a seed starter. On the other hand, the starter culture of yeasts, *Saccharomyces shaohsing* I and II, is prepared by fermenting a mixture of cooked glutinous rice, water, and seed starter. To make wheat chu, raw wheat kernels are coarsely crushed, moistened, shaped into round and flat cakes or packed in a basket, and cultured mainly with *Rhizopus* molds for 1-3 weeks. The wheat chu, yeast starter culture, and water are mixed and fermented for 1-2 months at 20°C in a large jar, followed by pressing, pasteurizing, clarifying, and bottling in a jar with a tight seal for aging (Fig. 4). The product is brown colored and contains 10-15% ethanol (v/v). The microbes isolated from wheat-chu were *Rhizopus japonicus*, *R. hangchon*, *R. chinesis*, *Absidia*, *Mucor*, *Monilia*, *Aspergillus*, *Lactobacillus*, and *Acetobacter* [62-64].

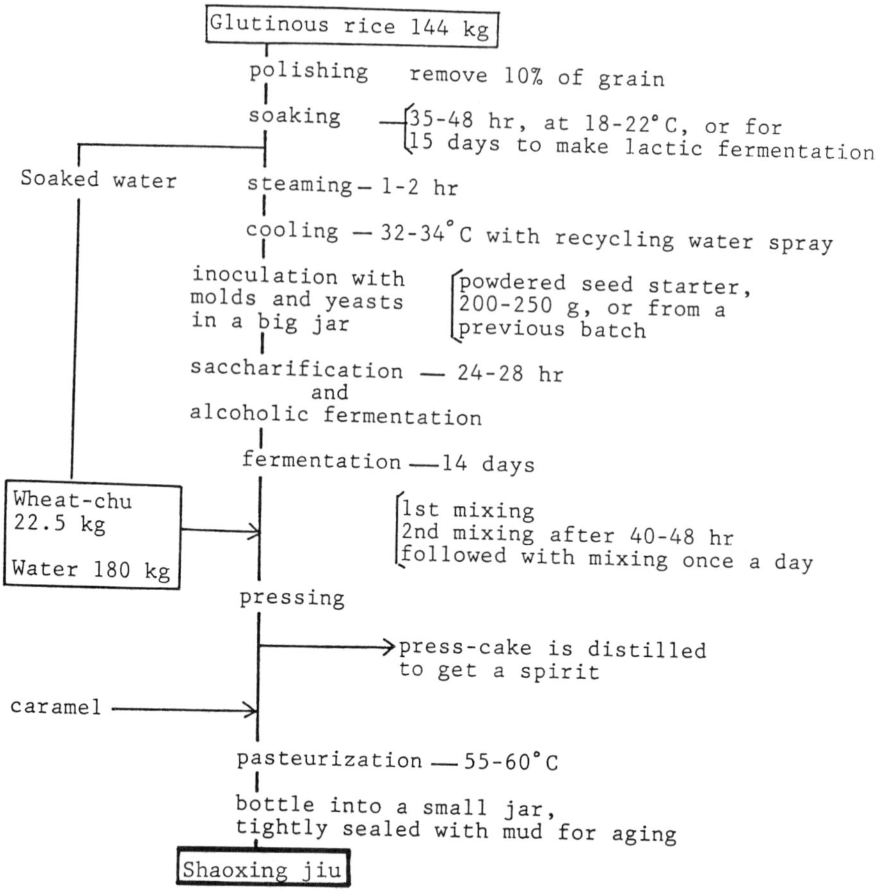

FIGURE 4. Manufacture of Shaoxing jiu. (Adapted from Ref. 62.)

Huan-chiew in Shuntung sheng is made by fermenting a mixture of cooked glutinous sorghum or glutinous millet and wheat-chu cultured with black or yellow *Aspergillus*. The fermented materials are mixed with water and pressed to get a light yellow-colored liquid containing 9-13% ethanol [65].

In making red wine or antyu, glutinous rice is inoculated with *Monascus* sp. to make ang-kak or red-koji, which is mixed with steamed glutinous brown rice and water and fermented with yeasts. The representative species are *Monascus anka, M. purpreus,* and *M. barkeri* [66,67].

Enzymes produced by *Monascus* include amylase, protease, ribonuclease, lipase, and α-glucosidase. This genus produces an edible red pigment called anka. Six chemical structures of anka are known. Anka production in Taiwan is about 100 tons per year [68]. Similar products are also popular in southeast China. Antyu contains 13-15% ethanol and is red but turns yellow during storage. Distilled rice wine is called Bi-chiew, is characteristic to Taiwan, and is often added to antyu mash. Bi-chiew is distilled from a mash prepared by fermenting a mixture of steamed, polished rice, ball-starter made of rice powder and cultured with *Rhizopus* and yeasts, and water [69].

Although Aristotle (384-322 B.C.) knew the process of distillation, it was not until the thirteenth or fourteenth century that humans consumed distilled liquor in China and in Scotland [70]. Kaoliang liquor [71] is a typical distilled wine produced in China and has different names. It is made from sorghum and corn through a characteristic solid-state fermentation. Semi-solid balls, cakes, or cubes made from cooked, crushed sorghum mixed with seed starter is stored underground, or more recently in a box sealed with plastic film, for about 10 days, where the mass is concurrently saccharified with mold enzymes and fermented with yeasts. The fermented material is steamed for distillation to get 40-60% ethanol (v/v) in the final product [72]. Earlier the seed starter was prepared naturally by inoculating *Rhizopus* molds onto balls or cakes of pulverized, moistened barley and red beans. However, recently the starter has been replaced by wheat-bran chu cultured with black *Aspergillus* molds.

According to Chiao [73], Chinese quality white spirits are generally classified into three types according to their distinguishing flavor components: ethyl acetate, ethyl capronate, and unknown fragrance components designated as Moutai, which is considered to consist of guaiacol, vanillic acid, and vanillate. There is a growing trend in China to produce white spirit with the liquid process in place of solid-substrate fermentation. In fact, about one-half of the white spirit factories in China have already adopted the liquid process, which greatly contributes to reduction of the fermentation period and increases the yield of alcohol. The key microbe which generates the characteristic flavor was isolated from the mud of old cellars and found to be *Clostridium*. This spore-forming bacterium can synthesize rather large amounts of caproic acid which is transformed into ethyl capronate in the fermenting mash [73]. One example of volatile flavor constituents of kaoliang liquor includes *n*-propylalcohol 270 ppm, isobutylalcohol 140, isoamylalcohol 360, ethyl acetate 1,280, ethyl peragonate 80, isoamyl acetate 840, acetaldehyde 330, and acetoin 280 [74].

Production of yellow wine and white wine in China amounted to 620,000 tons and 2,900,000 tons, respectively, in 1983.

E. Yakju and Takju in Korea

The traditional Korean alcoholic beverages are Yakju and Takju [75-77], which were originally made from rice but are now produced from wheat, barley, corn, or millet. They are also known as Maggally. They are made by saccharifying starch with *Aspergillus* mold enzymes and fermenting sugars to ethanol with *Saccharomyces* yeasts. The starter is named Nuruk, which is made from wheat, but historically was made from rice by a natural inoculum of molds, bacteria, and yeasts. However, *Aspergillus usami* is often used instead of the natural mixed-culture starter. Nuruk is the source of amylase produced by molds, and also of *S. cerevisiae* and *Hansenula* spp. that produce ethanol and a characteristic flavor. Concurrent saccharification of raw materials and lactic and alcoholic fermentation occur in mash prepared from cooked rice, nuruk, and water.

F. *Rhizopus* vs. *Aspergillus* as Enzyme Producers in Alcoholic Beverage Fermentations

In the saccharifying process for Japanese rice wine making, the following points are emphasized: (a) pure cultures of *Aspergillus* molds such as *A. oryzae*, *A. awamori*, or *A. usami* are predominantly used to produce amylolytic enzymes to degrade starch in raw materials; (b) *Aspergillus* mold is inoculated onto steamed rice grains in granular form to make koji. This is in sharp contrast to the chu used in alcoholic beverage making in China, Korea, and Taiwan, in which *Rhizopus* or *Mucor* is principally inoculated onto ball- or brick-shaped dough made from moistened, raw flour from rice or wheat.

It is worthwhile to consider why strains of these different molds are cultured in various ways in chu-making in China and koji-making in Japan. It is thought that these differences might have been caused by differences in the staple of diet: bread made of rice or wheat flour in China and cooked rice grains in Japan. It is likely that the different environmental conditions of temperature and humidity resulted in the difference in microbes naturally present. In fact, when cooked rice is placed in the open in Japan and subjected to spontaneous microbial contamination, growth of yellow *Aspergillus* molds in most instances predominates on the cooked rice, whereas the dominant microbes isolated from naturally contaminated ball or brick dough starter in China are *Rhizopus*, *Mucor*, and yeasts, although *Aspergillus*, *Penicillium*, and bacteria are present in lesser numbers. *Rhizopus* and *Mucor* molds and yeasts grow better at lower temperature, in more acidic, more humid, and less aerobic conditions than do *Aspergillus oryzae* and *A. sojae*.

Tanaka and Okazaki [78] and Tanaka [79] compared growth of *Aspergillus* (*A. oryzae* var. *viridis* Murakami RIB 128) and *Rhizopus* (*R. jawanicus* Takeda RIB 5501) when they are inoculated onto raw or steamed rice grains (25% polished) in granular form at 35°C and 95% RH. Growth was determined by oxygen consumption per unit of dry matter. The best growth was observed when *Rhizopus* was cultured on raw rice grains, followed by *Aspergillus* on raw rice, *Aspergillus* on steamed rice, and *Rhizopus* on steamed rice, in descending order. Retarded growth caused by steaming rice grains was greater with *Rhizopus* than with *Aspergillus*. Enzyme activities of these mold-cultured materials are listed in Table 1. The amylolytic activities of both steamed and raw brown rice grains cultured with *Rhizopus* are lower than those of *Aspergillus*. The inferior growth of *Rhizopus* but not *Aspergillus* on steamed rice is considered to be caused by the shortage of free amino acids available for growth due to less acid carboxypeptidase (ACPase) activity by *Rhizopus* than by *Aspergillus*. Molds in the genus *Rhizopus* grew well on polished rice steeped in a solution of amino acids (Ala, Glu, Lys, Tyr, Val) and steamed. It was concluded that the amino acids were important factors affecting *Rhizopus* growth [130].

This is even more evident when rice protein becomes less digestible by heat denaturation. Growth and enzyme formation by *Aspergillus* and *Rhizopus* were compared when they were cultured on tablets made from powdered and moistened brown rice (19.9 × 5.6 mm). Optimal amounts of moisture to be added to the powder to give the highest enzyme activities

TABLE 1. Enzyme Activity of Steamed or Unsteamed Glutinous Rice-Koji Inoculated with *Aspergillus oryzae* or *Rhizopus jawanicus*

	Mold strain	AAase	GAase	APase	ACPase
Steamed rice koji	A	1527	408	3961	12643
	B	100	100	4502	456
Nonsteamed rice koji	A	1255	331	4518	12360
	B	100	59	4184	466

AAase: α-amylase U/g, dry matter; GAase: glucoamylase dry matter; APase: acid protease dry matter; ACPase: acid carboxypeptidase dry matter.

Source: Ref. 79.

of α-amylase (AAase), glucoamylase (GAase), and acid protease (APase) were 40% for *Rhizopus* and 30% for *Aspergillus*. *Rhizopus* exhibited better growth and better formation of APase but lower AAase, GAase, and acid carboxypeptidase activities than did *Aspergillus*. Growth of seven different species of *Rhizopus* was better, and amylolytic enzyme formation was less when they were inoculated onto powdered rice grains than powdered brown wheat grains. Growth of *Aspergillus* was less but amylolytic enzyme formation was two to five times greater than that of *Rhizopus* when the substrate was powdered raw wheat grains [78,79]. It is reported that a semisolid microbial starter of *A. oryzae* cultured on powdered rice or wheat is currently used for opaque rice wine brewing in China [71], but *A. oryzae* is a better producer of amylolytic enzymes when grown on any form of steamed or nonsteamed rice or wheat.

G. Sake [80-87] and Shochu in Japan

The traditional alcoholic beverages in Japan are nonglutinous rice wine named sake or seishu and a distilled sake called shochu. Shochu is made from press-residue of sake mash or from an alcoholic mash made from sweet potatoes. The annual production of sake was 1.54 million kL and that of shochu 0.63 million kL in 1985.

1. Sake

Although the earliest mention of sake brewing in Japan is found in a record of the eighth century, its current production technology was established in the early sixteenth century. Sake brewing has been monopolized by the government ever since 1897. Japanese sake is characterized by its method for production of amylolytic enzymes by *A. oryzae* and vigorous lactic and yeast fermentations in the mash, and also by its very high contents of ethanol (as high as 20%) and free amino acids and lower peptides reaching 0.1% as total nitrogen. Moreover, it is often served warm.

Production of sake comprises rice-koji cultivation, preparation of starter mash of yeast, mash fermentation, pressing mash, and refining, as explained below.

Rice koji cultivation: Brown nonglutinous rice grains are polished to remove about 25% of the outer layer, which is rich in proteins and oils. The polished rice grains are soaked in water to absorb 27-30% moisture and then are steamed for 30-50 minutes. The steamed rice is inoculated with the starter mold or Tane-koji, *A. oryzae*, and incubated at about 25°C for 43-48 hours to make rice-koji.

A typical strain used for sake brewing is *A. oryzae* var. *globosas* Sakaguchi et Yamada. Functions required of *A. oryzae* when used for sake brewing include: (a) good growth on steamed rice, (b) good production of α-amylase and glucoamylase, (c) production of small amount of tyrosine to prevent the press-cake of sake mash from turning brownish, (d) no production of pherichrome, which causes color deepening when combined with Fe, (e) good spore formation, (f) production of good flavor, and (g) no production of mevalonic acid, which functions as a growth factor for the Hiochi bacterium, a lactic acid bacterium which deteriorates sake [88].

In making seed mold starter or tane-koji, generally two or three different strains of *A. oryzae* are inoculated on steamed, coarsely polished brown rice mixed with wooden ash, and cultured until abundant spore formation is obtained, followed by drying to 10-12% moisture.

Rice koji is cultured in the following two ways:

1. Conventional method (rice-chamber method): Steamed rice is cooled to 35°C and is mixed well with seed starter of *A. oryzae*, piled on the floor of a koji room at 28-30°C for about 10 hours, and the material is mixed. One and one-half kilograms of each of the materials (out of 100 kg) is placed in the center of a wooden tray about 30 × 45 × 5 cm; the resultant 270-300 trays are piled in the room for about 48 hours, during which time the trays are repiled two times. Materials are hand-mixed three times during cultivation when the temperature of materials rises to 35°C.
2. Mechanical rice-koji cultivation: A mixture of steamed rice and seed mold is spread on a perforated stainless steel plate with a thickness of 20-30 cm, and the molds are cultured by passing air with automatically controlled temperature and moisture through bottom holes and into materials according to conditions listed in Table 2 [86]. The enzyme composition of the rice-koji thus prepared is given in Table 3 [89] and is characterized by high activities of acid protease and carboxypeptidase and low activity of alkaline protease.

Since rice-koji cultivation is done in open air, the cleanest operation possible is required. Nevertheless, microbial contamination sometimes occurs: sake-yeasts 10^2-10^5/g, film-forming yeasts 10^2-10^5/g, lactobacilli 10^1-10^6/g, micrococci 10^4-10^6/g, and *Bacillus* 10^7/g. Substantial bacterial contamination of koji damages the quality of the final product.

Cultivation of seed mash of yeast or Moto: To ferment a sake mash as purely as possible with sake yeasts, a seed mash of yeast is prepared as the first step of mash fermentation. Historically, yeasts were naturally or spontaneously selected and grown in the mixture of rice-koji and water, simultaneously saccharifying starch into glucose and keeping the mash

TABLE 2. Temperature and Humidity Changes During Mechanical Cultivation of Rice-Koji for Sake Brewing by Through-Flow Aeration

Cultivation (h)	Temp. of materials[a]	Relative humidity (%)
0	32°C	90-94
20[b]	Elevate up to 35°C	80
25	Elevate up to 38°C	80
30	Elevate up to 38°C	80
30-40	Keep at 38-40°C[c]	80
40 (finish)		

[a]Thickness of materials = 30 cm.

[b]Mixing is conducted.

[c]Intermittent aeration.

Source: Ref. 86.

TABLE 3. Enzyme Composition of Rice-Koji for Sake Brewing

Enzyme	Average activity	Desired activity
α-Amylase[a]	1225	1000
Glucoamylase[b]	201	250
Transglucosidase[b]	286	—
Acid protease[c]	3674	4000
Acid carboxypeptidase[c]	5100	—

[a]Wholegemuth value, D_{30}^{40}/g koji.

[b]mg Glucose formed/h/g koji.

[c]μg Tyrosine found/h/g koji.

[d]Calculated when 20% of rice as raw materials is made into koji.

Source: Ref. 89.

weakly acidic by lactic acid fermentation, which suppresses growth of nonessential microorganisms. Eda in 1911 simplified and shortened the traditional method for cultivation of yeast starter mash by adding lactic acid to the mixture of rice-koji and water and propagated yeasts in the mash.

Good strains of *Saccharomyces cerevisiae* have been selected for aroma formation, nutrient requirement, pH, foam formation, color pigments, coagulation with lactobacilli, absorption by foams and fibers, film formation and a yeast-killing factor. Representative strains of sake yeast are supplied by the Brewing Society of Japan by the names of Nos. 6, 7, and 8 Kyokai Yeasts. Yeasts forming less foam, which reduce the increase in volume of mash so it can be contained in one tank have been introduced into the industry [90].

For conventional cultivation of yeast seed mash, 5-7% of rice, which is the raw material of sake brewing, is used for preparing seed mash of yeast. Of the rice, 30% is cultured into koji, and the koji is mixed with the remaining 70% steamed rice and water equivalent to 110% in volume of total rice used for seed mash. The mixture is kept at less than 10°C for 3-5 days followed by gradual elevation of temperature up to about 20°C over approximately 10 days. During this time, saccharification of koji proceeds, and growth of wild yeast is suppressed by the synergistic action between about 1% of lactic acid produced by lactobacilli such as *Leuconostoc mesentroides* var. *sake* and *Lactobacillus sake* and nitrite produced by nitrate-reducing bacteria dominated by *Pseudomonas* [91-93]. When sufficient glucose and lactic acid are produced and the nitrite disappears from the mash after some 15 days, seed starter is added and the mash is kept at 20-23°C until the yeast fermentation ends, taking about 30 days from the beginning. Microflora change during conventional yeast starter mash fermentation is shown in Figure 5.

Rapid cultivation of starter mash of yeast is done as follows. To the mixture of rice-koji and water, seed yeast and lactic acid equivalent to 0.5% of the water are added and fermented for about 2 weeks. The temperature of mash is initially 10-12°C for 2 hours, gradually elevated to 20°C after 7 days, kept at 20°C until the alcohol content of the mash is 10%, and then the mash is cooled. Almost no difference in quality of the final products was observed between the traditional and rapid cultivation of starter mash of yeast.

Sake mash fermentation: From 5 to 7% of the rice used as raw material serves for preparation of seed mash of yeast, and the remaining rice is divided into three portions in the ratio of 1:2:3-4. Each portion of rice is steamed, 30% of which is cultured into koji, and the mixture of koji plus steamed rice and water to give 110% volume of that of rice is added to the starter mash of yeast in three steps taking 3 days. The temperature of mash starts at 7-9°C, followed by gradual elevation to 15-17°C. The mash is fermented for a total of 15-25 days. Because of this stepwise addition of materials during sake mash fermentation, the pH of the mash is kept quite high, and the viable count of yeasts of the mash is kept at more than 10^7/g, which contributes to formation of ethanol as high as 20%. Hayashida et al. found that the proteolipid produced by koji mold promoted growth and ethanol tolerance of sake yeasts in the sake mash fermentation [94]. Instead of the conventional starter mash of yeast, abundant cells of

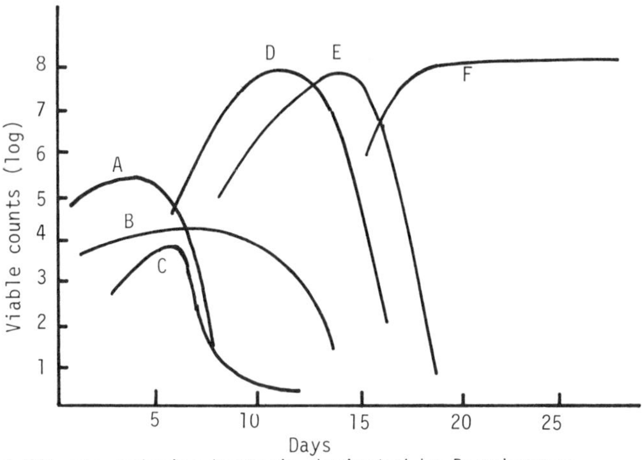

A:Nitrate-reducing bacteria dominated by Pseudomonas, B: Wild yeasts, C:Film-forming yeasts, D:Leuconostoc mesentroides, E:Lactobacillus sake, F:Sake-yeasts (Saccharomyces cerevisiae).

FIGURE 5. Microflora changes during conventional fermentation of starter mash of sake yeasts. (From Ref. 93.)

sake yeast, usually pure-cultured in a molasses medium, have come to be used in industrial sake mash fermentation (Fig. 6).

Refining process and product: The aged mash is press-filtered through cloth. The liquid thus separated is polished by cotton cloth filtration and then is pasteurized at 60°C to produce the final product. One ton of polished rice and water can make 1.8 kL of genuine sake containing 20% ethanol [86]. It is legally acceptable to increase this volume to 5.4 kL by adding fermented ethanol and legally permissible chemical constituents of sake into the genuine sake mash when the yeast fermentation is almost finished. Sake is pale yellow in color and its average chemical composition is as follows: ethanol, 15-16%; reducing sugar, 3-4% (80% glucose); total nitrogen, 0.1% (amino acids and lower peptides); and organic acid, 1-2% (35% lactic acid, 25% succinic acid). The composition of alcohols and esters of typical sake is given in Table 4 [95]. An attempt was made to improve the aroma of sake by increasing the isoamyl acetate content through addition of yeast able to produce a high level of isoamyl acetate [96].

The color of sake is largely formed and further deepened through the so-called Maillard reactions. Tadenuma and Sato [97] isolated a pigment from sake which was identified as ferrichrysine conjugated between deferrichrysine produced by koji mold and Fe^{3+} from water.

Years ago sake manufacturers experienced deterioration of sake mash and sake during storage. This phenomenon was known as "Hiochi" and was accompanied by turbidity and diacetyl formation. This was found to be caused by contamination with a wild lactobacillus, *Lactobacillus hiochi*. Tamura identified a growth factor for this bacterium, named "Hiochic acid"

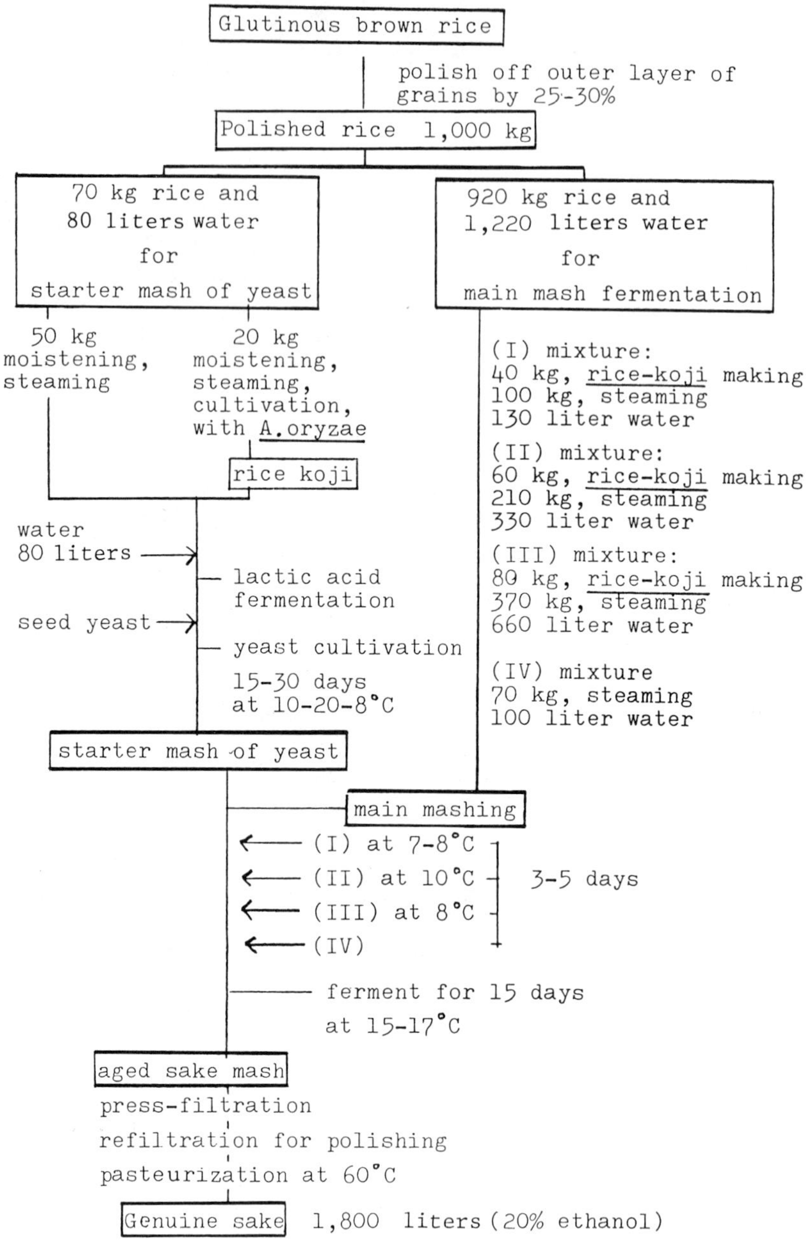

FIGURE 6. Japanese sake (rice wine) fermentation. (From Ref. 86.)

TABLE 4. Composition of Alcohols and Esters of Japanese Sake

Ethanol	15-17%	Ethylacetate	30-120 ppm
Glycerol	0.1-1.2%	Isobutyracetate	0.5-1.5
Isobutyralalcohol	30-70 ppm	Isoamyacetate	3-15
Act. amyalcohol	30-80	Ethylcapronate	3-10
Isoamylalcohol	100-200	Ethyllaurynate	2-11
Tyrosol	10-50	Ph. Et. acetate	8-7

Source: Ref. 95.

[98,99]. This compound was recognized to be identical with mevalonic acid, $HOCH_2CH_2C(CH_3)(OH)CH_2COOH$, which was found by Folkers, who isolated it as a growth-promoting factor of lactobacilli [100]. *Lactobacillus hiochi* can grow in the presence of 25% ethanol and at pH 4.0-5.5, but not at pH 6-7, and it is killed by heating at 60°C for 10 minutes. The hiochic bacteria were classified by Kitahara et al. into four types based on their sugar metabolic schemes and hiochic acid requirements [101].

Pasteurized sake sometimes becomes turbid, caused by the suspension of many white small particles found to be coagulated saccharogenic enzyme proteins. Sake exhibits some blue fluorescence under UV light, which is caused by harman photochemically produced from acetaldehyde and tryptophan in sun light [102].

2. Shochu [103-105]

Shochu of Japan is divided into two categories: (1) genuine shochu made by pot-still distillation of shochu mash, which is made from rice-koji, starchy raw materials such as rice, barley, sweet potatoes, and water; and (2) the new type of shochu made by mixing diluted fermented ethanol and some ingredients. The ethanol content of genuine shochu and the new type of shochu recognized by law are less than 45% and 36%, respectively. Total production of shochu in Japan in 1985 was about 600,000 kL, and about 40% was genuine shochu [106]. The production technology of shochu is said to have been introduced during the sixteenth century from Thailand to the southernmost parts of Japan, Ryukyu, and Kagoshima. Historically, genuine shochu was made only from rice-koji and water; however, since the seventeenth century when sweet potatoes were imported into Japan, much shochu came to be produced from rice-koji and steamed sweet potatoes. The latter type of shochu currently makes up 60% of total shochu produced in Japan. Currently, shochu is usually prepared in two steps: A starter mash or first mash is prepared from rice-koji and water, and then the second or main mash is prepared by adding steamed starchy raw materials such as rice, sweet potatoes, barley, buckwheat, and others to the first mash and fermented. As an example, rice-koji made from 100 kg of polished rice grains and 120 liters of water are mixed and fermented for 7 days at 25-30°C, then 200 kg of cooked rice and 440 liters of water are added and fermented for 15-20 days at 25-35°C to obtain a final alcohol content of

17-19% in the mash. When 500 kg of sweet potatoes and 300 liters of water are added to the first rice-koji mash instead of cooked rice and water, sweet potato-shochu mash is prepared which contains 13-15% alcohol after 15 days of fermentation. Press-cake of sake mash as is, or after enzymatically saccharifying its residual starch, is steam-distilled to get a shochu named Kasutori-shochu.

The strains of mold for shochu-koji are the so-called black koji molds, *A. awamori* and *A. usami*, used in the southernmost islands of Japan, Ryukyu, and Amami, and the so-called yellow koji molds, *A. oryzae*, used until 1907 in Kyushu Island, which is north of Amami Isle. Use of black koji molds for shochu-koji spread rapidly in the southern part of Kyushu, Kagoshima during 1907-1910, by which time the production yield of shochu was increased by 20-30% [108,108]. Currently, a white mutant strain of *A. usami* is used predominantly for shochu-koji, although yellow koji molds are still popular in northern Kyushu including Kumamoto.

Black koji molds are characterized by high productivity of citric acid, which keeps koji at an acidic level and prevents development of acid-forming bacterial contamination which is liable to occur in warmer districts.

Preparation procedures and facilities for black rice koji are similar to those for yellow rice koji. Both take about 45 hours, but the temperature changes during growth are markedly different. In culturing rice-koji with *A. oryzae* for sake and miso (fermented soybean paste), the temperature of materials is kept at ca. 30°C during the first half of the incubation. It is elevated to about 40°C during the second half to ensure higher activity of α-amylase. While culturing rice-koji with *A. awamori* or *A. usami*, the temperature begins at and is held at 43°C during the first half of the incubation to avoid growth of harmful bacteria and held at 30°C during the second half of the incubation. These are shown in Figure 7 [104]. Black rice koji is currently widely cultivated using a rotary-drum.

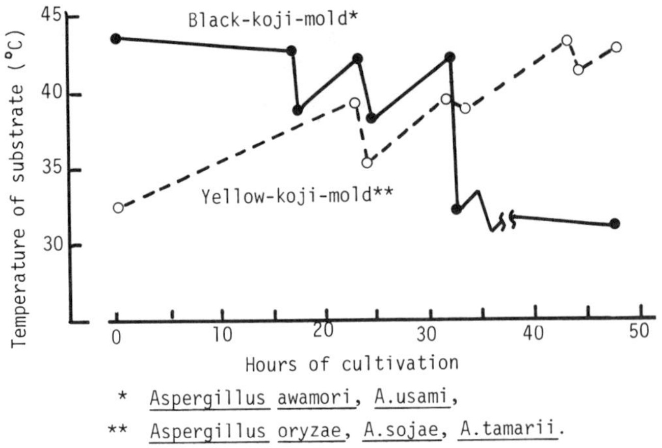

FIGURE 7. Temperature change of substrate during rice-koji cultivation with black koji mold and yellow koji mold. (From Ref. 104.)

Genuine shochu is manufactured by pot-still distillation of alcohol mash made from different raw materials; hence, numerous volatile ingredients, which contribute to sensory characteristics of commercial products, are present and consist of *n*-propylalcohol, 100-200 ppm; isobutylalcohol, 200-300 ppm; isoamylalcohol, 400-500 ppm; ethylacetate, 10-30 ppm; and aetaldehyde, 20-150 ppm [109,110].

H. Mirin, Amasake, and Lao-chao

Mirin manufacture began about 400 years ago in Japan and involved saccharifying steamed glutinous rice with nonglutinous rice koji in the presence of about 20% ethanol or in shochu. Its current annual production is about 70,000 kL. Mirin is used mainly for cooking as a condiment, and a small portion is flavored with several kinds of spice and then is consumed as "Toso" to celebrate the New Year. It is a transparent, pale yellow, very sweet and viscous liquid containing 12-14% ethanol, 35-45% glucose, and has a pH of 5.3-5.5. An example of its manufacturing procedure is as follows: steamed glutinous rice, 2500 kg; glutinous rice koji, 500 kg; and 1800 liters of 40% ethanol or shochu are mixed and the temperature adjusted to less than 30°C. Then the mixture is stored for 40-60 days at 20°C with agitation about once a week to allow saccharification and extraction of flavor compounds of koji. Because of the high alcohol content, no yeast or bacterial growth takes place during storage. Well-aged mash is press-filtered and the liquid thus separated is stored for 1-3 months for clarification. Gluterinlike protein derived from rice sometimes causes cloudiness of mirin, which can be avoided by steaming glutinous rice under pressure and dissolubilizing the gluterinlike protein.

Mirin mash is sometimes ground to make a white-colored viscous alcoholic drink named Shiro-sake or white sake, which is served on Girls' Day Festival on March 3 in Japan. Shiro-sake is very sweet and different from Nigori-sake or opaque-sake, which is a nonclarified sake. The mixture of nonglutinous rice koji and steamed glutinous rice is often saccharified at 55-60°C and diluted with water to make a sweet nonalcoholic drink named Ama-sake. According to Wang and Hesseltine, Lao-chao in China is a sweet gruel made from steamed glutinous rice and a commercial starter of yeasts and filamentous molds which saccharify the starch and ferment a portion of the starch to yield 1-2% ethanol [126].

IV. VINEGAR

A. Technological Development of Vinegar Manufacture in Western Countries

Vinegar has been used as a condiment, a preservative, and a medicine since ancient times. The vinegar made from date juice was known to Babylonians in 5000 B.C. [111-112]. Home vinegar production from all varieties of alcoholic substrates is practiced worldwide. Industrial vinegar production was established around the seventeenth century and included wine vinegar in Italy, Spain, and France and malt vinegar in Britain. Pasteur in 1864 found the mechanism of vinegar fermentation by *Acetobacter*. Vinegar can be prepared from any sugar-containing substrates and hydrolyzed starchy materials through the alcoholic fermentation followed by the acetic acid fermentation as follows:

$$C_2H_5OH + O_2 \xrightarrow[\textit{Acetobacter}]{} CH_3COOH + H_2O + 114.6 \text{ cal}$$

Acetic acid formation from ethanol by *Acetobacter* is an oxidative reaction generating heat. Almost all technological improvements associated with vinegar fermentation have been centered on how to increase the oxygen supply to the culture medium during fermentation. Until the Middle Ages, vinegar was produced by holding an alcoholic solution quiescently in kegs. This was followed by the so-called slow process, such as the fielding, Orleans, or French process, used until the seventeenth century to produce vinegar by surface culture in horizontal kegs and requiring several months. The so-called quick process or German process which used generators was invented by Boerhaave in 1670 and Schüzenbach in 1923. The generators (large wooden tanks) are packed with beechwood shavings, coke, birch twigs, or other materials and the alcoholic liquid medium acidified with about 20% fresh vinegar contacts the packing materials either by circulating the liquid through the generators or by alternating the fermenter until the typical slimy *Acetobacter* grows. The Frings Co. in 1932 developed a highly sophisticated rapid vinegar generator with automatic control of temperature, aeration, and recirculation of liquid substrate, which upgraded the yield of vinegar to 85-90%, reduced recycling time to 10 days, and increased the acetic acid content of finished vinegar to 12%. After World War II, introduction of the submerged aeration process for vinegar production was influenced by the submerged penicillin fermentation, which increased the fermentation speed to 10 times that of the quick surface generator process, increased yield to 94%, and reduced the recycling time of liquid to 25-35 hours [113-114].

Various vinegars are consumed worldwide, including wine vinegar, malt vinegar, fruit vinegars, grain vinegar, spirit vinegar, whey vinegar, honey vinegar, and others. Ethanol has been produced from molasses abundantly and economically since the beginning of the twentieth century, and so alcohol is commonly used as raw material for the vinegar fermentation. Moreover, in the United States edible acetic acid is legally fermented from synthesized ethanol. The acetic acid thus prepared promoted mass-production of vinegar, which was accelerated by the remarkable technological progress of vinegar production.

The annual per capita consumption of vinegar as 10% acetic acid in industrialized countries is ca. 1-2 liters. The acetic acid content of vinegar varies from 4 to 15% [113].

B. Rice Vinegar in Japan

A typical vinegar fermented with koji mold is rice vinegar, which is popular in China, Japan, Thailand, and other Oriental countries where rice wine is produced with koji mold. Rice vinegar making in Japan is said to have started together with rice wine making, which was first introduced from abroad to the southern part of Kyushu Island. Rice vinegar is still produced domestically by farmers in this district, and Fukuyama-city, located in southern Kyushu, has been the center of traditional rice vinegar brewing.

TABLE 5. Comparison of Chemical Analyses Between Rice and Alcohol Vinegar

	Rice vinegar	Alcohol vinegar
Specific gravity	1.0283	1.0134
Total acid	4.60%	4.29%
Total sugar	3.753	0.934
Total nitrogen	0.0320	0.0120
Amino nitrogen	0.0171	0.0071
NaCl	0.318	0.253
EX.	6.114	2.104

Analytical averages of 15 kinds of vinegar on the Japanese market are described.

Source: Ref. 116.

In the traditional method of rice vinegar production, polished rice (1.8 liter) is steamed, put into a china jar and mixed with 18 L water and 10.4 L rice koji, and the mixture is stored in the sealed jar for about one year to complete the fermentation. The vinegar is then press-filtered and pasteurized. Raw materials used to make vinegar in Japan have been rice, rice wine, and rice wine press-cake, a greater part of which is currently being replaced by ethanol.

The Japanese Agricultural Standard for vinegar issued in 1979 classifies vinegar into two categories: fermented and synthesized. Fermented vinegar includes (a) grain vinegar, (b) fruit vinegar, and (c) fermented vinegar. Grain vinegar is divided into rice vinegar and grain vinegar other than rice vinegar; one liter of the former must be prepared from more than 40 grams of rice and the latter from 40 grams of other grains including saccharified rice wine press-cake or malt; both types of vinegar must contain more than 4.2% acetic acid [115].

The annual production of vinegar in Japan was 343,000 kiloliters (based on acetic acid content of 5%) in 1985, and consisted of rice vinegar, 15%; grain vinegar other than rice vinegar, 42%; fruit vinegar, 5%; fermented vinegar, mainly from ethanol, 36%; and synthesized vinegar, 2% [106]. The average acetic acid content of commercial vinegars is 4-5%, but a vinegar containing 10% or more acetic acid can be produced under extremely aerobic conditions. Fifteen kinds of rice and alcohol vinegar available on the Japanese market are compared in Table 5 [116]

Rice vinegar and rice wine press-cake vinegar are produced by traditional still surface culture of *Acetobacter aceti* taking 30-50 days to oxidize ethanol by the film of living cells formed on the surface, or by the quick method with *Acetobacter suboxydans* (*Gluconobacter*) and vigorous stirring of the liquid in a submerged fermentation taking 4-15 days at about 30°C [117].

In culturing rice-koji used for genuine rice vinegar, the temperature of materials in the last stage after the second cooling is 45-47°C and 72

TABLE 6. Preferable C/N Ratio for High-Acidity Vinegar Fermentation by Submerged Cultivation

Carbon source	Ratio
G/TN	6-7
A/TN	100
G/Ala + Asp + Glu	4-5
A/Ala + Asp + Glu	70-80

G, glucose; TN, total nitrogen; A, ethanol; Ala, alanine; Asp, aspertic acid; Glu, glutamic acid. Rice-wine (sake) or rice wine press-cake was used as carbon source.

Source: Ref. 122.

hours are needed for total cultivation; this is 3-5°C higher and 24 hours longer than for ordinary rice-koji cultivation and yields a koji with high proteolytic and amylolytic enzyme activities. One part of rice-koji and 2.5-3.00 parts of cooked rice are mixed with 2-3 volumes of water based on total rice, and the mixture is saccharified at 60°C. The saccharified rice-koji is subjected to alcoholic fermentation at 28-30°C for 1-3 months, and then is filtered and stored for aging for about 3 months [120,121].

Entani and Masai selected *Acetobacter pasteurianus* for surface fermentation [118,119], *A. xylinum* having no cellulase-forming ability for intermediate submerged fermentation, and *A. polyoxygenes* for submerged fermentation as superior for vinegar production. The preferable C/N ratios for manufacturing high-acidity vinegar (10-20% acetic acid) by submerged fermentation are shown in Table 6 [122].

Fresh vinegar equivalent to 2-3% acetic acid is usually mixed with vinegar mash to prevent microbial deterioration. According to Entani and Masai [119], microbes other than *Acetobacter* that can grow in vinegar mash of more than 3% acetic acid are limited to *Lactobacillus fructivorans*, *L. acetotolerans*, and *Moniliella acetobutans*. These bacteria can be controlled by the use of sterilizing agents or by removing their growth factor, Ca^{2+}, from mash or other products.

Acetobacter cannot grow in a medium of only ethanol. In vinegar production in Japan, such nutrients as rice wine press-cake, rice-koji extract, or malt extract have been added to the fermenting medium to promote fermentation and provide a good flavor. The major factor for promoting growth of *Acetobacter* was found to be glycerol; some organic acids and amino acids were also associated with its growth [123].

The microflora change during rice vinegar fermentation was described by Entani [118] and is shown in Figure 8. Koji molds were rapidly destroyed after mashing; no molds other than koji molds were isolated. Viable counts of yeasts and lactobacilli reached maxima after 7 days and that of *Acetobacter* after 14 days. Major yeasts identified were *Saccharomyces cerevisiae* (6 strains), *Candida krusei* (2), *Zygosaccharomyces bailii* (1), and *Candida famata* (1). Major lactobacilli were *Lactobacillus casei* subsp.

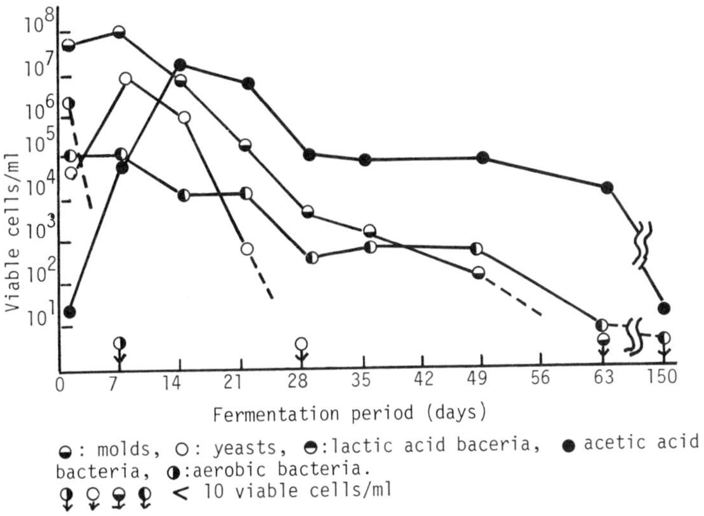

FIGURE 8. Change of microbial flora during Fukuyama traditional rice vinegar manufacture. (From Ref. 118.)

casei (4), *L. brevis* (3), *L. casei* subsp. *alactosus* (2), and *L. plantarum* (1); major acetobacters were *A. pasteurianus* (5), *A. xylinum* (4), and *A. aceti* (1). These microbes, other than lactobacilli and *Z. bailii* were less acid-tolerant and could not grow at less than pH 3.5 (1% acetic acid).

In formation of one mole of acetic acid from one mole of ethanol in vinegar fermentation, 114.6 cal of heat are generated. In the submerged process of vinegar production, 10-20 times the amount of acetic acid formed in the still process is produced, which is accompanied by heat generation, which makes this a technological problem. Omori et al. developed heat-resistant strains of *Acetobacter aceti* which gave better yields and better quality vinegar in submerged fermentation at 37°C than at 30°C, thereby greatly reducing the cost for cooling the fermentation medium [124].

The characteristic flavors of rice vinegar and malt vinegar result from their content of diacetyl reaching 10 ppm (Table 7) [125]. Too much diacetyl in vinegar is objectionable to some consumers. Diacetyl is fermented from acetoin, which is derived from lactic acid, succinic acid, and malic acid produced in the alcoholic fermentation by yeasts. Methods to reduce the diacetyl content of vinegar have been studied. In alcoholic fermentation, lactic acid is often added to mash to reduce its pH value and to control microbial contamination. Addition of enough acetic acid in place of lactic acid to reduce the pH of mash at 4.0 in alcoholic fermentation reduced formation of acetoin by 2.4. In apple vinegar fermentation, apple juice is first fermented with yeast to make ethanol, and then alcoholic apple juice is subjected to acetic acid fermentation. In malt vinegar fermentation, malt is saccharified and then fermented with yeast to make ethanol. Addition of ethanol to apple juice or to saccharified malt instead of producing it with yeast fermentation reduced diacetyl formation in the vinegar fermentation by more than one half. Similarly, in making rice vinegar, use of rice and

TABLE 7. Acetoin and Diacetyl Contents of Vinegars Compared with Rice Wine and Beer

Product	Raw materials	Acetoin (mg%)	Diacetyl (mg%)
Rice vinegar	A. Rice	191.8	22.5
	B. Rice, ethanol, press-cake of rice of rice wine	27.6	2.06
Malt vinegar	A. Barley, malt, rye	358.2	11.20
	B. Malt	85.0	1.62
Apple vinegar	A. Apple juice, ethanol	50.2	2.30
	B. Apple juice, ethanol	43.6	1.10
Wine vinegar	A. Grape juice, ethanol	73.2	0.52
	B. Grape juice, ethanol	38.6	2.32
Alcohol vinegar	A. Ethanol, rice	2.3	0.18
	B. Ethanol, press-cake of rice wine	1.1	0.12
Rice wine		1.4	0.70
Beer		0.2	0.01

Source: Ref. 125.

α-amylase in place of rice-koji was proposed to reduce the diacetyl content of rice vinegar.

Vinegar has been used to preserve foods, in pickle fermentation, for example, because of its high acidity. Nevertheless, deterioration of commercial vinegars sometimes is a serious problem for the manufacturers; such deterioration includes formation of turbidity, surface film, color change, bad odor and taste, and decrease in acidity. Proper conditions for pasteurization of vinegar were reported to be heating at 60-66°C for 30 minutes, or at 80°C for more than 5 minutes based on the heat tolerance of *A. aceti* [127,128].

C. Vinegar in China

Vinegar production in China is believed to have started more than 3000-3500 years ago, presumably with rice wine as the raw material. Since ancient times, vinegar has been prepared using both *Aspergillus oryzae* and *Rhizopus* sp. as is true for alcoholic beverage fermentation. The koji used to prepare vinegar, which is described in *Chi-min-yao-shu* of the sixth century, was most often prepared by culturing *Aspergillus oryzae* on steamed and crushed wheat; there were few descriptions of dough koji cultured with *Rhizopus* sp. on balls or cakes made from raw or steamed crushed wheat and water. It is presumed that the former has greater protease activity than the latter koji, which is favorable to increase the free amino acid and peptide contents and to improve the flavor of the product.

Crushed wheat gives a deeper color and better flavor to vinegar; however, crushed barley, wheat bran, and rice grains also were used as raw materials for the *Aspergillus oryzae* koji needed for vinegar production.

In traditional vinegar brewing, saccharification of starch (A), alcoholic fermentation of sugars (B), and acetic fermentation of alcohol (C) are conducted as follows: (I) all in parallel, (II) A and B in parallel and followed by C, or (III) all in succession in this order. Method (I) is considered to be the method for obtaining the best quality of vinegar, and it is conducted more by solid fermentation than by liquid fermentation. The same grains and other materials used for alcoholic beverage preparation, i.e., rice, wheat, barley, sorghum, millet, wheat bran, grain wine, fruit wine, presscake of wine mash, and honey, have been used as raw materials for vinegar making. Stepwise addition of starchy raw materials to solid mash has long been practiced, which is necessary to supply enough oxygen and to properly reduce the temperature of vinegar mash during fermentation.

Some famous traditional vinegars currently produced in China are the following:

1. Barley and sorghum vinegar in Shanxi sheng (near Taiyuan). Barley and peas (70:30) are crushed and moistened with 55% water, pressed into cakes, and cultured with *Rhizopus* sp. to make koji. Sorghum grains are crushed and moistened with 120% water, to which crushed koji (60% in weight) and water (180%) are added and mixed to make mash. The mash is subjected to the alcoholic fermentation at 27-28°C for 12-14 days. The finished alcoholic mash is mixed with millet bran (83% in weight) to make a solid mash in which the acetic fermentation is carried out at 45°C for 9-10 days. The acetic fermentation is stopped by adding 3% sodium chloride to the mash, and then vinegar is removed from the bottom, followed by washing of the mash twice with water to get the second crop.

2. Rice wine-cake and glutinous rice vinegar in Jiangsu sheng (Zhenjiang). Steamed glutinous rice is saccharified with barley dough koji cultured with *Rhizopus*, wheat koji in granular form cultured with *Aspergillus oryzae* is mixed with it, and the mixture undergoes alcoholic fermentation. The finished alcoholic mash is mixed with wheat bran and rice bran along with vinegar starter to make a solid mash in which the acetic fermentation is carried out. The vinegar starter is prepared by fermenting a mixture of yellow wine cake and water.

3. Wheat bran vinegar in Sichuan sheng. Yeast mash is prepared from steamed glutinous rice (39 kg), water (100 kg), and seed starter (0.5 kg) along with an extract of some spices (1-1.5 kg); this is mixed with raw wheat bran (650 kg) and water (40 kg) to make a solid mash. The mash is subjected to vinegar fermentation for 14 days, and the finished mash is covered with sodium chloride 3 cm thick, sealed, and aged for one year. About 1200 kg of vinegar is obtained from 650 kg of wheat bran and 30 kg of glutinous rice.

4. Anka-vinegar in Fujian sheng. Steamed red rice is saccharified with red-koji, and then subjected to alcoholic fermentation with addition of boiled nonglutinous rice. The alcoholic mash is subjected to vinegar fermentation for 3 years to give a product with 8% acetic acid content. Cane sugar (2%) is added in making the final product. The anka-vinegar fermentation is characterized by liquid fermentation throughout.

The greatest drawback of the traditional solid vinegar fermentation is the low conversion of starch into product and the long time needed to make the product. A new method for vinegar production using solid mash was developed in 1958, in which wheat bran koji cultured in black *Aspergillus* is used to saccharify dried sweet potatoes. The alcoholic fermentation is done with addition of pure cultured yeast starter mash, and then it is followed by the acetic fermentation. The complete vinegar fermentation is finished within one month with a higher yield of acetic acid than before. Incubation of wheat bran black-mold koji in wooden trays has been replaced by a mechanical through-flow aeration process on a perforated plate.

Modernized vinegar production in China is directed toward the submerged aeration system for liquid fermentation, as is true elsewhere in the world, but the product is not always preferred by consumers because of its poor flavor as compared with that of traditional products [131,132].

V. CONCLUSION

Fermented foods from around the world are divided for discussion into two categories: (a) foods made from nonproteinaceous, starchy raw materials and (b) foods made from proteinaceous substrates. Emphasis was on fermented foods, beverages, and condiments made from carbohydrate and plant proteinaceous materials which are manufactured using koji as a source of enzymes. The mold-cultured plant materials are called koji in Japanese and chu in Chinese, and many other names in East and Southeast Asian countries. Molds involved include the genera *Aspergillus*, *Rhizopus*, *Mucor*, and *Monascus*. As for starchy raw materials, many kinds of cereal grains have been and are used for food fermentation.

To make alcoholic beverages and vinegars from starchy raw materials, starch must be converted into sugars before it is subjected to alcoholic fermentation by yeasts. Amylolytic enzymes in germinated grains or malt, mold-cultured grains, or koji have been used for the saccharification of starch since ancient times.

The malt process, originated in the inland areas of Eurasian continent and Africa where the amount of yearly rainfall is limited, evolved into the beer, whiskey, and malt vinegar industries of today.

The koji process, on the other hand, original to East and Southeast Asia where the people took advantage of abundant rainfall and climate suitable for the growth of molds, evolved into industries producing alcoholic beverages, grain vinegars, and fermented soybean foods and condiments in the Orient. The malt process also existed in ancient China, but it was not as extensively developed later as the koji process in East and Southern Asian countries, presumably because of the superiority of the koji process.

Filamentous molds began to be applied to alcoholic beverage, vinegar, and proteinaceous food fermentation 3000 years ago in China. It is believed that yellow *Aspergillus* and white or grey *Rhizopus* and or *Mucor* were involved; they are still widely used in alcoholic beverage and vinegar fermentations in China: *Rhizopus* is grown on ball- or brick-shaped dough made from moistened flour of such cereal grains as wheat, barley, sorghum, or rice to make chu, and the chu is mixed with cereal grains and water for saccharification and then the mixture is subjected to alcoholic fermentation by yeasts.

In Japan, *Aspergillus* has been used for making rice wine, or sake, distilled rice wine, or shochu and vinegar. Moreover, the *Aspergillus* is grown on steamed rice in granular form to make koji, and the koji is mixed with cooked cereals and water for concurrent saccharification and alcoholic fermentation. The differences can be attributed to the difference in the form of staple foods of the respective countries, namely, boiled rice in granular form in Japan, and bread made of cereal flour in China. Furthermore, the exclusive use of the genus *Aspergillus* in Japan is said to be influenced by the rice culture introduced to Japan from Southeast Asia in the third century B.C., which preceded introduction of fermentation technology for alcoholic beverages and soybean foods from the northern part of China. In fact, *Aspergillus* molds can be easily isolated from the rice plant.

It was made clear that *Rhizopus* cannot grow on heat-treated rice or wheat because of its weak proteolytic enzyme activity, which cannot degrade heat-denatured protein. On the other hand, *Aspergillus* molds having strong proteolytic enzyme activity can grow better on heat-treated cereal grains. Moreover, the growth of *Aspergillus* molds is more oxygen dependent than that of *Rhizopus*, and hence a granular form of substrate having greater surface area is preferable for growth of *Aspergillus* molds.

REFERENCES

1. Ainsworth, G. C. *Ainsworth and Bisby's Dictionary of the Fungi*, 7th ed., Hawksworth, Sutton, and Ainsworth, CMI, Kew, 1983.
2. Lodder, J. *The Yeasts*. North-Holland Pub. Co., Amsterdam, 1970.
3. Kikkoman Co. *Thirty-five Years in the History of the Noda-shoyu Company*. Chibaken, Japan, 1955.
4. Kikkoman Co. *The History of Kikkoman*. Chibaken, Japan, 1977.
5. Sakaguchi Kinichiro. Search for the route of shoyu. *Sekai, 398*, Iwanami-shoten, Japan, 1979, pp. 252-266.
6. Wang, H. L., and Fuang, S. F. *History of Chinese Fermented Foods. Indigenous Fermented Foods of Non-Western Origin* (C. W. Hesseltine and Hwal Wang, eds.), J. Cramer, Berlin-Stuttgart, 1986, pp. 23-35.
7. Bo,Thi-an. Origin of jiang and jiang-yu and their production technology (1 and 2). *Brew Soc. Japan, 77*:365-371, 439-445, 1982.
8. Bo-Thi-an. Origin of douchi and its production technology (1 and 2) *Brew Soc. Japan, 79*:221-223, 395-402, 1984.
9. Lotong, N. *Koji. Microbiology of Fermented Foods*, Vol. 2 (Brian J. B. Wood, ed.), Elsevier Applied Sci. Publisher, London and New York, pp. 237-270.
10. Vogel, S., and Gobezie, A. Ethiopian tej. Symposium on indigenous fermented foods, Bangkok, Thailand. *Handbook of Indigenous Fermented Foods* (K. H. Steinkraus, ed.), Marcel Dekker, New York, 1977, pp. 306-312.
11. Merican, Z. Malaysian coconut palm toddy. In *Handbook of Indigenous Foods* (K. H. Steinkraus, ed.), Marcel Dekker, New York, 1977, pp. 315-328.

12. Okafor, N. Palm-mine. In *Handbook of Indigenous Fermented Foods* (K. H. Steinkraus, ed.), Marcel Dekker, New York, 1977, pp. 315-328.
13. Odeyemi, F. Ogogoro industry in Nigeria. In *Handbook of Indigenous Fermented Foods* (K. H. Steinkraus, ed.), Marcel Dekker, New York, 1977, pp. 315-328.
14. Faparusi. Nigerian palm-wine-emu. In *Handbook of Indigenous Fermented Foods* (K. H. Steinkraus, ed.), Marcel Dekker, New York, 1977, pp. 315-328.
15. Theivendirarajah, K., Jeyaseelan, K., and Puviraiasingam, V. Improvements in coconut toddy fermentation. In *Handbook of Indigenous Fermented Foods* (K. H. Steinkraus, ed.), Marcel Dekker, New York, pp. 315-328.
16. Theivendirarajah, K., Dassanayake, M. D., and Jeyaseelan, K. Studies on the fermentation of kitul (Caryyota urens) sap. In *Handbook of Indigenous Fermented Foods* (K. H. Steinkraus, ed.), Marcel Dekker, New York, 1977, pp. 315-328.
17. Samarajeewa, U. Fermentation of coconut sap (toddy). In *Handbook of Indigenous Fermented Foods* (K. H. Steinkraus, ed.), Marcel Dekker, New York, pp. 315-328.
18. Nyako, K. O. Palm-wine-an, alcoholic beverage of Ghana. In *Handbook of Indigenous Fermented Foods* (K. H. Steinkraus, ed.), Marcel Dekker, New York, 1977, pp. 315-328.
19. Shuaib, A. C., and Azumey, M. S. Pol-ra-coconut toddy of Sri Lanka. In *Handbook of Indigenous Fermented Foods* (K. H. Steinkraus, ed.), Marcel Dekker, New York, 1977, pp. 315-328.
20. Wong, P. W., and Jackson, H. Fermented foods of Sabah (Malaysia). In *Handbook of Indigenous Fermented Foods* (K. H. Steinkraus, ed.), Marcel Dekker, New York, 1977, pp. 315-328.
21. Ekmon, T. D., and Nagodawithana, T. W. Coconut toddy. In *Handbook of Indigenous Fermented Foods* (K. H. Steinkraus, ed.), Marcel Dekker, New York, 1977, pp. 315-328.
22. Harkishor, K. M. Kenyan sugarcane wine-muratina. In *Handbook of Indigenous Fermented Foods* (K. H. Steinkraus, ed.), Marcel Dekker, New York, 1977, 312-315.
23. Dahiya, D. S., and Prabhu, K. A. Indian jackfruit wine. In *Handbook of Indigenous Fermented Foods* (K. H. Steinkraus, ed.), Marcel Dekker, New York, 1977, pp. 337-338.
24. Sanchez-Marroquin, A. Mexican pulque, a fermented drink from Agave juice. In *Handbook of Indigenous Fermented Foods* (K. H. Steinkraus, ed.), Marcel Dekker, New York, 1977, pp. 328-335.
25. Herrera, T., Ullos, M., and Toboada, J. Microbiological studies on pulque. In *Handbook of Indigenous Fermented Foods* (K. H. Steinkraus, ed.), Marcel Dekker, New York, 1977, pp. 328-335.
26. Goncalves de Lima, O. Pulque, balche and pajauaru. In *Handbook of Indigenous Fermented Foods* (K. H. Steinkraus, ed.), Marcel Dekker, New York, 1977, pp. 328-335.
27. Escobar, A., Gardner, A., and Steinkraus, K. H. Studies of South American chicha. In *Handbook of Indigenous Fermented Foods* (K. H. Steinkraus, ed.), Marcel Dekker, New York, 1977, pp. 340-344.

28. Novilie, L. Sorghum beer and related fermentations in Southern Africa. In *Indigenous Fermented Food of Non-Western Origin* (C. W. Hesseltine and Hwal Wang, eds.), J. Cramer, Berlin-Stuttgart, 1986, pp. 219-235.
29. Ekundayo, J. A. Nigerian Pito. Symposium on indigenous fermented foods. In *Handbook of Indigenous Fermented Foods* (K. H. Steinkraus, ed.), Marcel Dekker, New York, 1977, pp. 358-363.
30. Harkishor, K. M. Kenian sugarcane wine-muratins. In *Handbook of Indigenous Fermented Foods* (K. H. Steinkraus, ed.), Marcel Dekker, New York, 1977, pp. 365-369.
31. Taboada, J., Ulloa, M., and Herrera, T. Microbiological studies on tesguino. In *Handbook of Indigenous Fermented Foods* (K. H. Steinkraus, ed.), Marcel Dekker, New York, 1977, pp. 352-357.
32. Marcos, S. R. Egyptian bouza. In *Handbook of Indigenous Fermented Foods* (K. H. Steinkraus, ed.), Marcel Dekker, New York, 1977, pp. 357-358.
33. Ulloa-Sosa, M., and Herrera, T. Fermented corn products of Mexico. *Advances in Biochemistry*, Vol. II, *Proc. 6th International Fermented Symposium*, July 20-25, London, Canada, 1980, pp. 534-540.
34. Boedijn, K. B. *Sydovia, 12*:321, 1958.
35. Dwidjoseputro, D., and Wolf, F. T. *Mycopathologia et Mycologia Applicata, 41*:211, 1970.
36. Ko, Swan Djien. *Research di Indonesia* (M. Makagiansar and R. M. Soemanti, eds.), Vol. 2, P. N. Balai Pustaka, Djakarta, Indonesia, 1965, p. 209.
37. Ko, Swan Djien. Tape fermentation. *Applied Microbiology, 23*:976-978, 1972.
38. Ko, Swan Djien. Indonesian ragi. In *Handbook of Indigenous Fermented Foods* (K. H. Steinkraus, ed.), Marcel Dekker, New York, 1977, pp. 381-409.
39. Ko, Swan Djien, and Hesseltine, C. W. Tempe and related foods. In *Economic Microbiology*, Vol. 4 (A. H. Rose, ed.), Academic Press, London, 1979, pp. 115-140.
40. Ko, Swan Djien. Indonesian Fermented foods not based on soybeans. In *Indigenous Fermented Food of Non-Western Origin* (C. W. Hesseltine and Hwal Wang, eds.), J. Cramer, Berlin-Stuttgart, 1986, pp. 68-84.
41. Ko, Swan Djien. Indigenous fermented foods. In *Economic Microbiology*, Vol. 7 (A. H. Rose, ed.), Academic Press, New York, 1982, pp. 16-36.
42. Saono, S., and Jenny, K. D. Microflora of ragi. In *Traditional Food Fermentation as Industrial Resources in ASCA Countries* (S. Saono et al., eds.), The Indonesian Institute of Science, Jakarta, Indonesia, 1982, pp. 241-249.
43. Ellis, J. J., Rhodes, L. J., and Hesseltine, C. W. *Mycologia, 68*:131, 1976.
44. Saono, S., and Basuki, T. B. *Annales Bogoriensis, 6*:207, 1976.
45. Saono, S., and Basuki, T. *Annales Bogoriensis, 7*:11, 1978.
46. Saono, S., Basuki, T., and Sastraatmadja, D. Indonesian ragi. In *Handbook of Indigenous Fermented Foods* (K. H. Steinkraus, ed.), Marcel Dekker, New York, 1977, pp. 381-409.

47. Yeoh, Quee Lan. Malaysian ragi. Malaysian tapai. *Handbook of Indigenous Fermented Foods* (K. H. Steinkraus, ed.), Marcel Dekker, New York, 1977, pp. 381-409.
48. Wong, P. W., and Jackson, H. Fermented foods of Sabah (Malaysia). In *Handbook of Indigenous Fermented Foods* (K. H. Steinkraus, ed.), Marcel Dekker, New York, 1977, 381-409.
49. Uyenco, F. R., and Gacutan, R. Z. Philippine bubdod. In *Handbook of Indigenous Fermented Foods* (K. H. Steinkraus, ed.), Marcel Dekker, New York, 1977, pp. 381-409.
50. Pichyangkura, S., and Kulprecha, K. Survey of mycelial molds in loopang from various sources in Thailand. In *Handbook of Indigenous Fermented Foods* (K. H. Steinkraus, ed.), Marcel Dekker, New York, 1977, pp. 381-409.
51. Cronk, T. C., Steinkraus, K. H., Hackler, L. R., and Mattick, L. R. *Applied and Environmental Microbiology, 33*:1067, 1977.
52. Cronk, T. C., Mattick, L. R., Steinkraus, K. H., and Hackler, L. R. *Applied and Environmental Microbiology, 37*:892, 1979.
53. Dhamcharee, B. Traditional fermented foods in Thailand. In *Traditional Food Fermentation as Industrial Resources in ASKA Countries* (S. Saono et al., eds.), The Indonesian Institute of Science, Jakarta, Indonesia, 1982, pp. 89-92.
54. Kozaki, M. Fermented foods and related microorganism in Southeast Asia. *J. A. Mycotoxicology, 2*:1, 1976.
55. Kozaki, M. Alcoholic beverages in Southeast Asia. *Shokuno-kagaku, 47*:31-36, 1979.
56. Sumikawa, T., Uchimura, Y., Okada, S., Obara, N., and Kozaki, M. Asian dough koji. *J. Research Institute on Freezing and Drying, 30*: 7, 1984.
57. Dahiya, D. S., and Prabhu, K. A. Indian jackfruit wine. In *Handbook of Indigenous Fermented Foods* (K. H. Steinkraus, ed.), Marcel Dekker, New York, 1977, pp. 405-408.
58. Batra, L. R. Fermented cereals and grain legumes of India and vicinity. In *Advances in Biotechnology*, Vol. II, Proc. 6th International Fermentation Symposium, July 20-25, London, Canada, 1980, pp. 547-554.
59. Steinkraus, K. H. Primitive Thai rice wine. In *Handbook of Indigenous Fermented Foods* (K. H. Steinkraus, ed.), Marcel Dekker, New York, 1977, pp. 404-405.
60. Nakao, S. Wine culture in Southeast Asia. *Shokuno-kagaku, 47*:22-29, 1979.
61. Komoda, K. Alcoholic beverages in Nepal. *J. Brew. Soc. Japan, 60*: 60-64, 1965.
62. Otani, S. Alcoholic beverages in China. II. Fermented wine-yellow-wine. *J. Brew. Soc. Japan, 68*:579-586, 1973.
63. Lin, Gen Nian. *Brewing, 66*, 1972.
64. Iizuka, H. Art of mold in wine making in the Orient. *Shokuno-kagaku, 47*:30-38, 1979.
65. Wang, H. H., and Hsieh, T. C. Kao-Liang brewing by pure cultures. *Proc. 4th International Fermentation Symposium*, Kyoto, Japan, 1972, pp. 651-658.

66. Nakazawa, R., and Sato, K. *J. Agric. Chem. Soc. Japan, 6*:252, 1930.
67. Iizuka, H., and Lin, F. On the genus *Monascus* of Asia and its specific characteristics. In *Advances in Biotechnology*, Vol. II, Proc. 6th International Fermentation Symposium, July 20-25, London, Canada, 1980, pp. 555-561.
68. Su, Yuan-Chi. Traditional fermented foods in Taiwan. In *Processing of the Oriental Fermented Foods*, Taipei, Taiwan, 1980, pp. 15-30.
69. Akiyama, T. Rice-wine in Taiwan. In *Brewing* (K. Ohtsuka, ed.), Yokendo, Tokyo, Japan, 1981, p. 64.
70. Sugama, S. Origin of genuine shochu and its technology development. *Proc. 14th Symposium on Brewing*, Brew. Soc. Japan, Sept. 9, 1982, pp. 76-80.
71. Otani, S. Alcoholic beverages in China. I. Distilled wine. *J. Brew. Soc. Japan, 68*:423-428, 1973.
72. Tonoike, R. Distilled wine in China. Shokuno-kagaku. Marunouchi Pub. Co., Tokyo, Japan, 1979, pp. 39-46.
73. Chiao, J. S. Modernization of traditional Chinese fermented foods and beverages. *Advances in Biotechnology*, Vol. II, *Proc. 6th International Fermentation Symposium*, July 20-25, London, Canada, 1980, pp. 511-516.
74. Otsuka, K. Shochu. In *Jozogaku* (K. Otsuka, ed.), Yokendo, Tokyo, 1981, p. 183.
75. Park, K. I., Mheen, T. I., Lee, K. H., Chang, C. H., Lee, S. R., and Kwon, T. W. Korean yakju and takju. In *Handbook of Indigenous Fermented Foods* (K. H. Steinkraus, ed.), Marcel Dekker, New York, 1977, pp. 379-381.
76. Yu, Ju-Hyun, and Yu Ryang Pyun. *Korean Fermented Foods. Processing of the Oriental Fermented Foods*, Taiwan, 1980, pp. 46-57.
77. Mheen, T. I., Kwon, T. W., and Lee, C. H. Traditional fermented food products in Korea. In *Traditional Food Fermentation as Industrial Resources in ASKA Countries* (S. Saono et al., eds.), The Indonesian Institute of Science, Jakarta, Indonesia, 1982, pp. 63-81.
78. Tanaka, T., and Okazaki, N. Growth of mold on uncooked grains. *Ferm. Technol., 60*:11-17, 1982.
79. Tanaka, T. Microbial starter cultured with granular substrate or ball or cake shaped dough (Bara-koji and Mochi-koji), *Proc. 14th Symposium on Brewing*, Brew. Soc. Japan, Sept. 9-10, Tokyo, Japan, 1982, pp. 4-8.
80. Imayasu, S. The progress of production method of Japanese sake. *International Symposium on Conversion and Manufacture of Foodstuffs by Microorganisms*, Dec. 5-9, Kyoto, Japan, 1971, pp. 141-145.
81. Murakami, H. Some problems in sake brewing. *Proc. 4th International Fermentation Symposium*, Mar. 19-25, Kyoto, Japan, 1972, pp. 639-641.
82. Kodama, K., and Yoshizawa, K. Sake. In *Economic Microbiology*, Vol. I (Rose et al., eds.), Academic Press, New York, 1977, pp. 432-475.
83. Yoshizawa, K. Japanese sake. In *Handbook of Indigenous Fermented Foods* (K. H. Steinkraus, ed.), Marcel Dekker, New York, 1977, pp. 373-379.

84. Nojiro, K. Recent advances in microbiology of sake brewing. In *Recent Advances in Food Science and Technology*, Vol. 2, International Symposium, Jan. 9-11, Taipei, Taiwan, 1980, pp. 100-108.
85. Nunokawa, Y. Enzymes affecting the fermentation of sake moromimash. In *Advances in Biotechnology*, 6th International Fermentation Symposium, July 20-25, London, Canada, 1980, pp. 563-568.
86. Akiyama, T. *Seishu* (K. Otsuka, ed.), Jozogaku, Yokendo, Tokyo, Japan, 1981, pp. 10-58.
87. Yoshizawa, K. Traditional alcoholic beverage industry in Japan. In *Traditional Food Fermentation as Industrial Resources in ASKA Countries* (S. Saono et al., eds.), The Indonesian Inst. of Sci., Jakarta, Indonesia, 1982, pp. 63-81.
88. Suginami, T., and Imayasu, S. Cell fusion between *Asp. awamori* and *Asp. oryzae* and its application. *Proc. 16th Symposium on Brewing*, Brew. Soc. Japan, Sept. 11-12, Tokyo, Japan, 1984, pp. 21-24.
89. Iwano, K., and Nunokawa, Y. Studies on the enzymes related to Japanese rice-wine (sake) brewing. *J. Brew. Soc. Japan, 73*:555-557, 1979.
90. Ouchi, K. Selection of non-forming sake-yeasts. *Proc. Symposium on Brewing*, Brew. Soc. Japan, Sept. 8-9, Tokyo, Japan, 1971, p. 1.
91. Saito, K. Microbiology in sake brewing, 1940.
92. Kodama, K. Some consideration on the cultivation of traditional sake-yeast starter mash. *Nikkyoshi, 54*:770, 1959.
93. Kitahara, S. *Research on Lactobacilli*, Todai Pub. Co., Tokyo, Japan, 1960.
94. Hayashida, S., Nanri, S., Tei, T., and Hongo, M. The high concentration alcohol-producing factor in koji. *J. Agric. Chem. Soc. Japan, 48*:529-535, 1974.
95. Yoshizawa, K. Volatile flavor components of alcoholic beverages. *J. Brew. Soc. Japan, 61*:481-485, 1966.
96. Akita, O., Hasuno, T., Takahashi, K., Miyano, H., and Yoshizawa, K. *J. Brew. Soc. Japan, 82*:703-708, 1987.
97. Sato, S., Takahashi, Y., and Tadenuma, M. The color of mirin. *J. Brew. Soc. Japan, 62*:657-659, 1967.
98. Tamura, G. Hiochic acid, a new growth factor for *Lactobacillus homohiochi* and *Lactobacillus heterohiochi*. *J. General and Applied Microbiology, 2*:431-434, 1956.
99. Tamura, G. The path from hiotic-acid to cell physiology. *J. Brew. Soc. Japan, 79*:760-765, 1984.
100. Tamura, G., and Folkers, F. Identity of mevalonic and biochic acids. *J. Org. Chem., 23*:772, 1958.
101. Kitahara, K., Kaneko, T., and Goto, O. Taxonomic studies on hiochi bacteria. *J. Agric. Chem. Soc. Japan, 31*:556-564, 1957.
102. Takase, S., and Murakami, H. Studies on the fluorescence of sake. I, II. *Agric. Biol. Chem., 30*:869-876, 1966; *31*:142-149, 1967.
103. Sugama, S. Shochu. In *Sogo-Shokuryo-kogyo* (Sakurai et al., eds.), Koseisha Pub., Tokyo, Japan, 1978, pp. 468-473.
104. Otsuka, K. Shochu. In *Jozogaku*, Yokendo Pub., Tokyo, Japan, 1981, pp. 177-183.

105. Sugama, S. Origin of genuine shochu and its technological developments. *Proc. 14th Symposium on Brewing*, Brew. Soc. Japan, Sept. 9-10, Tokyo, Japan, 1982, pp. 76-80.
106. Annual statistics of alcoholic beverages and foods. Agric. Forest. and Fish, Japan, 1985.
107. Kawaguchi, G. *J. Brew. Soc. Japan*, *435*:20, 1911.
108. Kawaguchi, G. *J. Brew. Soc. Japan*, *525*:87, 1919.
109. Omori, T. Shochu. *Encyclopedia of Food, Agriculture and Nutrition*, Kodansha, Tokyo, Japan, 1977, pp. 308-309.
110. Miyano, N. Quality and components of genuine shochu. *Proc. 10th Symposium on Brewing*, Brew. Soc. Japan, Sept. 13-14, Tokyo, Japan, 1978, pp. 105-111.
111. Huber, E. *Dtsch. Essigind.*, *31*:12-15, 18-30, 1927.
112. Conner, H. A. Vinegar: Its history and development. In *Applied Microbiology* (D. Perlman, ed.), Academic Press, New York, 1976, pp. 81-129.
113. Mori, A. History of vinegar. *Shokuno-kagaku*, *63*:22-32, 1981.
114. Omori, A. The way from Orlean method to submerged fermentation of vinegar. *Proc. 14th Symposium on Brewing*, Brew. Soc. Japan, Sept. 9-10, Tokyo, Japan, 1982, pp. 40-44.
115. Ebine, H. Quality of vinegar. *Shokuno-kagaku*, *63*:60-67, 1981.
116. Masai, H. Chemical analysis and sensory tests of vinegar. *Proc. 17th Symposium on Brewing*, Brew. Soc. Japan, Sept. 9-10, Tokyo, Japan, 1975, pp. 35-39.
117. Ito, H. Flavor components of vinegar. *Proc. 5th Symposium on Brewing*, Brew. Soc. Japan, Sept. 12-13, Tokyo, Japan, 1973, pp. 22-26.
118. Entani, E., and Masai, H. Morphology and prevention of acetic acid tolerant microbes in vinegar fermentation. *Proc. 17th Symposium on Brewing*, Brew. Soc. Japan, Sept. 10-11, Tokyo, Japan, 1985, pp. 19-23.
119. Entani, E., and Masai, H. Studies on microbes in vinegar fermentation. *Proc. 19th Symposium on Brewing*, Brew. Soc. Japan, Sept. 10-11, Tokyo, Japan, 1987, p. 58.
120. Itagaki, T. Vinegar. In *Fermented Foods* (Tomoda et al., eds.), Kyoritsu Pub. Co., Tokyo, Japan, 1966, pp. 275-306.
121. Higashi, K. Fukuyama-rice-vinegar. *Proc. 15th Symposium on Brewing*, Brew. Soc. Japan, Sept. 8-9, Tokyo, Japan, 1983, pp. 5-8.
122. Nakayama, S. On C/N ratio in high-acidity vinegar fermentation by submerged cultivation. *Proc. 6th Symposium on Brewing*, Brew. Soc. Japan, Sept. 11-12, Tokyo, Japan, 1974, pp. 8-10.
123. Namba, T., and Takeuchi, N. Promoting factor for the growth of acetobacter. *Proc. 10th Symposium on Brewing*, Brew. Soc. Japan, Sept. 13-14, Tokyo, Japan, 1978, pp. 48-53.
124. Omori, S., Okumura, H., Beppu, T., Arima, K., and Masai, H. Production technology of vinegar by submerged fermentation at high temperatures. *Proc. 10th Symposium on Brewing*, Brew. Soc. Japan, Sept. 13-14, Tokyo, Japan, 1978, pp. 89-91.

125. Yanagida, T., and Koizumi, K. Production method of vinegar of less diacetyl contents. *Proc. 12th Symposium on Brewing*, Brew. Soc. Japan, Sept. 11-12, Tokyo, Japan, 1980, pp. 79-82.
126. Wang, H. L., and Hesseltine, C. W. Sufu and Lao-Chao. *J. Agric. Food Chem.*, *18*:572, 1970.
127. Alwood, M. C., and Russel, A. D. *J. Appl. Bact.*, *32*:79, 1969.
128. Kokuto, K. Sterilization and preservation of vinegar. *Proc. 8th Symposium on Brewing*, Brew. Soc. Japan, Sept. 9-10, Tokyo, Japan, 1976, pp. 56-58.
129. Tachi, H., et al. Identification of microbes isolated from Chinese koji. *Proc. Annual Meeting*, Brew. Soc. Japan, Sept. 8-9, Tokyo, Japan, 1988, p. 9.
130. Taukioka, M., Hiroi, T., and Suzuki, T. Alcoholic beverage production with genus *Rhizopus* (Part 1), *Nippon Nogeikagaku Kaishi*, *62*: 1643-1647, 1988.
131. Bo,Thi-an. Brewing technology of Chinese vinegar (1-3). *J. Brew. Soc. Japan*, *7*:462-470; *8*:534-542, 681-688, 1988.
132. Shinoda, O. *Chinese History of Foods*. Shibata Shoten, Tokyo, Japan, 1981.
133. Mimmannitaya, S. Miang. *Traditional Foods and Their Processing in Asia*, NODAI Res. Inst., Tokyo, Japan, 1988, pp. 105-109.
134. Hesseltine, C. W., Ruth Rogers, and Winarno, F. G. Microbial studies on amylolytic oriental fermentation starters. *Mycopathologia*, *101*:141-155, 1988.
135. Hesseltine, C. W., and Ray, M. L. Lactic acid bacteria in murcha and ragi. *J. Appl. Bact.*, *64*:395-401.

11

PROTEINACEOUS FERMENTED FOODS AND CONDIMENTS PREPARED WITH KOJI MOLDS

TAMOTSU YOKOTSUKA *Kikkoman Corporation, Noda City, Chiba, Japan*

I. INTRODUCTION

Protein, carbohydrate, and fat, the three fundamental nutrients for the human body, are tasteless in themselves. Their degradation products do exhibit certain tastes and flavors, and they can be absorbed by the body through intestines. When a protein food is degraded into peptides and free amino acids, the products are more tasty and digestible than the original material and thus are used as condiments, side dishes, or staple foods. Since ancient times, several proteinaceous fermented foods have been made from milk. Milks from various animals have been fermented with lactobacilli to make cultured milk products, for example, yogurt in the Middle East and the Balkans, kaffir (kefir) in the Caucasus, koumis in Central Asia, dahi in India, kishk in Egypt, bramo in Bulgaria, acidophilus milk in the United States, and many other similar products in the USSR, Scandinavia, China, Japan, and other countries [1,2]. Lactobacilli in these fermentations serve not only to produce lactic acid but also to degrade protein into peptides and free amino acids to improve the palatability of products. In the West, proteins and fat involved in milks are enzymatically hydrolyzed by some *Penicillium* molds in blue cheese making to enrich the product with free amino acids and volatile flavor components. Examples are Stilton in England, Roquefort and Bress in France, Gorgonzola in Italy, Danablu in Denmark, and so on [3].

Enzymatic degradation of proteinaceous foods into digestible foods or tasty condiments in East and Southeast Asian countries is generally done with koji molds. The raw materials for these foods and condiments first mentioned in historical writings 3000 years ago were nonplant, e.g., gutted fish, bird, or meat. They were mixed with molded millet and salt and stored. In addition, another method was used for preparing a fermented fish sauce: Whole fish was digested by the indigenous enzymes in the presence of high salt concentrations. The latter type of sauce was transplanted to some of the Southeast Asian countries. They are patis in the

Philippines, nam-pla in Thailand, ngam-pya-ye in Burma, nuoc-mum in Vietnam, sambalikan in Singapore and Malaysia, nam-pa in Laos, ketjapikan in Kalimantan, terosi in Indonesia, tuk-trey in Cambodia, and, in the West, garos in Greece. In addition to fish sauce, fermented fish and shrimp pastes are equally popular in the same areas, as well as in China, Korea, and Japan. In Japan, the fish sauce shottsuru and fish paste shiokara are produced locally in small amounts, preparation of which is characterized by adding rice-koji or *Aspergillus*-cultured steamed rice to the fermented fish mash according to the ancient Chinese custom. Rice-koji is also mixed with vegetable and fish mixed pickles in Japan and Korea.

Katsuo-bushi or dried bonito is a characteristic solid seasoning material in Japan which is made mostly from a bonito. A bonito is stripped by removing skin, intestines, and bones, and then is cut into three pieces, cooked in water at 90-95°C for about one hour, dried in smoke at 85°C repeatedly until the moisture content is reduced to about 20%. After removal of the outer layer of oily substances, the fish is sun dried until the moisture content is 13-15% and stored. During storage, *Penicillium* molds grow first on the surface, the growth is brushed off, followed by *Aspergillus glaucus* (katsuobushi mold), *A. ruber*, and *A. rebens*. These molds degrade fat and protein into higher fatty acids, glycerol, and free amino acids. The major tasty ingredient of katsuo-bushi was found to be the hystidine-salt of 5'-inosinic acid. The product is flaked before using as a condiment. Its annual production in Japan is about 150,000 tons. A product similar to katsuo-bushi also is found in some southern Pacific Islands.

Soybeans, consisting of 35% protein, 28% carbohydrate, and 19% oil, are an excellent source of protein and edible oil for mankind worldwide. Three varieties of soybean, in terms of color of seeds, are known; yellow, green, and black. The yellow, and perhaps green, soybean *Glicine max* (L) Merrill, is native to eastern Asia. The cultivated soybean is thought by many investigators to have been derived from *Glycine ussuriensis* Regel and Maack, which grow wild throughout much of eastern Asia including Japan. Cultivation of yellow soybeans is considered to have spread to central China around 7000 B.C. In contrast, black soybeans originated independently from yellow beans and were popular in the southern part of China earlier than yellow beans. Soybeans were highly valued as food as well as medicine in China. Soybean cultivation was introduced to Korea and Japan a few thousand years B.C. judging from the remains found in Japan.

In the countries where soybeans were readily available, early in their history, nonplant protein foodstuffs as raw materials of fermented foods and condiments were replaced and surpassed by soybeans and cereals. The replacement occurred some 2000 years ago in China, and in Japan some 1000 years ago, when Buddhism was introduced from China. The major reason for the replacement is generally attributed to Buddhism and its tradition of abstinence from animal foods of all kinds. Thailand and Indonesia also have long been good soybean-producing countries.

The coincidence of soybean cultivation and application of koji molds in food fermentation prompted the development of numerous fermented soybean food products. In practice, soybeans were cultured directly with *Aspergillus*, *Rhizopus*, or *Mucor* molds or *Bacillus subtilis* to yield more digestive and/or tasty foods. Also, cooked soybeans were mixed with some cereal grains cultured with *Aspergillus* molds, and the mixture was further digested in the presence of salt with or without additional lactic acid and yeast fermentations. These foods and condiments have greatly enriched the dietary life of mankind both gastronomically and nutritionally.

II. FERMENTED SOYBEAN FOODS IN EAST AND SOUTHEAST ASIA

A. Douchi (China), Hama-natto (Japan), and In-yu (Taiwan)

Ancient Chinese literature reveals that two types of fermented proteinaceous food were popular 2000 years ago in China. They were chi (shi) (豉) and jiang (Chiang) (醬). Chi refers to proteinaceous plant foods cultured with edible microbes, which jiang designates proteinaceous plant or animal foods mixed with mold-cultured cereals or chu (麴) and salt and stored for fermentation [4-6]. The most popular chi, known as douchi, was mentioned first in *Shi-ji*, or Historical Records (written by Si-Ma-Qian in the second century B.C.), which states that at that time chi was a popular seasoning second only to salt.

The method of preparing chi is described in detail in the sixth century book *Chi-min-yao-shu*. Yellow or black soybeans are cooked and placed in a pile on a straw mat, which is covered with straw. The pile is turned frequently to disperse heat generated by microbial growth. When the beans are covered with white mold growth, aeration is increased as the temperature decreases. Beans covered with yellow mold mycelia are washed to remove yellow spores, which taste bitter; after washing, the beans are transferred to a pit, and covered tightly with straw. After 10-15 days, the beans turn black, and then are sun-dried to make dried douchi. Salt can be added before or during mold cultivation, or molded chi is stored in salt water to ferment and is then dried. A more detailed description of douchi than that of jiang is found in *Chi-min-yao-shu*, indicating that at that time chi was more common and important than jiang. Chi was consumed as a side dish or as a seasoning as is, or as chizhi after extracting it with boiling salt water.

In modern douchi preparation, the substrate inoculated with a pure culture starter of *A. oryzae* is cultivated at 25°C in wooden barrels. Molded beans in some instances are washed and mixed with 16-18% salt and fermented at 35°C for 30 days [6]. Nonsalted douchi is often stored in salt water together with some vegetables.

Mucor mold (*M. racemosus*) instead of *A. oryzae* is commonly used in making douchi in Sichuan sheng in China. Cooked beans packed in trays about 2-3 cm deep are incubated at 10-15°C for 1-2 weeks. When the white mold turns grey, the molded beans are mixed with salt and distilled liquor, tightly sealed into earthen jars, and fermented for 6 months in a shady place. Another soybean product fermented with *Mucor* in Sichuan sheng is known as labaci, which is made in winter when a relatively low temperature favors growth of *Mucor*. Cooked soybeans are wrapped with straw and placed in a room for about 15 days. The white mycelia of *Mucor* bind the beans into a cake, which is sliced, sun-dried, and used as a seasoning or is consumed fresh [6].

The Taiho Laws, enforced in Japan in 701, mention an office in the Imperial court dealing with several fermented soybean foods including douchi (豆豉) or kuki (豉); in Japanese, miso and jiang (醬) or hishio (醬). According to the book *Engishiki* (906), 360 ml of doujiang and 180 ml of douchi constituted a part of the monthly allowance supplied to government officials along with rice, soybeans, and redbeans among others. The method of preparing salted and nonsalted douchi is described in the book *Yoshufushi* (1686) and elsewhere. According to the oldest record about douchi in Korea, in 683, the product apparently was not an important food commodity in that country.

Salted douchi has appeared in the middle part of Japan every since with such names as Hama-natto, Daitokuji-natto, and others, and in Taiwan as In-si. It should be noted that these products are totally different from natto, which is much more popular in Japan in terms of microbes involved and method of preparation. Historically and literally, "natto" in Japanese meant any soybean products prepared in a kitchen of Buddhist temples. In preparing hama-natto, boiled soybeans, are cultured with yellow koji mold, sun-dried, mixed with salt water or shoyu, and stored for 6-12 months. Well-aged molded beans are sun-dried to make the final product.

B. Shuidouchi (China), Thua-nao (Thailand), Kinema (Nepal), and Natto (Itohiki-natto) (Japan)

Boiled soybean products cultured with *Bacillus subtilis* or *Bacillus natto* are widely distributed in China, eastern Nepal, Darjieeling, Sikkim, Bhutan, northern Thailand, Java, Korea, and Japan.

Boiled soybeans are naturally inoculated with *B. subtilis* and held at high humidity and at 30-40°C or higher. The optimum temperature for mycelial growth by *Aspergillus sojae* or *A. oryzae* is about 30-35°C, and the optimum humidity is moderate. Under such conditions, cultivation of cooked soybeans with *Aspergillus* molds free of bacterial contamination is technically more difficult than expected in practice. This is why in the book *Chi-min-yao-shu* of the sixth century, a precaution to keep the temperature rather low for douchi cultivation is mentioned. However, this may account for the prevalence in Asia of fermented soybean foods cultured with *B. subtilis*.

Representative douchi cultured with *B. subtilis* are shuidouchi in Shandong sheng in China and natto or, more specifically, Itohiki-natto in Japan. In making salted shuidouchi, soaked and cooked soybeans are placed in a cloth bag and covered with straw, which is the best natural source of *B. subtilis*. After 1-2 days at 25-30°C, the beans are covered with viscous substances. The sticky beans are mixed with minced ginger and salt and then tightly packed into jars. After aging for one week, they are ready for consumption [5,6].

Thua-nao is prepared as follows in the northern district of Thailand: Soaked and boiled soybeans are wrapped with banana leaves and fermented outdorrs at about 40°C for 2-3 days. The fermented soybeans are mashed into a paste along with salt and some flavoring agents such as garlic, onion, and red pepper, wrapped in small portions with banana leaves, and sun-dried. They are cooked before selling or eating [7-9].

Kinema in Nepal is prepared by culturing *B. subtilis* on cooked, mashed black or brown soybeans mixed with wood ash and wrapped in banana leaves [10]. Unlike Japanese natto, the product is not viscous and an almost equal number of cells (10^8-10^9/g) of *Pediococcus pentosaceus* and *Streptomyces fecalis* as well as of *Bacillus* were identified; no yeasts were found. The more lactobacilli found in kinema, the lower its pH value and the better its flavor. Katmandu, the capital of Nepal, is located 1300 m above sea level, has a temperature of 25-28°C in summer and 2-5°C in winter, and has a rainy season influenced by monsoons from May to September. Kinema prepared during the winter is consumed cooked in soups along with green vegetables or in curry as a substitute for meat with addition of pepper. The chemical compositions of kinema, thua-nao, natto, and douchi are given in Table 1.

Japanese itohiki-natto, or more commonly natto, has a more than 2000-year history in Japan. Its annual production was 170,000 tons in 1983, which was made from 85,000 tons of soybeans by about 1000 manufacturers. More than one ton of natto per day is being produced by each of 20 large producers in Japan by applying modernized automatic facilities in highly sanitary conditions. The largest 20 producers account for 20% of the total market of itohiki-natto in Japan [11].

Natto has a rich flavor, high nutritional value, and is highly digestible. It is consumed in Japan mostly as a side dish with a dab of mustard and shoyu as flavoring agents and Welsh onion as garnish. Natto exhibits a characteristic mucilaginous consistency. Sawamura identified *Bacillus natto;* Sawamura in 1905 was responsible for producing natto, and a pure culture of *B. natto* has been used in industrial natto manufacture since 1919 [12]. Extensive research on natto in terms of taxonomy, mucilage, enzymes, antibiotics, nutrition, bacteriophage, and genetics have been reported.

In natto manufacture, whole soybeans are soaked to 2.1-2.3 times their original weight, autoclaved at 1.0-1.5 kg/cm^2 for 20-30 minutes, cooled to 80-90°C, and sprayed with a pure culture seed starter of *B. natto*. Then 100 g is packed into polyethylene or polystylene packages, with 3000-4000 packages per lot, and incubated starting at 30°C followed by elevating the temperature to 40-50°C for 16-18 hours total. A high temperature of 50-52°C for the last 4-6 hours of incubation is necessary to obtain a highly mucilaginous product. The product is stored at 5-10°C, or -5°C in summer, before sale. Changes in nitrogenous compounds during natto fermentation are shown in Table 2 [13].

Instead of plastic containers used for incubation of natto, a wrapping made of straw is often employed according to the traditional method, and this often is a major source of microbial contaminants. The chemical composition of natto is H_2O, 59.5%; protein, 16.5% (of which 50% is water soluble and 20% is free amino acids); fat, 10%; fiber, 1.9%; reducing sugar, 9.8%; and vitamin B_2, 0.85 mg%. The mucilage of natto is equivalent to 0.1-0.8% on a dry solids basis and is a mixture of glutamic acid polypeptide and fructan, a polymer of fructose, in a ratio by weight of 90-81%: 25-47% [13].

The flavor of natto is influenced by content of dipeptides, free amino acids, diacetyl, and tetramethylpyrazine. Too much acetic acid and isovaleric acid in natto is not favored by consumers [14]. Yukiwari-natto is made by mixing itohiki-natto with rice-koji and salt.

TABLE 1. Chemical Composition of Kinema, Thua-nao, Natto, and Douchi

Product	Country of production	NaCl (g/100 g)	Total N (g/100 g)	Formol N[a]	Glutamic acid (mg/100 g)	pH	Acidity[a]	Acetic acid (g/100 g)
Kinema, wet type	Nepal	0.1	3.5	1.3	37	5.4	0.20	3.5
Kinema, dry type	Nepal	0.4	7.1	0.8	608	6.3	0.05	1.6
Thua-nao, wet type	Thailand	3.0	2.9	0.7	130	6.1	0.05	0.7
Thua-nao, dry type	Thailand	0.3	6.0	0.3	193	6.0	0.05	0.2
Natto	Japan	0.3	2.7	0.6	288	6.7	0.05	0.2

[a]ml N/10 NaOH/10 ml. Douchi made in Taiwan (in-si) was analyzed to be NaCl, 12.7%; acidity, 1.2; formol N, 0.98%; ammonia N, 0.36%; total N, 3.22%; FN/TN, 30.4%; and AN/TN, 11.2%.

Source: Refs. 8, 40.

TABLE 2. Changes in Nitrogenous Compounds During Natto Fermentation

Fermentation (h)	Total N in dry matter	Percent total nitrogen		
		Water-soluble N	Amino N	Ammonia N
0	7.36	17.1	0.9	0.3
4	7.45	16.9	0.9	0.3
6	7.45	36.9	2.3	0.3
8	7.29	44.2	2.7	0.3
12	7.29	51.9	5.9	2.0
16	7.41	53.9	8.1	2.2
18	7.23	54.9	8.3	2.8

Source: Ref. 13.

C. Tempe and Oncom (Indonesia)

Tempe is a fermented soybean food characteristically found in Indonesia. Tempe is a type of soybean-chi primarily cultured with *Rhizopus* molds and is thought to have been introduced into Indonesia centuries ago by the Chinese along with soybean cultivation. Central and east Java are the major tempe-producing areas. Tempe is a key source of protein for millions of Indonesian people who enjoy it daily in quantities of 20-129 g. The estimated nationwide annual production is about 325,000 tons. This means more than 35% of the total annual Indonesian soybean production is transformed into tempe. In Indonesia, tempe producers are still on a household scale.

Tempe is produced in small quantities in Malaysia, Scerinam, Thailand, Canada, and the Netherlands (since 1969) by immigrants from Indonesia. It has become a popular food in the United States since 1980 and in Japan since 1983. Traditionally, in Indonesia tempe is consumed on the day it is made. The tempe cake is cut into thin strips, sun-dried, and deep fat fried. Extensive studies on tempe in terms of microbiology, biochemistry, nutrition, and production technology were conducted in the United States after World War II [15-18].

Several kinds of tempe are distinguished by their raw materials, but *tempe kedelai*, or more commonly tempe, is most popular and is made from soybeans. *Tempe genbus* is made from tofu residue, *tempe bonkrek* from coconut press-cake, and *tempe benguk* from legumes boro benguk. All these tempe are prepared by culturing the substrate with *Rhizopus oligosporus*.

Tempe kedelai (soybean tempe) making: The technology of traditional tempe making is simple and rapid with an extremely low cost of production. Soybeans are soaked, dehulled, lightly boiled, and inoculated with molds of the genus *Rhizopus*. Incubation is at 30-37°C for about 1-3 days until the soybean cotyledons are combined into a compact cake with mycelia of the filamentous mold. A broad range of tempe-making processes has been described for different localities and countries (Fig. 1).

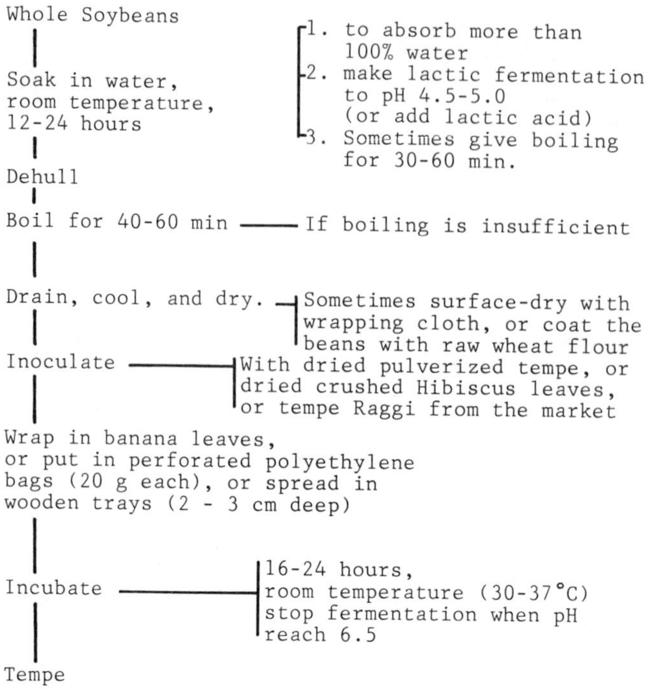

FIGURE 1. Flow sheet of Tempe making. (Compiled from Saono et al., Yeoh and Merican, Ijas and Peng, and Bate et al., Symposium on "Indigenous Fermented Foods," Bangkok, Thailand, 1977.)

Soaking soybeans and discarding the soak water is essential for the tempe making [19]. The mold cannot grow well on the intact soybeans. Moreover, removal of water-soluble substances from soybeans by soaking in water is effective to reduce bacterial contamination during the mold cultivation that follows. It also was found by Hesseltine et al. that soybeans contain a heat-stable water-soluble compound that inhibits growth of *R. oligosporus* [20].

Boiled and surface-dried soybean cotyledons are inoculated with tempe molds from a previous batch of tempe, or with the natural starter commonly known as "usar," or with ragi-tempe inocula sold on the Indonesian market. Usually 1-3 g of dried pulverized tempe is used to inoculate 1000 g of dehulled precooked soybean cotyledons. Ragi-tempe inocular are flattened, dry round rice cakes (2.5 cm in diameter) containing a variety of microorganisms. Usars prepared from leaves of *Hibiscus similis* are thought to be the best inoculum. Usar is still used by many tempe producers, and many of them claim that usar gives a better product than does an inoculum consisting of a single strain of *Rhizopus* [21]. The microbial populations of usar consist of molds dominated by *Rhizopus* spp., yeasts, and bacteria. Numbers of filamentous molds per gram are $7.2\text{-}9.5 \times 10^5$; yeasts, $0\text{-}3.2 \times 10^3$; and aerobic bacteria, $1.2\text{-}1.5 \times 10^5$ [21].

A comparative study revealed that the best results for tempe making were obtained with *R. oligosporus*, followed by *R. stolonifer*, *R. oryzae*, *R. microsporus*, *R. arrhizus*, and *R. chinensis* [16,22-25]. *R. oligosporus* NRRL 2710 was selected to be the best and has an optimal temperature of 30-42°C. It is highly proteolytic and lipolytic but has low amylase activity and no detectable pectinase activity [26]. It is a good producer of antioxidant substances and a pleasant flavor. Two kinds of protease are produced by *R. oligosporus*, both with optimal pH at 3.0, and yields are greatest at 50-55°C [27].

All commercial tempe examined so far had vitamin B_{12} activity. A tempe containing 30 ng vitamin B_{12}/g would provide 60% of the daily requirement of the vitamin [28]. Tempe made with the pure mold under aseptic conditions has no B_{12} activity. The bacterium responsible had been identified as *Klebsiella pneumoniae*, which is a common organism on plant materials.

Most commonly incubation of tempe begins at 31°C, and the temperature of materials reaches a maximum after 30 hours and then drops. Use of 37°C for tempe production is often recommended partly because if coconut press-cake is an ingredient, toxigenic *Pseudomonas cocovenans* will not grow, even if it is present. The best growth of *R. oligosporus* was observed at RH 75-85% [29]. Sixty-six percent of nonnitrogenous compounds, 50% of crude proteins, and 50% of fat are solubilized during tempe fermentation [30]. The most critical issue appears to be loss of protein during tempe preparation, which is generally considered to be more than 25%.

Until 25 years ago, most tempe prepared for sale in Indonesia was made with the traditional process, in which inoculated cotyledons are wrapped in small packets using wilted banana or other large leaves and are incubated in a warm place for 2-3 days. Today, several cottage industries have adopted the tray or plastic bag method. The tray method involves spreading inoculated beans on a tray covered with layers of banana leaves or wax paper and incubated at room temperature. A new method involves incubating tempe in plastic bags or plastic tubes with perforations at 0.25-1.3-cm intervals to allow access of oxygen [31].

Volatile flavor of tempe: Tempe should be harvested as soon as the cotyledons are knitted together with mold mycelia into a firm cake and the pH has risen above 6.5. Fresh tempe prepared in clean conditions has a pleasant, sweet, and bland odor, which changes into an ammonia, diacetyl-like (ethyl acetate-like) odor with storage at more than 10°C for one week. Rapidly frozen at below -35°C and stored at below -18°C, tempe remains fresh for more than one year. If the tempe is sterilized and stored at 5-10°C, it can be enjoyed at least for 10 days [29].

Moroe et al. [32] studied the volatile flavor components of tempe fermented at 31 and 38°C. The major volatile flavor components of cooked soybeans were maltol (6.7 ppm) and higher fatty acids. By fermenting tempe at 30°C, a top note was given, which consisted of isopentanal, isobutanol, isopentanol, acetoin, acetic acid, 2,3-butanediol, isovaleric acid, and pyrazines.

By fermenting at 38°C, four pyrazines and diacetyl (1.1 ppm) were produced in addition to the top note components developed by fermenting at 31°C. Contents of maltol and higher fatty acids of tempe fermented at

38°C were 2.6 and 1.4 times of those of boiled soybeans, respectively. Total volatile compounds of tempe fermented at 38°C were 4.6 times that of tempe fermented at 31°C. Fermenting tempe at higher temperatures promotes activity by microbes other than the tempe mold thus generating the fermented odor and natto-like odor associated with diacetyl (ethyl acetate-like odor), acetoin, pyrazines, and isovaleric acid. Tempe prepared by fermenting only with the tempe mold had a sweet odor resembling that of cooked soybeans as represented by maltol and higher fatty acids [32]. The characteristic beany odor or green note of cooked soybeans is totally lost through their incubation with *R. oligosporus*.

Chemical composition and nutritional value of tempe: The chemical composition of tempe was reported to be protein, 18.3%; fat, 4.0%; carbohydrate, 12.7%; vitamin A, 50.0 IU; vitamin B_1, 0.17 mg/100 g; vitamin B_{12}, 29 ± 5.0 ng/g on a dry solid basis [28]; or H_2O, 10.76%; fat, 27.5%; sugar, 17.75%; fiber, 3.98%; and ash, 3.64% [33]. Degradation of protein in tempe making into water-soluble forms and amino acids was calculated to be 21.2 and 2.7%, respectively.

According to Winarno [28], the average protein efficiency ratio (PER) of soy tempe is 2.4 as compared to casein, which is 2.5. The biological value (BV) of tempe is 58.7 as compared to that of meat at 80. The net protein utilization (NPU) of soy tempe is 56 as compared to that of chicken meat at 65. The digestibility quotient of tempe is 86.1%. Cereal grains, particularly rice, are deficient in lysine but well endowed with methionine-cystein, whereas lysine is not the limiting amino acid in tempe. Thus, serving these two foods at the same meal in the proper proportion (4 parts of rice and 1 part of tempe) will substantially increase the protein quality of the resulting combination [28].

Wang et al. found that the nutritive value of tempe made from a mixture of soybeans and wheat was comparable to that of milk casein [34]. Wang published a good review article on the nutritional quality of fermented foods including tempe in terms of the effect of fermentation on compositional change of substrates, protein quality, digestibility, and other factors [35]. Biotin and total folate compounds were, respectively, 2.3 and 4-5 times higher in tempe than in unfermented soybeans. The increase in riboflavin, niacin, pyridoxine, and vitamin B_{12} activity is of considerable importance nutritionally [36]. Peroxide values were determined for lipids extracted with ether from samples of soybeans and tempe stored for several months. Peroxide numbers of tempe samples ranged from 0 to 1.1, while those of soybeans ranged from 18.3 to 201.9 [37]. Gyorgy et al. attributed the improved nutritive value of tempe to stabilization of the oil by antioxidants produced during fermentation and to synthesis of B vitamins [18].

Tempe bonkrek: Bonkrek is a special type of tempe produced by fermenting freshly shredded coconut residue using the tempe mold *R. oligosporus*. The serious concern with bonkrek is that it sometimes develops toxins produced by the extraneous bacterium *Pseudomonas cocovenans* [meaning toxin (venom) producer from coconut] which overgrows the mold. Consumption of such tempe can be fatal. The bonkrek toxin consists of two substances: toxoflavin, $C_7H_7N_5O_2$ [38], and bonkrek acid, $C_{28}H_{38}O_7$ [39]. Manufacture of tempe bonkrek has been prohibited by local authorities since 1975 [34].

Oncom: Oncom is a food made by fermenting peanut press-cake with the tempe mold, *R. oligosporus*, or with *Neurospora sitophila*, and is called black oncom and red oncom, respectively. Oncom is produced mainly in West Java, and is an important ingredient of the daily menu. The average consumption of oncom is 255 g per capita per year. Red oncom produced in the Bogor area is usually made solely from tahu (tofu) residue. Oncom made from peanut press-cake is composed of protein, 10-13%; fat, 4-6%; carbohydrate, 22%; and moisture, 59%. Nutritional composition of oncom is: calories, 187.0; protein, 13.0%; fat, 6.0%; carbohydrate, 22.6%; vitamin B_1, 0.09 mg/100 g; and vitamin B_{12}, 31 ± 7.0 ng/g [28].

D. Fermented Tou-fu (Soybean Curd) Products: Sufu (China and Taiwan) and Tofu-yo (Japan)

Fermented soybean curd (tou-fu or tofu) products made with a mold belonging to genus *Actinomucor, Rhizopus, Mucor,* or *Aspergillus* are consumed in East Asian countries, particularly in China and Taiwan. Sufu is the Chinese name that first appeared in the literature. Since there are numerous dialects used in China, there are many names for the product. Major districts of sufu production in China are Jiangsu, Shaoxing, Zhejiang, Fujian, and Guangdong shengs. Sufu also is produced and consumed abundantly in Taiwan. Consumption of sufu in Mainland China is not known, but the annual production of tou-fu-ju (sufu) in Taiwan reached about 10,000 tons in 1977, which is equivalent to about 12 g per person per week [40].

Sufu tastes like creamy cheese and is thus called soybean cheese in the West. Protein and oil of soybean curd are hydrolyzed by mold enzymes into tasty free amino acids, glycerol, and higher fatty acids. In the original method of preparing nyu-fu (same as sufu) which appeared in literature of the fifth century, dried tou-fu is salted and stored in a fermented alcoholic mash or in salty soybean mash. Tofu-yo prepared in Okinawa in Japan even today is manufactured by this old method. The name of nyu-fu is found in *Chi-min-yao-snu*, published in the sixth century; tou-fu was prepared by coagulating soy milk protein with acetic acid (vinegar), and not by calcium sulfate as is practiced today. The most common method for preparing sufu consists of five procedures: preparation of tou-fu, cultivation of mold on the surface of cubed tou-fu (pehtze), fermenting pehtze in salt rice wine or soy sauce mash or the like, and packaging. A flow sheet for preparing sufu is shown in Figure 2.

The water content of common tou-fu is about 85% with 10% protein and 4% lipid, but the water content of tou-fu for sufu preparation is preferably less than 70%. Tou-fu is cut into small cubes of about 3 cm. Historically, rice straw was the source of mold used for inoculation. Tou-fu cubes are placed on straw arranged on a large bamboo trays, and the trays are piled. Cubes are separated from one another to permit air circulation on all surfaces of the cubes and to ensure total coverage of cubes with mold mycelia.

Cubes of tou-fu are surface-dried in a hot air oven at 100°C for approximately 15 minutes [41], or before the hot air treatment, tou-fu is immersed in an acidic saline solution composed of 6 g of NaCl and 2.5 g of citric acid dissolved in 100 ml of water [42]. The organisms must develop enzyme systems having high proteolytic activity, since the mold grows in a protein- and lipid-rich medium. *Actinomucor elegans* was confirmed to be the best organism for sufu making.

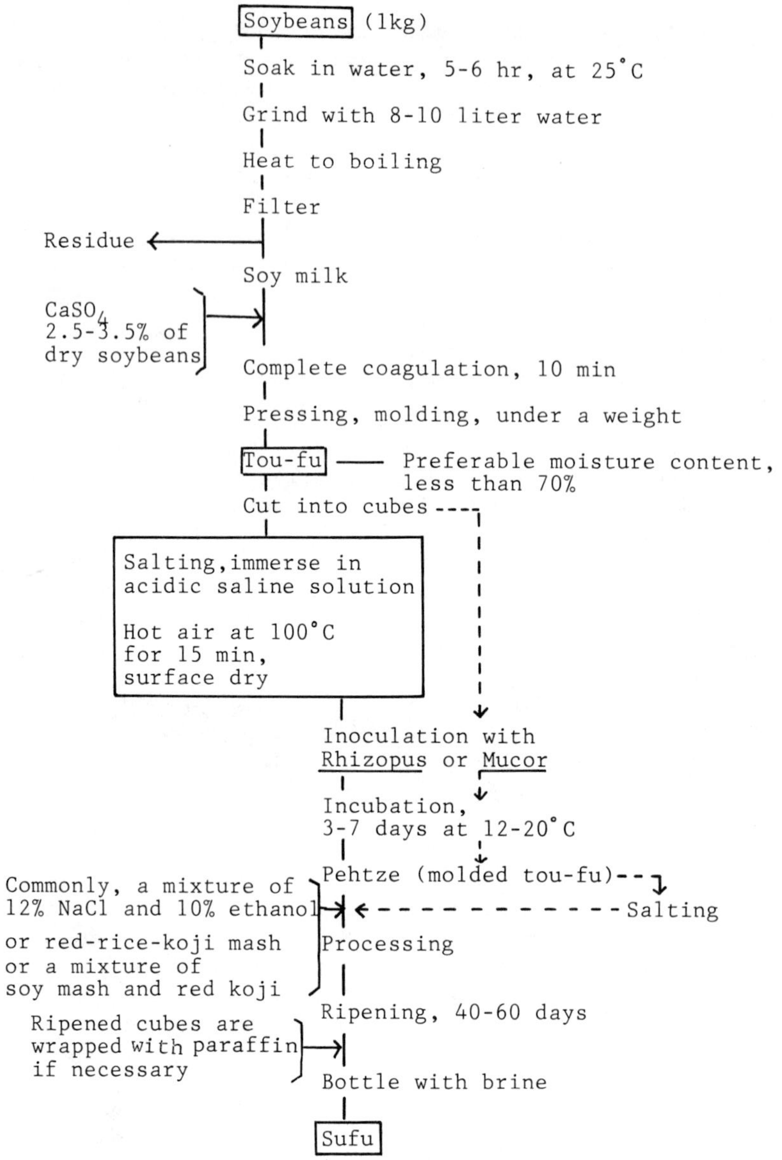

FIGURE 2. Flow diagram of Sufu making. (Compiled from Hesseltine 1965, Su-Yuan-chi 1980, Wai 1968, and Wang and Hesseltine 1970.) (From Refs. 22, 40, 41, 43.)

TABLE 3. Changes of Nitrogen Compounds in Sufu Making

N compounds	Tou-fu (%)	Tou-fu-ju (%) (Sufu)
Protein nitrogen	99.12	79.58
Nonprotein nitrogen	0.88	16.54
Formol nitrogen	1.37	17.82
Ammonia nitrogen	0.04	0.76

Source: Ref. 40.

Formerly, molded tou-fu (pehtze) was placed in an earthen jar in such a way that one layer of pehtze is followed by one layer of salt. After 3-4 days when salt has been absorbed, pehtze was taken out, washed with water, and put into another jar for processing. The molded cubes can be placed in various types of brine solution. A basic and most common brine solution is one containing 12% NaCl and rice wine containing 10% ethanol [43]. The mixture is allowed to age for about 40-60 days or longer. The product is then bottled with brine and marketed as sufu.

To make red sufu, red koji (anka) and soy mash are added; to make tsao-sufu, fermented rice mash is added; to make Guangdong sufu, red pepper and anise are added as are salt and red koji. The most common recipe in Taiwan is as follows: table salt, 1 pounds; soy mash, 1 pound; red koji, 0.6 pound; and raw sugar, 0.6 pound, thoroughly mixed with 6 pounds of water [40].

The loss of crude protein in molded tou-fu (pehtze) and sufu as compared to that of tou-fu are 5 and 20%, respectively. Conversion of protein into formol nitrogenous compounds and ammonia occurs during sufu ripening. The approximate amounts of products in sufu manufacture are as follows: soybeans, 1 kg; tou-fu, 3 kg; molded tou-fu (pehtze), 1.7 kg; and matured sufu, 1.9 kg [44]. Changes in nitrogenous compounds during sufu making are given in Table 3 [40].

Tofu-yo is a fermented soybean curd product made locally in Okinawa, the southernmost island of Japan. It originated in the southern part of China in the eighteenth and nineteenth centuries. The method of preparing tofu-yo differs from that of sufu in that dried tou-fu is stored in a fermented mixture of red rice koji (or yellow rice koji), a small amount of salt, and Awamori equal to 1.0-1.2 volumes of koji. Awamori is a distilled rice wine indigenous to Okinawa containing 43% ethanol. The fermentation period of dried tou-fu mash is 2-3 months. Red rice koji cultured with *Monascus purpurens*, *M. anka*, or *M. barkeri* is more commonly used for tofu-yo preparation than the yellow product.

III. FERMENTED SALTY CONDIMENTS IN A SLURRY OR PASTE MADE FROM SOYBEANS AND CEREALS

Current fermented salty condiments called jiang differ from that of the original terminology, and they appear to have been derived from chi. The condiments are prepared by fermenting a mixture of molded beans or cereals, salt, and water.

A. Doujiang (Touchiang) (China) and Tauco (Southeast Asia)

Doujiang are presumed to have originally been prepared by mixing soybean-chi (nonsalted soybeans grown with yellow koji mold such as *Aspergillus oryzae*), salt, and water to obtain chijhi or soy sauce, but the mixture was then stored for enzymatic digestion and lactic acid and yeast fermentations. Around the sixteenth century boiled soybeans were coated with raw wheat flour before mold cultivation. Aged soy mash, as such, was used as a seasoning agent, but often the liquid was separated from the mash as chijhi or tao-yu or soy sauce to be used as liquid seasoning. The extracted or filtered residue of the mash having the consistency of soft paste or porridge was used as a second class of seasoning agents. This pattern of production is still popular in China, Thailand, Singapore, Malaysia, Indonesia, and the Philippines (taosi or taoco).

In Indonesia, tauco is mainly produced in western Java and is used to prepare soups and other side dishes. The method of making tauco is greatly influenced by that of tempe. Soybeans are dehulled before culturing with mold. The soybean cotyledons with or without added wheat, tapioca, rice, or maize flour undergo spontaneous overgrowth by molds from the environment; alternatively, they may be inoculated with ragi tempe [45,46]. The mold-cultured material or koji is dried in the sun, mixed with salt water, and fermented and concentrated in the sun for 3-7 weeks. Tauco is a slurry containing 10% protein and sold in glass bottles, whereas Japanese miso is a paste containing 25% protein [47].

B. Doubanjiang (Toupanchiang)

Doubanjiang is very popular in Central China; particularly in Sichuan sheng and in the basin of the Zhujiang (Pearl) River. Soybeans and broad beans are used for doubanjiang production. In some instances, spices such as chili, pepper, fennel, etc. also are added. Broad beans are washed, soaked in water for one day, and then allowed to sprout. When the length of the sprout becomes one third or one fourth the length of beans, the bean coat is removed. Then the beans are cooked for 2-3 hours, cooled to 40°C, and mixed with 10% roasted wheat flour. Beans coated with wheat flour are cultured with *Aspergillus oryzae* to make koji. The broad bean koji is mixed with an equal volume of brine (Bè 20), put in an earthen jar, and fermented for about 2-5 months. The matured doubanjiang is transferred to glass bottles, boiled for 20 minutes, and then the bottles are sealed. The Chinese National Standard (Taiwan) for doubanjiang requires a moisture content of $<50\%$, salt 10-18%, crude protein $>8\%$, and ash (include salt) $<19\%$ [40].

C. Tianmianjiang (Tienmienchiang)

The tianmianjiang product which originated in north China contains much starch in its raw materials and tastes sweet. The product of the Hopei Province is the most famous one. Tianmianjiang has become popular as the seasoning agent for roasted duck. The major raw material is wheat flour. Wheat flour is mixed with water to make dough, then it is cut into bricks and steamed in a kettle. After cooling, it is inoculated with *Aspergillus oryzae* koji mold and incubated at 30°C for one week. The koji (200 liters) and 120 liters of brine (Bè 19) are mixed and stored for 3-4 months for fermentation.

D. Gochoojang and Doenjang (Korea)

Gochoojang and doenjang are fermented soybean paste products in Korea mainly made at home and used to make soup and for cooking in general. In their preparation, cooked and mashed soybeans are suspended in rice straw until overgrown with mold and bacteria during 2-3 months of winter storage. Mold-cultured material is called meju. In preparing gochoojang, 10 liters of sun-dried, crushed meju is mixed with 2 liters of cooked glutinous rice flour, 0.5-0.7 liters of hot pepper powder, kanjang or soy sauce and salt, and the mixture is fermented for 2-6 months [48]. For making doenjang, meju (1), salt (1), and water (4) are mixed and stored for fermentation for 6-10 months. Then the aged mash is separated into kanjang (liquid) and doenjang (paste); the latter is often mixed with hot pepper [49]. Commercial retail meju has been industrially prepared in recent years using *Aspergillus oryzae* as inoculum and a controlled fermentation room. Per capita daily consumption of kanjang, doenjang, and gochoojang was 23.3, 10.4, and 13.9 g, respectively, in 1971 [48].

E. Hishio (Japan)

Shoyu (liquid) and miso (paste) constitute the two major fermented condiments made in Japan from soybeans and such cereals as wheat, rice, or barley. Added to them, is another fermented condiment named hishio, which is a slurry paste made from the same raw materials. While the total production of hishio is very limited, numerous kinds of hishio are sold locally across the country and others very often are made at home. Kinzanji-miso or -hishio is most famous among them. Hishio is prepared by fermenting a mixture of vegetables, soybean-and-wheat koji, and salt in a slurry paste. Dehulled barley or peanuts is sometimes used in place of wheat or soybeans, respectively. Rice-koji, sugar, and/or shoyu are often added to the mash, which is then fermented with lactobacilli and yeasts. Vegetables commonly used for hishio include eggplant, cucumber, ginger, and red pepper.

F. Miso (Japan)

Miso is a general name for a semi-solid salty food made by fermenting soybeans and rice or barley. The original type of miso, introduced to Japan from Korea and China at least 1300 years ago, was called sho, miso, and shi in the Taiho-Law in 702. Although no mention is made of its production method, it is estimated that soybean chiang in China and doenjang in

Korea resembled each other and that both share a common process of shaping cooked soybeans into balls or bricks before incubation to develop mold. This particular process was used in farmers' households in prewar Japan. Molded materials, douchi in China and meju in Korea, were stored with salt for fermentation. Cooked soybean balls are presently made by a motor-driven extruder and inoculated with a mold-starter mixed with roasted barley flour. This modified method for making soybean meju is widely practiced in the prefectures of Aichi, Mie, and Gifu in the central part of Japan. But the amount of soybean-miso produced by this method is as little as about 8% of the total.

Miso is classified into three types on the basis of the raw materials used: (1) kome(rice)-miso made of rice-koji, cooked soybeans, and salt; (2) mugi(barley)-miso made from barley- or rye-koji, cooked soybeans, and salt; and (3) mame(soybean)-miso from soybean-koji, cooked soybeans, and salt. These three types of miso are further classified into sweet, medium, and salty, depending on their salt content, and white, light yellow, and red, depending on their color. About 80% of the miso consumed in Japan is rice-miso of such varieties as White, Edo, Shinshu, and Sendai. The amounts of mugi- and soybean-miso consumed are 12 and 8%, respectively, of the total. Constituents of some typical misos are indicated in Table 4 [50].

Production and consumption of miso: The amount of miso produced in Japan was 572,000 tons in 1985, but it has been decreasing during the last decade. Per capita per day consumption of miso is about 17 g, which is one-half that of shoyu in Japan. Raw materials for miso used in 1983 were soybeans, 181,000 tons (domestic 19,000 tons and imported 162,000 tons); milled rice, 105,000 tons; dehulled barley and wheat, 25,000 tons; and salt, 71,000 tons. There are about 2000 miso producers in Japan. The 10 primary producers supply 36% of the total amount made, and the market share of the biggest producer was 4.7% in 1980. Miso is largely consumed as an ingredient in miso soup, and it is also used as a condiment in cooking in general. Miso or tauco made from soybeans and a small amount of raw wheat flour is very popular in Asian countries, but in most instances, the soybean-miso produced in countries other than Japan is a slurry rather than semi-solid and is used much more as a condiment than as a soup base.

Rice-miso and barley-miso making: Whole yellow soybeans are almost always used to prepare typical miso. Soybeans are soaked in water until saturated and then cooked for 30-60 minutes at normal pressure, or are cooked in four volumes of water for 20-30 minutes at a pressure of 0.5-0.7 kg/cm^2, or are steamed for 20 minutes at a pressure of 0.7 kg/cm^2 (115°C), either batchwise or continuously. Cooked soybean granules preferably are pressed using less than 0.5 kg/cm^2 of pressure. Milled rice or barley is soaked in water and then steamed batchwise in an open cooker for 40 minutes or continuously on a net conveyor in a closed autoclave for 30-60 minutes.

The koji cultivation of *A. oryzae* on rice or barley is done at 35-38°C for 40-48 hr, sometimes with an increase in temperature of up to almost 40°C in the final stage. Finished koji is mixed with salt or sometimes cooled to stop further mold growth and to minimize inactivation of enzymes.

TABLE 4. Constituents of Some Types of Miso

Miso[a]	H_2O (%)	pH	NaCl (%)	Protein (%)	Fat (%)	Rs[b] (%)
White miso	45	5.3	4.5	7.9	3.8	38
Edo sweet miso	49	5.4	5.8	11.0	4.5	15
Shinshu-miso	48	5.2	12.0	11.4	5.5	12
Sendai-miso	50	5.1	12.8	12.0	5.3	11
Mugi-miso (salty)	48	5.1	12.0	12.7	5.5	11
Mugi-miso (sweet)	47	5.2	9.5	10.0	4.9	17
Soybean-miso	46	5.0	11.5	20.0	10.5	4

[a]The ratio between soybeans and rice as raw materials in white miso is 10:20-25; edo-miso, 10:13; shinshu-miso, 10:6-10; and sendai-miso, 10:3-5. The ratio between soybeans and barley or rye as raw materials in mugi-miso is 10:10-20. The barley or rye is milled to make it 60-85% by weight.

[b]Reducing sugars.

Source: Ref. 50.

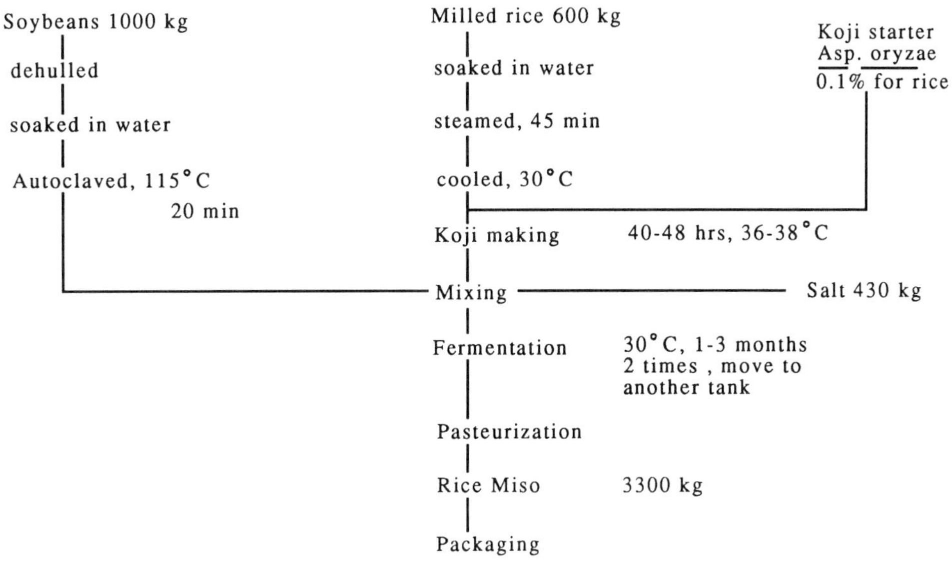

FIGURE 3. Diagram of rice-miso fermentation. (From Ref. 50.)

The amount of salt used is about 30% of the koji by weight. Various types of koji fermenters are currently being employed. Cooked soybeans are mixed with salted rice- or barley-koji, a small amount of water, and an inoculum of yeast and lactic acid bacteria, if necessary. It is important to mix these materials uniformly so that the variation in salt concentration in the mash is less than 0.5%. The mixture is packed in a fermentation tank; it is moved from one tank to another at least twice during the fermentation period to mix the contents and to provide aerobic conditions suitable for microbial growth. Fermentation is done at around 30°C for 1-3 months depending upon the type of miso. Well-ripened miso is then blended and mashed if necessary and pasteurized using a tube heater. About 2% alcohol is added to the product to stop growth of yeasts. The production process for manufacturing salty rice-miso is shown in Figure 3.

IV. FERMENTED SALTY LIQUID CONDIMENTS MADE FROM SOYBEANS AND CEREALS

There are many varieties of fermented salty liquid condiments produced in East and Southeast Asian countries. They are made from a mixture of whole soybeans or defatted soybean grits, and wheat kernels, or wheat flour (raw or roasted) or wheat bran in different ratios. The kind of microbe used and the conditions of preparation determine the characteristics of each fermented salty condiment. However, fish sauce is more popular in some Southeast Asian countries such as the Philippines, Vietnam, Thailand, and Malaysia. The 1982 statistics report that 10 liters of shoyu (soy sauce) is the annual per capita consumption in Japan, and the counterpart in other Asian countries is as follows: 13 L, Korea; 12 L, Beijing and Shanghai; 9 L, Taiwan; 6 L, Singapore; and 4-5 L, Indonesia.

A. Japanese Shoyu

Total annual production of five kinds of shoyu recognized by JAS (Japan Agricultural Standard) is about 1.2 million kL. Koikuchi is deep in color and makes up 85% of the total, whereas usukuchi is light in color and makes up 13%. Both are made from almost equal parts of soybeans and wheat kernels, and are characterized by thorough fermentations with lactobacilli and yeasts. Beside koikuchi and usukuchi, small amounts of tamari, shiro, and saishikomi are produced. Fish sauce is not recognized as shoyu by JAS. Tamari is made mostly from soybeans with only a small amount of wheat, and resembles Inyu in Taiwan. Shiro shoyu is very light in color and is made from wheat with a small amount of soybeans. Saishikomi shoyu is made by digesting shoyu koji in raw shoyu instead of salt water.

The JAS established three grades for each variety of shoyu; special, upper, and standard. The special grade of koikuchi shoyu contains more than 1.5% TN and 1.0% alcohol, and must be made by the genuine fermentation method. Of the shoyu in Japan, 65% was special grade in 1985. Three production methods of shoyu are also recognized by JAS: genuine fermentative, semi-fermented or HVP-mixed and fermented, and just HVP-mixed. About 74% of Japanese shoyu was made by the genuine fermentation method in 1985. The typical chemical composition of good quality genuine fermented shoyu is shown in Table 5.

TABLE 5. Chemical Composition of Various Kinds of Genuine Fermented Shoyu in Japan (Feb. 1988)

Kind	No. of Samples	Baume	NaCl[a]	TN[a]	FN[a]	RS[a]	Alc[b]	pH	Color[c]
Koikuchi	5	21.7	16.9	1.57	0.84	3.11	2.16	4.82	11
	3	22.5	17.0	1.69	0.92	3.89	2.12	4.79	11
d	5	20.0	13.5	1.57	0.84	3.54	3.34	4.85	11
e	2	15.7	8.6	1.54	0.86	3.12	5.40	4.75	7
Usukuchi	5	21.9	18.9	1.19	0.68	4.36	2.60	4.86	28 <
	1	22.2	18.0	1.49	0.85	4.03	2.57	4.85	27
Tamari	1	27.6	17.2	2.39	1.13	8.80	0.26	4.93	2 >
Shiro	1	25.1	17.6	0.53	0.28	16.6	0.21	4.70	46 <
Saishikomi	1	28.0	12.4	1.96	0.91	11.13	1.53	4.82	2 >

[a]% (w/v); TN (total nitrogen); FN (formol nitrogen); RS (reducing sugar).
[b]% (v/v); Alc (alcohol).
[c]Number of standard color of shoyu.
[d]Salt reduced by 20%.
[e]Salt reduced by 50%.
Source: Ref. 116.

Good quality genuine fermented koikuchi shoyu contains 1.5-1.8% (g/volume) total nitrogen, 3-5% reducing sugar (mainly glucose), 2-2.5% ethanol, 1-1.5% polyalcohol (primarily glycerol), 1-2% organic acid (predominantly lactic acid, pH 4.7-4.8), and 16-17% sodium chloride. To obtain a palatable product, about one-half of the nitrogenous compounds present must be free amino acids, and more than 10% free glutamic acid.

Shoyu manufacturers in Japan are assumed to approximate 2700. The five largest manufacturers produce 50%, and the next 50 account for 25% of the total production. The largest producers are Kikkoman, Yamasa, Higashimaru, Higeta, and Marukin, all of which produce genuine fermented koikuchi and usukuchi shoyu of the JAS special grade. Kikkoman is the largest, with recent annual production of 360,000 kL. The Japanese population of 250 years ago (1721) was 30 million, and 1.2 million people in Edo (now renamed Tokyo) consumed about 20,000 kL of shoyu a year. The koikuchi-type shoyu was consumed, and most of it was produced in Noda and Choshi, east of Tokyo. It was at about that time that the production method of koikuchi shoyu was established. In contrast, usukuchi shoyu originated in the western part of Japan (Tatsuno-city), and tamari in the central part of Japan (Nagoya-city). The Japanese shoyu was first exported to Europe from Nagasaki, Kyushu island in 1668 by Dutch traders.

Manufacture of koikuchi and usukuchi shoyu: Koikuchi shoyu is prepared using five main processes: treatment of raw materials, koji making,

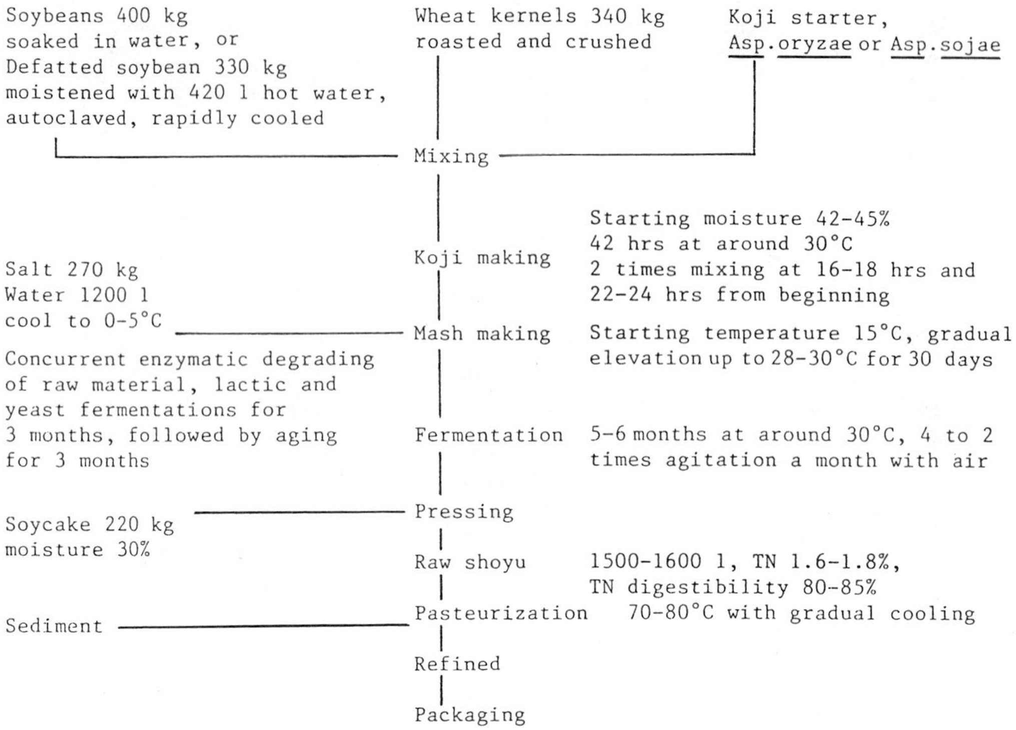

FIGURE 4. Diagram of koikuchi shoyu fermentation.

mash making and aging, pressing, and refining. An example of koikuchi shoyu manufacture is shown in Figure 4. While soybeans or, more commonly, defatted soybean grits are moistened and cooked with steam under pressure. Wheat kernels, the other half of the raw material, are roasted at 170-180°C for a few minutes, then coarsely crushed. The mixture of these two materials is inoculated with a small amount of seed mold or with a pure cultured *Aspergillus* mold and cultured for 42 hours in small wooden trays or, more recently, on a perforated stainless steel plate, allowing temperature- and moisture-controlled air to come up through the materials. The mold-cultured material or koji is mixed with strong salt water to make a mash, and the mashes are stored in wooden kegs or in larger tanks made of concrete, resin-coated iron, or FRP. It takes more than one year at the natural temperature, or about 6 months with adequately controlled temperature, for microbes in the mash to finish the fermentation. During this time, enzymatic degradation of raw materials, lactic acid and yeast fermentations, and interreaction of chemical components of the mash proceed concurrently. The aged mash is press-filtered through cloth, and the liquid portion obtained is pasteurized to make the shoyu. Production of usukuchi shoyu is similar to koikuchi except for use of 10% more salt water with higher salt concentrations, shorter fermentation time, and a lower temperature for mash fermentation and pasteurization.

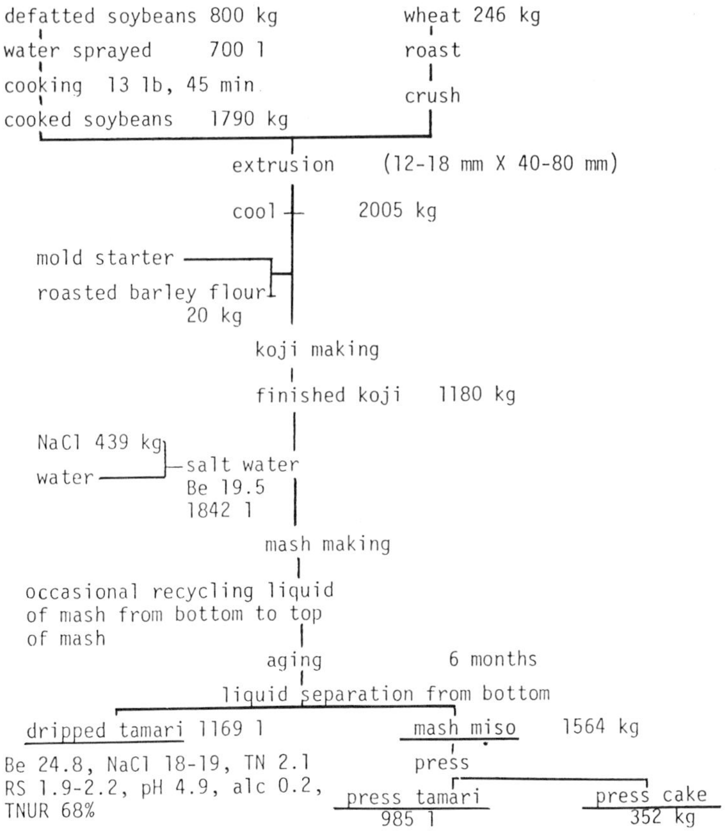

FIGURE 5. Tamari-shoyu fermentation. (From Ref. 118.)

Manufacture of tamari shoyu: A mixture of cooked soybeans or defatted soybean grits and roasted wheat (20:3) is extruded to form granules 12-16 mm in diameter. These granulated soybeans are inoculated with a mixture of the seed molds, *A. oryzae*, *A. sojae*, or *A. tamarii*, and roasted powdered barley in amounts less than 1.5% of the total. These materials are incubated at 26-28°C for about 45 hours to make koji. The moisture content of the koji preferably is about 35%, and its weight is about 120% that of the raw materials. This tamari-koji is sometimes dried to decrease its weight by 7-8% before making the mash. The koji is mixed with 0.5-1.3 volumes of salt water to make the mash. The mash is usually too solid to stir, so the liquid portion of the mash is repeatedly removed from the bottom and spread over the top of the mash. This is done instead of agitating the mash with compressed air as for koikuchi mash. Genuine tamari shoyu is rich in soluble solids and there is little contamination from film-forming yeasts, so the final product does not need pasteurization (Figure 5).

B. Soy Sauce Produced in East and Southeast Asian Countries Other than Japan

Domestic and industrial soy sauce in Korea: The amount of soy sauce produced in Korea in 1971 was 223,000 kL, which included 107,000 kL of industrially produced soy sauce and 116,000 kL produced in homes. The domestic soy sauce, kanjang, is produced with the meju process as described in the previous section. Industrial soy sauce is made largely by blending with HVP or with the semi-chemical method. The Senpyo Company is the largest shoyu manufacturer in Korea supplying about 50% of the industrial product.

Shoyu in Taiwan: Production of soy sauce in Taiwan in 1985 was estimated to be about 130,000 kL. Near 90% of the product is the Japanese koikuchi-type shoyu. There were 433 soy sauce factories in Taiwan in 1978; the eight largest producers supplied at least 45% of the local market. Soy sauce products include genuine fermented soy sauce (25%), chemically hydrolyzed soy sauce (5%) and blended soy sauce (75%) [40]. From 5 to 10% of Taiwan soy sauce is estimated to be in-yu, which is made only from black soybeans. There are three national standard grades of soy sauce in Taiwan, and their total percentages of nitrogen in 1980 were 1.4, 1.2, and 1.0 g/100 mL.

Soy sauce in Southeast Asian countries: Chijhi or soybean sauce is currently the most popular soy sauce in Hong Kong, Singapore, Malaysia, Indonesia, Thailand, and other Southeast Asian countries, where it is made from a mixture of whole soybeans and a small amount of raw wheat flour, and is cultured with yellow mold in granular form without making it into dough. The soybean sauce in these areas is usually fermented in 150-L earthen jars; they are replaced by 10-kL FRP tanks.

In Singapore, Malaysia, and Indonesia, two types of soy sauce are manufactured: a pale soy sauce (in Cantonese sung-show or in Hokkien shiew-cheng) and a dark soy sauce mixed with caramel, cane molasses, and/or palm sugar (in Cantonese low-chow or in Hokkien tau-iu).

In Indonesia, soy sause is called ketjap or kecap; two types of kecap are popular: kecap-asin, which has a salty taste, and kecap-manis, which is very sweet. The kecap mash is often washed more than twice with salt water, and the mixed extract is concentrated in the sun. The residue of the extraction is used for tauco or miso of an economical grade.

Soy sauce in the People's Republic of China: Historically China, the large country where fermented soy sauce originated, has many varieties of fermented soy sauces. A national standardized method of soy sauce manufacture, however, has generally been followed for only about two decades. From the author's observations and information provided by Chinese authorities in Beijing and Shanghai during a visit in 1979, it is clear that the method is based on a high-temperature short-time digestion of shoyu-koji to meet the pressing demand for shoyu in great quantity. Furthermore, koji is prepared by large-scale cultivation of *Aspergillus oryzae* on a mixture of steamed defatted soybeans and wheat or wheat bran (6:4). The koji is mixed with salt water to make hard mash with a moisture content of about 80% and a salt concentration of about 6-8%. The hard, low-salt mash is kept at 45-50°C for about 3 weeks to effect enzymatic digestion. The digested mash is extracted with hot salt water and then with hot water; the salt-free residue may be used for animal and poultry feed. There is no alcoholic fermentation or pressing of the mash as there is in Japanese shoyu manufacture. The yield of soy sauce on a nitrogen basis in 1979 was 75-80%. The highest government standard of soy sauce requires: total nitrogen, 1.6%; reducing sugar, 4%; and sodium chloride, 19%. Acid hydrolysis of plant protein for soy sauce manufacture was illegal in China until recently.

Chijhi or whole soybean soy sauce is still being widely distributed at home or in factories in the basins of the Zhujiang (Pearl) River and the Huanghe (Yellow) River in China. A factory located in Sichuan sheng, for instance, produced more than 3000 tons of chijhi a year [8].

V. BIOCHEMISTRY INVOLVED IN SHOYU AND MISO MANUFACTURE

A. Selection of Raw Materials

Before World War II, whole soybeans were used for shoyu and miso. However, now in Japan, defatted soybean grits are primarily used for shoyu production, whereas whole beans continue to be used for almost all kinds of miso. The amount of whole soybeans used for shoyu in 1983 was only 3% of the total. In contrast, in Southeast Asian countries whole beans are still the major raw material of fermented soy sauce.

A few decades ago whole soybeans had some advantages over defatted soybeans as raw material for shoyu in terms of ease of koji making and of fermenting the mash, higher glycerol content, greater stability of color, and superior sensory quality of shoyu. However, the remarkable technological progress achieved during the past 30 years in production of Japanese shoyu made it possible to produce a product with superior quality and low cost by using defatted soybean grits [51-53]. The chemical composition of whole and defatted soybeans is as follows: H_2O, 10% and 11%; total nitrogen, 6% and 8%; invert sugar, 18-20% and 21-22%; and crude fat, 17-19% and 1-2%; respectively. Soybeans contain about 6% sucrose, 4% stachiose, and less than 1% raffinose.

Of the nitrogen contained in koikuchi shoyu, 75% is derived from soybeans and the remainder from wheat kernels. The ratio of soybeans to wheat as raw materials for koikuchi shoyu ranges from 6:4 to 4:6. The glutamic acid content of soybeans and wheat is 20 and 30% of the total amino acids, respectively. Proteins present in wheat kernels are good sources of glutamic acid, which is an important flavor component of shoyu. The chemical analyses of wheat and wheat bran used for shoyu production are as follows: H_2O (%), 11.0 and 11.5; total nitrogen (%), 2.0 and 2.4; and starch (%), 65-63 and 45; respectively.

Wheat bran is sometimes used instead of wheat kernels; this decreases the alcohol content of shoyu and makes the color darker but its stability is inferior because of the increased amount of pentoses in the shoyu.

Milled rice is used for a wide variety of rice-misos in amounts ranging from 6 to 25% that of soybeans. The chemical compositions of starchy raw materials used for miso including refined rice, barley, and rye are as follows: H_2O (%), 15, 15, and 13; protein (%), 6-6.5, 13-14, and 13-14;

carbohydrate (%), 77, 60, and 62; and crude fat (%), 0.8, 6.0, and 4.0; respectively.

B. Contribution of Improved Cooking Methods of Raw Materials to the Increase of Enzymatic Protein Digestibility

The degree of enzymatic digestibility of proteins involved in raw materials is associated with both quality and quantity of the final product, and simultaneously with the rate of fermentation of salty mash.

High-temperature, short-time (HTST) heat treatment of raw materials, selection and improvement of koji molds in terms of protease formation and other functions, improved conditions for koji cultivation by use of mechanical apparatus, and improved control of the salty mash fermentation have contributed to the increase in enzymatic digestibility of proteins in shoyu manufacture from 65 to over 90% during the past 30 years.

Tateno et al. upgraded soybean protein digestibility from 69 to 73% by rapidly cooling cooked soybeans immediately after their autoclaving at 0.8 kg/cm^2 g for one hour [54]. Yokotsuka et al. further upgraded enzymatic digestion of defatted soybean and wheat kernel protein by 10 and 2-3%, respectively, by autoclaving at 6-7 kg/cm^2 g for 5-30 sec followed by rapid cooling accompanied with explosion, as shown in Table 6 [55,56]. Several HTST continuous cookers of soybeans and wheat kernels have been devised, and simultaneously the NK batch-type rotary cooker for soybeans was improved to more rapidly cool cooked soybeans than before.

TABLE 6. Effect of Cooking Conditions of Thoroughly Moistened Defatted Soybean Grits on the Enzymatic Digestibility of Protein

Steam pressure (kg/cm^2)	Cooking time (min)	Digestibility of protein[a]
0.9	45	86%
1.2	10	91
1.8	8	91
2.0	5	92
3.0	3	93
4.0	2	94
5.0	1	95
6.0	1/2	95
7.0	1/4	95

[a]In enzyme solution of 0% salt at 37°C for 7 days.

Source: Ref. 55.

TABLE 7. Differences Between *A. oryzae* and *A. sojae* Used for Shoyu Fermentation

Shoyu koji cultured with *A. oryzae*: lower pH value, lower carbohydrate content, higher activity of α-amylase, higher activity of acid protease, high activity of acid carboxypeptidase, lower activity of polygalacturonase
Shoyu koji cultured with *A. sojae*: higher pH value (less content of citric acid), more carbohydrate (lower consumption during koji cultivation)
Shoyu made from koji cultured with *A. sojae*: lower viscosity of mash, higher content of sugar, lower pH value of raw shoyu, higher content of lactic acid, less coagulant by pasteurization (fewer enzymes remained)

Source: Refs. 58, 59.

C. Selection and Improvement of Koji Molds

Strains of koji mold used for shoyu should be selected from various viewpoints, which include good spore formation, strong growth activity, absence of toxin production, high ability for enzyme formation, especially proteases and plant tissue-degrading enzymes, smaller consumption of starch and greater consumption of pentosan during growth, adaptability to the through-flow system of mechanical koji cultivation, and production of desirable and stable color and flavor in the final product.

About 75% of the shoyu producers in Japan are using *Aspergillus oryzae*, and most of the reat use *Aspergillus sojae*. Of miso producers, 88% use *A. oryzae*, and for sake almost 100% use *A. oryzae* [57]. *Aspergillus oryzae* and *A. sojae* have their own physiological characteristics, but generally, good amylase producers tend to be found among strains of *A. oryzae* and good protease producers among strains of *A. sojae*. The pH value, enzyme composition, and properties of mash of shoyu koji cultured with *A. oryzae* and *A. sojae* are compared in Table 7 [58,59].

More than 80% of protease produced by *A. sojae* is alkaline protease, but in addition, semi-alkaline protease, neutral protease I and II, and acid protease I, II, and III have been identified (Table 8). Three kinds of aminopeptidase and four kinds of carboxypeptidase also have been isolated [60]. Among exopeptidases or peptidases of koji mold, leucine aminopeptidase having optimum pH 7-8 is strongly associated with enzymatic formation of formol nitrogen and glutamic acid in the shoyu mash. Protein digestibility in shoyu fermentation is not always proportional to the protease activity of koji. The positive effect of cellulase or pectinases to increase the digestibility of soybean and wheat proteins by proteases and other enzymes of koji molds has been reported [61].

Proteolytic activity of 36 strains of *Rhizopus* was compared with that of *Aspergillus* molds. *Rhizopus tamarii* and *R. thermosus* grew well on wheat and soybeans and exhibited very high proteolytic activity at pH 3.0, but almost none at pH 6.0. In this regard, the *Rhizopus* molds were distinctly different from *Aspergillus* molds in that they gave lower protein

TABLE 8. Proteinases Produced by *Aspergillus sojae*

Proteinase	Molecular weight	Units/g koji
Acid I	39,000	41.1[a]
Acid II	100,000	10.0[a]
Acid III	31,000	4.6[a]
Neutral I	41,000	80.0[b]
Neutral II	19,300	8.7[b]
Semialkali	32,000	55.4[b]
Alkali	23,000	929.0[b]

[a]Activity on casein at pH 3.0.
[b]Activity on casein at pH 7.0.
Source: Ref. 60.

digestibility than did the *Aspergillis* molds during experimental brewing [62]. Rice miso was prepared by using *Rhizopus oryzae*, *R. japonicus*, and *Aspergillus oryzae*. The ratio between solubilized protein and total protein and that between amino acids and total protein of *Rhizopus* miso were 70 and 40% those of *Aspergillus* miso, respectively [63].

Improvement of proteolytic activity by koji molds has been achieved by induced mutation, and crossing or cell fusion. A 2-6% increase in protein digestion was reported in an experimental shoyu brewing test by using an induced mutant of *A. sojae*, which had protease activity six times greater than that of the original strain [64]. It is rather difficult to find a strain of koji mold which has strong activities for both proteinase and glutaminase. Thus a solution has been sought with protoplast fusion which has been applied to strains of *A. sojae* [65]. The high proteolytic activities of a mutant strain of *A. oryzae* induced by UV treatment were introduced into the mother strain through protoplast fusion [66]. A new strain of *A. oryzae* was obtained by UV radiation, and it consumed one-half as much starch during koji cultivation as did the original strain without changing other functions of the original strain [67].

D. Improvement in Koji Making

Mechanized preparation has greatly contributed to increased enzymatic activities and to reduced bacterial contamination of koji. This has served to increase the yield of shoyu and improve the chemical and sensory qualities of both shoyu and miso. Heat-treated raw materials spread to a depth of 30-40 cm on a large perforated stainless steel plate are aerated for 2 days with temperature- and moisture-controlled air, which comes up from the bottom through holes and into the materials to give proper conditions for mold growth and enzyme formation. During the first half of miso or shoyu koji cultivation, about 35°C is preferable for mycelial growth of koji molds, but in the second half, about 38°C is best for amylase formation in miso koji, or <30°C for protease formation and to avoid the inactivation of protease, especially of glutaminase in shoyu koji [68]. It takes about 42 hours to finish koji cultivation both for shoyu and miso. Materials are usually cooled twice by mixing during koji cultivation.

The pH values of the starting materials and those of the finished koji are between 6.0 and 7.0 for shoyu koji composed of soybeans and wheat, and soybean koji, and around 5.5 for milled rice or barley koji for miso. Higher amounts of ammonia are formed with smaller C/N ratios of the raw materials; the ammonia shifts the pH value from acidic toward neutral or alkaline, thus increasing the alkaline protease content. More acidic protease is formed with neutral to acidic raw materials (Table 9).

E. Microbial and Chemical Control of Salty Mash Fermentation

The temperature and length of shoyu mash fermentation are usually 15-30°C and 3-6 months, respectively, while those of miso mash are 30-50°C and 1-3 months. The shoyu koji is mixed with saline water of 0-5°C; storage of salty mash begins at about 15°C with a gradual increase to about 30°C over one month or more, and then maintaining this temperature to the end of the fermentation. Effects of this so-called cold-mash making are to restrain a rapid pH decrease by a lactobacillus fermentation and a too rapid alcoholic fermentation by yeasts at the beginning of mash fermentation; both reduce the activity of alkaline protease.

The pH value of shoyu mash starts at 6.5-7.0, depending on the pH value of the koji, which is greatly affected by acid-forming bacterial contaminants such as *Micrococcus*. The water activity (a_w) of 18% salt water is 0.88 in the new mash, and that of an average well-aged mash is 0.80. The types of microbes which can be grown in the shoyu mash are limited by the 16-18% salt concentrations, and only salt-tolerant lactobacilli and yeasts can grow in shoyu mash. The major lactic acid bacterium in shoyu mash has been identified as *Pediococcus halophylus*, which has an optimum pH of 5.5-9.0 a_w of >0.81 (opt. 0.99-0.94), and can grow in a 24% salt solution (opt. 5-10%) at 20-42°C (opt. 25-30°C) [69].

The initial pH value of shoyu mash decreases rapidly because of enzymatic degradation of proteins and lactic acid fermentation. When the pH value decreases to ≤5.5, growth of *S. rouxii* begins, replacing lactic acid bacteria, and reaching a viable count of 10^6-10^7/ml. The dominant yeast found in shoyu mash is *S. rouxii*, but sometimes *Torulopsis* yeasts such as *T. versatilis* and *T. etchellsii* are found along with *S. rouxii*. The a_w values of *S. rouxii* and *Torulopsis* yeasts are 0.78-0.81 and 0.98-0.84, respectively, and both can grow in salt concentrations of less than 24-26% [69]. These yeasts can grow at pH 3-7 in salt-free media, but this range is narrowed to 4-5 in an 18% salt solution [70]. Changes in microflora of shoyu mash are illustrated in Figure 6 [71,79].

Pediococcus halophylus and *Pediococcus acidi lactici* are the most dominant lactic bacteria in the miso mash fermentation. Since the acid tolerance of these bacteria is comparatively weak, they cannot survive below pH 5.0. *Pediococcus acidi lactici* is more acid tolerant and less salt tolerant than is *P. halophylus*.

TABLE 9. Enzyme Composition of Koji as Influenced by the Difference of Material

Enzyme	Substrate for assay	pH for assay	Shoyu koji[a]	Rice koji[a]	Wheat-bran koji[a]	Submerged culture[b]
Total proteinase	Casein	7.0	1500	73	2540	665
Acid proteinase	Casein	3.0	295	417	3020	304
α-Amylase	Starch	5.0	3920	3100	11700	257
Acid carboxypeptidase I	Cbz-Ala-Glu	4.0	0.456	2.153	5.650	0.665
Acid carboxypeptidase II, III, IV	Cbz-Glu-Tyr	3.0	0.708	1.491	7.430	0.842
Leucine aminopeptidase II, III	Leu-Gly-Gly	8.0	0.360	0.075	0.705	0.430
Leucine aminopeptidase I	Leu-β-NA	8.0	1.405	0.132	4.660	1.624
CM-cellulase	CM-cellulose	5.0	21.60	6.71	59.10	0.595
Pectin transeliminase	Pectin	5.5	12.33	1.84	12.88	0.036

[a]Units/g koji.
[b]Units/ml culture filtrate.
Source: Ref. 60.

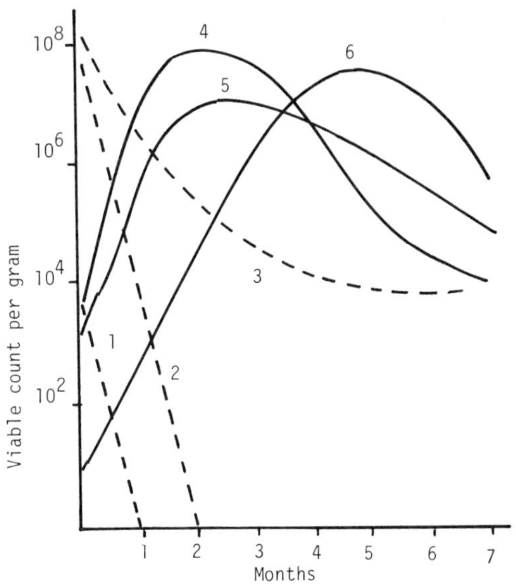

FIGURE 6. Microflora changes in shoyu mash fermentation. 1, wild yeasts; 2, *Micrococcus*; 3, *Bacillus*; 4, *Lactobacillus*; 5, *Saccharomyces rouxii*; 6, *Torulopsis* yeasts. (From Ref. 71.)

The fermentation of shoyu mash is controlled so that final contents of lactic acid and ethanol are 0.5-1.0% and 2-3%, respectively, and those of miso mash, 0.1-0.3% and more than 0.3%, respectively. To promote lactic acid and yeast fermentations concurrently with enzymatic degradation of raw materials, pure cultures of lactic acid bacteria and yeasts are usually added to shoyu and miso mash. Proper selection of strains of lactic acid bacteria to be added to shoyu mash is essential, because the diversity of shoyu lactics has been described in terms of metabolisms of sugars, organic acids, and some amino acids such as arginine, histidine, tyrosine, and aspartic acid. They influence aroma, pH value, color degree, and contents of some organic acids, amino acids, and amines of shoyu (Table 10) [72-75]. Several kinds of bacteriophages for *P. halophylus* in shoyu mash have been found recently [76].

Glutaminase in koji is most rapidly inactivated in shoyu mash, and amylase is most stable. Adding heat and salt-tolerant glutaminase produced by some yeasts to new mash effectively increases the glutamic acid content of shoyu regardless of the temperature [77,78]. Digestion of protein in current industrial shoyu, miso, natto, and tempe fermentations in Japan is indicated in Table 11.

F. Flavor Evaluation of Koikuchi Shoyu

The role of koikuchi shoyu in the diet is to provide a source of salt, flavor, and color. It increases the appetite and adds a delicious and spicy flavor to meals to promote digestion. Moreover, there are some who believe in the pharmaceutical benefits of shoyu and miso. Some physicians find shoyu comparable to caffeine in promoting secretion of gastric juice.

TABLE 10. Various Metabolic Patterns by Lactobacilli in Shoyu Mash

1. Homofermentation: Glucose → 2 mol lactic acid
2. Heterofermentation: Glucose → 1 mol lactic acid, ethanol, acetic acid, CO_2, H_2, acetone, butanol
3. 67 patterns of metabolic manners for arabinose, lactose, melibiose, manitol, and sorbitol
4. Metabolic manners for amino acids and citric acid:
   - Histidine → Histamine + CO_2
   - Tyrosine → Tyramine + CO_2
   - Arginine → Ornithine + 2 NH_3 + CO_2
   - Citric acid → Acetic acid + malic acid → lactic acid + CO_2
   - Aspartic acid → Alanine + CO_2

Source: Adapted from Refs. 72-75.

Fragrance and alcoholic odor were most favored among the components of the odor of fermented shoyu, whereas the butyric acid of chemically hydrolyzed shoyu and the odor produced when the fermentation temperature is too high were undesirable. As for characteristics of taste, a good aftertaste, a pure taste, a palatable taste, and moderate saltiness are important [80,81].

The importance of the volatile components of shoyu was indicated when the organoleptic ranking of two samples of shoyu could be reversed by exchanging their volatile fractions. Japanese investigators have identified nearly 300 volatile flavor compounds in the odor of koikuchi shoyu. One example of results of quantitative analyses of the major volatile flavor constituents of pasteurized koikuchi shoyu is given in Table 12 [51,82,83].

The most important part of aroma that is characteristic to koikuchi shoyu seems to exist in its weakly acidic fraction, from which the following important flavor compounds have been isolated: methionol (3-methylthio-1-propanol) [84], 4-ethylguaiacol and *p*-ethylphenol [85,86], maltol

TABLE 11. Digestibilities of Protein in Shoyu, Miso, Natto, and Tempe Fermentations

Product	Water-soluble nitrogen/ Total nitrogen (%)	Amino acid nitrogen/ Total nitrogen (%)
Shoyu mash	80-90	40-50
Miso	60-65	20-30
Natto	50	20
Tempe[a]	21	3

[a]From Ref. 33.

TABLE 12. Results of Quantitative Analysis of Flavor Constituents in Koikuchi Shoyu

Compound	Amount present (ppm)	Compound	Amount present (ppm)
Ethanol	31 501.10	Furfuryl alcohol	11.93
Lactic acid	14 346.57	Isopentyl alcohol	10.01
Glycerol	10 208.95	Acetoin	9.78
Acetic acid	2 107.74	*n*-Butyl alcohol	8.69
HMMF	256.36	HDMF	4.83
2,3-Butanediol	238.59	Acetaldehyde	4.63
Isovaleraldehyde	233.10	2-Phenylethanol	4.28
HEMF	239.04	*n*-Propyl alcohol	3.96
Methanol	62.37	Acetone	3.88
Acetol	24.60	Methionol	3.65
Ethyl lactate	24.29	2-Acetylpyrrole	2.86
2,6-Dimethoxyphenol	16.21	4-Ethylguaiacol	2.77
Ethyl acetate	15.13	Ethyl formate	2.63
Isobutyraldehyde	14.64	γ-Butyrolactone	2.02
Methyl acetate	13.84	4-Ethylphenol	Trace
Isobutyl alcohol	11.96		

Source: From Refs. 51, 82, 83.

and 5-hydroxymaltol [87,88], hydroxyfuranones, namely HEMF (4-hydroxy-2(or 5)-ethyl-5(or 2)-methylfuran-3(2H)-one), HMMF (4-hydroxy-5-methylfuran-3(2H)-one) and HDMF (4-hydroxy-2,5-dimethylfuran-3(2H)-one) [87-89], cyclothen (2-hydroxy-3-methylcyclopent-2-en-1-one) and several lactones [90].

Methionol (3-methylthio-1-propanol) is the metabolite of methional produced by yeasts, and methional is produced by Strecker degradation of methionine [84]. 4-Ethyl-guaiacol was isolated and is the metabolite of ferulic acid produced by *Candida* yeasts. Ferulic acid is produced by koji-mold from wheat bran, presumably from glucoside. About 25% of 50-70 samples of shoyu tested in 1964 contained 0.5-2.0 ppm 4-ethyl guaiacol (4EG) [91,92]. The correlation coefficient between 27 major odor components and the sensory evaluation of koikuchi shoyu was 0.313 at the highest, but the combination between methionol and 4EG was the most significant variable among all combinations of each two among 27 components, which can chiefly influence the variation of sensory data of shoyu. The optimum contents of 4EG and methionol were determined to be 0.8 ppm and 3.9 ppm, respectively [93,94].

The HMMF and HDMF are produced by the so-called browning reaction, and HMMF increases remarkably during pasteurization of shoyu, reaching

200 ppm. HEMF can be called the character impact component of fermented shoyu because 100-400 ppm is common in Japanese fermented shoyu and because it has a very low threshold value which is below 0.04 ppb in water [95]. It is important that HEMF is produced by yeast fermentation in shoyu mash; accordingly it is not found in tamari, and its content in usukuchi is smaller than in koikuchi.

Presence of substantial amounts of four of 70 pyrazines isolated from shoyu and their appreciable increase during pasteurization has been pointed out. Major top-note flavor constituents isolated from shoyu include ethanol, isovaleraldehyde, isobutyraldehyde, diethylacetals of these aldehydes, isoamylalcohol, and trace amounts of dimethylsulfide. Among the relative odor units of 14 constituents of headspace gas, isovaleraldehyde was highest, followed by ethanol [96].

G. Stability of Color of Pasteurized Shoyu

Color formation during mash fermentation and pasteurization is largely prompted by the heat-dependent Maillard reaction and the Strecker degradation carried at reducing and weakly acidic conditions. Color of pasteurized shoyu is fairly unstable toward oxidation, and exposure to air causes its reddish tone to darken and its flavor to deteriorate.

Oxidative browning of pasteurized shoyu is highly proportional to its initial degree of color. Factors which prevent the increase of color intensity in shoyu are: (a) reductive yeast fermentation of mash, (b) lactic fermentation with lactic acid bacteria having strong reducing power, (c) low pH and RH values, and (d) high salt concentrations [97,98]. Several Amadori compounds were isolated from shoyu, and they caused remarkable browning of shoyu in the presence of oxygen with distinct acceleration by iron [99,100].

H. Nutritional Concern About Shoyu and Miso

Salt contents of koikuchi shoyu were classified into three categories: ordinary salt, 16% (w/v), less salt, 13% (20% reduction), and reduced salt, 8-9% (50% reduction). In preparing shoyu with a 50% reduction, the salt content of ordinary shoyu is reduced by electric dialysis, or by a salt-retaining resin treatment. An addition to mash of 10^6 *Saccharomyces rouxii*/g in the early stage is done to promote rapid alcohol fermentation and antagonistically suppressing acid-forming bacteria during preparation of "less salt" miso [101].

I. Safety of Koji Molds and Shoyu

Many investigators have checked for aflatoxin production by koji molds, but so far all results indicate that koji molds are not aflatoxigenic [102,103]. In the course of this research, several pyrazine compounds were found to exhibit R_f values similar to those of aflatoxin B or G depending on solvent systems used in tests [104]. Some strains of *Aspergillus* mold were found to produce aspergillic acid, but the most prolific producer among them did not synthesize it on a solid substrate composed of soybeans and wheat within 2 days, the usual period for koji cultivation [105].

FIGURE 7. Classification of Aspergilli

1. Sakaguchi and Yamada (1944)

   Koji molds
   - *Aspergillus oryzae* (Ahlburg) Corn
   - *A. sojae*, Sakaguchi et Yamada (prominently echinulate conidia and smooth-walled conidiophore)

2. Raper and Fennel (1965)

   A. flavus group
   - *A. flavus* L.
   - *A. flavus* var. *colummaris* R. et F.
   - *A. parasiticus* A (include *A. sojae*)
   - *A. oryzae* (A) C.
   - *A. tamarii* K.
   - A. ----------

3. Murakami (1971)

   A. oryzae group
   - *A. oryzae* S. et Y. — *A. sojae* series
   - *A. tamarii* K. — *A. sojae* series
   - *A. oryzae* (A) C. — *A. oryzae* series
   - *A. oryzae* var. *viride* M. — *A. oryzae* series
   - *A. oryzae* var. *brunneus* M. — *A. oryzae* series

   A. flavus group
   - *A. parasiticus* S.
   - *A. toxicallius* M.
   - *A. flavus* L.

4. American Type Culture Collection (1982)

   Recognized *A. sojae* S. et Y. as a new species and separate from *A. parasiticus*.

5. Kurtzman (1983)

   A 90% or more of relatedness of *A. flavus*, *A. oryzae*, *A. parasiticus*, and *A. sojae* regarding DNA structures.

Source: From Refs. 107-109, 117.

Synthesis of aflatoxin, sterigmatosystine, ochratoxin, patulin, cyclopiazonic acid, and penicillic acid was checked for 33 kinds of industrial *Aspergillus* molds. Results indicate none of the strains tested produced these compounds, with the exception of a few strains, which produced cyclopiazonic acid [105]. Cyclopiazonic acid and kojic acid are easily degraded by yeasts in the shoyu mash fermentation [106].

Several classifications of aspergilli are summarized in Figure 7 [107-109]. In fact, it is sometimes difficult to definitely classify these molds using only their morphological features. Therefore, it becomes extremely important to confirm by means of chemical analysis that the molds to be used do not produce aflatoxin.

Shoyu itself is not a mutagen, although mutagenicity of heated products of some amino acids or proteins with or without added sugars is generally recognized. Existence of nitrosable compounds produced from more than 2000 ppm of nitrite at an elevated temperature in shoyu were identified as 1-methyl-1,2,3,4,tetrahydro-carboline-3-carboxylic acid (MTCA), tyramine, and acetylpyrole [110-112]. However, when shoyu was heated with 100, 500, 1000, and 2000 ppm of sodium nitrite for 30 minutes at 80°C and pH 3.0, only at 2000 ppm was mutagenicity exhibited in the Ames test [113]. However, it is generally recognized that the nitrite content remaining in the human stomach after a meal is about 5-15 ppm or 50 ppm at most. Nitrite at 2300 ppm and shoyu both at pH 3.0 and 37°C were introduced into the stomach of rats in vivo. Nitrite rapidly disappeared within 3 minutes and no mutagenicity was evident. The same reaction when it proceeded for 60 minutes in vitro showed some carcinogenicity with gradual consumption of 97% of added nitrite [114].

Rats in Canada fed Japanese shoyu (Kikkoman) for 33 months were somewhat smaller, but they were healthier, more active, and longer lived than controls. In fact, breast tumors developed in control rats, but none appeared in rats given shoyu. There was no evidence that shoyu is carcinogenic [115].

VI. CONCLUSION

One of the outstanding characteristics in preparation of traditional proteinaceous fermented plant foods in the Orient is application of mold cultivation on solid substrates, and particularly in granular form. This form of cultivation enables molds to grow in the extremely aerobic condition where their aerobic metabolism and characteristic enzyme formation can function fully. This is not technically feasible in an ordinary liquid fermentation.

Tempe in Indonesia is unique in that it is a highly degraded fermented soybean food with characteristically bland taste, which is a prerequisite for a staple food and, to some extent, for a side dish. Tempe serves as a nutritive and economical protein source in the Indonesian diet.

Technological excellence stands out in its manufacturing process and involves selection of *Rhizopus oligosporus* on the basis of its dominant weak acidic proteinase formation, and ability to grow the mold directly on boiled and dehulled soybeans, which are already acidified by natural lactic fermentation during soaking in water, thereby preventing growth of molds derived from undesirable contamination.

One of the technological problems of commercial tempe is its failure of maintaining freshness of flavor for more than a few days. Another technological problem of tempe is the great loss (as much as 25%) of proteins found in the raw materials during its manufacture. Wasting the water-soluble protein and thus sacrificing economy and nutrition while soaking soybeans in water is considered unavoidable from the viewpoint of preventing bacterial contamination during mold cultivation to ensure good flavor of the final products.

Japanese shoyu and miso industries experienced similar technological problems. However, discarding such waste water of soybeans was first challenged when the environmental pollution law prohibited it without further expensive treatment. To solve this problem, the soaking of soybeans or defatted soybean grits was replaced by spraying the proper amount of water on them before steaming. The problem of bacterial contamination during mold cultivation was reduced by applying pure-cultured starter mold and using thoroughly sanitized production facilities.

The chemical and physical properties of boiled or steamed soybeans make them most susceptible to natural contamination with *Bacillus subtilis*. It is justifiable to say that natto and the like are the most natural and primitive fermented soybean foods for various Asian peoples. The characteristic odor and mucilaginous properties of such products seem to limit the number of their consumers.

The first fermented soybean food that appeared in early history was douchi, boiled soybeans cultured with a yellow mold, presumably *Aspergillus oryzae*. In fact, it must have been difficult to grow *Aspergillus* yellow molds directly on boiled soybeans at room temperature while keeping the materials free from *Bacillus* contamination. Viewed scientifically, suggestions made in Chinese literature, in the sixth century on fermenting douchi at low temperature to avoid bacterial contamination are appropriate. The yellow koji molds, *Aspergillus oryzae* or *A. sojae*, are the best enzyme producers so soybean and/or wheat tissues are degraded into lower peptides and especially into amino acids. These molds are characterized by formation of amylase, proteases, and other enzymes in great amounts and mostly extracellularly. Some of these enzymes, peptidases including glutaminase, for example, are best produced when yellow koji molds are cultured aerobically on solid substrates having a great surface area, proper C/N ratio, and the right pH value. The technological progress in culturing yellow koji molds on solid plant materials has been achieved in three ways as follows:

1. Boiled soybeans or pelleted boiled soybeans are coated with raw wheat or barley flour to reduce the moisture content of the surface of soybean particles and to increase the C/N ratio of substrate, which is suitable for the growth of koji molds and thus avoids the impact of *Bacillus* contamination. This method is being applied to production of soybean sauce in China and Southeast Asian countries and to production of inyu in Taiwan and tamari in Japan.
2. Koji molds are cultured on cereal grains such as rice, wheat, or barley to make koji, and the koji is mixed with boiled soybeans along with salt and stored to digest the raw materials. Miso in Japan is made by this method.
3. A mixture of soybeans, or more commonly defatted soybean grits and wheat in a ratio of about 1:1 (volume:volume), is completely cultured with mold to make koji for manufacturing the Japanese shoyu, koikuchi, and usukuchi. Koji made by this method has stronger enzymatic, especially proteolytic, activities than that of the koji made from whole soybeans and a small amount of raw wheat flour. Enzymatic digestion of protein is greater in shoyu mash than that in miso mash because of the

greater amount of enzymes present in shoyu mash derived from the koji cultured with the total amount of raw materials.

The percentage of water-soluble proteinaceous compounds in fermented food products in Japan is in the following order: tempe < natto < miso < shoyu. Enzymatic liquefaction of protein in shoyu mash is important because it determines the yield of shoyu. A higher percentage of free amino acids in relation to total proteins is generally desirable for proteinaceous fermented condiments. As long as there are proper total and amino nitrogen contents in proteinaceous salty condiments, amounts of macro-chemical constituents such as salt, sugar, organic acids, and pH are of secondary importance as criteria for sensory quality of the products. If all these components are above standard, the final key to determine the consumer's preference is the volatile fraction of the products. A few bad odor components, resulting from microbial contamination or from some adulterants, for instance, are usually enough to reduce the total value of the products.

Control of effects of undesirable contaminants in traditional fermented foods has been achieved with an adequate salt concentration, by acidification of fermenting substrates by microbially produced lactic, acetic, or citric acid, or by presence of a few percent of ethanol formed by yeasts in the mash or derived from added alcoholic liquor, or by a combination of these approaches. Extremely clean operations and production facilities are vital to the manufacture of quality fermented foods.

For hygienic safety in household or industrial production of fermented food, application of pure culture microbial starters in place of inoculum from the previous batch or the traditional mixed culture dough or granular starters is highly recommended. To avoid possible harmful microbial contamination from the air, use of an overwhelming number of pure culture microbial spores or living cells as inocula, and rapid growth of the microbes to overpower contaminants must be done efficiently. Extended storage of raw materials in areas of high humidity and temperature is usually hazardous.

Japanese shoyu is characterized by the following features, which distinguish it from other similar products: use of soybeans and wheat kernels as raw materials in an approximate ratio of 1:1; up to 80-90% enzymatic digestion of protein; vigorous lactic and alcoholic fermentation of salty mash; and, in most instances, pasteurization of fermented mash-liquid at rather high temperatures to obtain an adequate color intensity and "brown flavor." This fundamental technology was essentially established some 350 years ago along with that of the rice wine or sake industry, utilizing *Aspergillus* koji molds in both processes.

Alcoholic beverages are consumed the world over as drinks, but their use as a condiment in cooking to make dishes more delicious must not be undervalued. Indeed, grape wine, sherry, brandy, whisky, vodka, rice wine, mirin, and the like which are indigenous to respective countries are used in their cookery. Generally, to degrade the protein of raw materials into water-soluble and tasty fragments is of utmost importance in soy sauce brewing. In Japan, equally important is the introduction of a vigorous yeast fermentation to shoyu mash, thereby adding abundant wheat kernels as a raw material primarily as carbohydrate source. Contribution of wheat to the sensory quality of the product is remarkable. This effect is evident from presence of such key flavor ingredients in shoyu as the vanilla series compounds including 4EG and vanillates, HEMF, 4-ethylphenol, methionol, and many kinds of alcohol and esters produced by yeasts.

Furthermore, the proper lactic fermentation concurrent with yeast fermentation in shoyu mash qualifies the product as an "all-purpose seasoning." Lactic acid fermented foods such as pickles and fermented dairy products are often used in cooking to provide extra flavors. In other words, Japanese koikuchi and usukuchi shoyu are the liquid condiments of a fermentative mixture and contain amino acids, alcohol, lactic acid, sugar, salt, and characteristic volatile flavors.

Recently, chemical reactions occurring between amino compounds and carbohydrates in foods have been widely discussed in terms of color, flavor, or biological effects of end products. It is amazing to realize that shoyu manufacturers 350 years ago in Japan knew that appropriate pasteurization of raw shoyu was indispensable for producing a product with an appetite-stimulating flavor. Pasteurization also serves such purposes as inactivation of microbial and enzymatic activities, color control, and clarification of liquid.

Japanese koikuchi shoyu, one of the most traditional of fermented condiments in East Asia, through research and development has come to be recognized as an industrial fermented food in the late twentieth century.

Remarkable research and technological progress has been achieved in the Japanese shoyu industry during the past 40 years. This progress includes:

1. Whole soybeans as raw material were replaced by defatted soybean grits, resulting in better yield and quality of final products and reduction in cost of their production.
2. Enzymatic digestion of protein in raw materials was upgraded from 65% to over 90% resulting from steaming raw materials through continuous HTST heating and from improvements in koji molds and koji making, which simultaneously contributed to shorten the fermentation period from over one year to less than 6 months.
3. The mechanical throughflow aeration system of the automatic koji cultivator definitely increased the proteolytic and plant-tissue-degrading enzymatic activities of koji and also greatly reduced bacterial contamination during koji cultivation.
4. Addition of pure cultures of essential lactobacilli and yeasts during the salty mash fermentation proved effective in improving the sensory quality of the final product and reducing the fermentation time. Almost all other production facilities were mechanized with automatic control, enlarged scale, and improved hygienic conditions.
5. The safety of koji molds and of final shoyu was confirmed by both chemical and biological methods.
6. More than 300 volatile flavor constituents of koikuchi shoyu were identified, and the key flavor components and their formation mechanisms were elucidated. The salt concentration of commercial shoyu was reduced by 10% and 20%.

Fermented foods and condiments made from soybeans with or without added cereal grains or flour are given in Figure 8.

No.	Raw materials	Kinds of mold	Form of koji, use of salt, mash fermentation	Products
1.	Soybeans	*Bacillus subtilis*	granular form salted or non-salted	Shuidouchi, douchi, Tua-nao, Kinema, Natto
		Aspergillus oryzae (*Mucor racemosus*)	granular form salted or non-salted	Douchi, Chizhi In-si, Hamanatto
		Rhizopus oligisporus	granular form non-salted	Tempe
2.	Soybeans — Toufu	*Actinomucor elegans* (*Mucor*, *Rhizopus*)	molded toufu is digested in salted ethanol solution	Sufu
		Monascus purpurens (*Aspergillus awamori*)	molded toufu is digested with rice-koji and distilled rice-wine	Tofu-yo
3.	Soybeans (10) + Wheat or barley flour (0.5-4)	*Aspergillus oryzae*	granular koji (Tou-shi) — mixed with salt-water and ferment	Douchi, Chizhi, Tau-co, Soybean-sauce
			ball or brick koji — mixed with salt-water and ferment	Jiangyu Soybean-sauce, Soybean-miso Doenjang, Kanjang
4.	Soybeans (10); Rice or barley (5-20)	*Aspergillus oryzae* (on rice or barley)	granular koji — mix with salt, and ferment with lacto-bacilli and yeasts	Miso (rice-miso, barley-miso)
5.	Soybeans (10) (defatted grits) + Wheat kernels (10)	*Aspergillus oryzae* or *Asp. sojae*	granular koji — mix with salt water, and ferment with lactobacilli and yeasts, filter	Shoyu (koikuchi, usukuchi)

FIGURE 8. Fermented foods and condiments made from soybeans mixed with or without cereal grains or flour. Note: Soybeans are boiled or steamed, wheat kernels are roasted and crushed, and wheat or barley flour is usually used raw. (From Yokotsuka 1989, original.)

REFERENCES

1. Batra, L. R. Fermented cereals and grain legumes of India and vicinity. In *Adv. in Biotech.*, Vol. II (M. Moo-Young, ed.), Pergamon Press, Canada, 1981, pp. 547-554.
2. Oberman, H. Fermented milk. In *Microbiology of Fermented Foods*, Vol. 1 (Brian J. B. Wood, ed.), Elsevier Appl. Sci. Pub., London, 1985, pp. 167-195.
3. Galloway, J. H., and Crawford, R. J. Cheese fermentation. In *Microbiology of Fermented Foods*, Vol. 1 (Brian J. B. Wood, ed.), Elsevier, London, 1985, pp. 111-165.
4. Bo, T. A. Origin of jiang and jiang-yu and their production technology (1 and 2). *Brew. Soc. Japan*, *77*:365-371, 439-445, 1982.
5. Bo, T. A. Origin of douchi and its production technology (1 and 2). *Brew. Soc. Japan*, *79*:221-223, 395-402, 1984.
6. Wang, H. L., and Suang, S. F. History of Chinese fermented foods. In *Indigenous Fermented Food of Non-Western Origin* (C. W. Hesseltine and Hwal Wang, eds.), J. Cramer, Berlin-Stuttgart, 1986, pp. 23-35.
7. Kubo, N. Fermented foods and cuisine in Thailand. *Science and Technology of Miso*, *34*(7):216-220, 1986.
8. Saito, K., Hasuo, T., Tadenuma, M., and Akiyama, H. Soybean fermented foods in India, Thailand and Japan. *J. Brew. Soc. Japan*, *78*(1):60-72, 1983.
9. Hesseltine, C. W., and Hwal Wang. Traditional fermented foods. *Biotechnol. Bioengng.*, *9*:275-288, 1967.
10. Karki, T. Microbiology of kinema. *Proc. Asian Symp. on Non-Salted Soybean Fermentation*, Tsukuba, Japan, 1986, pp. 39-41.
11. Toyota, M. Natto as a local industry and its hygienic control. *Proc. Asian Symp. on Non-Salted Soybean Fermentation*, Tsukuba, Japan, 1986, pp. 263-267.
12. Sawamura, S. On *Bacillus natto*. *J. of the College of Agriculture, Imperial University of Tokyo*, *5*(2):180-192, 1913.
13. Ota, T. Natto. In *Sogo-shokuryo-kogyo* (Sakurai et al., eds.), Koseisha-Koseikaku Pub. Co., Tokyo, 1978, pp. 576-579.
14. Fujii, H. On the formation of mucilage by *Bacillus natto*, I-VII. *J. Agric. Chem. Soc. Japan*, *36*:1000-1004, 1963; *37*:346-350, 407-411, 474-477, 615-618, 619-611, 1967; *41*:39-43, 1985.
15. Steinkraus, K. H. Historical perspective of tempe. In *Handbook of Indigenous Fermented Foods*, Marcel Dekker, New York, 1986, pp. 6-8.
16. Steinkraus, K. H., Yap, B. H., Van Buren, J. P., Providenti, M. L., and Hand, D. B. Studies on tempeh. *Food Res.*, *25*:777-788, 1960.
17. Hesseltine, C. W. Research at Northern Research Laboratory on fermented foods. *Proc. Conference on Soybean Products for Protein in Human Foods*, USDA, Peoria, IL, 1961, pp. 67-74.
18. Gyorgy, P., Murata, K., and Ikehata, H. Antioxidants isolated from fermented soybeans, tempeh. *Nature*, *203*:870-892, 1964.
19. Wang, H. L., and Hesseltine, C. W. Mold-modified foods. In *Microbiol. Technology*, 2nd ed., Vol. 2 (H. J. Peppler and D. Perlman, eds.), Academic Press, New York, 1979, pp. 96-129.

20. Hesseltine, C. W., deCamargo, R., and Rakis, J. J. A mould inhibitor in soybeans. *Nature, 200*:1226-1227, 1963.
21. Jutono, . The microbiology of Usar, a traditional tempe inoculum. *Proc. The Asian Symposium on Non-Salted Soybean Fermentation*, Tsukuba, Japan, 1986, pp. 50-59.
22. Hesseltine, C. W. A millenium of fungi, food, and fermentation. *Mycologia, 57*:149-157, 1965.
23. Hesseltine, C. W., Smith, M., Bradle, B., and Ko Swan Djien. Investigations of tempeh, an Indonesian food. *Developm. in Industr. Microbiol., 4*:275-287, 1963.
24. Ko Swan Djien and Hesseltine, C. W. Indonesian fermented foods. *Soybean Digest, 22*:14-15, 1961.
25. Renu Sharma and Sarbhoy, A. K. Tempe, a fermented food from soybean. *Current Science, 53*:325-326, 1984.
26. Wang, H. L., and Hesseltine, C. W. Oriental fermented foods. In *Prescott and Dunn's Industrial Microbiology*, 4th ed. (G. Reed, ed.), AVI, 1982, pp. 492-538.
27. Wang, H. L. Research on Oriental fermented foods at NRRL, processing of the Oriental fermented foods, Taipei, Taiwan, ROC, 1980, pp. 1-14.
28. Winarno, F. G. The nutritional potential of fermented foods in Indonesia. *Traditional Food Fermentation as Industrial Resources in ASKA Countries*, 1982, pp. 31-40.
29. Takamine, K. Fundamental technology of tempe making. Shokuhin-to-kagaku. Additional publication, 1987, pp. 151-152.
30. Van Buren, J. P., Hackler, L. R., and Steinkraus, K. H. Solubilization of soybean tempe constituents during fermentation. *Cereal Chem., 49*:208-211, 1972.
31. Martinelli, A., and Hesseltine, C. W. Tempeh fermentation: Package and tray fermentations. *Food Technol., 18*:167-171, 1964.
32. Moroe, M., Sato, A., and Yoshida, T. Flavor of tempe. *Proc. The Asian Symposium on Non-Salted Soybean Fermentation*, Tsukuba, Japan, 1986, pp. 210-218.
33. Ota, T., Ebine, H., and Nakano, M. *Report of Food Research*, Japan, No. 18, 1964, p. 67.
34. Wang, H. L., Rattle, C. I. R., and Hesseltine, C. W. Protein quality of wheat and soybeans after *R. oligosporus* fermentation. *J. Nutr., 96*:109-114, 1968.
35. Wang, H. L. Nutritional quality of fermented foods. In *Indigenous Fermented Foods of Non-Western Origin* (C. W. Hesseltine and Hwal Wang, eds.), J. Cramer, Berlin and Stuttgart, 1986, pp. 289-301.
36. Murata, K. Formation of antioxidants and nutrients in tempe. *Proc. The Asian Symposium on Non-Salted Soybean Fermentation*, Taukuba, Japan, 1986, pp. 186-198.
37. Steinkraus, K. H., Hand, D. B., Van Buren, J. P., and Hackler, L. R. Pilot plant studies on tempe. *Proc. Conference on Soybean Products for Protein in Human Foods*, USDA, 1961, pp. 75-84.
38. Van Damme, P. A., Johannes, A. G., Con, H. C., and Berends, W. On toxoflavin, the yellow poison of *Pseudomonus cocovenans*. *Rec. Trav. Chim. Pays-Bas, 79*:255-257, 1960.

39. de Bruin, J., Frost, D. J., and Nugteren, C. H. *The Structure of Bonkrek Acid,* Tetrahedron, 1973, pp. 1541-1547.
40. Su, Y.-C. Traditional fermented foods in Taiwan. *Processing of the Oriental Fermented Foods,* Taipei, Taiwan, 1980, pp. 15-30.
41. Wai, N. S. Investigation of the various processes used in preparing Chinese cheese by the fermentation of soybean curd with *Mucor* and other fungi. Final technical report, USDA, Public Law 480, Project UR-A6-(40)-1, 1968, p. 89.
42. Hesseltine, C. W. A millenium of fungi, food and fermentation. *Mycologia, 57*:149-197, 1965.
43. Wang, H., and Hesseltine, C. W. Sufu and Lao-Chao. *J. Agric. Food Chem., 18*(4):572-575, 1970.
44. Watanabe, T., Ebine, H., and Ota, T. Nyu-fu. Daizu-shokuhim (Soybean foods). Korin-Shoin, Tokyo, 1971, pp. 196-202.
45. Saono, S., Brotonegoro, S., Baski, T., Sastraatmadja, D. D., Jutono, I., Bakjre, G. P., and Ganjar, I. Indonesian tauco. In *Handbook of Indigenous Fermented Foods* (K. H. Steinkraus, ed.), Marcel Dekker, New York, 1977, pp. 479-482.
46. Winarno, F. G., Muchatadi, D., Laksani, B. S., Rahman, A., Swastomo, W., Zainnuddin, D., and Santaso, S. N. Indonesian tauco. In *Handbook of Indigenous Fermented Foods* (K. H. Steinkraus, ed.), Marcel Dekker, New York, 1977, pp. 479-482.
47. Nurhajati, S., Winarno, F. G., and Laksami, B. S. Studies on the effect of *Rhizopus oligosporus* and *R. oryzae*, and fermentation time on the quality of tauco. IFS Research Project, Bogor Agricultural University, Bogor, Indonesia, 1975.
48. Chang, C. H., Lee, S. R., Lee, K. H., Mheen, T. I., Kwon, T. W., and Park, K. I. Fermented soybean foods. In *Handbook of Indigenous Fermented Foods* (K. H. Steinkraus, ed.), Marcel Dekker, New York, 1977, pp. 482-487.
49. Lee, C. H., and Mogens, J. The effect of Korean soysauce fermentation on the protein quality of soybean. *Traditional Food Fermentation as Industrial Resources in ASCA Countries* (S. Saono, F. G. Winarno, and D. Kariade, eds.), The Indonesian Inst. of Sci., Jakarta, Indonesia, 1982, pp. 209-220.
50. Ebine, H. Miso. *Shokuno-kagaku, 56*:59-63, 1980.
51. Yokotsuka, T. Soy sauce biochemistry. *Adv. in Food Res., 30*:196-329, 1986.
52. Yokotsuka, T. Traditional fermented soybean foods. In *Comprehensive Biotechnology*, Vol. 3 (Moo-Young, ed.), Pergamon Press, Oxford, 1986, pp. 395-427.
53. Okuhara, A., and Yokotsuka, T. Difference of shoyu made from whole and defatted soybean. *J. Agric. Chem. Soc. Japan, 37*:255-261, 1963.
54. Tateno, M., and Umeda, I. Cooking method of soybeans and soybean cake as raw material for brewing. Japanese patent 204,858, Kikkoman Shoyu Co. Ltd. (1955).
55. Yokotsuka, T., Mogi, K., Fukuchima, D., and Yasuda, A. Dealing method of proteinous raw materials for brewing. Japanese patent 929,910, Kikkoman Shoyu Co. Ltd. (1966).

56. Yasuda, A., Mogi, K., and Yokotsuka, T. Studies on cooking method of proteinous materials for soy sauce brewing (I). *Seasoning Sci.*, *20*:20-24, 1973.
57. Murakami, H. Classification of koji molds and its meaning. *Seasoning Sci.*, *20*:2-14, 1973.
58. Terada, M., Hayashi, K., and Mizunuma, T. Distinction between *Aspergillus oryzae* and *Aspergillus sojae* by the productivity of some hydrolytic enzymes. *J. Japan Soy Sauce Res. Inst.*, *6*:75-81, 1980.
59. Terada, M., Hayashi, K., and Mizunuma, T. Enzyme productivity of species of koji-molds for soy sauce distributed in Japan. *J. Japan Soy Sauce Res. Inst.*, *7*:158-165, 1981.
60. Nakadai, T. The roles of enzymes produced by shoyu-koji-molds. *J. Japan Soy Sauce Res. Inst.*, *11*:67-79, 1985.
61. Ishii, S., Ogami, T., and Yokotsuka, T. Effect of degrading enzymes of plant tissue on shoyu production. *J. Agric. Chem. Soc. Japan*, *46*:340-354, 1972.
62. Ebine, H. Application of *Rhizopus* for shoyu manufacturing. *Seasoning Sci.*, *15*:10-18, 1968.
63. Harayama, F., and Yasuhira, H. Application of the genus *Rhizopus* for miso manufacturing (1). *Brew. Soc. Japan*, *80*:281-286, 1985.
64. Nasuno, S., and Ohara, T. Improvement of koji-mold for soy sauce production (2-3). *Seasoning Sci.*, *19*:32-39, 41-47, 1972.
65. Ushijima, S., and Nakadai, T. Breeding of *Aspergillus sojae*, a shoyu koji mold, by protoplast fusion. *Pro. Brew Soc. Ferment. Technol. Japan*, 1983, p. 260.
66. Furuya, T., Ishige, M., Uchida, K., and Yoshino, H. Koji mold breeding by protoplast fusion for soy sauce production (I). *J. Agric. Chem. Soc. Japan*, *57*:1-8, 1983.
67. Furuya, T., Noguchi, K., Miyauchi, K., and Uchida, K. Derivation of mutants with low α-amylase activity from *Aspergillus oryzae*. *Nippon Nogeikagaku Kaishi*, *59*:605-611, 1985.
68. Yamamoto, K. Studies on koji (3). *Bull. Agric. Chem. Soc. Japan*, *21*:319-324, 1957.
69. Yoshii, T. Fermented food and water activity. *J. Ferment. Assoc.*, *74*:213-218, 1979.
70. Ohnishi, H. Shoyu brewing and microorganisms (1-3). *J. Japan Soy Sauce Res. Inst.*, *5*:28-34, 78-82, 129-134, 1979.
71. Tamagawa, Y., Yamade, K., Kodama, K., and Suga, T. *Proc. Ann. Meeting on Ferment. Technol.*, 1975, p. 212.
72. Terasawa, M., Kadowski, K., Fujimoto, H., and Coan, M. Studies on lactic acid bacteria in shoyu-mash (2). *J. Japan Soy Sauce Res. Inst.*, *5*:15-20, 1979.
73. Fujimoto, H., Aiba, T., and Goan, M. Studies on lactic acid bacteria in shoyu-mash (3). *J. Japan Soy Sauce Res. Inst.*, *6*:5-9, 1980.
74. Uchida, K. Multiplicity of soy pediococci carbohydrate fermentation. *J. Gen. Appl. Microbiol.*, *28*:215-223, 1982.
75. Kanbe, C., and Uchida, K. Selection of lactobacilli having strong reducing power. *Proc. 16th Symp. on Brew., Brew. Soc. Japan*, 1984, p. 37.

76. Uchida, K., and Kanbe, C. Isolation of bacteriophase for shoyu-lactobacilli, *Pediococcus halophylus*. *Ann. Meet. Agric. Chem. Soc. Japan. Proc.*, 1987, p. 138.
77. Yokotsuka, T., Iwasa, S., Fujii, S., and Kakinuma, T. The role of glutaminase in shoyu brewing. *Proc. Ann. Meet. Agric. Chem. Soc. Japan*, Sendai, Japan, April 1, 1972, p. 333.
78. Yokotsuka, T., Iwasa, T., and Fujii, S. Studies on high temperature digestion of shoyu koji. *J. Japan Shoyu Res. Inst.*, *13*:18-25, 1987.
79. Yokotsuka, T. Microbial interactions in the Japanese shoyu and miso fermentations. *Proc. 4th International Symposium on Microbial Ecology*, Ljubljana, Yugoslavia, 1986, pp. 309-318.
80. Tanaka, T., Saito, N., and Yokotsuka, T. Relation between chemical factors and preference of soy sauce. *Seasoning Sci.*, *16*:21-26, 1969.
81. Tanaka, T., Saito, N., Nakajima, T., and Yokotsuka, T. Studies on sensory evaluation of soy sauce (4). *J. Ferment. Technol.*, *47*:137-145, 1968.
82. Yokotsuka, T., Sasaki, M., Nunomura, N., and Asao, Y. Flavor of shoyu (1,2). *J. Brew. Soc. Japan*, *75*:516-522, 717-728, 1980.
83. Yokotsuka, T. Recent advances in shoyu research. In *The Quality of Foods and Beverages*, Vol. 2 (G. Charalambous and G. Ingrett, eds.), Academic Press, New York, 1981, pp. 171-196.
84. Akabori, S., and Kaneko, T. On the flavor components of shoyu (2). *J. Chem. Soc. Japan*, *57*:832-836, 1936.
85. Yokotsuka, T. Studies on flavor substances in soy (9). *J. Agric. Chem. Soc. Japan*, *27*:276-281, 1953.
86. Asao, Y., and Yokotsuka, T. Studies on flavor substances in soy (17). *J. Agric. Chem. Soc. Japan*, *32*:622-628, 1958.
87. Nunomura, N., Sasaki, M., Asao, Y., and Yokotsuka, T. Isolation and identification of 4-hydroxy-2(or 5)-ethyl-5(or 2)-methyl-3(2H)-furanone, as a flavor component in shoyu. *Agric. Biol. Chem.*, *40*:491-495, 1976.
88. Nunomura, N., Sasaki, M., and Yokotsuka, T. Shoyu flavor components; acidic fractions and the characteristic flavor component. *Agric. Biol. Chem.*, *44*:339-351, 1980.
89. Nunomura, N., Sasaki, M., Asao, Y., and Yokotsuka, T. Shoyu volatile flavor components; basic fraction. *Agric. Biol. Chem.*, *42*:2123-2128, 1980.
90. Linardon, R., and Philippossian, G. Volatile components of soya hydrolyzate (1). Identification of some lactones. *Z. Lebensm. Unters.*, *167*:180-185.
91. Yokotsuka, T., Sakasai, T., and Asao, T. Studies on flavor substances in shoyu (25). *J. Agric. Chem. Soc. Japan*, *41*:428-433, 1967.
92. Asao, Y., Sakasai, T., and Yokotsuka, T. Studies on flavor substances in shoyu (26). *J. Agric. Chem. Soc. Japan*, *41*:434-441, 1967.
93. Mori, S., Nunomura, N., and Sasaki, M. A specific combination of flavor components influencing the odor preference of soysauce assessed by sensory evaluation. *Proc. Ann. Meet. Agric. Chem. Soc. Japan*, 1982, p. 36.

94. Mori, S., Nunomura, N., and Sasaki, M. The influence of 4-ethyl-2-methoxyphenol and 3-(methylthio)-1-propanol on the flavor of soy sauce. *Proc. Ann. Meet. Agric. Chem. Soc. Japan*, 1973, p. 236.
95. Ohloff, G. Importance of minor components in flavors and fragrance. *Perfume. Flavor*, *3*:11-22, 1978.
96. Sasaki, M., and Nunomura, N. Flavor components of topnote of shoyu. *J. Chem. Soc. Japan*, *55*:736-745, 1981.
97. Okuhara, A., Saito, N., and Yokotsuka, T. Color of soy sauce (6, 7), *J. Ferment. Technol.*, *49*:272-287, 1971; *50*:264-272, 1972.
98. Yokotsuka, T. Chemical and microbial stability of shoyu. In *Handbook of Food and Beverage Stability* (G. Charalambous, ed.), Academic Press, New York, 1986, pp. 518-619.
99. Hashiba, H. Oxidative browning of soy sauce. *J. Japan Soy Sauce Res. Inst.*, *7*:19-23, 116-120, 121-124, 1981.
100. Hashiba, H., and Abe, K. Oxidative browning and its control with the emphasis on shoyu. *Proc. Ann. Meet. Agric. Chem. Japan*, April 4, 1984, p. 708.
101. Imai, S. Current application of microbes in miso fermentation. *Sci. and Technol. of Miso*, *35*:334-348, 1987.
102. Hesseltine, C. W., Shotwell, O. L., Ellis, J. I., and Stublefield, R. D. Aflatoxin formation by *Aspergillus flavus*. *Bacteriol. Rev.*, *30*: 795-805, 1966.
103. Aibara, K., and Miyaki, K. Qualitative and quantitative analysis of aflatoxin. *J. Agric. Chem. Soc. Japan*, *39*:86, 1965.
104. Yokotsuka, T., Sasaki, M., Kikuchi, T., Asao, Y., and Nobuhara, A. Production of fluorescent compounds other than aflatoxins by Japanese industrial molds. In *Biochemistry of Some Foodborne Microbial Toxins* (R. Mateles and G. N. Wogan, eds.), MIT Press, Cambridge, 1966, pp. 131-152.
105. Yokotsuka, T., and Sasaki, M. Risks of mycotoxin in fermented foods. In *Indigenous Fermented Food of Non-Western Origin* (C. W. Hesseltine and Hwal Wang, eds.), J. Cramer, Berlin, Stuttgart, 1986, pp. 259-287.
106. Shinshi, E., Manabe, M., Goto, T., Misawa, S., Tanaka, K., and Matsuura, S. Studies on the fluorescent compound in fermented foods (7). *J. Japan Soy Sauce Res. Inst.*, *10*:151-155, 1984, and *Proc. 21st Ann. Meet.*, 1985, p. 9.
107. Raper, K. B., and Fennel, D. I. *The Genus Aspergillus*, Williams & Wilkins, Baltimore, 1965, p. 36.
108. Murakami, H. Classification of koji molds. *J. Soc. Brew. Japan*, *66*:658-662, 759-762, 859-863, 866-969, 1042-1045, 1150-1153, 1971.
109. Curtzman, C. *Int. Mycol. Conf. 34d*, Tokyo, Sept., 1983, p. 153.
110. Lin, J. Y., Wang, H. I., and Yen, Y. C. Mutagenicity of soybean sauce. *Consumer Toxicol.*, *17*:329-331, 1978.
111. Nagao, M. Identification of new premutagenic compound isolated from shoyu. Keidanren-kaikan, 1984, Jan. 17.
112. Yen, G. C., and Lee, T. C., Ann. Meet., IFT, 1984.
113. Shibamoto, T. Possible mutagenic constituents in nitrite-treated soy sauce. *Food Chem. Toxicol.*, *21*:745-747, 1983.

114. Nagahara, A., Ohshita, K., and Nasuno, S. Relation of nitrite concentration to mutagen formation in soy sauce. *Food Chem. Toxic.*, *25*:13-15, 1986.
115. MacDonald, W. C., and Dueck, J. W. Long term effect of shoyu on the gastric mucusa of the rat. *J. Natl. Cancer Inst.*, *56*:1143-1147, 1976.
116. *J. Japan Soy Sauce Res. Inst.*, *14*:112, 1988.
117. Sakaguchi, K., and Yamada, K. Morphology and classification of koji molds. *J. Agric. Chem. Soc. Japan*, *20*:66-73, 141-154, 1944.
118. Maeda, H. *Tamari-shoyu and soybean-miso.* Food Res. Inst., Ministry of Agric., Forest., and Fishery, Japan, 1977, p. 2.

12

FUNGI AND DAIRY PRODUCTS

ELMER H. MARTH and AHMED E. YOUSEF* *University of Wisconsin—Madison, Madison, Wisconsin*

I. INTRODUCTION

Molds and yeasts are widely distributed in the environment. Some dairy products, particularly cheese, are such favorable substrates for many molds that production of mold-free products is an important challenge for the dairy industry. This suggests that the first mold-ripened cheese was probably a discovery rather than an invention. Desperate attempts by humans to keep molds and yeasts away from cheese and reliance on those same fungi to produce some popular varieties of cheese remind us of the wisdom of the saying, "if you cannot beat them, join them."

Several species of molds and yeasts are used to manufacture certain varieties of cheese and fermented milks, whereas some of these fungi together with numerous others can spoil ripened cheeses, fresh cheeses, yogurt, butter, and some other dairy products. Additionally, some molds can produce toxic metabolites which can get into milk (and hence dairy products) via the cow that has consumed moldy feed, or which can get into dairy products directly when molds grow on these foods.

This chapter will first consider uses of molds and yeasts in the dairy industry; included are production of rennet-substitutes, manufacture of some varieties of cheese, and use of dairy by-products. This will be followed by a discussion of spoilage and health hazards associated with consumption of dairy foods and that are caused by molds and yeast.

II. USEFUL APPLICATIONS OF FUNGI IN THE CHEESE INDUSTRY

A. Production of Rennet Substitutes and Other Fungal Enzymes

The milk coagulant obtained from the fourth stomach of young animals (calves in particular) has long been used in the manufacture of cheese. The preparation is called rennet, and the active ingredient is the proteolytic enzyme

Present affiliation: Ohio State University, Columbus, Ohio rennin (or chymosin). A worldwide shortage of rennet prompted researchers to look for and find substitutes for the animal-derived rennet. Although enzymes from numerous sources can coagulate milk, only a few are suitable for making cheese commercially. Currently, much of the cheese is manufactured using a microbial rennet. Molds of the genera *Mucor* (*M. miehei* and *M. pusillus*) and *Endothia* (*E. parasitica*) are the major sources for commercial microbial rennet [42]. The active ingredients in fungal rennets are acid proteases. In general, these rennet substitutes give cheese with organoleptic properties and chemical composition comparable to that made using calf-rennet.

Fungal enzymes used to prepare rennet substitutes differ from rennin in (a) sensitivity to changes in environmental factors such as pH, temperature, and concentration of calcium ions in milk, (b) proteolytic activity, (c) extent of retention in the curd, and (d) resistance to heat [93]. During cheese manufacture, these enzyme preparations, compared with calf-rennet, give (a) milk coagula with different rheological properties, (b) curd that expels whey at a faster rate, and (c) faster development of acidity during cheese making. Therefore, use of fungal rennet substitutes necessitates modifications in the traditional procedures used to make cheese with calf-rennet.

Enzyme preparations from fungi may be used to accelerate cheese ripening. Numerous studies were conducted to test proteolytic enzymes from fungi, and some of these attempts are reviewed. Grieve [43] used crude extracts of the yeasts *Kluyveromyces lactis* and *Saccharomyces cerevisae* to manufacture Cheddar cheese. The researcher reported an increase in the rate of proteolysis and flavor enhancement in the resulting cheese. In another study [72], crude enzyme preparation obtained from *Penicillium roqueforti, Geotrichum candidum,* and *Streptococcus faecalis* var. *liquefaciens* were used to prepare Crescenza cheese. Cheese made with each of the two fungal extracts had a stronger flavor and better overall ripening than did cheese made with the bacterial extract. Other researchers, however, found extracts from *P. roqueforti* produced bitterness in cheese during ripening [45].

Lipase, another enzyme used to manufacture certain (mostly Italian) cheeses, is obtained mainly from animal sources. However, several commercial lipase preparations from fungal sources are currently available. Attempts to employ fungal lipases to accelerate ripening of certain cheeses were not always successful. For example, Long and Harper [65] found that fungal lipases improved the flavor of Italian cheeses. Sood and Kosikowski [108] made Cheddar cheese using a mixture of fungal protease and lipase. After 3 months of ripening, cheese made with fungal enzymes had a better flavor and contained more soluble proteins and free fatty acids than did the control cheese. In another study, a commercial preparation of lipase from *Aspergillus niger* produced flavor and texture defects in Ras cheese during its ripening [30].

B. Cheese Ripening

1. Surface-Ripened Cheeses

Camembert cheese: Camembert is an example of cheese that is made with a mold developing only on the surface rather than throughout the mass of cheese as happens with blue cheese. Apparently, cheese ripened through the action of mold growth on the surface has been produced in France for centuries. According to Kosikowski [58], it was in 1791 that Marie Harel, who lived in the village of Camembert in Normandy, developed a product similar to the present Camembert cheese.

The typical Camembert cheese is about 11 cm in diameter, 2.5-3.8 cm thick, and weighs 225-250 g. The interior is light yellow and waxy, and creamy or almost fluid in consistency, depending on the degree of ripening [122]. The rind is a thin feltlike layer of mold mycelium and dried cheese. The mold is gray-white in color; sometimes bacterial growth occurs on the surface of the cheese and results in development of areas that are reddish-yellow in color.

The basic procedure to manufacture Camembert cheese includes the following steps.

1. Pasteurized whole milk with about 3.5% milkfat is adjusted to about 32°C and is inoculated with 2% of an active lactic starter culture (*Streptococcus lactis* and/or *S. cremoris*) plus a sporulated culture of *Penicillium camamberti* (or *P. caseicolum*). Alternatively, spores of the mold can be applied to the surface of the cheese later in the manufacturing process.
2. Annatto (yellow coloring) may be added to the milk.
3. Inoculated milk is allowed to "ripen" for 15-30 minutes so that a titratable acid of 0.22% develops. Rennet extract (or other suitable coagulant) is added, and the milk is stirred and then held quiescently until a firm curd develops.
4. The curd is cut into cubes with 1.6-cm knives. Alternatively, uncut curd can be ladled into hoops.
5. Curd is not cooked but is placed into open-ended, round, perforated, stainless steel molds or hoops. Filled hoops are allowed to drain for about 3 hours at about 22°C; no pressure is applied to the cheese during draining.
6. Hoops of cheese are turned and draining continues. The turning process is repeated 3-4 times at 30-minute intervals.
7. Both flat sides of curd in hoops may now be inoculated by spraying the surface with a fine mist of *P. camemberti* (or *P. caseicolum*) spores suspended in water.
8. After an hour, cheese is removed from hoops, placed on a drain table, and held at 22°C for 5-6 hours. Weights are generally not placed on the cheese.
9. Dry salt is applied to the surface of cheese, which is then held overnight at about 22°C.
10. Cheese is held for 1 or 2 weeks at 10-15°C and 95-98% relative humidity. It may be turned once during storage to facilitate uniform development of mold on the surface.
11. Cheese is moved to storage at 4-10°C after being wrapped in foil. Storage under these conditions may last several weeks before the cheese is packaged and moved into distribution channels. Final ripening occurs during distribution.

Camembert cheese should be consumed within 6-7 weeks after it is made. The process used to make Camembert cheese has been mechanized in the United States and Europe.

Common defects associated with Camembert cheese include (a) early gas production during drainage of the cheese, and (b) growth of undesirable "wild" molds on the surface. The first defect can be minimized or eliminated by adequate sanitation and use of high quality milk. The second can be controlled by maintaining adequate humidity in the room where cheese is ripening; wild molds tend to develop on cheese when its surface becomes too dry for normal development of *P. caseicolum.*

Cheeses similar to Camembert: Walter and Hargrove [122] and Weigmann [132] described several kinds of cheese other than Camembert that are ripened largely through the activity of surface mold. Brie is probably the best known of these Camembert-like cheeses. Brie is made in three sizes: (a) large—about 40 cm in diameter and 3.8-4.2 cm thick, about 2.7 kg; (b) medium—about 30 cm in diameter and somewhat thinner than the large size, about 1.6 kg; and (c) small—14 to 20 cm in diameter and 3.2 cm thick, about 0.45 kg. The difference in size between Brie and Camembert causes differences in the ripening process of the two cheeses. This, together with variations in the manufacturing process, cause the flavor and aroma of Brie to differ from those of Camembert.

Coulommiers is a cheese similar to the small Brie but is ripened for less time. Another cheese similar to Brie is Monthery, which is made in two sizes roughly equivalent to the large-sized and medium-sized Brie. Monthery can be made from whole or partially skimmed milk. Melum is similar to the small Brie but has a firmer body and sharper flavor. This cheese also is designated as Brie de Melum. Other cheeses similar to Camembert and primarily produced in France include Olivet and Vendome. The latter is sometimes buried in ashes in a cool, moist cellar during ripening.

Brick and similar cheeses: Elsewhere in this chapter it has been indicated that yeasts develop on the surface of some mold-ripened cheeses. Additionally, yeasts are an important component of the microflora on the surface of some types of cheese that ripen without the aid of molds. Principal examples of such cheeses include brick and Limburger. Brick cheese will be discussed first, and then other similar kinds of surface-ripened cheese will be mentioned.

Olson [84] has described two methods that are generally used to make brick cheese. The following steps are involved in the first method.

1. Pasteurized whole milk at 32°C is inoculated with *Streptococcus thermophilus.* Alternatively, a combination of *S. thermophilus* and *S. cremoris* or *Lactobacillus bulgaricus* may be used.
2. After brief incubation, rennet or another suitable coagulant is added to the milk; the resultant curd is cut into 0.64-cm or 0.95-cm cubes and cooked at 38-45°C.
3. After the curd is sufficiently firm, enough whey is drained so that about 2.5 cm remains above the curd surface.
4. Curd and whey are dipped or pumped into rectangular hoops held on perforated screens.
5. Hoops of curd are allowed to drain for 6-18 hours and are turned at intervals. Weights can be placed on the cheese during the draining process.

6. Blocks of cheese are removed from hoops and immersed in brine containing 22% sodium chloride. Alternatively, salt can be applied directly to the exterior of cheese.
7. After 24-36 hours in brine, cheese is removed and placed in a room at 15°C for 4-10 days. During this time the "smear" develops on the surface of the cheese.
8. The smear is washed off, the cheese is waxed or packaged in plastic wrap, and is ripened for 4-8 weeks at about 4°C. More flavor can be obtained by leaving the smear intact for the entire ripening period. Some cheese marketed as brick is mild in flavor because little or no surface smear is allowed to develop on the product.

The second or "sweet curd" method for making brick cheese employs *S. lactis* and/or *S. cremoris* as the starter culture and addition of water to the curd-whey slurry to control development of acid in the cheese. Some other modifications in manufacturing must be made to ensure that the pH is 5.1-5.2 in 3-day-old cheese. Salting and ripening proceed as outlined in the preceding description for making brick cheese.

Brick cheese contains not more than 44% moisture, and at least 50% of the total dry matter must be milk fat. A typical brick cheese is about 12 cm wide, 25 cm long, and 7.5 cm thick; it weighs about 2.25 kg.

Olson [84] described the major defects that can appear in brick and other surface-ripened cheeses. Included are defects in flavor, body and texture, and surface microflora.

Flavor defects include:

1. Sour or acid—caused by excessive fermentation of lactose of inadequate washing of curd, or both. Too much acid retards development of surface microflora, and too little acid results in fruity and gassy cheese.
2. Bitterness—caused by abnormal protein degradation when the starter culture contains undesirable lactic acid bacteria such as *S. faecalis* var. *liquefaciens*.
3. Flat—caused by insufficient growth of surface microflora.
4. Fruity and fermented—caused when the pH of cheese is high and the salt content is low so that anaerobic spore-forming bacteria can grow.

Defects in body and texture include

1. Corky—caused by inadequate acid development or excessive washing of curd, or both.
2. Weak or pasty—caused by a combination of excessive moisture, too much or too little acid, and inadequate salt.
3. Mealy—too much acid.
4. Openness—caused by whey being trapped between firm cubes of curd; openings remain when whey drains from cheese.
5. Gassiness—caused by growth in the cheese of coliform bacteria, yeasts, certain strains of lactic acid bacteria, *Bacillus polymyxa*, or anaerobic spore-forming bacteria.
6. Split cheese—caused by gas from anaerobic sporeformers.

Defects in the surface microflora include

1. Lack of growth—caused by low temperatures during ripening, too much salt in cheese, or drying of the surface of cheese.
2. Mold growth—results when the surface smear fails to develop because the surface of cheese is too dry.

Other varieties of surface-ripened cheese: Limburger cheese has up to 50% moisture and is made by the procedures used for brick cheese. The initial ripening at 16°C and at a high relative humidity is longer than for brick cheese, thus allowing extensive growth of *Brevibacterium linens*. Surface growth is not removed when cheese is wrapped and moved to storage at 4-10°C. Extensive growth of *B. linens* on relatively small pieces of cheese accounts for the strong, pungent flavor and aroma of Limburger.

Port du Salut, Trappist, and Oka are wheel-shaped cheeses developed by Trappist monks and made by procedures somewhat similar to those for brick cheese. *Geotrichum* may appear in the surface microflora and contribute a distinctive flavor to the cheese. Other varieties of surface-ripened cheese that are more common in Europe than in the United States include Saint Paulin, Bel Paese, Königkäse, Bella Alpina, Vittoria, Fleur des Alpes, Butter, and Tilsit.

2. Blue-Veined Cheese

Blue cheese: Blue is an example of cheese that is ripened primarily by growth and activity of mold throughout the cheese mass rather than on the surface only, as is true of Camembert and related varieties. Blue cheese was not made successfully in the United States until about 1918; information on appropriate procedures for making this cheese was not available earlier [122]. A blue cheese is about 19 cm in diameter and weighs about 2-2.3 kg; it is round with a flat top and bottom.

The following steps [48] are involved in producing blue cheese.

1. Whole milk from cows is separated into cream and skim milk fractions, and the skim milk is pasteurized.
2. Cream is bleached by adding benzoyl peroxide (maximum 0.002% of the weight of milk), pasteurized, and homogenized. Bleaching is done so that the finished cheese is white (except for mold growth) in color and homogenization increases the surface area of milk fat globules and thus facilitates lipolytic action that occurs during ripening of the cheese.
3. Cream and skim milk are combined and 0.5% of an active lactic (*S. lactis* and/or *S. cremoris*) starter culture is added.
4. Inoculated milk is held at 30°C for 1 hour to allow some acid production.
5. A suitable coagulant is added, milk is allowed to coagulate, and the resultant curd is cut into cubes with 1.6-cm wire knives.
6. Curds remain in whey for about 1 hour while additional acid develops. Curds and whey are then heated to 33°C, held briefly, and whey is drained from the curds.
7. Curd is trenched and inoculated with *P. roqueforti* spores; salt also is added and then the curd is stirred. Spores can be obtained in powdered form from commercial culture firms.
8. Curd is placed into stainless steel blue cheese hoops.

9. Hoops of curd are turned once every 15 minutes for 2 hours and then are allowed to drain overnight at 22°C.
10. The next day curd is removed from the hoop and salt is applied to the surfaces of the cheese. Cheese is then stored at 16°C and 85% relative humidity and is salted once daily for 4 more days.
11. After salting is completed, each flat surface of the cheese is pierced about 50 times with a suitable needlelike steel rod. This facilitates escape of carbon dioxide from the cheese and entrance of air so that growth of the mold is encouraged.
12. Pierced cheese is then stored at 10-13°C and 95% relative humidity.
13. After 1 month, surfaces of cheese are cleaned; cheeses are wrapped in foil and then stored at about 2°C for 3-4 months to allow additional ripening.

Improper development of *P. roqueforti* can cause an assortment of defects. Too much growth can result in a musty, unclean flavor or in loss of the typical flavor, whereas too little growth is accompanied by defects in color, a body that is too firm, and insufficient flavor. Growth of unwanted molds can cause defects on the surface of blue cheese.

Cheeses similar to blue: Roquefort is the original blue cheese. The designation "Roquefort" is applicable only to cheese made from ewe's milk in the Roquefort area of France. A similar product made elsewhere in France is called bleu cheese. A peculiarity of Roquefort cheese is that it ripens in a network of caves and grottoes where cool moist air moves briskly, the temperature never exceeds 10°C, and relative humidity remains at about 95% throughout the year [122].

Gorgonzola is the principal blue-mold cheese of Italy where it is claimed to have been made in the Po Valley since A.D. 879. The English blue-mold cheese is Stilton, which has been made since about 1750. Stilton is milder than Roquefort or Gorgonzola. The texture of Stilton is sufficiently open so that the cheese usually does not need piercing to facilitate mold growth.

C. Changes in Cheese During Ripening

1. Changes in Fungal Flora

Several cheese varieties have surface microflora that usually include molds, yeasts, and some aerobic bacteria such as micrococci and coryneforms. This microflora gives surface-ripened cheese its characteristic appearance and participates in cheese ripening by developing the rheological and organoleptic properties of the cheese [59]. *Penicillium camemberti* and *P. caseicolum* are used to manufacture Camembert-type cheeses. These were commonly considered as different species; the former forms a pale gray-green mycelium, whereas that of the latter remains white. Currently, there is a tendency to consider *P. caseicolum* as a white mutant of *P. camemberti*. Another mold of importance to cheese is *G. candidum*, which forms a grayish-white crust on the surface of some cheeses. Its filaments develop forked branching and form chains of arthrospores by dearticulation of laterial hyphae at the level of septa. Yeast flora in cheese is diverse, but in many cheeses *Kluyveromyces* spp. predominate. Since the environment is the source of yeasts in cheese, the yeast flora can vary with each lot of the same kind of cheese. In addition to fungal flora, several bacteria play important roles in ripening of blue-veined and surface-ripened cheeses. These include lactic acid bacteria (from milk or added starter), micrococci (from cheesemilk and brine), and corynebacteria (from cheesemilk and soil). *Brevibacterium linens* is a corynebacterium of particular importance in cheese ripening [19].

Flora of Camembert cheese: Using scanning electron microscopy, Rousseau [95] studied the surface flora of traditional Camembert. Yeast cells appeared on 2-day-old cheese, and they covered the entire surface by day 6. They formed a layer close to the cheese itself; the thickness of this layer became maximum after ca. 10 or 11 days of ripening, and then remained unchanged until the end of ripening. *Geotrichum* conidiophores were detected in 2-day-old cheese, but developed at a slower rate than yeasts. By days 5 and 6, *Geotrichum* formed a highly ramified mycelium. During the third week of ripening, *Geotrichum* could be observed in the mass of degenerated *Penicillium* mycelium. Although *Geotrichum* seems to disappear from the surface by the third week of ripening, a layer (200-400 μm thick) of mold mycelium and arthrospores appears between the yeast layer (next to cheese curd) and the *Penicillium* layer (on the rind surface). A thick mycelium of *Penicillium* was observed after 6 days of ripening. Bundles or cords of mycelium were formed during the second and third weeks of ripening. After 3 weeks of ripening, a thick surface layer (450-600 μm) of *Penicillium* mycelium was observed, with conidiophores embedded in cheese curd. Some degenerating *Penicillium* was seen at this stage of ripening. Further degeneration was observed during the fourth and fifth weeks of cheese ripening. After 3 weeks of ripening, micrococci and *B. linens*-like bacteria from the environment were attached to the surface of mycelia and then migrated to the curd surface.

Examination of 500 yeast isolates from Camembert cheese obtained from five factories in Normandy, France, showed that the basic yeast flora is composed of *Kluyveromyces* sometimes in association with *Debaromyces* and *Saccharomyces*. Numbers of yeasts on the cheese surface were 10^6-10^7/g on the third day of ripening and reached a maximum of 5×10^8-10^9 by day 10. Yeast numbers in the cheese center were ca. 100-fold smaller but changed in a pattern similar to those on the surface [59].

For the first few days after manufacture of Camembert cheese, lactic acid bacteria represent almost the entire flora of the curd. Their numbers reach ca. 10^9/g and remain almost constant throughout ripening [19].

Flora of brick cheese: Yeasts predominate in the surface microflora of brick cheese during the initial stages of ripening [84]. This is because of their ability to grow at the temperature and relative humidity (approximately 95%) used for ripening as well as the low pH and high concentration of salt at the surface of the cheese. Depending on water activity of the cheese, yeasts present may be from one or more of the following genera: *Debaryomyces*, *Rhodotorula*, *Trichosporon*, and *Candida*.

Growth of yeasts serves to modify the surface of cheese so that *B. linens* and micrococci can grow. This is accomplished by metabolizing lactic acid and thus raising the pH of cheese at its surface above the minimum for growth of the bacteria. Additionally, yeasts produce vitamins which may enhance growth of the bacteria. Growth of yeasts also may contribute to the final flavor of brick cheese.

Brevibacterium linens and micrococci (*Micrococcus varians, M. caseolyticus*, and *M. freudenreichii*) develop after sufficient growth of yeasts has occurred. These bacteria release proteolytic enzymes that are largely responsible for producing the characteristic flavor of brick cheese that has had a surface smear develop during ripening.

Flora of blue cheese: In blue cheese, the high acidity and the amount of salt in the cheese cause a rapid demise of the lactic starter bacteria so that only a few viable cells remain in cheese that is 2-3 weeks old [35]. Growth of microorganisms can occur on blue cheese after 2-3 weeks of ripening. This growth consists of yeasts, micrococci, and *Brevibacterium* spp. and contributes to the final flavor of the cheese. If cheese is waxed, microorganisms will not grow to produce the surface slime.

Washam et al. [121] used scanning electron microscopy to monitor changes in the body structure and growth of mold during maturation of blue cheese. Five-day-old cheese had a porous, coral-like structure that changed to a globular appearance in 13-day-old cheese. Crevices and holes of various sizes that were present throughout the body of the 5-day-old cheese could serve as channels for development of blue veins and passages for diffusion of oxygen and carbon dioxide. Germinating mold conidia were seen in 5-day-old cheese. At 13 days of ripening, hyphae were evident along the walls of large crevices and holes, but tended not to penetrate into curd particles. This illustrates the importance of cheese structure in governing the veining of the cheese. Few conidia were detected in 13-day-old cheese, but they were abundant in 21-day-old cheese. In the latter cheese, conidia were oriented toward the center of crevices and cracks and were supported by a dense mat of hyphae that covered curd surfaces. Blue-green veining was seen in 21-day-old cheese. Cheeses after 55 and 65 days of ripening contained detached conidia that were in contact with the cheese matrix. Some of the hyphae were fragmented and detached from the cheese body.

Flora of other mold-ripened cheeses: Another example illustrating the diversity and complexity of the flora of mold-ripened cheese comes from a study on the Spanish mold-ripened Cabrales cheese [81]. The composition of the yeast and mold flora during making and ripening of that cheese was investigated. Species isolated from milk and curd were *Pichia fermentans, Saccharomyces unisporus, Trichosporon capitatum*, and *G. candidum*. During the early ripening period (5-15 days), the predominant yeasts were *Pichia membraneafaciens* in the interior and *Torulopsis candida* on the cheese surface. These yeasts metabolized lactic acid, but not lactose, so the pH of the cheese increased and thus yeasts contributed to the ripening process. Predominant molds during this period were *G. candidum* and *P. roqueforti*. During the rest of the ripening period (cave ripening, 16-120 days), most of the yeast flora consisted of *P. membraneafaciens* and *P. fermentans* in the interior, and the salt-tolerant *Debaryomyces hansenii* and *T. candida* on the cheese surface. Predominant molds were *P. roqueforti* inside, and *P. roqueforti* and *P. frequentans* on the cheese surface.

2. Chemical Changes

Some of the important biochemical capabilities of molds and yeasts used to manufacture cheese and manifestations of these capabilities during cheese ripening are described in the following paragraphs [19,59].

P. camemberti and P. roqueforti: When the proteolytic activity of *P. camemberti* was investigated, two extracellular endopeptidases were identified. One of these enzymes was a metaloprotease, and the other was an acid protease.

The metaloprotease is the principal proteolytic enzyme in cultures grown at pH 6.5. The optimum pH for activity on casein is ca. 6.0; α_{s1}-casein was preferentially degraded rather than β- or κ-casein. The enzyme is similar to the metaloproteases of *P. roqueforti* and *Aspergillus sojae*.

Acid protease is the principal proteolytic enzyme in cultures grown at pH 4.0. The optimum pH for activity on casein is ca. 5.0. α_{s1}-, β-, and κ-casein are degraded by this enzyme, but the enzyme has different pH optima for different casein fractions. At pH 5.0, α_{s1}-casein was preferentially degraded rather than β- or κ-casein. The enzyme is very similar to acid proteases of *P. roqueforti* and *P. janthinellum*.

The two enzymes are more active on isoelectric than on rennet or native casein. *Penicillium camemberti* also possesses exopeptidases including aminopeptidase (optimum pH 4-7) and carboxypeptidase (optimum pH 7.5-8.5) activities. Similar extracellular exopeptidases were produced by *P. camemberti*.

A study on *P. camemberti* showed that strains varied considerably in their capacity to produce extracellular lipolytic enzymes, but all strains had the same alkaline lipase system. Although the lipase system had an optimum pH of 8.5-9.5, it still retained 45% of its maximum activity at pH 5.5. The optimum temperature for activity was 35°C, but the enzyme retained 50% of its maximum activity at 1°C. *Penicillium roqueforti* produces an alkaline (optimum pH 7.5-8.0) and an acid (optimum pH 6.5) lipase. Contrary to what was found with *P. camemberti*, proteolytic and lipolytic activities of *P. roqueforti* are negatively correlated. Both *P. camemberti* and *P. roqueforti* can oxidatively degrade fatty acids. A positive correlation was found between the oxidative and lipolytic activity of *P. camemberti*.

P. camemberti uses lactic acid as soon as it starts growing on Camembert cheese (on the fifth and sixth day of ripening). This increases the pH of the cheese surface, while changes in pH of the cheese interior lag behind.

The following are possible manifestations of the biochemical activities of *P. camemberti* and *P. roqueforti* during ripening of cheese.

Although some researchers believe that proteolytic enzymes of *P. camemberti* contribute only indirectly to the ripening of Camembert cheese (see later discussion), others (e.g., Ref. 59) suggest a major role for these enzymes in cheese ripening. Evidence for such involvement includes:

Differences in the profile of nitrogenous matter in the internal and external parts of cheese.

Correlation between growth of mold on cheese and proteolytic activity at the surface of cheese compared with that at the center.
Degradation of β-casein seems to be related mainly to the activity of the fungal acid protease.

P. camemberti and *P. roqueforti* may be considered the principal agent in the pronounced lipolysis of milk fat that occurs in mold-ripened cheeses. Oxidation of milk fat by enzymes of these molds yields methyl ketones, which contribute to the aroma of this type of cheese.

P. camemberti consumes lactose and lactic acid on the surface of Camembert cheese. This raises the pH of the cheese surface and thus affects cheese appearance and texture, increases activity of enzymes involved in ripening (proteolytic, lipolytic, and oxidative), and enhances development of acid-sensitive bacterial flora on the surface (e.g., micrococci and coryne-bacteria) that are important for production of cheese aroma.

G. candidum: This mold is part of the surface flora of several varieties of cheese, including Camembert. The primary biochemical capabilities of this mold include metabolism of lactic acid, proteolysis, and lipolysis. In a study that included 80 strains from various sources, isolates from soft cheeses generally grew more rapidly and resulted in greater proteolytic activity than those from hard cheeses. Proteolytic activity of the mold is comparable to that of *P. camemberti*. Two extracellular endopeptidases capable of acting in cheese and intracellular endopeptidase activity were identified.

In a study Camembert cheese was made with *Penicillium* (control), *Geotrichum*, or both. Cheese made with *Geotrichum* alone was slower in ripening, had less proteolysis, and a smaller increase in pH than did the control and had a pronounced yeast taste. Cheese made with both molds (compared with the control) ripened faster and had a greater increase in proteolysis and pH at the surface.

Yeasts: Yeasts may use lactose during cheese making and the first few days of ripening and produce CO_2 that forms small holes in the curd. Formation of these holes in Roquefort cheese may be desirable since they assist the implantation of the mold in the cheese matrix. Yeasts also can metabolize lactic acid, thus increasing the pH of the cheese surface. In Camembert, brick, and Limburger, this deacidification of the surface permits growth of acid-sensitive bacteria such as micrococci and coryneforms.

Proteolytic and lipolytic activities were detected in yeasts isolated from cheeses. Isolates of *Candida valida* and *D. hansenii* produced lipases with optimum pH of 4.5. *Candida lipolytica* is known for its high lipolytic activity. Endocellular proteolytic activity was detected in several yeasts including *Debaryomyces*, *Kluyveromyces*, and *Candida*. High proteolytic activity was detected in *Kluyveromyces marxianus* var. *marxianus*. This yeast produces an acid and two alkaline proteases that retain more than 50% of their maximum activity at pH 5.5. Exopeptidase (especially aminopeptidase) activity was high in yeasts of importance to cheese making.

3. Texture and Flavor Development

Many researchers believe that textural development in surface mold-ripened cheese is caused by the proteolytic enzymes produced by the mold and other microflora involved in ripening [24,57,59]. Recently, evidence that this theory may not be accurate was reported. Alkaline phosphatase in high moisture (60%) cheese of the Meshanger type had very limited ability to migrate into the cheese. The diffusion coefficient (D') was ca. 0.003 cm^2/day [79]. The researcher found that proteolytic enzymes from *P. caseicolum* penetrated only ca. 6 mm into Camembert cheese in 40 days. This corresponds to a D' value of 0.001 cm^2/day, which is 100 times smaller than D' for diffusion of NaCl in a similar cheese.

During ripening of Camembert cheese, softening starts at the outside and gradually extends into the interior of the cheese. Noonan [79] explains this observation on the bases of (a) diffusion of ammonia formed by surface mold into the cheese body, and (b) action of residual rennet on α_{s1}-casein. To prove these hypotheses, the researcher treated acid cheese (made without rennet) with ammonia. Cheese became rubbery, with no signs of softening. Another cheese that was made with rennet and similarly treated with ammonia completely liquified in a few hours after the treatment. In this second cheese, α_{s1}-casein was almost completely degraded by residual rennet. Similar observations were reported by de Jong [22]. In the soft Meshanger cheese, softening occurs only in cheeses with degradation of α_{s1}-casein, which can be attributed to action of residual rennet in the cheese [22]. The researcher found a linear relationship between the concentration of residual rennet in the cheese and decomposition of α_{s1}-casein during ripening of Meshanger cheese.

Spread of mold mycelium throughout some cheeses (e.g., blue) may enhance the role of fungal proteolytic enzymes in ripening of such cheeses. Marth [67] indicated that proteolytic enzymes from the mold act to soften the curd and thus to produce the desired body of blue cheese. Hewedi and Fox [49] observed that during the first 10 weeks of ripening of blue cheese, proteolysis was limited and appeared to be caused by chymosin of the rennet. More proteolysis occurred thereafter, which may be attributed to action of proteinases/peptidases of *P. roqueforti*.

Production of numerous flavor compounds in Camembert cheese are attributed to the enzymatic activity of *P. camemberti* [82]. Methyl ketones, fatty acids, alkanols, and esters are derived from degradation of milk lipids; amines, sulfur compounds, short chain aldehydes, and amino acids are derived from the breakdown of proteins. In an experiment using a model system containing milk fat, Okumura and Kinsella [82] showed that *P. camemberti* hydrolyzed milk fat, then free fatty acids were oxidized to carbonyl compounds (mainly methyl ketones) such as 2-nonanone and 2-undecanone.

In blue and similar cheeses, *P. roqueforti* produces water-soluble lipases which hydrolyze milk fat to free fatty acids. Included are caproic, caprylic, and capric acids, which together with their salts are responsible for the sharp peppery flavor of blue cheese [32]. *Penicillium roqueforti* forms 2-heptanone from caprylic acid, and this ketone is an important component of blue cheese flavor. Other ketones (2-pentanone and 2-nonanone) have been recovered from blue cheese and probably contribute to its flavor [32]. Additionally, *P. roqueforti* reduces methyl ketones to form secondary alcohols (2-pentanol, 2-heptanol, and 2-nonanol), which also contribute to the flavor of the cheese [51]. The pH of blue cheese initially is 4.5-4.7 and increases to 6.0-6.25 after 2-3 months of ripening.

Tsugo and Matsuoka [120] detected small amounts of hydrogen sulfide and dimethyl disulfide during ripening of semi-soft white mold cheese; however, greater amounts of methyl mercaptan (0.542 ppm) were formed in this instance after 3 weeks of ripening than in Gouda-type cheese. It is generally believed that methyl mercaptan is an important flavor compound in such a cheese. The precursor of methyl mercaptan is methionine, which is produced by proteolysis during ripening of semi-soft white mold cheese. The suggested reaction for formation of methyl mercaptan from methionine is as follows:

D-methionine ↔ L-methionine → methyl mercaptan + ammonia
+ α-ketobutyric acid

Integrated models: A few investigators proposed integrated models that explain the interrelated physical, chemical, and microbial events that occur during ripening of mold-ripened cheeses. During ripening, cheese microflora and chemical composition change; at the same time texture and flavor characteristic of the fully ripened cheese develop gradually. Some of the early proposed models date back to the first decade of this century, when Dox [24] assumed that ripening of Camembert cheese is mainly the act of proteolytic enzymes produced by *P. camemberti* that grows on the cheese surface. Since ripening begins at the surface and proceeds toward the center, the author concluded that enzymes are produced by the mycelium of the mold and diffuse inward. Recent studies cast doubt on this hypothesis since enzymes produced on the cheese surface have limited capabilities to diffuse into the cheese [79].

Karahadian and Lindsay [52] investigated several processes that may contribute to the overall development of texture in mold surface-ripened cheeses. When present in a pure culture, *P. caseicolum* rapidly metabolized lactic acid and elaborated ammonia from an acidified milk-agar medium. Under similar cultural conditions, *G. candidum* also depleted lactic acid from the medium, but produced ammonia more slowly. In contrast, *B. linens* showed little or no metabolic activity, probably because of the low pH value of this medium. In Brie cheese, lactic acid decreased at a faster rate on the surface than inside the cheese wheels, but ammonia was produced and the pH increased at a faster rate on the surface than inside the cheese wheels. During ripening, calcium and phosphorus concentrations increased near the surface and decreased inside the cheese.

The model proposed by Karahadian and Lindsay [52] for textural development of surface-ripened cheese is as follows. The starter culture produces lactic acid, which decreases the solubility of casein and causes the firm, crumbly texture of fresh unripened cheese. During the first few days of ripening, *Geotrichum* and yeasts grow on the cheese surface, uninhibited by other microflora, and metabolize lactic acid but do not produce ammonia at this stage. Consequently, the cheese interior remains sour and crumbly. After about one week of ripening, *P. caseicolum* becomes dominant on the cheese surface, thus rapidly metabolizes lactic acid and produces ammonia. This causes the pH to increase at the surface, but that of the cheese interior remains unaltered. After about 2 weeks of ripening, the outer surface of cheese becomes somewhat softer and more pasty than the center, which remains firm and crumbly. The low pH inside cheese wheels favors

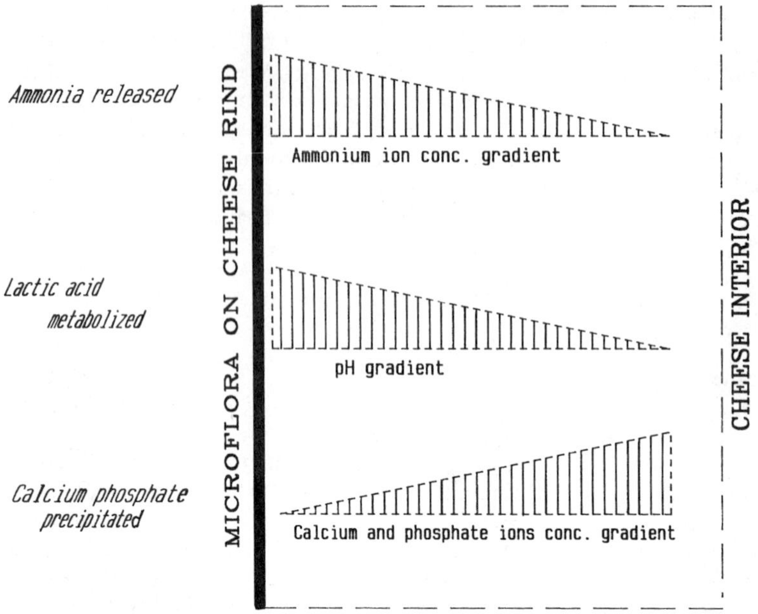

FIGURE 1. Cross section of a surface mold-ripened cheese showing the interrelated physical, chemical, and microbial factors that contribute to the ripening and textural development of this cheese. (Adapted from Ref. 52.)

solubilization of calcium salts. The pH gradient across the cheese wheel, caused by depletion of lactic acid and production of ammonia on the surface, results in a similar gradient of soluble calcium. This causes migration of soluble calcium forms from inside to the rind of the cheese. Ammonia also migrates from the surface to the center of the cheese, gradually increasing its pH. After 3 weeks of ripening, the cheese interior tastes slightly sour, and its texture becomes gel-like and soft. An increase in pH of cheese toward the center increases the solubility of casein, and solubility and migration of calcium from the center decrease calcium crosslinkings, thus softens cheese on the interior. It is likely that an increase in the pH of cheese activates residual proteolysis by milk plasmin and added rennet. If *B. linens* is excluded during cheese making, *P. casiecolum* continues to grow slowly during cheese ripening, forming a thick mat of mycelium on the cheese surface at 2-3 months of age. Surfaces of cheese sometimes are gritty from migration and precipitation of calcium near the surfaces. The mold does not produce enough ammonia to cause the body of cheese to become fluid and hence it remains gel-like. Exposure to ammonia in the head space softens the body of cheese and turns it fluid. When *B. linens* is included in cheese making, the microorganism grows and produces methyl mercaptan, which inhibits *P. casiecolum*, and also produces large amounts of ammonia, which results in a soft, runny texture in fully ripened cheese. The crust of the cheese becomes firm and gritty and contains a small amount of mold mycelia. An illustration of this model is shown in Figure 1.

In their study using scanning electron microscopy, Washam et al. [123] correlated the morphological changes of the mold flora with some of the characteristic ripening events of blue cheeses. Loaves of blue cheese are skewered (pierced) at an early stage of ripening to connect internal passages with atmospheric air, so that germinating spores can grow. At a later stage (ca. 3 weeks) mycelia proliferate, which is accompanied by a decrease in curdiness and an increase in soluble nitrogen, fatty acids, and methyl ketones. At this stage, also, the flavor of 2-alkanones is perceptible. Also at this stage, different manufacturing procedures may be followed depending on the extent of veining achieved. Some manufacturers transfer 3-week-old cheese to cool ripening rooms to slow growth of mold, whereas others skewer this cheese to increase growth of mold. After cheese reaches the legal minimum age of 60 days, it may be cut into wedges and left exposed to air for a short time before packaging. This causes the cheese to "bloom," i.e, mold color will improve and flavor will increase.

III. FUNGI AND FERMENTED MILKS

Fermented milks such as yogurt and cultured buttermilk are produced through the action of lactic acid bacteria. Since fungi are not involved in the fermentations, they will not be described in this chapter. However, some fermented milks are made with yeasts in addition to lactic acid bacteria. Kefir and kumiss are such products and will be described in the following paragraphs.

A. Kefir

This cultured milk originated in the Caucasus mountains and should not be confused with kaffir, a beer fermented from kaffir corn. Kefir can be made from the milk of goats, sheep, or cows. Although the product is not common in the United States, it is distributed widely in Europe. The finished kefir contains about 0.8% lactic acid, 1% ethanol, and carbon dioxide [58].

Kefir is made from whole milk after it has been held at 85°C for 30 minutes. The heated milk is cooled to 22°C, inoculated with kefir grains, and incubated overnight at 22°C. The resultant smooth curd is strained through a wire sieve to recover the kefir grains. The product is then cooled, after which it is ready for consumption.

Kefir grains are whitish or yellowish irregular granules that contain the mixture of microorganisms needed to culture the milk. Included are yeasts (*Candida* and *Saccharomyces* spp.), streptococci, micrococci, and bacilli; all are not necessarily useful or required for making kefir. Kefir grains recovered from a batch of kefir can be washed in clean water and stored moist at 4°C. Alternatively, they can be dried at room temperature for 36-48 hours. Dried kefir grains retain activity for 12-18 months, whereas moist grains retain activity for only 8-10 days [58].

B. Kumiss (Koumiss)

Kumiss is a fermented milk product found in Russia where it is commonly made from the milk of mares. Because of the composition of mare's milk, finished kumiss has a uniform consistency with no tendency to whey-off

[58]. Russians believe kumiss to be of value in treating tuberculosis, and a typical medical dose is 1.5 liters of kumiss daily for 2 months. Finished kumiss is naturally carbonated and contains 0.6-1.8% lactic acid and 0.7-2.5% ethanol.

Kumiss is made from fresh mare's milk warmed to 28°C. The milk is then inoculated with 30% of a starter consisting of lactic acid bacteria and *Torula kefyr* (*K. marxianus*). The culture is prepared by growing each organism separately in cow's milk, then adding both to mare's milk and after suitable incubation adding more mare's milk so that after 4 days the volume of culture is sufficient to serve as inoculum.

The inoculated mare's milk is agitated to facilitate growth of yeast. Approximately 2 hours of incubation yields a product with sufficient acidity. Kumiss is bottled warm, capped, and incubated for an additional 2 hours at 20°C. It is then stored at 4°C and distributed promptly.

A procedure to produce kumiss from cow's milk has been described by Puhan [91]. Skim milk or partially skimmed milk is pasteurized and then cooled to 26-28°C. The milk is inoculated with 30% of the yeast-lactic acid bacteria culture and then is stirred for 15 minutes so the milk is saturated with air. Stirring is necessary to facilitate growth of the yeast. Incubation at 26-28°C continues until the pH of the milk reaches 4.5-4.7, when the coagulum is broken through stirring, with the stirring being continued until the fermented milk is again saturated with air. At this time the product is filled into bottles with narrow necks, and each bottle is sealed with a crown cap. The bottled kumiss is held at 20°C for 2 hours, then is cooled to 4-6°C, and held refrigerated until consumed. The finished product is claimed to have characteristics similar to those of kumiss made from mare's milk.

IV. FUNGI IN THE USE OF DAIRY BY-PRODUCTS

A. Production of Yeast from Whey

Whey can serve as the substrate for production of food- and feed-grade yeast and yeast-whey products. These products, if properly produced, are highly nutritious, nontoxic sources of protein and vitamins that, under the right circumstances, can find application in human and animal nutrition. Worldwide protein deficits make production of yeast from whey particularly attractive since a much-needed food supplement could be made from a substrate that is often used inefficiently or wasted. Unfortunately, available whey is seldom located in those areas of the world that suffer from acute protein shortages. In addition to transportation problems, difficulties associated with acceptance and palatability of yeast products have not been resolved. Furthermore, consumption of excessive amounts of purines and pyrimidines from yeast can lead to high concentrations of uric acid in blood and thus to gout.

1. The Organism

Undoubtedly *K. marxianus* var. *lactis* (*K. fragilis*) is most widely recognized as the organism of choice for producing yeast from whey, although others have been suggested. Myers and Weisberg [76] in their early work found *K. marxianus* var. *lactis* suitable to make delactosed whey. Other investigators who have found *K. marxianus* var. *lactis* suitable for growth in whey include Amundson [3,4], Naiditch and Dikansky [78], Simek et al. [106], Porges et al. [90], Stimpson and Young [115], Wasserman [124], and Stuiber [117].

Graham et al. [39,40] compared *Candida krusei*, *C. utilis* (*Pichia jadinii*), *C. utilis* var. *thermophilus*, and *Torula cremoris* (*C. pseudotropicalis*) for their ability to grow in whey and concluded that *T. cremoris* was suitable for use to increase the value of whey as an animal feed. Successful growth in whey was attributed to *T. utilis* (*P. jadinii*), *Torula casei* (*C. pseudotropicalis*), and *T. cremoris* by Tomisek and Gregr [118]. Other yeasts suggested as suitable for growth in whey include *Candida tropicalis* [21], *C. utilis* [23], *Torulopsis sphaerica* (*Kluyveromyces lactis*), *Torula lactosa* (*Candida kefyr*) [31], and *Torula* spp. [77]. (The genera *Torulopsis* and *Torula* are no longer recognized.)

2. The Process

Numerous procedures have been suggested for growth of yeast in whey. The most detailed studies on methods for maximum yield of yeast cells (and protein) in whey are those of Wasserman and his associates [124-131]. The process developed through these studies will be described first, and other options will be considered later.

Medium: Wheys resulting from cottage, Cheddar, ricotta, Italian, or other cheeses made by similar procedures are satisfactory, although some (especially those resulting from the manufacture of Italian hard cheeses) contain some milk fat, which may make drying of the finished product difficult.

Supplementation of whey with added nutrients is necessary to obtain maximal yields of yeast. Phosphorus must be added in the amount of 0.225% and, since only 25% of whey nitrogen is available to yeasts, additional nitrogen must be supplied. Ammonium sulfate can be added in the amount of 0.85% to compensate for the nitrogen deficiency. Yeast extract or dried brewer's yeast is another necessary additive.

The optimum pH for maximal yeast production is in the range of 5.0 to 5.7, but it may rise during incubation. If it is allowed to reach 8.5, growth of yeast is impaired, and the pH must be reduced by addition of acid or by an interruption in aeration so that the yeast produces its own acid under anaerobic conditions.

Heating of whey alters the medium to make it more suitable for yeast growth. If a short growth cycle is used (as described in the following paragraphs), sterile conditions are not necessary since the yeast outgrows contaminating microorganisms.

Temperature: The optimum temperature for growth of *K. marxianus* var. *lactis* in whey ranges from 31 to 33°C, although good yields have been obtained up to 43°C. The wide temperature range in which growth is possible reduces the need for cooling since proliferation of yeast is accompanied by liberation of heat.

Inoculum: The time required to attain the yield limit depends to a great extent on the size of the inoculum. For example, when 3.5 g of yeast are added per liter, the maximum yield can be expected in 8 hours, but when 23.8 g of yeast are used, the maximum yield is reached in 3-4 hours. The former condition involves a ninefold increase in number of cells, whereas the latter requires only a twofold increase. Inoculation with yeast equivalent in dry weight to approximately 30% of the weight of lactose in the medium results in a 4-hour incubation for maximum yield.

Aeration: Sufficient oxygen must be provided to enable yeast to oxidize approximately 35% of the lactose in whey to carbon dioxide and water. The remainder of the carbohydrate is used to produce new cell materials or nonoxidizable metabolic products. From 13 to 15 liters of oxygen are needed per liter of whey, although the calculated requirement for whey with 4% lactose is only 10.5 liters. The excess oxygen is required for endogenous cell respiration and for oxidizing small amounts of protein and lactic acid.

The demand for oxygen rises to a peak midway during the 4-hour incubation period, hence information on the total amount of oxygen required is not enough. If the total amount of oxygen is supplied in uniform increments, there will be both excessive and inadequate amounts at various times during the fermentation. Anaerobic conditions occur quickly and growth becomes limited if the supply of oxygen is inadequate. The peak requirement, occurring after approximately 2 hours of incubation in a 4-hour fermentation, is 100-120 mL of oxygen per liter per minute. The increase in oxygen demand rises virtually in a straight line from approximately 30 mL per liter per minute initially to approximately 110 mL per liter per minute after 2 hours and then decreases sharply, and nearly in a straight line, to less than 20 mL per liter per minute at the end of 4 hours. The work of Stuiber [117] further emphasizes the importance of adequate aeration in this fermentation. He observed a marked decrease in yield of yeast (33.1 versus 46.9 g of cells per 100 g of lactose) under conditions of oxygen starvation. A higher proportion of carbon in the medium was converted to carbon dioxide when oxygen was limited than when the supply was ample (63 vs. 47.5%).

Oxygen absorption by the medium is dependent on (a) quantity of air passing through the fermenter, (b) type of impeller used for agitation (turbine impeller is preferable to Rheinhütte or propeller types), and (c) speed of the agitator (slower speed is associated with lower absorption). Use of the Waldhof propagator, as opposed to the conventional stirred fermenter, results in good yields of yeast with lower oxygen input. The Waldhof fermenter employs both an agitator (turbine-type) and a draft tube. Demmler [23] also reported successful use of the Waldhof fermenter for yeast propagation, and Stimpson and Trebler [113] suggested modifying this type of fermenter by allowing air to enter from the bottom to minimize foaming.

Harvesting: The yeast suspension is concentrated by centrifugation so that a slurry with 15-18% solids is produced. Yeast may be washed at this point so that the final product is bland in flavor. The yeast cream can be spray-dried or fed to a drum dryer operating with 85 psi (about 8300 g/cm^2) steam pressure and rotating at 12 rpm. Dried yeast may be pulverized with a hammer mill before bagging.

Yield: The yield will be variable, depending on fermentation conditions. Reports suggest that approximately 0.42 kg of yeast (dry) can be obtained per kilogram of lactose. This crop represents a yield of 75% of that which could theoretically be expected, a considerable improvement over the 24-45% yield reported by Porges et al. [90] or the 50% yield obtained by Müller [73].

3. Suggested Modifications in the Process

Numerous investigators other than Wasserman and his associates have studied production of yeast from whey, and some have suggested procedures which differ from those just outlined. These procedures will be described in the following paragraphs.

Medium: Several suggestions have been made regarding the heat treatment given to whey before fermentation. Myers and Weisberg [76] suggest pasteurization (flash at 74-85°C or 30 minutes at 63°C), whereas Stuiber [117] and Amundson [3,4] recommend heating to 93°C and holding for 5 minutes. Other recommendations include heating to 60°C and holding at that temperature, followed by heating to 77°C [115], and heating to 85-90°C [73].

Some researchers have claimed improved yields from whey fortified with compounds other than those listed earlier. For example, Siman and Mergl [105] suggested that nitrogen and phosphorus be supplied in the form of ammonium nitrate, potassium nitrate, sodium nitrate, ammonium chloride, ammonium sulfate, or ammonium (mono- and di-) phosphate. Another combination of ingredients suggested by Müller [73] consists of 9 g of ammonium sulfate, 1 g of diammonium phosphate, 0.4 g of potassium sulfate, 0.2 g of magnesium sulfate, and several drops of a 5% ferric chloride solution (all on a per liter basis). Eichelbaum [28] recommended the acid or enzyme hydrolysis of whey protein before fermentation. Other additives that have been found useful are corn steep liquor [4], ammonia [115], and byproducts of the manufacture of glutamic acid and sodium glutamate [107]. The suggested pH for this fermentation ranges from 3.5 obtained with added phosphoric acid [3,4,117] to 8.0 [46], although most reports recomment a pH of 5.0 or below.

Temperature: Although various temperatures of incubation have been suggested, most of them are close to 30°C. Specifically, the following temperatures have been found suitable: 26°C [6], 28-31°C [5], 25-30°C [40, 76], and 32-35°C [73].

Inoculum: The volume of inoculum suggested by Wasserman and coworkers is quite high but is necessary for rapid fermentation. Smaller inocula and correspondingly slower fermentations have been found satisfactory by some workers. For example, Graham et al. [40] reported use of a 5-10% inoculum of *T. cremoris* (*P. jadinii*), and Stimpson and Young [115] suggested that inoculated whey should contain 8×10^8-1×10^9 yeast cells per milliliter.

Harvesting: Several procedures different from those described earlier have been reported and may be useful in some applications. Myers and Weisberg [76] suggested that yeast and whey solids be concentrated in a vacuum pan before drying. An alternative procedure also was mentioned in which the fermented whey was first filtered or centrifuged to recover yeast cells and precipitated protein. This was followed by adjusting the pH of the liquor to 7.0, heating to precipitate soluble albumin, and

TABLE 1. Composition (%) of Various Dried Yeast Products Derived from Whey

Component	Myers and Weisberg [76]	Simek et al. [106]	Stimpson and Young [115][a]	Davidov et al. [21]	Naiditch and Dikansky [78]	Graham et al. [78]	Mergl and Siman [70b]
Protein	31.70	54.0	40.0	46.0	25-36	39	19
Ash	30.42	22.0	—	—	16-20	30	14
Lactic acid	13.63	—	—	—	—	—	—
Moisture	7.28	—	—	—	—	—	5
Lactose	0	0	—	—	17-30	—	—
Reducing substances	0.59	—	—	—	—	—	—
Fat	—[b]	4.0	—	—	—	19	—
Nitrogen-free extract	—	19.0	—	—	30-40	12	60
Riboflavin	—	—	p[c]	—	—	p	—
Thiamin	—	—	p	—	—	p	—

[a]Niacin, pyridoxine, pantothenic acid and folic acid also reported as present.

[b]Not reported.

[c]Reported as present.

filtering to recover the protein. The filtrate was then condensed in a vacuum pan, calcium lactate was allowed to crystallize, and it was removed by centrifugation. The resultant liquor contained vitamins, and further concentration was suggested. Stuiber [117] and Amundson [3,4] proposed harvesting of yeast and whey protein by means of a self-cleaning clarifier which produced a sludge with 30% solids. The sludge was then pasteurized and spray-dried.

Composition of dried products: The composition of various yeast products derived from whey by different investigators is given in Tables 1 and 2. Examination of the data reveals that the dried products are high in protein, although there is considerable variation ranging from a low of 19% to a high of 54.4%. Manufacturing procedures, including the degree of washing given the product, influence the final protein content as well as the content of other constitutents (Table 2).

It is also apparent that the mineral (ash) content of dried products is high, ranging from 14 to 30.42% (Table 1). This condition may limit their use in diets of some animals, although washing (Table 2) can reduce the ash content to reasonable levels. A comment is appropriate concerning the relatively high fat content of the product described by Graham et al. [41] and listed in Table 1. The fat in the dried product undoubtedly is attributable to either or both of the following: (a) the whey contained a substantial amount of fat, or (b) the yeast used was able to produce this material.

TABLE 2. Composition (%) of Dried Unwashed and Washed Yeast-Whey and Dried Whey

Component	Unwashed yeast-whey[a]	Washed yeast-whey[a]	Dried whey[b]
Protein	25.9	54.4	12-14
Fat	8.6	10.3	0.7-1.3
Lactose	5.8	2.4	78-81
Moisture	2.0	2.6	—
Ash	18.6	5.7	7.7-8.7
Lactic acid	0.8	—[c]	0.5-2.0

[a]From Amundson [4].

[b]From Olling [83].

[c]Not reported.

The amino acid composition of protein produced by *K. marxianus* var. *lactis* is given in Table 3. These data reveal the presence of a substantial amount of lysine (an amino acid usually deficient in cereals) and a shortage or absence of sulfur-containing amino acids.

Table 4 summarizes information on the vitamin content of dried yeast-whey products, whey, and *K. marxianus* var. *lactis*. The dried yeast-whey products are uniformly good sources of the B vitamins.

TABLE 3. Amino Acid Composition (g/16 g N) of *Kluyveromyces fragilis* (*K. marxianus* var. *lactis*)

Essential		Nonessential	
Lysine	10.20	Arginine	7.08
Threonine	6.46	Histidine	1.87
Valine	7.78	Asparagine	11.16
Methionine	1.25	Serine	6.96
Isoleucine	6.0	Glutamic acid	13.26
Leucine	9.60	Proline	4.31
Aromatic amino acids		Glycine	4.63
Tyrosine	3.42	Alanine	8.17
Phenylalanine	5.39		

Source: Adapted from Ref. 124.

TABLE 4. Vitamin Content (μg/g) of Unwashed and Washed Yeast-Whey, Whey, and *Kluyveromyces fragilis* (*K. marxianus* var. *lactis*) (dry weight basis)

Vitamin	Unwashed yeast-whey[a]	Washed yeast-whey	Whey[b]	*Kluyveromyces fragilis*[c]
Pantothenic acid	94.8	65.2	22-90	67.2
Riboflavin	57.5	42.5	5-68	36.0
Pyridoxine	12	15	0.5-15	13.6
Niacin	61	104	3-22	280
Thiamin	32.8	29.7	4-15	24.1
Folic acid	3.1	2.06	0-1	6.83
p-Aminobenzoic acid	—[d]	—	—	24.2
Biotin	—	—	—	1.96
Choline	—	—	—	6670
Inositol	—	—	—	3000

[a]Adapted from Amundson [4].
[b]Adapted from Wasserman [127].
[c]Adapted from Wasserman [124].
[d]Not reported.

B. Production of Ethanol from Whey

Ethyl alcohol can be produced from whey by one of the lactose-fermenting yeasts. Although ethanol was a by-product of the whey fermentation proposed by Myers and Weisberg [76], these investigators did not concern themselves with this product in their patent issued in 1938. One year later a patent describing a procedure for production of ethanol from whey was issued to Kauffmann and Van der Lee [53]. They employed pasteurized whey, a lactose-fermenting yeast, and a 3-day incubation. During this time, 65% of the lactose was converted to ethanol.

Some years later Rogosa et al. [94] compared different lactose-fermenting yeasts for their ability to produce ethanol in whey and concluded that *T. cremoris* (*K. marxianus*) was the most efficient of the yeasts tested. Results of their experiments led them to suggest the following procedure to produce ethanol from whey.

1. Residual milk fat is removed from whey.
2. Whey is heated to 100°C and the pH is adjusted to 4.7-5.0 using sulfuric acid.
3. Hot whey is filtered to recover the precipitated protein and is then cooled to 34°C.
4. An inoculum of 0.45 kg per 454 liters of whey is added, and the fermentation is allowed to proceed for 48-72 hours.
5. Fermented whey is centrifuged to recover yeast cells.
6. The liquid is distilled to recover ethanol.

According to Rogosa et al. [94], 84-91% of the theoretical ethanol yield was obtained with *T. cremoris* (*K. marxianus*).

Wilharm and Sack [135] increased the lactose content of whey to 10, 20, and 30% and the fortified wheys underwent an alcoholic fermentation. Their results indicated that up to 90, 75-80, 60-75, and 37-40% of the lactose was fermented in ordinary whey, and in wheys containing 10, 20, and 30% lactose, respectively. These data suggest that some fortification (or concentration) of single-strength whey might be possible to increase the yield of alcohol and that excessive supplementation with lactose apparently reduces the efficiency of the fermentation.

Treatment of whey by ultrafiltration results in permeate (liquid free of whey proteins) which can be fermented. A process claimed to be 5-30 times more productive than batch fermentations and employing a continuous fermentor has been described by Mehaia et al. [70a] as a means of producing alcohol from permeate. Regardless of the process used, ethanol derived from whey by fermentation may be used for conversion to acetic acid (vinegar) or in other food applications.

C. Production of Lactase from Whey

Lactase is an enzyme which catalyzes hydrolysis of the galactosidic linkage of lactose. Since the linkage between the hexoses of lactose is of the β-D configuration, the enzyme is also termed β-D-galactosidase [88]. Lactose occurs naturally in some plants and in the intestines of various animals. Additionally, it is produced by numerous molds, yeasts, and bacteria [88].

Although a widespread application has not been found for lactase in the food industry, limited uses accompanied by beneficial results have been reported. Pomeranz [89] summarized this information and listed the following possibilities: (a) minimize age-thickening of frozen, concentrated milk [111], (b) permit an increase in nonfat milk solids content of ice cream by hydrolyzing lactose to prevent it from crystallizing, (c) hydrolyze lactose in animal feeds, since many animals cannot tolerate high concentrations of this sugar in their diets, (d) improve the color of some fried foods by making hexoses available, and (e) produce fermentable carbohydrates from nonfat dry milk in bread dough. The last suggested use has not proven to be very successful [89]. Lactase has been found useful for treating milk so it can be consumed by persons suffering from lactose intolerance.

Use of a lactose-fermenting yeast as a source of lactase and growth of the yeast in whey have been suggested. Most of the early work in this area was done in the research laboratories of National Dairy Products Corp. (now Kraft-General Foods) and results are recorded in patents issued to Stimpson [109,110], Morgan [71], Myers [74], Myers and Stimpson [75], Young and Healey [142], Connors and Sfortunato [20], Sfortunato and Connors [102], Stimpson and Stamberg [112], and Stimpson and Whitaker [114]. Wendorff [133] and Wendorff et al. [134] at the University of Wisconsin also have studied lactase production.

According to Morgan [71], lactase may be derived from *K. marxianus* var. *lactis*, *Zygosaccharomyces lactis* (*K. lactis*), and *P. jadinii*, although *K. marxianus* var. *lactis* appears to be the organism of choice for large-scale production of the enzyme. The process for lactase production, as recorded by Morgan [71], Stimpson [109], and Myers and Stimpson [75], involves the following.

1. Whey containing 2-8% solids is adjusted to pH 4.5 and heated to 85°C for 30 minutes to precipitate whey proteins.
2. The liquid fraction is fortified with 0.1% corn steep liquor and with a source of nitrogen (0.2% urea, 0.14% ammonia, or 0.4% dibasic ammonium phosphate).
3. The fortified liquid is cooled to 30°C and inoculated with 10% of an actively growing culture of *K. marxianus* var. *lactis* so that the freshly inoculated whey will contain 1×10^7-6×10^7 cells per milliliter.
4. The fermentation is completed in 2-8 hours with, or in 30 hours without, aeration. According to Young and Healey [142], optimum lactase production was obtained when the medium contained 0.06% available nitrogen and was aerated at the rate of 1 volume of air per volume of medium per minute.
5. Yeast cells are separated from whey and washed if a product with good flavor is desired. Washing is not necessary if the lactase preparation is to be used in animal feeds.
6. The concentrated yeast is dried and stored at 4°C until used.

Drying of the yeast accomplishes two purposes. The first is the obvious one of removing water, and the second is that of destroying "zymase" activity without markedly affecting the lactase activity. Zymase is a collective term referring to the group of enzymes in yeast which catalyze reactions involved in converting glucose and galactose into ethanol and carbon dioxide. Procedures which must be followed in drying have been outlined by Stimpson [110], Myers [74], and Myers and Stimpson [75]. The following methods have been found to yield a satisfactory product.

1. A yeast cream (10-18% yeast solids) is fed into a spray drier with an inlet air temperature of 93-154°C and an outlet air temperature of 54-96°C. The optima are 121°C for inlet air and 77°C for outlet air.
2. Tray-dry at 66°C or below for no more than 4 hours unless a vacuum is applied; in this event the period may be extended to 8 hours. Alternatively, drying at 46°C for 8 hours without vacuum is satisfactory.
3. Freeze yeast at -43 to -18°C followed by freeze-drying, with the yeast remaining in the frozen state during drying. The temperature during freezing is not critical, although rapid freezing results in retention of more enzyme activity. Temperature during drying is not critical so long as the yeast remains frozen.

Although roller drying inactivates lactase as well as zymase [110], yeast can be treated with toluene, chloroform, or ethyl ether, or heated to 52°C at pH 7 to destroy zymase activity without untoward damage to lactase [75]. An additional treatment to kill yeast cells as well as chance contaminants without inactivating lactase was proposed by Morgan [71]. To accomplish these purposes he suspended wet yeast cake in 8-17.5% aqueous ethanol and allowed the mixture to stand for 1-5 hours. After this treatment yeasts were removed from the alcoholic solution and dried by procedures described earlier.

Wendorff et al. [134] reported the following conditions were needed for maximum lactase production by *K. marxianus* var. *lactis*: (a) whey containing 10-15% lactose, (b) addition to whey of corn steep liquor or a casein digest, (c) pH of 4.0-4.5, and (d) incubation at 28°C. Yeast grown under optimal conditions yielded 175 units of lactase per gram or 1300 units per liter of whey. A unit of lactase was defined by Young and Healey [142] as the grams of lactose hydrolyzed by 1 g of lactase preparation when done in 30% skim milk solids at 51°C and pH 6.8 for 4 hours.

Additional studies by Wendorff [133] characterized the lactase enzyme produced by *K. marxianus* var. *lactis*. He found the enzyme to be stable at pH 6.0-7.0 and that optimum lactose hydrolysis occurred at 37°C and pH 6.5. Activation of the enzyme was attributed to potassium ions, whereas manganese ions served as cofactors for lactase. Inactivation of lactase was associated with urea and inhibition with heavy metals, *p*-chloromercuribenzoate, iodoacetate, cysteine, galactose, glucose, and various amines. Milk products contain a naturally occurring lactase inhibitor which can be inactivated by heating the products at 74°C for 30 minutes.

V. FUNGUS-RELATED PROBLEMS IN THE DAIRY INDUSTRY

A. Spoilage of Dairy Products by Fungi

1. Spoilage by Molds

Of all the dairy products, cheese is probably most often spoiled by growth of molds. This led some to suggest that cheese is either nonvisibly or visibly moldy [101]. Yousef and Marth [145] compared the ability of cheeses to sustain the growth of molds. Results show that cheeses varied in their susceptibility to mold growth. Aged Cheddar was less susceptible to mold growth than was mild Cheddar. Natural cheeses were more resistant to growth of mold than were processed cheeses made therefrom.

The duration and conditions of ripening and storage (e.g., temperature, relative humidity, packaging, sanitation) and composition (especially water activity, a_w) of cheese are major factors affecting the susceptibility of cheese to mold growth. Cheese normally has an a_w in the range of 0.94-0.89, depending upon its age and type [119]. In general, yeasts and molds are more tolerant to reduced water activity than bacteria. Minimum water activity at which bacteria can grow is 0.75-0.98, yeasts 0.62-0.92, and molds 0.605-0.94 [119]. As a result, cheeses are subject to growth of mold (and sometimes yeasts), particularly on their surfaces, and sometimes in the air pockets caused by mechanical holes in the cheese body. Growth of molds on cheese may produce off-flavors (mostly because of lipid hydrolysis) and/or discoloration of the surface.

As might be expected, a variety of molds can grow on and spoil cheese. In one study Gaddi [38] isolated 144 mold cultures from 14 varieties of cheese made or distributed by three United States food processors. Of the isolates, 69% were *Penicillium* spp., 9% *Aspergillus* spp., 8% *Scopulariopsis* spp., 3% *Mucor* spp., and 2% *Cephalosporium* spp. Some isolates of *Cladosporium* spp. and *Syncephalastrum* spp. also were encountered. In the Netherlands, cheese samples with visible mold growth were collected from shops, households, and warehouses [80]. The predominant species of molds found on cheese from shops and households was *Penicillium verrucosum* var. *cyclopium*, and those from warehouses were *P. verrucosum* var. *cyclopium*, *Aspergillus versicolor*, *Aspergillus repens*, and *P. verrucosum* var. *verrucosum*.

Growth of mold can occur during ripening of cheese as well as later when cheese is cut and packaged for the consumer. Processed cheese, if contaminated after heating, also can be spoiled by molds provided the cheese is not packaged to exclude virtually all oxygen. Molds also have been found responsible for spoilage of farm-separated cream (*G. candidum*), butter (several genera), cottage cheese, yogurt, and sweetened condensed milk.

2. Spoilage by Yeasts

Spoilage of dairy products by yeasts has been discussed by Walker and Ayres [121]. Problems caused by yeasts include (a) production of gas and off-flavors in improperly handled cream, (b) production of gas in sweetened condensed milk, (c) production of gas and off-flavors in cheese, (d) growth, often on the surface, in cottage cheese with accompanying off-flavors and reduction in shelf life, and (e) production of rancidity and other flavor defects in butter, particularly if the yeast population is large.

3. Control Measures

Control of molds and yeasts in dairy products generally involves one or more of the following:

1. Use of accepted hygienic practices before, during, and after manufacture of the product. Processing areas and machinery should be checked regularly for freedom from contamination. Machinery mold (*G. candidum*) is often used to assess the sanitation of processing operations.
2. Use of proper heat treatments and refrigerated storage. Doyle and Marth [25,26] found that fungal spores are readily inactivated by heat processes common to the dairy industry, and hence it is especially important for heated products to be handled in a sanitary manner.
3. Since many mold spores are transmitted by air, special attention should be given to the quality of air in the dairy-processing facility.
4. Packaging to essentially eliminate oxygen from the atmosphere surrounding a product also is useful to control mold growth.
5. Acceptable antimycotic agents such as sorbic acid (sorbates), propionic acid (propionates), or natamycin (also known as pimaracin) may be added to some dairy products to inhibit yeast and mold growth and thus extend the shelf life of the products. Natamycin has been approved for use in the United States on certain dairy products.

If sorbates are used, it must be remembered that some penicillia can grow in the presence of concentrations of the chemical greater than the amount (3000 ppm) commonly used on foods. When that happens, the penicillia are likely to decarboxylate sorbic acid and thus produce 1,3-pentadiene, a compound which imparts a hydrocarbon-like (kerosene) odor and flavor to cheese [68]. Although more toxic than sorbic acid, according to results of the Ames test, 1,3-pentadiene is not mutagenic [41].

When Marth et al. [68] first studied degradation of sorbic acid, their isolates of penicillia (all obtained from cheese) initiated growth in the presence of 1800-7100 ppm potassium sorbate (reduce by about 30% to convert to ppm sorbic acid). Later Finol et al. [34] isolated penicillia also from cheese, that initiated growth in the presence of 3,000-12,000 ppm potassium sorbate. Growth of penicillia in the presence of sorbate is accompanied by a loss of the chemical from the substrate, and so a loss of antifungal activity.

Liewen and Marth [63,64] compared sorbate-sensitive and sorbate-resistant strains of *P. roqueforti*. They observed that supplementing the broth medium (YES, YM, or mycological broth) with maltose (rather than glucose) increased mycelial growth (over the control) of the resistant strain of the mold in the presence of sorbate. Furthermore, uptake of sorbate by mycelia of the resistant strain was considerably less than by mycelia of the sensitive strain. Conidia of the two strains lost ATP rapidly after exposure to sorbate, but conidia of the resistant strain recovered ATP upon continued incubation. It is possible that some penicillia are resistant to sorbate because they can maintain an internal ionic balance and pH value high enough to reduce the effectiveness of sorbic acid. Additional information about growth and inhibition of microorganisms in the presence of sorbic acid can be found in a review by Liewen and Marth [62].

B. Health Hazards Associated with Presence of Molds in Dairy Products

Some molds, when they grow on food or feed, can produce hazardous metabolites that are collectively known as mycotoxins. When such molds grow on dairy products, particularly cheese, they may produce mycotoxins, which then can pose a direct hazard to consumers. Toxigenic molds also may grow on animal feed and produce these mycotoxins. Such feed might be consumed by dairy cows, and the mycotoxin, especially aflatoxin, can be carried over into milk and subsequently into dairy products. Other health hazards are indirectly related to growth of molds in mold-ripened cheeses. Growth of such molds may provide favorable environments for proliferation of pathogenic bacteria. Details about these types of hazards are included in the following discussion.

1. Toxigenic Molds in Dairy Products

Toxigenic molds were occasionally isolated from cheeses [101]. Some reports indicate that such molds cannot grow and produce mycotoxins on cheese during normal ripening conditions. In other studies, however, molds produced mycotoxins when allowed to grow on cheese. That toxigenic aspergilli can produce aflatoxin during growth on Cheddar and brick cheese was demonstrated by Lie and Marth [60] and Shih and Marth [103]. Furthermore, these authors demonstrated that aflatoxin can penetrate into cheese to a depth of up to 4 cm from the surface. In Greece, 22 *Aspergillus* species, isolated from Teleme cheese or the cheese factory environment, were grown on the cheese and their ability to produce aflatoxin was tested [147]. One isolate produced aflatoxin when grown directly on Teleme cheese.

Kiermeier and Böhm [56] examined 19 varieties of German cheese and found aflatoxin B_1 in samples of Tilsit, Edelpilzkäse, butter cheese, smoked cheese, Parmesan, and Romadur. Additionally, aflatoxin G_1 was found in samples of Emmentaler, Gouda, Tilsit, and Romadur. A combination of B_1 and G_1 occurred in some samples of still other varieties of cheese. Northolt et al. [80] collected cheese samples with visible mold growth from shops, households, and warehouses. Nine of 39 cheeses contaminated with *A.*

versicolor contained sterigmatocystin in the surface layer in concentrations ranging from 5 to 600 μg/kg.

Under certain cultural conditions, some strains of mold normally used to manufacture cheese may produce mycotoxins. Sieber [104] found that *P. roqueforti* cultures can produce PR-toxin on special media. The toxin, however, was not produced in cheeses ripened with this mold. No carcinogenic mycotoxins have been discovered in these types of cheeses, except aflatoxin M_1, which might originate from contaminated milk.

According to other reports, *P. roqueforti* and *P. caseicolum* (used to manufacture blue and Camembert cheeses, respectively) can produce toxic substances [55,100]. The known metabolites of *P. roqueforti* are penicillic acid, roquefortine, isofumigaclavines A and B, PR-toxin and related metabolites, mycophenolic acid, and siderophores. Penicillic acid and PR-toxin are unstable in cheese. However, roquefortine (intraperitoneal LD_{50} of 15-20 mg/kg for the mouse), isofumigaclavine A (intraperitoneal LD_{50} value of 340 mg/kg for the mouse), mycophenolic acid (oral LD_{50} values of 2500 and 700 mg/kg, respectively, for the mouse and rat), and the siderophore ferrichrome, all in low ppm concentrations, have been found in blue cheese [100]. The metabolite cyclopiazonic acid (oral LD_{50} of 36-63 mg/kg for rats) is produced by *P. caseicolum*. Low concentrations of this mycotoxin have been detected in crusts of cheese. At this time no acute health hazard is associated with the presence of the metabolites produced in cheese by either *P. roqueforti* or *P. caseicolum*.

2. Aflatoxin in Milk and Milk Products

In 1963, Allcroft and Carnaghan [1,2] first reported that a factor toxic for ducklings was contained in milk from cows whose rations contained peanut meals contaminated with aflatoxin. They also demonstrated that the toxic factor remained with the casein fraction that precipitated when milk was treated with rennet. According to Kiermeier [54] various workers fed controlled amounts of aflatoxin to cows and found that milk generally contained less than 1% of the amount of toxin fed to the cows. Allcroft et al. [2] designated the milk toxin as aflatoxin M, and Holzapfel et al. [50] found aflatoxins M_1 and M_2. The latter researchers also determined chemical structures and concluded that M_1 was hydroxyaflatoxin B_1 and M_2 was hydroxyaflatoxin B_2.

Kiermeier [55] fed cows with peanut meal that contained 10.9 mg of aflatoxin per kilogram. In one experiment the cow received a single dose of 850 g of feed, in another 415 g were given to a cow on each of 2 successive days, and in a third, two cows each received 914 g daily for 5 days. Results indicated that (a) aflatoxin appeared in milk within 12 hours after toxin was consumed by the cow, (b) the amount of aflatoxin in milk varied with the individual cow, (c) excretion of aflatoxin in milk decreased markedly about 1 day after feeding of toxin had stopped, although small amounts appeared in milk for 2-3 additional days, and (d) from 0.18 to 0.39% of the amount consumed was excreted in milk.

Applebaum et al. [12] gave daily doses, for 7 days, of 13 mg of aflatoxin B_1 to each of 10 fistulated Holstein cows in midlactation. Six of the cows received pure aflatoxin B_1, whereas four received aflatoxin B_1 plus other aflatoxins and metabolites as produced by *A. parasiticus*. Toxin was administered to each animal twice daily, one-half of the total dose each time, via the rumen orifice. Aflatoxin M_1 appeared in milk in 4 but not 2 hours after aflatoxin B_1 was administered to the cows. Milk contained from 1.05 ppb to 10.58 ppb aflatoxin M_1, with greatest amounts appearing in milk of cows that received the impure preparation of toxin. Somatic cell counts and Standard Plate Counts of milk from two treated cows were not affected by administration of aflatoxin, but milk production by cows receiving the impure toxin preparation was reduced. The amount of toxin (13 mg/day) administered to cows in these trials was based on the amount likely to be ingested if the cow consumed aflatoxin-contaminated feed. The amount of aflatoxin in milk of these experiments regularly exceeded 0.5 ppb, the maximum amount, according to guidelines of the U.S. Food and Drug Administration, that can occur in fluid milk that enters interstate commerce. The biosynthesis and occurrence of aflatoxin M_1 in milk is discussed in greater detail in a review by Applebaum et al. [11].

Several surveys have shown that aflatoxin sometimes occurs in milk produced under normal farm conditions. During 1972 Kiermeier [55] sampled raw milk produced in the vicinity of Freising, Federal Republic of Germany. Twelve of 36 samples taken from individual farms during February to April contained from 0.03 to 0.25 μg of aflatoxin M_1 per liter. In May, 9 of 12 samples taken from tank trucks (6000-9000 liters per truck) contained 0.01-0.08 μg per liter. Small amounts of aflatoxin M_1 also appeared in milks taken during February from the holding tank in the factory, but none was found when milks were tested in March and June, suggesting that the diet of cows was less contaminated with aflatoxin in spring than winter. In 1973, the U.S. Food and Drug Administration surveyed milk products for presence of aflatoxin M_1. The toxin (0.05-0.5 μg per liter) was detected in all 16 samples from one milk shed where use of contaminated feed was suspected [116]. Major problems with aflatoxin M_1 in milk occurred in the southeastern United States in 1976 because in that year much corn produced in that region and fed to dairy cows contained aflatoxin. In 1978 problems occurred in the southwestern part of the United States because contaminated cottonseed was consumed by dairy cows. Widespread contamination of corn with aflatoxin occurred in much of the United States during 1988. Some of this corn was consumed by dairy cows and led to production of some contaminated milk.

Initially, processing of milk was thought to reduce its content of detectable aflatoxin M_1. For example, Purchase et al. [92] reported a reduction in amount of aflatoxin M_1 of 32% when milk was pasteurized at 62°C for 30 min, of 45% when pasteurized at 72°C for 45 seconds, of 64% when pasteurized at 80°C for 45 seconds, of 64% when processed into condensed milk, of 61% when roller dried at low pressure, of 76% when roller dried at 4.9 kg/cm^2, of 81% when sterilized at 115°C, and of 86% when spray-dried.

Wiseman et al. [141] and Wiseman and Marth [138,139] studied heat resistance of aflatoxin M_1 and stability of the toxin when dry dairy products were made. Naturally contaminated skim milk and cream were heated at 64°C for 30 minutes, and there was no evidence of a decrease in aflatoxin M_1 content as a consequence of the heat treatment. Heating milk at 64, 84, or 100°C for up to 120 minutes also failed to reduce the amount of aflatoxin M_1 that was present. Samples heated for 120 minutes were held overnight at 4°C and tested again. Such holding caused no appreciable change in content of aflatoxin M_1 in the heated milks. Wiseman and Marth [139] produced both nonfat dry milk and dried buttermilk (residue from making butter) by freeze-drying naturally contaminated starting materials. Both products were stored at 22°C for up to 4 months. There was no evidence of loss of aflatoxin M_1 during manufacture and subsequent storage of the products.

Wiseman and Marth [137] and Applebaum and Marth [8] made yogurt, buttermilk, kefir, and cottage cheese from milk naturally contaminated with aflatoxin M_1. Toxin content was not affected appreciably by the fermentations involved in making these products or by refrigerated storage (6 weeks for yogurt, 14 days each for buttermilk, kefir, and cottage cheese) of these foods. Wiseman and Marth [138] demonstrated that aflatoxin M_1 was stable for 4 days at 28°C in McIlvaine buffer at pH 4.0. Hence, it is not surprising that aflatoxin M_1 was stable in cultured milks and cottage cheese. Aflatoxin M_1 has been found in commercial dried milks, cottage cheese curd, and evaporated milk [47,116].

Ice cream and sherbet were made by Wiseman and Marth [140] from milk naturally contaminated with aflatoxin M_1. The toxin remained stable in both products during 8 months of frozen storage. Brackett et al. [17] and Brackett and Marth [15,16] studied behavior of aflatoxin M_1 in various kinds of cheese made from milk naturally contaminated with the toxin. The toxin was stable in Cheddar cheese during ripening at 7°C for about 1 year. Converting the cheese into process cheese spread did nothing to inactivate the aflatoxin, but appeared to improve recovery of toxin during analytical procedures. Aflatoxin M_1 also was stable in brick and Limburger-like cheese, except for the rind of the Limburger where some loss of aflatoxin occurred during the ripening process. Aflatoxin M_1 was stable in mozzarella cheese held for 19 weeks at 7°C. In contrast to this, in Parmesan cheese there was an initial decrease in amount of detectable aflatoxin M_1 followed by stability through 42 weeks of ripening at 10°C. Wiseman and Marth [136] made Queso Blanco and baker's cheese from milk naturally contaminated with aflatoxin. Again, the aflatoxin was stable in these cheeses, both during 1 or 2 months of refrigerated (4°C) or 2 months of frozen (-23°C) storage. Other work by Brackett and Marth [14] demonstrated the affinity of aflatoxin for casein. Consequently, the presence of the toxin in cheese, or other casein-containing products, is to be expected if milk contaminated with aflatoxin M_1 is used as a starting material for these foods. Additional information on aflatoxin M_1 in milk and milk products appears in a review by Applebaum et al. [11].

Methods to reduce the amount of or eliminate aflatoxin M_1 from milk have been studied by Applebaum and Marth [9,10] and Yousef and Marth [143,144]. They found the following to be effective in eliminating 50% or more of aflatoxin M_1 present from milk: (a) 1% hydrogen peroxide plus 0.5 mM riboflavin, (b) 0.1% hydrogen peroxide plus 5 units of lactoperoxidase added per milliliter, (c) 0.1-0.4 g of bentonite added per 20 ml of milk and then removed (bentonite adsorbs aflatoxin M_1), (d) treatment of milk with ultraviolet irradiation, or (e) addition of hydrogen peroxide to milk followed by treatment with ultraviolet irradiation. Further information on degradation of mycotoxins in foods and agricultural commodities is available in other publications [27,146].

3. Pathogenic Bacteria in Fungus-Ripened Cheeses

Occasionally, raw milk harbors pathogenic bacteria such as *Listeria monocytogenes* [48,66]. Normal pasteurization commonly is sufficient to inactivate such a pathogen in cheesemilk [18,29]. Several cheeses are traditionally made from raw or minimally heated milks. Some mold-ripened cheeses are made from raw milk [19,44,59]. During ripening of these cheeses, yeasts and molds utilize lactic acid, thus raising the pH of the cheese. In addition, most mold-ripened cheeses are high in moisture content. All these factors make mold-ripened cheeses favorable media for proliferation of some pathogenic bacteria. This may explain the frequency and severity of disease outbreaks associated with consumption of mold-ripened cheeses.

Late in 1971 at least 227 persons in eight states of the United States became ill with acute gastroenteritis about 24 hours after consuming French Camembert or Brie cheese [13,99]. The illness was attributed to the presence in cheese of enteropathogenic *Escherichia coli*. Park et al. [87] investigated the fate of enteropathogenic *E. coli* during the manufacture and ripening of Camembert cheese. They added *E. coli* to cheese milk and observed that growth of the bacterium was minimal until after the curd was cut and hooped, when a population in excess of 10^4/g had developed. Overnight storage of cheese in hoops was accompanied by a decrease in the number of viable *E. coli*, and this trend continued during the ripening period. When production of acid was inadequate during cheese making, approximately 10^9 *E. coli* developed per gram of 24-hour-old cheese, and 10^7/g persisted after 9 weeks of ripening.

Frank et al. [36] also studied survival of enteropathogenic strains of *E. coli* during manufacture and subsequent storage of Camembert cheese. They observed that one pathogenic (B_2C) and two nonpathogenic strains (H-52 and B) survived in cheese beyond 4 weeks of age at populations ranging from 10^0 to 10^3/g. Hence the pathogen could have been present in cheese when it would have been consumed. Growth and survival curves of the pathogen were nearly parallel when inocula of 10^2, 10^3, or 10^4 *E. coli* B_2C/mL of cheesemilk were used. Reducing the amount (0.25% instead of 2.0%, v/v) of lactic starter culture used to make the cheese resulted in increased (4 instead of 2 orders of magnitude) growth of *E. coli* B_2C during the cheese-making process. During ripening, because the decrease in initial pH of the cheese was less, numbers of *E. coli* B_2C decreased less in cheese made with the smaller rather than larger inoculum of lactic acid bacteria. Because ripening of this variety of cheese proceeds from the outside in, the authors determined survival of *E. coli* at different places in the cheese wheel during ripening. From the bactericidal unripened core outward, conditions became more favorable for survival, and growth of *E. coli* occurred at the outer surface of the cheese. Rutzinski et al. [96] did similar experiments but used *Hafnia* sp. instead of *E. coli*. Their work indicated that *Hafnia* sp. increased from about 10^2/g to about 10^8/g in Camembert cheese with normal pH values during ripening. Of the coliforms studied by Park et al. [87], Frank et al. [36], and Rutzinski et al. [96], only *Hafnia* sp. (two strains) was able to initiate growth and then to attain substantial populations in Camembert cheese.

Late in 1985, Brie cheese, a variety similar to Camembert, imported into the United States from France was found to contain *L. monocytogenes*, the cause of listeriosis. French Brie cheese contaminated with this pathogen also appeared in Canada and the United Kingdom. Surveys of Swiss and German mold-ripened soft cheese also revealed the occasional presence of *L. monocytogenes* [7].

Ryser and Marth [97] studied behavior of *L. monocytogenes* during the manufacture and ripening of Camembert cheese. Their results indicate that the bacterium survived but did not grow during the initial stage of ripening before there was a substantial increase in pH of cheese caused by the metabolic activity of *P. caseicolum*. When the pH had increased from about 4.7 to about 5.5, *L. monocytogenes* initiated growth, and at the end of ripening had attained a population of about 10^7/g of cheese. *Listeria monocytogenes* is a psychrotroph and hence grew in the ripening cheese even though it was at 6°C. In another study, Papageorgiou and Marth [85] investigated behavior of *L. monocytogenes* during the manufacture and ripening of blue cheese. During ripening, numbers of the pathogen decreased at a variable rate that appears to be correlated with changes in pH caused by growth of *P. roqueforti*. An initial inoculum in cheese milk of ca. 10^2 cfu/mL gave a cheese that contained *L. monocytogenes* after up to 120 days of ripening. Dependence of growth and survival of *L. monocytogenes* upon the pH of medium was demonstrated in several studies [33,69,86].

Liederkranz is a trade name for surface-ripened cheese once made in the United States by procedures similar to those for Limburger. In 1985, Liederkranz cheese was found to contain *L. monocytogenes*. Contaminated cheese was removed from market channels before it could be consumed and cause illness. Apparently, a *Listeria*-contaminated culture applied to the surface of the cheese to facilitate its ripening was the source of the infection. Currently (1990) Liederkranz cheese is not being made in the United States.

Behavior of enteropathogenic strains of *E. coli* during the manufacture and subsequent storage of brick cheese was studied by Frank et al. [37]. Growth of two of three test strains of *E. coli* was about 10 times greater during the initial hours of the manufacture of brick than of Camembert cheese, probably because in making the former the temperature of milk was higher and the decrease in pH was slower than in the latter kind of cheese. After 7 weeks of ripening (2 weeks at 15.5°C and 5 weeks at 7°C), populations of *E. coli* ranged from 7×10^2/g to 2×10^4/g. Growth of *E. coli* on the surface of brick cheese was much less than on the surface of Camembert cheese.

Ryser and Marth [98] made brick cheese from milk that was inoculated with one of four strains of *L. monocytogenes* to contain ca. 10^2-10^3 cfu/mL. Strains of *L. monocytogenes* behaved differently during smear development (2-4 weeks at 15°C) and ripening (22 weeks at 10°C) of brick cheese. Strains Scott A and Ohio grew during smear development, and then their numbers remained nearly constant during the rest of ripening period. In contrast, numbers of strains V7 and California decreased rapidly during smear development. During ripening, these two strains were detected in cheese samples only by cold enrichment and occasionally by direct plating.

ACKNOWLEDGMENTS

A contribution from the College of Agricultural and Life Sciences, University of Wisconsin-Madison. Preparation of this chapter was supported in part by the National Dairy Promotion and Research Board.

REFERENCES

1. Allcroft, R., and Carnaghan, R. B. A. Groundnut toxicity: An examination for toxin in human food products from animals fed toxic groundnut meal. *Vet. Rec.*, *75*:259-263, 1963.
2. Allcroft, R., Rogers, H., Lewis, G., and Nabney, J. Metabolism of aflatoxin in sheep: Excretion of the "milk toxin." *Nature* (London), *209*:154-155, 1966.
3. Amundson, C. H. Increasing the protein content of whey through fermentation. *Proc. 33rd Wash. State Univ. Inst. Dairy*, 1966, pp. 23-30.
4. Amundson, C. H. Increasing protein content of whey. *Am. Dairy Rev.*, *29*(7):22-23, 96-99, 1967.
5. Anonymous. Baker's yeast from whey. *Südd, Molkereiztg.*, *67*:114-115, 1946.
6. Anonymous. On the utilization of whey from cheese. *Technician Lait*, *13*:13-14, 1961.
7. Anonymous. Second World Congress on food infections and intoxications. *Deut. Molkerei-Zeitung*, *107*:874-875, 1986.
8. Applebaum, R. S., and Marth, E. H. Fate of aflatoxin M_1 in cottage cheese. *J. Food Prot.*, *45*:903-904, 1982.
9. Applebaum, R. S., and Marth, E. H. Inactivation of aflatoxin M_1 in milk using hydrogen peroxide and hydrogen peroxide plus riboflavin or lactoperoxidase. *J. Food Prot.*, *45*:557-560, 1982.
10. Applebaum, R. S., and Marth, E. H. Use of sulphite or bentonite to eliminate aflatoxin M_1 from naturally contaminated raw whole milk. *Z. Lebensm. Unters. Forsch.*, *174*:303-305, 1982.
11. Applebaum, R. S., Brackett, R. E., Wiseman, D. W., and Marth, E. H. Aflatoxin: Toxicity to dairy cattle and occurrence in milk and milk products—A review. *J. Food Prot.*, *45*:752-777, 1982.
12. Applebaum, R. S., Brackett, R. E., Wiseman, D. W., and Marth, E. H. Responses of dairy cows to dietary aflatoxin: Feed intake and yield, toxin content and quality of milk of cows treated with pure and impure aflatoxin. *J. Dairy Sci.*, *65*:1503-1508, 1982.
13. Barnard, R., and Callahan, W. Follow-up on gastroenteritis attributed to French cheese. *Morbidity Mortality Rep.*, *25*:445, 1971.
14. Brackett, R. E., and Marth, E. H. Association of aflatoxin M_1 with casein. *Z. Lebensm. Unters. Forsch.*, *174*:439-441, 1982.
15. Brackett, R. E., and Marth, E. H. Fate of aflatoxin M_1 in Cheddar cheese and in process cheese spread. *J. Food Prot.*, *45*:549-552, 1982.
16. Brackett, R. E., and Marth, E. H. Fate of aflatoxin M_1 in Parmesan and mozzarella cheese. *J. Food Prot.*, *45*:597-600, 1982.

17. Brackett, R. E., Applebaum, R. S., Wiseman, D. W., and Marth, E. H. Fate of aflatoxin M_1 in brick and Limburger-like cheese. *J. Food Prot.*, *45*:553-556, 1982.
18. Bunning, V. K., Donnelly, C. W., Peeler, J. T., Briggs, E. H., Bradshaw, J. G., Crawford, R. G., Beliveau, C. M., and Tierney, J. T. Thermal inactivation of *Listeria monocytogenes* within bovine milk phagocytes. *Appl. Environ. Microbiol.*, *54*:364-370, 1988.
19. Choisy, C., Gueguen, M., Lenoir, J., Schmidt, J. L., and Tourner, C. The ripening of cheese: Microbiological aspects. In *Cheesemaking—Science and Technology* (A. Eck, ed.), Lavoisier Pub. Inc., New York, 1987.
20. Connors, W. M., and Sfortunato, T. Purification of lactase enzyme and spray-drying with sucrose. U.S. Pat. 2,773,002 (1956).
21. Davidov, R. B., Gul'ko, L. E., and Fainger, B. I. Enrichment of whey with protein and vitamins. Izv. Timiryazevsk. Skh. Akad. 1963, No. 5, 166 171 (1963).
22. de Jong, L. Protein breakdown in soft cheese and its relation to consistency. 2. The influence of the rennet concentration. *Neth. Milk Dairy J.*, *31*:314-327, 1977.
23. Demmler, G. Production of yeast from whey using the Waldhof method. *Milchwissenschaft*, 5:11-17, 1950.
24. Dox, A. W. Proteolytic changes in the ripening of Camembert cheese. U.S. Dept. of Agric., Bur. Anim. Ind., Bul. 109, 1908, pp. 1-24.
25. Doyle, M. P., and Marth, E. H. Thermal inactivation of conidia from *Aspergillus flavus* and *Aspergillus parasiticus*. I. Effects of moist heat, age of conidia, and sporulation medium. *J. Milk Food Technol.*, *38*:678-682, 1975.
26. Doyle, M. P., and Marth, E. Thermal inactivation of conidia from *Aspergillus flavus* and *Aspergillus parasiticus*. II. Effects of pH and buffers, glucose, sucrose, and sodium chloride. *J. Milk Food Technol.*, *38*:750-758, 1975.
27. Doyle, M. P., Applebaum, R. S., Brackett, R. E., and Marth, E. H. Physical, chemical and biological degradation of mycotoxins in foods and agricultural commodities. *J. Food Prot.*, *45*:964-971, 1982.
28. Eichelbaum, G. Process of obtaining food extracts. U.S. Pat. 708,330 (1902).
29. El-Shenawy, M. A., Yousef, A. E., and Marth, E. H. Thermal inactivation and injury of *Listeria monocytogenes* in reconstituted nonfat dry milk. *Milchwissenschaft.*, *44*:741-745, 1989.
30. El Soda, M., Hussein, S., and Ezzat, N. Acceleration of Ras cheese ripening with commercial enzyme preparation. *J. Dairy Sci.* (Suppl. 1), *68*:71 (Abstr.), 1985.
31. Enebo, L., Lundin, H., and Myrbäck, K. Yeasts from whey. *Sven. Kem. Tidskr.*, *53*:137-147, 1941.
32. Ernstrom, C. A., and Wong, N. P. Milk-clotting enzymes and cheese chemistry. In *Fundamentals of Dairy Chemistry*, 2nd ed. (B. H. Webb, A. H. Johnson, and J. A. Alford, eds.), AVI Publishing, Westport, CT, 1974.
33. Farber, J. M., Sanders, G. W., Dunfield, S., and Prescott, R. The effect of various acidulants on the growth of *Listeria monocytogenes*. *Letters Appl. Microbiol.*, *9*:181-183, 1989.

34. Finol, M. L., Marth, E. H., and Lindsay, R. C. Depletion of sorbate from different media during growth of *Penicillium* species. *J. Food Prot.*, *45*:398-404, 1982.
35. Foster, E. M., Nelson, F. E., Speck, M. L., Doetsch, R. N., and Olson, J. C., Jr. *Dairy Microbiology*, Prentice-Hall, Englewood Cliffs, NJ, 1957.
36. Frank, J. F., Marth, E. H., and Olson, N. F. Survival of enteropathogenic and non-pathogenic *Escherichia coli* during the manufacture of Camembert cheese. *J. Food Prot.*, *40*:835-842, 1977.
37. Frank, J. F., Marth, E. H., and Olson, N. F. Behavior of enteropathogenic *Escherichia coli* during manufacture and ripening of brick cheese. *J. Food Prot.*, *41*:111-115, 1978.
38. Gaddi, B. L. Mycotoxin-producing potential of fungi isolated from cheese. Ph.D. thesis, University of Wisconsin, Madison, 1973.
39. Graham, V. E., Gibson, D. L., and Klemmer, H. W. Increasing the food value of whey by yeast fermentation. II. Investigations with small scale fermenters. *Can. J. Technol.*, *31*:92-97, 1953.
40. Graham, V. E., Gibson, D. L., Klemmer, H. W., and Naylor, J. M. Increasing the food value of whey by yeast fermentation. I. Preliminary studies on the suitability of various yeasts. *Can. J. Technol.*, *31*:85-91, 1953.
41. Graham, V. E., Gibson, D. L., and Lawton, W. C. Increasing the food value of whey by yeast fermentation. III. Pilot plant studies. *Can. J. Technol.*, *31*:109-113, 1953.
42. Green, M. L. Review of the progress of dairy science: Milk coagulants. *J. Dairy Res.*, *44*:159-188, 1977.
43. Grieve, P. Use of yeast protease to accelerate Cheddar cheese ripening. *21st Int. Dairy Congress*, Vol. 1, Book 2, 1982, p. 491.
44. Gripon, J. C. Mold-ripened cheeses. In *Cheesemaking—Science and Technology* (A. Eck, ed.), Lavoisier Pub. Inc., New York, 1987.
45. Gripon, J., Le Bars, D., and Vassal, L. Addition of protease from *Penicillium roqueforti* in hard cheese. *21st Int. Dairy Congress*, Vol. 1, Book 2, 1982, pp. 482-483.
46. Hanson, A. M., Rodgers, N. E., and Meade, R. E. Method of enhancing the yield of yeast in a whey medium. U.S. Pat. 2,465,870 (1949).
47. Hanssen, E., and Jung, M. On the occurrence of aflatoxins in foods that are not moldy and suggestions for sampling. *Z. Lebensm. Unters.-Forsch.*, *150*:141-145, 1972.
48. Hayes, P. S., Feeley, J. C., Graves, L. M., Ajello, G. W., and Fleming, D. W. Isolation of *Listeria monocytogenes* from raw milk. *Appl. Environ. Microbiol.*, *51*:438-440, 1986.
49. Hewedi, M. M., and Fox, P. F. Ripening of blue cheese: Characterization of proteolysis. *Milchwissenschaft*, *39*:198-201, 1984.
50. Holzapfel, C. W., Steyn, P. S., and Purchase, I. F. H. Isolation and structure of aflatoxins M_1 and M_2. *Tetrahedron Lett.*, *25*:2799-2803, 1966.
51. Jackson, H. W., and Hussong, R. V. Secondary alcohols in blue cheese and their relation to methyl ketones. *J. Dairy Sci.*, *41*:920-924, 1958.

52. Karahadian, C., and Lindsay, R. C. Integrated role of lactate, ammonia and calcium in texture development of mold surface-ripened cheese. *J. Dairy Sci.*, *70*:909-918, 1987.
53. Kauffmann, W., and Van der Lee, P. J. Method of fermenting whey to produce alcohol. U.S. Pat. 2,183,141 (1939).
54. Kiermeier, F. Aflatoxin M excretion in cow's milk in relation to the quantity of aflatoxin B_1 ingested. *Milchwissenschaft*, *28*:683-685, 1973.
55. Kiermeier, F. The significance of aflatoxins in the dairy industry. Int. Dairy Fed. Doc. No. 30, 1974.
56. Kiermeier, F., and Böhm, S. On aflatoxin formation in milk and milk products. V. Application of the chick embryo test for the affirmation of thin-layer chromatographic determination of aflatoxin in cheese. *Z. Lebensm. Unters.-Forsch.*, *147*:61-64, 1971.
57. Knoop, A. M., and Peters, K. H. Submikroskopische Strukturveränderungen im Camembert-Käse während der Reifung. *Milchwissenschaft*, *26*:193, 1971.
58. Kosikowski, F. *Cheese and Fermented Milk Foods*, Edwards Brothers, Ann Arbor, MI, 1970.
59. Lenoir, J. The surface flora and its role in the ripening of cheese. *Int. Dairy Fed. Bull.*, *171*:3-20, 1984.
60. Lie, J. L., and Marth, E. H. Formation of aflatoxin in Cheddar cheese by *Aspergillus flavus* and *Aspergillus parasiticus*. *J. Dairy Sci.*, *50*:1708-1710, 1967.
61. Liewen, M. B., and Marth, E. H. Evaluation of 1,3-pentadiene for mutagenicity by the Salmonella/mammalian microsome assay. *Mut. Res.*, *157*:49-52, 1985.
62. Liewen, M. B., and Marth, E. H. Growth and inhibition of microorganisms in the presence of sorbic acid: A review. *J. Food Prot.*, *48*:364-375, 1985.
63. Liewen, M. B., and Marth, E. H. Growth of sorbate-resistant and sensitive strains of *Penicillium roqueforti* in the presence of sorbate. *J. Food Prot.*, *48*:525-529, 1985.
64. Liewen, M. B., and Marth, E. H. Viability and ATP content of conidia of sorbic acid-sensitive and -resistant strains of *Penicillium roqueforti* after exposure to sorbic acid. *Appl. Microbiol. Biotechnol.*, *21*:113-117, 1985.
65. Long, J. E., and Harper, W. J. Italian cheese ripening: VI. Effect of different types of lipolytic enzyme preparations on the accumulation of various free fatty and free amino acids and the development of flavor in Provolone and Romano cheese. *J. Dairy Sci.*, *39*:245-252, 1956.
66. Lovett, J., Francis, D. W., and Hunt, J. M. *Listeria monocytogenes* in raw milk: Detection, incidence, and pathogenicity. *J. Food Prot.*, *50*:188-192, 1987.
67. Marth, E. H. Fermentations. In *Fundamentals of Dairy Chemistry*, 2nd ed. (B. H. Webb, A. H. Johnson, and J. A. Alford, eds.), AVI Publishing, Westport, CT, 1974.
68. Marth, E. H., Capp, C. M., Hasenzahl, L., Jackson, H. W., and Hussong, R. V. Degradation of potassium sorbate by *Penicillium* species. *J. Dairy Sci.*, *49*:1197-1205, 1966.

69. McClure, P. J., Roberts, T. A., and Otto Oguru, P. Comparison of the effects of sodium chloride, pH and temperature on the growth of *Listeria monocytogenes* on gradient plates and in liquid medium. *Letters Appl. Microbiol., 9*:95-99, 1989.

70a. Mehaia, M. A., Cheryan, M., and Argondellis, A. Conversion of whey permeate to ethanol. Improvement of fermentor productivity using membrane reactors. *Cult. Dairy Prod. J., 20*:9-12, 1985.

70b. Mergl, M., and Siman, J. Dried yeast feeding stuffs from fermented whey. *Drubeznictvi, 11*(9):136, 1963.

71. Morgan, E. R. Lactase enzyme preparation. U.S. Pat. 2,715,601 (1955).

72. Mucchetti, G., Neviani, E., Yoo Yok, J., and Cabrini, A. Enzymatic systems in cheese making. II. *Latte, 8*:17-25, 1983.

73. Müller, W. R. Growth of yeast in whey. *Milchwissenschaft, 4*:147-153, 1949.

74. Myers, R. P. Drying of yeast to inactivate zymase and preserve lactase. U.S. Pat. 2,762,748 (1956).

75. Myers, R. P., and Stimpson, E. G. Production of lactase. U.S. Pat. 2,762,749 (1956).

76. Myers, R. P., and Weisberg, S. M. Treatment of milk products. U.S. Pat. 2,128,845 (1938).

77. Naiditch, V. Method and plant for treating whey. French Pat. 86,177 (1965).

78. Naiditch, V., and Dikansky, S. Method and equipment for the treatment of whey. French Pat. 1,2.35,978 (1960).

79. Noomen, A. The role of the surface flora in the softening of cheeses with low initial pH. *Neth. Milk Dairy J., 37*:229-232, 1983.

80. Northolt, M. D., Van Egmond, H. P., Soentoro, P., and Deijll, E. Fungal growth and the presence of sterigmatosystin in hard cheese. *J. Assoc. Off. Anal. Chem., 63*:115-119, 1980.

81. Nuñez, M., Medina, M., Gaya, P., and Dias-Amado, C. The yeasts and molds of Spanish mould-ripened Cabrales cheese. *Le Lait, 61*:62-79, 1981.

82. Okumura, J., and Kinsella, J. E. Methyl ketone formation by *Penicillium camemberti* in mold systems. *J. Dairy Sci., 68*:11-15, 1985.

83. Olling, C. H. J. Composition of Fresian whey. *Neth. Milk Dairy J., 17*:176, 1963.

84. Olson, N. F. *Ripened Semisoft Cheese*, Chas. Pfizer & Co., New York, 1969.

85. Papageorgiou, D. K., and Marth, E. H. Fate of *Listeria monocytogenes* during the manufacture and ripening of blue cheese. *J. Food Prot., 52*:459-465, 1989.

86. Parish, M. E., and Higgins, D. P. Survival of *Listeria monocytogenes* in low pH model broth systems. *J. Food Prot., 52*:144, 147, 1989.

87. Park, H. S., Marth, E. H., and Olson, N. F. Fate of enteropathogenic strains of *Escherichia coli* during the manufacture and ripening of Camembert cheese. *J. Milk Food Technol., 36*:543-546, 1973.

88. Pomeranz, Y. Lactase (beta-D-galactosidase). I. Occurrence and properties. *Food Technol., 18*:682-687, 1964.

89. Pomeranz, Y. Lactase (beta-D-galactosidase). II. Possibilities in the food industry. *Food Technol.*, *18*:690-697, 1964.
90. Porges, N., Pepinski, J. B., and Jasewicz, L. Feed yeast from dairy byproducts. *J. Dairy Sci.*, *34*:615-621, 1951.
91. Puhan, Z. Several foreign sour milk products. In *Sour Milk Products*, Dairy Technology Institute, Swiss Federal Institute of Technology, Zürich, Switzerland, 1973.
92. Purchase, I. F. H., Steyn, M., Rinsma, R., and Tustin, R. C. Reduction of the aflatoxin M content of milk by processing. *Food Cosmet. Toxicol.*, *10*:383-387, 1972.
93. Ramet, J. P. Preparation of the curd: The agents of milk conversion. In *Cheesemaking—Science and Technology* (A. Eck, ed.), Lavoisier Pub. Inc., New York, 1987.
94. Rogosa, M., Browne, H. H., and Whittier, E. O. Ethyl alcohol from whey. *J. Dairy Sci.*, *30*:263-269, 1947.
95. Rousseau, M. Study of the surface flora of traditional Camembert cheese by scanning electron microscopy. *Milchwissenschaft*, *39*:129-134, 1984.
96. Rutzinski, J. L., Marth, E. H., and Olson, N. F. Behavior of *Enterobacter aerogenes* and *Hafnia* species during the manufacture and ripening of Camembert cheese. *J. Food Prot.*, *42*:790-793, 1979.
97. Ryser, E. T., and Marth, E. H. Fate of *Listeria monocytogenes* during the manufacture and ripening of Camembert cheese. *J. Food Prot.*, *50*:372-378, 1987.
98. Ryser, E. T., and Marth, E. H. Behavior of *Listeria monocytogenes* during manufacture and ripening of brick cheese. *J. Dairy Sci.*, *72*: 838-853, 1989.
99. Schnurrenberger, L. W., Beck, R., and Pate, J. Gastroenteritis attributed to imported French cheese. *Morbidity Mortality Weekly Rep.* *20*:427-428, 1971.
100. Scott, P. M. Toxins of *Penicillium* species used in cheese manufacture. *J. Food Prot.*, *44*:702-710, 1981.
101. Scott, P. M. Mycotoxigenic fungal contaminants of cheese and other dairy products. In *Mycotoxins in Dairy Products* (H. P. Van Egmond, ed.), Elsevier Appl. Sci., New York, 1989.
102. Sfortunato, T., and Connors, W. M. Conversion of lactose to glucose and galactose with a minimum production of oligosaccharides. U.S. Pat. 2,826,502 (1958).
103. Shih, C. N., and Marth, E. H. Experimental production of aflatoxin on brick cheese. *J. Milk Food Technol.*, *35*:585-587, 1972.
104. Sieber, R. Zur Frage der gesundheitlichen Unbedenklichkeit von in der Käsefabrikation verwendeten Schimmelpilzkulturen. *Z. Ernährungswissenschaft*, *17*:112-123, 1978.
105. Siman, J., and Mergl, M. Growth of strains of *Torulopsis* and *Candida* in whey with added sources of inorganic N and P. *Sb. Vys. Sk. Chem. Technol. v Praze Potraviny Tech.*, *6*:127-138, 1962.
106. Simek, F., Kovacs, J., and Sarkany, I. Production of fodder yeast using dairy by-products. *Tejipar*, *13*:75-78, 1964.
107. Société des alcools du vexin. Method and equipment for the treatment of whey. French Pat. 80,198 (1963).

108. Sood, V., and Kosikowski, F. Accelerated Cheddar cheese ripening by added microbial enzymes. *J. Dairy Sci.*, *62*:1865-1872, 1979.
109. Stimpson, E. G. Conversion of lactose to glucose and galactose. U.S. Pat. 2,681,858 (1954).
110. Stimpson, E. G. Drying of yeast to inactivate zymase and preserve lactase. U.S. Pat. 2,693,440 (1954).
111. Stimpson, E. G. Frozen concentrated milk products. U.S. Pat. 2,668,765 (1954).
112. Stimpson, E. G., and Stamberg, O. E. Conversion of lactose to glucose, galactose, and other sugars in the presence of lactase activators. U.S. Pat. 2,749,242 (1956).
113. Stimpson, E. G., and Trebler, H. A. Aerating method and apparatus. U.S. Pat. 2,750,328 (1956).
114. Stimpson, E. G., and Whitaker, R. Ice cream concentrate. U.S. Pat. 2,738,279 (1956).
115. Stimpson, E. G., and Young, H. Increasing the protein content of milk products. U.S. Pat. 2,809,113 (1957).
116. Stoloff, L. Occurrence of mycotoxins in foods and feeds. In *Mycotoxins and Other Fungal Related Problems* (J. V. Rodricks, ed.), Adv. Chem. Ser. No. 149, Am. Chem. Soc., Washington, D.C., 1976, pp. 23-50.
117. Stuiber, D. A. Whey Fermentation by *Saccharomyces fragilis*. M.S. thesis, University of Wisconsin, Madison, 1966.
118. Tomisek, J., and Gregr, V. Yeast protein manufacture from whey. *Kvasny Prum.*, *7*:130-133, 1961.
119. Troller, J. A. Food spoilage organisms tolerating low-a_w. *Food Technol.*, *54*:72-75, 1979.
120. Tsugo, T., and Matsuoka, H. The formation of volatile sulfur compounds during the ripening of the semi-soft white mold cheese. *16th Int. Dairy Congress Proc.*, Sect. B, 1962, pp. 385-392.
121. Walker, H. W., and Ayres, J. C. Yeasts as spoilage organisms. In *The Yeasts*, Vol. 3, *Yeast Technology* (A. H. Rose and J. S. Harrison, eds.), Academic Press, London, 1970.
122. Walter, H. E., and Hargrove, R. C. Cheese varieties and descriptions. U.S. Dept. Agric. Handbook 54, 1969.
123. Washam, C. J., Kerr, T. J., and Todd, R. L. Scanning electron microscopy of blue cheese: Mold growth during maturation. *J. Dairy Sci.*, *62*:1384-1389, 1979.
124. Wasserman, A. E. The rapid conversion of whey to yeast. *Dairy Eng.*, *77*:374-379, 1960.
125. Wasserman, A. E. Whey utilization. II. Oxygen requirements of *Saccharomyces fragilis* growing in whey medium. *Appl. Microbiol.*, *8*:291-293, 1960.
126. Wasserman, A. E. Whey utilization. IV. Availability of whey nitrogen for the growth of *Saccharomyces fragilis*. *J. Dairy Sci.*, *43*: 1231-1234, 1960.
127. Wasserman, A. E. Amino acid and vitamin composition of *Saccharomyces fragilis* grown in whey. *J. Dairy Sci.*, *44*:379-386, 1961.
128. Wasserman, A. E., and Hampson, J. W. Whey utilization. III. Oxygen absorption rates and the growth of *Saccharomyces fragilis* in several propagators. *Appl. Microbiol.*, *8*:293-297, 1960.

129. Wasserman, A. E., Hampson, J., Alvare, N. F., and Alvare, N. J. Whey utilization. V. Growth of *Saccharomyces fragilis* in whey in a pilot plant. *J. Dairy Sci.*, *44*:387-392, 1961.
130. Wasserman, A. E., Hopkins, W. J., and Porges, N. Whey utilization—growth conditions for *Saccharomyces fragilis*. *Sewage Ind. Wastes*, *30*:913-920, 1958.
131. Wasserman, A. E., Hopkins, W. J., and Porges, N. Rapid conversion of whey to yeast. *Int. Dairy Congr. 1959*, *2*:1241-1247, 1959.
132. Weigmann, H. *Handbook of Practical Cheesemaking*, 4th ed., Verlag Paul Parey, Berlin, 1933.
133. Wendorff, W. L. Studies on the β-galactosidase activity of *Saccharomyces fragilis* and effect of substrate preparation. Ph.D. thesis, University of Wisconsin, Madison, 1969.
134. Wendorff, W. L., Amundson, C. H., and Olson, N. F. Production of lactase by yeast fermentation of whey. *J. Dairy Sci.*, *48*:769-770, 1965.
135. Wilharm, G., and Sack, U. The properties of several lactose-fermenting yeasts. *Milchwissenschaft*, *2*:382-389, 1947.
136. Wiseman, D. W., and Marth, E. H. Behavior of aflatoxin M_1 during manufacture and storage of Queso Blanco and bakers' cheese. *J. Food Prot.*, *46*:910-913, 1983.
137. Wiseman, D. W., and Marth, E. H. Behavior of aflatoxin M_1 in yogurt, buttermilk and kefir. *J. Food Prot.*, *46*:115-118, 1983.
138. Wiseman, D. W., and Marth, E. H. Heat and acid stability of aflatoxin M_1 in naturally and artificially contaminated milk. *Milchwissenschaft*, *38*:464-466, 1983.
139. Wiseman, D. W., and Marth, E. H. Stability of aflatoxin M_1 during manufacture and storage of a butter-like spread, non-fat dried milk and dried buttermilk. *J. Food Prot.*, *46*:633-636, 1983.
140. Wiseman, D. W., and Marth, E. H. Stability of aflatoxin M_1 during manufacture and storage of ice-cream and sherbet. *Z. Lebensm. Unters.-Forsch.*, *177*:22-24, 1983.
141. Wiseman, D. W., Applebaum, R. S., Brackett, R. E., and Marth, E. H. Distribution and resistance to pasteurization of aflatoxin M_1 in naturally contaminated whole milk, cream and skim milk. *J. Food Prot.*, *46*:530-532, 1983.
142. Young, H., and Healey, R. P. Production of *Saccharomyces fragilis* with an optimum yield of lactase. U.S. Pat. 2,776,928 (1957).
143. Yousef, A. E., and Marth, E. H. Degradation of aflatoxin M_1 in milk by ultraviolet energy. *J. Food Prot.*, *48*:697-698, 1985.
144. Yousef, A. E., and Marth, E. H. Use of ultraviolet energy to degrade aflatoxin M_1 in raw or heated milk with and without added peroxide. *J. Dairy Sci.*, *69*:2243-2247, 1986.
145. Yousef, A. E., and Marth, E. H. Quantitation of growth of mold on cheese. *J. Food Prot.*, *50*:337-341, 1987.
146. Yousef, A. E., and Marth, E. H. Stability and degradation of aflatoxin M_1. In *Mycotoxins in Dairy Products* (H. P. Van Egmond, ed.), Elsevier Appl. Sci., New York, 1989.
147. Zerfiridis, G. K. Potential aflatoxin hazards to human health from direct mold growth on Teleme cheese. *J. Dairy Sci.*, *68*:2184-2188, 1985.

13

FUNGAL METABOLITES IN FOOD PROCESSING

RAMUNAS BIGELIS *Amoco Research Center, Amoco Technology Company, Naperville, Illinois*

I. INTRODUCTION

Fungal metabolites play an important role in food processing. They are produced with fermentation or enzyme technology and, in many cases, there are no competitive means for their production. This review discusses fungal metabolites that are used in large volumes by the food industry and briefly considers some compounds that show promise for future food applications.

II. FUNGAL METABOLITES USED IN FOODS

A. Organic Acids

Fungi are producers of a variety of organic acids. However, citric acid, gluconic acid, and malic acid are the only organic acids currently made via large-scale industrial fungal processes. Some organic acids formerly made by fungal fermentation are now produced by direct chemical synthesis. Elevated prices for the precursors of chemical synthesis and new or expanded markets could make fungal fermentation an alternative for the production of these organic acids. The availability of new organisms or genetically engineered fungal strains could also increase the economic feasibility of biotechnological methods for their production.

Organic acids may be natural components of foods, they made be added to foods, or they may be introduced by food fermentation. Citric acid, gluconic acid, and malic acid fit all three categories. Their roles in food processing are considered below.

1. Citric Acid

Industrial output of citric acid exceeds that of most other fermentation-derived primary metabolites. Worldwide production is estimated at more than 350,000 tons [1]. The traditional organisms for citric acid fermentation are selected strains of *Aspergillus niger*, though citric acid has also been produced with several yeasts, mainly species of *Candida*, especially *Candida lipolytica* or its sexual form *Yarrowia* (*Saccharomycopsis*) *lipolytica*.

Many of the citric acid production organisms and much of the fungal fermentation technology are proprietary. Nevertheless, articles that provide details on the citric acid fermentation are numerous and informative [1-14].

Citric acid has many uses [1,8,15-20]. About 70% of all citric acid is used in the food industry, and about 10% of the total output is used in cosmetics and pharmaceuticals. The remainder is employed for diverse industrial purposes, including an increasing use in liquid wash products.

Citric acid is widely used in the food industry owing to its versatility and multifunctional nature as a food additive. It has unrestricted GRAS (generally recognized as safe) status. Citric acid occurs naturally in almost all natural foods and is a component of almost all animal and plant cells. In foods and beverages citric acid serves as an acidulant, preservative, pH regulator, flavor enhancer, chelating agent, stabilizer, and antioxidant. It is highly soluble, permitting use in concentrated syrups. It has extremely low toxicity. And, being a product of fermentation and a native component of many foods, it can be termed "natural," a property of food ingredients that is perceived as healthful and preferred by many consumers.

Citric acid and other organic acids are used extensively to adjust the acid flavor of soft drinks, fruit and vegetable juices, wine and wine-based drinks, ciders, and canned fruit. Citric acid provides a pleasant tartness and complements fruit, berry, and other flavors. Acidulant levels of most fruit-flavored carbonated drinks range from 0.1% to 0.25%, those of uncarbonated drinks range from 0.25 to 0.40%, and those of soft drink mixes vary from 1.5 to 5.0%, depending on flavor. Sufficient citric acid is added to yield a final pH of 2.5-4.5. Being a natural ingredient in many fruits and juices, citric acid effectively brings out flavors and blends well with flavor systems. Sodium citrate plays a similar role in beverages, especially in lemon-lime drinks. Blends of citric acid combined with malic acid, lactic acid, or fumaric acid are also used in beverage formulations. Other functions of citric acid in these applications rely on its ability to prevent metal-catalyzed off-flavors and the deterioration of color. This important food additive also serves as an antimicrobial preservative, retarding the growth of spoilage organisms.

Fruits and vegetables are often treated with acidulants like citric acid during canning. Citric acid lowers the pH of the food, altering processing parameters. Thus, low acid foods can be converted to high acid foods, changing the cook/retort time and temperature. The lowering of pH prevents the growth of certain dangerous organisms, especially *Clostridium botulinum*. The addition of citric acid during processing preserves the natural color of fruit and vegetable pigments. It protects the activity of ascorbic acid by chelating the trace metals that react with it, serving as an antioxidant in this manner. Citric acid also retards the action of naturally occurring enzymes in food, preventing their effect on color, e.g., enzymatic browning, and flavor. Citric acid levels of 0.1-0.3% inhibit color and flavor deterioration during the processing of frozen fruits and vegetables and during canning.

Citric acid, combined with erythorbic or ascorbic acid, is suitable as a sulfite replacer to prevent browning of fresh fruits and vegetables. Citric acid acts as a chelating agent, removing the copper prosthetic group from polyphenol oxidases, which are responsible for the browning reaction. In addition, such acidification lowers the pH below the optimum (~6-7) for oxidative enzymes. Fresh food material dipped in a citric acid/ascorbic acid bath for 30 seconds, packaged in oxygen-permeable bags, and then refrigerated resists discoloration as well as or better than sulfite-treated material. Immersion of fresh vegetables in a 1-2% citric acid bath for 30 seconds prevents browning for 2-4 hours at room temperature. A solution of 0.3-1.0% citric acid and 0.05-0.5% ascorbic or erythorbic acid used for 30-45 seconds permits even longer preservation if the food item is vacuum-packaged and then refrigerated [21-23].

In jams, jellies, preserves, and pie fillings citric acid serves as a means of adjusting flavor. It also maintains the correct pH in the range of 3.0-3.5 for pectin setting. During processing, citric acid is added as a 50% solution to permit even distribution throughout the batch. Citric acid plays a comparable role in the acidification of gelatin desserts, where it maintains the pH near the isoelectric point of gelatin, permitting proper gelation as well as tartness.

In candies, citric acid is added mainly for tartness and a pleasant sour taste. It also prevents graining, that is, the crystallization of sucrose in the candy product. The confectionary industry uses concentrations of 0.5-1.0% in its products, although sour candies may contain up to 2% citric acid. Citric acid is usually added with color and flavor to molten candy glass after the cook to avoid acid-catalyzed sucrose hydrolysis. When added to the candy batch during the cooking process, it prevents graining by promoting conversion of sucrose to glucose and fructose, both of which are more soluble than sucrose. Agar- and starch-containing jellies in confectioneries contain citric acid as a flavoring agent. Citric acid is an important addition to pectin jellies since it affects not only flavor, but also the gelling properties of the pectin.

Sodium citrate finds wide use in dairy products. The citrate salt aids in the emulsification of processed cheeses and cheese foods. Levels of up to 3% are permitted in such cheese products in the United States. It prevents fat separation, maintains the flexibility and texture of cheese slices, prevents adherence of sliced cheese, and promotes favorable and uniform cheese-melting properties, all without detracting from cheese flavor. Sodium citrate is a stabilizer in whipping cream products and dairy substitutes derived from vegetables. It also promotes the whippability of ice cream and custard products.

The addition of citric acid to wines and ciders is used to dissolve the tannin-iron and phosphate-iron complexes that cause turbidity. Furthermore, citric acid can prevent an increase in the pH of white wine and the accompanying browning reaction that sometimes occurs during storage.

Citric acid at concentrations in the range of 0.005-0.2% is used to chelate metals in fat-containing foods, thereby retarding the development of rancidity. Monostearyl citrate, a fat-soluble food additive, is especially effective in sequestering heavy metals which promote oxidation reactions. This citrate derivative is suitable for use in solid fats, plant and animal oils, and fish products.

Citric acid plays an important role in seafood processing [24,25,26]. Citric acid helps to maintain the stability and flavor of seafood products by inactivating endogenous enzymes and enhancing the action of antioxidants. Frozen fish and shellfish are dipped in a solution of 0.25% citric acid (also containing 0.25% erythorbic or ascorbic acid) to inactivate certain fish enzymes and sequester trace metals that lead to rancidity and discoloration. Citric acid is used to chelate iron and copper ions, which catalyze the oxidative formation of off-flavors and fishy odors associated with dimethylamine. Dimethylamine is formed by natural fish enzymes or bacteria from trimethylamine, a decomposition product of phospholipids. Citric acid, used in conjunction with other organic acids, retards the formation of dimethylamine. Citric acid also deters the metal-catalyzed oxidation of unsaturated fatty acids. High temperature, light, and lipases can generate such fatty acids that can act as precursors of off-flavor and malodorous compounds. Citric acid is used in shellfish processing as well. It is an effective inhibitor, when used at levels of 0.1-0.5%, of tyrosinases and phenol oxidases that cause the browning of shrimp and the blueing of crabmeat. Off-odor formation can also be retarded by citric acid. Other uses of citric acid in the seafood industry have been reviewed by Porter [24], including its use in the prevention of struvite (magnesium ammonium phosphate) formation during canning.

Fermented foods may contain citric acid that was produced during the microbial fermentation stage. Sauerkraut, fermented sausages and fish, pickles, olives, and condiments owe their flavor, in part, to the presence of citric acid.

Citric acid derivatives can deliver essential nutrients in dietary supplements and are used to fortify foods. Ferric ammonium citrate, calcium citrate, and magnesium citrate are examples of such derivatives.

2. Gluconic Acid

Aspergillus niger, the organism first observed to overproduce gluconic acid, is still used today to make this metabolite from glucose by submerged fermentation. The high yields approach 95% and make the process competitive with electrochemical synthesis. More than 45,000 tons of gluconic acid are produced annually worldwide, mainly as the sodium salt, which is used as a sequestering agent for calcium or iron. Iron (II) gluconate serves as an iron source in treatment of anemia, whereas calcium gluconate is used to treat calcium deficiencies via oral or intravenous therapy. Gluconic acid itself is employed as a cleaning agent. A portion of the industrial output of gluconic acid is converted to δ-gluconolactone, a slow-acting acidulant which slowly hydrolyzes to gluconic acid [6,26-30]. It is used with sodium bicarbonate for controlled leavening of baked goods. The lactone is also used in the dairy industry for cheese curd formation and for the improvement of the heat stability of milk. In addition, it serves as an additive in cured meat products, luncheon meats, tofu, and packet dessert mixes.

3. L-Malic Acid

Fungal fermentations for L-malic acid have been investigated, but DL-malic acid is currently made by chemical synthesis in Western countries [31,32]. In China, some L-malic acid is manufactured from fumaric acid with

TABLE 1. Fungal Processes for Amino Acid Production

Amino acid	Organism and enzyme(s)	Process	Reference
L-Lysine[a]	*Cryptococcus laurentii* L-aminocaprolactam hydrolase and *Achromobacter obae* aminocaprolactam racemase	Conversion of synthetic DL-α-amino-ε-caprolactam to L-lysine by two enzymes	44
L-Methionine[b]	*Aspergillus oryzae*	Conversion of synthetic DL-methionine to *N*-acetyl derivative, then asymmetric hydrolysis with immobilized aminoacylase	42,43
L-Phenylalanine[c]	*Rhodotorula rubra* phenylalanine ammonia lyase	Conversion of *trans*-cinnamic acid to L-phenylalanine with immobilized cells in presence of 7.85 M ammonia	45

[a]L-Lysine is also produced by bacterial fermentation.

[b]DL-Methionine is also synthesized by chemical means; aminoacylase has been used to synthesize L-valine, L-phenylalanine, and other L-amino acids.

[c]Most L-phenylalanine is produced by bacterial fermentation.

immobilized *Candida rugosa* cells at an overall yield of 82-85% [33] and processes in Japan employ fumarase from immobilized or intact cells of *Brevibacterium* spp. [34-36].

Malic acid plays a role similar to that of citric acid as a food ingredient. As an acidulant, it has a tart, smooth, long-lasting flavor profile. Malic acid exhibits a synergy with aspartame permitting a 10% reduction in the usage of this sweetener in beverages [37].

4. Other Organic Acids

Fermentation processes employing fungi have been used for the production of organic acids other than citric acid and gluconic acid, but have been replaced by chemical methods. Thus, fumaric and lactic acids are now prepared more economically by chemical means, or, in the case of the latter compound, also by bacterial fermentation. Tartaric acid, succinic acid, and erythorbic acid (D-araboascorbic acid) are obtained chemically, though fungal sources for these food additives have been described [11,31].

B. Amino Acids

Amino acids are used primarily as nutritional supplements, as food flavoring agents, and as starting materials for pharmaceutical production. Almost all of the 20 standard amino acids are manufactured by bacterial processes, though fungal technology is also used in a few cases [38-41].

Some amino acids are synthesized with immobilized fungal cells or enzymes (Table 1). L-Methionine is synthesized with immobilized *Aspergillus oryzae* aminoacylase [42,43]; L-lysine is produced with L-amino-caprolactam hydrolase from *Cryptococcus laurentii* [44]; and L-phenylalanine is produced with immobilized *Rhodotorula rubra* cells bearing phenylalanine ammonia lyase [45]. Phenylalanine is in demand for the synthesis of aspartame, a low-calorie, artificial sweetener used in soft drinks and desserts [46-47].

Tryptophan, lysine, and methionine, all essential amino acids in the human diet, are used to enrich foods. Certain fungi are able to produce substantial amounts of these amino acids but are surpassed by improved industrial strains of bacteria. Strains of *Pichia anomala* are prolific producers of tryptophan, making up to 6-14 g/L, respectively, from the added precursors anthranilic acid or indole [48]. Strains of the yeasts *Candida utilis* and *Saccharomyces cerevisiae* [49] and the molds *Ustilago maydis* and *Gliocladium* spp. [50-51] produce significant quantities of lysine. Mutants of the yeasts *Candida tropicalis* [52], *Y. lipolytica* [53], *Pichia pastoris* [54], and *C. utilis* [55] have an increased methionine content and may have potential as food supplements.

C. Nucleotides and Related Substances

Industrial production of flavor nucleotides originated in Japan. Today over 2000 tons of inosinic acid (5'-IMP) and 1000 tons of guanylic acid (5'-GMP) are manufactured there every year, primarily by direct fermentation with improved strains of *Bacillus* or *Brevibacterium* species. Both nucleotides are also produced in other countries and used as food additives. Review articles have summarized progress on the microbial production of these and other nucleic acid-related compounds [56-66].

1. 5'-IMP and 5'-GMP

The nucleotides IMP and GMP, and also monosodium-L-glutamate (MSG), are termed *umami* compounds in Japan and flavor enhancers or potentiators in English-speaking countries [67-69]. Umami compounds appear to be chemically related to substances that give rise to the four basic tastes (salty, sweet, sour, and bitter), though their taste properties are independent. The three primary umami compounds mentioned above have been used for centuries in Japan in the form of seasonings derived from kelp (*konbu*). dried bonito fish (*katsuobushi*), and Japanese mushroom (*shiitake*). They were first identified as the principal umami substances in these sources and then isolated and characterized.

GMP, IMP, and to a lesser extent xanthylic acid (XMP) cause the development of a fuller flavor and impart a "meaty" or "mouth-filling" taste to foods, acting alone or synergistically with MSG (Fig. 1). Besides developing a sense of greater smoothness, body, or viscosity in certain foods, they also have the capacity to suppress undesirable flavors and aromas [57,69-74]. The molecular basis for the flavor activity of umami compounds has undergone considerable investigation and has been reviewed elsewhere [58,69].

Industrial production of 5'-IMP and 5'-GMP began in 1961 and involved the enzymatic degradation of ribonucleic acids with nucleolytic enzymes from selected strains of *Penicillium citrinum* or *Streptomyces aureus*. These and

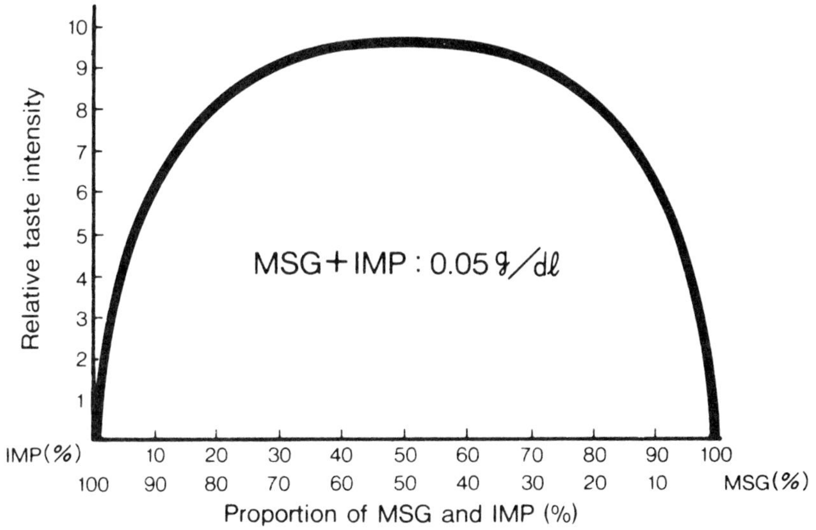

FIGURE 1. Synergistic effect of MSG and IMP on the relative taste intensity of umami. (From Ref. 70.)

other methods are still used today, though bacterial fermentation is much more common (Fig. 2). The RNA is derived from yeast, usually improved mutants of *Candida utilis* which are grown under conditions that maximize RNA content. RNA levels can reach 10-15% of the cell dry weight. Adenylic acid deaminase from *A. oryzae* is used to convert 5'-adenosine monophosphate (5'-AMP) to 5'-IMP. In 1967 an alternate process was introduced that employs alkaline hydrolysis of yeast RNA followed by chemical phosphorylation of the nucleosides. Adenosine is then converted to inosine by deamination with nitrous acid. In both cases, 5'-IMP and 5'-GMP are purified by ion exchange chromatography, treated with charcoal, and then precipitated with ethanol [65,67].

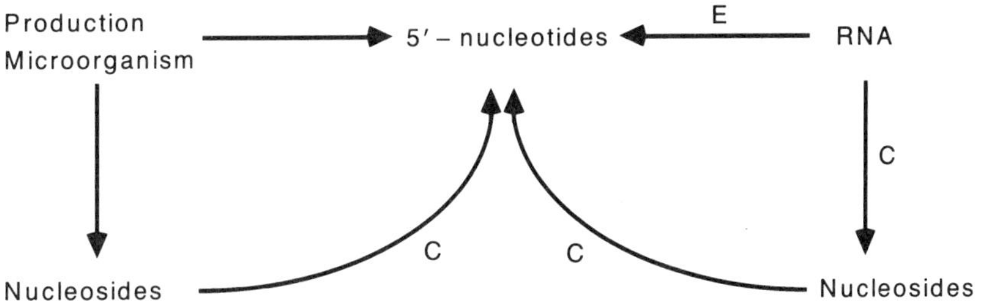

FIGURE 2. Industrial production of 5'-nucleotides via four routes involving direct fermentation, chemical (C) processing, or enzymatic (E) processing. The nucleotide 5'-GMP is also made via the intermediate AICAR.

The production of 5'-mononucleotides from yeast RNA using both immobilized 5'-phosphodiesterase from *Penicillium* and 5'-AMP deaminase from *Aspergillus* has been investigated by Samejima et al. [75] but has not replaced the batch system that uses native enzymes. Marginal savings in enzyme costs preclude commercialization of the process, whose economics are determined primarily by the cost of the RNA substrate. Newer, improved methods with immobilized enzymes may be more efficient, however [76].

The flavor nucleotides are usually produced by direct fermentation with bacterial species to yield the nucleotide or its precursor (Fig. 2). Improved industrial strains of *Brevibacterium ammoniagenes* accumulate high levels of 5'-IMP. Alternatively, the nucleosides inosine and guanosine are excreted by industrial production organisms, typically mutants of *B. ammoniagenes* or *Bacillus subtilis*. The nucleosides are phosphorylated chemically or enzymatically, yielding 5'-IMP and 5'-GMP. Another method for the production of 5'-GMP involves chemical conversion of 5-amino-4-imidazolecarboxamide (AICAR) excreted by *Bacillus megaterium* to the 5'-nucleotide [62,65, 66].

2. Nucleic Acid-Related Substances

Other nucleic acid-related substances are produced on a large scale, primarily in Japan. Some processes rely on enzymatic or chemical degradation of yeast RNA or fungal fermentation. These compounds are used primarily as starting materials for pharmaceutical production or as nutritional supplements [64-65,67,77-78].

D. Vitamins and Nutritional Supplements

Vitamins are essential mediators of metabolism and play an important nutritional role in the human diet. They are often supplements in foods and animal feeds. Microbiological methods for the production of most of the vitamins have been described, yet the majority are still extracted from biological materials or are chemically synthesized. Only riboflavin (vitamin B_2) and cyanocobalamin (vitamin B_{12}) are produced by fermentation on a commercial scale. Industrial production of ascorbic acid involves chemical synthesis and uses a single microbiological step; biotechnological process improvements for vitamin C production that rely on microbial fermentation are anticipated. The journal and patent literature documents these and many other processes for potential production of vitamins by microbial fermentation [11,79-85].

The following discusses the production of riboflavin, vitamin D, vitamin A, and nutritional supplements using mycological methods.

1. Riboflavin

Riboflavin is used as a nutritional supplement to food products, and a derivative, riboflavin-5'-phosphate, is used as a food coloring. Approximately 1.25 million kg of riboflavin are manufactured worldwide every year by chemical synthesis, a semi-synthetic process, or by direct fermentation with fungi or bacteria [84-86]. About half of all commercial riboflavin is made by the semi-synthetic method, in which microbially derived D-ribose is chemically converted to riboflavin. The D-ribose is obtained from hyper-producing mutants of *Bacillus subtilis* or *B. pumilis* [87].

Many fungi and bacteria can produce riboflavin directly. The two ascomycetes *Eremothecium ashbyii* and *Ashbya gossypii* have attracted the most attention since their levels of riboflavin are suitable for industrial production [79,82,84,88-90]. Industrial fermentation with *A. gossypii* is now preferred and generates about 30% of world output. The yield of riboflavin is 10-15 g/L, compared to 2.5 g/L for *E. ashbyii* [91]. Mutants of *A. gossypii* obtained after mutagenesis and screening are grown at 28°C in aerated medium containing glucose, sucrose, or maltose as the carbon source and a protein hydrolyzate as the nitrogen source. Glycine supplements enhance riboflavin production significantly, and the addition of plant oils, or their use as a carbon source instead of the carbohydrate, reportedly boosts yields even more [86,92].

During the fungal fermentation *A. ashbyii* is allowed to proceed to the end of the growth phase, at which point the size of a vegetative mycelia and the number of vacuoles increase. Riboflavin crystallizes in cellular vacuoles and also accumulates in the medium. Riboflavin solutions destined for pharmaceutical use or food supplements are treated with reducing agent to cause specific precipitation of reduced riboflavin, reoxidized, and then crystallized. On the other hand, the culture broth can be concentrated, dried, and added to animal feeds [84,91].

2. Vitamin D

Sterols are widely distributed among the fungi [93,94]. Ergosterol, a commercially important lipid, can be extracted from dehydrated yeast, irradiated with ultraviolet light, and converted to vitamin D [95]. Efficient ergosterol-producing yeast strains have been isolated by screening natural isolates [96], by hybridization and polyploidization [97], or by mutagenesis followed by enrichment with echinocandin and nystatin [98]. Such yeast strains may serve as alternative sources of vitamin D in the future. At present, synthetic vitamin D is added to dairy products, often milk and margarine, and sometimes to breakfast cereals and baby foods.

3. Vitamin A

Carotenoid pigments are often precursors of vitamin A when ingested by animals and are added to some foods and animal feeds as coloring agents. Most commercial β-carotene is synthetic, though algal material is also available. Fungal sources are being investigated. Interest has focused on *Blakeslea trispora*, a heterothallic fungus of the Mucorales group which is extremely rich in mycelial β-carotene, as a potential source of carotenoids [99,100]. The β-carotene yield is as much as 1 g per liter of medium or 20 mg per gram of dry mycelia when both sexual forms are grown together [101]. *Phycomyces blakesleeanus* also produces substantial amounts of β-carotene. A process has been patented for the formation of intersexual heterokaryons of *P. blakesleeanus*, strains having nuclei of both the + and - sexual types, from mutants that overproduce β-carotene [102]. The heterokaryon strains produce 0.5 mg of β-carotene per gram dry weight cells, but are unstable. Superproducing strains of *P. blakesleeanus* have been isolated which synthesize 25 mg of β-carotene per gram dry weight of

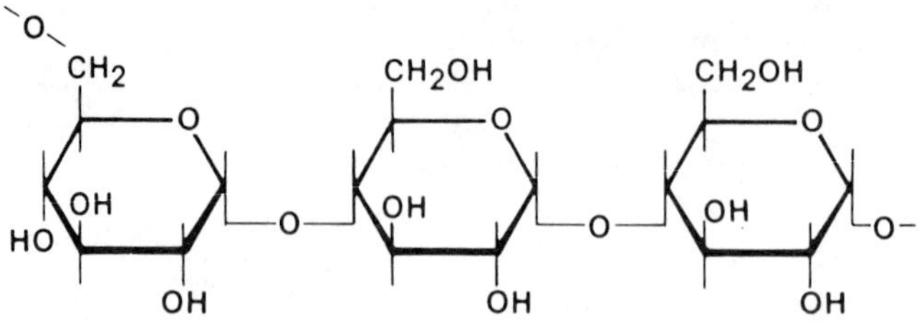

FIGURE 3. Repeating unit of pullulan.

of cells [103,104]. Further research could lead to commercial production of carotenoid coloring agents and vitamin A precursors derived from fungi.

4. Nutritional Supplements

Dried brewer's, baker's, and torula yeasts, as well as species of *Candida*, have been used to fortify foods, vitamin supplements, and animal feed. Microbial protein used for this purpose is termed single-cell protein (SCP). Single-cell protein is obtained by large-scale fermentations that often use cheap raw material or waste biomass from food processors. Whole cells of suitable fungi obtained in this way may become an important future source of food, food ingredients, or nutritional supplements [105,106].

E. Polysaccharides

Pullulan and scleroglucan have attracted considerable interest as industrial fungal polymers with unique properties. Numerous other fungal polysaccharides have been investigated, but food applications have been proposed for only a few of them.

1. Pullulan

Pullalan is a neutral, linear homopolysaccharide composed of maltotriose units, i.e., three α-1,4-linked glucose molecules which are held together by α-1,6 linkages (Fig. 3). Occasional, randomly inserted maltotetraose units and some α-1,3 linkages may also be present [107]. The dimorphic black yeast *Aureobasidium* (*Pullularia*) *pullulans* is used for industrial production of this α-D-glucan. The molecular weight of the polysaccharide can range from 10^4 to 10^6 daltons, depending on strain, substrate, phosphate supplements, pH, and the length of the fermentation. Pullulan synthesis is induced by glucose, fructose, or certain disaccharides, and more than 70% of the sugar substrate can be converted to polymer. Cells are grown at room temperature in a fermentation medium which contains thiamine and other nutrients under slightly acidic conditions. Cell growth precedes pullulan production and then elaboration of the exopolymer coincides with the depletion of nitrogen in the medium. Reduced availability of nitrogen favors a transition of *A. pullulans* from a mycelial form to the yeast-like form which produces the polysaccharide [108-110].

Pullulan does not form viscous solutions at low concentrations, but, owing to its unique physical properties, it can be molded into a variety of shapes, in addition to serving as an adhesive. Its ability to form strong

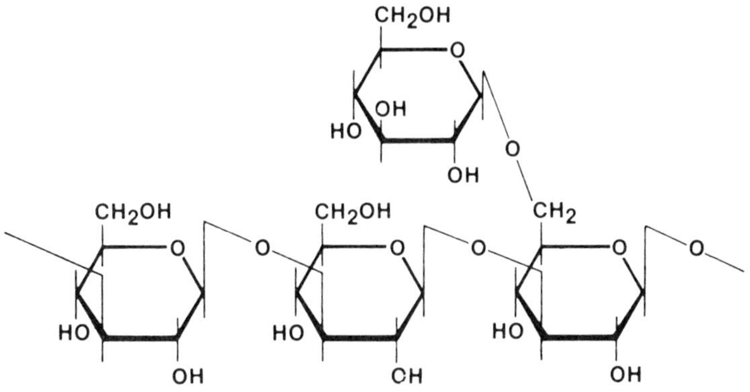

FIGURE 4. Repeating unit of scleroglucan.

transparent films with low oxygen permeability enhances its usefulness as a food coating and packaging agent. The properties of these films can be modified by blending with starch, gelatin, or plasticizers. If approved for such use, pullulan may also find application as a tasteless, odorless dietetic food ingredient, and perhaps also serve as a food texturizer and preservative. It is apparently non-toxic and non-digestible. Pullulan products in a range of molecular weights, and their esterified derivatives, have been manufactured in Japan since 1974 [111-113].

2. Scleroglucan

Scleroglucan (Polytran®, Actigum CS®) is produced by industrial fermentation with *Sclerotium glucanicum* or with *S. rolfsii*. As a commercial product, it has competed with xanthan biopolymer for use in enhanced oil recovery [114]. The polymer is a highly branched, neutral homopolysaccharide with a backbone of about 90 D-glucose units in β-1,3 linkage bearing single D-glucose side groups attached in β-1,6 linkage to every third or fourth backbone residue [115] (Fig. 4). Molecular weight and the extent and length of branching varies with each species of *Sclerotium*. The high molecular weight of scleroglucan and its extreme molecular rigidity in water make the purified form an excellent viscosity enhancer in hot or cold water even at low polymer concentrations [116]. X-ray diffraction data and conformational studies indicate that scleroglucan is triple-stranded in solid fibers [115,117].

Scleroglucan is produced by submerged fermentation in batch or continuous culture with a yield of 50% based on input glucose. The mold grows with a pellet morphology under these conditions. The pH drops from 4.5 to 2.0 during the fermentation as a result of oxalic acid accumulation. Once the fermentation is complete, the fermentation liquor is heated to inactivate glucanases. Homogenization detaches glucan adhering to the mycelia. The exopolymer is then precipitated with alcohol, dried, and milled [114,118].

Scleroglucan exhibits pseudo-plasticity over a broad range of temperature, pH, and salt conditions, making it suitable for use as a food additive when approved by regulatory agencies. It may serve as a suspending, coating, and gelling agent [113]. Studies with chicks and dogs indicate it has the caloric equivalence of starch and lowers cholesterol levels [119].

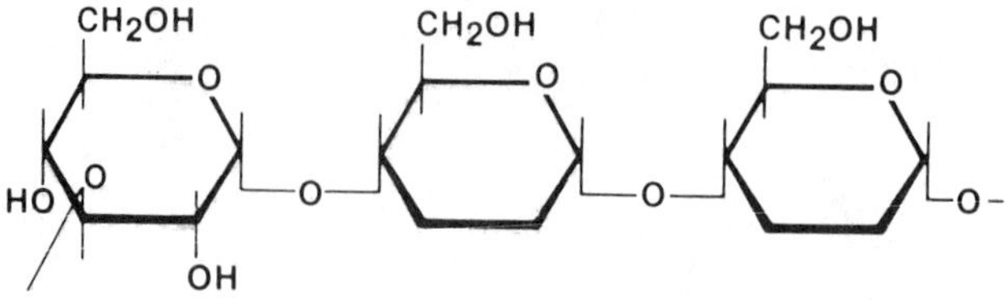

FIGURE 5. Repeating unit of elsinan.

Scleroglucan may also find application in the cosmetics and pharmaceutical industries [120].

3. Other Fungal Polysaccharides

Other fungal polysaccharides may be considered for food use once applications research and safety studies have been completed. Among these, elsinan, an extracellular glucan (Fig. 5) produced by the mold *Elsinoe leucospila* and related species, may be suitable for packaging or coating foods [113,121]. Baker's yeast glycan (BYG®) is a biopolymer with projected use in the food industry as a thickening agent. This glycan has the mouthfeel of a fat or oil and may serve as a lipid substitute and thickener in certain diet foods, frozen desserts, cheese products, sauces, soups, and dairy drinks (Table 2) [113,122]. Other fungal polysaccharides, including various mannans and phosphohexans, have not been examined closely by food scientists and may also have potential as food ingredients [110,113,118,123-128].

F. Lipid Substances

About 70% of the world's supply of fats and oils is derived from plants. Currently no food lipids are produced by microbiological means. Advances in agricultural genetics and enzyme engineering, as well as economic considerations, could influence the production and supply of food lipids in the future. A dwindling global food supply or periods of hardship may also be factors. It is notable that processes for fat production with *Candida* and *Fusarium* species were developed in Germany and seriously considered as a means to alleviate food shortages during both world wars [129,130].

Fat-producing yeasts and molds have been thoroughly surveyed [94, 129-137]. The potential for commercialization of lipids and fatty acids from yeasts and molds has been considered as well [138-141].

1. Lipids

Research on microbial lipid production has focused on the evaluation of microorganisms with a high fat content. The yeasts *Rhodotorula gracilis*, *Lipomyces lipofer*, *Lipomyces starkeyi*, *Endomycopsis vernalis*, *Cryptococcus terricolus*, and *Candida* sp. no. 107 accumulate significant levels of fat. The molds *Mortierella vinacea*, *Mucor circinelloides*, *Aspergillus ochraceus*, *A. terreus*, and *Penicillium lilacinum* are also good producers of fat, as are certain fungi of the genera *Hansenula*, *Chaetomium*, *Cladosporium*, *Malbranchea*, *Rhizopus*, and *Pythium*. According to Ratledge, ideal lipid-producing fungi

TABLE 2. Baker's Yeast Glycan—Potential Applications

Frozen desserts:	
Chocolate pudding	Ice cream
Cheese cake	Sherbet
Cheese spread and dips:	Dairy drinks:
Sour cream dip	Thick milk shake
Cream cheese spread or dip	Skim milk
Cheddar cheese spread or dip	Hot chocolate
Low calorie margarine	Soups:
Sauces:	New England clam chowder
Canned white sauce	
Others:	
Salad dressings (pourable and spoonable)	Extruded french fried potatoes
	Cake frosting
Fruit-gel dessert	Yellow cake
Batter breading for fish	"No-cook" fudge
Coating agent on corn flakes	Coffee creamer

Source: Ref. 113.

should be rapidly growing organisms that efficiently produce concentrated nontoxic lipids that are easily extracted and handled in bulk, in addition to being suitable alternatives to animal or plant fats. The high lipid content of numerous oleaginous fungi approaches 60% of the cell dry weight and makes them candidates for food use. As a rule about 80% of the total lipid content is represented by triglycerides. The saturated and unsaturated fatty acid composition is quite variable. The remaining 20% of the content is composed of phospholipids, sterols, and sterol esters [130,139].

2. Fatty Acids

Fatty acids for the food, medical, and cosmetics industries could be standard products of fungal fermentation in coming years. A process in Great Britain is already used to yield an oil rich in γ-linolenic acid from food-approved *Mucor* sp. [141], and a method that employs *Mortierella* sp. for the production of this fatty acid is expected to be operational in Japan [142]. The refined oil from *Mucor* sp. contains twice as much γ-linolenic acid as the best plant oils and is of a more consistent quality. The fungal mass of *Mortierella* sp. contains 37-53% by weight of lipid and yields 30-83 g/L, of which γ-linolenic acid is 8-11% by weight. The fungal oils will be an attractive alternative to material that is laboriously extracted from seeds of evening primrose (*Oenothera binnis*) at present and used in dietary supplements.

γ-Linolenic acid is an essential acid in human nutrition and has been shown to be an effective and practical means of lowering human plasma cholesterol levels. Mold oil (containing 6% γ-linolenic acid and 10% linoleic acid) from *Mortierella ramanniana* var. *angulispora* lowers blood cholesterol in rats fed a cholesterol-enriched diet and has an effect similar to that of safflower oil (containing 75% linoleic acid) [143]. Mold oil may have a comparable hypocholerolemic activity in humans.

Some polyunsaturated fatty acids exhibit unique biological activities and serve as natural precursors of compounds such as prostaglandins, thromboxanes, and leucotrienes. Recent data suggest that certain fungi may serve as practical and reliable natural sources of these and other polyunsaturated fatty acids and may replace marine fish oils as the main source. For example, *Mortierella alpina* [144,145] and *Mortierella elongata* [146] produce large amounts of arachidonic acid; *Mortierella* spp. produce significant quantities of 5,8,11,14,17-*cis*-eicosapentaenoic acid (EPA) [147,148]; *M. alpina* is a rich source of 8,11,14-*cis*-eicosatrienoic acid [149]; and immobilized *Mucor ambiguus* secretes γ-linolenic acid in the presence of nonionic surfactants [150]. An *M. alpina* strain can even convert various natural oils, especially linseed oil, containing α-linolenic acid as 60% of the total fatty acids, to an oil rich in EPA and arachidonic acid [151].

Many fungi are known to accumulate fatty acids and their derivatives both intra- and extracellularly. Filamentous fungal and yeast fatty acids have been comprehensively reviewed by Lösel [137] and Rattray [94], respectively.

G. Ethanol

Ethanol is a traditional product of biotechnology [152,153]. Significant advances have been made in the improvement of yeast strains used in the brewing and industrial fermentation ethanol industries. Yeast molecular genetics now permits the application of recombinant DNA technology to strain and process improvement [154-158]. The application of gene technology is attracting considerable interest and undoubtedly will have impact on the beverage alcohol industry. A substantial literature on ethanol manufacture has appeared in recent years and should be consulted for information on the latest developments.

H. Other Metabolites

Other classes of fungal metabolites have been investigated for their potential as food additives and processing aids. They are discussed only briefly below, and references are provided instead.

Polyhydric alcohols have many food applications [159]. Fungal polyhydric alcohols may be developed as humectants, sweetening agents, or preservatives with food uses. Fungi are known to produce a variety of such compounds [11,160-164].

Fungal extracellular glycolipids which act as biosurfactants have been proposed as food ingredients. Such compounds could act as emulsifiers; wetting, spreading, and penetrating agents; solubilizers and dispersants; foaming or defoaming agents; and detergents in food applications [165,166]. For example, sophorolipids (Fig. 6) from *Candida bombicola* have been proposed as bread dough conditioners [167].

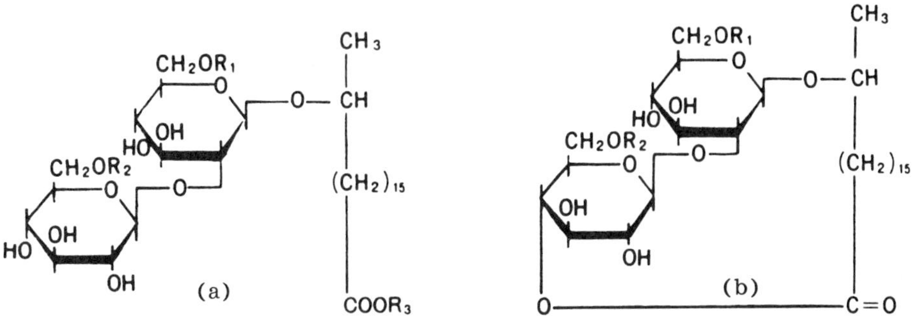

FIGURE 6. Major sophorolipids produced by *Candida bombicola* shown with 17-hydroxyoctadecenoic acid. (a) Type I is a 17-L-[(2'-O-β-D-glucopyranosyl-β-D-glucopyranosyl)oxy]octadecanoic acid 6',6"-diacetate, where R_1 = R_2 = acetyl and R_3 = H. In the acid form R_1 = R_2 = R_3 = H, and in the methyl form R_1 = R_2 = H and R_3 = methyl. (b) Type II is the 1,4"-lactone derivative of type I. Type III is the 6'-deacetylated derivative of type II.

The phytopathogenic fungus *Gibberella fujikuroi*, which causes rice disease in the Orient and pink ear-rot of corn in the United States, is used to make gibberellins, an important group of plant hormones (Fig. 7) [168]. More than 60 compounds of the group are known. Microgram amounts of gibberellins have marked effects on plant germination, growth, flowering, and ripening, making them valuable as agricultural chemicals. Gibberellins are also sometimes added to barley to reduce the time needed for malting and to induce amylase production during beer brewing. They are applied to cotton and certain fruit and vegetable crops as a growth regulator. For example, application delays citrus fruit ripening, increases biennial plant development, improves the yield of seedless grapes, and prolongs the harvesting period for artichokes. The major commercial use is to increase fruit size and sugar cane yield [169-172].

Zearalenone and its natural or chemical derivatives are fed to cattle as a growth promoter (Fig. 8). These polyketides are manufactured by submerged fermentation using high-yielding mutants of *Gibberella zeae* (*Fusarium graminearum*), the causative agent of corn red-ear rot. In fungi, zearalenone acts as an essential sex hormone regulating perithecia formation and thus reproduction. Based on studies with mammals, in the liver it is metabolized to α-zearalenol, a compound with estrogenic activity. When used as a growth promoter, zearalanol is administered to steers and calves

FIGURE 7. Classes of gibberellins produced by *Gibberella fujikuroi*. (a) Gibberellin A7, R = H; Gibberellin A3, R = OH. (b) Gibberellin A4, R^1 = OH, R^2 = H, R^3 = H.

FIGURE 8. (a) Zearalenone. (b) α-Zearalanol.

via an ear implant, which releases a growth-promoting activity comparable to that of diethylstilbesterol, but without its undesirable side effects [173-176].

Many fungi produce pigments. Food scientists have examined some as potential sources of alternatives to synthetic red food colorants. Among these fungi are species of *Monascus* and *Phaffia rhodozyma*. Species of *Monascus*, especially *M. purpureus* and *M. anka*, have been traditionally grown on rice in the Orient since ancient times to produce a red mass which can be added to foods and beverages or dried to a powder and used as needed. The four main *Monascus* pigments (Fig. 9) are heat stable and unchanged over a pH range of 2-10, besides having solubility properties adaptable to water- or oil-based systems. The colors range from yellow to orange to red and even purple, depending on culture conditions, species and strain, means of pigment preparation and storage, and possible chemical substitution on the polyketide structure. In the last 15 years 38 patents pertaining to *Monascus* colorants have appeared. They have been proposed for use with meat products, meat substitutes, vegetable protein, fish, candies, cakes, milk products, and various beverages. In cured meats their addition yields a color similar to that developed by nitrite treatment [177,178].

Phaffia rhodozyma accumulates the red pigment astaxanthin, also known as 3,3'-dihydroxy-β,β-carotene-4,4'-dione [179,180] (Fig. 10). Mutants with increased astaxanthin content have been isolated [181]. The pigment is found naturally in marine animals or poultry, depending on diet. Fungal astaxanthin has been proposed by Johnson et al. [182,183] as a dietary supplement for poultry, and also for crustaceans and fish grown by aquaculture. High levels of deposition can be achieved in egg yolks, lobsters, and rainbow trout. Astaxanthin, when fed as a yeast preparation to commercial pen-reared salmonids, could yield fish with improved flavor and color more efficiently than the synthetic carotenoid.

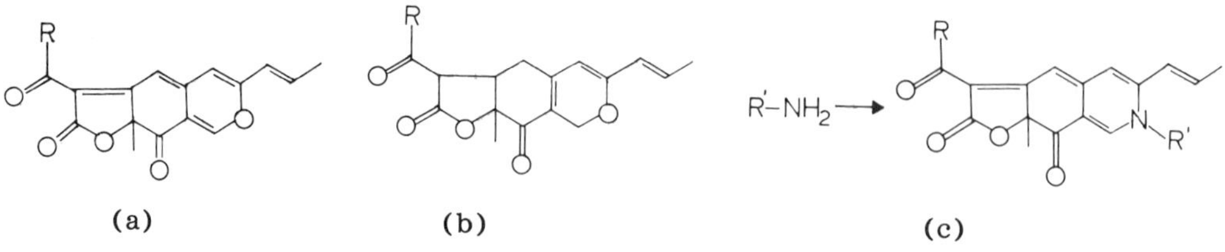

FIGURE 9. *Monascus* pigments. (a) Monascin, R = C_5H_{11}; ankaflavin, R = C_7H_{15}. (b) Rubropunctatin, R = C_5H_{11}; monascorubin, R = C_7H_{15}. (c) Colorant formed by reaction with protein via the amino group.

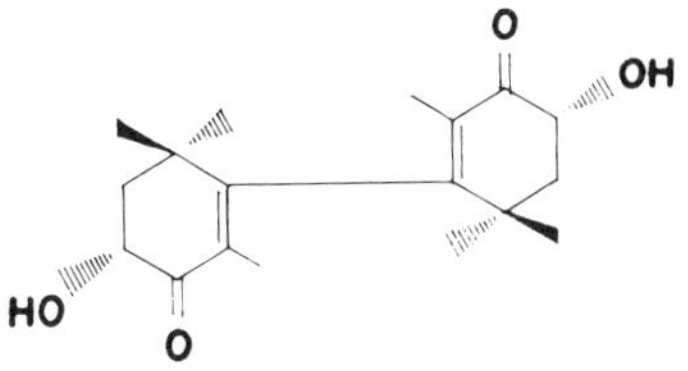

FIGURE 10. (3R,3'R)-Astaxanthin from *Phaffia rhodozyma*.

It would be impractical to review here the many flavor and aroma compounds produced by fungi, some of which have potential as food ingredients. References 184 through 196 discuss these interesting primary and secondary metabolites and the organisms that produce them.

Other reviews have discussed fungal metabolites and considered some aspects of food applications [11,31,85,197-204]. Miller [205], Shibata et al. [206], Turner [207], and Laskin and Lechevalier [208] have compiled comprehensive surveys of microbial metabolites.

III. CONCLUSIONS

Food biotechnology is classical food technology wedded to new techniques and new approaches. What does it offer food science and food process engineering from a mycological perspective? It offers *gene technology* as a means to improve industrial fungi that produce metabolites and enzymes used as food ingredients and processing aids. Fungal gene manipulation also provides methods to develop better strains for food fermentations, both traditional and novel. It offers *protein engineering* as a means of adapting food processing enzymes to new conditions or new tasks, both in vivo and in vitro. It offers unique approaches to *enzyme and fermentation engineering*, that is novel ways to use cells and enzymes in fermenters, bioreactors, or reaction vessels to achieve better yields and rates of production with cheaper raw materials. And, perhaps most notably, it raises the possibility of making new substances for food technology, substances that can be made economically and used safely to enhance food quality and nutrition. It may even allow production of valuable metabolites that are made only in trace quantities by plants, animals, or microbes which are too difficult or impossible to handle on a commercial scale.

How could fungal technology be used to make new classes of commercially promising primary and secondary metabolites for food use? A number of methods that rely on recent advances could play a role. New screening methods may identify productive organisms for diverse metabolites. Pathway engineering may combine genes for biochemical pathways from different sources, resulting in novel catabolic or anabolic routes. Bioconversions with growing cells, resting cells, or spores may permit processing of natural or synthetic compounds. Fungal biotransformations may permit stereospecific conversion of readily available starting materials to valuable products. Immobilized sequential enzymes, perhaps covalently linked and associated with cofactor regeneration elements, could lead to useful reaction sequences unknown in nature. Cofermentation systems may combine two metabolic capabilities in a fermenter and achieve similar results. The introduction into fungi of genes that encode peptides with food preservative properties, like cecropin perhaps [209], could generate safe bacteriostatic agents and natural, specific preservatives effective against foodborne pathogens. Fungal production of plant peptides with food functionality, like the sweetener thaumatin [210], could also be important to food technology.

These and other approaches will permit food biotechnology to benefit the food industry and the consumer. The benefits will have impact on the way food is harvested, processed, preserved, and prepared, as well as on human diet and nutrition.

ACKNOWLEDGMENT

The author thanks Kathy Hardy for help in preparing the manuscript.

REFERENCES

1. Kubicek, C. P., and Röhr, M. Citric acid fermentation. *CRC Crit. Rev. Biotechnol.*, *3*:331-373, 1986.
2. Smith, J. E., Nowakowski-Waszcuk, A., and Anderson, J. G. Organic acid production by mycelial fungi. In *Industrial Aspects of Biochemistry* (B. Spencer, ed.), Elsevier, Amsterdam, 1974, pp. 297-317.
3. Berry, D. R., Chmiel, A., and Obaidi, Z. A. Citric acid production by *Aspergillus niger*. In *Genetics and Physiology of Aspergillus* (J. E. Smith and J. A. Pateman, eds.), Academic Press, London, 1977, pp. 405-426.
4. Miall, L. M. Organic acids. In *The Filamentous Fungi. Industrial Mycology*, Vol. 1 (J. E. Smith and D. R. Berry, eds.), Arnold, London, 1975, pp. 104-121.
5. Lockwood, L. B. Organic acid production. In *The Filamentous Fungi. Industrial Mycology*, Vol. 1 (J. E. Smith and D. R. Berry, eds.), Edward Arnold Publishers, London, 1975, pp. 140-157.
6. Lockwood, L. B. Production of organic acids by fermentation. In *Microbial Technology. Microbial Processes*, 2nd ed., Vol. 1 (H. J. Peppler and D. Perlman, eds.), Academic Press, New York, 1979, pp. 353-387.
7. Bouchard, E. F., and Merritt, E. G. Citric acid. In *Kirk-Othmer Encyclopedia of Chemical Technology*, 3rd ed., Vol. 6 (M. Grayson and D. Eckroth, eds.), John Wiley and Sons, New York, 1979, pp. 150-179.
8. Kapoor, K. K., Chaudhary, K., and Tauro, P. Citric acid. In *Prescott and Dunn's Industrial Microbiology*, 4th ed. (G. Reed, ed.), AVI Publ., Westport, CT, 1982, pp. 709-747.
9. Röhr, M., Kubicek, C. P., and Kominek, J. Citric acid. In *Biotechnology. Microbial Products, Biomass and Primary Products*, Vol. 3 (H. Delweg, ed.), Verlag Chemie, Weinhem, 1983, pp. 419-454.
10. Abou-Zeid, A.-A. A., and Ashy, M. A. Production of citric acid: A review. *Agric. Wastes*, *9*:51-76, 1984.

11. Bigelis, R. Primary metabolism and industrial fermentations. In *Gene Manipulations in Fungi* (J. W. Bennett and L. L. Lasure, eds.), Academic Press, New York, 1985, pp. 357-401.
12. Milsom, P. E. Organic acids by fermentation, especially citric acid. In *Food Biotechnology-1* (R. D. King and P. S. J. Cheetham, eds.), Elsevier Applied Science, London, 1987, pp. 273-307.
13. Milsom, P. E., and Meers, J. L. Citric acid. In *Comprehensive Biotechnology. The Principles, Applications, and Regulations of Biotechnology in Industry, Agriculture and Medicine*, Vol. 3 (H. W. Blanch, S. Drew, and D. I. C. Wang, eds.), Pergamon Press, Oxford, 1985, pp. 665-680.
14. Bigelis, R., and Arora, D. K. Organic Acids of fungi. In *Handbook of Applied Mycology*, Vol. 4, *Fungal Biotechnology* (D. K. Arora, R. P. Elander, and K. G. Mukerji, eds.), Marcel Dekker, New York (in press).
15. Atticus. Citric acid forges ahead as an industrial chemical. *Chem. Age India, 26*:49-54, 1975.
16. Fricke, H., and Jensen, S. B. Citric acid—a versatile food additive. *Process. Ind., 44*(528):37, 39-40, 44, 1975.
17. Phillips, G. F., and Woodruff, J. G. Beverage acids, flavors, colors, and emulsifiers. In *Beverages: Carbonated and Non-carbonated* (J. G. Woodruff and G. F. Phillips, eds.), AVI, Westport, CT, 1981, pp. 152-207.
18. Irwin, W. E. *The Use of Citric Acid in the Beverage Industry*, Miles Inc., Elkhart, IN, 1983.
19. Schmidt, T. R. *The Use of Citric Acid in the Canned Fruit and Vegetable Industry*, Miles Inc., Elkhart, IN, 1983.
20. Andres, C. Acidulants. Flavor, preserve, texturize, and leaven foods. *Food Process., 46*(5):52-54, 1985.
21. Anonymous. Acidulants find growth niches in mature market. *Food Eng., 58*(10):83-86, 87, 1986.
22. Langdon, T. T. Preventing of browning in fresh prepared potatoes without the use of sulfiting agents. *Food Technol., 41*(5):64, 66-67, 1987.
23. Andres, C. Ingredient/package/process system is alternative to sulfites. *Food Process., 48*(5):54-55, 1987.
24. Porter, A. F. *The Use of Citric Acid in the Seafood Industry*, Miles Inc., Elkhart, IN, 1984.
25. Shenouda, S. Y. K., Montecalvo, J., Jhaveri, S., and Constantinides, S. M. Technological studies on ocean pout, an unexploited fish species, for direct human consumption. *J. Foods Sci., 44*:164-168, 1979.
26. Milsom, P. E., and Meers, J. L. Gluconic and itaconic acids. In *Comprehensive Biotechnology. The Principles, Applications, and Regulations of Biotechnology in Industry, Agriculture, and Medicine*, Vol. 3 (M. W. Blanch, S. Drew, and D. I. C. Wang, eds.), Pergamon Press, Oxford, 1980, pp. 681-700.
27. Underkofler, L. A. Gluconic acid. In *Industrial Fermentations* (L. A. Underkofler and R. J. Hickey, eds.), Chemical Publishing, New York, 1954, pp. 446-469.

28. Feldberg, C. Gluco-delta-lactone. *Cereal Sci. Today*, *4*:96-99, 1959.
29. Ward, G. E. Production of gluconic acid, glucose oxidase, fructose, and sorbose. In *Microbial Technology* (H. J. Peppler, ed.), Reinhold Publishing, New York, 1967, pp. 200-221.
30. Röhr, M., Kubicek, C. P., and Kominek, J. Gluconic acid. In *Biotechnology. Microbial Products, Biomass and Primary Products*, Vol. 3, (H. Dellweg, ed.), Verlag Chemie, Weinheim, 1983, pp. 456-465.
31. Buchta, K. Organic acids of minor importance. In *Biotechnology. Microbial Products, Biomass, and Primary Products*, Vol. 3 (H. Dellweg, ed.), Verlag Chemie, Weinheim, 1983, pp. 467-478.
32. Irwin, W. E., Lockwood, L. B., and Zienty, M. F. Malic acid. In *Kirk-Othmer Encyclopedia of Chemical Technology* (A. Standen, H. F. Mark, J. J. McKetta, Jr., and D. F. Othmer, eds.), John Wiley and Sons, New York, 1967, pp. 837-849.
33. Zhang, S. Z. Industrial applications of immobilized biomaterials in China. In *Enzyme Engineering* (I. Chibata, S. Fukui, and L. B. Wingard, Jr., eds.), Plenum Press, New York, 1982, pp. 265-279.
34. Yamamoto, K., Tosa, T., Yamashita, K., and Chibata, I. Continuous production of L-malic acid by immobilized *Brevibacterium ammoniagenes* cells. *Eur. J. Appl. Microbiol.*, *3*:169-183, 1976.
35. Takata, I., Yamashita, K., Tosa, T., and Chibata, I. Immobilization of *Brevibacterium flavum* with carrageenan and its application for continuous production of L-malic acid. *Enzyme Microb. Technol.*, *2*:30-36, 1980.
36. Yukawa, H., Yamagata, H., and Terasawa, W. Production of L-malic acid by the cell reusing process. *Proc. Biochem.*, *21*(5):164-166, 1986.
37. Duxbury, D. D. Malic acid/aspartame synergy reduces sweetener usage 10% in diet drinks. *Food Process.*, *47*(7):42, 1986.
38. Hirose, Y., and Okada, H. Microbial production of amino acids. In *Microbial Technology. Microbial Processes*, 2nd ed., Vol. 1 (H. J. Peppler and D. Perlman, eds.), Academic Press, New York, 1979, pp. 211-240.
39. Soda, K., Tanaka, H., and Esaki, N. Amino acids. In *Biochemistry. Microbial Products, Biomass, and Primary Products*, Vol. 3 (H. Dellweg, ed.), Verlag Chemie, Weinheim, 1983, pp. 467-478.
40. Yoshinaga, F., and Nakamori, S. Production of amino acids. In *Amino Acids Biosynthesis and Regulation* (K. H. Herrman and R. L. Somerville, eds.), Addison-Wesley, Reading, MA, 1983, pp. 405-429.
41. Aida, K., Chibata, I., Nakayama, K., Takinami, K., and Yamada, H. *Biotechnology of Amino Acid Production*, Elsevier, Amsterdam, 1986.
42. Chibata, I. *Immobilized Enzymes*. John Wiley, New York, 1978.
43. Tanaka, H., and Soda, K. Methionine. *Prog. Ind. Microbiol.*, *24*:183-187, 1986.
44. Fukumura, T. Conversion of D- and DL-α-amino-ε-caprolactam into L-lysine using both yeast cells and bacterial cells. *Agric. Biol. Chem.*, *41*:1327-1330, 1977.
45. Hamilton, B. K., Hsiao, H. W., Swann, W. E., Anderson, D. M., and Delente, J. J. Manufacture of L-amino acids with bioreactors. *Trends Biotechnol.*, *3*:64-68, 1985.

46. Vetsch, W. Aspartame: Technical considerations and predicted use. *Food Chem.*, *16*:245-258, 1985.
47. Klausner, A. Building for success in phenylalanine. *Bio/Technology*, *3*:301-307, 1985.
48. Terui, G. Tryptophan. In *Microbial Production of Amino Acids* (K. Yamada, S. Kinoshita, T. Tsunoda, and K. Aida, eds.), Kodansha, Tokyo, 1973, pp. 515-531.
49. Broquist, H. P., Stiffey, A. V., and Albrecht, A. M. Biosynthesis of lysine from α-ketoadipic acid and α-aminoadipic acid in yeast. *Appl. Microbiol.*, *9*:1-5, 1961.
50. Dulaney, E. L., Stapley, E. O., and Simpf, K. Studies on ergosterol production by yeasts. *Appl. Microbiol.*, *2*:371-378, 1954.
51. Sanchez-Marroquin, A., Ledezma, M., and Barreiro, J. Oxygen transfer and scale-up in lysine production by *Ustilago maydis* mutant. *Biotechnol. Bioeng.*, *13*:419-429, 1971.
52. Okanishi, M., and Gregory, K. F. Isolation of mutants of *Candida tropicalis* with increased methionine content. *Can. J. Microbiol.*, *16*:1139-1143, 1970.
53. Morzycka, E., Sawnor-Korszynska, D., Paszewski, A., Grabski, J., and Raczynska-Bojanowska, K. Methionine overproduction by *Saccharomycopsis lipolytica*. *Appl. Environ. Microbiol.*, *32*:125-130, 1976.
54. Shay, L. K., and Wegner, E. H. High methionine content *Pichia pastoris* yeasts. U.S. Patent 4,439,525 (1984).
55. Dunyak, S. A., and Cook, T. M. Continuous fermenter growth of a methionine-overproducing mutant of *Candida utilis*. *Appl. Microbiol. Biotechnol.*, *21*:182-186, 1985.
56. Kuninaka, A., Kibi, M., and Sakaguchi, K. History and development of flavor nucleotides. *Food Technol.*, *18*:221-223, 1964.
57. Shimazono, H. Distribution of 5'-ribonucleotides in food and their application to foods. *Food Technol.*, *18*:36-45, 1964.
58. Kuninaka, A. Recent studies of 5'-nucleotides as new flavor enhancers. In *Flavor Chemistry* (I. Hornstein, ed.), American Chemical Society, Washington, DC, 1966, pp. 261-274.
59. Kuninaka, A. Nucleic acids, nucleotides, and related compounds. In *Biotechnology. Microbial Products II*, Vol. 4 (H. Pape and H.-J. Rehm, eds.), VCH Verlagsgesellschaft, Weinheim, 1986, pp. 71-114.
60. Demain, A. L. Production of purine nucleotides by fermentation. *Prog. Ind. Microbiol.*, *8*:35-72, 1968.
61. Ogata, K. Industrial production of nucleotides, nucleosides and related substances. In *Biochemical and Industrial Aspects of Fermentation* (K. Sakaguchi, T. Uemura, and S. Kinoshita, eds.), Kodansha, Tokyo, 1971, pp. 37-59.
62. Ogata, K. The microbial production of nucleic acid-related compounds. *Adv. Appl. Microbiol.*, *19*:209-247, 1975.
63. Shibai, H., Enei, H., and Hirose, Y. Purine nucleoside fermentations. *Proc. Biochem.*, *13*(11):6-8, 32, 1978.
64. Hirose, Y., Enei, H., and Shibai, H. Nucleosides and nucleotides. *Annu. Report Ferment. Process.*, *3*:253-273, 1979.
65. Nakao, Y. Microbial production of nucleosides and nucleotides. In *Microbial Technology*, 2nd ed., Vol. 1 (H. J. Peppler and D. Perlman, eds.), Academic Press, New York, 1979, pp. 311-354.

66. Enei, H., Shibai, H., and Hirose, Y. 5'-Guanosine monophosphate. In *Comprehensive Biotechnology. The Principles, Applications and Regulations of Biotechnology in Industry, Agriculture and Medicine*, Vol. 3 (H. W. Blanch, S. Drew, and D. I. C. Wang, eds.), Pergamon Press, Oxford, 1985, pp. 653-658.
67. Kuninaka, A. Taste and flavor enhancers. In *Flavor Research. Recent Advances* (C. R. Teranishi, R. A. Flath, and H. Sugisawa, eds.), Marcel Dekker, New York, 1981, pp. 305-353.
68. Anonymous. *A Handbook for the World of Umami*, Umami Information Center, Tokyo, 1986.
69. Kawamura, Y., and Kare, M. R. *Umami: A Basic Taste*, Marcel Dekker, New York, 1987.
70. Yamaguchi, S. The synergistic taste effect of monosodium glutamate and disodium 5'-inosinate. *J. Food Sci.*, *32*(4):473-478, 1967.
71. Wagner, J. R., Titus, D. S., and Schade, J. E. New opportunities for flavor modification. *Food Technol.*, *17*:52-57, 1963.
72. Kurtzman, C. H., and Sjöström, L. B. The flavor-modifying properties of disodium inosinate. *Food Technol.*, *18*:221-223, 1964.
73. Caul, J. F., and Raymond, S. A. Home-use test by consumers of the flavor effects of disodium inosinate in dried soup. *Food Technol.*, *18*: 95-99, 1964.
74. Sjöström, L. B. Flavor potentiators. In *The Handbook of Food Activities* (T. E. Furia, ed.), The Chemical Rubber Co., Cleveland, 1968, pp. 493-500.
75. Samejima, H., Kimura, K., Noguchi, S., and Shimura, G. Production of 5'-mononucleotides using immobilized 5'-phosphodiesterase and 5'-AMP deaminase. In *Enzyme Engineering*, Vol. 3 (E. K. Pye and H. H. Weetall, eds.), Plenum Press, New York, 1975, pp. 469-475.
76. Hoechst. U.S. Patent 4,649,111 (1987).
77. Sakai, T. Microbial production of coenzymes. *Biotechnol. Bioeng. 22* (Suppl.) *1*:143-162, 1980.
78. Shimizu, S., and Yamada, H. Coenzymes. In *Biotechnology. Microbial Products II*, Vol. 4 (H. Pape and H.-J. Rehm, eds.), VCH Verlagsgesellschaft, Weinheim, 1986, pp. 157-184.
79. Goodwin, T. W. Production and biosynthesis of riboflavin in microorganisms. *Prog. Ind. Microbiol.*, *1*:137-177, 1959.
80. Prescott, S. C., and Dunn, C. G. Vitamin production by yeasts. In *Industrial Microbiology*, 3rd ed., McGraw-Hill, New York, 1959, pp. 218-235.
81. Hanson, A. Microbial production of pigments and vitamins. In *Microbial Technology* (H. J. Peppler, ed.), Reinhold Publishing, New York, 1967, pp. 222-250.
82. Yamada, K., Nakahara, T., and Fukui, S. Petroleum microbiology and vitamin production. In *Biochemical and Industrial Aspects of Fermentation* (K. Sakaguchi, T. Uemura, and S. Kinoshita, eds.), Kodansha, Tokyo, 1971, pp. 62-99.
83. Perlman, D. Vitamins. In *Economic Microbiology, Primary Products of Metabolism*, Vol. 2 (A. H. Rose, ed.), Academic Press, New York, 1978, pp. 303-326.

84. Florent, J. Vitamins. In *Biotechnology. Microbial Products II,* Vol. 4 (H. Pape and H.-J. Rehm, eds.), Verlag Chemie, Weinheim, 1986, pp. 115-158.
85. Bigelis, R. Industrial products of biotechnology: Application of gene technology. In *Biotechnology. A Comprehensive Treatise. Gene Technology,* Vol. 7b (S. Jolly and G. Jacobson, eds.), Verlag-Chemie, Weinheim, 1989, pp. 229-259.
86. Lago, B. D., and Kaplan, L. Vitamin fermentations: B_2 and B_{12}. In *Advances in Biotechnology. Fermentation Products* (C. Vezina and K. Singh, eds.), Pergamon Press, New York, 1981, pp. 241-257.
87. Takeda Chemical Industries, Ltd. Ger. Offen. 2,454,931 (1975).
88. Demain, A. L. Riboflavin oversynthesis. *Annu. Rev. Microbiol., 26*: 369-388, 1972.
89. Yoneda, F. Riboflavin (B_2). In *Kirk-Othmer Encyclopedia of Chemical Technology* (M. Grayson and D. Ekroth, eds.), John Wiley, New York, 1984, pp. 108-124.
90. Nakajima, K. Riboflavin biosynthesis. *Koshein Daigaku Kijo., 13*:1-11, 1985.
91. Perlman, D. Microbial process for riboflavin production. In *Microbial Technology,* Vol. 1 (H. J. Peppler and D. Perlman, eds.), Academic Press, New York, 1979, pp. 521-527.
92. Ozbas, T., and Kutsal, T. Comparative study of riboflavin production from two microorganisms: *Eremothecium ashbyii* and *Asbya gossypii. Enzyme Microb. Technol., 8*:593-596, 1986.
93. Weete, J. D. Sterols of the fungi: Distribution and biosynthesis. *Phytochemistry, 12*:1843-1864, 1973.
94. Rattray, J. B. M. Yeasts. In *Microbial Lipids,* Vol. 1, Academic Press, London, 1988, pp. 555-697.
95. Harrison, J. S. Yeast as a source of biochemicals. *Proc. Biochem., 3*:467-476, 1957.
96. Dulaney, E. L. Formation of extracellular lysine by *Ustilago maydis* and *Gliocladium* sp. *Can. J. Microbiol., 3*:467-476, 1957.
97. Kosikov, K. K., Lyapunova, T. S., Raevskaya, O. G., Semiknatova, N. M., Kochkina, I. B., and Meisel, M. N. Ergosterol synthesis in yeast hybrids and strains of the genus *Saccharomyces* of different ploidy. *Mikrobiologiya, 46*:86-91, 1977.
98. Parks, L. W., Rodriguez, R. J., and McCammon, M. T. Sterols of yeast: A model for biotechnology in the production of fats and oils. *J. Am. Oil. Chem. Soc., 59*:294A-295A, 1982.
99. Ciegler, A. Microbial carotenogenesis. *Adv. Appl. Microbiol., 7*:1-33, 1965.
100. Goodwin, T. W. Carotenoids in fungi and non-photosynthetic bacteria. *Prog. Ind. Microbiol., 11*:29-88, 1972.
101. Ninet, L., and Renaut, J. Carotenoids. In *Microbial Technology. Microbial Processes,* 2nd ed., Vol. 2 (H. J. Peppler and D. Perlman, eds.), Academic Press, New York, 1979, pp. 529-544.
102. Araujo, F. J. M., Calderon, I. M., Diaz, I. L., and Olmedo, E. C. U.S. Patent 4,318,987 (1982).
103. Davies, B. H. Carotene biosynthesis in fungi. *Pure Appl. Chem., 35*:1-28, 1973.

104. Murillo, F. J., Calderon, I. L., Lopez-Diaz, I., and Cerda-Olmeda, E. Carotene super-producing strains of *Phycomyces*. *Appl. Environ. Microbiol.*, *36*:639-642, 1983.
105. Peppler, H. J. Production of yeasts and yeast products. In *Microbial Technology*, 2nd ed., Vol. 1 (H. J. Peppler and D. Perlman, eds.), Academic Press, New York, 1979, pp. 157-185.
106. Solomons, G. L. Single cell protein. *Crit. Rev. Biotechnol.*, *1*:21-58, 1983.
107. Catley, B. J., and Whelan, W. J. Observations on the structure of pullulan. *Arch. Biochem. Biophys.*, *143*:138-142, 1971.
108. Catley, B. J. The rate of elaboration of the extracellular polysaccharide, pullulan, during growth of *Pullularia pullulans*. *J. Gen. Microbiol.*, *78*:33-38, 1973.
109. Catley, B. J. Pullulan synthesis by *Aureobasidium pullulans*. In *Microbial Polysaccharides and Polysaccharases* (R. C. W. Berkeley, G. W. Gooday, and D. C. Ellwood, eds.), Academic Press, London, 1979, pp. 69-84.
110. Slodki, M. E., and Cadmus, M. C. Production of microbial polysaccharides. *Adv. Appl. Microbiol.*, *23*:19-54, 1978.
111. Yuen, S. Pullulan and its applications. *Proc. Biochem.*, *9*(9):7-9, 22, 1974.
112. Jeanes, A. Dextrans and pullulans: Industrially significant α-D-glucans. In *Extracellular Microbial Polysaccharides* (P. A. Sandford and A. Laskin, eds.), American Chemical Society, Washington, DC, 1977, pp. 284-298.
113. Sandford, P. A. Potentially important microbial gums. In *Food Hydrocolloids*, Vol. 1 (M. Glicksman, ed.), CRC Press, Boca Raton, FL, 1982, pp. 167-202.
114. Compere, A. L., and Griffith, W. L. Scleroglucan biopolymer production, properties and economics. *Advances in Biotechnology. Fermentation Products* (C. Vezina and K. Singh, eds.), Pergamon, New York, 1983, pp. 441-446.
115. Bluhm, T. L., Deslandes, Y., and Marchessault, R. H. Solid-state and solution conformation of scleroglucan. *Carbohydr. Res.*, *100*:117-130, 1982.
116. Holzwarth, G. Xanthan and scleroglucan: Structure and use in enhanced oil recovery. *Dev. Ind. Microbiol.*, *26*:271-280, 1985.
117. Marchessault, R. H., Deslandes, Y., Ogawa, K., and Sundararajan, P. R. X-ray diffraction data for β-(1→3)-D-glucan. *Can. J. Chem.*, 55:300-303, 1977.
118. Kang, K. S., and Cottrell, I. W. Polysaccharides. In *Microbial Technology*, 2nd ed., Vol. 1 (H. J. Peppler and D. Perlman, eds.), Academic Press, New York, 1979, pp. 417-481.
119. Griminger, P., and Fisher, H. Antihypercholesterolemic action of scleroglucan and pectin in chickens. *Proc. Soc. Exp. Biol. Med.*, *122*:551-553, 1966.
120. Rodgers, N. E. Scleroglucan. In *Industrial Gums. Polysaccharides and Their Derivatives* (R. L. Whistler and J. N. BeMiller, eds.), Academic Press, New York, 1973, pp. 499-511.

121. Misaki, I. S., and Tsumuraya, Y. Structure and enzymatic degradation of Elsinan, a new α-D-glucan produced by *Elsinoe leucospila*. In *Chemistry and Biochemistry of Fungal Polysaccharides* (P. A. Sandford and K. Matsuda, eds.), American Chemical Society, Washington, DC, 1980, pp. 197-220.
122. Seeley, R. D. Fractionation and utilization of baker's yeast. *MBAA Tech. Quart.*, *14*:35-39, 1977.
123. Gorin, P. A. J., and Spencer, J. F. T. Structure chemistry of fungal polysaccharides. *Adv. Carbohydr. Chem.*, *23*:367-417, 1968.
124. Slodki, M. E., and Boundy, J. A. Yeast phosphohexans. *Dev. Ind. Microbiol.*, *11*:86-91, 1970.
125. Siodki, M. E., Ward, R. M., and Cadmus, M. C. Extracellular mannans from yeasts. *Dev. Ind. Microbiol.*, *13*:428-435, 1972.
126. McNeely, W. M., and Kang, K. S. Xanthan and some other biosynthetic gums. In *Industrial Gums*, 2nd ed. (R. L. Whistler and J. N. BeMiller, eds.), Academic Press, New York, 1973, pp. 473-497.
127. Slodki, M. E. Structural aspects of exocellular yeast polysaccharides. In *Chemistry and Biochemistry of Fungal Polysaccharides* (P. A. Sandford and K. Matsuda, eds.), American Chemical Society, Washington, DC, 1980, pp. 183-196.
128. Cottrell, I. W. Industrial potential of fungal and bacterial polysaccharides. In *Chemistry and Biochemistry of Fungal Polysaccharides* (P. A. Sandford and K. Matsuda, eds.), American Chemical Society, Washington, DC, 1980, pp. 251-270.
129. Prescott, S. C., and Dunn, C. G. *Industrial Microbiology*, 3rd ed., McGraw-Hill, New York, 1959.
130. Ratledge, C. Lipids and fatty acids. In *Primary Products of Metabolism. Economic Microbiology*, Vol. 2 (A. H. Rose, ed.), Academic Press, New York, 1978, pp. 263-302.
131. Lundin, H. Fat synthesis by microorganisms and its possible applications in industry. *J. Inst. Brew.*, *56*:17-28, 1950.
132. Woodbine, M. Microbial fat: Microorganisms as potential fat producers. *Prog. Ind. Microbiol.*, *1*:179-245, 1959.
133. Stodola, F. H., Deinema, M. H., and Spencer, J. F. T. Extracellular lipids of yeasts. *Bacteriol. Rev.*, *31*:194-213, 1967.
134. Hunter, K., and Rose, A. H. Yeast lipids and membranes. In *The Yeasts*, Vol. 2 (A. H. Rose and J. S. Harrison, eds.), Academic Press, New York, 1971, pp. 211-270.
135. Rattray, J. B. M., Schibeci, A., and Kidby, D. K. Lipids of yeasts. *Bacteriol. Rev.*, *39*:197-231, 1975.
136. Ratledge, C., and Wilkinson, S. G. *Microbial Lipids*, Vol. 1, Academic Press, London, 1988.
137. Lösel, D. M. Fungal lipids. In *Microbial Lipids*, Vol. 1, Academic Press, London, 1988, pp. 699-805.
138. Ratledge, C. Microbial conversions of n-alkanes to fatty acids: A new attempt to obtain economical microbial fats and fatty acids. *Chem. Ind.* (London) *(June)*:843-854, 1970.
139. Whitworth, D. A., and Ratledge, C. Microorganisms as a potential source of oils and fats. *Proc. Biochem.*, *9*:14-22, 1974.

140. Rattray, J. B. M. Biotechnology of the fats and oils industry—an overview. *J. Am. Oil Chem. Soc.*, *61*:1701-1712, 1984.
141. Sinden, K. W. The production of lipids by fermentation within the EEC. *Enzyme Microb. Technol.*, *9*:124-125, 1987.
142. Suzuki, O. Production of γ-linolenic acid by fungi. *J. Am. Oil Chem. Soc.*, *64*:1251, 1987.
143. Sugano, M., Ishida, T., Yoshida, K., Tanaha, K., Niwa, M., Arima, M., and Morita, A. Effects of mold oil containing γ-linolenic acid on the blood cholesterol and eicosanoid levels in rats. *Agric. Biol. Chem.*, *50*:2483-2491, 1986.
144. Totani, N., and Oba, K. The filamentous fungus *Mortierella alpina*, high in arachidonic acid. *Lipids*, *22*:1060-1062, 1987.
145. Totani, N., and Oba, K. A simple method for production of arachidonic acid by *Mortierella alpina*. *Appl. Microbiol. Biotechnol.*, *28*: 135-137, 1988.
146. Yamada, H., Shimizu, S., and Shinmen, Y. Production of arachidonic acid by *Mortierella elongata* 1S-5. *Agric. Biol. Chem.*, *51*:785-790, 1987.
147. Shimizu, S., Kawashima, H., Shinmen, Y., Akimoto, K., and Yamada, H. Production of eicosapentaenoic acid by *Mortierella* fungi. *J. Am. Oil Chem. Soc.*, *65*:1455-1459, 1988.
148. Shimizu, S., Shinmen, Y., Kawashima, H., Akimoto, K., and Yamada, H. Fungal mycelia as a novel source of eicosapentaenoic acid. *Biochem. Biophys. Res. Commun.*, *150*:335-341, 1988.
149. Shimizu, S., Akimoto, K., Kawashima, H., Shinmen, Y., and Yamada, H. Production of dihomo-γ-linolenic acid by *Mortierella alpina* 15-4. *J. Am. Oil Chem. Soc.*, *66*:237-241, 1989.
150. Fukuda, H., and Morikawa, H. Secretive fermentation of γ-linolenic acid production using cells immobilized in biomass support particles. In *Bioreactors and Biotransformations* (G. W. Moody and P. B. Baker, eds.), Elsevier Applied Science, London, 1987, pp. 386-394.
151. Shimizu, S., Kawashima, H., Akimoto, K., Shinmen, Y., and Yamada, H. Microbial conversion of an oil containing α-linolenic acid to an oil containing eicosapentaenoic acid. *J. Am. Oil Chem. Soc.*, *66*:342-347, 1989.
152. Kosaric, N., Wieczorek, A., Cosentino, G. P., Magee, R. J., and Prenosil, J. E. Ethanol fermentation. In *Biotechnology. A Comprehensive Treatise*, Vol. 3 (H. Dellweg, ed.), Verlag-Chemie, Weinhem, 1983, pp. 257-385.
153. Maiorella, B. L. Ethanol. In *Comprehensive Biotechnology. The Principles, Applications and Regulations of Biotechnology in Industry, Agriculture and Medicine*, Vol. 3 (H. W. Blanch, S. Drew, and D. I. C. Wang, eds.), Pergamon Press, Oxford, 1985, pp. 861-914.
154. Stewart, G. G. The genetic manipulation of industrial yeast strains. *Can. J. Microbiol.*, *27*:973-990, 1981.
155. Stewart, G. G., and Russell, I. One hundred years of research and development in the brewing industry. *J. Inst. Brew.*, *92*:537-448, 1986.
156. Struhl, K. The new yeast genetics. *Nature*, *305*:391-397, 1983.
157. Spencer, J. F. T., and Spencer, D. M. Genetic improvement of industrial yeasts. *Annu. Rev. Microbiol.*, *37*:121-142, 1983.

158. Tubb, R. S. Genetics of ethanol-producing microorganisms. *CRC Crit. Rev. Biotechnol.*, *1*:241-261, 1983.
159. Griffin, W. C., and Lynch, M. J. Polyhydric alcohols. In *CRC Handbook of Food Additives*, 2nd ed. (T. E. Furia, ed.), CRC Press, Cleveland, 1972, pp. 431-455.
160. Spencer, J. F. T., Roxburgh, J. M., and Sallans, H. R. Factors influencing the production of polyhydric alcohols by osmophilic yeasts. *J. Agric. Food Chem.*, *5*:64-67, 1957.
161. Nickerson, W. J., and Brown, R. G. Uses and products of yeasts and yeastlike fungi. *Adv. Appl. Microbiol.*, *7*:225-272, 1965.
162. Lewis, D. H., and Smith, D. C. Sugar alcohols (polyols) in fungi and green plants. *New Phytol.*, *66*:143-184, 1967.
163. Spencer, J. F. T. Production of polyhydric yeasts. *Prog. Ind. Microbiol.*, *7*:1-42, 1968.
164. Spencer, J. F. T., and Spencer, D. M. Production of polyhydric alcohols by osmotolerant yeasts. In *Primary Products of Metabolism. Economic Microbiology*, Vol. 2 (A. H. Rose, ed.), Academic Press, New York, 1978, pp. 392-425.
165. Kosaric, N., Gray, N. C. C., and Cairns, W. L. Microbial emulsifiers and de-emulsifiers. In *Biotechnology. A Comprehensive Treatise*, Vol. 3 (H. Dellweg, ed.), Verlag-Chemie, Weinheim, 1983, pp. 575-592.
166. Kosaric, N., Cairns, W. L., and Gray, N. C. C. *Biosurfactants and Biotechnology*, Marcel Dekker, New York, 1987.
167. Kao Soap. Japn. Pat. 61-205499 (1986).
168. Borrow, A., Jefferys, E. G., and Nixon, J. S. Process of producing gibberellic acid by cultivation of *Gibberella fujikuroi*. U.S. Patent 2,906,671 (1959).
169. Graebe, J. E., Hedden, P., and Rademacher, W. Gibberellin biosynthesis. In *Gibberellins—Chemistry, Physiology and Use*, Monograph 6 (J. R. Lenton, ed.), Wantage, British Plant Regulator Group, 1980, pp. 31-47.
170. Nickell, L. G. *Plant Growth Regulators*, Springer-Verlag, Berlin, 1982.
171. Hanson, J. R. Aspects of diterpenoid and gibberellin biosynthesis in *Gibberella fujikuroi*. *Biochem. Soc. Trans.*, *11*:522-527, 1983.
172. Crozier, A. *The Biochemistry and Physiology of Gibberellins*, Vols. 1, 2, Praeger, New York, 1983.
173. Shipchandler, M. T. Chemistry of zearalenone and some of its derivatives. *Heterocycles*, *3*:471-520, 1975.
174. Mirocha, C. J., Pathre, S. V., and Christensen, C. M. Mycotoxins. In *Economic Microbiology. Secondary Products of Metabolism*, Vol. 3A (A. H. Rose, ed.), Academic Press, London, 1979, pp. 488-494.
175. Leslie, J. F. Some genetic techniques for *Gibberella zeae*. *Phytopathol.*, *73*:1005-1008, 1983.
176. Anke, T. Further secondary metabolites of biotechnological interest. In *Biotechnology. Microbial Products*, Vol. 4 (H. Pape and H.-J. Rehm, eds.), Verlagsgessellschaft, Weinheim, 1986, pp. 611-628.
177. Francis, F. J. *Handbook of Food Colorant Patents*, Food and Nutrition Press, Westport, CT, 1986.

178. Francis, F. J. Lesser-known food colorants. *Food Technol.*, *41*(4): 62-68, 1987.
179. Johnson, E. A., and Lewis, M. J. Astaxanthin formation by the yeast *Phaffia rhodozyma*. *J. Gen. Microbiol.*, *115*:173-183, 1979.
180. Andrewer, A. G., and Starr, M. P. (3R,-3'R)-Astaxanthin from the yeast *Phaffia rhodozyma*. *Phytochem.*, *15*:1009-1011, 1976.
181. Gil-Hwan, A., Schuman, D. B., and Johnson, E. A. Isolation of *Phaffia rhodozyma* mutants with increased astaxanthin content. *Appl. Environ. Microbiol.*, *55*:116-124, 1989.
182. Johnson, E. A., Lewis, M. J., and Grau, C. R. Pigmentation of egg yolks with astaxanthin from the yeast *Phaffia rhodozyma*. *Poultry Sci.*, *59*:1777-1782, 1980.
183. Johnson, E. A., Villa, T. G., and Lewis, M. J. *Phaffia rhodozyma* as an astaxanthin source in salmonid diets. *Aquaculture*, *20*:123-134, 1980.
184. Webb, A. D., and Ingraham, J. L. Fusel oil. *Adv. Appl. Microbiol.*, *5*:317-353, 1963.
185. Collins, E. B. Biosynthesis of flavor compounds by microorganisms. *J. Dairy Sci.*, *55*:1022-1028, 1972.
186. Maga, J. A. The potential of certain fungi as sources for natural flavor compounds. *Chem. Sens. Flav.*, *2*:255-262, 1976.
187. Tressl, R., Apetz, M., Arrieta, R., and Grunewald, K. G. Formation of lactones and terpenoids by microorganisms. In *Flavor of Foods and Beverages* (G. Charalambous and G. Inglett, eds.), Academic Press, New York, 1978, pp. 145-168.
188. Schreir, P. The role of micro-organisms in flavour formation. In *Process in Flavour Research* (D. G. Land and H. E. Nursten, eds.), Applied Science Publishers, London, 1978, pp. 175-196.
189. Margalith, P. Z. *Flavor Microbiology*, Charles C. Thomas, Springfield, IL, 1981.
190. Hanssen, H.-P., and Sprecher, E. Aroma-producing fungi: Influence of strain specificity and culture conditions on aroma production. In *Flavour '81* (P. Schreier, ed.), W. de Gruyter, Berlin, 1981, pp. 547-556.
191. Schindler, J., and Schmid, R. D. Fragrance of aroma chemicals—microbial synthesis and enzymatic transformation—a review. *Proc. Biochem.*, *17*:2, 4-6, 8, 1982.
192. Schindler, J. Terpenoids by microbial fermentation. *Ind. Eng. Chem. Prod. Res. Dev.*, *21*:537-539, 1982.
193. Kempler, G. M. Production of flavor compounds by microorganisms. *Adv. Appl. Microbiol.*, *29*:29-51, 1983.
194. Sharpell, F. H. Microbial flavors and fragrances. In *Comprehensive Biotechnology. The Principles, Applications and Regulations of Biotechnology in Industry, Agriculture and Medicine*, Vol. 3 (H. W. Blanch, S. Drew, and D. I. C. Wang, eds.), Pergamon Press, Oxford, 1985, pp. 965-981.
195. Armstrong, D. W., and Yamazaki, H. Natural flavours production: A biotechnological approach. *Trends Biotechnol.*, *4*:264-268, 1986.
196. Gatfield, I. L. Production of flavor and aroma compounds by biotechnology. *Food Technol.*, *42*(10):110, 112,118, 120-122, 169, 1988.

197. Bothast, R. J., and Smiley, K. L. Metabolites of fungi used in food processing. In *Food and Beverage Mycology* (L. R. Beuchat, ed.), AVI Publishing, Westport, CT, 1978, pp. 368-396.
198. Demain, A. L. Industrial microbiology. *Science, 214*:987-995, 1981.
199. Bennett, J. W. Differentiation and secondary metabolism in mycelial fungi. In *Secondary Metabolism and Differentiation in Fungi* (J. W. Bennett and A. Ciegler, eds.), Marcel Dekker, New York, 1983, pp. 1-32.
200. Bigelis, R. Microbial production of Primary metabolites. In *Trends in Food Biotechnology, Proceedings of the 7th World Congress of Food Science and Technology 1987,* Singapore Institute of Food Science and Technology (in press).
201. Fleet, G. H. Biotechnology and the food industry. *Food Technol.,* (Australia), *36*:464-471, 1986.
202. Lin, Y.-L. Genetic engineering and process development for production of food processing enzymes and additives. *Food Technol., 40*(10): 104-112, 1986.
203. Ruttloff, H. Impact of biotechnology on food and nutrition. In *Food Biotechnology* (D. Knorr, ed.), Marcel Dekker, New York, 1987, pp. 37-94.
204. Bigelis, R., and Lasure, L. L. Fungal enzymes and primary metabolites used in food processing. In *Food and Beverage Mycology,* 2nd ed.. (L. R. Beuchat, ed.), AVI/Van Nostrand, Westport, CT, 1987, pp. 473-516.
205. Miller, M. W. *The Pfizer Handbook of Microbial Metabolites*, McGraw-Hill, New York, 1961.
206. Shibata, S., Natori, S., and Udagawa, S. *List of Fungal Products.* University of Tokyo Press, Tokyo, 1964.
207. Turner, W. B. *Fungal Metabolites*, Academic Press, London, 1971.
208. Laskin, A. I., and Lechevalier, H. A. *Handbook of Microbiology. Microbial Products,* Vol. 3, CRC Press, Cleveland, OH, 1973.
209. Xanthopoulos, K. G., Lee, J.-Y., Gan, R., Kockum, K., Faye, I., and Boman, H. G. The structure of the gene for cecropin B, and antibacterial immune protein from *Hyalophora cecropis. Eur. J. Biochem., 172*:371-376, 1988.
210. Edens, L., Bom, I., Ledeboer, A. M., Maat, J., Toonen, M. Y., Visser, C., and Verrips, C. T. Synthesis and processing of plant protein thaumatin in yeast. *Cell, 37*:629-633, 1984.

14

FUNGAL ENZYMES IN FOOD PROCESSING

RAMUNAS BIGELIS *Amoco Research Center, Amoco Technology Company, Naperville, Illinois*

I. INTRODUCTION

Microorganisms use enzymes to process specific biological molecules during metabolism and growth. Some of the same enzymes can be used to process biological molecules in ways that are valuable in food technology. The food industry exploits these catalysts and often uses fungal enzymes. When used in food processing, the enzymes may be indispensable to the formation of the final product; the enzymes may improve the quality of the final product; or the enzymes may improve the efficiency of the food-processing steps. This review will discuss specific applications of fungal enzymes that involve all of these roles. It will also consider new developments in their application and new means for their production. Other articles cover topics relevant to this review [1-25]. A number of excellent books discuss the role of fungal enzymes in food processing [26-47].

II. FOOD PROCESSES THAT USE FUNGAL ENZYMES

A. Baking

Wheat flour used in baking contains a number of enzyme activities. However, it lacks the necessary levels of enzymes that determine some of the functional properties of doughs, especially critical for automated manufacturing processes. The key enzymes that are deficient in doughs are α-amylase and protease, and therefore both are often supplemented. Malted wheat, barley, bacteria, and fungi are common sources of α-amylase, while fungi and bacteria are the traditional sources of proteases [32,48-52]. The following discussion of the use of fungal α-amylase and protease emphasizes their importance in the baking industry.

1. α-Amylase

α-Amylases from *Aspergillus oryzae, A. niger,* and *A. awamori,* or *Rhizopus* spp. are used to supplement the amylolytic activity in flour. The enzymes

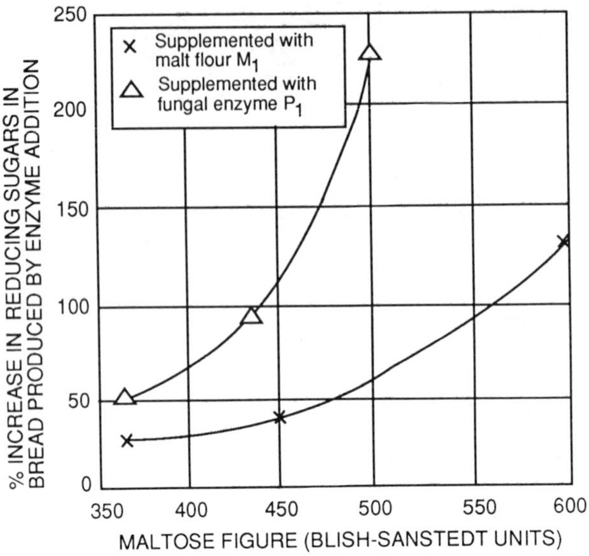

FIGURE 1. Increase in sugar content of bread made from flour with an enzyme supplement from *Aspergillus oryzae*. (From Ref. 48.)

elevate the levels of fermentable monosaccharides and disaccharides from 0.5% of the dough to concentrations that activate the yeast (Fig. 1). A sustained release of glucose and maltose by added and endogenous enzymes provides the nutrients essential for yeast growth and gas production during panary fermentation. An α-amylase supplementation may be beneficial for another reason. The enzyme degrades starch granules usually present in bread flour much more effectively than does wheat β-amylase. Since white bread flours contain 6.7-10.5% damaged starch, addition of enzyme at the mill or the bakery is often essential to maintain the yeast fermentation in such doughs [49]. The *A. oryzae* α-amylase is often favored for baking applications over the bacterial enzyme since it is heat-labile at 60-70°C and thus does not survive the baking process. Its thermolability prevents action on the gelatinized starch in the finished loaf and the production of a soft or sticky crumb.

Amylase supplementation also improves other aspects of bread quality. Amylase lowers dough viscosity and influences its softness, improving the ease of manipulation by hands or machines (Fig. 2). The treatment improves the volume (Fig. 3), taste, crust, and toasting qualities. The volume of the bread is improved probably because α-amylase reduces the viscosity of the gelling starch, allowing greater expansion during baking before protein denaturation and enzyme inactivation fix the loaf volume. The storage characteristics of breads are changed also, yielding a product with a softer, more compressible crumb that firms more slowly. Thus, it keeps better. However, despite having a softer crumb, fungal amylase-treated doughs do not yield fragile loaves. Amylolytic activity also increases the sugar concentration in bread. This outcome is beneficial in many instances

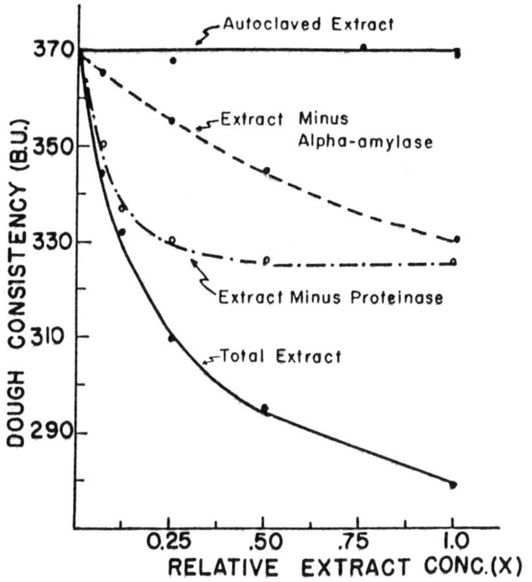

FIGURE 2. Effect of normal fungal extract, extract without α-amylase, extract without proteinase, and autoclaved extract on dough consistency. Dough consistency is expressed in arbitrary Brabender units, and enzyme concentration is also expressed in arbitrary units. (From Ref. 54.)

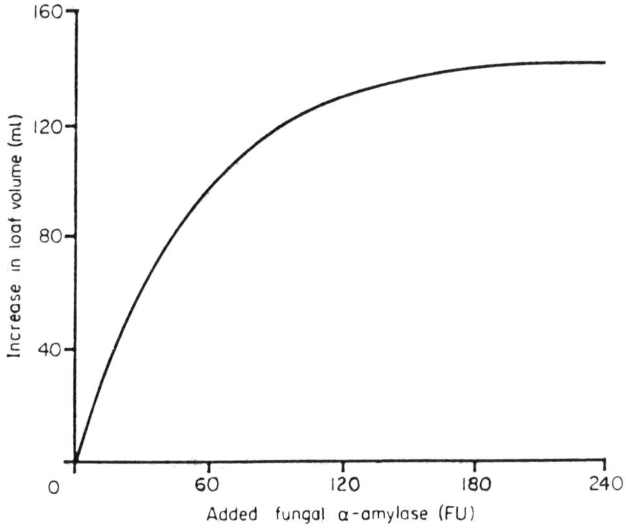

FIGURE 3. The response of loaf volume to added *Aspergillus oryzae* α-amylase. The average response curve is fitted to 174 data points obtained with 58 commercially milled white flours. (From Ref. 55.)

since an elevated sugar content in bread is often preferred. In addition to its beneficial effects on breads, amylase treatment enhances the desirable qualities of rolls, buns, and crackers [49,51,53-58].

Fungal α-amylase is added to doughs in the form of diluted powders and prepacked doses or water-dispersible tablets. Preparations have been added to flours at the bakery or, in some countries, at the mill itself. Malted wheat or barley can also serve as a source of enzyme by blending with wheat flour at the mill.

Though more heat-stable, bacterial α-amylase from species of *Bacillus* is also used in baking. Its properties make it suitable for the production of coffee cake, fruit cake, brownies, cookies, snacks, and crackers.

2. Protease

In addition to α-amylase, wheat flour also contains proteinases. However, the native proteases of flour do not play a role in the preparation of doughs and breads since their levels are so low. For this reason perhaps, the addition of fungal proteases may be more critical for dough processing than α-amylase supplements. About two-thirds of the bread in the United States is treated with proteolytic enzyme from *A. oryzae*. Strains with only protease or only amylase activity are available, allowing control of the specific enzyme formulation. Fungal protease is added to the bread dough during the fermentation stage. It releases amino acids and peptides necessary for yeast growth and proliferation. The enzyme also improves the handling and machining characteristics of the dough (Fig. 2), reduces mixing time, increases dough volume, and enhances the texture and elasticity of gluten. The appearance of the crust and the internal grain and texture are improved, as is the aroma and taste of the loaf. Fungal protease is also used to produce strong, extensible doughs for crackers and snack foods [32,49,52-53, 59].

The molecular mechanisms for dough improvement by proteases are not fully understood. Since gluten proteins play a major role in determining the viscoelastic properties of dough, it is likely that limited proteolysis affects this protein. Disruption of gluten molecules and their disulfide links may cause aggregation of the shortened gluten peptides. The subsequent formation of microfibrillar sheets may then introduce desirable changes in the elastic properties of the dough and permit it to be distended by the pressure of gases released during fermentation. Release of water bound to gluten after proteolysis and then its transfer to starch molecules may also cause favorable alterations in dough characteristics. Such a process could enhance the action of amylases on starch [32,52,60].

3. Other Enzymes

Several other fungal enzymes have been examined in baking applications. Lactase from *Aspergillus* species has been tested as a means of generating fermentable sugars from lactose; but the approach has proved to be uneconomic [53]. Fungal phytase has been explored as a means of reducing the phytate content of wheat flours. Phytase removes the phosphate groups from phytic acid (inositol hexaphosphoric acid) and prevents phytic acid from interfering with mineral absorption [61]. Fungal pentosanase has been used to degrade hemicellulose, a wheat constituent which may lead to a coarser bread crumb. Treatment of bread dough with purified fungal xylanase results in a dramatic decrease in dough strength, yielding loaves with an open crumb structure. Bread made with wheat and guar flour treated with xylanase from *Trichoderma viride* has 12% greater loaf volume and height than control loaves. In addition, crumb texture is finer and crumb color is superior. Preparations of *T. viride* pentosanase and xylanase are also useful in degrading wheat gums and fibers during wheat starch and gluten processing [62-64].

B. Beer Brewing

Beer brewing relies on yeast cultures and quite often on microbial enzymes that are added at several stages of the brewing process [65-69]. The enzymes that are used most commonly are amylase (Table 1), protease, and glucanase. The enzymes are obtained as the extracellular products of selected industrial strains of fungi and bacteria and are used as crude enzyme preparations. They may constitute only several percent of the dry weight of the preparation. They have characteristic temperature and pH optima which are important to their application as brewing enzymes (Table 2). The application of these enzymes during the early (Fig. 4A) and the late stages (Fig. 4B) of brewing is discussed below, as is the application of amylases to distilled alcoholic beverage production.

1. Enzyme Use During the Early Stages of Brewing

Enzymes are often used in the early stages of brewing, that is, during the wort production process that generates the constituents needed for the subsequent fermentation. They are added during the mashing process, during barley brewing, during the production of low carbohydrate beers, or during cereal cooking [67,68,71-77].

Enzymes are often added at the brewery during mashing, a step which involves the liberation of fermentable sugar. The enzymes supplement the action of endogenous barley β-amylase and proteases. They are especially important when nonmalted cereal grains (adjuncts) such as corn and rice are used, since these grains are deficient in carbohydrase and proteases. Added fungal α-amylase (and sometimes glucoamylase) from *Aspergillus* species such as *A. niger* or *A. oryzae* increases starch digestion, reduces the proportion of unmalted grain, and improves the reliability of mashing. The enzyme solubilizes barley amylose and amylopectin, exposing these substrates to further degradation by β-amylase. Ultimately, the levels of maltose and small dextrins are elevated, yielding the wort ingredients essential for the yeast fermentation. Amylase preparations with low transglucosidase activity are preferred since this enzyme is responsible for the synthesis of isomaltose and panose, both of which are nonfermentable [78]. *Aspergillus* protease may also be supplemented during mashing to solubilize protein and release amino acids, thereby making nitrogen available for yeast growth.

Fungal amylases may be used in barley brewing, a method of brewing that uses unmalted barley and a mixture of enzymes in the mash tun. The added enzymes are typically α-amylase, β-glucanase, and proteinase, all chosen for their heat stability properties. The enzymes are available in various degrees of purity. In general, additions of 0.05-0.10% of the grist are made to cooled fermenters since β-glucanase and proteinase act in the

TABLE 1. Properties of Amylases Used in Brewing of Malt Beverages

	α-Amylase from *Aspergillus* spp.[a]	Glucoamylase from *Aspergillus* spp.	β-Amylase from barley grain
Bond cleavage	α-1,4 glycosidic linkages	α-1,4 and α-1,6 glycosidic linkages, slow α-1,3 cleavage	α-1,4 glycosidic linkages
Substrate specificity	Endoamylase: random	Exoamylase: sequential hydrolysis from nonreducing end	Endoamylase: random
Branchpoint	Traverse α-1,6 bonds	Traverse and hydrolyze α-1,6 branch-points	Do not traverse or hydrolyze α-1,6 bonds
Main hydrolysis products	Dextrins	Glucose	β-Maltose
Glucose production	Slow	Fast	None
Maltose production	None	None	Fast
Dextrin production	Fast	Slow	Slow
Viscosity reduction	Fast	Slow	Slow

[a]Thermostable *Bacillus licheniformis* α-amylase is used during cooking of some unmalted cereals.

Source: Refs. 70, 71.

range of 45-50°C. After an incubation period, the temperature is raised to 65-68°C, a range which is optimal for starch hydrolysis. Care must be taken to maintain these temperatures, since excessive heating will reduce the levels of fermentable sugars and thus damage the wort. Finally, in barley brewing the enzymes are inactivated during the boiling step to avoid hydrolytic activity in the fermenting wort. In effect, the barley brewer simulates a malt wort, but avoids the mashing step by using exogenous enzymes to generate a fermentation medium. The use of cheaper barley plus added enzymes, rather than malt, yields economic benefits for the brewer and permits more freedom in selecting various enzyme combinations and mashing conditions.

Fungal enzymes are critical to the production of low carbohydrate "light" beers. Worts normally contain 4% (w/v) of their carbohydrates as unfermentable dextrins. These are carried over to the beer and contribute to its caloric content since yeast ferments maltose well, maltotriose poorly, and larger oligosaccharides, the dextrins, not at all. Digestion of wort dextrins with glucoamylase added directly to the fermenter eliminates the unfermentable carbohydrate fraction almost completely and yields a "light" beer. The alcohol content of the beer remains the same if the

TABLE 2. Temperature and pH Optima of Fungal and Bacterial Amylases

	Fungal α-amylase from *Aspergillus* spp.	Fungal gluco-amylase from *Aspergillus* spp.	Bacterial α-amylase from *Bacillus licheniformis*
Temperature optimum	50°C	55°C	75°C
pH optimum	5.5	4.5	6.5

Source: Refs. 72, 73.

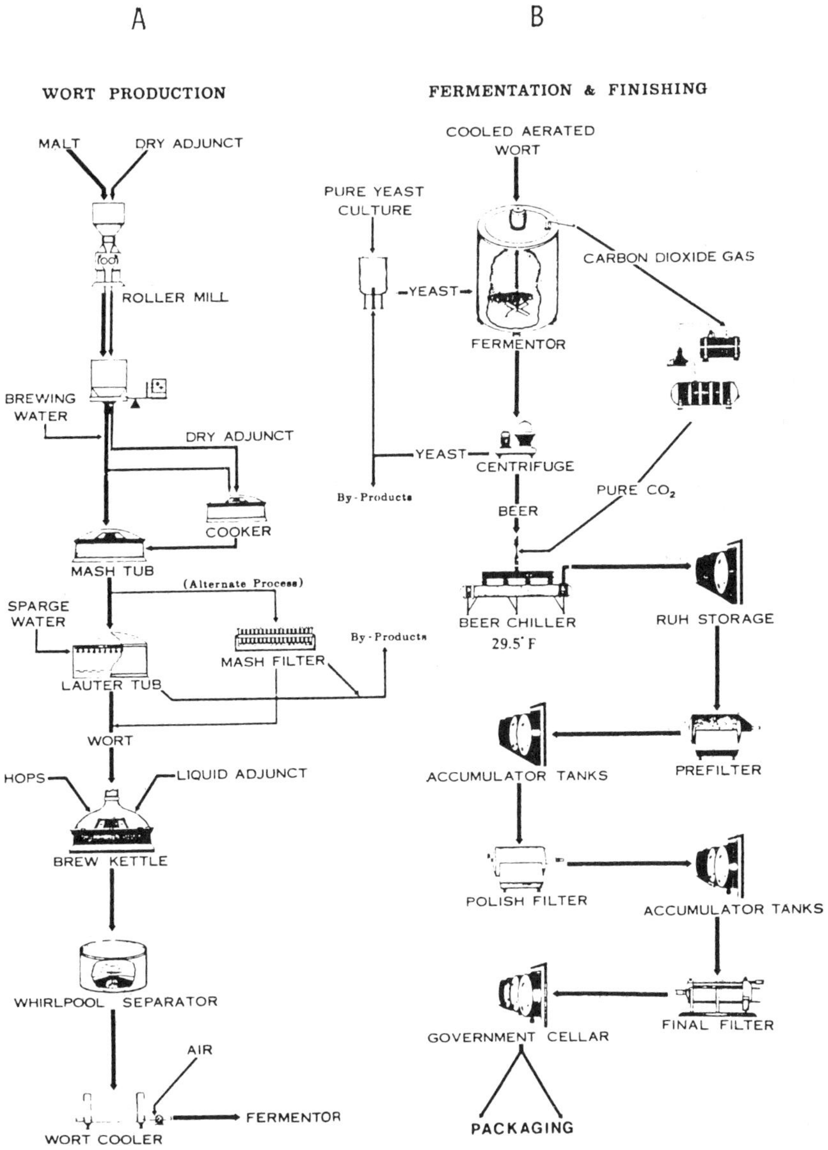

FIGURE 4. Flow chart of the brewing process. (A) The early stage of brewing involves wort production. (B) The late stage of brewing involves fermentation and finishing. (From Ref. 66.)

gravity of the starting material is adjusted. Glucoamylase hydrolysis of the α-1,4 linkages and α-1,6 branchpoints yields glucose, which also must be metabolized if a sweet, reduced-calorie beer is to be avoided. An alternative approach to glucoamylase treatment is the use of a combination of fungal α-amylase and bacterial pullulanase in the fermenter [20,79]. Since pullulanase can hydrolyze α-1,6 linkages, both enzymes working together can convert dextrins to fermentable sugars.

In the United States, enzymes are also used in cereal cooking for brewing processes that employ unmalted cereal such as rice, maize, or sorghum in the mash. The adjunct starch from these cereals is only partially gelatinized under mashing conditions. Therefore, an additional cereal cooking step is necessary to increase the exposure of adjunct starch to malt enzymes and promote liquefaction. The boiling time in the cereal cooker depends on the nature of the adjunct cereal. Enzyme treatment of the mash with thermostable bacterial α-amylase allows a reduction of the cereal cooker temperature, leading to energy savings and greater uniformity of cooker operations. This enzymatic treatment also facilitates gelatinization and liquefaction of the starch and permits shorter cooker times. Flavor and color may be improved by avoidance of harsh temperature treatments that may extract undesirable compounds. *Bacillus licheniformis* α-amylase with an optimum temperature of 85-87°C is commonly used during cereal cooking. Fungal α-amylase may also be added afterwards to the cooled mash or fermenter to ensure proper fermentability of the wort.

Barley β-glucans are unbranched polymers of β-linked D-glucosyl residues and are referred to as barley gums, mixed linkage β-glucans, or (1→3)(1→4)-β-D-glucans [80,81]. These β-glucans are major constituents of barley endosperm cell walls, representing about 75% of their total carbohydrate. During the malting of barley grain, β-glucans may be fully degraded by barley β-glucanases to glucose. If they are only partially degraded, a high molecular weight, viscous polymeric material is released into solution and incompletely digested cell wall material may act as a physical barrier to starch digestion. Enzyme supplements may be needed to degrade the residual β-glucans, especially if the malting process is accelerated or if unmodified barley adjuncts are used. Bacterial β-glucanase, typically from *Bacillus subtilis*, may be added to the wort to help break down the cell walls of barley endosperm cells and facilitate exposure of starch granules to amylases. *Aspergillus niger, Penicillium emersonii, Trichoderma reesei*, and *T. viride* are other sources of this enzyme. Together with proteinases, the β-glucanase acts to loosen the endosperm structure. Since high levels of residual high molecular weight β-glucans may remain in the wort, the enzyme is also used to reduce wort viscosity and degrade the viscous barley β-glucans that may clog pumps and filters during transfer operations. Preparations of *T. viride* xylanase can perform a similar function by degrading xylans that cause viscosity problems [81-83].

2. Enzyme Use During the Late Stages of Brewing

After production of the wort, commercial fungal enzymes are also used in the late stages of the brewing process, during maturation, processing, and finishing. The enzymes play a role in chillproofing, natural conditioning, the production of sweet beers, and the improvement of viscosity [67,68,71,74,75,77].

A common practice in brewing is chillproofing. Chillproofing is the addition of papain or sometimes fungal protease from strains of *Aspergillus*, mainly *A. oryzae*, to reduce a colloidal haze. The haze, which consists of a polyphenolic complex and polypeptides, is intensified by chilling. Prevention of this cloudiness is important since most beers are served cold. The proteolytic enzyme preparation is added after fermentation and clarification, but before pasteurization. Careful maintenance of the temperature at 60°C permits the enzyme to survive heating and then function after pasteurization. Fungal α-amylase may be blended with chillproofing proteases to solubilize both the remaining starch and protein that may contribute to beer turbidity.

Glucoamylase or α-amylase can be added along with yeast to promote a process termed natural conditioning. This process requires a second fermentation and is fueled by residual dextrins which increase the carbonation of draught beer.

Enzymes can also play a role in the manufacture of sweet beers. Glucoamylase from *Aspergillus* species is added directly to the fermenter. The enzymatic breakdown of residual dextrins generates the sweetness and obviates the need for supplemented glucose.

In the late stages of brewing, viscosity problems can arise during filtration and transfer of beer [80,81,84-87]. As during wort production, the addition of β-glucanase from *B. subtilis*, *A. niger*, or *T. reesei* alleviates these problems, especially in processes that employ unmalted barley containing higher levels of particulate matter composed of nonstarch polysaccharides, primarily hemicelluloses. Filtration throughput is increased, filter aid requirements are reduced, and beer clarity and stability are improved. Moreover, β-glucanase treatment shortens lautering times and increases brew-house yield [74-88].

3. Other Beer-Brewing Enzymes

Other microbial enzymes have been used in manufacturing beer, while the potential of some has been investigated. Glucose oxidase from *A. niger* has been used to remove oxygen from bottled and canned beer (see Sec. II.J). Anthocyanase from *A. niger*, *A. oryzae*, or other aspergilli has been shown to be effective in degrading anthocyanins in beer [89]. Diacetyl reductase from the bacterium *Aerobacter aerogenes* has been successfully applied to the removal of diacetyl from beer, as has added α-acetolactate decarboxylase to the removal of α-acetolactate and α-aceto-α-hydroxybutyrate [90,91]. However, the availability of diacetyl-less brewer's yeasts and related strains may eliminate the need for enzymatic removal of acetyl compounds [92].

Additional proposed uses of microbial enzymes in the brewing industry include: tannases for removal of polyphenolic compounds, phosphatases for hydrolysis of phosphorylated compounds, nucleases for the generation of flavor enhancers, cellulases and macerating enzymes for yeast cell wall degradation, and esterases for lowering the ester content of beer [93]. Carbohydrase preparations from *T. reesei* have been shown to be effective in the recovery of fermentable sugars from brewer's spent grains [94].

A number of novel brewing strains, some genetically engineered, that produce enzymes beneficial to the brewing process are already available or awaiting regulatory approval. Production of such enzymes directly by the brewer's yeast could obviate the need for exogenous enzymes [25,95-103].

4. Distilled Alcoholic Beverage Production

Alcoholic beverages are produced from starch-bearing plant materials or sugar-containing raw materials worldwide. Corn, milo, barley, malt, and rye are commonly used as starch-bearing substrates in the United States; potatoes, barley, malt, maize, and rye are used in Europe; potatoes, rye, and wheat are sources in the U.S.S.R.; rice and sweet potatoes are common substrates in the Orient; and cassava is used in tropical countries.

Examples of distilled spirits and the associated starch-containing raw materials are: barley—whisky; maize and rye—bourbon whisky; potatoes and barley—aquavit; rice—Chinese brandies. The starch in the raw materials must first be hydrolyzed to fermentable sugars to fuel the ethanolic fermentation. Enzymes of vegetable origin, such as those in malt, or enzymes from microbial surface culture, such as those in koji, have been and still are used for this purpose. However, today industrial microbial enzymes are preferred for starch conversion processes for potable alcohol production. Distilled spirits are also produced from sugar-containing raw materials: rum and cachaca from molasses; cognac and pisco from grape wines; tequila from agave cactus juice; kirsch from cherries; brandy from pears; slivovice from plums; and ogogoro from palm juice. Commonly, species of the yeasts *Saccharomyces* or *Schizosaccharomyces* are employed to produce alcohol from both types of raw materials—those rich in starch and those rich in sugar. Bacteria such as *Zymomonas* spp. are also used [104,105].

The starch conversion processes that fuel the yeast fermentation are often similar to those used in brewing beer. Both malt and microbial amylases are used in the mashing step. As in beer brewing, local regulations may specify certain beverage alcohol production conditions, especially the amounts and proportions of certain mashing enzymes. The use of malt treated with fungal gibberellic acid can reduce malt requirements by 30%, and the addition of mold enzymes can further reduce the need for malt. Thermostable *Bacillus* α-amylases have been employed in the mashing step but are not favored since they are less efficient at the low pH conditions typical of mashes. Instead, glucoamylases from strains of *Aspergillus* spp. or *Rhizopus* spp. are utilized more commonly and can reduce malt usage to as little as 2% of the weight of the grain. The raw material is first cooked in batch, yielding a gelatinized starch that can be enzymatically degraded to sugars. The cooked mash is cooled to about 20-25°C and then saccharified with enzymes, often directly in fermenters. The fungal enzymes are stable under distillery fermentation conditions and show no loss of activity after 96 hours in fermenters, even at pH 3.5 or less. In contrast, malt enzymes become totally inactive after such a treatment. Furthermore, fungal glucoamylase saccharifies the starch more rapidly and completely than malt. It generates less maltose, isomaltose, and oligosaccharides, boosting yields of fermentable sugars for the yeast fermentation. And it increases the rate of the fermentation itself, as well as the number of proof gallons per bushel of grain. Thus, glucoamylase and α-amylase treatment with *Aspergillus* enzyme preparations, usually from *A. niger* or *A. awamori*, in conjunction with malt can convert an inexpensive grain mash into an excellent medium for distiller's yeast growth and alcohol production. Proteases can also play a beneficial role in the mashing step. Fungal proteolytic activity in the amylase preparation can degrade grain protein and elevate the levels of available nitrogen essential for yeast propagation. Once the mashing step is complete, distiller's yeast equal to 2-3% of the total grain is added and optimal conditions for alcohol production are maintained. The fermentation is complete in 40-72 hours and yields fermented mash containing 6-8% ethanol by volume [104-108].

Recombinant strains of distiller's yeast could be important producers of ethanol or alcoholic beverages from starch-containing raw materials. A strain of *S. cerevisiae* able to secrete a glucoamylase derived from the *A. awamori* gene has been constructed. This strain utilizes 95% of the carbohydrate in soluble starch at a 25% (w/v) concentration in growth medium and produces high yields of ethanol. Though the yeast shows poor ethanol tolerance, glucoamylase expression, and plasmid stability, it is a major step toward the construction of industrial distiller's yeast strains able to convert starch to ethanol [109].

Mold cultures are used for starch conversion in the Orient. In the two-stage fermentation process, aerated steamed rice or grain is inoculated with mold, incubated to promote starch digestion, and then inoculated once again with yeast. In another approach that combines both stages, the starch is degraded by molds during the yeast fermentation. Both methods are used to produce a variety of distilled alcoholic beverages [104].

C. Fruit and Juice Processing

A wide variety of sparkling clear juices and also cloud juices are produced by the food industry. Juice production has relied on the use of fungal enzymes to facilitate processing of the raw material and to change the properties of the final product [110-116].

1. Pectic Enzymes

The most commonly used juice-processing enzymes are pectic enzymes. They are often collectively termed "pectinases." Pectic substances are acidic heteropolysaccharides with molecular weights of 30,000-300,000. They are composed primarily of pectin, which consists of polygalacturonic acid, where at least 75% of the D-galacturonic acid monomers are esterified with methanol, as well as rhamnogalacturonans, galacturonans, galactans, arabinogalactans, and arabinans. These polysaccharides play a structural role in the plant cell wall or middle lamella [111,116,117].

Several key enzymes usually make up commercial fungal pectic enzyme mixtures. Pectinases, also termed polygalacturonases, break down polygalacturonic acid and other polymers bearing D-galacturonic acid to soluble oligosaccharides. They are endoenzymes that prefer low-methoxyl pectin or completely deesterified pectate. Pectases, also called pectin methylesterases or pectinesterases, split methanol from esterified carboxyl groups and convert pectins to low methoxyl pectins and eventually to completely deesterified pectate. Pectin lyases randomly break glycosidic linkages next to a methyl ester group. They are endoenzymes which prefer high methoxyl pectins. Commercial pectic enzyme preparations contain the above-mentioned and many other enzymes, including other polysaccharide-degrading enzymes,

TABLE 3. Enzymes Produced by *Aspergillus niger* that Hydrolyze Polymers

Substrate	Enzymes
Arabinans	α-L-Arabinofuranosidase
Cellulose	C_1,C_x-type cellulases
Dextran	Dextranase
DNA, RNA	Deoxyribonuclease, ribonuclease
β-Glucans	β-Glucanase
Hemicellulose	Hemicellulases
Inulin	Inulinase
Mannans	β-Mannanase
Pectic substances	Pectin methylesterase, pectate lyase, polygalacturonase
Proteins	Proteases
Starch	α-Amylase, glucoamylase
Xylans	Xylanase

Source: Ref. 116.

which also may act on pectic substances. For example, more than 64 enzymes have been detected in *A. niger* culture media, and many of these degrade polymers (Table 3). These enzymes can influence the results obtained with pectinolytic enzyme preparations [114,116,118,119].

Pectic enzymes play a useful role in the processing of fruits when used to treat the juice. After processing, juices can contain considerable amounts of solids, consisting mainly of pectic substances which may affect juice appearance and texture. This material may constitute 5-10% of the fresh weight of treeborne fruit. The pectic substances may interfere with further processing or detract from the desired characteristics of the finished product. In many countries, including the United States, pectic enzyme mixtures from *A. niger* (or its close relatives) are used in apple and grape juice processing to reduce the cloudiness formed by colloidal suspension of pulp, to increase the filterability of the juice, and to prevent gelling of pectin in concentrated juice products. The supplemented fungal enzymes partially hydrolyze the soluble pectin to smaller particles, which can then flocculate and be removed. A nonenzymatic process that involves neutralization of electrostatic charges on the partially degraded particles also plays a role in pectin removal. In addition to improving the clarity and viscosity of sparkling juices such as grape and apple juices, depectinization also increases yield, usually 10-20%, when applied to the pulp before processing (Fig. 5). It can increase the efficiency of pressing and juice extraction and even enhance flavor and color release. Depectinization is especially valuable in continuous pressing operations [110,113,115,116,120].

Many noncitrus tree fruits, berry fruits, and tropical fruits are processed with the aid of pectinases. Among them are black currants, cherries,

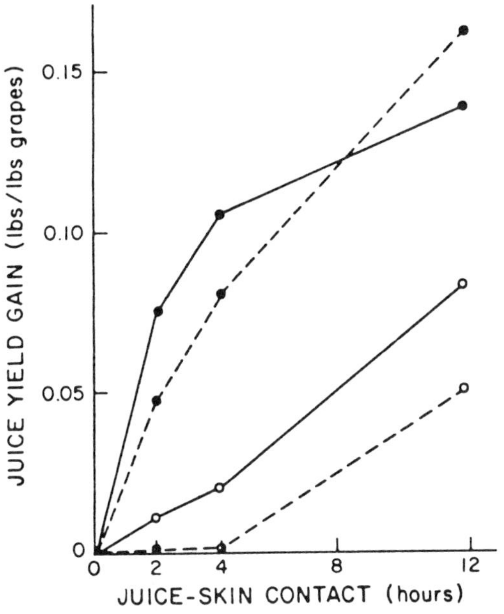

FIGURE 5. Effect of added pectic enzymes on juice yield. ○, No added enzymes; •, added enzymes; ——, Chenin blanc grapes; -----, Muscat of Alexandria grapes. (From Ref. 120.)

raspberies, strawberries, and bananas. The type and variety of fruit, and often its maturity, determine the content of pectic substances, the ratio of soluble to insoluble material, and the processing variables important for juice extraction. The type of fruit thereby also determines the nature of the enzyme treatment and the point in the manufacturing process where the treatment is especially important. Of course, adjustment of the enzyme concentration, the temperature, and the reaction time will also influence the efficiency of the enzyme application [110].

Pectinase treatment can play a specialized role in citrus fruit processing. It can improve the stability of the cloud, a desirable feature in many citrus juices [115,116] (Fig. 6). Recent studies by the U.S. Department of Agriculture report that pectinase injected into citrus fruit dissolves the albedo and loosens peels. Machines can then remove the peels, yielding high-quality sectioned fruit that can be canned or eaten fresh [122]. Citrus oils such as lemon oil may be made with the help of pectic enzymes. The oil is obtained from emulsified citrus peel extracts. Pectic enzymes destroy the emulsifying properties of pectins, which interfere with the collection of oil [2].

Pectic enzymes are produced on a large scale with selected strains of filamentous fungi. Species of *Aspergillus* are the most common sources of enzyme, while *Coniothyrium diplodiella*, *Sclerotinia libertiana*, and species of *Botrytis*, *Penicillium*, and *Rhizopus* are also used. The fungi are grown by semi-solid substrate fermentation or by submerged fermentation. The latter method is more convenient and is favored by many industrial producers.

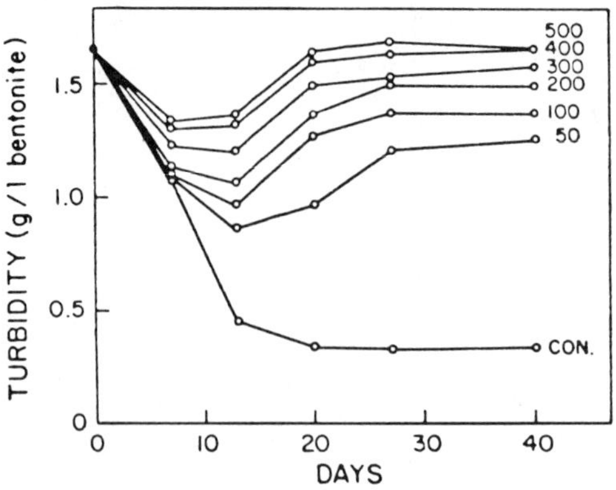

FIGURE 6. Stabilization of orange juice cloud at 4°C by addition of several concentrations of pectic enzymes. (From Ref. 121.)

The medium composition for commercial production is proprietary. Generally, an acid medium, about pH 3.5, is used and contains pectin and a carbon source containing sucrose, lactose, or glucose, or carbohydrate mixtures. The added pectin, which serves as an inducer of pectic enzymes, is typically processed fruit or vegetable waste, such as apple pomace, citrus peel, or beet pulp. Pectic enzyme production usually reaches a maximum when the carbohydrate is exhausted. The deep tank fermentation takes 3-6 days. Cell-free fermentation fluid from submerged culture, which bears extracellular pectic enzymes, is usually sold as is or processed, yielding a liquid concentrate. In contrast, after semi-solid fermentation, the culture mass is extracted to produce an enzyme solution. Enzyme from submerged or semi-solid substrate culture can be precipitated with inorganic salts or organic solvents and then dried, milled, and finally mixed with stabilizers or inert ingredients yielding a powder [112,123,124].

2. Other Juice Processing Enzymes

Enzymes other than pectinases are important to fruit juice technology (Table 4). Some of these enzymes are present in pectinase preparations (Table 3), while others are added to formulations. Proteases and amylases, often derived from fungal sources, may be used to prevent haze and sediment formation, and also to reduce viscosity that may interfere with procedures to concentrate or filter the juice. Glucoamylase may be needed to degrade starch colloid, which may be a cause of postbottling haze in apple and pear juice. Juices from unripe apples may require enzyme treatment since the starch content may approach 15% [110,113,116].

Some enzymes may be indispensable fruit- and vegetable-processing aids via the process of liquefaction. Cellulases and hemicellulases, usually from *A. niger* or *Trichoderma* spp., may be employed in conjunction with pectinases to reduce agricultural products to pulps, nectars, or purees. Such enzyme mixtures are used to process apples, pears, grapes, berries, citrus fruits, or vegetables. Juice production by enzymatic liquefaction is

TABLE 4. Haze Formation in Liquid Food Products

Product	Presumed cause	Treatment
Sake wine	Polyphenols-proteins	Proteases
Cider	Polyphenols-proteins	Proteases
Grape juice	Polyphenols-proteins	Proteases
Beer	Tannins-proteins	Papain
Apple juice	Starch (retrograded)	α-Amylase
Pear juice	Arabinan	—
Soluble tea	Gallotannin, theaflavins thearubigins with caffeine	Tannase
Raspberry juice	Pectin, polyphenols, polyvalent cations	—
Mandarin oranges	Hesperidin crystals	Hesperidinase
Grapefruit	Naringinin	Naringinase
Enzyme potentiation (several)	Polymerized phenolics	—

Source: Ref. 116.

especially useful with agricultural products for which no juice extraction equipment exists, e.g., tropical fruit, or for products which would lose important constituents in the pulp, e.g., the carotene in carrots. Another process of tissue disintegration, one which preserves intact cells, is called maceration, and the essential enzymes for the process, the pectin depolymerases, are termed macerases. Maceration also protects important juice constituents that might be lost by mechanical, enzymatic, or oxidation processes [110-113].

The enzymes naringinase and hesperidinase are used to treat citrus fruit juices. Naringinase reduces excessive levels of the intensely bitter flavonoid naringin, which is found in grapefruit and oranges. Naringin, the 7-(2-rhamnosido-β-glucoside) of naringenin, is converted to the nonbitter compound naringenin by rhamnosidase and β-glucosidase in naringinase preparations. The enzymes also prevent the crystallization of this compound in processed products. Naringinase is derived from *A. niger* or other fungi and is usually acid-treated to destroy pectinolytic activities. *Aspergillus niger* hesperidinase is usually used to reduce the levels of hesperidin, a tasteless flavonoid commonly found in oranges, lemons, and some mandarins. Excess hesperidin may crystallize, causing muddiness in citrus juice products or deposits on processing machinery or juice containers [90,110,115,125].

D. Winemaking

Winemaking often relies on enzyme treatment of grape juice, or juices of other fruits such as berries, peaches, apples, and pears [110,115,120,124, 126-128]. These enzymes and their modes of action have been discussed in Section C. Since grape juice is the most common starting material for winemaking, the application of enzymes to grape wines is the main focus of this discussion.

1. Pectic Enzymes

Pectic enzymes are used to reduce haze or gelling of grape juice at various stages of the winemaking process. The enzymes are derived from fungi, typically *A. niger, Penicillium notatum,* or *Botrytis cinerea.* Pectic enzymes are especially valuable to those American winemakers who use both thin-skinned grapes and thick-skinned grapes like the Concord variety. Thick-skinned grape varieties, which are typically native to northern America, possess a high pectin content, requiring higher levels of pectinases for the depectinization of juices made from them. During the processing of either type of grape, pectic enzymes may be added at any one of three stages during the winemaking process: stage 1—to the crushed grapes before pressing; stage 2—to the must (free-run juice) before fermentation or after; or stage 3—to the wine after fermentation.

Addition of pectic enzymes at stage 1 is favored since it increases the volume of the free-run juice and reduces pressing time, besides aiding in juice filtration and must clarification. The yield of juice per ton of grapes is improved; a 10% increase usually corresponds to an increase of 16 gallons of juice per ton of grapes. Furthermore, enzyme treatment promotes extraction of pigments when the grapes are heat extracted or fermented on the skin. The color from enzyme-treated juice is generally superior and the color of red wines is usually more intense. Pectic enzyme preparations from *A. niger* and *A. oryzae* may even contain an anthocyanase that hydrolyzes colored anthocyanins to colorless derivatives. The decolorization process may depend not only on the enzyme source but also on the pigment composition of the grape variety [120,129,130].

Treatment with pectic enzymes at stage 2 before or during fermentation causes settling of many of the suspended particles and often some undesirable microorganisms. A firmer yeast sediment and clearer wine is produced. Addition of pectic enzymes to the fermented wine at stage 3 increases the filtration rate and clarity. The enzyme mixture causes flocculation and precipitation of pectin particles, floating microorganisms, and protein. Elimination of the protein improves the stability of the wine and reduces the reliance on bentonite for protein removal. The level of enzyme supplement must be adjusted owing to the inhibitory effect of alcohol on pectinases. Use of pectic enzymes at all three stages promotes a faster aging of the wine. However, the effect on the flavor and bouquet is more difficult to assess. Some sensory tests indicate that wine quality is either enhanced or unchanged by pectic enzyme treatment [110,120].

2. Other Enzymes in Winemaking

Though most protein is removed in early processing steps, traces of remaining material from grapes or lysed yeast may cause a haze. Endogenous grape proteolytic activity or even enzymes from lysed yeast cells can play a role in reduction of this slight cloudiness. Treatment of wines with fungal protease from *A. oryzae* or other species of *Aspergillus* has been explored but not yet applied routinely by the wine industry [110].

A preparation of β-glucanase from a strain of *Trichoderma* may be added to aid in the clarification and filtration of wine. The enzyme is added to the wine between the first racking and filtration [20].

As mentioned, fungal anthocyanase can decolorize or lighten red wines or fruit juices. It has been shown to be effective in laboratory trials and is now commercially available [131].

The use of fungal glucose oxidase and catalase preparations has been explored for the removal of excess oxygen from wine, beer, and fruit juices. The addition of glucose necessary for the reaction and the instability of catalase are drawbacks to this procedure. Another drawback is an undesirable browning reaction that occurs in white table wines during enzyme treatment [132].

E. Starch Processing

Starch processing in the food industry converts starch-bearing raw materials to starches, starch derivatives, and starch saccharification products [24, 133-145]. These products have a multitude of uses (Table 5) [146-149] and some serve as the starting materials for a wide variety of industrial products [150]. The raw materials for starch processing are usually maize, wheat, or potatoes, though barley, cassava, and other sources may be important in the future. Projected corn yields indicate that this crop will continue to be of importance as a carbohydrate source in the United States, and the following will concentrate on the processing of this starch-rich raw material to industrial sweeteners.

1. Corn Starch

Corn starch is a mixture of polysaccharides composed of about 25% amylose, a linear homopolymer of glucose held together by α-1,4 glycosidic linkages, and 75% amylopectin, a branched glucose homopolymer which has α-1,6 branchpoints [151-153]. Microbial amylase plays an important role in biotechnological methods for starch saccharification [22,35,138,154-159]. Starch-processing protocols are variable, and Figure 7 illustrates the diversity. Typically, bacterial α-amylases are used as endoenzymes to hydrolyze internal α-1,4 linkages; fungal glucoamylases are used as exoenzymes to remove terminal glucose units; and bacterial debranching enzymes, e.g., pullulanases, are used to cleave α-1,6 branchpoints. The glucoamylases also have a weak debranching activity. Microbial glucose isomerase plays an important role in the conversion of glucose to another valuable product, fructose, especially in high-fructose corn syrup (HFCS) production. Thus, glucose and fructose, both of which are important sweeteners in the food industry, can be generated from starch by the use of the above-mentioned enzymes.

Inorganic acid and enzyme-catalyzed systems are used to hydrolyze starch to glucose. The latter method is more common at this time. The use of enzymes offers distinct advantages over acid hydrolysis. The need for acid neutralization is eliminated in enzymatic processes, reducing refining costs. Energy demands are lowered, and steps requiring expensive reactors operating under extreme conditions are unnecessary. Furthermore, enzymatic processes permit greater control of the reaction, its specificity, and the reactivity of its products. Lower temperatures and a near-neutral pH during enzymatic processing reduce unwanted side reactions and thus

TABLE 5. Properties and Industrial Applications of Hydrolyzed Starch Products

Type of syrup	DE[a]	Composition (%)	Properties	Application
Low DE maltodextrins	15-30	1-20 D-glucose 4-13 Maltose 6-22 Maltotriose 50-80 Higher oligomers	Low osmolarity	Clinical feed formulations; raw materials for enzymic saccharification; thickeners, fillers, stabilizers, glues, pastes
Maltose syrups	40-45	16-20 D-glucose 41-44 Maltose 36-43 Higher oligomers	High viscosity, reduced crystallization, moderately sweet	Confectionary, soft drinks, brewing and fermentation, jams, jellies, ice cream, conserves, sauces
High maltose syrups	48-55	2-9 D-glucose 48-55 Maltose 15-16 Maltotriose	Increased maltose content	Hard confectionary, brewing and fermentation
High DE syrups	56-68	25-35 D-glucose 40-48 Maltose	Increased moisture holding, increased sweetness, reduced content of higher sugars, reduced viscosity, higher fermentability	Confectionary, soft drinks, brewing and fermentation, jams, conserves, sauces
Glucose syrups	96-98	95-98 D-glucose 1-2 Maltose 0.5-2 Isomaltose	Commercial liquid "dextrose"	Soft drinks, caramel, baking, brewing and fermentation, raw material
Fructose syrups	98	48 D-glucose 52 D-fructose	Alternative industrial sweeteners to sucrose	Soft drinks, conserves, sauces, yogurts, canned fruits

[a]Dextrose equivalent.

Source: Ref. 146.

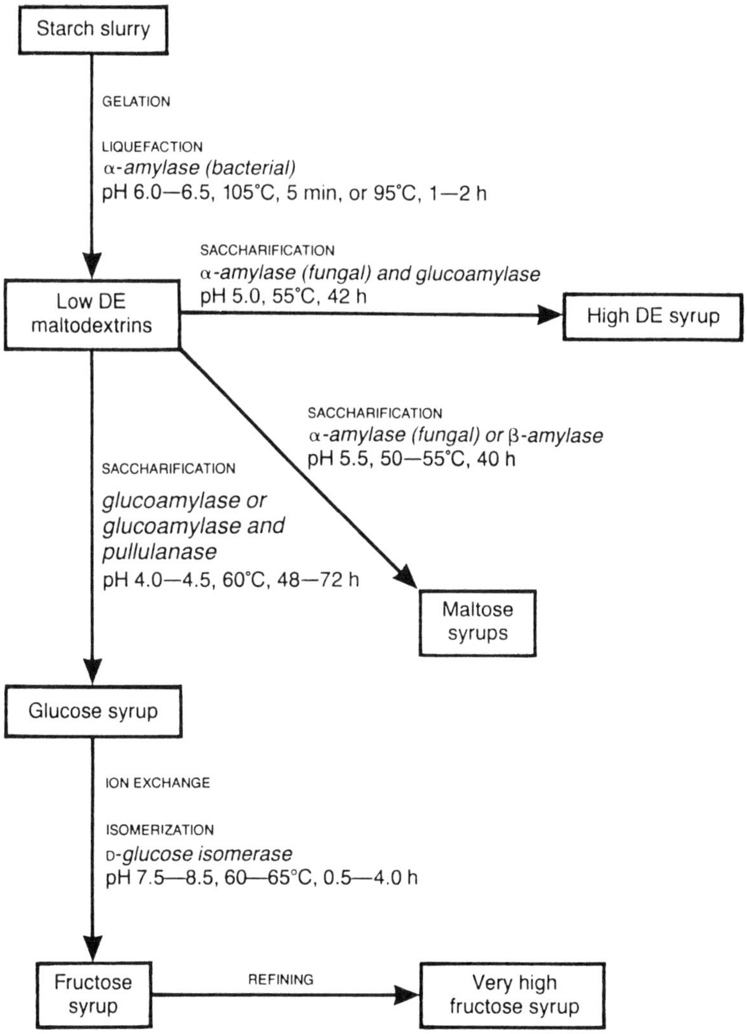

FIGURE 7. Starch processing using enzymes. (From Ref. 146.)

boost yields. Another benefit is the avoidance of color and deleterious flavor that are often generated by such side reactions. For example, 5-hydroxy-2-methylfurfuraldehyde and anhydroglucose compounds are absent, as are undesirable salts. Of course, these main advantages of enzymatic processing require that the production and use of highly active enzymes must be economic and that the final product, the corn-based sweetener, be competitive with sucrose. In both cases, these requirements are met.

2. Gelation and Liquefaction of Starch

After wet milling of the corn, existing starch-processing technology requires pretreatment of the starch before enzymatic hydrolysis. This pretreatment step gelatinizes the starch, which is present initially as granules 5-25 μm in diameter. Exposure to temperatures above 60°C swells and disrupts the particles. Addition of heat-stable α-amylase from *Bacillus subtilis* or *B. subtilis* var. *amyloliquifaciens* thins the starch slurry by causing random hydrolysis of glycosidic bonds. Both bacterial enzymes have a temperature optimum of about 70°C and easily tolerate the pretreatment temperature. The thinning step is essential to produce a slurry of manageable viscosity, which can then be treated further with enzymes.

The pretreated slurry, which is 30-40% starch solids, is then processed with heat-stable α-amylase, typically from *Bacillus licheniformis*, at pH 6.5. In the continuous process, which is preferred at present, the starch slurry is passed through a steam jet cooker. The temperature is maintained at 103-107°C for 5-10 minutes during liquefaction. Then, a 1-2-hour treatment at 95°C converts the thinned starch to a hydrolyzate of 0.5-1.5 DE (dextrose equivalent). (DE, simply stated, is the percentage of glucosidic bonds hydrolyzed.) The thermostable *B. licheniformis* α-amylase operates throughout this entire process and need not be re-added during the 95°C treatment. (Fungal α-amylases, being less thermostable, would not survive this procedure.) Enzymatic hydrolysis is allowed to continue until a DE of 10-15 is reached. Complete hydrolysis of starch, if desired, using a thermostable α-amylase and then fungal glucoamylase yields 95-97% dextrose. An added advantage of the *B. licheniformis* enzyme in this step is its reduced requirement of calcium for stability. Cost savings and convenience result since less reliance is placed on ion exchange columns to remove the calcium, which inhibits glucose isomerase used in HFCS production later on.

Liquefied starch derivatives of DE = ~10 obtained by degradation with thermostable bacterial α-amylase are dried and sold as maltodextrins (Table 5). Maltodextrins consist of 3% maltose, 4% maltotriose, and 93% of carbohydrate with a degree of polymerization higher than 4. They are used in the food industry as thickening agents and are useful additives for drying hygroscopic substances such as yeast extracts, molasses, malt, and certain food ingredients.

3. Saccharification of Starch

Saccharification is performed with fungal α-amylase alone or acting together with fungal glucoamylase. A pullulanase may also be added. A fungal glucoamylase preparation, typically obtained from *A. niger*, *A. awamori*, *A. oryzae*, or *Rhizopus oryzae*, is used to manufacture corn syrups of high DE and crystalline dextrose. Since the enzyme has exoamylase and debranching activity and since fungal glucoamylase preparations also contain α-amylase, liquefied starch can be hydrolyzed almost completely by these enzymes at about pH 4 and 60°C. The enzyme dose determines the exact treatment time, which may last up to 4 days. The pH and temperature conditions of the commercial saccharification process with fungal glucoamylase offer several distinct advantages. Acid pH greatly reduces isomerization reactions to fructose and other sugars, which lower the glucose yield. Also, low pH, high temperature, and high levels of solids restrict the growth of contaminating microbes.

Fungal α-amylases (Table 1) from species of *Aspergillus* are *thermolabile*, operating in the range of 50-55°C. They are used to saccharify starch to high DE corn syrups. Glucoamylase is a co-product in many commercial fungal α-amylase preparations unless the mixture is purified before sale or unless production strains or fermentation conditions are carefully chosen. As discussed earlier, *thermostable* α-amylases used in starch processing are obtained from industrial *Bacillus* strains.

At high concentrations of glucose, glucoamylase preparations synthesize maltose, isomaltose, and panose via a transglucosylation reaction. Industrial fungal strains with reduced levels of this activity have been selected or methods developed for its removal [78]. Upon achievement of maximum glucose levels, the glucoamylase is inactivated by heat treatment of the starch hydrolyzate to prevent disaccharide formation via the accompanying transglucosylation activity.

Debranching enzymes, such as pullulanases, are obtained from bacterial sources [20,160,161]. Commercial pullulanases, typically obtained from *Bacillus* species, are used to improve glucose yields and decrease reaction times during starch saccharification, and could be used to produce high maltose syrups. The pH and temperature profiles must match those of the other enzymes used in the saccharification process. The *Bacillus* enzyme is acidophilic and thermostable and thus can operate effectively under saccharification conditions. However, since *A. niger* glucoamylase possesses debranching activity, the fungal glucoamylase can still fulfill many of the needs of the sweetener industry.

4. Isomerization to Fructose

Glucose isomerase is obtained exclusively from bacterial sources. Immobilized glucose isomerase in a packed bed reactor is routinely used to convert glucose to fructose, a monosaccharide which is 1.2-1.8 times sweeter than sucrose on a dry weight basis. Corn starch is a preferred raw material for the glucoamylase-catalyzed production of glucose [134-137,143,158].

Today, HFCS is used as a caloric sweetener in a wide variety of foods and beverages. The worldwide market has grown 20% every year since 1980 and was estimated at 4.58 million metric tons in 1985 [162]. When appropriate in product formulations, HFCS is cheaper and its use results in lower calories than sucrose, while retaining the same level of sweetness. The use of 42% fructose HFCS in various food products and 55% fructose HFCS (instead of invert sugar) in soft drinks will continue to grow significantly in the future, as it did in the 1970s and 1980s. New forms of fructose products are being developed. For example, HFCS containing 90% fructose [162] is already available, as is crystalline fructose [163]. The technology for conversion of HFCS to a crystalline form by sonic drying is on the horizon [164,165].

5. Alternative Sweeteners

The full impact of alternative sweeteners on the consumption of cane-, beet-, and corn-based caloric sweeteners remains to be assessed. Depending in many cases on regulatory factors, cyclamate, thaumatin, monellin, stevioside, glycyrrhizin, phyllodulcin, hernandulcin, PS 99, PS 100, neosugar, miraculin, neohesperidin, trichlorogalactosucrose, L-aspartyl-3-((bicycloalkyl)-L-alanine alkyl esters, L-altrose, L-sugars, improved derivatives of these compounds, or new substances could find use in the food industry. Some could join aspartame, acesylfame K, saccharin, and xylitol as alternative sweeteners in the United States [144,166,167].

F. Sucrose Refining

1. α-Galactosidase

Beet sugar refining in Japan depends on α-galactosidase from *Mortierella vinaceae* var. *raffinose utilizer* for hydrolysis of raffinose to sucrose and galactose. The enzyme, also termed raffinase or melibiase, reduces the levels of the trisaccharide raffinose, which interferes with the crystallization of sucrose from molasses during refining [168,169]. Levels of raffinose as low as 0.05-1.5% can reduce sucrose yields since they induce the formation of fine needles of sucrose rather than large crystals. Levels of 8% raffinose are unmanageable without enzyme treatment [39,170,171].

α-Galactosidase is produced from a strain of *M. vinaceae* that has negligible invertase and gives rise to mycelial pellets with bound enzyme [172]. Enzyme activity can be induced most effectively by lactose, though raffinose, melibiose, or galactose can also act as inducers. Mycelial pellets are harvested and maintained as suspensions in chambered troughs while molasses containing raffinose is fed through the horizontal reactors. Treatment of the pellets with glutaraldehyde increases their stability. Continuous beet sugar processing with quasi-immobilized *M. vinaceae* α-galactosidase converts about 65% of the raffinose to sucrose. Thus, a sugar-processing plant that refines 600 tons of sucrose per day from 3000 tons of beets can transform up to 3.25 tons of raffinose into sucrose that is then available for crystallization [173-175].

2. Dextranase

Dextran is a glucan of 1.5×10^4 to 2×10^7 molecular weight or higher containing primarily α-1,6 linkages and branches consisting of α-1,3 and α-1,4 linkages. The degree of branching is variable. The polysaccharide is synthesized by *Leuconostoc mesenteroides* or *L. dextranicum*, which convert sucrose to fructose and dextran using the enzyme dextransucrase. Deteriorated sugar cane juice, especially juice from temperature-abused beets, can become contaminated with these organisms and contain dextrans which interfere with beet sugar refining and reduce the economic efficiency of the mill. The increased viscosity of the contaminated juice is correlated with slower heating, a reduction in the rate of sucrose crystallization, increased turbidity, slower filtration, and the presence of elongated sugar crystals [174,176-180]. The use of fungal dextranase at raw sugar factories alleviates these problems. Treatment of the juice with a dextranase preparation from *Penicillium funiculosum* or *P. lilacinum* for 20 minutes at 40°C and pH 5.4 results in degradation of 68% of the dextran to mainly isomaltose and isomaltotriose. These products do not interfere with sucrose crystallization. The specific viscosity of the cane juice can be reduced significantly by enzyme treatment, while the processing rate is increased. Dextranase production by *P. funiculosum* can be induced by growing the appropriate strain in medium containing dextran [181]. *Leuconostoc mesenteroides* cultures from which cells have been removed can serve as culture medium. After 4-5 days of growth at 30°C, the dextranase can be obtained from culture filtrates of *P. funiculosum*, processed, and then used to treat cane juice that has undergone postharvest biodeterioration [182].

G. Confectionery

Invertase from baker's yeast has been known since the last century and is used to manufacture confections, desserts, syrups, and artificial honey. The yeast enzyme hydrolyzes the terminal nonreducing portion of β-fructofuranosides and can be used to break down disaccharides, trisaccharides, and fructans to desired products. Its primary use is the conversion of sucrose to fructose and glucose [154,183].

Commercial invertase preparations from *S. cerevisiae* or *S. carlsbergensis* are obtained by growing yeast on molasses and then treatment of the cells with organic solvents to remove the enzyme from the periplasmic space [11]. Autolysis may be a superior method of enzyme extraction in the future. Purification is necessary to remove objectionable flavors originating from the yeast culture. The yeast enzyme activity has a high thermal stability and a broad pH optimum. Industrial invertases can be further stabilized with crosslinking agents and used as immobilized systems [39,183,184].

In the confectionery industry, yeast invertase is used in the manufacture of chocolate-coated candies with a soft center. A solid sucrose core containing a dose of invertase is coated; liquefaction occurs during storage. Invertase is also applied to the production of fondants and chocolate coatings [183,185,186].

H. Dairy

The dairy industry depends on microorganisms for the production of diverse milk products. The following section focuses on two applications of fungal enzymes to the dairy industry: whey processing with lactase and flavor development in cheeses. Fungal rennets are discussed in Chapter 12.

1. Lactase

Lactase, or β-galactosidase, is used to hydrolyze lactose, the main sugar (about 5% w/v) in milk and whey, to glucose and galactose [7,8,187-192]. These two monosaccharides are sweeter, more digestible, and more soluble than lactose. The enzymatic breakdown of lactose offers an opportunity for the dairy industry not only to develop new milk-derived products, but also to market the whey by-products of cheese manufacturing that are discarded, sometimes to the detriment of the environment. Lactases are already used to transform cheese whey into a sweet syrup and fermentation media (for ethanol, xanthan gum, wine, beer, and soy sauce production, for example), to produce low-lactose milk and other dairy products, and to reduce lactose crystallization in dairy products. Lactose-derived syrups have good humectancy, promote Maillard browning, and can serve as replacements for corn syrup or sucrose in a wide variety of dairy, bakery, soft drink, dessert, or confectionery products. When added to low pH dairy foods such as sour cream, yogurt, and buttermilk, lactases have been shown to improve the taste without significantly adding calories.

Lactase treatment of threefold concentrated milk before freezing reduces thickening and extends its storage life, and it improves the flavor of the reconstituted product. Lactase treatment of milk for yogurt manufacture accelerates acid development by the starter cultures and increases the viscosity, sweetness, and shelf life of the final products. Similarly, lactase-treated milk leads to faster acidification in cheese manufacture. It yields a firmer, more elastic cottage cheese curd, besides reducing set time. Of greater importance is the effect on ripened cheeses: Lactase treatment of milk significantly shortens the ripening process and reduces cost. Flavor development is also hastened and enhanced. Accompanying fungal proteases apparently contribute to the effects of added lactase on the ripening process [8,11,39,189,192-197]. Lactase can also serve as a digestive aid for lactose intolerance, a phenomenon characteristic of a large majority of the world's population [198].

Commercial lactase is produced as an extracellular enzyme by *A. niger* or *A. oryzae* or intracellularly by the yeasts *Kluyveromyces marxianus* var. *lactis*, *Kluyveromyces fragilis*, and *Candida pseudotropicalis*. The enzymes have been characterized (Table 6). The yeast enzyme has a pH optimum around neutrality and is used to treat milk or sweet whey. The more thermostable *A. niger* lactase has a broader but lower pH optimum of 4-5 and is used primarily to process acid wheys. It can still be used at pH 6.5, however. Since the production of fungal lactases is expensive, once-only use is uneconomic. Therefore, fungal lactases have been immobilized for large-scale industrial hydrolysis of lactose in milk or whey. Many processes and reactor designs have been investigated. Most immobilized systems are adapted to the quality of the substrate and are sensitive to the presence of milk proteins and macroscopic impurities. Pretreatment, especially pasteurization, ultrafiltration, and demineralization are usually necessary [189,191,192,199-201].

Commercial processes with the yeast or *A. niger* lactase are in operation. Lactase from *K. marxianus* var. *lactis* has been immobilized in triacetate fibers and used to produce up to 10 tons of low lactose milk per day [197]. Immobilized mycelia of *A. niger* have been used to convert cheese whey into a sweet syrup that can serve as an ingredient in baked goods, ice cream, yogurts, canned fruit, candies, and other foods [202]. In the Corning process, which uses lactase immobilized on porous glass beads, approximately 30,000 liters of whey yields 1.7 tons of hydrolyzed syrup and about 80% of input lactose is hydrolyzed. Each lactose hydrolysis unit operates continuously at pH 5.0 and 50°C and processes 360 to 500 liters of lactose per hour (Fig. 8) [192,194].

Other fungal lactases have also been immobilized for pilot studies of potential commercial processes [203].

2. Cheese Ripening Enzymes

Microorganisms and enzymes cause cheese ripening by their action on butterfat, casein, and lactose in the milk and curd. A multitude of breakdown products that determine flavor are released: fatty acids (especially short-chain fatty acids), acetic acid, lactic acid, ammonia, amines, amino acids, peptides, alcohols, aldehydes, ketones, esters, sulfides, and mercaptans. The characteristic flavor of cheese varieties is related to the levels and

TABLE 6. Properties of Microbial Lactases

Source	pH optimum	pH stability	Temperature optimum (°C)	K_m for lactose (mM)	Cofactors needed	Molecular weight	Subunits
Aspergillus niger	3.0–4.0	2.5–8.0	55–60	85	None	124,000	n.d.
Aspergillus oryzae	5.0	3.5–8.0	50–55	50	None	90,000	n.d.
Kluyveromyces fragilis	6.6	6.5–7.5	37	14	Mn^{2+}, K^+	201,000	2 or 10
Kluyveromyces lactis	6.9–7.3	7.0–7.5	35	12–17	Mn^{2+}, Na^+	135,000	1
Escherichia coli	7.2	6–8	40	2	Na^+, K^+	540,000	4
Lactobacillus thermophilus	6.2	n.d.	55	6	n.d.	540,000	n.d.
Leuconostoc citrovorum	6.5	n.d.	60	7.8	None	n.d.	n.d.

n.d. = not determined.

Source: Ref. 199.

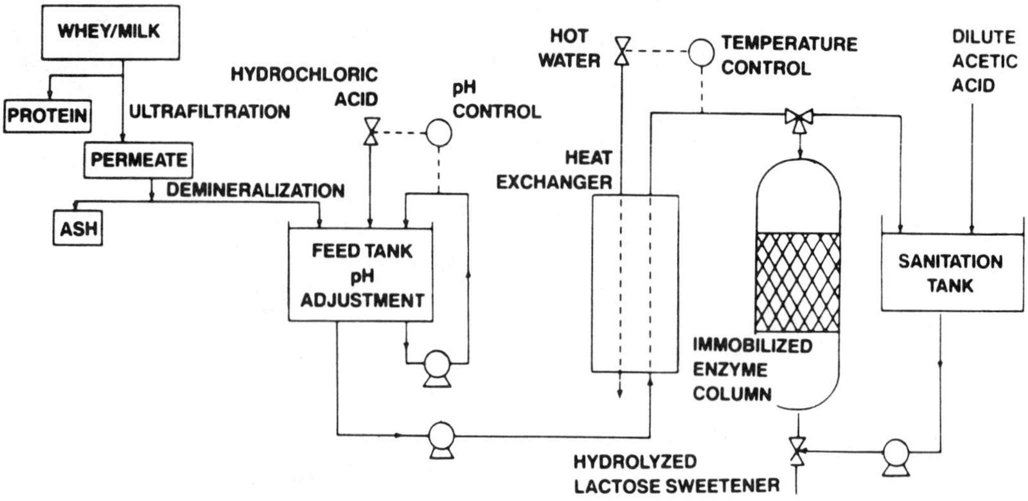

FIGURE 8. Commercial process with immobilized lactase that hydrolyzes lactose using whey, whey permeate, or milk permeate as starting material. (From Ref. 194.)

proportions of relatively few key components of these breakdown products. Thus, the influence of microbial or enzymatic processes on the balance of a few flavor components may be critical to success or failure in cheese making [204].

The addition of lipolytic and proteolytic enzymes to cheese as constituents of rennet pastes influences the rate of cheese ripening and flavor development [205-208]. Secreted mold enzymes also can be important in the development of flavor and body texture of certain cheeses. A number of mold cheeses, especially Roquefort, blue, and Stilton, depend on release of fatty acids by fungal lipases to impart a characteristic sharp flavor [206, 209-211] and desirable aroma and texture [23,212].

In modern times, the addition of exogenous microbial enzymes to cheese milk has been used to accelerate cheese ripening and flavor development. Lipases and neutral proteases from fungi and bacteria have been investigated most extensively. Early attempts to supplement the lipolytic activity of cheese cultures were unsuccessful since by-products with unpleasant tastes were formed. There is still no general agreement on the efficacy of exogenous microbial enzymes on cheese ripening and flavor. And there is a lack of reliable methods for the determination of free fatty acids and other key flavor compounds as they appear in maturing cheese [206,207,213].

Present day processes for flavor development and accelerated cheese ripening employ bacterial and fungal lipases. The enzymes are added in carefully measured amounts to avoid bitterness and rancidity, and the temperature is maintained at about 10°C. The enzyme preparations and treatments are chosen with a specific cheese variety in mind. Mainly crude enzyme preparations from *Micrococcus* spp., *Lactobacillus* spp., *Candida* spp., *Aspergillus* spp., *Penicillium* spp., or *R. miehei* are used. Esterases in the preparations, especially the esterase in *R. miehei* enzyme mixtures, contribute desirable flavor characteristics [7,37,39,208,214-216].

Results with *A. oryzae* or *Bacillus* proteinases have been inconsistent. Flavor defects, especially bitterness, are sometimes observed, as are textural flaws resulting from excessive breakdown of β-casein [206].

Microbial enzymes are added to cheese milk in soluble, insoluble, or even encapsulated forms. For example, soluble *B. subtilis* neutral protease in combination with insoluble *A. oryzae* lipase has been used to produce an excellent Cheddar cheese [217]. About 90% of the insoluble *A. oryzae* lipase becomes associated with the curd and little is lost in the whey. However, since uneven flavor distribution in the cheese curd may be evident with use of insoluble lipase, treatment with soluble enzyme may sometimes be preferable. Soluble food-grade lipase from *Candida cylindracea* together with soluble bacterial protease has also been used successfully [218].

An alternative approach to conventional methods of enzyme supplementation and accelerated ripening is the addition to cheese milk of encapsulated enzymes or substrates. One method, the use of liposomes in controlled ripening of cheese, shows promise for the future. Microencapsulation in liposomes entraps flavor enzymes in food-grade materials, stabilizes and protects the enzyme activities, prevents their premature action, and distributes the enzymes uniformly throughout the milk and eventually throughout the curd as the liposomes disintegrate. Liposome composition can be varied, permitting control of their permeability, stability, and affinity properties. The microcapsules are enmeshed in the curd matrix during coagulation and enzyme loss to the whey is less than 10%. Whereas the ripening process can take up to a year for mature Cheddar, the use of microencapsulated cheese maturing enzymes can produce good-quality, mature Cheddar in half the usual time [219-220].

Industrial cheese making relies on two enzyme processes for flavor and body texture transformation of cheese: These produce accelerated ripened cheese (ARC) and enzyme-modified cheese (EMC). As discussed in this section already, ARC production involves long-hold, natural ripening which aims to reduce ripening time by months and produce cheese with more intense flavor and smoother body. EMC production, which requires larger amounts of added enzymes than ARC, involves a rapid 4-8-day transformation that utilizes lipases and other enzymes. The EMC process is generally applied to the manufacture of processed cheese, cheese flavors, cheese-flavored snacks, baked goods, and various food mixtures. A similar process is used to produce enzyme-modified milk powder and butterfat. Both the ARC and EMC processes are becoming more common and point to an increasing role of microbial flavor enzymes in the food industry [205,208,221, 222].

I. Food Fermentations

Fermented foods are edible substances that have been subjected to the action of microorganisms or enzymes so that favorable biochemical processes can lead to significant modification of the food. As a consequence of this modification, the food may be more nutritious, palatable, or digestible, or it may be safer to eat [223]. Preceding sections have already considered the more familiar food fermentations.

TABLE 7. Production of Classes of Fermented Foods by Geographical Region

World production	Region	Importance: Major	Minor
High	Europe	Dairy; beverages, cereals; meat	Legumes; starch crops
	N. America	Beverages; dairy; meat	Fish; legumes; starch crops
	Africa S.	Starch crops; cereals; beverages	Dairy
Medium	S. America	Beverages; dairy	Legumes
	Middle East	Dairy	Legumes; meat
	Indian subcon.	Cereals; legumes	Meat
	E. Asia	Fish; legumes	Dairy
	S.E. Asia	Fish; legumes	Dairy
Low	Oceania	Dairy	Legumes
	N. Africa	Dairy	Legumes

Source: Ref. 223.

Fermented foods can be divided into nine classes: beverages, cereal products, dairy products, fish products, fruit and vegetable products, legumes, meat products, starch crop products, and miscellaneous products. Many of these classes of foods are specific to particular regions of the world, and, in many cases, they are important factors in the local diet (Table 7). Fermented foods include bread, yogurt, soy sauce, fish sauce, sausages, kimchi, gari, and mushrooms. Examples of fermented beverages include beer, wine, sake, brandy, whiskey, gin, liqueurs, and nonalcoholic tea, coffee, and cocoa [31,34,47,223-228].

In many traditional food fermentations, fungi that secrete enzymes are used to make desirable flavor, aroma, texture, and nutritional changes in edible substances. Since antiquity Western peoples have accented the flavor of cheese by fermenting it with *Penicillium* molds; and peoples in the Orient have used *Aspergillus* and *Rhizopus* molds that secrete amylases and proteases to enrich the flavor of beans, pulses, many cereals, various vegetables, fish, and meat. Today, added fungal enzymes can lead to changes similar to those that occur in traditional food fermentations, or they can exaggerate desirable characteristics, without the need for the whole organism. In the following sections, a few additional examples of the use of enzymes in present-day food fermentations illustrate these points.

1. Koji

The koji fermentation originated in China and is now widespread in the Orient. It produces an essential component of many food fermentations. "Koji," a Japanese term, refers to a molded cooked cereal which acts like a malt in various food fermentations such as soy sauce, miso, mirin, sake, and shochu. Alcoholic beverages are prepared by simultaneous saccharification with koji and fermentation of starchy materials from rice, corn, cassava, or potatoes. Though treatment with koji is a step in a two-stage fermentation, the koji process itself can also serve as a single-stage fermentation for the production of various fermented foods [229].

The koji process is a solid substrate fermentation [230,231] on soybeans, wheat, rice, or other cereals or seeds. It is a rich source of hydrolytic enzymes which break down solid raw material to soluble products. When added to a food fermentation, these enzyme preparations generate fermentable substrates for yeast or bacteria. Koji also provides various nutrients that may be essential, in addition to contributing flavor and aroma components. Enzyme classes that have been detected in soybean koji include amylase, invertase, cellulase, hemicellulase, glutaminase, lipase, pectinase, phosphatase, and protease. The proteases are acid, neutral, and alkaline enzymes and include aminopeptidases and carboxypeptidases [226,227,229].

A number of different fungi are used to produce koji. The organisms for koji production are commonly *A. oryzae*, *A. awamori*, or *A. sojae*. The koji is named after the color of mold spores or pigments in the mold or is named after the substrate or endproduct of the fermentation. *Rhizopus* spp., mainly *Rhizopus oryzae*, are used to produce a koji (pang-cheu) used in white spirit and whiskey distilleries. *Monascus purpureus* is used to produce ang-kok (red rice), a type of koji rich in pigments and flavor substances that impart coloring and flavoring to foods. This koji is also used as a starting material in the brewing of red rice wine [229].

The koji process also serves as a single-stage solid substrate fermentation for foods. For example, *Rhizopus oligosporus* produces tempeh on soybean (tempe kedala) or coconut meal (tempe bonkrek); *Neurospora sitophila* produces ontjom on peanut press cake; and *Mucor* spp. produce a fermented soy bean cheese (sufu) [224,229].

2. Vinegar Fermentation

In vinegar production a yeast fermentation generates the alcohol feedstock for the subsequent vinegar fermentation carried out by selected species of *Acetobacter*. Present-day vinegar fermentations depend on enzyme treatment of the waste fruit or vegetable biomass used as substrate for the first stage of the manufacturing process, the yeast fermentation. The starch in the pulp suspension is treated with fungal or bacterial α-amylase and glucoamylase to release sugars steadily for yeast growth. Acidophilic pectic enzymes from species of *Aspergillus* are also used to disrupt structural tissues of the plant material which serves as substrate. Such enzyme treatments permit a more efficient extraction of nutrients from plant pulp supplied for the yeast fermentation [232,233].

In Japan, *A. oryzae* is grown on steamed rice and then used as the source of koji for rice vinegar production. The koji is mixed with steamed rice, and its amylolytic enzymes promote the alcoholic yeast fermentation. Species of *Mucor* and *Rhizopus* are used in China as the source of saccharifying enzymes for rice vinegar production [233].

3. Coffee and Tea Fermentation

The processing of coffee depends on fermentation with pectinolytic microorganisms to remove the mucilage coat from the coffee beans. Exogenous fungal pectic enzymes are sometimes supplemented to the fermentation to remove this pulpy layer, three-fourths of which consists of pectic substances. A diluted commercial enzyme preparation is sprayed onto the cherries at a dose of 2-10 g per ton at 15-20°C. The subsequent fermentation stage of coffee processing is accelerated by the enzymatic digestion and is reduced from 40-80 hours to about 20 hours. Cellulase and hemicellulase present in the enzyme preparation promote hydrolysis of the mucilage. Enzyme in the decanted liquid can be reused for several additional treatments. However, such large-scale treatments with industrial pectinases are often costly and uneconomic. Instead, inoculated waste mucilage is fermented in tanks, and the fermentation liquor is washed and filtered. It is then sprayed onto the cherries. Ten kg of crude enzyme is sufficient to treat 1000 kg of washed ripe cherries. The subsequent coffee fermentation is reduced to about 12 hours after application of such a crude enzyme solution [232,234-237].

Fungal enzymes are also used in the manufacture of tea. Fungal pectinase treatment accelerates tea fermentation. However, the enzyme dose must be carefully adjusted to avoid damage to the tea leaf. Pectinase also improves the foam-forming property of instant tea powders by destroying tea pectins [237-239]. Fungal cellulase releases soluble compounds from tea leaves that enhance the quality of tea. Another enzyme, *A. niger* tannase, is used to treat insoluble material that forms during the manufacture of instant tea. It hydrolyzes tea flavonols esterified with gallic acid to galloyl derivatives that are then oxidized during tea fermentation [174,238, 240].

4. Other Food Fermentations that Use Exogenous Enzymes

Soybean fermentations have been performed with industrial enzymes. Fungal pectic enzymes, proteases, and hemicellulases have been added to improve soy flavor [241,242].

Fungal and bacterial proteases have been used to accelerate protein degradation and fermentation in the conversion of fish protein to pastes and sauces [243]. Proteolytic enzymes from *A. niger* are especially effective in reducing fermentation time and increasing yield and may be added in the form of koji [244].

Cheese whey has been treated with fungal lactase and then subjected to yeast fermentation to produce an intensely flavored product resembling soy sauce [245].

J. Canning, Bottling, and Packaging

1. Glucose Oxidase

Glucose oxidase, also known as notatin, is used in food processing to remove glucose or oxygen or to form gluconic acid or hydrogen peroxide [137,246-249]. Commercial preparations also contain catalase, which participates in the oxygen removal reaction. When added to canned or bottled fruit juices, carbonated soft drinks, beer, or some wines, glucose oxidase removes traces of oxygen and prevents oxidative rancidity and enzymatic browning. Glucose supplements may be necessary to promote oxygen removal via the glucose oxidase reaction. Removal of oxygen remaining in the headspace of cans containing soft drinks or juices can increase the shelf life of the products up to 6 months [249,250]. The enzyme treatment does not cause detectable flavor changes and reduces color fading and iron pickup by the beverage [249]. Glucose oxidase is obtained primarily from *A. niger*. *Penicillium amagasakiense* is the source of the enzyme in Japan, and *Penicillium vitale* is used in the U.S.S.R. Fermentation conditions are quite dependent on strain and organism. The enzyme is intracellular and is extracted from disrupted cells before partial purification, processing, and final preparation as a powder or stabilized liquid [8,11,246].

2. Lactase

An appealing concept in enzymology and food packaging is the proposal that functional enzymes be built into paperboard-based or plastic-based container constructions. Research on "process-in-package" functions is focusing on the incorporation of lactase (see Sec. H.2) into special milk cartons. The milk cartons would serve as enzyme reactors and would convert lactose into glucose and galactose during transit or storage. The lactase is retained in the container material and would not be consumed with the dairy product [251]. It is estimated that 15% of all Americans and a large majority of the world population are lactose-intolerant [198]. Such built-in enzyme components of food packaging would generate a lactose-free milk product and thus make available nutritional dairy products to many lactose-intolerant people, including lactase-deficient individuals.

Novel packaging systems could also be used to remove oxygen with glucose oxidase or other enzymes, and also to eliminate traces of complex sugars, starches, proteins, and pectin that give rise to turbidity in the food product. A process-in-package technology could preserve or stabilize liquid products, convert sugars, generate natural chelators, and even reduce or eliminate cholesterol in food products [251]. The concept is not totally new, however. For example, a process for the preservation of foods with plastic wrappers coated with a system that control-releases glucose oxidase and catalase was patented in 1956 [252].

An enzyme dosing system designed for individual packages of aseptically processed low-acid foods has been developed and used in many countries. Its use in the United States, especially for milk products, is expected in the near future. The system allows the introduction of measured doses of liquid enzyme into a sterilized milk line and piping directly into an aseptic filling machine. When lactase is introduced this way, hydrolysis of lactose occurs within the package in several days. Lactose does not alter the nutritional quality of the milk but does convert lactose to more digestible sugars and increases the sweetness of the milk. Post-process addition of the lactase to the package prevents possible inactivation of the enzyme by heat treatment and also avoids an increase in the caramelization of sugars [253].

K. Fungal Enzymes Used in Other Food Processes

Fungal enzymes, including many of those already considered, are used in other food-processing applications. These industrial applications may involve highly specialized or less-common uses, small-scale processes or small markets, or experimental studies. A few of these fungal enzymes and their actual or potential roles in food processing are discussed below.

1. Cellulases

Cellulases are multienzyme complexes bearing endo-1,4-β-glucanase, cellobiohydrolase, and β-glucosidase activity. They cleave the β-1,4 linkages of cellulose or its chemically modified forms and also cellodextrin or cellobiose [254,255]. Cellulases have not been used widely in the food industry or in large volumes since commercial preparations have low activity and few large-scale uses exist. Fungal cellulases are used occasionally in brewing, cereal processing, fruit processing, wine production, and alcohol production. Other examples of cellulase applications include uses to improve the palatability of low-quality vegetables, increase the flavor of mushrooms, promote the extraction of natural products, and alter the texture of foods. In some cases, cellulases are used to supplement the action of pectinases, β-glucanases, and starch-degrading enzymes. Cellulases are being investigated as means of converting food-processing wastes to food ingredients, single-cell protein (SCP), or substrates for fuel-producing microorganisms [16,254,256-265].

Industrial production of commercial cellulases generally involves the fermentation of *A. niger* to yield enzyme for food use and *T. viride* to yield enzyme for nonfood use [2]. Other fungi are also prolific producers of cellulases but are used less commonly [12,266]. *Trichoderma viride* is grown by submerged fermentation in the presence of inducers like cellulose, mixed β-1,4 glucans, or oligosaccharides. Large-scale cultivation of *A. niger* usually involves surface fermentation in trays containing wheat bran. Cellulase is extracted from the dried bran. Both *T. viride* and *A. niger* enzyme preparations are concentrated by precipitation, partially purified, and then mixed with sulfate salts or other stabilizing agents prior to vacuum drying [11,254,261].

2. Hemicellulases

Hemicelluloses are an indefinite group of polysaccharides classified according to the component sugar residues. They are alkali-soluble polysaccharides found in plant cell walls, but do not include cellulosic or pectic substances [267]. Diverse microorganisms produce enzymes that break down these polysaccharides. Fungi are the preferred sources for hemicellulases such as xylanases glucanases, galactomannanases, and pentosanases. Hemicellulases are by-products of commercial cellulase and pectinase production [267,268].

Fungal hemicellulases are used to process various foods. In the manufacture of instant coffee they degrade coffee mucilage and prevent gelling of the liquid coffee concentrate. As mentioned above, they are also used to degrade barley β-glucans that clog filters and pumps in the brewing industry [88]. Hemicellulases may be used to hydrolyze wheat flour pentosans to improve processes for the production of baked goods [269].

3. Lipases

Fungal lipases have few major food applications, though the potential for new uses exists [215,270,271]. Lipases from various sources have been used for flavor development in processed cheese products and chocolate crumb, production of butterfat flavors, flavor modification of cooking fats, antistaling action in baked goods, and improved whipping properties of egg white [1,6]. Other food uses include applications to the improvement of apple wine flavor, fish processing, meat product curing, degreasing bones for gelatin production, and vegetable fermentation [174,272]. Lipases are also used as digestive aids. Possible future applications involve the synthesis and modification of valuable food lipids or the production of pseudolipid food additives [216,248,270,273,274], perhaps using immobilized fungal cells or enzymes operating in organic solvents [274-279].

A number of fungi are rich sources of lipases. Important industrial producers include *A. niger*, *Rhizopus arrhizus*, *Rhizopus niveus*, *Rhizopus japonicus*, *Candida rugosa*, *Candida cylindracea*, *Thermomyces* (*Humicola*) *lanuginosa*, *Geotrichum candidum*, *Penicillium roqueforti*, *Penicillium* spp., *R. miehei*, and *R. pusillus*. Submerged fermentation in a complex medium typically involves the use of lipid material to induce the enzyme and elevated levels of nitrogen to promote lipase production [11,226,280].

4. Invertases

In addition to applications in the confectionery industry, yeast invertase is used to make syrups of invert sugars (glucose and fructose), D-fructose from inulin, melibiose from raffinose, and gentiobiose from gentianose. It is also used in the recovery of scrap sucrose products and for the hydrolysis of sucrose in molasses-based microbial fermentations [8,183,281,282].

Immobilized fungal invertases in various types of reactors have been studied for the commercial production of sugar mixtures from sucrose [39, 170,183,184]. The product syrups of glucose, fructose, and sucrose do not crystallize as readily as mixtures and are free of the colored by-products generated by acid hydrolysis of sucrose. High yields are possible. For example, conversions of 50-72% of the substrate can be observed in reactors containing invertase immobilized within ceramic membranes [283] or onto agricultural by-products [184], depending on sucrose concentration, flow rate, and enzyme activity. Though research results have been promising, the use of invertase has been limited owing to the success of processes that use cheaper raw materials and amylase/glucose isomerase treatment of starch for the production of invert sugar.

5. Proteases

Fungal protease from species of *Aspergillus*, *Mucor*, *Rhizopus*, or *Cryphonectria* (*Endothia*) are used in the baking, dairy, and soybean-processing industries or in food fermentation [3,16,22,284-288]. Applications in the brewing industry for chillproofing of beer have been explored, but plant and bacterial proteases are better suited for reducing beer cloudiness.

Fungal proteases have been used only infrequently to process and tenderize meat, since they digest both structural and connective proteins [211]. Nevertheless, the *A. oryzae* protease has been approved for such use in the United States and is commercially available [289,290]. The use of thermophilic microbial proteases prior to slaughter may be effective in meat tenderization. The injection and vascular diffusion of small doses of enzyme has been proposed as a means of improving the quality of meat. Other uses of fungal proteases involve the hydrolysis of meat scrap and fish protein to produce high-grade fat preparations and both soluble protein concentrates and insoluble protein fractions [16,291].

Fungal proteases have been used to modify proteins, change their functionality, upgrade their value, permit utilization of by-products, and lower processing costs [20,287,292-293].

6. Oxidation-Reduction Enzymes

The oxidation of food constituents such as lipids, vitamins and flavor compounds is a major problem in food processing [294]. Oxidation reduces the nutritional quality of foods, influences sensory properties, and may even produce toxic substances. The role of activated oxygen species in the initiation of oxidation in food systems has received limited attention. Recently, the use of superoxide dismutase from *S. cerevisiae* has been proposed as a possible antioxidative additive in foods. Purified yeast superoxide dismutase acts as an antioxidant in model oxidative systems of emulsified linolenic acid, emulsified cholesterol, and ascorbic acid. It very likely could play the same role in food systems, probably most effectively at alkaline pH. And, owing to its heat stability, it could be a potential antioxidant in heat-treated foods [7,295].

Glucose oxidase is used to prevent oxidative processes in foods. It is effective in removing residual oxygen from beer, wine, fruit juices, and high-fat products such as mayonnaise. Glucose oxidase from *A. niger* or or species of *Penicillium* is added to some foods such as processed eggs or dehydrated potatoes to remove residual glucose and thereby prevent Maillard browning or off-flavors. The enzyme converts the sugar to gluconic acid in the presence of oxygen [7,8,11,174,249,266,296-298].

Catalase breaks down hydrogen peroxide to water and oxygen and is employed to remove H_2O_2 that may be formed after food irradiation. Catalase may also be applied to the removal of excess H_2O_2 that may be added as an antimicrobial agent to milk and cheese [7,196] or formed during commercial cake baking [18]. The enzyme is obtained from the submerged fermentation of *A. niger* or *Penicillium* species, and bacterial sources are also available [266].

7. Other Enzymes

Enzymes from fungi have been used to break down certain oligosaccharides, such as stachyose, verbascose, and raffinose, found in legumes. These compounds cause flatulence when the vegetables are consumed by humans. For example, stachyase (an α-galactosidase) degrades stachyose, a flatulence-producing oligosaccharide found in soybeans. The enzyme can be obtained from species of *Aspergillus*. Soybean foods prepared by fermentation are devoid of flatus activity apparently as a result of enzymes produced by *Rhizopus* cultures used in the food fermentation process [61].

Egg white lysozyme is used as a food preservative in countries where it is permitted. It is effective as an antimicrobial in dairy products, especially certain hard cheeses, that are spoiled by late blowing caused by

Clostridium tyrobutyricum [299,300]. The availability of recombinant human lysozyme produced by yeast may have impact on lysozyme use by the food industry [301].

Fructosyl transferase from *Aureobasidium pullulans* breaks down sucrose and attaches the fructose to an acceptor sucrose molecule. The resulting fructose polymers have potential uses in food technology as bulking agents. *Aureobasidium pulullans* or another microorganism may serve as the source for this enzyme in large-scale synthesis [136].

Nucleases from *Penicillium* spp., usually *P. citrinum*, and 5'-AMP deaminase from *Aspergillus* spp., usually *A. oryzae*, have been used to produce nucleotide flavor enhancers [302-304].

III. APPLICATION OF NEW TECHNOLOGIES

The first microbial enzyme manufactured for commerce was α-amylase from *Aspergillus oryzae* almost 100 years ago. The patent for its production was awarded to Jokichi Takamine in 1894 [305]. Since then the number of enzymes produced by industrial fermentation and the diversity of their applications have increased significantly. Innovations for their production have matched this growth. Now, since the birth of the new biotechnology in the last decade, the pace of innovation has quickened. Disciplines that were new or unknown several decades ago are emerging as the driving forces for progress in industrial enzymology. Genetic engineering, protein engineering, enzyme immobilization and engineering, and fermentation engineering are changing the ways industrial enzymes are produced and the ways they are applied.

A. Genetic and Protein Engineering

The application of genetic engineering to food biotechnology holds promise for the future. Though to date no fungal products of recombinant DNA technology have received approval for food use, such products are expected in coming years once proven to be completely safe. A significant number of genes that encode enzymes of actual or potential importance to food technology have been expressed in fungi [25]. Many of the fungal hosts themselves are already approved for food use. Table 8 lists these recombinant enzymes and the fungi that produce them. The size of this list will increase as candidate enzymes for food use are identified and as genes that encode them are cloned and manipulated for maximum expression.

A major goal of the molecular mycologist will be the maximization of enzyme yield and rate of synthesis. Manipulation of the recombinant gene, the vector, or the host organism will facilitate commercial production of the enzyme and its recovery as a secreted product. The development of novel production organisms able to grow on cheap raw materials and adapted to specialized, large-scale fermenters or bioreactors will also contribute to the economic feasibility of the enzyme product. Recombinant approaches, coupled with classical methods for strain improvement, may become standard methods for development of efficient organisms for the production of enzymes useful to the food industry [22,306,307].

TABLE 8. Enzymes of Importance in Industrial Mycology Which Have Been Examined by Recombinant DNA Methods[a]

Industrial enzyme	Source	Recombinant fungal host
α-Amylase	Mouse salivary	*Saccharomyces cerevisiae*
	Mouse pancreas	*Saccharomyces cerevisiae*
	Human salivary	*Saccharomyces cerevisiae*
	Saccharomycopsis fibuligera	*Saccharomyces cerevisiae*
	Wheat	*Saccharomyces cerevisiae*
Glucoamylase	*Aspergillus awamori*	*Saccharomyces cerevisiae*
	Aspergillus niger	*Aspergillus niger*, *Aspergillus awamori*
	Rhizopus sp.	*Saccharomyces cerevisiae*
	Saccharomyces diastaticus	*Saccharomyces cerevisiae*
	Saccharomyces diastaticus	*Schizosaccharomyces pombe*
	Saccharomycopsis fibuligera	*Saccharomyces cerevisiae*
Protease (rennet)	Calf	*Saccharomyces cerevisiae*
	Calf	*Aspergillus nidulans*
	Rhizomucor miehei	*Aspergillus nidulans*
Cellulases		*Aspergillus oryzae*
Cellobiohydrolase	*Trichoderma reesei*	*Saccharomyces cerevisiae*
	Cellulomonas fimi	*Saccharomyces cerevisiae*
Endoglucanase	*Trichoderma reesei*	*Saccharomyces cerevisiae*
	Cellulomonas fimi	*Saccharomyces cerevisiae* *Aspergillus nidulans*
β-Glucosidase	*Aspergillus niger*	*Saccharomyces cerevisiae*
	Kluyveromyces fragilis[b]	*Saccharomyces cerevisiae*
	Candida pelliculosa	*Saccharomyces cerevisiae*
Lactase	*Kluyveromyces marxianus* var. *lactis*	*Saccharomyces cerevisiae*
Invertase	*Saccharomyces cerevisiae*[b]	*Saccharomyces cerevisiae*
β-Glucanase	*Bacillus subtilis*[b]	*Saccharomyces cerevisiae*
	Clostridium thermocellum[b]	*Saccharomyces cerevisiae*
	Barley	*Saccharomyces cerevisiae*
Ligninase	*Phanerochaete chysosporium*	*Saccharomyces cerevisiae*
Lysozyme	Human	*Saccharomyces cerevisiae*
	Bovine	*Pichia pastoris*

[a]References can be found in Bigelis and Das [25].

[b]Recombinant enzymes that are not secreted by the host.

Protein engineering is also expected to make substantial contributions to the development and improvement of enzymes for food use [308,309]. An understanding of the rules of enzyme structure-function relationships should permit the design of novel catalysts [310,311]. Such catalysts would permit the development of new food ingredients with unique functionalities, desirable organoleptic properties, and nutritional characteristics.

B. Enzyme Immobilization

Enzyme immobilization is the fixation of enzymes onto or within water-soluble materials and already has proved invaluable to the food industry in the production of HFCS and the breakdown of lactose in cheese whey. Many of the enzymes discussed in this chapter have been immobilized and applied to the production or processing of food-related substances on an experimental scale. Immobilization offers a number of advantages. It involves interaction of the enzyme, or even the cell itself, with an insoluble material and prevents diffusion and mixing of the enzyme with reactants and products. Thus, the enzyme can be easily recovered and reutilized, and even used in continuous processes. Enzyme immobilization lowers the amount of enzyme required and thereby cuts costs, besides also greatly reducing plant dimensions. Some disadvantages to using immobilized enzymes do exist. These primarily involve contamination problems and enzyme instability. Moreover, some enzymes play a multifunctional role in food processing, e.g., the action of rennet on milk, and cannot perform their tasks during short-term exposure to food substances in an enzyme reactor [5,15,30,190,312-317].

The pioneering successes with immobilized enzymes such as bacterial glucose isomerase, fungal lactase, and fungal α-galactosidase suggest that applications with other immobilized enzymes can be anticipated as enzyme engineers become more skilled in the art. Immobilized single enzymes, multiple enzyme systems, and multiple enzyme systems with cofactor regeneration elements can be expected to serve as valuable means for processing food-related substances. These immobilized constructs will bring together native, recombinant, or engineered enzymes from diverse plants, animals, and microbes. Such constructs will not only serve to process food materials, they will also serve as means, with few economic alternatives, for the production of primary and secondary metabolites that function as safe and effective food additives, including food ingredients that are not yet in our food technological vocabulary.

This chapter has considered the role of fungal enzymes in food processing. The role is clearly an important one. Fungi have served humans since ancient times helping to process food, to preserve it, to enrich it with nutrients, or to improve its taste. New technologies will extend the role of fungi in food production. Still, in the future, as in the past, fungi will continue to provide humans with staple foods and beverages—a loaf of bread, a jug of wine, ...

ACKNOWLEDGMENT

The author thanks Kathy Hardy for assistance in preparation of the manuscript.

REFERENCES

1. Wiseman, A. Enzyme utilization in industrial processes. In *Handbook of Enzyme Biotechnology* (A. Wiseman, ed.), Ellis Horwood, Chichester, 1975, pp. 111-124.

2. Scott, D. Enzymes, industrial. In *Kirk-Othmer Encyclopedia of Chemical Technology* (M. Grayson and D. Ekroth, eds.), John Wiley and Sons, New York, 1978, pp. 173-224.
3. Aunstrup, K. Enzymes of industrial interest. Traditional products. *Annu. Report Fermentation Process.*, *1*:181-204, 1977.
4. Taylor, M. J., and Richardson, T. Applications of microbial enzymes in food systems and in biotechnology. *Adv. Appl. Microbiol.*, *25*:7-35, 1979.
5. Kilara, A., and Shahani, K. M. The use of immobilized enzymes in the food industry: A review. *CRC Crit. Rev. Food Sci. Nutr.*, *12*: 161-198, 1979.
6. Fox, P. F., and Morrissey, P. A. Exogenous enzymes in food technology. In *Industrial and Clinical Enzymology* (L. Vitale and V. Simeon, eds.), Pergamon Press, Oxford, 1979, pp. 39-48.
7. Fox, P. F. Enzymes other than rennets in dairy technology. *J. Soc. Dairy Technol.*, *33*:118-128, 1980.
8. Barker, S. A., and Shirley, J. A. Glucose oxidase, glucose dehydrogenase, glucose isomerase, β-galactosidase, and invertase. In *Economic Microbiology. Microbial Enzymes and Bioconversions*, Vol. 5 (A. H. Rose, ed.), Academic Press, London, 1980, pp. 171-226.
9. Anonymous. *Food Chemicals Codex*, 3rd ed., National Research Council, Food and Nutrition Board, National Academy Press, Washington, DC, 1981.
10. Fogarty, W. M., and Kelly, C. T. Developments in microbial extracellular enzymes. *Top. Enzyme Ferment. Biotechnol.*, *3*:45-102, 1981.
11. Böing, J. T. P. Enzyme production. In *Prescott and Dunn's Industrial Microbiology* (G. Reed, ed.), AVI Publishing, Westport, CT, 1982, pp. 634-708.
12. Vanbelle, M., Meurens, M., and Crichton, R. R. Enzymes in foods and feeds. *Revue des Fermentations et des Industries Alimentaires*, *37*:124-135, 1982.
13. Lambert, P. W., and Meers, J. L. The production of industrial enzymes. *Phil. Trans. R. Soc. Lond.*, *B300*:263-282, 1983.
14. Cheetham, P. S. J. The applications of enzymes in industry. In *Handbook of Enzyme Biotechnology*, 2nd ed. (A. Wiseman, ed.), Ellis Horwood, Chichester, 1985, pp. 274-379.
15. Swaisgood, H. E. Immobilization of enzymes and some applications in the food industry. In *Enzymes and Immobilized Cells in Biotechnology* (A. I. Laskin, ed.), Benjamin Cummings, Menlo Park, CA, 1985, pp. 1-24.
16. Ward, O. P. Hydrolytic enzymes. In *Comprehensive Biotechnology. The Principles, Applications and Regulations of Biotechnology in Industry, Agriculture and Medicine*, Vol. 3 (H. W. Blanch, S. Drew, and D. I. C. Wang, eds.), Pergamon Press, Oxford, 1985, pp. 819-835.
17. Bigelis, R. Primary metabolism and industrial fermentations. In *Gene Manipulations in Fungi* (J. W. Bennett and L. L. Lasure, eds.), Academic Press, New York, 1985, pp. 157-401.
18. Bigelis, R. Industrial products of biotechnology: Application of gene technology. In *Biotechnology. A Comprehensive Treatise. Gene*

Technology, Vol. 7b (S. Jolly and G. Jacobson, eds.), Verlag Chemie, Weinheim, 1989, pp. 229-259.

19. Bigelis, R. Microbial production of primary metabolites. In *Food Biotechnology*, Proceedings of the 7th World Congress of Food Science and Technology—1987, Singapore, Singapore Institute of Food Science and Technology (in press).
20. Sheppard, G. The production and uses of microbial enzymes in food processing. *Prog. Ind. Microbiol.*, *23*:237-283, 1986.
21. Frost, G. M. Commercial production of enzymes. *Dev. Food Proteins*, *4*:57-134, 1986.
22. Bigelis, R., and Lasure, L. L. Fungal enzymes and primary metabolites used in food processing. In *Food and Beverage Mycology*, 2nd ed. (L. R. Beuchat, ed.), AVI/Van Nostrand, New York, 1987, pp. 473-516.
23. Rutloff, H. Impact of biotechnology on food and nutrition. In *Food Biotechnology* (D. Knorr, ed.), Marcel Dekker, New York, 1987, pp. 37-94.
24. Peppler, H. J., and Reed, G. Enzymes in food and feed processing. In *Biotechnology. Enzyme Technology*, Vol. 7a (J. F. Kennedy, ed.), VCH Verlagsgesellschaft, Weinheim, 1987, pp. 547-603.
25. Bigelis, R., and Das, R. C. Secretion research in industrial mycology. In *Protein Transfer and Organelle Biogenesis* (R. C. Das and P. W. Robbins, eds.), Academic Press, New York, 1988, pp. 771-808.
26. Reed, G. R. *Enzymes in Food Processing*, 2nd ed., Academic Press, New York, 1975.
27. Wiseman, A. *Handbook of Enzyme Biotechnology*, Ellis Horwood, Chichester, 1975.
28. Wingard, L. B., Katchalski-Katzir, E., and Goldstein, L. *Applied Biochemistry and Bioengineering. Enzyme Technology*, Vol. 2, Academic Press, New York, 1979.
29. Peppler, J., and Perlman, D. *Microbial Technology*, 2nd ed., Vols. 1-2, Academic Press, New York, 1979.
30. Pitcher, W. H. *Immobilized Enzymes for Food Processing*, CRC Press, Boca Raton, FL, 1980.
31. Rose, A. H. *Economic Microbiology*, Vols. 1-7, Academic Press, New York, 1978-1982.
32. Schwimmer, S. *Source Book of Food Enzymology*, AVI Publishing, Westport, CT, 1981.
33. Rehm, H.-J., and Reed, G. *Biotechnology. A Comprehensive Treatise*, Vols. 1-8, Verlag-Chemie, Weinheim, 1981-1989.
34. Reed, G. *Prescott and Dunn's Industrial Microbiology*, 4th ed., AVI, Westport, CT, 1982.
35. Fogarty, W. M. *Microbial Enzymes and Biotechnology*, Applied Science Publishers, London, 1983.
36. Dupuy, P. *Use of Enzymes in Food Technology*, Technique et Documentation Lavoisier, Paris, 1983.
37. Godfrey, T., and Reichelt, J. *Industrial Enzymology. The Application of Enzymes in Industry*, The Nature Press, New York, 1983.
38. Birch, G. G., Blakebrough, N., and Parker, K. J. *Enzymes and Food Processing*, Applied Science Publishers, London, 1981.

39. Crueger, A., and Crueger, W. *Biotechnology. A Textbook of Industrial Microbiology*, 2nd ed., Sinauer Associates, Sunderland, MA, 1990.
40. Wiseman, A. *Handbook of Enzyme Biotechnology*, 2nd ed., Ellis Horwood, Chichester, 1985.
41. Pomeranz, Y. *Functional Properties of Food Components*, Academic Press, New York, 1985.
42. Laskin, A. I. *Enzymes and Immobilized Cells in Biotechnology*, Benjamin Cummings, Menlo Park, CA, 1985.
43. Moo-Young, M. *Comprehensive Biotechnology*, Vols. 1-4, Pergammon Press, Oxford, 1985.
44. Harlander, S. K., and Labuza, T. P. *Biotechnology in Food Processing*, Noyes Publications, Park Ridge, NJ, 1986.
45. Knorr, D. *Food Biotechnology*, Marcel Dekker, New York, 1987.
46. Präve, P., Faust, U., Sittig, W., and Sukatsch, D. A. *Fundamentals of Biotechnology*, VCH Verlagagesellschaft, Weinheim, 1987.
47. Beuchat, L. R. *Food and Beverage Mycology*, 2nd ed., AVI/Van Nostrand, New York, 1987.
48. Amos, A. J. The use of enzymes in the baking industry. *J. Sci. Food Agric.*, *6*:489-495, 1955.
49. Barrett, F. F. Enzyme uses in the milling and baking industries. In *Enzymes in Food Processing*, 2nd ed. (G. Reed, ed.), Academic Press, New York, 1975, pp. 301-330.
50. Marston, P. E., and Wannan, T. L. Bread baking. The transformation of dough to bread. *Bakers Dig.*, *50*(4):24-28, 49, 1976.
51. Reichelt, J. R. Baking. In *Industrial Enzymology. The Application of Enzymes in Industry* (T. Godfrey and J. Reichelt, eds.), The Nature Press, New York, 1983, pp. 210-220.
52. Drapron, R., and Godon, B. Role of enzymes in baking. In *Enzymes and Their Role in Cereal Technology* (J. E. Kruger, D. Lineback, and C. E. Stauffer, eds.), American Assoc. of Cereal Chemists, Inc., St. Paul, MN, 1987, pp. 280-324.
53. ter Haseborg, E. Enzymatic treatment of flour. *Alimenta*, *27*:2-10, 1988.
54. Johnson, J. A., and Miller, B. S. Studies on the role of alpha-amylase and proteinase on breadmaking. *Cereal Chem.*, *26*:371-383, 1949.
55. Cauvain, S. P., and Chamberlain, N. The bread improving effect of a fungal α-amylase. *J. Cereal Sci.*, *8*:239-248, 1988.
56. Kaur, M., and Bains, G. S. Amylase supplementation of Indian wheat flours for improving bread potential. *Indian Miller*, *8*:38-41, 1978.
57. Brabender, M., and Seitz, W. Effect of fungal amylases and malt flour on baking properties of flour. *Muehle Mischfuttertechn.*, *116*:219-220, 223-224, 1979.
58. Maninder, K., and Jorgensen, O. B. Interrelations of starch and fungal alpha-amylase in bread making. *Starch/Starke*, *35*:419-426, 1983.
59. Lyons, T. P. Proteinase enzymes relevant to the baking industry. *Biochem. Soc. Trans.*, *10*:287-290, 1982.
60. Kasarda, D. D., Birnardin, J. E., and Nimmo, C. C. Wheat proteins. *Adv. Cereal Sci.*, *1*:158-236, 1976.

61. Liener, I. E. Removal of naturally occurring toxicants through enzymatic processing. In *Food Proteins. Improvement Through Chemical and Enzymatic Modification* (R. E. Feeney and J. R. Whitaker, eds.), American Chemical Society, Washington, DC, 1977, pp. 283-299.
62. McCleary, B. V., Gibson, T. S., Allen, H., and Gams, T. C. Enzymic hydrolysis and industrial importance of barley β-glucans and wheat flour pentosans. *Starch/Starke, 38*:433-437, 1986.
63. McCleary, B. V. Enzymatic modification of polysaccharides in brewing, baking and syrup manufacture. *Food Hydrocolloids, 1*:445-448, 1987.
64. Wong, K. K. Y., Tan, L. U. L., and Saddler, J. N. Multiplicity of β-1,4-xylanase in microorganisms: Functions and applications. *Microbiol. Rev., 52*:305-317, 1988.
65. Broderick, H. M. *The Practical Brewer—A Manual for the Brewing Industry*, Master Brewers Association of the Americas, Madison, WI, 1977.
66. Westermann, D. H., and Huige, N. J. Beer brewing. In *Microbial Technology, Fermentation Technology*, 2nd ed., Vol. 2 (H. J. Peppler and D. Perlman, eds.), Academic Press, New York, 1979, pp. 1-37.
67. Briggs, D. E., Hough, J. S., Stevens, R., and Young, T. W. Adjuncts, sugars, wort-syrups and industrial enzymes. In *Malting and Brewing Science*, 2nd ed., Vol. 1, Chapman and Hall, New York, 1981, pp. 222-253.
68. Godfrey, T. Brewing. In *Industrial Enzymology. The Application of Enzymes in Industry* (T. Godfrey, ed.), The Nature Press, New York, 1983, pp. 221-259.
69. Priest, F. G., and Campbell, I. *Brewing Microbiology*, Elsevier, Amsterdam, 1987.
70. Allen, W. G., and Spradlin, J. E. Amylases and their properties. *Brewers Dig., 48*:48-50, 52-53, 65, 1973.
71. Marshall, J. J., Allen, W. G., Denault, L. J., Glenister, P. R., and Power, J. Enzymes in brewing. *Brew. Dig., 57*(9):14-18, 1982.
72. Saletan, L. T. Carbohydrases of interest in brewing, with particular reference to amyloglucosidase. *Wallerstein Lab. Commun., 31*(104);33-44, 1968.
73. Woodward, J. D. Enzymes in practical brewing. *Brew. Dig., 53*(5); 38, 40, 42-44, 1978.
74. Bass, E. J., and Cayle, T. Beer. In *Enzymes in Food Processing*, 2nd ed. (G. Reed, ed.), Academic Press, New York, 1975, pp. 445-471.
75. Slaughter, J. C. Enzymes in the brewing industry. In *Alcoholic Beverages* (G. G. Birch and M. G. Lindley, eds.), Elsevier, Amsterdam, 1985, pp. 15-27.
76. Denault, L. J., Glenister, P. R., and Chau, S. Enzymology of the mashing step during beer production. *J. Am. Soc. Brew. Chem., 39*:46-52, 1981.
77. Beckerich, R. P., and Denault, L. J. Enzymes in the preparation of beer and fuel alcohol. In *Enzymes and Their Role in Cereal Technology* (J. E. Kruger, D. Lineback, and C. E. Stauffer, eds.), American Assoc. of Cereal Chemists, St. Paul, MN, 1987, pp. 335-355.

78. McCleary, B. V., Gibson, T. S., Sheehan, H., Casey, A., Horgan, L., and O'Flaherty, J. Purification, properties, and industrial significance of transglucosidase from *Aspergillus niger*. *Carbohydr. Res.*, *185*:147-162, 1989.
79. Willox, I. C., Rader, S. R., Riolo, J. M., and Stern, W. The addition of starch debranching enzymes to mashing and fermentation and their influence on attenuation. *M.B.A.A. Tech. Quart.*, *14*:105-110, 1977.
80. Bamforth, C. W. Barley β-glucans. Their role in malting and brewing. *Brewer's Dig.*, *57*:22-27, 1982.
81. McCleary, B. V. Problems caused by barley beta-glucans in the brewing industry. *Chemistry in Australia*, *53*:306-308, 1986.
82. Scott, R. W. The viscosity of worts in relation to their content of β-glucan. *J. Inst. Brew.*, *78*:179-186, 1972.
83. Bathgate, G. N., and Dalgliesh, C. E. The diversity of barley and malt β-glucans. *Proc. Am. Soc. Brew. Chem.*, *33*:32-36, 1975.
84. Takayanagi, S., Amaha, M., Satake, K., Kuroiwa, Y., Igarashi, H., and Murata, A. Studies on frozen beer precipitates. I. Formation and general characters. *J. Inst. Brew.*, *75*:284-292, 1969.
85. Leedham, P. A., Savage, D. J., Crabb, D., and Morgan, G. T. Materials and methods of wort production that influence beer filtration. *Proc. Eur. Brew. Conv. Congress*, Nice, 1975, pp. 201-216.
86. Bournes, D. T., Jones, M., and Pierce, J. S. Beta-glucan and beta glucanases in malting and brewing. *M.B.A.A. Tech. Quart.*, *13*:3-7, 1976.
87. Narziss, L. Beta-glucan and beta-glucanases. *Proc. Eur. Brew. Conv. Barley and Malt Symposium*, Helsinki, 1981, pp. 99-117.
88. Enkenlund, J. Externally added beta-glucanase. *Proc. Biochem.*, *7*(8):27-29, 1972.
89. Schneider, H. J. Enzymatic removal of anthocyanin components in beer. *Tech. Proc. Convention Master Brewers Assoc. of America*, Chicago, 1962, pp. 2-9.
90. Chase, T., Jr. Flavor enzymes. In *Food Related Enzymes* (J. R. Whitaker, ed.), American Chemical Society, Washington, DC, 1974, pp. 241-265.
91. Godtfredsen, S. E., and Ottesen, M. Maturation of beer with the α-acetolactate decarboxylase. *Carlsberg Res. Commun.*, *47*:93-102, 1982.
92. Gjermanssen, C., Nilsson-Tillgren, T., Petersen, J. G. L., Keilland-Brandt, M. C., Sigsgaard, P., and Holmberg, S. Towards diacetylless brewer's yeast. Influence of the *ilv2* and *ilv5* mutations. *J. Basic Microbiol.*, *3*:175-183, 1988.
93. Hysert, D. W. Recent advances in enzymology of relevance to brewing. *Amer. Soc. Brew. Chem., Proc.*, *33*:114-118, 1975.
94. Khan, A. W., Lamb, K. A., and Schneider, H. Recovery of fermentable sugars from the brewer's spent grains by the use of fungal enzymes. *Proc. Biochem.*, *23*(6):172-175, 1988.
95. Hinchliffe, E. β-Glucanase: The successful application of genetic engineering. *J. Inst. Brew.*, *91*:384-389, 1985.
96. Stewart, G. G., and Russell, I. One hundred years of yeast research and development in the brewing industry. *J. Inst. Brew.*, *92*:537-558, 1986.

97. Russell, I., Jones, R., and Stewart, G. The genetic modification of brewer's yeast and other industrial yeast strains. In *Biotechnology in Food Processing* (S. K. Harlander and T. P. Labuza, eds.), Noyes Publications, Park Ridge, NJ, 1986, pp. 171-195.
98. Tubb, R. S. Amylolytic yeasts for commercial applications. *Trends Biotechnol., 4*:98-105, 1986.
99. Enari, T. M. Prospects of biotechnology for brewers. *J. Inst. Brew., 93*:501-505, 1987.
100. Penttila, M. E., Suihko, M.-L., Lehtinen, U., Nikkola, M., and Knowles, J. K. C. Construction of brewer's yeasts secreting fungal endo-β-glucanase. *Curr. Genet. 12*:413-420, 1987.
101. Fenton, M. S., Kavanagh, T. E., and Clarke, B. J. Recent developments in brewing technology. *Food Technol.* (Australia), *40*:132-134, 148, 1988.
102. Sone, H., Fujii, T., Kondo, K., Shimizu, F., Tanaka, J.-I., and Inoue, T. Nucleotide sequence and expression of *Enterobacter aerogenes* alpha-acetolactate decarboxylase gene in brewer's yeast. *Appl. Environ. Microbiol., 54*:36-42, 1988.
103. Hammond, J. R. M. Brewery fermentation in the future. *J. Appl. Bacteriol., 65*:169-177, 1988.
104. Brandt, D. A. Distilled alcoholic beverages. In *Enzymes in Food Processing*, 2nd ed. (G. Reed, ed.), Academic Press, New York, 1975, pp. 443-453.
105. Poulson, P. B. Alcohol-potable. In *Industrial Enzymology. The Application of Enzymes in Industry* (T. Godfrey and J. Reichelt, eds.), The Nature Press, New York, 1983, pp. 170-178.
106. Maisch, W. F., Sobolov, M., and Petricola, A. J. Distilled beverages. In *Microbial Technology. Fermentation Technology*, 2nd ed., Vol. 2 (H. J. Peppler and D. Perlman, eds.), Academic Press, New York, 1979, pp. 79-94.
107. Berry, D. R. The physiology and microbiology of scotch whiskey production. *Prog. Ind. Microbiol., 19*:199-243, 1984.
108. Sobolov, M., Booth, D. M., and Aldi, R. G. Whiskey. In *Comprehensive Biotechnology. The Principles, Applications and Regulations of Biotechnology in Industry, Agriculture and Medicine*, Vol. 3 (H. W. Blanch, S. Drew, and D. I. C. Wang, eds.), Pergamon Press, Oxford, 1985, pp. 383-393.
109. Cole, G. E., McCabe, P. C., Inlow, D., Gelfand, D. H., Ben-Bassatt, A., and Innis, M. A. Stable expression of *Aspergillus awamori* glucoamylase in distiller's yeast. *Bio/Technology, 6*:417-421, 1988.
110. Neubeck, C. E. Fruit, fruit products. In *Enzymes in Food Processing*, 2nd ed. (G. Reed, ed.), Academic Press, New York, 1975, pp. 397-442.
111. Fogarty, W. M., and Ward, O. P. Pectinases and pectic polysaccharides. *Prog. Ind. Microbiol., 13*:59-113, 1974.
112. Rombouts, F. M., and Pilnik, W. Pectic enzymes. In *Economic Microbiology. Microbial Enzymes and Bioconversions*, Vol. 5 (A. H. Rose, ed.), Academic Press, London, 1980, pp. 227-282.
113. Baumann, J. W. Application of enzymes in fruit juice technology. In *Enzymes and Food Processing* (G. G. Birch, N. Blakebrough,

and K. J. Parker, eds.), Applied Science Publishers, London, 1981, pp. 129-147.

114. Pilnik, W., and Rombouts, F. M. Pectic enzymes. In *Enzymes and Food Processing* (G. G. Birch, N. Blakebrough, and K. J. Parker, eds.), Applied Science Publishers, London, 1981, pp. 105-128.
115. Pilnik, W. Enzymes in the beverage industry (fruit juices, nectar, wine, spirits, beer). In *Utilisation des Enzymes en Technologie Alimentaire* (P. Dupuy, ed.), Technique et Documentation Lavoisier, Paris, 1982, pp. 425-450.
116. Whitaker, J. R. Pectic substances, pectic enzymes and haze formation in fruit juices. *Enzyme Microb. Technol.*, *6*:341-349, 1984.
117. MacMillan, J. D., and Sheiman, M. I. Pectic enzymes. In *Food Related Enzymes* (J. R. Whitaker, ed.), American Chemical Society, Washington, DC, 1974, pp. 101-130.
118. Lindhardt, R. J., Galliher, P. M., and Cooney, C. L. Polysaccharide lyases. *Appl. Biochem. Biotechnol.*, *12*:135-176, 1986.
119. Ward, O. P. Enzymatic degradation of cell wall and related plant polysaccharides. *CRC Crit. Rev. Biotechnol.*, *8*:237-274, 1989.
120. Ough, C. S., and Crowell, E. A. Pectic-enzyme treatment of white grapes: Temperature, variety and skin-contact time factors. *Am. J. Enol. Vitic.*, *30*:22-27, 1979.
121. Baker, R. A., and Bruemmer, J. H. Pectinase stabilization of orange juice cloud. *J. Agric. Food Chem.*, *20*:1169-1173, 1972.
122. Anonymous. USDA develops vacuum infusion process for citrus. *Food Business*, Jan.:16-17, 1988.
123. Nyiri, L. Manufacture of pectinases. *Proc. Biochem.*, *4*(8):27-30, 1969.
124. Fogarty, W. M., and Kelly, C. T. Pectic enzymes. In *Microbial Enzymes and Biotechnology* (W. M. Fogarty, ed.), Applied Science Publishers, London, 1983, pp. 131-182.
125. Horowitz, R. M., and Gentilli, B. Taste and structure in phenolic glycosides. *J. Agr. Food Chem.*, *17*:696-700, 1969.
126. Robertson, G. L. Pectic enzymes and winemaking. *Food Technol. (New Zealand)*, *12*:32, 34-35, 1977.
127. Haupt, W. Pectolytic enzymes in wine production. *Weinwissenschaft*, *117*:1014-1017, 1981.
128. Felix, R., and Villettaz, J.-C. Wine. In *Industrial Enzymology. The Application of Enzymes in Industry* (T. Godfrey and J. Reichelt, eds.), MacMillan, New York, 1983, pp. 410-421.
129. Ough, C. S., and Berg, H. W. The effect of two commercial pectic enzymes on grape musts and wines. *Am. J. Enol. Vitic.*, *25*:208-211, 1974.
130. Ough, C. S., Noble, A. C., and Temple, D. Pectic enzyme effects on red grapes. *Am. J. Enol. Vitic.*, *26*:195-200, 1975.
131. Huang, H. T. Decolorization of anthocyanins by fungal enzymes. *J. Agric. Food Chem.*, *3*:141-146, 1955.
132. Ough, C. S. Chemicals used in making wine. *Chem. Eng. News*, *65*(1):19-28, 1987.
133. MacAllister, R. V., Wardrip, E. K., and Schnyder, B. J. Modified starches, corn syrups, containing glucose and maltose, corn syrups

containing glucose and fructose, and crystalline dextrose. In *Enzymes and Food Processing*, 2nd ed. (G. Reed, ed.), Academic Press, New York, 1975, pp. 331-359.
134. MacAllister, R. V. Nutritive sweeteners made from starch. *Adv. Carbohydr. Chem.*, *36*:15-56, 1979.
135. Bucke, C. Enzymes in fructose manufacture. In *Enzymes and Food Processing* (G. G. Birch, N. Blakebrough, and K. J. Parker, eds.), Applied Science Publishers, London, 1981, pp. 51-72.
136. Luesner, S. J. Microbial enzymes for industrial sweetener production. *Dev. Ind. Microbiol.*, *24*:79-96, 1983.
137. Bucke, C. Glucose-transforming enzymes. In *Microbial Enzymes and Biotechnology* (W. M. Fogarty, ed.), Applied Science Publishers, London, 1983, pp. 93-129.
138. Norman, B. The application of polysaccharide degrading enzymes in the starch industry. In *Microbial Polysaccharides and Polysaccharases* (R. C. W. Berkeley, G. W. Gooday, and D. C. Ellwood, eds.), Academic Press, London, 1979, pp. 339-376.
139. Norman, B. E. New developments in starch syrup technology. In *Enzymes and Food Processing* (G. G. Birch, N. Blakebrough, and K. J. Parker, eds.), Applied Science Publishers, London, 1981, pp. 15-50.
140. Reichelt, J. R. Starch. In *Industrial Enzymology. The Application of Enzymes in Industry* (T. Godfrey and J. Reichelt, eds.), The Nature Press, New York, 1983, pp. 375-396.
141. Ostergaard, J. Enzymes in the carbohydrate industry. In *Utilisation des Enzymes en Technologie Alimentaire* (P. Dupuy, ed.), Technique et Documentation Lavoisier, Paris, 1983, pp. 57-79.
142. Kempf, W. New possible outlets for starch and starch products in chemical and technical industries. *Food Technol.* (Australia), *37*:241-245, 1985.
143. Coker, L. E., and Venkatasubramanian, K. Starch conversion processes. In *Comprehensive Biotechnology. The Principles, Applications and Regulations of Biotechnology in Industry, Agriculture and Medicine*, Vol. 3 (H. W. Blanch, S. Drew, and D. I. C. Wang, eds.), Pergammon Press, Oxford, 1985, pp. 777-787.
144. Newsome, R. L. Sweeteners: Nutritive and non-nutritive. *Food Technol.*, *40*(8):195-206, 1986.
145. Luallen, T. E. Structure, characteristics and uses of some typical carbohydrate food ingredients. *Cereal Foods World*, *33*(11):924-927, 1988.
146. Kennedy, J. F., Cabalda, V. M., and White, C. A. Enzymic starch utilization in genetic engineering. *Trends Biotechnol.*, *6*:184-189, 1988.
147. Corn Refiners Association. *Nutritive Sweeteners from Corn*, Corn Refiners Association, Inc., Washington, DC, 1979.
148. Corn Refiners Association. *Corn Starch*, Corn Refiners Association, Inc., Washington, DC, 1979.
149. Lineback, D. R., and Inglett, G. E. *Food Carbohydrates*, AVI Publishing, Westport, CT, 1982.

150. Koch, H., and Roper, H. New industrial products from starch. *Starch/Starke, 40*:121-131, 1988.
151. Whistler, R. L., BeMiller, J. W., and Paschall, E. F. *Starch: Chemistry and Technology,* 2nd ed., Academic Press, New York, 1984.
152. Guilbot, A., and Mercier, C. Starch. In *Polysaccharides* (G. O. Aspinall, ed.), Academic Press, Orlando, FL, 1985, pp. 209-282.
153. Zobel, H. F. Molecules to granules: A comprehensive starch review. *Starch/Starke, 40*:44-50, 1988.
154. Kulp, K. Carbohydrases. In *Enzymes in Food Processing,* 2nd ed. (G. Reed, ed.), Academic Press, New York, 1975, pp. 53-122.
155. Fogarty, W. M., and Kelly, C. T. Starch-degrading enzymes of microbial origin, Part 1. *Prog. Ind. Microbiol., 15*:89-150, 1979.
156. Fogarty, W. M., and Kelly, C. T. Amylases, amyloglucosidases, and related glucanases. In *Economic Microbiology. Microbial Enzymes and Bioconversions,* Vol. 5 (A. H. Rose, ed.), Academic Press, London, 1980, pp. 115-170.
157. Meyrath, J., and Bayer, G. Environmental factors and cultivation in fungal α-amylase production. *Proc. FEBS Mtg., 61*:331-338, 1980.
158. Linko, P. Enzymes in the industrial utilization of cereals. In *Enzymes and Their Role in Cereal Technology* (J. E. Kruger, D. Lineback, and C. E. Stauffer, eds.), American Association of Cereal Chemists, Inc., St. Paul, MN, 1987, pp. 280-324.
159. Shetty, J. K., and Allen, W. G. An acid-stable thermostable alpha-amylase for starch liquefaction. *Cereal Foods World, 33*:929-934, 1988.
160. Takasaki, Y., and Yamanobe, T. Production of maltose by pullulanase and β-amylase. In *Enzymes and Food Processing* (G. G. Birch, N. Blakebrough, and K. J. Parker, eds.), Applied Science Publishers, London, 1981, pp. 73-88.
161. Boyce, C. O. L. *Novo's Handbook of Practical Biotechnology,* Bagsvaerd, Denmark, 1986.
162. Anonymous. High fructose corn syrup comes of age. *Dairy Field,* March:43-45, 1988.
163. Anonymouse. Crystalline fructose: A breakthrough in corn sweetener process technology. *Food Technol., 41*(1):66-67, 72, 1987.
164. Swientek, R. J. Sonic technology applied to food drying. *Food Process., 47*(7):62-63, 1986.
165. Davenport, R. Sonic drying offers new product potential. *Food Business,* Jan:20, 1988.
166. Anonymous. Sweeteners. *Food Technol., 40*(1):112-130, 1986.
167. Anonymous. Sweeteners: Nutritive and non-nutritive. *Food Technol., 40*(8):195-206, 1986.
168. Suzuki, H., Ozawa, Y., Oota, H., and Yoshida, H. Studies on the decomposition of raffinose by α-galactosidase of mold. *Agric. Biol. Chem., 33*:507-513, 1969.
169. Kobayashi, H., and Suzuki, H. Studies on the decomposition of raffinose by α-galactosidase of mold. *J. Ferment. Technol., 50*:625-632, 1972.
170. Reilly, P. J. Potential and uses of immobilized carbohydrates. In *Immobilized Enzymes for Food Processing* (W. H. Pitcher, Jr., ed.), CRC Press, Boca Raton, FL, 1980, pp. 113-151.

171. Lindley, M. G. Cellobiase, melibiase and other disaccharides. *Dev. Food Carbohydr.*, *3*:141-165, 1982.
172. Shimizu, J., and Kaga, T. Apparatus for continuous hydrolysis of raffinose. U.S. Patent 3,664,927 (1972).
173. Blanch, H. W. Immobilized microbial cells. *Annu. Reports Fermentation Process.*, *7*:81-105, 1984.
174. Scott, D. Miscellaneous applications of enzymes. In *Enzymes in Food Processing*, 2nd ed. (G. Reed, ed.), Academic Press, New York, 1975, pp. 493-517.
175. Obara, J., Hashimoto, S., and Suzuki, H. Enzyme applications in the sucrose industries. *Sugar Technol. Rev.* *4*:209-258, 1977.
176. Keniry, J. S., Lee, J. B., and Davis, C. W. Deterioration of mechanically harvested chopped-up cane. Part I. Dextran—A promising quantitative indicator of the processing quality of chopped-up cane. *Int. Sugar J.*, *69*:330-333, 1967.
177. Foster, D. H. Deterioration of chopped cane. *Proc. Queensl. Soc. Sugar Cane Technol.*, *36*:21-28, 1969.
178. Imrie, F. K. E., and Tilbury, R. H. Polysaccharides in sugar cane and its products. *Sugar Technol. Rev.*, *1*:291-361, 1972.
179. Abram, J. C., and Ramage, J. S. Sugar refining: Present technology and future developments. In *Sugar Science and Technology* (G. G. Birch and K. J. Parker, eds.), Applied Science Publishers, London, 1979, pp. 49-95.
180. Barfield, S., and Mollgaard, A. Dextranase solved dextran problems in DDS' beet sugar factory. *Zuckerind.*, *112*:391-395, 1987.
181. Kosaric, N., Yu, K., and Zajic, J. E. Dextranase production from *Penicillium funiculosum*. *Biotechnol. Bioeng.*, *15*:729-741, 1973.
182. Tilbury, R. H. Sucrose from sugar cane—addition of dextranase during processing to remove dextran. Brit. Pat. 1290694 (1972).
183. Wiseman, A. New and modified invertases—and their applications. In *Topics in Enzyme and Fermentation Biotechnology*, Vol. 3 (A. Wiseman, ed.), Ellis Horwood, Chichester, 1982, pp. 265-288.
184. Monsan, P., and Combes, D. Application of immobilized invertase to hydrolysis of concentrated sucrose solutions. *Biotechnol. Bioeng.*, *26*:347-351, 1984.
185. Ingleton, J. F. The use of invertase in the confectionery industry. *Confectionery Prod.*, *29*:773-774, 776-777, 790, 1963.
186. Szekely, P. Use of invertase enzyme in confectionery. *Edesipar*, *21*:80-83, 1970.
187. Shukla, T. P. Beta-galactosidase technology. A solution to the lactose problem. *CRC Crit. Rev. Biotechnol.*, *5*:325-356, 1975.
188. Miller, J. J., and Brand, J. C. Enzymic lactose hydrolysis. *Food Technol.* (Australia), *32*:144-146, 1980.
189. Coughlan, R. W., and Charles, M. Applications of lactase and immobilized lactase. In *Immobilized Enzymes for Food Processing* (W. H. Pitcher, ed.), CRC Press, Boca Raton, FL, 1980, pp. 153-173.
190. Linko, P., and Linko, Y.-Y. Applications of immobilized microbial cells. In *Immobilized Microbial Cells*, Vol. 4 (I. Chibata and L. B. Wingard, Jr., eds.), Academic Press, New York, 1983, pp. 53-151.

191. Gekas, V., and Lopez-Leiva, M. Hydrolysis of lactose: A literature review. *Proc. Biochem.*, *20*(2):2-12, 1985.
192. Mahoney, R. R. Modification of lactose and lactose-containing dairy products with beta-galactosidase. *Dev. Dairy Chem.*, *3*:69-109, 1985.
193. Woychik, J. H., and Holsinger, V. H. Use of lactose in the manufacture of dairy products. In *Enzymes in Food and Beverage Processing* (R. L. Ory and A. J. St. Angelo, eds.), American Chemical Society, Washington, DC, 1977, pp. 67-79.
194. Moore, K. Immobilized enzyme technology commercially hydrolyzes lactose. *Food Prod. Devel.*, *14*(1):50-51, 1980.
195. Nijpels, H. H. Lactases and their applications. In *Enzymes and Food Processing* (G. G. Birch, N. Blakebrough, and K. J. Parker, eds.), Applied Science Publishers, London, 1981, pp. 89-104.
196. Burgess, K., and Shaw, M. Dairy. In *Industrial Enzymology. The Application of Enzymes in Industry* (T. Godfrey and J. Reichelt, eds.), The Nature Press, New York, 1983, pp. 260-283.
197. Moulin, G., and Galzy, P. Whey, a potential substrate for biotechnology. In *Biotechnology and Genetic Engineering Reviews*, Vol. 1 (G. E. Russel, ed.), Intercept, Newcastle upon Tyne, 1984, pp. 347-374.
198. Houts, S. S. Lactose intolerance. *Food Technol.*, *42*(3):110-113, 1988.
199. Greenberg, N. A., and Mahoney, R. R. Immobilization of lactase (β-galactosidase) for use in dairy processing: A review. *Proc. Biochem.*, *16*(2):2-8, 1981.
200. Richmond, M. L., Gray, J. I., and Stine, C. M. β-Galactosidase: A review of recent research related to technological application, nutritional concerns, and immobilization. *J. Dairy Sci.*, *64*:1759-1771, 1981.
201. Van Griethuysen-Dilber, E., Flaschel, R., and Renken, A. Process development of the hydrolysis of lactose in whey by immobilized lactase of *Aspergillus oryzae*. *Proc. Biochem.*, *23*(2):55-59, 1988.
202. Finnocchario, T., Olson, N. F., and Richardson, T. Use of immobilized lactase in milk systems. *Adv. Biochem. Eng.*, *15*:71-88, 1980.
203. Fox, P. F. Proteolysis in milk and dairy products. *Biochem. Soc. Trans.*, *10*:282-284, 1982.
204. Kristofferson, T. Biogenesis of cheese flavor. *J. Agr. Food Chem.*, *21*:573-575, 1973.
205. Moskowitz, G. J., and LaBelle, G. G. Enzymatic flavor development in foods. In *The Quality of Foods and Beverages*, Vol. 2, Academic Press, New York, 1981, pp. 21-35.
206. Law, B. A. Microorganisms and their enzymes in the maturation of cheeses. *Prog. Ind. Microbiol.*, *19*:245-283, 1984.
207. Law, B. A. Flavor development in cheeses. In *Advances in the Microbiology and Biochemistry of Cheese and Fermented Milk* (F. L. Davies and B. A. Law, eds.), Elsevier Applied Science Publishers, London, 1984, pp. 187-208.
208. Kosikowski, F. V. Enzyme behavior and utilization in dairy technology. *J. Dairy Sci.*, *71*:557-573, 1988.

209. Dwivedi, B. K. The role of enzymes in food flavors. I. Dairy products. *CRC Crit. Rev. Food Technol.*, *3*:456-478, 1973.
210. Shahani, K. M. Lipases and esterases. In *Enzymes in Food Processing* (G. Reed, ed.), Academic Press, New York, 1975, pp. 181-217.
211. Shahani, K. M., Arnold, R. G., Kilara, A., and Dwivedi, B. K. Role of microbial enzymes in flavor development of foods. *Biotechnol. Bioeng.*, *23*:891-907, 1976.
212. Gatfield, I. L. Production of flavor and aroma compounds by biotechnology. *Food Technol.*, *42*(10):110, 112-118, 120-122, 169, 1988.
213. Law, B. A. The accelerated ripening of cheese by the use of nonconventional starters and enzymes. A preliminary assessment. *Bull. Int. Dairy Fed.*, No. 108:40-50, 1978.
214. Peppler, H. J., Dooley, J. G., and Huang, H. T. Flavor development in Fontina and Romano cheese by fungal esterase. *J. Dairy Soc.*, *59*:859-862, 1976.
215. Macrae, A. R. Extracellular microbial lipases. In *Microbial Enzymes and Biotechnology* (W. M. Fogarty, ed.), Applied Science Publishers, London, 1983, pp. 225-249.
216. Kilara, A. Enzyme-modified lipid food ingredients. *Proc. Biochem.*, *20*(2):35-45, 1985.
217. Arbige, M. V., Freud, P. R., Silver, S. C., and Zelko, J. T. Novel lipase for cheddar cheese flavor development. *Food Technol.*, *40*(4):91-96, 98, 1986.
218. Sood, V. K., and Kosikowski, F. V. Accelerated cheddar cheese ripening by added microbial enzymes. *J. Dairy Sci.*, *62*:1865-1872, 1979.
219. Magee, E. L., Jr., and Olson, N. F. Microencapsulation of cheese ripening systems: Formation of microcapsules. *J. Dairy Sci.*, *64*:600-610, 1981.
220. Kirby, C. J., and Law, B. A. Recent developments in cheese flavor technology application of enzyme microencapsulation. *Dairy Ind. Int.*, *52*(2):19, 21, 1987.
221. Law, B. A. The accelerated ripening of cheese. In *Advances in the Microbiology and Biochemistry of Cheese and Fermented Milk* (F. L. Davies and B. A. Law, eds.), Elsevier Applied Science Publishers, London, 1984, pp. 209-228.
222. Dziezak, J. D. Enzyme modification of dairy products. *Food Technol.*, *40*(4):114, 116, 118-120, 1986.
223. Campbell-Platt, G. *Fermented Foods of the World. A Dictionary and Guide*, Butterworths, London, 1987.
224. Hesseltine, C. W. Fungi, people, and soybeans. *Mycologia*, *77*:505-525, 1985.
225. Steinkraus, K. H. *Handbook of Indigenous Fermented Foods*, Marcel Dekker, New York, 1983.
226. Yokotsuka, T. Traditional fermented soybean foods. In *Comprehensive Biotechnology. The Principles, Applications and Regulations of Biotechnology in Industry, Agriculture and Medicine*, Vol. 3 (H. W. Blanch, S. Drew, and D. I. C. Wang, eds.), Pergamon Press, Oxford, 1985, pp. 395-427.

227. Yokotsuka, T. Fermented protein foods in the Orient with emphasis on shoyu and miso in Japan. In *Microbiology of Fermented Foods*, Vol. 2 (B. J. B. Wood, ed.), Elsevier Applied Science Publishers, London, 1985, pp. 197-248.
228. Wood, B. J. B. *Microbiology of Fermented Foods*, Vols. 1 and2, Elsevier Applied Science Publishers, London, 1985.
229. Lotong, N. Koji. In *Microbiology of Fermented Foods*, Vol. 2 (B. J. B. Wood, ed.), Elsevier Applied Science Publishers, London, 1985, pp. 237-270.
230. Aidoo, K. E., Hendy, R., and Wood, B. J. B. Solid substrate fermentations. *Adv. Appl. Microbiol.*, *28*:201-237, 1982.
231. Cannel, E., and Moo-Young, M. Solid substrate fermentation systems. *Proc. Biochem.*, *15*(5):2-7, 1980.
232. Godfrey, A. Production of industrial enzymes and some applications in fermented foods. In *Microbiology of Fermented Foods*, Vol. 2 (B. J. B. Wood, ed.), Elsevier Applied Science Publishers, London, 1985, pp. 345-371.
233. Adams, M. R. Vinegar. In *Microbiology of Fermented Foods*, Vol. 1 (B. J. B. Wood, ed.), Elsevier Applied Science Publishers, London, 1985, pp. 1-47.
234. Amorim, H. V., and Amorim, V. L. Coffee enzymes and coffee quality. In *Enzymes in Food and Beverage*, ACS Symposium Series No. 47 (R. L. Ory and A. J. St. Angelo, eds.), American Chemical Society, Washington, DC, 1977, p. 39.
235. Arunga, R. C. Coffee. In *Economic Microbiology. Fermented Foods*, Vol. 7 (A. H. Rose, ed.), Academic Press, London, 1982, pp. 259-292.
236. Jones, K. L., and Jones, S. E. Fermentations involved in the production of cocoa, coffee and tea. *Prog. Ind. Microbiol.*, *19*:411-456, 1984.
237. Carr, J. G. Tea, coffee, and cocoa. In *Microbiology of Fermented Foods*, Vol. 2 (B. J. B. Wood, ed.), Elsevier Applied Science Publishers, London, 1984, pp. 133-154.
238. Sanderson, G. W., and Coggon, P. The use of enzymes in the manufacture of black tea and instant tea. In *Enzymes in Food and Beverage Processing* (R. L. Ory and A. J. St. Angelo, eds.), American Chemical Society, Washington, DC, 1977, pp. 12-26.
239. Sanderson, G. W. Tea manufacture. In *Biotechnology. A Comprehensive Treatise*, Vol. 15 (H.-J. Rehm and G. Reed, eds.), Verlag-Chemie, Deerfield Beach, FL, 1983, pp. 577-586.
240. Thomas, R. L., and Murtagh, K. Characterization of tannase activity on tea extracts. *J. Food Sci.*, *50*:1126-1129, 1985.
241. Wood, B. J. B. Progress in soy sauce and related fermentations. *Prog. Ind. Microbiol.*, *19*:373-409, 1984.
242. MacLeod, G., and Ames, J. Soy flavor and its improvement. *CRC Crit. Rev. Food Sci. Nutr.*, *27*:219-400, 1988.
243. Beddows, C. G. Fermented fish and fish products. In *Microbiology of Fermented Foods*, Vol. 2 (B. J. B. Wood, ed.), Elsevier Applied Science Publishers, 1985, pp. 1-39.

244. Beddows, C. G., and Ardeshir, A. G. The production of soluble fish protein solution for use in fish sauce manufacture. I. The use of added enzymes. *J. Food Technol.*, *14*:603-612, 1979.
245. Luksas, A. J. Fermenting whey and producting soy sauce from fermented whey. U.S. Pat. 3,552,981 (1971).
246. Scott, D. Oxidoreductases. In *Enzymes in Food Processing*, 2nd ed. (G. Reed, ed.), Academic Press, New York, 1975, pp. 219-254.
247. Scott, D. Applications of glucose oxidase. In *Enzyme in Food Processing*, 2nd ed. (G. Reed, ed.), Academic Press, New York, 1975, pp. 519-547.
248. Schmid, R. D. Biotechnology: Application to oleochemistry. *J. Am. Oil Chem. Soc.*, *64*:563-570, 1987.
249. Richter, G. Glucose oxidase. In *Industrial Enzymology. The Application of Enzymes in Industry* (T. Godfrey and J. Reichelt, eds.), The Nature Press, New York, 1983, pp. 428-435.
250. Underkofler, L. A. Manufacture and uses of industrial microbial enzymes. *Chem. Eng. Proj. Symp. Ser.*, *62*:11-20, 1966.
251. Rice, J. Enzymology and food packaging. *Food Process.*, *49*(12): 188-189, 1988.
252. Sarett, B. L., and Scott, D. Enzyme-treated sheet product and article wrapped therewith. U.S. Pat. 2,765,233 (1956).
253. Duxbury, D. D. Liquid enzyme provides economical milk sugar sweetness replacement. *Food Process.*, *50*(2):77-78, 1989.
254. Enari, T. M. Microbial cellulases. In *Microbial Enzymes and Biotechnology* (W. M. Fogarty, ed.), Applied Science Publishers, London, 1983, pp. 183-223.
255. Wood, T. M. Properties of cellulolytic enzyme systems. *Biochem. Soc. Trans.*, *13*:407-410, 1985.
256. Ghose, T. K., and Pathak, A. N. Cellulases—2: Applications. *Proc. Biochem.*, *8*(5):20-21, 24, 1973.
257. Fox, P. F. Enzymes in food processing. In *Industrial Aspects of Biochemistry* (B. Spencer, ed.), North Holland Publishing, Amsterdam, 1974, pp. 213-239.
258. Emert, G. H., Gum, E. K., Jr., Lang, J. A., Liu, T. H., and Brown, R. D., Jr. Cellulases. In *Food Related Enzymes* (J. R. Whitaker, ed.), American Chemical Society, Washington, DC, 1974, pp. 79-100.
259. Halliwell, G. Microbial β-glucanases. *Prog. Ind. Microbiol.*, *15*:3-61, 1979.
260. Goksøyr, J., and Eriksen, J. Cellulases. In *Economic Microbiology. Microbial Enzymes and Bioconversions*, Vol. 5 (A. H. Rose, ed.), Academic Press, London, 1980, pp. 283-330.
261. Enari, T. M. Cellulases. In *Food Related Enzymes* (J. R. Whitaker, ed.), American Chemical Society, Washington, DC, 1983, pp. 183-224.
262. Coughlan, M. P. The properties of fungal and bacterial cellulases with comment on their production and application. *Biotechnol. Genet. Eng. Rev.*, *3*:39-111, 1985.
263. Mandels, M. Applications of cellulases. *Biochem. Soc. Trans.*, *13*:414-416, 1985.

264. Montencourt, B. S., and Eveleigh, D. E. Fungal carbohydrases: Amylases and cellulases. In *Gene Manipulations in Fungi* (J. W. Bennett and L. L. Lasure, eds.), Academic Press, New York, 1985, pp. 491-512.
265. Beguin, P., and Gilkes, N. R. Cloning of cellulase genes. *CRC Crit. Rev. Biotechnol.*, *6*:129-162, 1987.
266. Frost, G. M., and Moss, D. A. Production of enzymes by fermentation. In *Biotechnology. Enzyme Technology*, Vol. 7a (J. F. Kennedy, ed.), VCH Verlagsgesellschaft, Weinheim, 1987, pp. 65-211.
267. Dekker, R. F. H., and Richards, G. N. Hemicellulases: Their occurrence, purification, properties, and mode of action. *Adv. Carbohydr. Chem.*, *32*:277-352, 1976.
268. Woodward, J. Xylanases: Functions, properties and applications. In *Topics in Enzyme and Fermentation Biotechnology*, Vol. 8 (A. Wiseman, ed.), Ellis Horwood, Chichester, 1984, pp. 9-30.
269. Kulp, K. Enzymolysis of pentosans of wheat flour. *Cereal Chem.*, *45*:339-350, 1968.
270. Posorske, L. H. Industrial-scale application of enzymes to the fats and oils industry. *J. Am. Oil Chem. Soc.*, *61*:1758-1760, 1984.
271. Macrae, A. R., and Hammond, R. C. Present and future applications of lipases. *Biotechnol. Genet. Eng. Rev.*, *3*:193-217, 1985.
272. Seitz, E. W. Industrial application of microbial lipases: A review. *J. Am. Oil Chem. Soc.*, *51*:12-15, 1974.
273. Gillis, A. Research discovers new roles for lipases. *J. Am. Oil Chem. Soc.*, *65*:846-850, 1988.
274. Applewhite, T. H. *Proceedings. World Congress on Biotechnology for Fats and Oils Industry*, American Oil Chemists Society, Hamburg, FRG, 1988.
275. Macrae, A. R. Enzyme catalyzed modification of oils and fats. *Phil. Trans. Roy. Soc. Lond.*, *B310*:227-233, 1985.
276. Zaks, A., and Klibanov, A. M. Enzyme-catalyzed processes in organic solvents. *Proc. Natl. Acad. Sci. USA*, *82*:3192-3296, 1985.
277. Yamane, T. Enzyme technology for the lipids industry: An engineering overview. *J. Am. Oil Chem. Assoc.*, *64*:1657-1662, 1987.
278. Miller, C., Austin, H., Posorske, L., and Gonzalez, J. Characteristics of an immobilized lipase for the commercial synthesis of esters. *J. Am. Oil Chem. Soc.*, *65*:927-931, 1988.
279. Huge-Jensen, B., Galluzzo, D. R., and Jensen, R. G. Studies on free and immobilized lipases from *Mucor miehei*. *J. Am. Oil Chem. Soc.*, *65*:905-910, 1988.
280. Iwai, M., and Tsujisaka, Y. Fungal lipase. In *Lipases* (B. Borgstrom and H. L. Brockman, eds.), Elsevier, Amsterdam, 1984, pp. 443-469.
281. Woodward, J., and Wiseman, A. Invertase. *Dev. Food Carbohydr.*, *3*:1-21, 1982.
282. Buchholz, K., and Rabet, D. The effect of cell and microbial invertases on juice extraction from sugar beet. *Zuckerind.*, *112*:792-795, 1987.
283. Nakajima, M., Jimbo, N., Nishizawa, K., Nabetani, H., and Watanabe, A. Conversion of sucrose by immobilized invertase in an asymmetric membrane reactor. *Proc. Biochem.*, *23*(2):32-35, 1988.

284. Aunstrup, K. Proteinases. In *Economic Microbiology. Microbial Enzymes and Bioconversions*, Vol. 5 (A. H. Rose, ed.), Academic Press, London, 1980, pp. 49-114.
285. Weetall, H. H. Immobilized proteases—potential application. In *Immobilized Enzymes in Food Processing* (W. H. Pitcher, Jr., ed.), CRC Press, Boca Raton, FL, 1980, pp. 175-183.
286. Ward, O. P. Proteinases. In *Microbial Enzymes and Biotechnology* (W. M. Fogarty, ed.), Applied Science Publishers, London, 1983, pp. 251-317.
287. Kilara, A. Enzyme-modified protein food ingredients. *Proc. Biochem.*, *20*(5):149-157, 1985.
288. Loffler, A. Proteolytic enzymes: Sources and applications. *Food Technol.*, *40*(1):63-70, 1986.
289. Wieland, H. *Enzymes in Food Processing and Products*, Noyes Data Corp., Park Ridge, NJ, 1972.
290. Bernholdt, H. F. Meat and other proteinaceous foods. In *Enzymes and Food Processing*, 2nd ed. (G. Reed, ed.), Academic Press, New York, 1975, pp. 473-492.
291. Cowan, D., Daniel, R., and Morgan, H. Thermophilic proteases: Properties and potential applications. *Trends Biotechnol.*, *3*:68-72, 1985.
292. Feeney, R. E., and Whitaker, J. R. *Food Proteins. Improvement Through Chemical and Enzymatic Modification*, American Chemical Society, Washington, DC, 1977.
293. Cowan, D. Proteins. In *Industrial Enzymology. The Application of Enzymes in Industry* (T. Godfrey and J. Reichelt, eds.), The Nature Press, New York, 1983, pp. 352-374.
294. Korycka-Dahl, M. B., and Richardson, T. Activated oxygen species and oxidation of food constituents. *CRC Crit. Rev. Food Sci. Nutr.*, *9*:209-241, 1978.
295. Linghert, H., Akesson, G., and Eriksson, C. E. Antioxidative effect of superoxide dismutase from *Saccharomyces cerevisiae*. *J. Agric. Food Chem.*, *37*:23-28, 1989.
296. Baldwin, R. R., Campbell, H. A., Thiessen, R., and Lorant, C. J. The use of glucose oxidase in the processing of foods with special emphasis on desugaring of egg white. *Food Technol.*, *7*:275-282, 1953.
297. Schmid, R. D. Oxidoreductases—past and potential applications in technology. *Proc. Biochem.*, *14*(5):2, 4-6, 8, 35, 1979.
298. Sharma, B. P., and Messing, R. A. Application and potential of other enzymes in food processing: Aminoacylase, aspartase, fumarase, glucose oxidase-catalase, sulfhydryl oxidase, and controlled release enzymes. In *Immobilized Enzymes for Food Processing* (W. H. Pitcher, ed.), CRC Press, Boca Raton, FL, 1980, pp. 185-209.
299. Carini, S., Mucchetti, G., and Neviani, E. Lysozyme: Activity against clostridia and use in cheese production—a review. *Microbiologie Aliments Nutrition*, *3*:299-320, 1985.
300. Proctor, V. A., and Cunningham, F. E. The chemistry of lysozyme and its use as a food preservative and a pharmaceutical. *CRC Crit. Rev. Food Sci. Nutr.*, *26*:359-395, 1988.

301. Yoshimura, K., Toibana, A., Kikuchi, K., Kobayashi, M., Hayakawa, T., Nakahama, K., Kikuchi, M., and Ikehara, M. Differences between *Saccharomyces cerevisiae* and *Bacillus subtilis* in secretion of human lysozyme. *Biochem. Biophys. Res. Commun.*, *145*:712-718, 1987.
302. Samejima, H., Kimura, K., Noguchi, S., and Shimura, G. Production of 5'-mononucleotides used immobilized 5'-phosphodiesterase and 5'-AMP deaminase. In *Enzyme Engineering*, Vol. 3 (E. K. Pye and H. H. Weetall, eds.), Plenum Press, New York, 1975, pp. 469-475.
303. Nakao, Y. Microbial production of nucleosides and nucleotides. In *Microbial Technology*, 2nd ed., Vol. 1 (H. J. Peppler and D. Perlman, eds.), Academic Press, New York, 1979, pp. 311-354.
304. Kuninaka, A. Nucleic acids, nucleotides, and nucleic acid related compounds. In *Biotechnology. Microbial Products*, Vol. 4 (H. Pape and H.-J. Rehm, eds.), VCH Verlagsgesellschaft, Weinheim, 1986, pp. 71-114.
305. Takamine, J. Process for making diastatic enzyme. U.S. Pat. 525,823 (1984).
306. Lin, Y.-L. Genetic engineering and process development for production of food processing enzymes and additives. *Food Technol.*, *40*(11):104-112, 1986.
307. Meade, J. H., White, R. J., Shoemaker, S. P., Gelfand, D. H., Chang, S., and Innis, M. A. Molecular cloning of carbohydrases for the food industry. In *Food Biotechnology* (D. Knorr, ed.), Marcel Dekker, New York, 1987, pp. 393-411.
308. Wetzel, R. Protein engineering: Potential applications in food processing. In *Biotechnology in Food Processing* (S. K. Harlander and T. P. Labuza, eds.), Noyes Publications, Park Ridge, NJ, 1986, pp. 57-71.
309. Jiminez-Flores, R., Kang, Y. C., and Richardson, T. Genetic engineering of enzymes and food proteins. In *Protein Tailoring for Food and Medical Uses* (R. E. Feeney and J. R. Whitaker, eds.), Marcel Dekker, New York, 1986, pp. 155-180.
310. Ulmer, K. Protein engineering. *Science*, *219*:666-671, 1983.
311. Shaw, W. B. Protein engineering. *Biochem. J.*, *246*:1-17, 1987.
312. Brodelius, P. Industrial applications of immobilized biocatalysts. *Adv. Biochem. Eng.*, *10*:75-129, 1978.
313. Weetall, H. H., and Zelko, J. T. Application of microbial enzymes for production of food-related products. *Dev. Ind. Microbiol.*, *24*:71-77, 1982.
314. Katchalski-Katzir, E., and Freeman, A. Enzyme engineering reaching maturity. *Trends Biochem. Sci.*, *7*:427-431, 1982.
315. Hultin, H. O. Current and potential uses of immobilized enzymes. *Food Technol.*, *37*(10):66, 68, 72, 74, 76-78, 80, 82, 176, 1983.
316. Kennedy, J. F., and Cabral, J. M. S. Immobilized living cells and their applications. In *Immobilized Microbial Cells*, Vol. 4 (I. Chibata and L. B. Wingard, Jr., eds.), Academic Press, New York, 1983. pp. 189-280.
317. Chibata, I. *Immobilized Enzymes*, John Wiley, New York, 1978.

15

SINGLE-CELL PROTEIN FROM MOLDS AND HIGHER FUNGI

SURINDER SINGH KAHLON *Punjab Agricultural University, Ludhiana, Punjab, India*

I. INTRODUCTION

Protein is an essential component of the diet. One of the greatest problems in the world today is achieving efficient distribution of available protein foods. Without such foods, humans suffer from poor health, including mental health, and this problem is fast assuming a new dimension because of the population explosion, especially in traditional societies such as those in Asia and Africa. Having adequate quantities of food is only one aspect of the problem. Another is the supply of desirable elements in the form of balanced and nutritious food. An all-out effort is being made to increase production of calories by bringing more land under cultivation, introducing high yielding varieties of cereals and pulses, use of proper fertilizers, judicious irrigation, pest and disease control, and improved methods of farming. However, little attention is being given to protein requirements and production. Consumers, in general, are ignorant of protein technology, and there is a background of preference and prejudice which governs patterns of consumption. Notable is the emphasis on flesh protein with the invalid impression that only fish and meat provide good protein. The eating of flesh is primeval, dating from the hunting and scavenging period of man's development. It is directly related to the killer instinct and has, therefore, become subject to regional and sectarian habits, preferences, traditions, ceremonies, and convictions. Throughout human history, these social controls have influenced technological progress and have caused unnecessary famines. Many of the present restricting factors in the diet can be related to social problems. There is a need to educate all concerned about protein and to eradicate attitudes and activities which prevent effective use of available protein.

It is estimated that more than half of the world's population not only has too little food, but the food it does have does not meet the minimum quality requirements for health and survival. It is, therefore, necessary to differentiate between protein needed for survival and that needed for health. The significance of dietary protein during infancy for resistance to infection [1] and for physical growth and mental development has been recognized [2]. Approximately two-thirds of the world's population is suffering from the protein deficiency disease called "kwashiorkor." Children with inadequate protein in their diet are victims of this disease [3].

Usually cereals have low protein values, whereas pulses are not available in the desired quantity because of poor yields and high costs. The good quality protein available in fish, meat, poultry, and eggs is not available to some humans in required amounts because of poor economic living standards as well as varied dietary habits; with the current system of distribution and production, agriculture cannot be relied upon to feed the world population. Hence there is an urgent need to find protein sources which do not rely on climate and which can, in effect, provide protein foods not subjected to the confused pattern of social and religious prejudice. The most potential is seen in microbial protein, which is too new to be condemned on religious grounds; recent advances in microbiology and biotechnology has made microbial protein production feasible on an industrial scale. With the worldwide upsurge of interest in protein from both conventional and unconventional sources, emphasis has centered on biomass production by microbial fermentation of waste materials of agriculture origin. The unconventional substrates, through suitable microbial conversions, are upgraded into microbial proteins for use in animal feed and ultimately as human food [4-6].

Analysis of the cost of single-cell protein (SCP) production using various substrates indicates that carbohydrate substrates are the cheapest source of carbon except for methane (Table 1). The most expensive carbon source is ethanol. Oxygen transfer and heat removal are lowest for carbohydrates and most expensive for methane, followed by methanol, *n*-paraffins, and ethanol. Table 1 indicates that carbohydrates, as carbon sources, are the best and cheapest sources for single-cell protein production [7]. Moreover, some of the carbohydrates of agriculture and forestry residues and from food industries are amply available at lower costs than

TABLE 1. Effect of Substrate and Yield Coefficient on SCP Operating Costs[a]

Substrate	Substrate cost	Oxygen transfer	Heat removal	Combined
Glucose from molasses	3.9	0.23	0.54	4.7
n-Paraffins	4.0	0.97	1.4	6.4
Methanol	5.0	1.2	1.7	8.1
Methane	1.6	3.3	3.7	8.6
Ethanol	8.8	0.75	1.3	10.9

[a]Shown in cents per pound.

Source: Ref. 7.

that of hydrocarbons, methanol, and ethanol. Recently, Das [8] reported that fermentable sugars can be obtained from agricultural crop residues.

Many reviews and articles on microorganisms as sources of food or feed have been published in the last 10 years or so [7,9-15], giving a fairly good idea of the current status of the protein problem in the world. An attempt has been made in this chapter to show the potential that molds and higher fungi possess for contributing to the world protein food supply.

II. BRIEF HISTORY

In the beginning, the field now known as "single-cell protein" was concerned exclusively with the production of yeast on hydrocarbons. During the last 15 years, the field has expanded considerably and now includes bacteria, fungi, and algae grown on a variety of carbon substrates, including *n*-paraffins, methane, methanol, ethanol, acetate, CO_2, and carbohydrate and cellulosic materials from different byproducts and waste sources. Activity in the field has been truly global; no area in the world has ignored SCP [16]. There is nothing new in the consumption of microorganisms by humans, either by design or accident; it must have been going on for millenia rather than centuries, when we consider alcoholic beverages, cheese, such foods as yogurt, soy sauces, breads, and others [17]. However, the first effort to cultivate microbes for food on a large scale was started in Germany during World War I, where torula yeast was produced. During the period between the two world wars, production of microbial food continued and reached a level of 16,000 tons per year [16]. By 1954, the potential of hydrocarbon-grown microorganisms as a source of food protein was well recognized [18,19]. However, commercial production of microbial food on hydrocarbons started in the late 1950s [20]. Table 2 shows the production of SCP on an industrial scale, which has been undertaken by a range of companies, whose main financial strengths vary from food to petrochemicals [21]. The British Petroleum Company (BP) was first to develop SCP from hydrocarbons, followed by other petroleum and chemical companies, government agencies, and universities. The British Petroleum Company constructed a 16,000 tons/year factory in France. Also, Japanese chemical and U.S. oil companies started SCP projects, based on *n*-paraffin feed stocks. Gulf Research and Development Company also produced SCP from hydrocarbons in 1963 [22] but abandoned their plans for commercialization several years ago on economic grounds [16]. Japan also could not make much significant headway [23] in its *n*-paraffin-based SCP projects, due to a strong anti-SCP campaign by Tokyo newspapers urging consumers to demand halt to "petroprotein" development plans. Ultimately, upon advice of the government, the companies decided to withdraw voluntarily from all their hydrocarbon-based SCP projects [28]. In 1970, methanol, a petroleum-derived substrate, received great attention for microbial food products [25-27]. The Imperial Chemical Industries (ICI) and Mitshuishi Gas Chemical of Japan and many other organizations in other parts of the world began to develop a methanol process employing both yeast and bacteria for SCP production. Imperial Chemical Industries made great progress compared to others in the field of SCP from methanol, and it had planned to

TABLE 2. Organizations Involved in SCP Development

Type of substrate	Organism	Organizational location	Production	Use[a] A	Use[a] H	Commercial status
Starches, Sugars/Yeast						
Molasses	*C. utilis*	Cuban sugar	70000	+	?	Product being developed for H application
Lactose/whey	*Kluyveromyces fragilis*	Bellyeast/France	6000	+	+	High value application sought
Sulfite liquor	*C. utilis*	Pekilo/Finland	3000	+		
Starches, Sugars/Fungi						
Hydrolyzed starch	*Fusarium graminarium*	RHM/UK	Pilot plant		+	Awaiting scale-up
Alcohols/Yeasts						
Ethanol	*C. utilis*	Amoco/USA	5000		+	High value markers
Methanol	*Pichia pastoris*	IFP/France	Pilot plant	+		
Methanol	*Pichia* spp.	Phillips/USA	Pilot plant		+	Research phase
Alcohols/Bacteria						
Methanol	*Methylophilus methylotrophus*	ICI/UK	55000		+	Successful large-scale development
Methanol	*Methylomonas clara*	Hoechst/Germany	Pilot plant	?	+	Research phase
Hydrocarbons/Yeast						
Gas oil	*Candida tropicalis*	BP/France	15000	+		Project closed
	Candida spp.	VEB/DDR	100000	+		Commercial development
n-Alkanes	*Candida lipolytica*	BF/UK	4000	+	?	Project closed
	Candida lipolytica	ANIC/Italy	100000	+		Project closed
	Candida maltosa	Liquichemical/Italy	100000	+		Project closed
	Candida tropicalis	IFP/France	Pilot plant	+		Project closed

[a]A = animal; H = human.

Source: Ref. 43.

start methanol-based SCP projects in Italy, West Germany, Sweden, etc. The Shell company developed a process to produce SCP from methane with mixed symbiotic cultures of *Pseudomonas* sp. and *Hyphomicrobium* sp., but its plan for commercialization was abandoned because of low prices of soybeans, maize, and the problem of applying sophisticated processes in developing countries [16]. Production of SCP from ethanol, which has become very popular in recent years due to certain advantages over other substrates, has been started by a number of companies like Amoco in the United States, Mitsubishi Petrochemical Co. in Japan, the Instituto de Fermentaciones Industries of the Centro Superior in Spain, and Exxon-Nestlé's project in Switzerland. There are reports in the literature of SCP produced from ethanol in Czechoslovakia, Russia, and India [28-30]. However, ethanol-based SCP industries may not be able to survive for long because of the rising prices of ethanol.

The future of any new product to be used as food or feed depends on its nutritional quality, desirable food characteristics like texture, flavor, taste, etc., and its production cost compared with other foods and feeds. Production of SCP from *n*-paraffins, methanol, methane, and ethanol would not be successful because of the factors just mentioned. Therefore, the effort to produce microbial food or feed shifted to conversion of cheap carbohydrates like lignocellulosic crops, forest residues, food industry waste, etc. In fact, SCP production from lignocelluloses was the earliest approach practiced during World War I and II [18], when torula yeast was produced on wood hydrolysate and sulfite waste liquor in Germany. In the U.S.S.R., the hydrolysis industry was started in the mid-1930s and continues to the present time, when fodder yeast is producted on hydrolysates of plant materials [30].

III. ADVANTAGES AND DISADVANTAGES

The term single-cell protein (SCP) was coined at the Massachusetts Institute of Technology at the insistent prodding of Professor Wilson in May, 1966, as a substitute for names that implied derivation from microorganisms or from waste materials or petroleum compounds, arousing unpleasant connotations when applied to food ingredients. The wisdom of this decision has been confirmed not only by the prompt and wide acceptance of the term, but also by the unfortunate misunderstandings that have ensued in the public mind when terms such as "petroprotein" have been used. However, SCP is a misnomer, as all proteins come originally from single cells. For a layperson it is a neutral and exotic name that arouses curiosity or interest rather than revulsion [31]. The microorganisms used for SCP production are usually algae, bacteria, yeast, and filamentous fungi. In this chapter, filamentous fungi as a potential source of SCP, their specific advantages, and problems encountered in cultivation will be discussed.

The general advantages of microbes over plants and animals as protein producers are:

1. Bacteria and yeast can double their biomass within 2-4 h, whereas plants like soya take 1-2 weeks, birds like chickens 2-4 weeks, and cattle 2-4 months.

2. Most microorganisms contain 7-12% nitrogen on a dry weight basis, which is higher than that of most common foods.
3. SCP can be produced in a continuous culture independent of climate and with only a small area and water requirements.
4. The waste disposal problems are not great compared with other processes of food production.
5. SCP production can be based upon substrates locally available in large quantities.
6. Through genetic manipulation of the organisms, it is possible to obtain mutants of desired characteristics like rapid growth rate and heat resistance [32].
7. The amino acid composition of the organism can be altered with changes in the enzyme pattern [33].
8. Development of SCP can be a complementary route to augment food and fodder production [34]. This is in agreement with what has been established by the Protein Advisory Group of the UNO [35], that the protein from the microbes offer the best hope for a new source of major protein independent of agriculture.

Some advantages and drawbacks of filamentous fungi as SCP are that fungi can be readily filtered and then squeezed or extracted to give certain texture and form. The problems with mycelial fungi is that they are usually slow-growing, have high maintenance requirements, and require sterile processing conditions, whereas yeast can grow at low pH (4.5-5.5) and the chance of bacterial contamination is less. This is not true of the fungal species. They are generally susceptible to contamination and must be grown under sterile conditions. Due to their slow growth rate, fungi have comparatively less RNA than bacteria and yeasts [36]. Many fungi produce a wide range of undesirable metabolites like mycotoxins, which give cause for concern about their safety in use [37]. Also, some of them have the potential to produce oxalates and other undesirable compounds, whereas bacteria and yeast have no such drawbacks. Furthermore, fungi are genetically unstable [38].

IV. SUBSTRATES FOR FUNGAL PRODUCTION

To produce SCP from various organisms like yeast, bacteria, and fungi, many substrates can be used. Depending on availability, raw materials can be grouped into renewable and nonrenewable substrates (Table 3). Growth of bacteria and yeasts to produce SCP has been studied on a wide range of substrates. Some fungi like *Graphium* spp. and *Trichoderma lingorum* can grow on unusual carbon sources like methane, ethane, and methanol, respectively [39,40], but most investigations into the use of fungi for protein production have recently centered on the use of substrates of agricultural origin or food processing byproducts or wastes. Thus, for fungi, starch, cellulose wastes, milk or cheese whey, etc. feature prominently as carbon sources [4-6,41,42]. Another cheap substrate for fungal biomass is sulfite liquor [43]. Availability of carbohydrates and their cost, including collection, transport, storage, and pretreatment [44], should be considered before selecting a substrate for fungal biomass

TABLE 3. Substrates for SCP Production

- I. Renewable Source
  - A. Industrial Wastes
    1. Bagasse
    2. Molasses
    3. Animal manure
    4. Sulfite waste liquor
    5. Corn steep liquor
  - B. Derived from Plants
    1. Starch
    2. Sugar
    3. Cereal
    4. Cellulose
    5. Molasses
    6. Ethyl alcohol
- II. Nonrenewable Sources
    1. Methane
    2. Methanol
    3. Gas oil
    4. *n*-Parrafinic hydrocarbons

production. However, some fungi do not readily utilize cellulose, starch, etc.; hence, it becomes necessary to hydrolyze crude substrate to yield simple sugars. Some fungi degrade waste materials very slowly for the process to be economically feasible. Therefore, pretreatment of the substrates by physical means, namely, griding [45], ball milling [46], electron irradiation [47], steaming [48], or microbial cellulases is necessary before the substrates are used for SCP production. The most promising fungi suitable for production of cellulases identified to date are *T. reesei*, *T. viride* QM 9414, and *Chaetomium cellulolyticum* [49-52].

V. PRODUCTION OF SINGLE-CELL PROTEIN

A. Fleshy and Higher Fungi

The use of fungi as food is not new. There is a long history of human consumption of fungi called mushrooms, and today the sporophores (fruiting bodies) of the basidiomycete *Agaricus bisporus* are cultivated on a large scale and are sold in the normal food distribution system of Western countries. Many other species are commonly consumed elsewhere in the world. Early humans picked wild mushrooms from their natural habitat, mainly as a supplement to the diet and not as a source of protein. The idea of mass production of fungal mycelium as a possible protein source is relatively recent. The idea of microbiological conversion of cellulosic materials, e.g., various straws, wood, and plant residues, was first expressed in 1920 [53], and in the same year animals were fed with straw enriched with mycelium of *Aspergillus fumigatus* in Germany [54]. Again in Germany, during World War II, to meet the food shortage, mycelium of fungi was used as human food, and the fungi were grown in submerged culture [55-57]. Today, scientists [58,59] are of the opinion that in the near future, mushroom mycelium may become a major source of food or feed. Using the technique of culture inoculum, Humfeld [60] produced mycelium of *Agaricus bisporus* in submerged culture, using press juice from pear waste as a carbon and energy source. He forecasted that production of fungal mycelium as a protein source on a large scale has commercial potential, thereby encouraging others to explore different edible fungi for mycelium production. Mycelium of high temperature-loving edible fungi, e.g., *Agaricus bazei*, produced on various media containing citrus press water, orange juice, and corn steep liquor as carbon sources [61], contained 23.5% protein compared with 43% in the wild type. In Canada, Reusser et al. [62] screened growth of 10 fungi on molasses and sulfite waste liquor media. Mycelium of *Tricholoma nudum* from sulfite liquor contained 38.7% protein and, on the basis of sugar utilized, *T. nudum* produced 18.9 g of protein per 100 g of sugar. *Tricholoma nudum*, when grown in waste from beet molasses production, yielded a mycelium with higher tryptophan content than reported for yeast [63]. A process was patented to cultivate *Tricholoma nudum* on sulfite waste liquor [64]. Along with other fungi, *T. nudum* [65] was cultivated on an industrial waste called vinasee; only three fungi could grow, namely, *A. bisporus*, *Boletus indecisus*, and *T. nudum*. These three were further cultivated on soybean whey; *T. nudum* and *B. indecisus* grew fast and had a high protein content [66]. The wood-rotting basidiomycetes were grown in 6 L malt extract medium under forced aeration for 7 days [67]. Following fermentation, the harvested mycelium protein content ranged from 21.7% in *Lentinus tigrinus* to 40.8% in *Polyporus tulipiferns*, with an average of 30.6% for 17 species screened. To date no one has been able to cause *Morchella* to fruit under artificial conditions, whereas its mycelium has been successfully produced in submerged culture [29], resulting in two U.S. patents [68,69]. Some people [70,71] cultivated *Morchella* spp. in 2000-gallon fermenters with food-processing waste as the culture medium. On analysis of the morel mycelium, the protein value was found to be similar to those reported for fungi. Martin [72] grew *Morchella esculenta* NRRL 2603 in a peat acid hydrolysate in a bench fermenter at 24°C and at an initial pH 7.0. A yield of 0.37 g/L and a substrate consumption of 45-52% were achieved. Martin [73,74] also produced edible mushroom *A. compestris* biomass in submerged culture with peat extract medium. The optimum growth temperature and effect of carbon source on growth of *Morchella crassipes* NRRL 2686, *A. bisporus* IRJP, *L. edodes* IFRI 257/C, *Pleurotus ostreatus* IFRI 958, and *Polyporus sulphuieus* IFRI 954 were investigated [75,76]. The biomass yield of 0.40 and 0.45 and productivities of 1.66 and 1.31 g biomass/L/h of *A. bisporus* and *M. crassipes*, respectively, on citrus peel extracts supplemented with glucose and inorganic nitrogen were achieved [77], and growth of *A. compestris* and *M. hortensis* in dilute whey plus potato infusion medium for 3-6 days at 24°C yielded 20 g/L biomass [78]. Mycelium obtained from several *Morchella* spp. collected from the Aegean region was grown as a submerged culture for protein production. Cheese whey, olive black water, beet molasses, potatoes, pumpkin, and carob were used as substrates in submerged culture. The best mycelium developed was in the beet molasses medium. In every carbon source used, from the standpoint of mycelium development, *M. elata* was most productive [79]. Fruiting bodies of *Pleurotus sapidus* were produced at 20°C in alfalfa brown juice supplemented with 0.1% safflower oil and used in a submerged culture [80]. Growth of various filamentous and higher fungi was studied [81] using three agricultural wastes, namely, sugarcane bagasse, wheat straw, and cow dung added as suspensions at 15 g/L in Chahal's medium. A biomass of 8 g/L of *Pleurotus carnucopia, P. ostreatus*, and *P. pulmonarius* in protein-free potato juice plus glucose medium was recorded [82]. The effect of vitamins, hormones, and fatty acids on biomass production by *Termitomyces clypeatus, Panafolus papillionacens, Gymnopilus chrysimyces, Coprinus lagopus, Lentinus squarrosubis, Volvariella volvacea*, and *A. bisporus* was studied [83]. Litchfield and associates [70] advocated the use of morel mycelia as a flavor supplement in food and only incidental value in protein contribution. However, according to Gray [9] this view was an exceptionally short-sighted estimation by Litchfield and associates [70], because in a world where protein deficiency is so widespread, a much better use than as a flavoring agent can be made of the fungal mycelium containing about 50% protein. Today mushroom cultivation is a well-developed industry all over the world. In addition, since production of truffles in France and paddy and shiitake mushrooms in Japan, the demand for mushrooms as flavoring agents is no longer a problem.

B. Molds

The use of mushroom mycelium has an advantage since carpophores of many species are nontoxic, edible, and hence one can simply expect the mycelia also to be nontoxic. The drawback is that many of them grow slowly compared to the molds, and their production on a large scale involves use of equipment for a longer duration than might be economically feasible. Second, there are not as many species of fleshy or higher fungi as there are of the simpler molds. Moreover, the mycelium of molds is relatively rich in protein; however, such fungi are not suggested as human food except in developing countries, where there is a serious protein problem. The mycelia of the molds is intended for animal feed [9], and this use is not new [54]. Various fungi, e.g., *Aspergillus* sp., *Penicillium* sp., and *Trichoderma koningii*, were grown in synthetic culture media, and the mycelium, after harvest, was fed to rats, pigs, and cattle. Based on feeding trials, fungal protein was found inadequate as cattle protein feed [84,85]. But Chastukhin [86] was adamant about the possibilities of fungal mycelium as a source of fodder proteins in the near future. In 1960, Gray and his associates [87-92] initiated a program to explore the possibilities of producing protein from imperfect fungi. Gray [87] compared the yields of protein/acre of corn with the yield/acre when grain carbohydrate was used as food for livestock or used as a carbon and energy source in production of fungal protein. According to him, efficiency of microbial conversion was higher than that of animal conversion. Microbes are much more efficient in synthesizing protein from carbohydrates and wastes, which are unfit even for animal feed. The minimum daily protein requirements of humans can be met either by 150 g of algae or fungi, 3.6 kg of potatoes, or 14.5 kg of fresh sugarcane [93]. Fungi Imperfecti was selected as a possible source of edible protein [89], and in a series of articles [94-97], cheap carbohydrates like sweet potatoes, corn, rice, cassava roots, beet molasses, paper pulp, and waste products of the food-processing industries were used as substrate for conversion into fungal protein. Exploratory studies [98] using sugar cane juice for synthesis of fungal protein also were published. In India, Chahal and co-workers [99] used cellulosic materials, the cheapest source of organic carbon, as substrate for conversion into fungal protein. Forty-four cellulolytic fungi were screened for their ability to grow on wood pulp in culture medium; only four, namely, *Myrothecium verrucaria, Chaetomium globosum, Rhizoctonia* sp., and *Trichoderma* sp., produced the highest protein biomass. Also, all essential amino acids appeared in *Rhizoctonia solani, Trichoderma* sp., and *Chaetomium globosum* except that threonine was not detected in *Trichoderma* sp. when grown in acid hydrolysate of wood pump [100]. Single-cell protein also was produced from various cellulosic materials like filter paper, cellulose powder, para-aminobenzoic cellulose, carboxymethyl cellulose cotton fibers, paper towel, bagasse pith, and fibers of Sorgo bagasse, etc. [101]. It is only during the last decade or so that numerous papers began to appear on fungal protein production from different agro and starch industrial wastes and effluents. Rogers et al. [102] reported 13.3% crude protein (DW) in the final product obtained by growing *Aspergillus fumigatus* on alkali-treated cellulose. Peitersen [103] obtained 21.25% (DW) crude protein by growing *T. viride* on alkali-treated barley straw. Romanelli [104] reported 60% utilization of fine solka-flock powder in 3 days by *Sporotrichum thermophile*. Eriksson and Larsson [105] obtained a product with 6% (DW) crude protein from powdered cellulose, 13.8% from waste fibers, and 32% from highly amorphous cellulose by growing the lignocellulolytic organism *S. pulverulentum*. Miller and Srinivasan [106] used *A. terreus* for protein production from cellulose, and the end product containing 32.9% (DW) crude protein was harvested on pure cellulose (Solka-Floc) treated with 1 N NaOH. Updegraff [107] obtained a product containing 10% crude protein by growing *M. verrucaria* on ball-milled newspaper. When *Penicillium notatum* and *P. digitatum* were grown on potato-processing wastes, a biomass yield of 9-24 g/L was recorded [108]. Stakheev et al. [109] developed a new fungal protein feed "digitatum" from potato waste through the action of *P. digitatum* 24P. The fungus synthesized 10-17 g/L mycelium in a medium containing potato waste. The mycelium protein had a high content of essential amino acids and about 40-60% crude protein. Three cellulolytic fungi, namely, *M. verrucaria, Paecilomyces* sp., and *A. terreus*, were grown for 7 days at 28°C in a basal medium at pH 5.5 containing sugarbeet pulp [110]. *Myrothecium verrucaria* had the highest protein in the biomass. Hydrolyzed oat hulls were fermented with *T. viride* QM 6a to produce animal feed [111].

A high lipid concentration between 8.9 and 27.5% was noted in the biomass of *Fusarium oxysporum, A. luchuensis, A. niger,* and *Oospora* sp.

when grown in a carbon and nitrogen source [112]. Similar results were obtained [113] when metabolism of the different lipid fractions of *F. oxysporum* was studied. Ghai et al. [114] grew *C. cellulolyticum* 32319 in delignified canning "sag" waste in submerged culture and harvested biomass containing 20.6-24.6% crude protein. *Penicillium frequentans*, *A. nidulans*, and *Fusarium* sp. were cultivated in submerged culture for 4 days at 28°C. The first two yielded more protein and the last more fat [115]. Ek and Eriksson [116] grew *S. pulverulentum* ATCC 32629 in waste waters from a fiber board mill in a continuous 10-L fermenter and harvested 0.19-0.35 g/L/h fungal biomass. Milled wheat bran was selected as substrate to produce fungal protein using *A. terreus* and *A. niger* in shake flasks for 15 days; the resultant biomass contained 40% crude protein [117]. Garg and Neelakantan [118] isolated *A. terreus* GNI and cultivated it on 1% alkali-treated bagasse in shake culture for 7 days. The highest crude protein content was 20% at 30°C and pH 4.0.

Recently Miller and Srinivasan [119] grew *A. terreus* ATCC 20514 in alkali-treated Solka floc and sugarcane bagasse both by batch and continuous fermentation and achieved 80-89% and 78-84% cellulose degradation, respectively, at 35-45°C within 30-36 h. *Aspergillus carneus* proved best for synthesizing protein in alkali-treated straw [120], but the most efficient fungus for protein production on wheat straw was *Cochliobolus specifer* [121]. Sidhu and Sandhu [122] reported they succeeded in growing *T. longibrachiatum* Refai (IMI 228288) in NaOH delignified bagasse and obtained a higher yield of biomass. A white-rot fungus, *Phanerochaete chrysosporium*, yielded 12-16% crude protein on alkali-treated maple bark [123], which led to a process for protein production and water purification by using the white-rot fungus *S. pulverulentum* [116]. The capacity of white-rot fungi to degrade lignin is of special potential; however, use of agitated fermenters should be limited if the sensitivity of lignin degradation to culture agitation found with *P. chrysosporium* [124] applies to other white-rot fungi. The effect of nitrogen on lignin degradation by white-rot fungi also creates problems [125-127] as low concentrations are required for lignin breakdown, and under such conditions protein production is reduced. However, it is not certain whether all white-rot fungi require low nitrogen or not as *Ganoderma applanatum* and *P. florida* degraded lignin under conditions of high nitrogen [128-130]. Furthermore, Levonen-Munoz et al. [131] also found exceptions to the repressive effect of nitrogen on lignin degradation. Hatakka and Pirhonen [14] cultivated, in submerged culture, 10 wood-decaying fungi, 8 of which were white rot, on various agricultural wastes including straw, alkali-treated straw, straw-cob supplemented with molasses, and urea and grainary waste. The highest protein content, 29.2%, was achieved with the white-rot fungus *Fomes fomenterius* 80 on grainary waste. Protein content on alkali-treated straw increased 7.8-fold and 7.1-fold after 10 days with non-white-rot fungi *Cylindrobasidium evolvens* 58 and *Agaricus* sp. strain 153, respectively. In Egypt, Abdel-Monem et al. [132] isolated microorganisms with different capacities to degrade sugarbeet bagasse cellulose and to synthesize cellular protein. No consistent relationship was found between decomposition and SCP production. The highest cellulose degradation (>67%) with corresponding high protein yields (>43 mg/culture) were recorded for *A. fumigatus* IB and *P. funiculosum* 2B, 3B, and 6B.

The 6B fungus proved to be the best for SCP production (65 mg/50 ml medium).

Waste sludges from pulp and paper mills represent a potential source of lignocellulose material, which has been rendered accessible to enzymatic attack by pulping process [133]. Sludges from three pulp and paper mills in New York were characterized for potential use as substrate for SCP and hydrolytic enzyme production by growing the cellulolytic fungus *T. Reesei* DAUM 167654, which accumulated a product containing over 22% crude protein and caused a conversion of sludge to protein of about 15% in 3 days of growth in shake flasks [134].

Scanning of the literature shows that there has been little success in converting lignocelluloses from agricultural residues and industrial wastes into fungal biomass protein. When pure cellulose treated with high concentrations of alkali was used as substrate, a product containing 33% protein was achieved [106]. However, this process cannot be economically viable because pure cellulose was used and large quantities of NaOH were required for pretreatments; also, an efficient organism to convert or break down lignocelluloses into sugars was required. Sekhon [135], in her studies on utilization of wheat straw for production of cellulases and protein by thermophilic fungi, searched for thermophilic fungi in composting wheat straw. When wheat straw is composted with inorganic nitrogen fertilizer, phosphorus, and potassium, a large number of indigenous microorganisms develop. These convert the straw into a suitable substrate for growth of mushrooms [136]. Samples of degrading wheat straw were taken 5-10 cm deep in the composting heap where fungi and actinomycetes predominate. A large number of fungi were isolated, identified, and grouped into eight distinct species or strains based upon their morphological characteristics [137]. A fungus belonging to the genus *Chaetomium* different from all the known species was found, thus it was named as a new species *Chaetomium* sp. nov. When all the isolates were screened for their ability to convert wheat straw into fungal protein, *Chaetomium* sp. nov. was the highest protein producer. Chahal and Hawksworth [138] studied the taxonomic and morphological characteristics of this new *Chaetomium* sp. and assigned it the name *C. cellulolyticum.* Its growth rate was faster and over 80% more biomass protein was formed than by *T. viride* [139]. Later, the growth behavior and protein production of *C. cellulolyticum* were established on pure cellulose in the slurry state in a 14-L fermenter [140]. In initial trials on pure cellulose, *C. cellulolyticum* produced only 17.5% protein [137], but in subsequent experiments [140], with improved cultural techniques, 47.5% protein was achieved within 24 h. When grown on untreated wheat straw, the protein content was only 25.4% [137], but when the straw was delignified or treated with NaOH and washed, the protein content increased to over 40% [137,138,140,141]. Cattle manure pretreated with NaOH was an excellent substrate for production of *C. cellulolyticum* fungal feed rich in protein. The final product contained 37% protein, and conversion was achieved within 11 h [139]. Similarly, swine manure supplemented with glucose also proved to be an excellent substrate, and a product containing 50% protein was harvested, when supplemented swine manure was fermented by *C. cellulolyticum* in shake culture [142]. Hemicellulose obtained from steamed pretreated wood also was used for fungal protein production by

C. cellulolyticum [143]. Following fermentation for 22 h, a product containing 45% protein was harvested. Hemicellulose from acid- cum steam-treated wheat straw at pH 6.0 was fermented for enzyme production by *T. reesei* and for protein production by *C. cellulolyticum* and *P. sajor-caju*. The biomass produced by these two fungi contained 40-47% crude protein on a dry weight basis, which was most suitable for animal protein feed [144]. Recently, the "Waterloo process" [145] has been developed to produce fungal protein at the pilot plant level, using corn stover as the substrate for *C. cellulolyticum* to synthesize protein-rich biomass.

Both untreated and delignified wheat straw were used in submerged fermentation for SCP production using *C. globosum*, and the optimum cultural conditions for maximum conversion of wheat straw into protein were 5 days, 37°C, pH 5.0, and 400 mg N/L in the form of sodium nitrate [4]. Rosen and Schuegerl [146] compared cellulose conversion both in submerged and solid state reactors using pretreated wheat straw as substrate and *C. cellulolyticum* as the conversion agent. Both methods yielded equal cellulose conversions; however, in solid state culture, the duration of cultivation was 2-3 times longer than that of submerged cultures. Valdivia et al. [147] fermented cassava meal with *Rhizopus oligosporus* in stirred tanks under nonseptic conditions. The fungal product contained 26% true protein.

Submerged culture production of the antibiotic penicillin from 1944 onward also resulted in a substantial amount of fungal mycelium. The spent *Penicillium chrysogenum* was dried and used as animal feed supplement, most suitable for chickens [148]. However, the problem of antibiotic residues and acceptability caused a ban on the use of spent *P. chrysogenum* biomass in the West, although interest continued elsewhere [66,149]. A pilot scheme to treat corn and pea-canning wastes has been developed [150], where both *T. viride* and *Gliocladium deliquescens* were used to treat the waste effluent. Likewise, use of fungi as fixed films [151] appears to hold promise as a means of dealing with dilute effluent streams. In Guatemala, a method to treat coffee wastes and produce microbial biomass by using various fungi, especially *T. harzianum*, is described [152]. Peitersen [15] produced cells of *T. viride* on alkali-treated straw and observed that water-soluble inhibitors in the straw batch fermentation with 2 and 4% straw lasted 6-8 days. Addition of straw during fermentation made it possible to increase the straw concentration to 6% without prolonging the fermentation time of 6 days. The protein content of the product was 22%. A combined mild heat and biological treatment of wheat straw with four cellulolytic fungi was developed to improve digestibility of cellulose [153] in submerged cultivation. *Aspergillus japonicus* converted about half of the treated straw dry matter into fungal biomass containing 10% crude protein on a dry weight basis, and yielding 60% more protein than *T. harzianum* (8.6% protein). *Aspergillus japonicus* was found to be the most suitable of the four fungi screened for upgrading straw in submerged culture. Also, a novel two-stage bioreactor has been designed for combined submerged (SF) and solid substrate fermentation (SSF) of wheat straw [154]. The straw is pretreated with steam, and cellulases from the culture fluid of *T. reesei* are absorbed on it to increase bioconvertibility. SSF is conducted on the top part of the bioreactor by inoculating the straw with a 36-h mycelial culture of *T. reesei* or *Coriolus versicolor*. In the bottom part of the fermenter, *Endomycopsis fibuliger* is grown in SF. The SF liquor is recirculated through the SSF stage at 24-h intervals to remove glucose and other metabolites that may inhibit growth. Removed glucose and other metabolites provide nutrients for yeast on the SF stage. Two products are obtained, the SSF product having a protein content of 10.5% wet weight, increasing the digestibility of a feed ration of bull calves when mixed in the ration at the 10% level. The second product is the yeast generated in SF with 12.6% wet-weight protein content, which is an excellent feed supplement for monogastrics such as poultry and pigs. Bioconversion of waste paper to SCP by *Scytalidium acidophilom* has been reported [155]. Shake flasks and a pH of less than 1 were used, and the author claimed 97% sugar conversion and yields (dry weight of fungus produced and amount of sugar utilized) of 43-46% and 47%, respectively. Also, the crude protein had relatively low levels of nucleic acids. In dilute peat extract medium with 0.3% yeast extract, *Scytalidium acidophilum* produced maximum biomass protein of 42% in 10 days at 25°C with pH 2.0 [156] in shake culture. The important factors in SCP production are the aseptic operating condition and costs associated with harvesting the biomass. A cross-flow microscreen cultivation technique, particularly applicable for selective cultivation of microorganisms in dilute substrate solution, is successfully used to select and maintain an easily harvestable microbial culture, with a limited number of species under nonaseptic condition in dilute cheese whey [157]. Thus, it makes possible SCP production from dilute waste organic effluents. The microbial selective pressure exerted by the system could be manipulated by varying the hydraulic (θ) and mean cell (θn) residence times. The optimum system parameters were $\theta = 1$ h and $\theta n = 10$ h, resulting in selected microbial population comprising three species, namely, *Geotrichum candidum* a fungus, *Streptococcus cremoris*, and *Leuconostoc lactophilum*. The amino acid profile of the SCP produced compared favorably with other types of protein.

The search for fungi adequate for bioconversion of various agricultural wastes into SCP has been reported in a series of publications since 1974, and some of the important findings are presented in brief.

When Dhillon et al. [158] grew *C. globosum* on sucrose and untreated wheat straw, the fungal product contained 44 and 14% crude protein after 7 days at 28°C, respectively. The protein of *C. globosum* contained nearly all the essential amino acids and was at a par with the prescription of the FAO, except that two amino acids, namely methionine and tyrosine, were lacking.

From among the eight fungi screened for fungal protein production on molasses, glucose, and sucrose [159], six had higher mycelial growth and protein content on molasses than on glucose and sucrose, with *P. crustosum* having 33.7% crude protein. When proteins were analyzed for lysine, methionine, and tryptophan, five of eight fungi had a higher content of lysine with molasses as carbon source. All eight fungi screened were rich in lysine, tryptophan, and had more than adequate methionine content.

Penicillium crustosum was grown on rice flour in shake culture and used as a fungal protein supplement for wheat flour to improve its protein quality. However, the palatability for rats of the diet containing fungal protein was reduced, compared to whole wheat flour. The effect of the fungus on different organs, weight, liver, and plasma nitrogen was statistically nonsignificant [160].

Using wheat straw and groundnut hulls, six cellulolytic fungi were screened for fungal protein production [161]. *C. globosum* yielded maximum protein of 100.39 mg/g substrate, the minimum was produced by *M. verrucaria* on wheat straw, whereas protein biomass production ranged between 20.4-26.7 mg/g substrate for the six fungi growing on groundnut hulls. With corn cobs as substrate *M. verrucaria* yielded 204.6 mg biomass, and on spent grain waste it produced 24.7% biomass which was 8.8-fold more than the control [162-164]. Although *M. verrucaria* produced maximum biomass on corn cobs, rice stover, and wheat straw and its fractions, holocellulose and cellulose, the resultant protein was deficient in methionine [162, 165,166]. A waste from canneries, citrus pulp, was a more suitable substrate for *T. koningii* than for *P. ostreatus*. The former produced maximum protein (118.81 mg/g substrate), whereas the latter yielded only 7.25% biomass with a maximum fat content of 3.68% in 72 h [167].

The feed value of wheat straw was improved through solid substrate fermentation using *P. ostreatus*; the optimum incubation, temperature, and nitrogen source were 21 days, 30°C, and ammonium sulfate, respectively. Digestibility increased to 54.8% from 15.5% and protein enriched to 10.5% from 2.6% [168]. With *S. pulverulentum*, optimum conditions were 21 days, 35°C, and ammonium chloride, respectively, and protein content increased from 3 to 10%. In vitro digestibility of fermented straw was upgraded from 16 to 34.1% [169]. When rice straw was cultured with *S. pulverulentum*, protein of the fungus increased from 3.23 to 9.69% in 30 days at 35°C [170]. In vitro digestibility of *P. ostreatus*-treated rice straw increased from 39.4% to 60.2% [171]. When wheat straw was steamed under pressure for 20 minutes, treated with NH_4OH, and fermented with *C. cellulolyticum* in shake culture, maximum protein yield was 15.9% compared to 5.8% in untreated straw, in 5 days at 40°C [172].

Protein content of potato waste was enriched with *P. ostreatus* through solid state fermentation and then was used as poultry feed [173]. The crude and true protein of the substrate increased from 11 to 21% (1.88 times) and 7.5 to 14.6% (1.95 times), respectively, in 17 days at 30°C, with NH_4Cl as nitrogen source.

C. Treatment of Lignocellulosic Materials for Feed

The low digestibility of various straws due to high encrustation of cellulose and lignin is well known [30,174,175]. Various physical, chemical, and biological methods can depolymerize lignin to increase digestibility of lignocellulosic plant materials. Steam under pressure [176,177] makes the materials easily accessible to hydrolytic enzymes. Phenol-like compounds increase in the steam-treated bagasse [178], and such compounds plus furfurals are toxic to most microorganisms [179], thus rendering the materials unsuitable for SCP production [180]. Alkali treatment of straws to improve their nutritive value has been practiced for more than a century. Dilute alkali increases the fiber saturation point and the swelling capacity of lignocelluloses, which results from saponification of esters of 4.0 methylglucosomic acid attached to xylan chains. Alkali treatment also increases the availability of plant polysaccharides to microorganisms [181]. The feed efficiency of straws also is improved by alkali treatment [182]. Strong oxidizing agents like sodium chlorite ($NaClO_2$) have long been used to remove lignin [183,184]. Han and Anderson [185] pretreated rye grass with 0.5 N H_2SO_4 at 121°C and obtained a material containing 30% sugars. When fermented with *T. viride*, the substrate was enriched four times with fungal protein biomass. These various physical and chemical methods to increase digestibility and nutritional quality of plant residues, however, are tedious and costly, and further treatments to eliminate the side effects of the chemicals can make the process uneconomical.

Nutritional upgrading of agricultural residues and other lignocellulosic waste by use of biological methods has been extensively employed [6,168, 170,186-188]. Feeds so produced are better than those produced by chemical and physical means, as no side effects are noted in cattle. Moreover, feeds are enriched with microbial protein. The huge amount of lignocellulosic agricultural residue generated every year can be a suitable substrate for SCP production by employing various cellulolytic fungi either in submerged or solid state fermentation. Recently, greater attention has been given to bioconversion of these lignocellulosic materials in protein-rich feeds by solid state fermentation (SSF) because of its advantages.

D. Single-Cell Protein by Solid State Fermentation

The term SSF refers to any fermentation process in which the substrate is not a free liquid [189]. Solid substrate or solid state fermentation are terms used by different workers, but they are essentially the same when applied to processes for enriching marginally valuable crop residues, manure, fibers, or other agricultural residues with microbial proteins and/or increasing their digestibility. Solid state fermentation may be suitable due to its low cost technology, reduced reactor volume per unit substrate converted, and direct use of the fermented product rich in protein for animal feed [185,190,191]. The high product concentration and reduced cost of dewatering also makes SSF attractive for production of cellulase enzyme or for production of secondary metabolites [192-194]. Fermenters used in SSF systems vary from a simple open pan or tray to mechanized rotating drums. Studies [189] on some SSF with mechanized solid state fermenters revealed the ease of operating solid state fermenters. Their low energy requirements, in comparison with stirred tank reactors, suggests that these fermenters are of potentially great economic importance for both the industrial and developing countries. Advantages of the SSF process over the conventional stirred tank system [192] include use of a medium that is relatively simple, and space required for fermentation is small relative to yield of product because less water is used and the substrate is concentrated. The equipment, on a laboratory, pilot plant, or factory scale, is simple as compared to conventional fermentation equipment. Spores are directly used in the fermentation, hence seed tanks are not required. Conditions under which the fungus grows are more like those in its natural habitat. Aeration is easily obtained since there is air space between each particle of the substrate. The desired product may be readily extracted directly from the vessel by addition of solvent.

A limitation of SSF is that only those organisms can be employed which can grow at low moisture levels. Furthermore, heat becomes a problem when large quantities of moist substrate are fermented. It is difficult to control pH, and a large quantity of spore inoculum is needed. The most important limitation encountered in SSF is the difficulty in estimation of mycelial biomass [189], and biomass generation on lignocellulose substrate by SSF is severely limited by steric factors of fungal growth. Still, SSF is most attractive as low-level farm technology supplementing traditional methods, namely, ensiling and composting to improve digestibility and feed value of a marginally valuable agricultural residue while reducing its bulk [12].

Solid substrate fungal growth, in real life, is used widely to produce fermented foods and edible mushrooms [9] from lignocellulosic biomass. Asians have used various fungi for many fermented food processes [195,196]. These fermentations often do not produce an overall enrichment in protein, but significantly alter the taste of the substrate [197]. The Tropical Product Institute, using principles of Asian fermentation, produced SCP using cassava enriched with nitrogen and *Rhizopus arrhizus* or *R. oligosporus* [197,198]. *Pleurotus ostreatus* degraded labeled lignin [199] and *A. niger* and *R. nigricans* through SSF of moistened bagasse produced a product containing 3% protein, an increase of 3-4 times the raw material. *Sporotrichum thermophile* grown on ground newspaper in SSF for 6 days [201] yielded a product containing 6.5% protein on dry weight basis, while acid-pretreated (NH_3-neutralized) straw fermented with *Aureobasidium pullulans* and *Trichoderma* sp. yielded 14.0 and 10.9% protein, respectively, by SSF within 2-3 days [185]. Zadrazil [202] reported enhanced mycelium growth by *P. ostreatus*, *P. florida*, *P. cornucopiae*, and *P. eryngii* on wheat straw, with the addition of organic nitrogen and CO_2 concentration between 22 and 28. The soluble glucose concentration in wheat straw increased depending upon the phase of fungal growth and method of extraction. Also, the cellulose-lignin complex decomposed variably at different stages of mycelial growth releasing various quantities of water-soluble substances. Zadrazil [187] fermented wheat straw without any chemical treatment or nitrogen supplementation by SSF using *P. cornicopiae*, *P. florida*, *Agrocybe aegerita*, and *Stropharia rugosoannulata*. The highest rate of straw decomposition and release of metabolic energy was by *P. cornucopiae* and *S. rugosoannulata*. After 120 days at 30°C, *S. rugosoannulata* and *P. cornucopiae* decomposed 60% of OM, whereas *P. florida* and *A. aegerita* only 40 and 20%, respectively. Straw fermented with *S. rugosoannulata* (44 days at 30°C, 79 and 120 days at 22, 25, and 30°C), and with *P. cornucopiae* (120 days at 22 and 25°C) attained the digestibility (60-70%) of an average quality hay. Zadrazil [187] also used *Pleurotus*-cultured spent wheat straw as feed and found it had a feed quality equivalent to hay. An investigation into the influence of NH_4NO_3 [128] supplementation on degradation and in vitro digestibility of straw revealed that the rate of decomposition of OM by *P. salmonea stramineus* and *P. eryngii* decreased, whereas that of *L. edodus* and *P. florida* increased at a low NH_4NO_3 concentration but decreased at a higher concentration. *Pleurotus florida* had the highest lignin degradation rate, while lignin degradation by *L. edodus*, *P. eryngii*, and *P. salmonea stramineus* was depressed at different NH_4NO_3 levels. No correlation between lignin degradation and nitrogen supplementation was possible when *Ganoderma applanatum* and *P. florida* were used. The in vitro digestibility of the substrate with and without nitrogen supplementation decreased with *Agrocybe aegerita*, *Flammulina velutipes*, and *G. applantum*. All other fungi studied had enhanced in vitro digestibility when no NH_4NO_3 was added to the substrate. When important physical parameters for SSF of straw were studied [203], the highest decomposition rate at 22, 25, and 30°C was by *Abortiporus bienais* and *Trametes hirsuta*. A positive correlation between incubation temperature and the extent of substrate decomposition was observed with *T. hirsuta*. A reduced decomposition rate at 30°C was displayed by *A. bienais*, *G. applanatum*, and *P. serotinus*, whereas no influence of temperature was observed for *Lenzites betulina*, *P. ostreatus*, and *P. sajor-caju*. Less lignin was degraded at 30 than at 22°C by *G. applanatum*, *P. ostreatus*, and *P. serotinus*, whereas *A. bienais* and *T. hirsuta* decomposed lignin better at 30 than at 22°C. In vitro digestibility dropped below the value obtained with sterile straw as the control. When the water content in SSF of straw was increased, the gas phase was reduced and gas exchange was hindered. At low water content, because of high water tension and low degree of substrate swelling, growth of fungi was suboptimal. An intermediate range of water (75 ml/25 g substrate) was optimum for SSF, whereas a high water content effected increased formation of mycelium and degradation of lignin. Most fungi investigated degraded lignin best at water contents between 50 and 125 ml/25 g substrate, and an increase in incubation time increased degradation of the substrate with the slow-growing fungi *G. applanatum*, *L. betulina*, and *P. sajor-caju*. The faster growing species "Chille 88" and *T. hirsuta* decomposed the degradable fraction of the substrate within the first 60 days of incubation. Whereas *G. applanatum* produces a feed called "Palo-podrido" by degrading wood, the in vitro digestibility of the feed in cattle increased 20-fold [204]. Performance of two white-rot fungi was compared for degradation of wheat straw and other residues [205]; *S. pulverulentum* was faster than *Dichotinus squalens;* however, the latter after prolonged incubation preferentially degraded lignin and improved the in vitro digestibility of the straw. Rosenberg [206] found *Thielania thermophile* to be cellulolytic and *Chrysoporium prionosum* and *S. pulverulentum* highly lignocellulolytic. Optimum temperature and pH for efficient degradation of the lignocellulose complex by *S. pulverulentum* was 40°C and 4.9, respectively. The increases in nitrogen and protein of rice straw were 50% when fermented with *P. sajor-caju* [207] and indicates the possibility of the dual purpose of improving feed quality of paddy straw and mushroom production. When *A. bisporus* and *Volvariella diplasia* [208] were cultivated, respectively, on wheat and paddy straws, the crude protein, cell-soluble, and lignin contents increased in postharvested straw, unlike that in untreated straw. Use of spent straw rather than original straw as feed supplemented with oil cakes and/or nonprotein nitrogen was advocated for ruminants. Fermenting paddy straw with *P. ostreatus* and *S. pulverulentum* significantly degraded cellulose and lignin and increased the crude protein of fermented straw to 9.6 and 9.0%. In vitro dry matter digestibility (IVDRD) increased to 60.2% and 59.3%, respectively, within 15 days [6,170]. Spent wheat straw [209] obtained after cultivation of *A. bisporus* was fed to adult buffaloes as a 50% replacement for wheat straw. The washed spent straw diet, as compared to unwashed straw, resulted in better feed intake, which was equivalent to the control. This resulted from the reduction in total ash content and increase in crude protein and cell-soluble contents of spent straw. However, Bakshi et al. [210]

reported that increased crude protein content in *Pleurotus*-spent straw contributed very little toward nitrogen metabolism by buffaloes. Recently, a study on protein enrichment of wheat straw by *P. ostreatus* [211] revealed high neutral detergent fiber (NDF), cell-soluble contents, a minimum of 15% dry matter loss with a maximum of 44% in vitro digestibility of straw, and initiation of sporophore primordia by the 27th day of straw fermentation. The product was suitable for animal feed. Kokke [212] employed *R. oligosporus* CBS 324-35 or *Monascus ruber* to improve carob pods as feed. Chopped carob pods were autoclaved in an ammonium salt solution, inoculated, and incubated for 3 days at 30°C. Following fermentation, a cake-like structure containing 7% protein was obtained. The fermented product contained 73.8% of original carob sugars, whereas the amount of tannins was reduced considerably in the pods. Knapp and Howell [213], with the SSF technique, produced SCP from various fungi, and Gulati [214] improved the SSF process for fungal protein production employing unutilized cellulose and protein as carbon and nitrogen sources. Mukhopadhya and Sikyta [215] did SSF of barley straw for SCP, using *Trichoderma* sp. 39 and *F. oxysporum*, which showed good growth and cellulolytic activity after 5 days. Solid state fermentation by *Polyporus anceps* was done in three types of static trays at 30°C, using alkali-treated maple wood shavings as substrate [216]; 70% of substrate was utilized to produce 17% fungal crude protein in the final product. One percent treated hardwood sawdust processed by SSF [217] yielded a product containing 12.5% crude protein and in vitro digestibility of 30% after 48 h. *Chaetomium cellulolyticum* and *P. ostreatus* NRRL 2366 were used, and when they were grown on wheat alone or with addition of inorganic nitrogen or oat meal, they degraded lignin and cellulose from 10 to 45% and 15 to 55%, respectively, after 36 days. Free reducing sugars in the substrate and in vitro digestibility increased two- to threefold after 90 days [218]. A high percentage of fungal protein, 35%, was produced when banana peels were treated with hot water and fermented with *P. funiculosum* rather than when they were fermented naturally [219]. When cotton stalks, tur stalks, and groundnut shell were pretreated with water in a ratio of 3:1, the fermentation with *P. funiculosum* resulted in an increase of crude protein [220]. A continuous solid potato starch waste composting system was developed [221], which permitted complete fermentation within 7-14 days. When SSF of potato peels with *P. ostreatus* was done under optimum conditions, the organism increased crude protein of waste peels from 11.2 to 21.0% and true protein by 1.95 times [5,173]. Chahal et al. [222] grew thermophilic *C. cellulolyticum* ATCC 32319 in SSF at 37°C and pH 6.0 in alkali-pretreated wheat straw and corn stover and obtained a final product with 19 and 24% (DW) of protein and cellulose utilization of 63 and 48%, respectively. The same fungi in SSF of delignified canning "sag" waste produced 9-12% protein in the total biomass [114]. Ibrahim and Pearce [223] grew 11 rot fungi on barley and pea straw, sugarcane bagasse, and sunflower hulls for 21 days at 14-25°C; the fungi that reduced lignin and increased the in vitro dry matter digestibility were *Peniphora gigantea* in barley straw and sugarcane bagasse, and *G. lucidom* in pea straw and sunflower hulls. It was observed that 8 weeks growth of *P. anceps* on sterilized poplar shavings enriched with minerals increase the in vitro digestibility from 30 to 72%, 64% with *G. applanatum* and 61%

with *P. versicolor* after 3 weeks, and 42% with *Formitopsis ulmarius* after 4 weeks [224].

Poultry droppings were fermented with *P. ostreatus*, and their feed value as a replacement of maize was assessed for broiler production. The fungus contained 17.3% crude protein, and up to 10% maize could be replaced with fungus-treated poultry droppings [225]. Some workers [226] were of the view that the increase in nutritional value of poultry droppings probably resulted from reducing the uric acid and crude fiber contents by *P. ostreatus* growth.

Cassava meal fermented with *R. oligosporus* in packed columns or trays under nonaseptic conditions [147] resulted in a product containing 22.8% true protein in packed columns and 16.8% in trays. Amylolytic *R. oligosporus* produced a product with the appearance of meat, from partially gelatinized barley starch [227]. When growing on potato-processing waste, *Panus tigrinus* IBK-131 exhibited the highest protein productivity. However, protein synthesis was hindered when a nitrogen source was not provided [228]. Similarly, *A. niger* was the most efficient of three fungi, including *A. foetidus* and *A. awamori*, for protein production from potato-processing waste [229]. Mori et al. [230] inoculated heat-sterilized potato waste with TK_2 strain incubated for 4 days at 30°C, and the protein content of the product was increased to 14.5%. When potato waste was fermented with TK 41 and TK 42, the protein content in the biomass was 12.9 and 11.4%, respectively, and when fresh potato waste containing 5% carbohydrates was used as carbon source for *Cephalosporium cichharniae* ATCC 38255 at 45°C and pH 3.75, the final product was an excellent protein-rich animal feed [231]. *Rhizopus oligosporus* NRRL 2710 under nonseptic conditions produced 22.8 and 16.8% protein in packed column and trays, respectively, from cassava meal; the growth rate of the fungus was lower but protein productivity was higher in the solid culture [232].

Raimbault et al. [233] isolated 24 molds from traditional foods and screened them for ability to upgrade the protein content of cassava by SSF. They obtained 10-16% protein in the final product. Smith et al. [234] obtained the highest percentage of protein biomass from *Aspergillus* spp. in an aerated bench tray fermenter, whereas *S. pulverulentum* produced 30.4 g protein/100 g dry cassava root in 48 h at 45°C.

Steam-treated wheat straw containing 70% moisture was subjected to SSF with *T. reesei* or a mixture of *T. reesei* and *E. fibuliger* (R-574). The best protein productivity was obtained in a stationary layer fermenter, with product containing 13% protein [12]. When the various parameters for SSF of wheat straw were optimized, total sugar in the substrate available for bioconversion was 60.5% and the overall efficiency of biomass production was 42.7% [235]. Yadav [236] reported that SSF of wheat straw with *Coprinus* sp. was influenced by the levels of N, P, and S and free carbohydrates. Addition of free carbohydrates such as molasses and whey had a detrimental effect on biodegradation of lignin and on digestibility of degraded dry matter. However, protein production in supplemented straw was enhanced.

A simple method to produce SCP by a hypercellulolytic mutant of *T. viride* grown on NaOH-treated bagasse was described [237]. The fermentation is carried out under septic conditions in nylon bags at room temperature for 30 days. This treatment increased the protein content of the bagasse 10 times and reduced the lignin content by 50%. Also, the effect of nitrogen supplementation on fermentation of bagasse by *S. pulverulentum* was studied [238] and the in vitro dry matter rumen digestibility (IVRD) of bagasse increased from 27 to 36% when the fermentation medium was supplemented with peptone, yeast extract, NH_4NO_3 or NH_4Cl, but asparagine reduced IVDRD of bagasse below the control value. Similarly, alkali-treated sugar cane pith was protein-enriched through semi-solid culture with *C. cellulolyticum* ATCC 32319 [239]; the biomass obtained from the substrate treated with 1% and 4% NaOH contained 10 and 15% protein, respectively.

Sugarbeet residue supplemented with 1% nitrogen was protein-enriched with *T. album* through SSF [240]; 22% protein was in the final product after 4 days. Apple pomace also was protein-enriched [241] by fermenting the material with *Saccharomycopsis lipolytica* or *T. reesei* in single and mixed cultures. A product containing 13-15% protein on a dry weight basis was obtained, which was suitable for cattle feeding. The process also was applied to bioconversion of apple distillery slop. *Trichoderma* and *Phanerochaete* were found most suitable [242,243]; the fungi degraded raw fibers by 20% resulting in the filter cakes containing 17-22% raw protein.

Citrus peel when fermented with *A. niger* through SSF was enriched by more than 10% crude protein [244], and banana waste was protein-enriched by 6 to 18% by *A. niger* and SSF [245]. Conditions needed to enrich straws by *Tyromyces lacteus* and *Corilus versicolor* [246] are initial pH, 4-7; temperature, 25-30°C; and substrate humidity, 65-75%. Baraga et al. [247] enriched residues of Napier grass for animal feed using *Phanerochaete chrysosporium* through SSF. The total protein value of Napier grass increased from 10.2% to approximately 16.3 and 18.7% within 6 and 7 days for spore and mycelium inoculation, respectively.

Coffee pulp was subjected to SSF using *A. niger* [248], and maximum total amino acid (14.55%) content was obtained after 43 h, with a moisture content of 80%, pH of 3.5, temperature of 35°C, and aeration of 8 L/min/kg as optimum fermentation conditions. The fermented product, beside high total amino acids, had a lower cell wall constituent value than did the original pulp. A ration for growing chickens containing 10% of the fermented coffee pulp had a feed efficiency (2.14) similar to that of a standard ration (2.19) and was significantly better than that of a diet containing 10% of the original coffee pulp (2.53).

Improved digestibility of cellulose was obtained with a combination of mild heat and biological treatment of wheat straw [153]. All fungi screened converted hot water-treated components into biomass through SSF. *Corilus versicolor* decreased the cellulose content by 60%, whereas others used 35-40% of the cellulose. *Corilus versicolor* was the most efficient lignin degrader in solid state cultures and contained 11.2% protein. Steam-treated wheat straw at 70% (w/w) moisture was enriched through SSF with *T. reesei* (Riga, U.S.S.R.) in fermentation equipment of various design including some with mixing, some with stationary layers, and a mixed-layer 1.5 m^3 pilot plant scale fermenter. The best protein productivity was achieved in stationary layer fermenters, with a product containing 13% protein [12].

Mushrooms are the oldest SCP food and represent the first solid state fermentation. However, basic research on microbial technology has not been applied to any significant extent to mushroom cultivation. In fact, the only means for converting unmodified lignocellulosic material biologically is through production of various mushrooms, which are regarded as a great delicacy [9,249].

We can conclude that with development of SSF to enrich agricultural residues with microbial proteins and directly using the product for animal feed, there is hope that in the near future, farm-sized protein-generating units will be developed for converting straws, grass, hedge-trimmings, vegetable wastes, and the like into protein-rich materials for direct feeding. Perhaps two stages involving fungi and then yeast will be used.

VI. PROTEIN CONTENT AND NUTRITIONAL VALUE

The protein content of fungi is usually low when compared to yeast and bacteria. While this is true, fungi selected for biomass production do have a true protein content near that of yeasts (in the range of 33-45% on dry weight basis) and bacteria (60-65% protein, accompanied by high levels of RNA, which can be up to 15-25%) [250]. The distinguishing feature of fungal composition lies in the distribution of its nitrogen content. In determining the quality of protein, the most misleading expression used is "crude protein" content, meaning total nutrogen × 6.25. Fungi may have a high nitrogen content, without being rich in protein. Thus a better measure is α-amino nitrogen × 6.25. Anderson et al. [251] studied many fungi cultured by continuous fermentation in a 400-L fermenter, to determine total nitrogen (TN), which comprises α-amino nitrogen (AN) and nonprotein nitrogen (NPN) in their biomass. They found NPN accounts for a large portion of cell TN, thus the precaution in determining protein as TN × 6.25 [29] is justified. In practice, amino acids and small peptides account for 5-20% of the amino acid total, thus lowering the protein content. Litchfield [29] and Anderson et al. [251] independently reported a similar TN × 6.25 of 40-50% in *A. niger*, but the true protein was only 25.5%. Two major nonprotein sources in fungi are nucleic acids and chitin in the cell wall. Levels of RNA vary with growth rate, temperature, pH, etc. [251-253]. The distribution of TN in *F. gramineum* [251] grown in continuous culture at a growth rate of 0.1/h is given below:

1.	Free amino acid N	0.71%
2.	Protein N	6.34%
3.	Nucleotide N	0.15%
4.	RNA N	1.46%
5.	*n*-Acetylglucosamine N	1.00%
	Total N	9.66%
	TN by Kjeldahl's method	9.72%

However, the protein content of the mycelium of several species of mushrooms grown in submerged culture is generally less than that of the fruiting bodies. *Lentinus edodes* is an exception; its mycelium protein exceeds that of fruiting body protein content by more than 2.5 times [29]. Usually the protein value is calculated by multiplying the nitrogen content of the mycelium by the factor 6.25, which is based on the assumption that protein contains 16% nitrogen, as do cereal grains and feeds. However, there is a report [254] of 11.79% nitrogen in fruiting bodies of *A. compestris*. No one has reported a similar fraction of protein in mycelia of edible fungi [29]. Cultivation of fungi for protein biomass should never be done with nitrogen limitation; this results in an increase in lipid content of cells [23].

The nutritional quality of the SCP is generally determined by estimation of digestibility, biological value, and either net protein utilization (NPU) or protein efficiency ratio (PER) in animals and by nitrogen balance studies in humans [255]. The nutritional value of fungal protein is very satisfactory [256] and compares well with that of yeasts and bacteria.

The amino acid composition of a protein determines its biological value as a source of nitrogen for growth and maintenance of an organism. In general, various microorganisms have a well-balanced amino acid pattern but are low in content of the S-amino acids methionine and cystine [257, 258]. In conventional studies with experimental animals and human subjects fed at the deficient levels of protein required to obtain values comparable to those in animal studies, methionine was the limiting amino acid in SCPs as it is in legumes. When legume proteins are fed to humans at required levels, the deficiency of methionine is not detected, and addition of this amino acid does not improve nitrogen retention [259]. The same is true for SCPs [260], which means that digestibility and amount of amino acid nitrogen are the main limiting factors in use of SCP [258]. The amino acid analysis and pattern provides little information on the true nutritional usefulness of the SCP product [23]. Reported digestibility and biological value of *F. graminearum* supplemented with methionine is 78 and 73, respectively [261]. Protein digestibility is not greatly affected by the level of protein intake, but biological value and net protein utilization are high when the protein intake is adequate [260]. Fungi produce a wide range of undesirable metabolites, which cause concern as to their safety when used as food or feed. Mycotoxins, oxalates, and other undesirable compounds are produced by numerous fungi [22]. Therefore, toxicological testing of any SCP is imperative, and should be done with inbred laboratory rodents and dogs because of their consistency in response to diet when suitable procedures are used [43]. The health of workers involved with manufacture of SCP should be of prime importance. There may be adverse chemical reactions when humans are exposed to fine SCP particles. Clinical symptoms caused by SCP have been described [262] and apply to all types of SCP products. The simple physical control of fine SCP dust in the factory or treatment of material with edible oil creates a condition in which allergic reactions are obviated [263].

The nutritional value of SCP is not complete without nucelic acids (NA). The NA content of microorganisms is approximately 8-25 g/100 g of protein biomass, compared with liver, which contains about 4 g/100 g protein, and wheat flour with approximately 1 g/100 g protein [10]. Any rapidly growing cell contains relatively large amounts of nucleic acids. The medical importance of the NA content of SCP is that purines are metabolized to uric acid, which is eliminated as such by humans [264,265]. Thus RNA causes an increase in urinary uric acid of 100-150 mg/g RNA, which means that only a part of the purines in nucleic acid is absorbed and eventually excreted as uric acid via the kidneys. The risk of high nucleic acid intake is that since uric acid is sparingly soluble, an increase in plasma concentration may result in the precipitation of ureate in tissues and joints

TABLE 4. Permissible Level of Intake of Nucleic Acid from SCP for Different Age Groups and Sex

Age group (yr)	Sex	Body weight (kg)	Intake of nucleic acid (purine N × 9:g)
Adult	M	65	2.0
Adult	F	55	1.7
16–19	M	63	1.9
16–19	F	54	1.7
13–15	M	51	1.6
13–15	F	50	1.5
10–12	M	37	1.1
10–12	F	38	1.2
7–9	M.F.	28	0.9
4–6	M.F.	20	0.6
1–3	M.F.	13	0.4

Source: Ref. 266.

analogous to the condition in gout. Also, stones may be formed in the kidney and bladder. Sometimes during evolution, humans lost the enzyme uricase (urate oxidase), which oxidizes uric acid to allantoin. In contrast to humans, other mammals possess uricase and thus are able to convert uric acid into the soluble and easily excretable metabolite allantoin. Thus the purine content in untreated SCP, when used as fodder, does not cause a problem in animals. The advisory group of the UN system [20] suggested an additional 2 g NA/day in the usual mixed diet as an upper safe limit for healthy young adults. Normally the amount of dietary NA safely tolerated is based on body weight. Table 4 gives such calculations taken from the working group report [266].

VII. FUNGAL PROTEIN PROCESSES

A. Waterloo Process

The concept of the Waterloo process emerged at the University of Waterloo in Canada from an integrated scheme for bioconversion of agricultural and forestry wastes and animal manure into animal feed supplement [145]. A new thermophilic cellulolytic fungus, *Chaetomium cellulolyticum*, is used to synthesize protein from the pretreated cellulosic materials. The process can be operated either in a conventional aerated fermenter system or in a solid state system. In the SSF process, materials such as corn stover or paper mill sludge are used. The fungal product after fermentation contains up to 45% protein. There are plans to set up a commercial pilot plant based on the Waterloo process in British Columbia [267].

B. RHM Process

Rank, Hovis, and McDaugall (RHM) in association with Dupont, made use of *Fusarium graminearum* ATCC 20334 to produce protein from carbohydrate-containing wastes [257]. The fungus has a doubling time as low as 2.4 h. Their primary aim was to produce fungal protein for human consumption, hence they used all food grade components for the culture medium and operated the process under aseptic conditions. Recently, the U.K. government sanctioned its product called "Myco-protein" for sale as human food. This is the first fungal biomass so approved [268]. The fermented product containing 30% solids is used to fabricate a number of consumer products; the most important are meat analogues, which are difficult to distinguish from actual meat products [269].

C. Pekilo Process

A process based on *Paecilomyces variatii* grown on spent sulfite liquor has been developed in Finland and is called the Pekilo process [28]. At present two factories with 10,000-ton capacity per annum have been set up in Finland. Since spent sulfite waste liquor has long been used as a substrate for yeast production and is considered to be nontoxic, the Finnish government has recently allowed use of Pekilo protein in animal feeds [270]. The dried product, "Pekilo," contains 55-60% protein, of which 87% is digestible. The Pekilo product resembles animal protein more than vegetable protein. "Pekilo" is cream-colored, odorless, and its taste appeals to animals. The Pekilo protein has been produced in a pilot plant since 1971 [28].

D. Tate and Lyle Process

The Tate and Lyle process in England involves growing filamentous fungi, *Fusarium* sp. and *Aspergillus niger*, on carbohydrate-containing wastes. The objective of the Tate and Lyle process is to develop a simple, small-scale village-level technology which can be easily operated in developing countries. The process is SSF and is carried out nonaseptically in plastic vessels. Substrates used are such locally available agricultural wastes as carob papay wastes, water wastes from olives, palm oil, potatoes, dates, citrus and cassava, as well as molasses and sulfite waste liquor [271].

E. Heurtey Process

Heurtey isolated *Penicillium cylopium* and studied its growth characteristics [27]. The mold used lactose of milk whey as a carbon and energy source for SCP production. Existing processes are based on yeasts that can utilize lactose in whey. The SCP from *P. cylopium* has a good amino acid composition [272]. The optimum temperature and pH for efficient operation of the process are 28°C and 3.5, respectively.

F. Protein Enrichment Feeds

Starchy materials like banana, cassava, or potatoes are fermented with *Aspergillus niger* in France. The fermentation employs batch cultures, semi-solid media [273], at 38°C and pH 4.0. About 150-200 kg of product is obtained in 30 h. The process is most suitable at village or farm levels [274].

VIII. CONCLUSIONS

A survey of the literature indicates that production of single-cell protein (SCP) from nonconventional sources, namely, *n*-paraffin, methanol, methane, and ethanol, could not survive due to nonavailability and rising costs. Hence interest has shifted to the substrates that are derived from agriculture or food processing byproducts or wastes, which are abundantly available at lower costs, to produce SCP using higher fungi and molds. There is a long history of human consumption of higher fungi as SCP food, and today the fruiting bodies (mushrooms) of the basidiomycete *Agaricus bisporus* are cultivated on an industrial scale. However, there has been little success in converting lignocelluloses into fungal biomass, which indicates that there is a dire need for finding an organism that could convert lignocellulose efficiently and economically into a fungal biomass rich in essential amino acids. This may be possible by strain improvement through genetic engineering.

From the current research and development efforts on SCP production from fungi, one can draw the conclusion that fungal biomass production on a commercial scale is feasible. Indeed, some processes are being operated in the world on a pilot scale. However, introduction of SCP products depends on the economy and the market for the product. Thus, the future of large-scale SCP production appears to be limited to use as a protein supplement and functional protein ingredients, rather than as a primary source of protein in human diets, unless we realize that two great dangers facing the world today are a rapidly rising population and widespread protein manipulation. For animal feed, SCP production will be possible in those areas where low cost substrates such as carbohydrates are available in plenty throughout the year and conventional protein feed stuffs like soya and fish meal, etc., are in short supply.

REFERENCES

1. Scrimshaw, N. S. *Mammalian Protein Metabolism* (H. N. Munro and J. B. Allison, eds.), Academic Press, New York, 1964, pp. 568-592.
2. Malcolm, H. H. Diet and maintenance of mental health in the elderly. *Nutritional Rev.*, *46*:79-82, 1988.
3. Scrimshaw, N. S., and Behar, M. Protein malnutrition in young children. *Science*, *133*:2039-2047, 1961.
4. Kahlon, S. S., and Kalra, K. L. *Chaetomium globosum*. A non-toxic fungus: A potential source of protein (SCP). *Agric. Wastes*, *18*:207-213, 1986.
5. Kahlon, S. S., and Arora, M. Utilization of potato peels by fungi for protein production. *J. Fd. Sci. Technol.*, *23*:264-267, 1986.
6. Kahlon, S. S., and Das, S. K. Biological conversion of paddy straw into feed. *Biol. Wastes*, *22*:11-21, 1987.
7. Kharatyan, S. G. Microbes as food for humans. *Annu. Rev. Microbiol.*, *32*:301-327, 1978.
8. Das, S. K. Fermentation of rice straw with non-toxic edible fungi to increase its feed value and in vitro digestibility. M.Sc. thesis, Punjab Agricultural University, Ludhiana, India, 1983.

9. Gray, W. D. In *The Use of Fungi as Food and In Food Processing* (W. D. Gray, ed.), Butterworth and Co., New York, 1970, pp. 60-82.
10. Kihlberg, R. The microbes as a source of food. *Annu. Rev. Microbiol.*, *26*:427-466, 1972.
11. Litchfield, J. H. Single cell proteins. *Science*, *219*:740-746, 1983.
12. Laukevics, J. J., Apsite, A. F., Viesturs, U. E., and Tengerdy, R. P. Straw to fungal protein. *Biotechnol. Bioeng.*, *26*:1465-1474, 1984.
13. Martin, A. M., and White, M. D. Growth of the acid tolerant fungus *Scytalidium acidophilum* as a potential source of SCP. *J. Fd. Sci.*, *50*:197-200, 1985.
14. Hatakka, A. I., and Pirhonen, T. I. Cultivation of wood rotting fungi on agricultural lignocellulosic materials for the production of crude protein. *Agric. Wastes*, *12*:81-97, 1985.
15. Peitersen, N. Single cell protein from straw. *Proc. Symp. on Enzymatic Hydrolysis of Cellulose*, Aulanko, Finland, 1975, pp. 407-418.
16. Laskin, A. I. Single cell protein. *Ann. Rept. Fermt. Process* (D. Perlman, ed.), Academic Press, New York, 1977, pp. 151-175.
17. Bunker, H. J. Sources of single cell protein: Prospective and prospect. In *Single Cell Protein* (R. I. Mateles and S. R. Tannenbaum, eds.), MIT, Cambridge, MA, 1968, pp. 67-89.
18. Beerstecher, E. *Petroleum Microbiology*, Elsevier Press Inc., Houston, Texas, 1954.
19. Davis, J. B., and Updegraff, D. M. Microbiology in the petroleum industry. *Bacteriol. Rev.*, *18*:215-238, 1954.
20. Champagnat, A., Vernet, C., Laine, B., and Filosa, J. Biosynthesis of protein-vitamin concentrates from petroleum. *Nature*, *197*:13-14, 1963.
21. Stringer, D. A. Acceptance of single cell protein for animal feeds. In *Comprehensive Biotechnology. The Principles, Applications and Regulations of Biotechnology in Industry, Agriculture and Medicine* (M. Moo-Young, ed.), Pergamon, Oxford, 1985, pp. 685-693.
22. Cooper, P. G., Silver, R. S., and Boyle, J. P. Semi-commercial studies of a petro protein process based on *n*-paraffins. In *Single Cell Protein II* (S. R. Tannenbaum and D. I. C. Wang, eds.), MIT, Cambridge, MA, 1975, pp. 438-453.
23. Kanazawa, M. The production of yeast from *n*-paraffins. In *Single Cell Protein II* (S. R. Tannenbaum and D. I. C. Wang, MIT, Cambridge, MA, 1975, pp. 438-453.
24. Kotoh, K. In *Single Cell Protein* (P. Davis, ed.), Academic Press, London, 1974, pp. 223-232.
25. Cooney, C. L., and Levine, D. W. Microbial utilization of methanol. *Adv. Appl. Microbiol.*, *15*:337-351, 1972.
26. Cooney, C. L., and Levine, D. W. SCP production from methanol by yeast. In *Single Cell Protein II* (S. R. Tannenbaum and D. I. C. Wang, eds.), MIT, Cambridge, MA, 1975, pp. 402-423.
27. Cow, J. S., Littlehailes, J. D., Smith, S. R. L., and Walter, R. B. SCP production from methanol. In *Single Cell Protein II* (S. R. Tannenbaum and D. I. C. Wang, eds.), MIT, Cambridge, MA, 1975, pp. 370-401.

28. Maclaren, D. D. *Single Cell Protein. A Food for the Future.* First North American Chemical Congress, Mexico City, 1975.
29. Sista, V. R., and Strivastava, G. C. Ethanol—a renewable source of SCP. In *Proc. Recycling Residues of Agriculture and Industry Symp.* (M. S. Kalra, ed.), Punjab Agric. Univ., Ludhiana, India, 1980, pp. 115-123.
30. Pokrovsky, A. Some results of SCP. Medico-biological investigations. In *Single Cell Protein II* (S. R. Tannenbaum and D. I. C. Wang, eds.), MIT, Cambridge, MA, 1975, pp. 475-502.
31. Scrimshaw, N. S. Introduction. In *Single Cell Protein* (R. I. Mateles and S. R. Tannenbaum, eds.), MIT, Cambridge, MA, 1968, pp. 3-7.
32. DeZeeu, J. R. Genetic and environmental control of protein composition. In *Single Cell Protein* (R. I. Mateles and S. R. Tannenbaum, eds.), MIT, Cambridge, MA, 1968, pp. 181-191.
33. Bressani, R. The use of yeast in human foods. In *Single Cell Protein* (R. I. Mateles and S. R. Tannenbaum, eds.), MIT, Cambridge, MA, 1968, pp. 90-121.
34. Krishna, M. G. Socio-economic situations in developing countries and potential for the production of petro proteins. UNIDO, Vienna, 1973.
35. PAG. *Statement No. 4 on Single Cell Protein.* Protein Advisory Group of the U.N., New York, 1970.
36. Sinskey, A. J., and Tannenbaum, S. R. Removal of nucleic acid in SCP. In *Single Cell Protein II* (S. R. Tannenbaum and D. I. C. Wang, eds.), MIT, Cambridge, MA, 1975, pp. 158-178.
37. Litchfield, J. H. Production of single cell protein for use in food or feed. In *Microbial Technology* (H. J. Peppler and D. Perlman, eds.), Academic Press, New York, 1979, pp. 93-155.
38. Solomons, G. L. Production of biomass by filamentous fungi. In *Comprehensive Biotechnology, The Principles, Applications and Regulations of Biotechnology in Industry, Agriculture and Medicine* (M. Moo-Young, ed.), Pergamon, Oxford, 1985, pp. 483-505.
39. Volesky, B., and Zojic, J. E. Batch production of protein from ethane and ethane-methane mixtures. *Appl. Microbiol., 21*:614-622, 1971.
40. Tye, R., and Willets, A. Fungal growth on C_1 compounds: Quantitative aspects of growth of a methanol utilising strain of *T. ligorum* in batch culture. *Appl. Environ. Microbiol., 33*:758-761, 1977.
41. Rolz, C., and Humphrey, A. E. Microbial biomass from renewable: Review of alternatives. *Adv. Biochem. Eng., 21*:1-54, 1982.
42. Kim, J. H., and Lebeault, J. M. Protein production from whey using *Penicillium cyclopum*: Growth parameters and cellular composition. *Eur. J. Appl. Microbiol. Biotechnol., 13*:151-154, 1981.
43. Romantschuk, H. The Pekilo process: Protein from spent sulfite liquor. In *Single Cell Protein II* (S. R. Tannenbaum and D. I. C. Wang, eds.), MIT, Cambridge, MA, 1975, pp. 344-356.
44. Litchfield, J. H. The production of fungi. In *Single Cell Protein* (R. I. Mateles and S. R. Tannenbaum, eds.), MIT, Cambridge, MA, 1968, pp. 309-329.
45. Han, Y. W., and Callihan, C. D. Cellulose fermentation: Effect of substrate pretreatment on microbial growth. *Appl. Microbiol., 27*:159-165, 1974.

46. Andren, R. K., Erickson, R. J., and Medeiros, J. E. Cellulosic substrate for enzymatic saccharification. *Biotechnol. Bioeng. Symp.*, *6*: 177-203, 1976.
47. Millet, M. A., Baker, A. J., Salter, L. D., and Gaden, E. L. Physical and chemical pretreatment for enhancing cellulose saccharification. *Biotechnol. Bioeng. Symp.*, *6*:125-154, 1976.
48. Saddler, J. N., Mes-Hartree, M., Yu, E. K. C., and Brownell, H. H. Enzymatic hydrolysis of various pretreated lignocellulosic substrates and fermentation of liberated sugars to ethanol. *Biotechnol. Bioeng. Symp.*, *5*:225-238, 1984.
49. Katz, M., and Reese, E. T. Production of glucose by enzymatic hydrolysis of cellulose. *Appl. Microbiol.*, *16*:419-420, 1968.
50. Moo-Young, M., Chahal, D. S., and Viach, D. Single cell protein from various chemically pretreated wood substrates using *Chaetomium cellulolyticum*. *Biotechnol. Bioeng.*, *20*:107-118, 1978.
51. Kahlon, S. S., and Namrita, C. Production of ethanol by saccharification of saw dust. *J. Res.* (India), *23*:235-246, 1988.
52. Mandels, M., Hontz, L., and Nystrom, J. Enzymatic hydrolysis of waste cellulose. *Biotechnol. Bioeng.*, *16*:1471-1493, 1974.
53. Robertson, T. B. Principles of biochemistry. In *Single Cell Protein I* (R. I. Mateles and S. R. Tannenbaum, eds.), MIT, Cambridge, MA, 1968.
54. Pringsheium, H., and Lichtenstein, S. Versuche zur Erreicherung von Draftsroh mit Pilzeiweiss. *Cellulosechem.*, *1*:29-32, 1920.
55. Bunker, H. J. *Biochemistry of Industrial Microorganisms* (C. Rainbow and H. H. Rose, eds.), Academic, New York, 1963, pp. 34-67.
56. Bunker, H. J. *Global Impacts of Applied Microbiology* (M. P. Starr, ed.), John Wiley and Sons, New York, 1964, pp. 234-240.
57. Thatcher, F. S. Foods and feeds from fungi. *Annu. Rev. Microbiol.*, *8*:449-472, 1954.
58. Robinson, R. F., and Davidson, R. S. The large-scale growth of higher fungi. *Adv. Appl. Microbiol.*, *1*:261-278, 1959.
59. Gilbert, F. A., and Robinson, R. F. Food from fungi. *Econ. Bot.*, *11*:126-145, 1957.
60. Humfeld, H. The production of mushroom mycelium *Agaricus compestris* in submerged culture. *Science*, *107*:373-376, 1948.
61. Block, S. S., Stearns, T. W., Stephens, R. L., and McCandless, R. F. J. Mushroom mycelium experiments with submerged culture. *J. Agric. Fd. Chem.*, *1*:890-893, 1953.
62. Reuesser, F., Spencer, J. F. T., and Salians, H. R. Protein and fat content of some mushrooms grown in submerged culture. *Appl. Microbiol.*, *6*:1-4, 1958.
63. Prescott, S. C., and Dunn, C. G. *Industrial Microbiology*, 2nd ed., McGraw Hill, New York, 1974.
64. Cirillo, V. P., Crestwood, W. A., Hardwick, O., and Seeley, R. D. Fermentation process for producing edible mushroom mycelium. U.S. Patent 2928210 (1960).
65. Falanghe, H. Production of mushroom mycelium as a protein and fat source in submerged culture in medium of Vinasse. *Appl. Microbiol.*, *10*:572-576, 1962.

66. Falanghe, H., Smith, A. K., and Rackis, J. J. Production of fungal mycelial protein in submerged culture of soybean whey. *Appl. Microbiol.*, *12*:330-334, 1964.
67. Jennison, M. W., Richberg, C. G., and Krikszens, A. E. Physiology of wood rotting basidiomycetes II. Nutritive composition of mycelium grown in submerged culture. *Appl. Microbiol.*, *5*:87-95, 1957.
68. Szuecs, J. Mushroom culture. U.S. Patent 2761246 (1956).
69. Szuecs, J. Methods of growing mushroom mycelium and the resulting products. U.S. Patent 2850841 (1958).
70. Litchfield, J. H., and Overbeck, R. C. Submerged culture growth of *Morchella* species in food processing waste substrates. In *Proc. 1st Intern. Cong. Fd. Sci. Technol.*, London, Gordon and Breech B-6 Part II, New York, 1963.
71. Anderson, R. F., and Jackson, R. W. Essential amino acids in microbial proteins. *Appl. Microbiol.*, *6*:369-373, 1958.
72. Martin, A. M. Submerged growth of *Morchella esculenta* in peat hydrolysates. *Biotechnol. Lett.*, *4*:13-18, 1982.
73. Martin, A. M. Submerged production of *Agaricus compestris* mycelium in peat extracts. *J. Fd. Sci.*, *48*:206-210, 1983.
74. Martin, A. M. Submerged production of edible mushroom mycelium in peat extracts. *Can. Inst. Fd. Sci. Technol.*, *16*:215-219, 1983.
75. Labaneiah, M. E., Abou-Donia, S. A., Esmat, S. A., El-Zalaki, E. M., and Mohamed, M. S. In *Annu. Report on Fermentation Process*, Vol. 7 (G. S. Tsao, ed.), Academic Press, New York, 1984, pp. 213-356.
76. Labaneiah, M. E., Abou-Donia, S. A., Mohamed, M. S., and El-Zalaki, E. M. In *Annu. Report on Fermentation Process*, Vol. 7 (G. S. Tsao, ed.), Academic Press, New York, 1984, pp. 213-356.
77. Labaneiah, M. E., Abou-Donia, S. A., Mohamed, S. A., and El-Zalaki, E. M. Utilization of citrus wastes for production of fungal protein. *J. Fd. Technol.*, *14*:95-100, 1979.
78. Duvnjak, Z., Erick, M., and Tamburasev, G. Growth of higher fungi *Agaricus compestris* and *Morchella hortensis* in submerged culture using whey as a growth medium. *Mljekarstvo.*, *28*:38-42, 1978.
79. Dizbay, M., and Karaboz, I. Submerged culture and ascocarp production of *Morchella* spp. *Doga Biyol Serisi.*, *10*:326-330, 1986.
80. Kurtzman, R. H., Jr. Production of mushroom fruiting bodies on the surface of submerged cultures. *Mycologia*, *70*:179-184, 1978.
81. Jauhri, K. S., Kumari, M. L., and Sen, A. *Zbl. Bakt. II*, *133*:588-591, 1978.
82. Vecher, A. S., Solomko, E. F., Shachov, E. N., Dudka, I. A., Bukhalo, A. S., Paromchik, I. I., and Pchelintseva, R. H. Potato juice concentrate as a substrate for the cultivation of higher mushroom mycelium. *Dokl. Akad. Nauk. BSSR*, *23*:855-858, 1979.
83. Ghosh, A. K., and Sengupta, S. Influence of some growth factors on the production of mushroom mycelium in submerged culture. *J. Fd. Sci. Technol.*, *19*:57-60, 1982.
84. Skinner, J. T., Peterson, W. H., and Steenbock, H. Nahrwert von Schimmelpilzmycel. *Biochem. Zeit.*, *267*:169-172, 1933.

85. Woolley, D. W., Berger, J., Peterson, W. H., and Steenbock, H. Toxicity of *Aspergillus sydowi* and its correction. *J. Nutr., 16*: 465-469, 1938.
86. Chastukhin, V. Ya. Mass cultures of microscopic fungi. Nat. Reservations Headquarters Press, Moscow, 1948.
87. Gray, W. D. Microbial protein for the space age. *Dev. Ind. Microbiol., 3*:63-67, 1962.
88. Gray, W. D. Fungi as nutrient source in biologistics for space systems. *Symp. 6570th Aerospace Medical Research Laboratories*, Wright Petterson AFB, Ohio AMRL-TDR-62-116, 1962, p. 356.
89. Gray, W. D. Protein and world population increase. *Proc. 13th Ann. Meet. Agric. Res. Inst.*, 1964, p. 97.
90. Gray, W. D. Process for production of fungal protein. U.S. Patent 3151038 (1964).
91. Gray, W. D. Fungi as a potential source of edible protein. *Research and Dev. Associates Activities Report, 17*:1-18, 1965.
92. Gray, W. D. Fungal protein for food and feed I. Introduction. *Econ. Bot., 20*:89-93, 1966.
93. Bhattacharjee, J. K. Microorganisms as potential source of food. *Adv. Appl. Microbiol., 13*:139-161, 1970.
94. Gray, W. D., and Abou El-Seoud, M. Fungal protein for food and feed II. Whole sweet potato as a substrate. *Econ. Bot., 20*:119-126, 1966.
95. Gray, W. D., and Abou El-Seoud, M. Fungal protein for food and feed III. Monioc as a potential crude raw material for tropical areas. *Econ. Bot., 20*:251-255, 1966.
96. Gray, W. D., and Abou El-Seoud, M. Fungal protein for food and feed IV. Whole sugarbeets or beet pulp as a substrate. *Econ. Bot., 20*:372-376, 1966.
97. Gray, W. D., and Karve, M. D. Fungal protein for food and feed V. Rice as a source of carbohydrate for the production of fungal protein. *Econ. Bot., 21*:110-114, 1967.
98. Gray, W. D., and Paugh, R. Fungal protein for food and feed VI. Direct use of cane juice. *Econ. Bot., 21*:273-279, 1967.
99. Chahal, D. S., and Gray, W. D. Growth of cellulolytic fungi on wood pulp I. Screening of cellulolytic fungi for their growth on wood pulp. *Indian Phytopath., 22*:80-92, 1969.
100. Chahal, D. S., Gray, W. D., and Munshi, G. D. Amino acid composition of cellulolytic fungi grown on wood pulp. *Indian J. Microbiol., 10*:23-26, 1970.
101. Han, Y. W., Dunlap, C. E., and Callihan, C. D. Single cell protein from cellulosic wastes. *Fd. Technol., 25*:32-35, 1971.
102. Rogers, D. J., Coleman, E., Spino, D. F., and Purcell, T. C. Production of fungal protein from cellulose and waste cellulosic. *Environ. Sci. Technol., 6*:715-718, 1972.
103. Peitersen, N. Production of cellulase and protein from barley straw by *Trichoderma viride*. *Biotechnol. Bioeng., 17*:361-374, 1975.
104. Romanelli, R. A., Houston, C. W., and Barnett, S. M. Studies on thermophillic cellulolytic fungi. *Appl. Microbiol., 30*:276-281, 1975.

105. Eriksson, K. E., and Larsson, K. Fermentation of waste mechanical fibres from a newsprint mill by the rot fungus *Sporotrichum pulverulentum*. *Biotechnol. Bioeng.*, *17*:327-348, 1975.
106. Miller, T. F., and Srinivasan, V. R. Production of fungal protein from *Aspergillus terreus*. Presented at ACS Div. Microbiology and Biochemical Technology Meeting, Washington, DC, 1979, pp. 9-14.
107. Updegraff, D. M. Utilization of cellulose from waste paper by *Myrothecium verrucaria*. *Biotechnol. Bioeng.*, *13*:77-97, 1971.
108. Stakheev, I. V. Cultivation of protein producing yeasts and molds on potato processing waste (cited from Potato Abstract 1980), *5*, 1978.
109. Stakheev, I. V., Beker, V. F., Babitskaya, V. G., Krauze, R. Yu, Bertulite, O. V., Kolomiets, E. I., and Pitran, B. S. A new fungal protein in feeds for chickens. In *V asyvanie Obmen Veshchestv u Zhivotnykh*, Riga, Lativian, SSR, 1980, pp. 138-145.
110. Gupta, R. P., Singh, A., Kalra, M. S., Sharma, V. K., and Gupta, S. K. Single cell protein production from sugarbeet pulp. *Indian J. Nutr. Diet.*, *14*:302-307, 1977.
111. Rosenberg, H., Obrist, J., and Stobs, S. J. Production of fungal protein from oat hulls. *Econ. Bot.*, *32*:413-417, 1978.
112. Chahal, D. S., Singla, M., and Gupta, R. P. Comparison of lipid production by unicellular and filamentous fungi. *Proc. Indian Natl. Sci. Acad.*, *45*:18-26, 1979.
113. Bhatia, I. S., and Arneja, J. S. Lipid metabolism in *Fusarium oxysporum*. *J. Sci. Fd. Agric.*, *29*:619-626, 1978.
114. Ghai, S. K., Kahlon, S. S., and Chahal, D. S. Single cell protein from canning industry waste. Sag waste as substrate for thermotolerant fungi. *Indian J. Expt. Biol.*, *17*:789-793, 1979.
115. Abraham, M. J., and Srinivasan, R. A. Utilization of whey for production of microbial protein and lipid. *J. Fd. Sci. Technol.*, *16*:11-15, 1979.
116. Ek, M., and Eriksson, K. E. Utilization of white rot fungus *S. pulverulentum* for water purification and protein production on mixed lignocellulosic waste water. *Biotechnol. Bioeng.*, *22*:2273-2284, 1980.
117. Gibriel, A. Y., Mahmoud, R. M., Goma, M., and Abou-Zied, M. Production of single cell protein from cereal by-products. *Agri. Wastes*, *3*:229-240, 1981.
118. Garg, S. K., and Neelakantan, S. Effect of cultural factors on cellulase activity and protein production by *Aspergillus terreus*. *Biotechnol. Bioeng.*, *23*:1653-1659, 1981.
119. Miller, T. F., and Srinivasan, V. R. Production of single cell protein from cellulose by *Asperbillus terreus*. *Biotechnol. Bioeng.*, *25*:1509-1519, 1983.
120. Geethadevi, B. R., Sitaram, N., Kunhi, A. A. M., and Rao, T. N. R. Screening of fungi for single cell protein and cellulase production. *Indian J. Microbiol.*, *18*:85-89, 1979.
121. Chahal, D. S., Moo-Young, M., and Dhillon, G. S. Bioconversion of wheat straw components into single cell protein. *Can. J. Microbiol.*, *25*:793-797, 1979.

122. Sidhu, M. S., and Sandhu, D. K. Single cell protein by *Trichoderma longibrachiatum* on treated sugarcane bagasse. *Biotechnol. Bioeng.*, *22*:689-692, 1980.
123. Daugulis, A. J., and Bone, D. H. Production of microbial protein from tree bark by *Phanerochaete chrysosporium*. *Biotechnol. Bioeng.*, *20*:1639-1649, 1978.
124. Kirk, T. K., Schultz, E., Connors, W. J., Lorenz, L. F., and Zeikus, J. G. Influence of cultural parameters on lignin metabolism by *Phanerochaete chrysosporium*. *Arch. Microbiol.*, *117*:277-285, 1978.
125. Keyser, P., Kirk, T. K., and Zeikus, J. G. Lignolytic enzyme system of *P. chrysosporium* synthesized in the absence of lignin in response to nitrogen starvation. *J. Bacteriol.*, *135*:790-797, 1978.
126. Buswell, J. A., Ander, P., and Eriksson, K. E. Ligninolytic activity and levels of ammonia assimilating enzymes in *S. pulverulentum*. *Arch. Microbiol.*, *133*:165-171, 1982.
127. Hatakka, A. I., Buswell, J. A., Pirhonen, T. I., and Uugi-Rauver, A. K. Degradation of C-14 labelled lignins by white rot fungi. In *Recent Advances in Lignin Biodegradation* (T. Higuchi, H. M. Chang, and T. K. Kirk, eds.), UNI Publishers Co. Ltd., Tokyo, 1983, pp. 176-187.
128. Zadrazil, F., and Brunnert, H. The influence of ammonia nitrate supplementation in degradation and in vitro digestibility of straw colonized by higher fungi. *Eur. J. Appl. Microbiol. Biotechnol.*, *9*:37-44, 1980.
129. Freer, S. N., and Detroy, R. W. Biological delignification of C-14 labelled lignocellulosics by basidiomycetes fungi: Degradation and stabilization of lignin and cellulose components. *Mycologia, 74*:943-952, 1982.
130. Leatham, G. F., and Kirk, T. K. Regulation of ligninolytic activity of nutrient nitrogen in white rot basidiomycetes. *FEMS. Microbial Lett.*, *16*:65-67, 1983.
131. Levoren-Munoz, E., Bone, D. H., and Daugulis, A. J. Solid state fermentation and fractionation of oat straw by basidiomycetes. *Eur. J. Appl. Microbiol. Biotechnol.*, *18*:120-123, 1983.
132. Abdel-Monem, H. El. R., Mohamed, M. A., and Houssam, A. El. S. Microbial formation of cellulases and proteins from some cellulosic residues. *Agri. Wastes*, *11*:105-113, 1984.
133. Pamment, N., Robinson, C. W., and Moo-Young, M. Pulp and paper mill solid waste as substrate for SCP production. *Biotechnol. Bioeng.*, *21*:561-573, 1979.
134. John, C. R., and Nakas, J. P. Production of mycelial protein and hydrolytic enzymes from paper mill sludges by cellulolytic fungi. *J. Industrial Microbiol.*, *2*:9-13, 1987.
135. Sekhon, A. Studies on utilization of wheat straw for production of cellulase and protein by thermophilic fungi. M.Sc. thesis, Punjab Agric. Univ., Ludhiana, India, 1975.
136. Chahal, D. S. Mushroom growing for beginners, Punjab Agri. Univ. Publication, Ludhiana, India, 1974.

137. Chahal, D. S., Sekhon, A., and Dhaliwal, B. S. In *Proc. 3rd International Biodegradation Symp.* (J. M. Sharpley and A. M. Kaplan, eds.), Appl. Science Publishers, London, 1975, pp. 665-671.
138. Chahal, D. S., and Hawksworth, D. L. *Chaetomium cellulolyticum*, a new thermotolerant and cellulolytic chaetomium. *Mycologia, 68*:600-610, 1976.
139. Moo-Young, M., Chahal, D. S., Swan, J. E., and Robinson, C. W. Single cell protein production by *C. cellulolyticum*, a new thermotolerant cellulolytic fungus. *Biotechnol. Bioeng., 19*:527-538, 1977.
140. Chahal, D. S., and Wang, D. I. C. *Chaetomium cellulolyticum*, growth and behaviour on cellulose and protein production. *Mycologia, 70*:160-170, 1978.
141. Chahal, D. S., Swan, J. E., and Moo-Young, M. Cellulase and protein production by *Chaetomium cellulolyticum* on wheat straw. *Dev. Ind. Microbiol., 18*:443-442, 1977.
142. Chahal, D. S., Vlach, D., Stickeny, B., and Moo-Young, M. Swine manure as substrate for SCP. Paper presented at *IV International Fermentation Symp. London*, Ontario, Canada, 1980.
143. Chahal, D. S., McGuire, S., Pikor, H., and Noble C. Bioconversion of lignocellulosics into food and feed. In *Proc. 2nd World Cong. Chem. Eng.*, Montreal, Canada, 1981, pp. 245-248.
144. Chahal, D. S. Bioconversion of hemicellulose into useful products in an integrated process for food/feed and fuel (ethanol) production from biomass. *Biotechnol. Bioeng. Symp., 14*:421-432, 1984.
145. Moo-Young, M., Daugulis, A. J., Chahal, D. S., and Macdonald, D. C. The Waterloo process for SCP production from waste biomass. *Process Biochem., 14*:38-40, 1979.
146. Rosen, W., and Schuegerl, K. Pretreatment and conversion of straw into protein in a state culture. In *Production and Feeding of SCP* (M. P. Ferranti and A. Fiecater, eds.), Applied Science, Barking, Essex, 1983, pp. 87-89.
147. Valdivia, A., Ramos, A., Torre, M. de la, and Casascompallo, C. Solid state fermentation of cassava with *Rhizopus oligosporus* NRRL 2710. In *Production and Feeding of SCP* (M. P. Ferranti and A. Fiecater, eds.), Applied Science, Barking, Essex, 1983, pp. 104-111.
148. Fink, H. I., Schlie, I., and Ruge, U. Production of biomass from filamentous fungi. In *Comprehensive Biotechnology, the Principles, Application and Regulations of Biotechnology in Industry, Agriculture and Medicine* (M. Moo-Young, ed.), Pergamon, Oxford, 1985, pp. 483-505.
149. Doctor, V. M., and Kerve, L. Penicillium mycelium waste as protein supplement in animals. *Appl. Microbiol., 16*:1723-1726, 1968.
150. Church, B. D., Erickson, E. E., and Widmer, C. M. Fungal digestion of food processing wastes. *Fd. Technol., 27*:36-42, 1973.
151. Anderson, J. G., and Blain, J. A. Novel developments in microbial film reactors. In *Fungal Biotechnology* (J. E. Smith, D. R. Berry, and B. Kristiansen, eds.), Academic Press, New York, 1980, pp. 125-152.
152. Aguirre, F., Maldonado, O., Rolz, C., Menchu', J. F., Espinosa, R., and Cabrera, S. de. Protein from waste: Growing fungi on coffee waste. *Chem. Technol., 6*:636-642, 1976.

153. Milstein, O., Vered, Y., Sharma, A., Gressel, J., and Flowers, H. M. Heat and microbial treatments for nutritional upgrading of wheat straw. *Biotechnol. Bioeng.*, *28*:381-386, 1986.
154. Viesturs, U. E., Strikauska, S. V., Leite, M. P., and Berzins, A. J. Combined submerged and solid substrate fermentation for the bioconversion of lignocellulose. *Biotechnol. Bioeng.*, *30*:282-288, 1987.
155. Ivarson, K. C., and Morita, H. Single cell protein production by the acid tolerant fungus, *Scytalidium acidophilum* from acid hydrolysate of waste paper. *Appl. Environ. Microbiol.*, *43*:643-652, 1982.
156. Martin, A. M., and White, M. D. Growth of the acid tolerant fungus, *Scytalidium acidophilum* as a potential source of single cell protein. *J. Fd. Sci.*, *50*:197-200, 1985.
157. Kuhn, A. L., and Pretorius, W. A. Use of cross flow microscreen technique for SCP production from dilute substrates. *Appl. Microbiol. Biotechnol.*, *27*:593-600, 1988.
158. Dhillon, G. S., Gupta, R. P., and Gupta, S. K. Amino acid composition of *Chaetomium globosum* grown on sucrose and wheat straw. *Res. Dev. Report*, *4*:93-95, 1987.
159. Garcha, J. S., Kaur, C., Singh, A., and Raheja, R. K. Production of fungal proteins from molasses—a comparative study using three carbon sources. *Indian J. Anim. Res.*, *7*:19-22, 1973.
160. Sethi, R. P., Singh, A., Kalra, M. S., and Chahal, D. S. Nutritional evaluation of *Penicillium crustosum* protein. *Indian J. Nutr. Dietet.*, *13*:332-335, 1976.
161. Singh, A., Sethi, R. P., Kalra, M. S., and Chahal, D. S. Single cell protein production from wheat straw and groundnut hulls. *Indian J. Microbiol.*, *16*:37-39, 1976.
162. Singh, A., and Kalra, M. S. Single cell protein production from corn cobs. *J. Fd. Sci. Technol.*, *15*:249-252, 1978.
163. Singh, A., Dhillon, G. S., Gupta, R. P., and Kalra, M. S. Single cell protein production from spent grain waste. *Indian J. Expt. Biol.*, *16*:1317-1318, 1978.
164. Singh, A., Dhillon, G. S., and Kalra, M. S. Supplementary value of single cell protein from *Myrothecium verrucaria* to wheat protein. *Proc. Indian Acad. Sci.*, *90*:275-279, 1981.
165. Ghai, S. K., Singh, A., and Kalra, M. S. SCP production from rice stover. Solid vs. submerged fermentation. *Indian J. Expt. Biol.*, *19*:61-64, 1981.
166. Dhillon, G. S., Singh, A., and Kalra, M. S. Single cell protein production from wheat straw and its fractions by *Myrothecium verrucaria*. *J. Fd. Sci. Technol.*, *19*:74-78, 1982.
167. Chai, S. K., Kahlon, S. S., and Sood, S. M. SCP from canning industry wastes. 1. Citrus fruit pulp as substrate. *J. Res.* (India), *16*:64-68, 1979.
168. Kahlon, S. S., and Parveen, N. Protein enrichment of wheat straw with non-toxic fungus *Pleurotus ostreatus*. *J. Res.* (India), *20*:327-331, 1983.
169. Parveen, N., Kahlon, S. S., Sethi, R. P., and Chopra, A. K. Solid substrate fermentation of wheat straw into animal feed. *Indian J. Anim. Sci.*, *53*:1191-1194, 1983.

170. Kahlon, S. S., and Das, K. Improving the nutritive value of rice straw through fungal fermentation. *J. Res.* (India), *22*:527-533, 1985.
171. Kahlon, S. S., and Das, K. Paddy straw feed through solid substrate fermentation. *J. Res.* (India), *23*:466-472, 1986.
172. Kahlon, S. S. SCP production by *Chaetomium cellulolyticum* on treated waste cellulose. *J. Res.* (India), *23*:330-335, 1986.
173. Kahlon, S. S., and Arora, M. Protein enrichment of potato waste by solid substrate fermentation. *Indian J. Poultry Sci.*, *20*:46-53, 1985.
174. Feist, W. C., Baker, A. J., and Tarkow, H. Alkali requirement for improving digestibility of hard woods by rumen micro-organisms. *J. Anim. Sci.*, *30*:832-835, 1970.
175. Baker, A. J. Effect of lignin on the in vitro digestibility of wood pulp. *J. Anim. Sci.*, *36*:768-771, 1973.
176. Bender, F., Heaney, D. P., and Bowder, A. Potential of steamed wood as a feed for ruminant. *Forest Prod. J.*, *20*:36-41, 1970.
177. Heaney, D. P., and Bender, F. The feeding value of steamed aspen for sheep. *Forest Prod. J.*, *20*:99-102, 1970.
178. Campbell, C. M., Waynam, O., Stanley, R. W., Kamstra, L. D., Albrich, S. E., Hoa, E. B., Nakajama, T., Kohler, G. D., Walker, H. G., and Graham, R. Effect of pressure treatment of sugarcane bagasse upon nutrient utilization. *Proc. West. Soc. Anim. Soc. Animal Sci.*, *24*:173-184, 1973.
179. Harris, S. E., Hajny, G. H., Mannan, M., and Rogers, S. C. Fermentation of Douglas fir for hydrolysate by *S. cerevisiae*. *Ind. Eng. Chem.*, *38*:896-904, 1946.
180. Leonard, R. H., and Hajny, G. H. Fermentation of wood sugar to ethyl alcohol. *Ind. Eng. Chem.*, *37*:390-395, 1945.
181. Torkow, H., and Feist, W. D. A mechanism for improving the digestibility of lignocellulosic materials with dilute alkali and liquid ammonia. *Adv. Chem. Series*, *95*:197-218, 1969.
182. Han, Y. W., Yu, P. L., and Smith, S. K. Alkali treatment and fermentation of straw for animal feed. *Biotechnol. Bioeng.*, *20*:1015-1026, 1978.
183. Green, J. W. Wood cellulose. *Methods in Carbohydrate Chemistry*, *3*:9-20, 1963.
184. Goering, H. K., Smith, L. W., Van Soest, P. J., and Gordon, C. H. Digestibility of roughages materials ensiled with sodium chloride. *J. Dairy Sci.*, *56*:233-240, 1973.
185. Han, Y. W., and Anderson, A. W. Semi-solid fermentation of rye grass. *Appl. Microbiol.*, *30*:930-934, 1975.
186. Han, Y. W. Microbial fermentation of rice straw. Nutritive composition and in vitro digestibility of the fermented product. *Appl. Microbiol.*, *29*:510-514, 1974.
187. Zadrazil, F. The conversion of straw into feed by basidiomycetes. *Eur. J. Appl. Microbiol.*, *4*:273-278, 1977.
188. Zadrazil, F. Cultivation of *Pleurotus*. In *The Biology and Cultivation of Edible Mushrooms* (S. T. Chang and W. A. Hayes, eds.), Academic Press, New York, 1978, p. 521.

189. Aidoo, A. E., Hendry, R., and Wood, B. J. B. Solid substrate fermentation. *Adv. Appl. Microbiol.*, *28*:201-237, 1982.
190. Moo-Young, M., Moreira, A. R., and Tengerdy, R. P. In *Fungal Technology, Filamentous Fungi* (J. E. Smith, D. R. Berry, and B. Kristiansen, eds.), Arnold, London, 1983, pp. 117-142.
191. Han, Y. W., Cheeke, P. R., Anderson, A. W., and Lekprayoon, C. Growth of *Aureobasidium pullulans* on straw hydrolysate. *Appl. Environ. Microbiol.*, *32*:799-802, 1976.
192. Hesseltine, C. W. Solid state fermentation. *Biotechnol. Bioeng.*, *14*:517-532, 1972.
193. Shotwell, O. L., Hesseltine, C. W., Stubblefield, R. D., and Sorenson, W. G. Production of aflatoxin on rice. *Appl. Microbiol.*, *14*: 425-428, 1966.
194. Silman, R. W. Enzyme formation during solid-substrate fermentation in rotating vessels. *Biotechnol. Bioeng.*, *22*:411-420, 1980.
195. Hesseltine, C. W. A millenium of fungi, food and fermentation. *Mycologia*, *57*:149-197, 1965.
196. Stanton, W. R., and Wallbridge, A. W. Fermented food processes. *Proc. Biochem.*, April, *45*:1-9, 1969.
197. Vankeen, A. G., and Steinkraus, K. H. Nutritive value and wholesomeness of fermented foods. *J. Agric. Fd. Chem.*, *18*:576-592, 1970.
198. Brook, E. J., Stanton, W. R., and Wallbridge, A. W. Fermentation methods for protein enrichment of cassava. *Biotechnol. Bioeng.*, *11*: 1271-1284, 1969.
199. Haider, K., and Trojanowski, J. Decomposition of specifically labelled phenols and dehydropolymer of coniferyl alcohol as model for lignin degradation by soft and white rot fungi. *Arch. Microbiol.*, *105*:33-41, 1975.
200. Cruz, R. A., Malasia, R. E., Cruz, T. J., and Pusag, C. C. Biological treatment of bagasse enhances its value as a soil conditioner or feed. *Sugar News*, *43*:15-18, 1967.
201. Barnes, T. A., Eggins, H. O. W., and Smith, E. L. Preliminary stages in the development of a process for the microbial upgrading of waste paper. *Int. Biodet. Bulletin*, *8*:112-117, 1972.
202. Zadrazil, F. The ecology and industrial production of *Pleurotus ostreatus*, *P. florida*, *P. cornucopial* and *P. eryngii*. *Mush. Sci.*, *9*: 621-652, 1974.
203. Zadrazil, F., and Brunnert, H. Investigation of physical parameters important for solid substrate fermentation of straw by white rot fungi. *Eur. J. Appl. Microbiol. Biotechnol.*, *11*:183-188, 1981.
204. Zadrazil, F., Grinbergs, J., and Gonzales, A. 'Palo-podrido' decomposed wood used as feed. *Eur. J. Appl. Microbiol. Biotechnol.*, *15*: 167-171, 1982.
205. Zadrazil, F., and Brunnert, H. Solid state fermentation of lignocellulose containing plant residues with *Sporotrichum pulverulentum* Nov. and *Diochomitus squalens*. *Eur. J. Appl. Microbiol. Biotechnol.*, *161*: 13-18, 1982.
206. Rosenberg, S. L. Cellulose and lignocellulose degradation by thermophillic and thermotolerant fungi. *Mycologia*, *70*:1-13, 1978.

207. Shetty, K. S., and Krishnamurthy, V. Possibility of protein enrichment of paddy straw by mushroom, *Pleurotus sajor-caju*. In *Recycling Residues of Agriculture and Industry* (M. S. Kalra, ed.), Punjab Agric. Univ., Ludhiana, India, 1980, pp. 363-367.
208. Langer, P. N., Sehgal, J. P., and Garcha, H. S. Chemical changes in wheat and paddy straws after fungal cultivation. *Indian J. Anim. Sci.*, *50*:942-946, 1979.
209. Langer, P. N., Sehgal, J. P., Rana, V. K., and Garcha, H. S. Utilization of fungal treated spent straw in the ruminant diets. In *Recycling of Residues of Agriculture and Industry* (M. S. Kalra, ed.), Punjab Agric. Univ., Ludhiana, India, 1980, pp. 173-180.
210. Bakshi, M. P. S., Gupta, V. K., and Langer, P. N. Acceptability and nutritive evaluation of *Pleurotus* harvested spent straw in buffaloes. *Agric. Wastes*, *13*:55-57, 1985.
211. Gupta, V. K., and Langer, P. N. *Pleurotus florida* for upgrading the nutritive value of wheat straw. *Biol. Wastes*, *23*:57-64, 1988.
212. Kokke, R. Improvement of carob pods as feed by solid substrate fermentation. *J. Appl. Bacteriol.*, *43*:303-307, 1977.
213. Knapp, J. S., and Howell, J. A. Solid substrate fermentation. *Tropical Enz. Ferment. Biotechnol.*, *4*:85-143, 1980.
214. Gulati, S. L. Improvement of production of fungal protein by solid substrate fermentation. *Zentralbl. Bakteriol. Parasitenkd Infektionsker Umweltschutzes*, *135*:413-417, 1980.
215. Mukhopadhyaya, A. K., and Sikyta, B. Solid state fermentation of barley straw. *Bakteriol. Parasitenkd. Infektionsker. Hyg.*, *135*: 682-684, 1980.
216. Matteau, P. P., and Bone, D. H. Solid state fermentation of maple wood by *Polyporus anceps*. *Biotechnol. Lett.*, *2*:127-132, 1980.
217. Pamment, N., Moo-Young, M., Hsieh, F. H., and Robinson, C. W. Growth of *C. cellulolyticum* on alkali pretreated hard wood sawdust solids and pretreatment liquor. *Appl. Environ. Microbiol.*, *36*:284-290, 1978.
218. Lindensfelser, L. A., Detroy, R. W., Ramstack, J. M., and Waiden, K. A. Biological modification of the lignin and cellulose components of wheat straw by *Pleurotus ostreatus*. *Dev. Ind. Microbiol.*, *20*: 541-551, 1979.
219. Sethi, R. P., and Sood, S. M. Utilization of banana peels by fungi for protein and CMCase production. *Indian J. Mycol. Pl. Path.*, *7*: 39-42, 1977.
220. Balasubramanya, R. H., and Bhatawdekar, S. P. Utilization of agricultural waste as feed for ruminants. *Indian J. Microbiol.*, *21*:14-16, 1981.
221. Hokkao Kakohi, K. K. Composting system for potato starch wastes. *Japan Kokai. Tokyo Koho.*, *8169*:292-299, 1981.
222. Chahal, D. S., Vlach, D., and Moo-Young, M. In *Adv. in Biotechnology* (M. Moo-Young and C. W. Robinson, eds.), Pergamon Press, Toronto, 1981, pp. 327-333.
223. Ibrahim, M. N. M., and Pearce, G. R. Effect of gamma irradiation on the composition and in vitro digestibility of crop by-products. *Agric. Wastes*, *2*:199-205, 1980.

224. Reade, A. E., and McQueen, R. E. Investigation of white rot fungi for the conversion of poplar into a potential feedstuff for ruminants. *Can. J. Microbiol., 29*:457-463, 1983.
225. Virk, R. S., Sethi, R. P., and Garcha, H. S. Note on the conversion of poultry droppings by *P. ostreatus* into feed. *Indian J. Anim. Sci., 50*:293-295, 1980.
226. Sethi, R. P., Garcha, H. S., and Virk, R. S. Effect of edible fungus treated poultry droppings on broiler chicks. *Indian J. Poultry Sci., 16*:138-141, 1981.
227. MacLennan, M., and Lawson, M. Protein containing food materials. U.S. Patent 4265915 (1981).
228. Stakhyeyew, I. V., Kapicw, A. M., and Babitskaya, V. R. Cultivation of mycelium of wood decomposing basidiomycetes on the waste of potato processing. *Navuk kaz Syer Biyal Navuk., 0*:55-59, 1984.
229. Abouzied, M. M., and Mostafa, M. M. Production of single cell protein on media containing potato processing waste starch. *Egypt. J. Fd. Sci., 13*:63-74, 1985.
230. Mori, K., Yanagimoto, M., Okada, N., and Yanai, S. Protein enrichment of potato waste by solid substrate fermentation. *Rep. Natl. Fd. Res. Inst., 0*:15-20, 1986.
231. Stevens, C. A., and Gregory, K. F. Production of microbial biomass protein from potato processing wastes by *Cephalosporium eichhorniae. Appl. Environ. Microbiol., 53*:284-291, 1987.
232. Senez, J. C. Protein enrichment of starchy materials by solid state fermentation. In *Production and Feeding of Single Cell Protein* (M. P. ferranti and A. Fiecher, eds.), Appl. Sci., Barking, Essex, 1984, pp. 104-111.
233. Raimbault, M., Revah, S., Pina, F., and Villalobos, P. Protein enrichment of cassava by solid substrate fermentation using mold isolated from traditional foods. *J. Ferment. Technol., 63*:395-400, 1985.
234. Smith, R. E., Osothsilp, C., Bicho, P., and Gregory, K. F. Improvement of the protein content of cassava by *Sporotrichum pulverulentum* in solid state culture. *Biotechnol. Lett., 8*:31-36, 1986.
235. Abdullah, A. L., Tengerdy, R. P., and Murphy, V. G. Optimization of solid substrate fermentation of wheat straw. *Biotechnol. Bioeng., 27*:20-27, 1985.
236. Yadav, J. S. Influence on nutritional supplementation on solid substrate fermentation of wheat straw with an alkaliphilic white rot fungus (*Coprinus*). *Appl. Microbiol. Biotechnol., 26*:474-478, 1987.
237. Hamissa, F. A., El. A., Wang, D., Shaker, H. M., and El-Rafai, A. M. H. Improvement of nutritional value of sugarcane bagasse by simple solid state fermentation. *Microbiol. Lett., 26*:129-134, 1984.
238. Faiez, A. A., and Smith, J. E. Effect of different nitrogen supplementation on fermentation of bagasse by *Sporotrichum pulverulentum* wild type and mutant. *Trans. Br. Mycol. Soc., 86*:165-168, 1986.
239. Carrizales, V., and Saenz, D. Protein enrichment of sugarcane pith using semisolid culture of *Chaetomium cellulolyticum. Acta Cient. Venez., 37*:580-586, 1986.
240. Yang, S., Durand, A., and Blachere, H. Protein enrichment of sugarbeet residue with the inoculation of conidia of *Trichoderma*

album by solid state fermentation. *Chin. J. Microbiol. Immunol.*, *19*:69-80, 1986.

241. Hours, R. A., Alberto, E. M., and Ertola, R. J. Microbial biomass product from apple pomace in batch and fed batch cultures. *Appl. Microbiol. Biotechnol.*, *23*:33-37, 1985.
242. Friedrich, J., Cimerman, A., and Perdih, A. Comparison of different cellulolytic fungi for bioconversion of apple distillery waste. *Appl. Microbiol. Biotechnol.*, *24*:432-434, 1986.
243. Friedrich, J., Cimerman, A., and Perdih, A. Mixed culture of *Aspergillus awamori* and *T. reesei* for bioconversion of apple distillery waste. *Appl. Microbiol. Biotechnol.*, *26*:299-303, 1987.
244. Rodriquez, J. A., Echevarria, J., Rodriquez, F. J., Sierra, N., Daniel, A., and Martinez, O. Solid state fermentation of dried citrus peel by *A. niger*. *Biotechnol. Lett.*, *7*:577-580, 1985.
245. Baldensperger, J., Lewer, J., Hannibal, L., and Quinto, A. J. Solid state fermentation of banana wastes using a strain of *A. niger*. *Biotechnol. Lett.*, *7*:743-748, 1985.
246. Babitaskaya, V. G., Stakheev, I. V., Schcherba, V. V., and Vadetskii, B. Yu. Bioconversion of lignocellulosics substrates by mycelial fungi in solid state culture. *Prikl. Biokhim. Mikrobiol.*, *22*:531-539, 1986.
247. Baraga, V., Martin, M., and Nicoli, J. R. Influence of spore and mycelial inoculation of *Phanerochaete chrysosporium* on SSF of Napier grass. *Rev. Microbiol.*, *18*:58-66, 1987.
248. Walter, P., Molina, M. R., Brenes, R. G., and Bressani, R. Solid state fermentation: An alternative to improve the nutritive value of coffee pulp. *Appl. Environ. Microbiol.*, *49*:388-393, 1985.
249. Ragini, B., and Madan, M. Mushrooms: Potential protein source from cellulosic residues. *Enz. Microb. Technol.*, *5*:251-259, 1983.
250. Sinskey, A. J., and Tannenbaum, S. R. Removal of nucleic acids in SCP. In *Single Cell Protein II* (S. R. Tannenbaum and D. I. C. Wang, eds.), MIT, Cambridge, MA, 1975, pp. 158-178.
251. Anderson, C. J., Longton, C., Maddix, G. W., and Scammell Solomans, G. L. The growth of microfungi on carbohydrates. In *Single Cell Protein II* (S. R. Tannenbaum and D. I. C. Wang, eds.), MIT, Cambridge, MA, 1975, pp. 314-329.
252. Mateles, R. I. The physiology of SCP production. In *Microbial Technology. Current State, Future Prospects* (A. T. Bull, D. C. Ellwood, and C. Ratledge, eds.), Cambridge Univ. Press, London, 1979, pp. 29-52.
253. Righelato, R. C. Growth kinetics of mycelial fungi. In *Filamentous Fungi* (J. E. Smith and D. R. Berry, eds.), Edward Arnold, London, 1975, pp. 79-103.
254. Fitzpatrick, W. H., Essenlen, W. B., Jr., and Wier, E. Production of fungi. In *Microbial Technology* (H. J. Peppler and D. Perlman, eds.), Academic, New York, 1946, pp. 93-155.
255. PAG. Guidelines on the production of single cell protein for human consumption. In *Single Cell Protein II* (S. R. Tannenbaum and D. I. C. Wang, eds.), MIT, Cambridge, MA, 1975, pp. 670-673.

256. Duthie, I. F. Animal feeding trials with microfungal protein. In *Single Cell Protein II* (S. R. Tannenbaum and D. I. C. Wang, eds.), MIT, Cambridge, MA, 1975, pp. 505-544.
257. Young, V. R., and Scrimshaw, N. S. Clinical studies on the nutritional value of SCP. In *Single Cell Protein II* (S. R. Tannenbaum and D. I. C. Wang, eds.), MIT, Cambridge, MA, 1975, pp. 564-586.
258. Scrimshaw, N. S. Acceptance of SCP for human food applications. In *Comprehensive Biotechnology, The Principles, Application and Regulations of Biotechnology in Industry, Agriculture and Medicine* (M. Moo-Young, ed.), Pergamon, Oxford, 1985, pp. 673-683.
259. Scrimshaw, N. S., and Young, V. R. Soy protein in adult human nutrition: A review with new data. In *Soy Protein and Human Nutrition* (H. K. Wilcke, D. T. Hopkins, and D. H. Waggle, eds.), Academic Press, New York, 1979, p. 121.
260. Pellett, P. L., and Young, V. R. Nutritional evaluation of protein foods. *The U.N. Univ. World Hunger Programme. Food and Nutrition Bulletin Supplement 4.* United Nations, Univ. Tokyo, Japan (WHTR-3/UNDP-29), 1980.
261. Udall, J. N., Lo, C., Young, V. R., and Scrimshaw, N. S. The tolerance and nutritional value of two microfungal foods in human subjects. *Am. Clin. Nutr., 40*:285-289, 1984.
262. Ekenvall, L., Bolling, B., Gothe, C. J., Ebbinghaus, L., Von-Stedingk, L. V., and Wasserman, J. SCP as an occupational hazard. *Br. J. Ind. Med., 40*:212-215, 1983.
263. Mayes, R. W. Lack of allergic reactions in workers exposed to 'Pruteen' (Bacterial SCP). *Br. J. Ind. Med., 39*:183-186, 1982.
264. Keilin, J. The biological significance of uric acid and guanine excretion. *Biol. Rev., 34*:265-280, 1959.
265. Christen, P., Peacock, W. C., Chisten, A. E., and Walker, W. E. C. Urate oxidase in primate phylogenesis. *Eur. J. Biochem., 12*:3-12, 1970.
266. Anon. PAG adhoc working group meeting on clinical evaluation and acceptable nucleic acid levels of SCP for human consumption. *PAG Bull., 54*:17-33, 1975.
267. Anon. Canadian plant will test single cell protein process. *Feed Stuffs, 54*:6-9, 1982.
268. Gellender, M. Microbes upgrade starch to high protein foods. *Chem. Int., 1*:21-25, 1981.
269. Anon. Food from a fermenter looks and tastes like meat. *Fd. Eng.*, May:117-118, 1981.
270. Forss, K. G., Gadd, G. O., Lundell, R. O., and Williamson, H. W. Process for the manufacture of protein containing substances for fodder, food stuffs and technical applications. U.S. Patent 3809614 (1974).
271. Imrie, F. K. E., and Vlitos, A. J. Production of fungal protein from carob (*Ceratoma siliqua* L.). In *Single Cell Protein II* (S. R. Tannenbaum and D. I. C. Wang, eds.), MIT, Cambridge, MA, 1975, pp. 223-243.
272. Kosman, W. A better way to make protein from whey. *Chem. Eng.*, March 13th 36C and D, 1978.

273. Deschamps, F. Production de proteines d'organismes unicellulaires a partir d'amidon. *Ann. Cong. Int.* (Paris), *1*:49-63, 1979.
274. Senez, J. C., Raimboult, M., and Deschamps, F. Protein enriched fermented feeds (PEFF). *VI International Fermentation Symposium*, London, Ontario, 1980, pp. 205-210.

16

ANTIFUNGAL FOOD ADDITIVES

MICHAEL B. LIEWEN *General Mills, Inc., Minneapolis, Minnesota*

I. INTRODUCTION

The problem of food spoilage has plagued humans throughout history. As early humans evolved from a gathering and hunting life style to raising crops and keeping animals, they were forced to store their own foods. Early attempts to preserve foods involved use of sugars, spices, salts, and wood smoke. Today, however, preservation has used such factors as temperature, water activity, pH, gases, organic acids, salts, antibiotics, irradiation, packaging, and various combinations of these factors. No matter which factors are selected, use of the proper antimicrobial is dependent on the chemical properties of the antimicrobial; properties and composition of the food product; type of preservation system, other than the chemical, used in the food; type, characteristics, and number of microorganisms; safety of the antimicrobial; and cost-effectiveness of the antimicrobial [7].

Fungi are among the most challenging organisms to inhibit in foods given their ability to grow under a diverse range of environmental conditions. Environmental conditions such as water activity, pH, temperature, and atmosphere can be manipulated to control fungal growth. However, these conditions often need to be taken to extremes to control fungi since subgroups exist that have become adapted to extreme environmental conditions. Using water activity as an example, most molds are inhibited by an a_w of 0.80 or lower, although some xerophilic molds can grow at a_w values as low as 0.65 [22,28,42]. Most yeasts are inhibited by a_w values of 0.87, but some osmophilic yeasts can grow at a_w values as low as 0.60 [25].

Most fungi are little affected by pH over a broad range, commonly 3 to 8. Some molds can grow down to pH 2.0, and yeasts down to pH 1.5 [22]. However, as pH moves away from an organism's optimum growth range, typically about pH 5.0 for fungi, the effect of other growth-limiting factors becomes more apparent.

Fungi are generally easily inactivated by heat treatments such as pasteurization, although some heat-resistant molds associated with fruits and fruit products can survive rather severe heat treatments and spoil products such as pasteurized juices and canned fruits. In addition, molds can grow over a wide range of temperatures. Some molds can grow at temperatures less than 0°C while others can grow at 50-55°C.

Molds have an absolute requirement for oxygen. Many species, however, are efficient oxygen scavengers and can grow in atmospheres containing less than 1.0% O_2 [9]. From a practical standpoint, it can be difficult to inhibit mold growth in foods solely by exclusion of oxygen from the package. Foods often contain dissolved oxygen, which slowly equilibrates with the package headspace, and oxygen can leach through all but the most impermeable packages. While mold growth can be delayed, it is not inhibited over long-term storage. Yeasts have no requirement for oxygen and can grow in its complete absence.

Antifungal food additives are an efficient, cost-effective, and often the only successful way to control fungal growth in foods. Antifungal food additives are basically chemicals that prevent or interfere with mold growth. These chemicals may be found naturally occurring in certain foods, such as some organic acids and essential oils, or may be added to foods during processing [6]. Antifungal food additives should never be used as a substitute for good manufacturing practices or proper sanitation procedures. Their proper use is as a processing aid to complement the practices just mentioned. Obviously, antifungal food additives must be safe for human consumption, and their use is limited by law in most countries to relatively low levels and to specific foods. The various antifungal food additives are briefly reviewed in this chapter.

II. ORGANIC ACIDS

Organic acids have been used for years to control fungal spoilage of foods. They find wide use because of solubility, taste, and low toxicity. The mode of action of organic acids is attributed to depression of intracellular pH by ionization of the undissociated acid molecule or disruption of substrate transport by alteration of cell membrane permeability [30,44]. In addition to inhibiting substrate transport, organic acids may inhibit NADH oxidation, thus eliminating supplies of reducing agents to electron transport systems [18].

Since the undissociated portion of the acid molecule is primarily responsible for antifungal activity, effectiveness is dependent upon the dissociation constant of the acid and pH of the food to be preserved. Because the dissociation constant of most organic acids is between pH 3 and 5, organic acids are generally most effective at low pH values [6]. This along with solubility properties determines the foods in which organic acids may be effectively used.

A few fungal species possess mechanisms of resistance to organic acid preservatives. *Saccharomyces bailii* is resistant to high concentrations of sorbic and benzoic acids [18,33,44]. Some molds in the genus *Penicillium* can grow in the presence of high concentrations of sorbic acid and decarboxylate sorbic acid to 1,3-pentadiene, a volatile compound with an extremely strong kerosene-like odor [29,30,41]. When resistance to or metabolism of an organic acid is a problem, other preservative systems must be used.

A. Sorbic Acid

Sorbic acid and its potassium salt are the most widely used forms of this compound and are collectively known as sorbates. The salt forms are highly soluble in water, as is true for all organic acids. Their most common use is preservation of food, animal feed, cosmetic and pharmaceutical products, as well as technical preparations that come in contact with the human body. Methods of application include direct addition into the product; dipping, spraying, or dusting of the product; or incorporation into the wrapper [12,30].

Typical use levels in foods range from 0.02% in wine and dried fruits to 0.3% in some cheeses (Table 1). Foods in which sorbate has commercially useful antimicrobial activity include baked goods (cakes and cake mixes, pies and pie fillings, doughnuts, baking mixes, fudges, icings), dairy products (natural and processed cheeses, cottage cheese, sour cream), fruit and berry products (artificially sweetened confections, dried fruits, fruit drinks, jams, jellies, wine), vegetable products (olives, pickles, relishes, salads), and other miscellaneous food products (certain fish and meat products, mayonnaise, margarine, salad dressings).

Environmental factors such as pH, water activity, temperature, atmosphere, type of microbial flora, initial microbial load, and certain food components, singly or in combination, can influence the activity of sorbate. Together with preservatives such as sorbic acid, they often act to broaden antimicrobial action or increase it synergistically. Use of other preservatives in combination with sorbate can broaden or intensify antimicrobial action. Length and temperature of storage are other important considerations. If growth of spoilage or pathogenic organisms is inhibited, but the microorganisms is not killed, growth will eventually resume under proper conditions. The length of inhibition will vary with storage temperature as well as with any of the other factors discussed.

TABLE 1. Typical Concentration (%) of Sorbic Acid Used in Various Food Products

Cheeses	0.2-3.0
Beverages	0.03-0.10
Cakes and pies	0.05-0.10
Dried fruits	0.02-0.05
Margarine (unsalted)	0.05-0.10
Mayonnaise	0.10
Fermented vegetables	0.05-0.20
Jams and jellies	0.05
Fish	0.03-0.15
Semi-moist pet food	0.1-0.3
Wine	0.02-0.20
Fruit juices	0.05-0.20

Source: Ref. 30.

TABLE 2. Minimum Concentrations (%) of Preservative Required for Inhibition of Mold Growth at pH 5.0

	Species		
Compound	*A. soini*	*P. citrinum*	*A. niger*
Benzoic acid	0.15	0.20	0.20
Propionic acid	0.06	0.08	0.08
Sorbic acid	0.02	0.08	0.08

Source: Ref. 34.

Sorbic acid is a broad-spectrum antimycotic that is effective against yeasts and molds. The antifungal effect of sorbate is greater at pH 5.0 or lower than at higher pH values. Sorbic acid has little antifungal activity at pH values higher than 5.5 or 6.0. Above this pH range, little of the acid is in the antimicrobially active undissociated form. However, sorbic acid has a relatively high dissociation constant compared to benzoic or propionic acids and is therefore usually the most effective of the organic acids at pH levels of 5.0 and higher. This is illustrated in Table 2.

B. Benzoic Acid

Benzoic acid also has widespread use in the food industry. It occurs naturally in raspberries, cranberries, plums, prunes, cinnamon, and cloves [6]. As an antifungal food additive, the water-soluble sodium and potassium salts and the fat-soluble acid form are suitable for foods and beverages with pH below 4.5. Benzoates have little effect at neutral pH values. They are not as effective as sorbates at pH 5.0 (Table 2), but effectiveness increases at lower pH values.

Benzoic acid is active against yeasts and molds, including aflatoxin-forming microorganisms [34]. The acid form is often added to the fat phase or the sodium salt to the water phase of products such as salad dressings, mayonnaise, pickled vegetables, fruit products, and fruit drinks. Because benzoate can impart a fairly strong bitter off-flavor, it is frequently used in combination with sorbate. This mixture is often more effective in inhibiting yeasts and molds than a comparable level of either preservative alone. In addition, the mixture is less offensive organoleptically than benzoate alone.

C. Propionic Acid

This organic acid inhibits molds but not yeasts. It occurs in some foods as a result of natural processing. It is present in Swiss cheese at concentrations up to 1%, where it is produced by the bacterium *Propionibacterium shermanii* [6]. Since yeasts are typically unaffected, the acid can be added to bread dough without interfering with leavening.

In the food industry, propionic acid is often used as a sodium or calcium salt [36]. Propionates are used primarily to inhibit molds in bakery goods. In addition to their antimycotic properties, propionates will inhibit

Bacillus mesentericus, the rope-causing bacterium. Propionates also are used to a limited extent to inhibit mold growth in processed cheeses.

The antifungal activity of propionic acid is weak compared to the other organic acids. Therefore, propionates must be used in relatively high concentrations to be effective. As with other organic acids, the pH value of the food to be preserved affects antimicrobial activity. Because of its low dissociation constant, propionic acid is active in a pH range similar to that of sorbic acid.

III. MEDIUM-CHAIN FATTY ACIDS

Generally, fatty acids are most effective as inhibitors of gram-positive bacteria and yeasts, although some fatty acids exhibit antimycotic activity [6]. Chipley et al. [11] observed that fatty acid derivatives reduced growth and aflatoxin production by *Aspergillus* spp.

Polyhydric alcohol fatty acid esters have great potential for use as emulsifiers in food formuations. They also possess antifungal properties and therefore may exert a preservative effect in foods. Kato and Shibasaki [27] demonstrated strong fungistatic activity of glycerol monocaprate and glycerol monolaurate toward *Aspergillus niger*, *Penicillium citrinum*, *Candida utilis*, and *Saccharomyces cerevisiae*. Sucrose monocaprate and sucrose monolaurate are slightly inhibitory to a spoilage film-forming yeast inoculated into a soy sauce substrate [26]. Six sucrose esters substituted to different degrees with a mixture of palmitic and stearic acids were examined by Marshall and Bullerman [31] for antifungal properties. Growth of *Aspergillus*, *Penicillium*, *Cladosporium*, and *Alternaria* spp. was inhibited in media containing 1% of sucrose esters.

IV. ANTIBIOTICS

A. Natamycin

Natamycin, formerly called pimaricin, is an antibiotic which possesses strong antifungal properties, yet is not active against bacteria. Its use is currently allowed in several countries. Commercially it is used primarily in the preservation of cheese as a surface coating.

Researchers have demonstrated that natamycin is active at very low concentrations against many of the fungi known to cause food spoilage. Levels of 1 to 100 ppm added to laboratory media have been reported to be effective for inhibiting mold growth [20]. Similar concentrations have been found effective in cottage and Cheddar cheese, while solutions of 1000-2000 ppm are effective as dips for cheese. Natamycin inhibits aflatoxin formation by molds only when growth was completely inhibited [34].

B. Nisin

Nisin is active against gram-positive bacteria exclusively according to most reports [34]. However, Yousef et al. [46] reported that nisin at levels of 5 and 125 ppm delayed growth of *Aspergillus parasiticus* in culture media, but after long periods of incubation, the mold grew faster and produced more aflatoxin in nisin-containing cultures than in cultures without it.

V. METABOLITES FROM LACTIC ACID BACTERIA

A wide variety of raw foods are preserved by lactic acid fermentation, including milk, meat, fruits, and vegetables. Reduction of pH and removal of large amounts of carbohydrates by fermentation are the primary preserving actions that lactic acid bacteria (LAB) provide to a fermented food. These actions are largely ineffective in preventing growth of fungi in foods. However, it has also been recognized that LAB can produce inhibitory substances other than organic acids (lactate and acetate) that are antagonistic toward other microorganisms [16]. The antibacterial properties of LAB are well documented. Several LAB, typically of the genera *Streptococcus* (*Lactococcus*) and *Lactobacillus*, produce antibacterial substances. Antifungal properties of LAB have received little attention; however, several metabolites of LAB have been reported to have antifungal activity [5,16].

Batish et al. [5] screened different lactic starter cultures for their antifungal activity with the goal of commercially exploiting their antifungal potentials. They found several strains of *Streptococcus* that inhibited a wide variety of molds. While the antifungal substances produced by their LAB were not identified or characterized, maximum production occurred at 30°C and pH 6.8. Several specific LAB metabolites have been reported to have antifungal activity.

A. Diacetyl

Diacetyl is a metabolic end product produced by some species of LAB. It is best known for the buttery aroma that it imparts to cultured dairy products [16]. Its antimicrobial action has been investigated by Jay [23,24], who reported that a concentration of 200 ppm was inhibitory to yeasts and 300 ppm was inhibitory to molds. Acidity of the growth medium was shown to have a direct effect on the antimicrobial activity of diacetyl. The compound was clearly more effective as an antifungal agent below pH 7.0 than above this value. Reasons for pH-associated antifungal activity is not clear.

Since effective concentrations of diacetyl impart a sharp odor of butter, potential for use in foods as an antifungal agent is limited. However, its use as a utensil sanitizer and in wash or rinse water for certain products is more feasible [24].

B. Microgard

Microgard (Wesman Foods, Inc., Beaverton, OR) is grade A skim milk that has been fermented by *Propionibacterium shermanii* and then pasteurized [45]. The product prolongs the shelf life of cottage cheese by inhibiting psychrotrophic spoilage bacteria. The product is also antagonistic toward some yeasts and molds [16]. Microgard consists of propionic acid, diacetyl, acetic acid, and lactic acid [2].

C. Reuterin

Reuterin is a low molecular weight, nonproteinaceous, highly soluble, pH-neutral metabolite produced by *Lactobacillus reuterii*. The compound is a broad-spectrum antimicrobial with activity that encompasses yeasts and molds as well as bacteria. It may have application in preservation of foods by reducing populations of pathogenic and spoilage microorganisms [16].

VI. HERBS AND SPICES

Herbs and spices are widely used to impart flavor to foods. It is generally accepted that certain herbs and spices have antimicrobial activity and may influence the keeping quality of foods to which they have been added. However, they are not currently used with the primary purpose of providing a preservative effect.

Hoffman and Evans [21] were among the earliest to describe the preservative action of cinnamon, cloves, mustard, allspice, nutmeg, ginger, black pepper, and cayenne pepper. They found that cinnamon, cloves, and mustard were most effective and ginger, black pepper, and cayenne pepper were least effective.

Bachmann [4] studied the effect of spices and their essential oils on growth of several test organisms, including *Aspergillus* and *Penicillium* species, and concluded that spices used in amounts as employed normally for ordinary foods were insufficient as preservatives. However, when used in larger amounts, cinnamon, cloves, and allspice retarded mold growth. Bullerman [8] reported that cinnamon in concentrations as low as 0.02% inhibited mold growth and aflatoxin production in culture media and cinnamon bread.

Cloves, cinnamon, mustard, allspice, garlic, and oregano at the 2% level in potato dextrose agar completely inhibited mycotoxigenic molds for up to 21 days [3]. Combinations of different levels of potassium sorbate with cloves showed an enhanced or possibly synergistic inhibitory effect on the growth of molds, indicating the possibility of using spices and commercial antifungal agents together in small amounts to obtain antifungal activity.

In most instances, herbs and spices are not effective antifungal agents when used alone in amounts normally added to foods. However, when used in combination with other preservative systems, they can be valuable contributors to an antifungal system consisting of interacting physical and chemical preservatives.

VII. ESSENTIAL OILS

The antimicrobial activities of extracts from several types of plants and plant parts used as flavoring agents in foods and beverages have been recognized for many years. Certain of these essential oils have antifungal properties. Conner and Beuchat [14] documented the effects of garlic and onion against yeasts, and other investigators have shown these extracts to be inhibitory to molds [13,32,38,40]. Alderman and Marth [1] examined the effects of lemon and orange oils on *Aspergillus flavus* and found that when the citrus oils were added to grapefruit juice or glucose-yeast extract medium at a concentration of 3000-3500 ppm, growth and aflatoxin production were suppressed. When orange oil was added to either medium at concentrations up to 7000 ppm, growth and aflatoxin production were greatly reduced although still evident. Connor and Beuchat [14] showed that essential oils of allspice, cinnamon, clove, oregano, savory, and thyme were inhibitory to food and industrial yeasts.

VIII. PHENOLIC ANTIOXIDANTS

Phenolic antioxidants have been shown by several researchers to possess antifungal activity. Chang and Branen [10] demonstrated that in a glucose salt medium, 1000 ppm butylated hydroxyanisole (BHA) inhibited growth and aflatoxin production of *Aspergillus parasiticus* spores, and >250 ppm inhibited growth and aflatoxin of *A. parasiticus* mycelia. However, they found that at 10 ppm BHA, total aflatoxin production was more than twice that of the control, with virtually no effect on mycelial weight. Their results indicate that at high levels, BHA may serve as an effective antifungal agent; however, at low levels BHA may actually stimulate aflatoxin production. Fung et al. [19] tested effects of BHA and butylated hydroxytoluene (BHT) on several strains of *A. flavus*. On solid agar, BHT was not inhibitory at any levels tested. BHA inhibited growth and aflatoxin production.

Since the primary use of these compounds in foods is as antioxidants, their effectiveness as antifungal agents in food systems has not been adequately studied. While results of experiments in growth media indicate that these compounds exert antifungal effects, extrapolation of these results to food systems should be done with caution. Interaction of these compounds with food components will undoubtedly affect their antifungal properties.

IX. GASES AND MODIFIED ATMOSPHERES

Elimination of oxygen is often used as a control measure for inhibiting growth of molds. Exclusion of oxygen will not prevent growth of yeasts. Studies on bakery products have demonstrated that atmospheric O_2 levels must be reduced to 0.1-1.0% to effectively inhibit growth of molds. In studies on toasted bread, Cerny [9] demonstrated that visible mold would occur in 3 days in air; 5 days in 99% N_2—1% O_2; and >100 days in 99.9% N_2—0.1% O_2, 99% CO_2—1.0% O_2, 99.8% CO_2—0.2% O_2, and 100% CO_2. This study demonstrates that although molds are considered to be aerobic organisms, certain species have the ability to grow at very low O_2 concentrations. Effectively controlling molds by simple gas flushing can be difficult in practice. Chemical oxygen scavengers can be used in place of or to supplement gas flushing [39]. Oxygen scavengers will also give protection against package leaks and infiltration of O_2 through the package.

Carbon dioxide exerts antifungal action that supplements simple exclusion of O_2 and thus is more effective than inert gases such as nitrogen. The gas probably exerts antifungal activity by altering intracellular pH levels [15].

Recent research has shown CO to have potential use with foods. Carbon monoxide inhibits yeasts and molds which cause postharvest decay in fruits and vegetables [43]. The potential toxicity of this compound to workers requires special handling procedures.

Sulfur dioxide is broadly effective against yeasts and molds [17]. It is used extensively to control growth of undesirable microorganisms in fruits, fruit drinks, wines, sausages, fresh shrimp, and pickles. The antimicrobial activity of SO_2 is associated with the unionized form of the molecule. Therefore, it is most effective at pH values < 4.0, where this form predominates [43].

Ethylene oxide has been widely used to reduce microbial contamination and kill insects in various dried foods. The gas has been used to treat gums, spices, dried fruits, corn, wheat, barley, dried egg, and gelatin [22]. Concern over toxicity of residues has limited use of this gas in recent years.

Propylene oxide has been less studied than ethylene oxide. However, it appears that its antifungal effects are similar [43]. Yeasts and molds are more sensitive to the gas than bacteria. Propylene oxide has been used as a fumigant for control of microorganisms and insects in bulk quantities of goods such as cocoa, gums, processed spices, starch, and processed nut meats [22].

X. INTERACTION OF FACTORS

Many of the antifungal agents reviewed in this chapter need to be used at extreme concentrations or levels in a food to be effective when used alone. However, a variety of factors can prevent growth of fungi. While fungi tend to be more tolerant to adverse environmental conditions than bacteria, combining inhibitory factors such as temperature, water activity, or pH with antifungal agents can result in considerable improvement of the microbial stability of foods. Suitable combinations of growth-limiting factors at subinhibitory levels can be devised so that certain microorganisms can no longer proliferate.

Sorbic acid at 1000 ppm and pH 7.0 will not inhibit mold growth. However, if the pH is lowered to 5.0, growth of most molds will be inhibited [29]. Antioxidants such as BHA and BHT have been shown to potentiate the action of sorbic acid [37]. In general, antifungal food additives become more effective as environmental conditions move away from the optimum for a particular organism.

The level of a single growth-limiting factor that will inhibit a microorganism is usually determined under conditions in which all other factors are optimum. In preserving foods, more than one factor is usually relied upon to control microbial growth. Addition of a substance which in itself does not give full inhibition can effectively preserve products in the presence of other subinhibitory factors. The effect of superimposing limiting factors is known as the "hurdles concept" [37].

Little information is currently available on combining subinhibitory factors to preserve foods. It is very time consuming and expensive to design preservative systems using the hurdles concept by random design. Predictive modeling can be used to test the consequences of a number of factors changing at the same time. With proper design and interpretation, preservative systems can be designed rapidly and efficiently [35]. With greater emphasis on development and marketing of refrigerated foods by the food industry, these methods will become more widely used and accepted. However, product challenge studies should be conducted to verify the effectiveness of a combination of subinhibitory factors.

REFERENCES

1. Alderman, G. G., and Marth, E. H. Inhibition of growth and aflatoxin production of *Aspergillus parasiticus* by citrus oils. *Z. Lebensm. Unters.-Forsch.*, *160*:353-358, 1976.
2. Al-Zoreky, N. Microbiological control of food spoilage and pathogenic microorganisms in refrigerated foods. M.S. thesis, Oregon State University, Corvallis, 1988.
3. Azzouz, M. A., and Bullerman, L. B. Comparative antimycotic effects of selected herbs, spices, plant components and commercial antifungal agents. *J. Food Prot.*, *45*:1298-1301, 1982.
4. Bachman, F. M. The inhibitory action of certain spices on the growth of microorganisms. *J. Ind. Eng. Chem.*, *8*:620, 1916.
5. Batish, V. K., Grover, S., and Lal, R. Screening lactic starter cultures for antifungal activity. *Cultured Dairy Prod. J.*, *24*(2):21-25, 1989.
6. Beuchat, L. R., and Golden, D. A. Antimicrobials occurring naturally in foods. *Food Technol.*, *43*:134-142, 1989.
7. Branen, A. L. Introduction to the use of antimicrobials. In *Antimicrobials in Foods* (A. L. Branen and P. M. Davidson, eds.), Marcel Dekker, Inc., New York, 1983, p. 1.
8. Bullerman, L. B. Inhibition of aflatoxin production by cinnamon. *J. Food Sci.*, *39*:1163-1165, 1974.
9. Cerny, G. Retardation of toast bread by gassing. *Chem., Mikrobiol., Technol. Lebensm.*, *6*(1):8-10, 1979.
10. Chang, H. C., and Branen, A. L. Antimicrobial effect of butylated hydroxianisole (BHA) and butylated hydrotoluene (BHT). *J. Food Sci.*, *40*:349-351, 1975.
11. Chipley, J. R., Story, L. D., and Kabara, J. J. Inhibition of *Aspergillus* growth and extracellular aflatoxin accumulation by sorbic acid and derivatives of fatty acids. *J. Food Safety*, *2*:109-115, 1981.
12. Cirilli, G. Preservation of bread with sorbic acid. *Industrie Alimentari*, *15*:67-68, 1976.
13. Collins, M. A., and Charles, H. P. Antimicrobial activity of carnosol and ursolic acid: Two antioxidant constituents of *Rosmarinus-officinalis*. *Food Microbiol.*, *4*:311-316, 1987.
14. Conner, D. E., and Beuchat, L. R. Effects of essential oils from plants on growth of spoilage yeasts. *J. Food Sci.*, *49*:429-434, 1984.
15. Corral, L. G., Post, L. S., and Montville, T. J. Antimicrobial activity of sodium bicarbonate. *J. Food Sci.*, *53*:981-982, 1988.
16. Daeschel, M. A. Antimicrobial substances from lactic acid bacteria for use as food preservatives. *Food Technol.*, *43*(1):164-166, 1989.
17. Dziezak, J. D. Preservatives: Antimicrobial agents. *Food Technol.*, *40*(9):104-107, 1986.
18. Fresse, E., Sheu, C. W., and Galliers, E. Function of lipophilic acids as antimicrobial food additives. *Nature*, *241*:321-325, 1973.

19. Fung, D. Y. C., Taylor, S., and Kahan, J. Effects of butylated hydroxianisole (BHA) and butylated hydrotoluene (BHT) on growth and aflatoxin production of *Aspergillus flavus*. *J. Food Safety*, *1*: 39-51, 1977.
20. Gourama, H., and Bullerman, L. B. Effects of potassium sorbate and natamycin on growth and penicillic acid production by *Aspergillus ochraceus*. *J. Food Prot.*, *51*:139-144, 155, 1988.
21. Hoffman, C., and Evans, A. C. Use of spices as preservatives. *J. Ind. Eng. Chem.*, *3*:835, 1911.
22. ICMSF. *Microbial Ecology of Foods*, Vol. 1. *Factors Affecting Life and Death of Microorganisms*, Intl. Commission of Microbiological Specifications for Foods, Academic Press, New York, 1980.
23. Jay, J. M. Antimicrobial properties of diacetyl. *Appl. Environ. Microbiol.*, *44*:525-532, 1982.
24. Jay, J. M. Effect of diacetyl on foodborne microorganisms. *J. Food Sci.*, *47*:1829-1831, 1982.
25. Jermini, M. F. G., and Schmidt-Lorenz, W. Growth of osmotolerant yeasts at different water activity values. *J. Food Prot.*, *50*:404-410, 1987.
26. Kato, N. Antimicrobial activity of fatty acids and their esters against a film-forming yeast in soy sauce. *J. Food Safety*, *3*:321-325, 1981.
27. Kato, N., and Shibasaki, I. Comparison of antimicrobial activities of fatty acids and their esters. *J. Ferment. Technol.*, *53*:793-795, 1975.
28. Leistner, L., and Rodel, W. Inhibition of microorganisms in foods by water activity. In *Inhibition and Inactivation of Vegetative Microbes* (F. A. Skinner and W. B. Hugo, eds.), Academic Press, London, 1976, pp. 219-237.
29. Liewen, M. B., and Marth, E. H. Inhibition of penicillia and aspergilli by potassium sorbate.
30. Liewen, M. B., and Marth, E. H. Growth and inactivation of microorganisms in the presence of sorbic acid: A review. *J. Food Prot.*, *48*:364-375, 1985.
31. Marshall, D. L., and Bullerman, L. B. Antimicrobial activity of sucrose fatty acid ester emulsifiers. *J. Food Sci.*, *51*:468-470, 1986.
32. Moleyar, V., and Narasimham, P. Fungitoxicity of binary mixtures of citral, cinnamic aldehyde, menthol and lemon grass oil against *Aspergillus niger* and *Rhizopus stolonifer*. *Lebensm. Wiss.-Technol.*, *21*:100-102, 1988.
33. Pitt, J. I., and Richardson, K. C. Spoilage by preservative-resistant yeasts. *CSIRO Food Res. Q.*, *33*:80-85, 1973.
34. Ray, L. L., and Bullerman, L. B. Preventing growth of potentially toxic molds using antifungal agents. *J. Food Prot.*, *45*:953-963, 1982.
35. Roberts, T. A. Combinations of antimicrobial and processing methods. *Food Technol.*, *43*(1):156-163, 1989.
36. Sauer, F. Control of yeasts and molds with preservatives. *Food Technol.*, *31*(2):66-67, 1977.
37. Scott, V. N. Interaction of factors to control microbial spoilage of refrigerated foods. *J. Food Prot.*, *52*:431-435, 1989.
38. Sharma, A., Tewari, G. M., Shrikhande, A. J., Padwal-Desai, S. R., and Bandyopadhyay, C. Inhibition of aflatoxin-producing fungi by onion extracts. *J. Food Sci.*, *44*:1545-1547, 1979.

39. Smith, J. P., Orraikv, B., Koersen, W. J., Jackson, E. D., and Lawrence, R. A. Novel approach to oxygen control in modified atmosphere packaging of bakery products. *Food Microbiol.*, *3*:315-320, 1986.
40. Tansey, M. R., and Appelton, J. A. Inhibition of fungal growth by garlic extract. *Mycologia*, *67*:409-415, 1975.
41. Tsai, W. J., Liewen, M. B., and Bullerman, L. B. Toxicity and sorbate sensitivity of molds isolated from surplus commodity cheeses. *J. Food Prot.*, *51*:457-462, 1988.
42. Trucksess, M. W., Stoloff, L., and Mislivec, P. B. Effect of temperature, water activity and other toxigenic mold species on the growth of *Aspergillus flavus* and aflatoxin production on corn, pinto beans and soybeans. *J. Food Prot.*, *51*:361-363, 1988.
43. Wagner, M. K., and Moberg, L. J. Present and future use of traditional antimicrobials. *Food Technol.*, *43*(1):134-142, 1989.
44. Warth, A. D. Mechanism of resistance of *Saccharomyces balii* to benzoic, sorbic and other weak acids used as food preservatives. *J. Appl. Bacteriol.*, *43*:215-230, 1977.
45. Weber, G. H., and Broich, W. A. Shelf life extension of cultured dairy foods. *Cultured Dairy Prod. J.*, *21*(4):19-23, 1986.
46. Yousef, A. E., El-Gendy, S. M., and Marth, E. H. Growth and biosynthesis of aflatoxin by *Aspergillus parasiticus* in cultures containing nisin. *Z. Lebensm. Unters.-Forsch.*, *171*:341-343, 1980.

17

PRODUCTS AND USES OF YEAST AND YEASTLIKE FUNGI

TILAK W. NAGODAWITHANA *Technical Center, Universal Foods Corporation, Milwaukee, Wisconsin*

I. INTRODUCTION

The splendors of ancient Egypt not only included an abundance of pyramidal and tomb texts, stone and wooden sculptures, but also numerous scenes depicting life that existed during the time of the Pharoahs nearly 4000 years ago. Among these were scenes of bakeries, brew houses, and vintage scenes showing gathering of grapes from a cultivated arched vine, pressing of the grapes, collection of juice, and storage in amphorals presumably for fermentation [1]. The remarkable skills of this civilization stretch back even further. As the records indicate, a menu presumably of an aristocrat of ancient Egypt dating back approximately 6000 years included 16 different kinds of bread and cakes, 6 kinds of wines, and 4 types of beer [2]. After these early beginnings, fermenting of beverages like beer and wine developed as an art and a craft and began to spread to other civilizations in complete absence of scientific understanding of the fermentation process.

In studying the development of the brewing, baking, and distilling processes during the Middle Ages, it is important to consider the influence of the Christian clergy. Originally, making of beer and wine was a part of the work of the household, and such chores were done primarily by women. It took a new turn toward the end of the ninth century when monasteries developed brewing into an industry to meet the needs of taverns and wayside inns. Two centuries later, brew houses were found not only in monasteries but also in all the important courts and castles of rulers at the time. The excess yeast produced in such brewing operations was often used for bread baking.

As the baking industry expanded, the supply of yeast from the brewing industry was found inadequate to meet the total requirement for baking. This prompted initiation of a separate industry to produce baker's yeast. In the meantime, the brewing industry gradually migrated from monasteries into the cities, eventually culminating into a lucrative business. The progress of winemaking closely paralleled the beer industry with the original know-how of winemaking predominantly in the hands of the Christian clergy. Less wine than beer was produced, and hence winemaking was expensive and was considered the beverage of the aristocrats who lived at the time. It is, thus, understandable how such crops as barley and grapes became so important to these early civilizations whose records illustrate how their arts and crafts were so closely interwoven with agriculture and religion.

Antonie van Leeuwenhoek, a Dutch lens grinder, made the first recorded observation of the yeast cell in 1680. Despite this discovery, the exploration of the microbial world was not appreciably extended for nearly two centuries. During the middle of the eighteenth century, several scientists began to realize the causal relationship between microorganisms living in certain sugary substrates and the chemical changes that take place in such infusions. Finally, in 1866, Louis Pasteur, through his brilliant studies, concluded that viable yeast cells cause fermentation under anaerobic conditions and that sugar is converted to carbon dioxide and ethanol by the yeast. These findings provided to the world the basic principles of fermentation technology, thus making the beginning of exciting new opportunities for improvement of processes related to baking, brewing, and winemaking.

Today, more than ever, yeast and yeastlike fungi have become almost inseparable from the routine day-to-day activities of mankind. Although society has largely been benefited by this group of organisms, an undesirable group of yeasts in the wrong environment can mean a difference between profit and loss from an industrial point of view or excessive suffering from disease when the yeast in question is pathogenic.

The most recent classification of yeast is found in the treatise edited by Kreger-Van Rij [9]. This classification is based primarily on sexual characteristics, physiological and biochemical properties such as the ability to ferment or assimilate certain sugars, resistance to antibiotics, vitamin requirements, temperature tolerance for growth, and osmotolerance. Numerous surveys of the distribution of yeast and yeastlike fungi have revealed that this group appears to be far less ubiquitous than most bacterial species.

Like molds, the term "yeast" is commonly used, but a simple precise definition for it is difficult because many unicellular yeast forms sometimes have highly complicated life cycles. The true yeasts belong to the class Hemiascomycetes, and these reproduce sexually, resulting in production of ascospores. Formation of ascospores follows conjugation of two cells in most true yeasts, but some may produce ascospores without conjugation, followed by conjugation of ascospores in the daughter cells. False yeasts, in contrast, do not have a sexual cycle and hence do not produce ascospores or other sexual spores. These yeasts are grouped in the subdivision Deuteromycotine or fungi imperfecti.

Most yeasts used industrially belong to the class Hemiascomycetes and predominantly are in the genus *Saccharomyces*. A special emphasis will be placed in this chapter on yeasts in this genus. However, an attempt also will be made to include a number of other species from ascomycetes, basidomycetes, and asporogenous genera which are, in some ways, associated with the food industry.

II. YEASTS IMPORTANT FOR INDUSTRIAL FERMENTATIONS

A. Baker's Yeast

From time immemorial, bakers have relied on yeast for leavening of dough. Such modifications in flour doughs improve palatability and stability of the baked products. Certain rheological changes take place as a result of fermentative metabolism by the yeast in their quest for energy for their sustenance. In this process the byproduct, carbon dioxide, becomes entrapped in the dough matrix, bringing about a honeycomb-like appearance in a typical bread product.

In the mid-nineteenth century, with the rapid expansion of the baking industry, the supply of surplus brewer's yeast became limited, which eventually led to initiation of the baker's yeast industry. These large-scale yeast propagations were done originally using malted grain mashes as the sugar substrates for the yeast. Although this practice was continued for a considerable period, a grain shortage became apparent in the early 1920s. This prompted yeast manufacturers to search for cheaper sugar substrates to manufacture baker's yeast. Blackstrap molasses, which was abundantly available at the time, was an obvious choice for investigative purposes. The extensive research that followed resulted in successful transfer to molasses substrates for baker's yeast production. Currently, blends of beet and cane molasses are being used for this purpose based on economy and general availability of the two molasses types.

A yeast propagation initiated from the test tube takes approximately 6-8 days to produce a commercial batch of baker's yeast suitable for distribution. At least six different fermentation stages (Pasteur flask, PC1, PC2, F1, F2, and F3) (Fig. 1) are needed to build up an inoculum large enough to meet the pitching volume for a 20,000-50,000-gallon fermenter [4]. The success of a commercial propagation is largely dictated by the condition of the pitching yeast. Hence, to ensure rapid growth characteristics in the commercial propagation, the seed yeast is grown to contain approximately 9% nitrogen, 3% phosphate as P_2O_5 with a 17-20% final bud count.

Final commercial propagations generally last 12-20 hours with an aeration rate of 0.5-1 VVM (air volume/fermenter volume/min). To prevent wasteful production of ethanol and maximize biomass production, wort is added incrementally to the fermenter so that the sugar fed at any given time is equivalent to the sugar assimilated by the growing yeast population. Under these conditions, residual sugar in the medium is maintained at a minimal level, thereby minimizing or completely eliminating the Crabtree effect throughout the propagation. This is sometimes referred to as the "fed batch" or "Zulauf" process.

For successful propagation, yeast requires an adequate supply of nitrogen and phosphorus. This is generally fed into the fermenter as ammonia and phosphoric acid, respectively. In addition to requirements supplied by the molasses blend, yeast also needs additional minerals such as magnesium and zinc and a supply of vitamins such as thiamine and pantothenic acid and, in certain instances, biotin if the mash is made to contain more than 80% beet molasses. The temperature is maintained at 30°C with pH controlled at approximately 5.2. Aeration without feed is generally applied

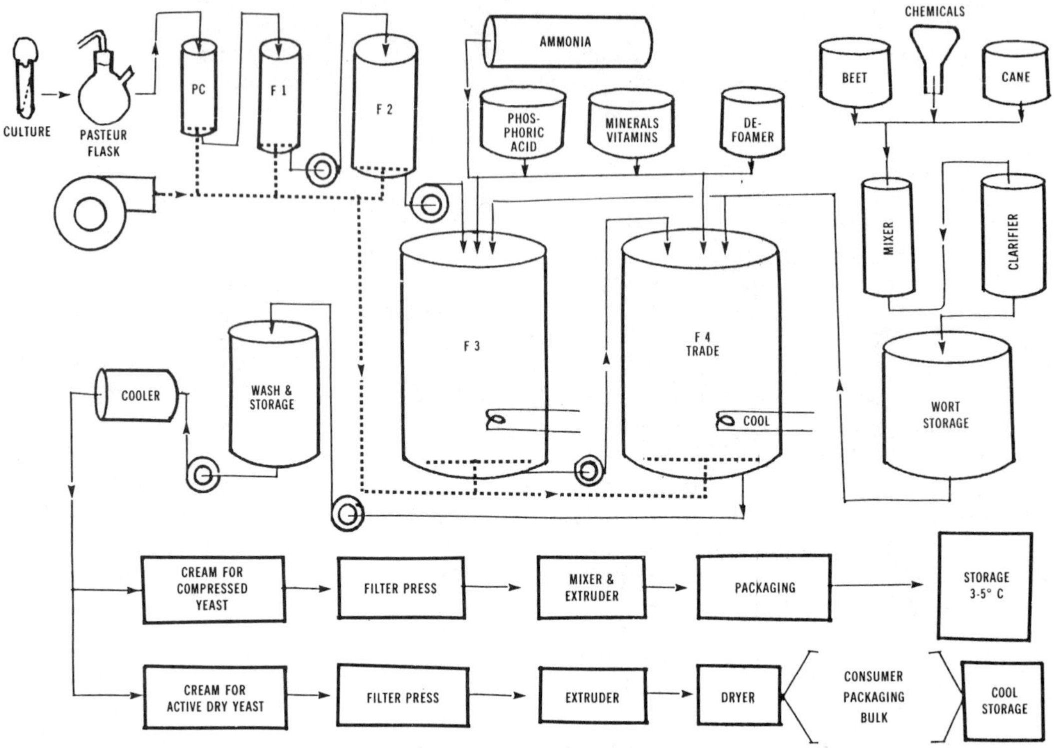

FIGURE 1. Schematic flow diagram for the production of baker's yeast. PC, pure culture; F, fermentation series 1-4.

during the last half hour of the propagation to ensure that yeast is adequately matured and ripened for improved stability during storage.

The optimal level of aeration is critical for success of a baker's yeast propagation. This can be used as a control parameter to maintain the desired alcohol pattern in the exhaust air from the fermenter. Although hydrocarbon analyzers are presently used to maintain this control, more modern equipment like mass spectrometer-type gas analyzers coupled to computers are beginning to provide more advanced control systems for baker's yeast propagations.

Most compressed yeast, or cake yeast (CY), disperses readily when mixed with water. Occasionally a batch of CY may fail to disperse in water readily and could settle out in large clumps. This is sometimes the result of a phenomenon referred to as "grit," which may be due to cell-to-cell interaction with no apparent relationship to flocculation common to brewer's yeast. Grit has caused serious problems, especially to those bakers who use this yeast in water brew systems for production of buns. Formation of grit is dependent on both strain and growth conditions. Gritty strains are less likely to produce grit when they are subjected to conditions that result in catabolite repression or intense aeration that prevents ethanol accumulation during the crucial part of the fermentation [5].

Yeast that is grown to full maturity in commercial propagations is then separated using continuous nozzle separators, and then the yeast is washed several times to improve the color of the final product. Yeast cream is cooled to refrigeration temperature and is stored in clean holding tanks until it is ready for pressing. Removal of extracellular water in yeast cream is achieved using plate and frame presses or, in the continuous mode, by use of rotary vacuum filters. Pressed cake which has a moisture content of approximately 70% is used to make several baker's yeast products. The major types include: (a) compressed yeast (cake yeast), (b) crumbled yeast, (c) active dry yeast (ADY), and (d) instant active dry yeast (IADY). In addition to these products, there is also some interest on the part of some bakers to use yeast cream in their processes, primarily due to convenience and easy handling.

Compressed yeast or cake yeast has a moisture content of approximately 70% and a protein level of 48-60% on a dry solids basis. It also has an emulsifier added to it. Due to the high moisture content, the product is perishable and requires refrigeration during storage. This product is primarily used to produce buns, rolls, bread, etc. Crumbled yeast has characteristics similar to compressed yeast except that it is processed without the emulsifier and is sold in 50-lb bags in the crumbled form. The product is used by large bakeries to produce buns, rolls, bread, etc.

Specially selected strains of baker's yeast have been used to produce ADY. These strains are grown to contain a final nitrogen content of 6.5-7% on a dry solids basis with the final bud index less than 1%. Compressed yeast made with this strain is extruded and noodles or strands are generally dried in continuous tunnel or rotolouver dryers. The products have a moisture content of approximately 6-8% and a protein content of 40-45%. This product is relatively stable at room temperature. Active dry yeast must be rehydrated when used by slurrying it in water at 41-45°C for 5-10 minutes. The product can be used in any of the dough systems where compressed or crumbled yeast is used. Another dry yeast, which is a relatively new product in the market, is instant active dry yeast (IADY). There are also specially selected or constructed strains of yeast which can retain maximum bake activity through the specially designed drying process. The protein content may vary from 38-60% with a moisture content of 4-6%. The drying process provides a porous particle for better rehydration characteristics. Accordingly, the product must be hermetically packaged in inert gas or vacuum to prevent rapid destruction through oxidative process. A special advantage of this product to the baker is that it can be directly added to dry ingredients without prior rehydration. This product is used by housewives for home baking, small bakeries for making bread and buns, etc., and pizza manufacturers for controlled leavening of the pizza dough. The reader desiring more extensive information of these different products is encouraged to refer to Refs, 4, 6, and 7.

B. Brewer's Yeast

Beer is a carbonated beverage made by fermenting an aqueous extract of malt and hops using a special strain of brewer's yeast. Starchy adjuncts are now being used as a partial replacement for the more expensive barley malt to reduce the production cost of beer.

Two species of the genus *Saccharomyces* are important for brewing. These are *Saccharomyces cerevisiae*, a top-fermenting yeast, and *Saccharomyces uvarum*, known for its bottom-fermenting character. Both species are now grouped under *S. cerevisiae* in the latest classification [3]. Top-fermenting brewer's yeasts are used to brew ale-type beers. In such fermentations the yeast tends to rise to the top of the fermenting medium and form a yeast head, which can easily be skimmed off at the end of the fermentation. The bottom-fermenting yeast stays dispersed in the suspension during the active phase of the fermentation and finally settles to the bottom of the fermenting vessel when the fermentation is near completion. Beer thus produced is termed lager beer, which is the most widely consumed beer in most parts of the world (Fig. 2).

In addition to the strain differences already mentioned, ale fermentations differ from lager fermentations in several ways. Most often, worts prepared by the infusion method are used in the fermentations, whereas for lager beer production the accepted method is the decoction or double mash system. The top-fermenting ale strains provide the optimum flavor profile at 15-20°C, although the optimum is 25-26°C. Lager fermentations are generally conducted at 15°C to achieve the desired flavor profile. The differences in strains and fermentation conditions make the two major types of beers—ale and lager—uniquely different in their flavor characteristics.

The wort medium which has been pasteurized by heating before fermentation should contain no viable yeast cells. As the first step in the fermentation process, a culture of the selected yeast strain is inoculated into the wort medium so that initially it contains 10-15 million cells/mL. Typically, a part of the yeast grown in a previous fermentation is used to pitch (inoculate) subsequent fermentations. Most brewers periodically isolate and maintain their most effective cultures in pure form. However, a few who have inadequate facilities generally depend on outside help to achieve these objectives. Different brewing cultures also are available from several institutes throughout the world.

The primary fermentation can be thought of as consisting of two different overlapping phases, one being the initial aerobic growth phase which prevails until all the dissolved oxygen in the wort is depleted, and the other the anaerobic phase in which most of the biochemical reactions take place to provide the characteristic flavor profile to the beer.

The first visible sign of fermentation appears 12-24 hours after the wort is pitched. At this stage the fermenting wort becomes saturated with carbon dioxide and small bubbles begin to appear on the surface, and a white creamy head known as "Kräusen" is formed. Lager-type fermentations are generally conducted at temperatures that do not exceed 15°C at pH 5.5. As the fermentation progresses, a significant amount of heat is liberated, and, for these reasons, fermenters are equipped with cooling devices to maintain optimum fermentation conditions. At 100-120 hours, most of the fermentable sugars which were originally present have been fermented. Accordingly, the yeast tends to become less active and begins to settle.

Completion of a typical lager fermentation takes place in 7-9 days, at which time the fermented beer has a relatively thin surface cover of fine foam. During this phase of the process, most flavor and aroma components, such as higher alcohols, esters, sulfur compounds, and organic acids, are already formed. Such a beer is highly immature and is referred to as "green beer." This beer is generally transferred to lager cellars for maturation. This phase is referred to as lagering.

In most breweries, beer is matured in closed vessels at 3-6°C in the presence of 6-10 million yeast cells/mL initially suspended in the medium. Under these conditions, diacetyl produced during the primary fermentation is reduced through assimilation by the yeast and the sulfury character is slowly purged out through the bubbling action of carbon dioxide. A certain degree of yeast cell autolysis occurs during this period contributing certain characteristic flavor components to the beer.

A part of carbon dioxide saturation occurs during the primary fermentation. Additional carbonation is done at high pressure and low temperature in the conditioning tanks by injecting purified carbon dioxide under pressure. However, the traditional method of carbonation is by use of a secondary fermentation termed "Kräusening." This is a natural carbonation process carried out in the conditioning tanks by use of some wort in the presence of wood chips so yeast generates the carbon dioxide necessary for the carbonation.

During the conditioning period, beer is treated with chill-proofing agents such as papain to enhance chill stability. The finished product is then filtered to make it sparkling clear and transferred to surge tanks for final bottling and packaging.

During recent years, the beer-drinking public has become increasingly calorie conscious. In response to this demand, low-calorie beer production has now become an important segment of the brewing business. Several methods for light beer production have been described in the literature, and almost all such processes have involved addition of fungal glucoamylase to reduce the dextrin content of beer, which would otherwise have contributed additional calories. Several attempts have been made to use *Saccharomyces diastaticus*, a starch-utilizing yeast, to make light beer, but in all such attempts fermentation was inadequately attenuated, and the beer was unpalatable due to the medicinal phenolic flavor.

Genetic manipulation of brewing strains by protoplast fusion or transformation techniques has shown promise [8]. For low calorie or light beer production, selection should be made of strains that can utilize dextrin or starch without introducing phenolic off-flavors to the beer. Since expression of traits as a result of protoplast fusion is random, use of this technique to selectively introduce a single trait into a yeast strain without introducing other undesirable characters is difficult. Transformation is a technique widely used to achieve genetic recombination while overcoming the nonspecificity inherent in fusion. Hence use of plasmids for transfer of DEX genes to brewing strains is presently being investigated in several laboratories. Despite extensive studies throughout the world, there is no

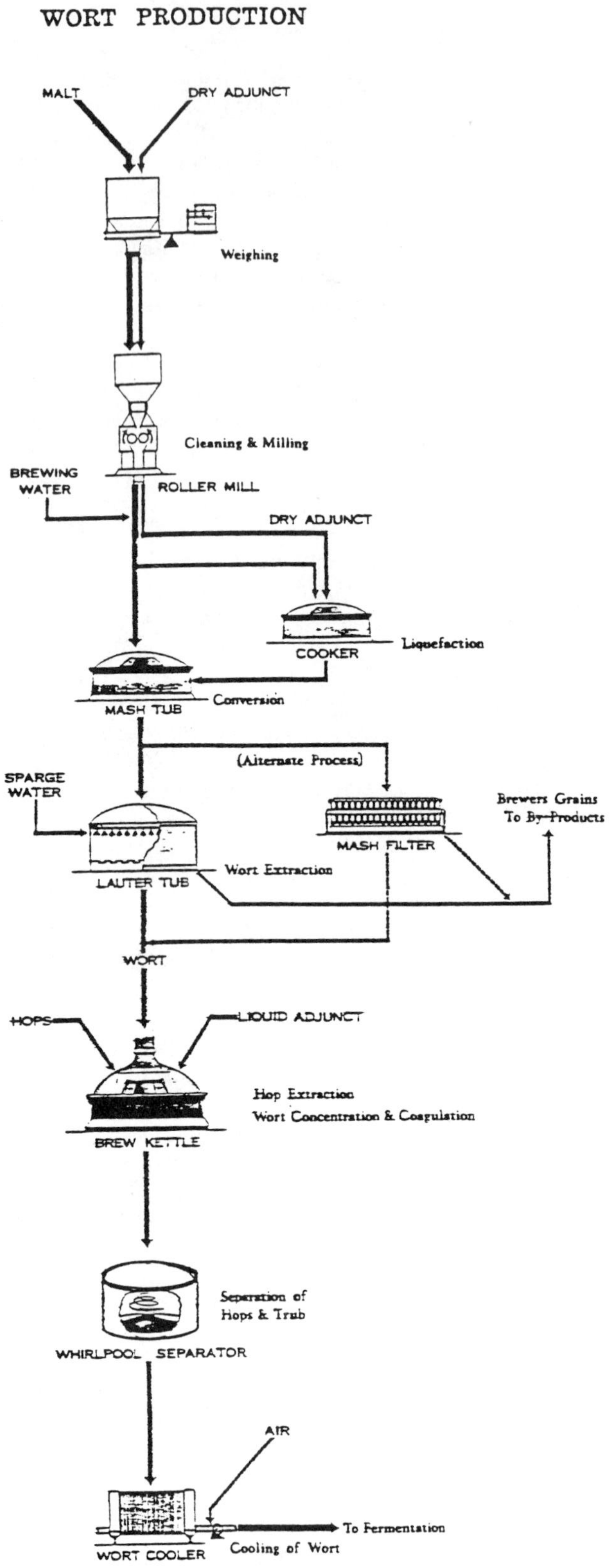

FIGURE 2. Schematic flow diagram for the production of beer.

FERMENTATION & FINISHING

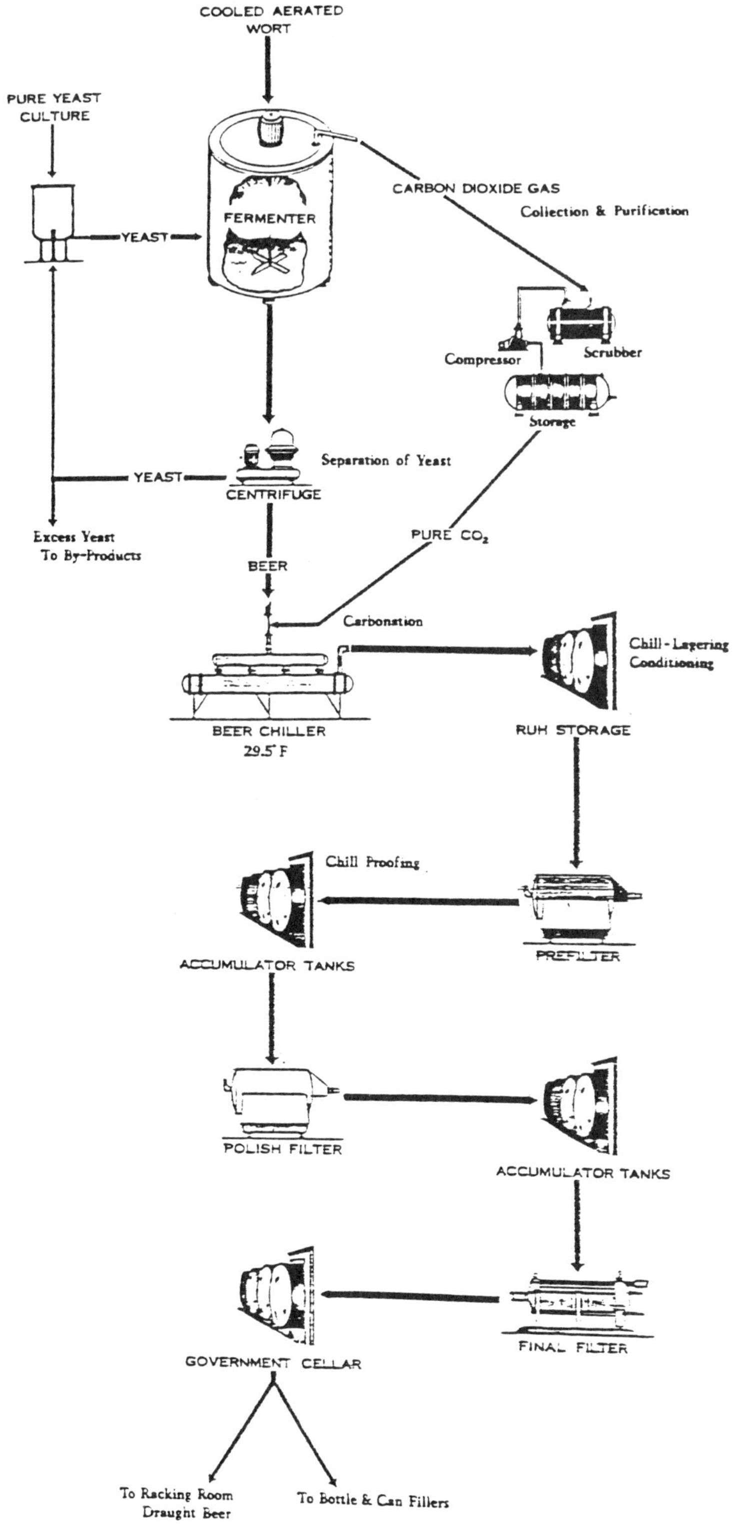

evidence to indicate that a suitable dextrin-utilizing brewing strain is available for commercial production of light beer of acceptable quality. However, advances made to date should eventually result in construction of strains with desired characteristics needed to produce light beer more economically [9].

Wild yeast contamination has been a problem for many brewers. Certain wild yeast strains that excrete a protein toxic to other sensitive yeast strains were first recognized by Bevan and Markower [10]. These yeasts, termed killers, were not effective against other fungi and bacteria. Contamination by killer yeast in both batch and continuous fermentations has resulted in loss of viability and consequent displacement of the brewery strain with production of an unpalatable beer. Manipulation of brewing strains so that they acquire killer characteristics would provide two significant advantages. First, they would be immune to the toxins of their own and related killer yeasts, and second, production of toxins would prevent growth of many wild yeasts, thereby giving protection against wild yeast contamination. Several procedures have been shown to transfer the cytoplasmically inherited killer character from laboratory strains of *S. cerevisiae* to brewing strains. To accomplish this, the rare mating technique has proven successful. Despite these developments, commercial use of such killer strains in brewery fermentation is seldom reported in the literature.

C. Distiller's Yeast

Unlike brewers and winemakers, distilled beverage manufacturers have less dependence on the strain of yeast in their manufacturing practices. One important reason is that distillers can adjust the final flavor profile of their distilled products to a large degree by appropriately controlling conditions of the distillation setup. Such distilled beverages include whiskey, brandy, and grain neutral spirits used to produce such beverages as gin and vodka. There is yet another type of distilled spirit known as industrial alcohol which is commonly produced in large quantities because of its increasing importance as a liquid fuel. These developments have enabled many developing countries to drastically reduce their dependency on imported oil. Although flavors generated by yeast are of some importance to the distilled beverage manufacturer, this character in yeast is of no significance to the manufacturer of industrial alcohol. Accordingly, strains of yeast used in fuel alcohol manufacture possess high alcohol tolerance and a high rate of alcohol production while exhibiting an acceptable level of tolerance to temperature and osmotic pressure.

Whiskey is an alcoholic product distilled from grain mashes such as barley, rye, corn, or wheat at a concentration lower than 190 proof so that the product will have flavor and aroma characteristics of whiskey. A clear distinction can be made among various whiskey products based on differences in their flavor profile resulting from variations in their manufacturing procedures and their corresponding mash bills. For example, a specially prepared barley malt, having been kilned with burning peat, is used as the only substrate to produce Scotch malt whiskey. In production of Scotch grain whiskey, fermentables are derived from unpeated malt, green malt, and corn grit. Fermentation is done at 20-22°C for 36-100 hours, depending on initial wort concentration and level of pitch. A strain of *S. cerevisiae* is used in these fermentations.

Distilled spirits for both malt and grain whiskeys are reduced to 110 proof and stored in oak casks for a minimum of 3 years. When the products are fully matured, the malt whiskey is often blended with grain whiskey and sold as blended Scotch whiskey.

In the manufacture of neutral spirits, the distiller's primary objective is to produce a high grade ethanol, free of congeners, using a highly cost-effective process. Fermentations are designed to achieve maximum productivity using cheap carbohydrate sources. The product is usually distilled at a strength exceeding 190 proof. It is most often produced from grain mashes, although molasses and potatoes are used in certain countries for economic reasons. A more detailed account of production of these distilled beverages may be found in Nagodawithana [4].

With the ever-increasing depletion of economically recoverable petroleum reserves, production of ethanol from vegetable sources as a partial or complete replacement for conventional fossil-based liquid fuels has, in the past, become more attractive. In such fermentation processes, yeasts play an important role in the conversion process. Hence, selection of yeasts for industrial fermentations is based on high ethanol productivity and tolerance to high concentrations of ethanol. Accordingly, the yeast is geared to produce a uniform, rapid fermentation and maximum alcohol yields at a range of temperatures and pH values.

Almost all the yeasts currently in use are strains of *S. cerevisiae*. Industrial alcohol is produced by fermentation of various sugar sources such as corn, molasses, and other starchy substrates after saccharification. For most fuel alcohol production in the United States, corn is employed as the principal raw material (Fig. 3).

Starch in the corn is solubilized and saccharified to glucose as whole mash or after the various purification steps as practiced in the corn wet milling industry. It is not necessary that complete conversion of the dextrins occur before pitching with yeast. As soon as sufficient sugars are available to support a yeast population, the fermentation process can be initiated. At this stage, the mash should be about 8-12% solids (33 gallons liquid/bushel).

Yeast strains specially selected for distillery fermentations with high ethanol productivity and exceptionally high ethanol tolerance are now commercially available, most often as distiller's active dry yeast (DADY). These dry products are first rehydrated with three to five times their weight of warm (41-45°C) water for 5-10 minutes. The recommended 2-4 pounds of dry yeast per 1000 gallons of mash will result in an inoculum of 5-10 million cells/mL. Optimal fermentation conditions are 30°C and pH 4-5 with intermittent aeration at very low levels [1]. When grains are left in the wort, or when backset is used, the buffering capacity of these components assists in maintaining the required pH. The time required for the fermentation, usually 48-72 hours, is dependent on temperature, pitching rate, and strain of yeast used.

The final step in the process is distillation of the beer and ultimate disposal of the stillage. Ethanol in the beer is 9-12% (v/v), and this is concentrated to 30-40% in the beer still. After a series of columns to

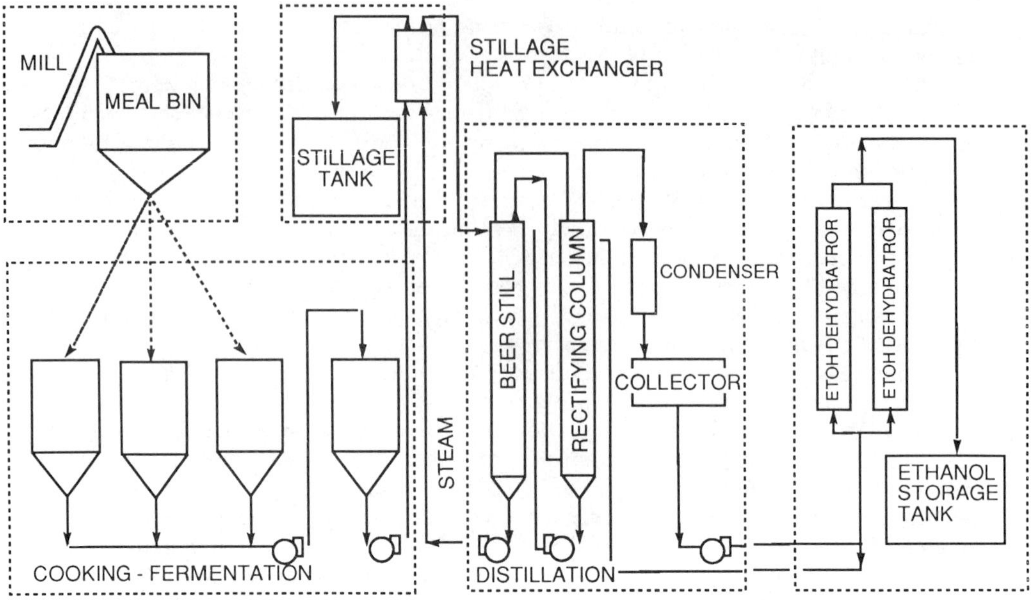

FIGURE 3. Schematic flow diagram for the production of fuel alcohol.

remove other organic compounds formed during the fermentation, ethanol of 95% concentration is produced. This is dried to exceed 99.5% purity by an azeotropic distillation. The dried ethanol is blended with denaturants, such as 5% unleaded gasoline, and shipped out as fuel ethanol.

D. Wine Yeast

The character and quality of a wine is determined by (a) composition of the raw material from which it is made, (b) nature of the fermentation process, and (c) changes that occur naturally, or are made to occur, during the postfermentation period. The fermentation process with emphasis on the fermenting yeast will be dealt with in some detail in this report.

Traditionally, most of the large wine-producing countries have relied almost entirely on spontaneous fermentation. This means fermentation is induced by indigenous microorganisms already found on grapes and equipment used for wine making. While most of the world's most acclaimed wines are made by spontaneous fermentations, there are insufficient data to support that their superior quality is a result of the adoption of this fermentation method. Nevertheless, wines produced by spontaneous fermentations often have a more rounded flavor [12].

Yeasts responsible for natural (spontaneous) fermentation of must include 18 genera and in excess of 150 species. These strains could be roughly divided into three groups: those which exert little or no influence, those which can carry out a normal wine fermentation, and those able to cause spoilage. Progress of the fermentation and quality of the finished wine depend largely on prevalence of the "normal" yeast, which in turn depends on the climatic conditions, vineyard environment, grape and must handling, and cellar sanitation. Winkler [13] found a natural yeast count of approximately 10^5-10^6 cells/mL in fresh grape juice.

During the early stages of the must fermentation, asporogenous yeasts predominate, to be replaced entirely by sporogenous yeasts during the latter stages of the fermentation. The fermentation is usually initiated by yeast with a low tolerance for ethanol such as *Kloeckera apiculata, Hanseniaspora guillermondi,* and species of *Torulopsis* and *Candida.* These are superseded by more alcohol-tolerant *Saccharomyces* yeasts. This phenomenon has been described extensively over the last 20 years [14-18]. It is not, however, clear whether superior quality wines require such sequential fermentation by naturally occurring microbial flora as previously described.

Use of selected cultures was started earlier in this century. It is based on the idea of using pure culture strains of a suitable wine yeast. These strains are usually available in pure culture form in test tubes and can be grown in sterile musts, first in flasks and then gradually increasing the cell mass by passing from one vessel to a vessel of increasing size until a sufficient amount of yeast has been produced to inoculate a commercial-size fermenter.

The procedure mentioned above does not, however, lead to true pure culture fermentations since, on a large commercial scale, the must is not sterilized. Sulfur dioxide is used to suppress some of the normal flora. Despite this treatment, some yeasts survive, and these together with the inoculated yeast culture make it possible to produce a clean-tasting wine of consistent quality. These pure culture strains are generally available to the winemaker from such places as the University of California, Davis, the American Type Culture Collection in Washington, DC, Geisenheim Institute in West Germany, the South African Research Institute in Stellenbusch, the Australian Wine Institute, etc.

Interest in having commercial yeast manufacturers produce compressed wine yeast or active dry wine yeast (WADY) began in the late 1950s. Pioneering investigations were done independently [19-22].

Wine yeast is grown aerobically at carbohydrate-limiting conditions and dried like active dry yeast (ADY), as described elsewhere in this chapter. Active dry yeast usually contains 20-30 billion cells/g of dry product. Currently, active dry wine yeasts are used extensively in Germany, Italy, the United States, and, to a minor extent, in France. Introduction into Spain and South American countries has been quite slow.

Several strains of wine yeasts are now available in active dry form, but Montrachet (UCD 522) and Pasteur Champagne (UCD 595) constitute the most used. Montrachet's strain of *S. cerevisiae* is a fast-starting, vigorous, fermenting strain with high resistance to SO_2. It is well suited to a variety of red and white table wine types. Epernay 2, which is a strain of *S. cerevisiae,* is now beginning to receive attention from many winemakers. Pasteur champagne is a strain of *Saccharomyces bayanus.* It has a moderate fermentation rate and strong finishing characteristics which may be beneficial in the presence of high CO_2 and alcohol levels at higher than normal temperatures. Other strains available in an active dry form are used for specialized applications. California champagne (UCD 505), a strain of *S. bayanus*, is used for sparkling wine fermentation because of its rapid and compact flocculation. Flor Sherry (UCD 519), which is a strain of *Saccharomyces fermentati*, is used for submerged culture sherry fermentations because of its high aldehyde production.

In the cooler wine-producing regions of the world, there is a higher concentration of malic acid in the must which is reflected by lower pH and poor quality. An efficient malo-lactic fermentation can reduce acidity of wine by conversion of a strong acid (malic) into a weak acid (lactic), thereby improving quality and stability through action of lactic acid bacteria.

Although the malolactic fermentation is highly beneficial to red wines, it occurs less frequently in white wines. Several species of lactic acid bacteria occur in wines. Of these, a species of heterofermentative cocci, *Leuconostoc oenos*, has been considered the most effective organism for the malolactic fermentation. This fermentation is not, however, important for musts low in malic acid concentration. Many commercial preparations of *L. oenos* are now available in many parts of the world.

The pathways suggested earlier involve mediation by enzymes such as malate dehydrogenase or malic enzyme for conversion of malic acid to an intermediary compound, pyruvic acid, and then reduction of pyruvic acid to lactic acid through action of the enzyme lactic acid dehydrogenase. More recent studies have shown existence of a complex bacterial enzyme system designated by Radler et al. [22] as "malolactic enzyme" (EML), which can convert malic acid to lactic acid without mediation of lactic acid dehydrogenase. Although the precise composition of the EML complex is not determined, it is activated by NAD.

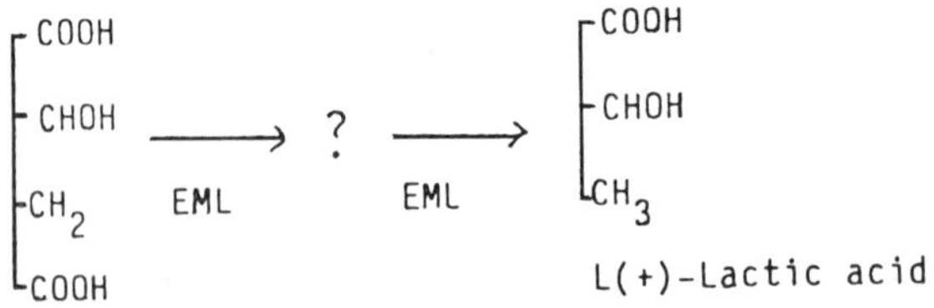

Malic acid

As with brewery fermentations, the presence in the must of killer yeasts has resulted in reduced fermentation activity as a result of loss of viability by the fermenting yeast. Recent studies have shown that the killer characteristic can be introduced into sensitive wine strains by hybridization or cytoduction. Such manipulations would give the wine yeast two significant advantages. First, it would be immune to its own toxin and to other related killer yeasts, and second, production of the toxin would prevent growth of many wild yeasts that would otherwise have grown, thereby giving protection against wild yeast contamination.

Thus it would be highly desirable to have strains for winemaking which can ferment musts to relatively high ethanol concentrations and, at the same time, produce fermentation products with just the desired concentrations of fusel oil and other congenerics to provide the balanced flavor profile characteristic for a particular wine. Certain leucine-requiring yeast mutants have already been developed which produce only trace quantities of isoamyl alcohol (a component of fusel oil) with other characteristics associated with winemaking unaffected. It is also known that some valine-requiring mutants produce markedly less isobutyl alcohol than do their parent strains. Although there are several publications on this subject [24,25], currently there is no evidence that these strains are used for commercial winemaking.

Recombinant techniques are presently being applied to strain improvement of wine yeast, but there is a need for thorough understanding of the precise genetics of yeast associated with the important characteristics of winemaking. However, despite overcoming such technical problems, there is still the difficult task ahead to obtain federal regulatory approval and, above all, acceptance of such products by the consumer.

III. YEASTS IMPORTANT FOR PRODUCTION OF SINGLE-CELL PROTEINS (SCP)

Difficulties of providing adequate food for the world's population have become increasingly apparent since the early 1960s. These problems are likely to intensify with time. While people in many parts of the world are already experiencing malnutrition, reports of starvation and death from the worst affected areas are common. Protein-rich food supplies from conventional sources are finding it difficult to keep pace with the increasing population and also are becoming unaffordable to a large segment of the human race. Accordingly, considerable effort has been expended to develop processes to produce low-cost protein-rich foods from conventional sources such as algae, bacteria, and yeasts. The dried products made from such organisms are collectively referred to as "single-cell proteins" (SCP), which was originally proposed in 1966 by C. L. Wilson of the Massachusetts Institute of Technology [26]. In general, yeast and yeastlike fungi are preferred because their cells are larger than those of bacteria and thus permit easy recovery by conventional methods with low energy input. Above all, yeast is accepted as a safe food ingredient as a result of its usage throughout the centuries.

The widely available single-cell protein is a byproduct of the brewing industry. Yeast grown during a beer fermentation exceeds the amount that is required for pitching fresh wort. Roughly three-quarters of a pound of excess yeast solids is recovered per barrel of beer brewed under standard conditions. This surplus yeast is concentrated into a cream and may be pasteurized and often drum-dried as feed yeast, mixed with spent brewer's grains for use as fodder, debittered and dried as food yeast, or converted to value-added products with nutritional or flavor-enhancing characteristics by further processing. The bitterness in raw brewer's yeasts results from adsorbed hop resins (isohumulones) on cell surfaces of yeast. They are commonly removed by solubilizing the hop components under alkaline conditions and recovering the yeast by centrifugation. The final suspension is dried after neutralization with an acid such as hydrochloric acid. The product is available as powder, flakes, or tablets, often supplemented with additional minerals and/or vitamins.

The two most widely used species for SCP production in the family Saccharomycetaceae are *Saccharomyces cerevisiae* and *Kluyveromyces marxianus*. The corresponding asporogenous species in the family Cryptococcaceae is *Candida utilis*. These three strains can use cheap carbohydrate substrates which are essentially difficult to treat waste streams of other industries, and also have high growth rates and yield coefficients which have a favorable impact on production cost of SCP.

A. *Saccharomyces cerevisiae*

Baker's yeast, *Saccharomyces cerevisiae*, which is grown as a primary yeast using cane and/or beet molasses, was used as a food or feed yeast in Germany during World War I. This yeast requires certain B vitamins, mainly thiamine, inositol, pyridoxine, and pantothenic acid, which are commonly derived from blends of cane and beet molasses. The detailed procedure for the propagation has already been described in this chapter.

B. *Candida utilis* (Torula Yeast)

The carbon sources for this species are hexoses, pentoses, ethanol, and many organic acids. This species was propagated as a source of food or feed yeast in Germany during World War II using wood hydrolysates. It can use pentoses in wood hydrolysates. Unlike *S. cerevisiae*, *C. utilis* does not require any amino acids or B vitamins for growth. However, such propagations should be supplemented with phosphoric acid and ammonia as sources of phosphorus and nitrogen, respectively. This combination, when fed appropriately, should provide the necessary pH control. When sulfite waste liquor is used, it is first sterilized by passing through heat exchangers. Nevertheless, sterile conditions are not maintained during growth. Control of contaminants is achieved by maintaining an acidic (pH 4.5) pH, aeration with sterile air, and by maintaining a high concentration of yeast cells within the fermenter. The temperature is maintained at 35-37°C.

Candida utilis also can be propagated on ethanol [27]. In comparison with other waste materials employed as substrates in many commercial fermentations, ethanol is sufficiently expensive to require that it be used most efficiently if selected as a substrate. Operation of the closed propagation system is generally continuous under highly aerated conditions with the incoming sterile feed containing 100-500 ppm ethanol. Several macronutrients are generally fed along with the ethanol substrate. The micronutrients are fed separately to the fermenter. Ammonia is added continuously as a nitrogen source and also to maintain a pH within the 4-4.5 range. Temperature should not exceed 37°C, and the headspace pressure within the fermenter is maintained at approximately 10 psig to increase the oxygen transfer rate and to assist in preserving aseptic conditions. Fermented broth is continuously withdrawn when the suspended yeast in the liquor reaches approximately 3% on a solids basis. The withdrawal rate maintaining this cell concentration should provide an average residence time of approximately 3 hours with a dilution rate of 0.33/hour. The fermentation broth is separated by centrifugation, washed two or three times, pasteurized in plate and frame heat exchangers at 95°C, and the yeast cream is spray-dried to give a food grade yeast product.

When supplies of crude oil were cheap and plentiful and the cost of protein from conventional sources was high, there was considerable interest in using hydrocarbon fractions from the petroleum industry as substrates for production of SCP. Several *Candida* sp. were used for large-scale biomass production. Of the petrochemical substrates available, *n*-alkanes and gas oil were found suitable for use in biomass production. Air lift-type fermenters with their economical way of achieving the required oxygen transfer rate can make the process highly cost effective. Yeast grown on substrates like *n*-alkane have shown yields of 0.88-1.11 based on the weight of the *n*-alkane used [28]. These yields are significantly higher than those for sugar substrates which generally fall within the 48-50% range. There are, however, several setbacks which prevent extensive use of hydrocarbon-based substrates for production of SCP for food use. A major obstacle is to achieve a clean SCP free of residual hydrocarbons. Presently, there are clinical studies in progress aimed at confirming the safety of these protein-rich products for human use.

C. *Kluyveromyces fragilis (marxianus)*

Approximately 18 million tons of cheese whey are produced annually in the United States, of which about 4% is lactose. This represents a supply of 700,000 tons of whey sugar annually, which can be used as a substrate for production of SCP using a lactose-fermenting yeast. Although there are a few yeast strains with lactose-utilizing ability, *K. fragilis* (*marxianus*) is used in the United States to produce SCP for food use.

Currently, there are three factories in the United States able to produce food grade *K. fragilis*, the newest being at Juneau, Wisconsin, and administered by Universal Foods Corporation, Milwaukee. This factory has a production capability of 10,000 tons/year. The process described by Bernstein and Plantz [29,30] includes an initial addition of ammonia, phosphoric acid, mineral, and yeast extracts to whey. This is followed by an adjustment of pH to 4.5, pasteurization (80°C for 45 minutes), cooling to 30°C and transferring to 15,000-gallon fermenters equipped with aeration, agitation, and cooling to maintain the fermentation temperature of 35°C. Eight to twelve hours following inoculation with *K. fragilis*, a very low level of fermentable sugar can be expected. This marks the beginning of the continuous mode with initiation of continuous feed at 1250 gallons/hour. Rapid monitoring of the fermenting medium for lactose and glucose concentration can help in optimization of the feed, thereby improving production efficiency. For feed yeast manufacture, the whole fermented mash can be concentrated and dried. Primary grown yeast intended for food use requires extensive cleaning. Fed batch protocols as used for baker's yeast production also can be used to propagate *K. fragilis* without hampering either product quality or productivity.

Several reviews have been published on nutritive value of SCP [6,26]. Accordingly, both biological value (BV) and net protein utilization of SCP products of yeast origin are comparable to vegetable protein but significantly lower than for animal proteins. This is primarily due to deficiency in methionine, although yeast proteins are rich in lysine. Supplementation with methionine improved the biological value of yeast proteins.

Yeast contains 6-15% nucleic acid as compared to 2% in meat products. High nucleic acid intake in the human diet increases the uric acid level at physiological pH. It has a tendency to precipitate or crystallize in joints, causing gout or gout-arthritis. Formation of renal stones in the urinary tract is also common with high nucleic acid diets. The safe level of nucleic acid intake for humans has been established at 2 g/day [31]. This corresponds to 20 g SCP/day, which would account for only one-sixth of the recommended daily allowance of 65 g of protein per day for a 70-kg adult male. Hence, it is clear that nucleic acid content of yeast must be reduced before it can be treated as a major source of protein in the human diet.

IV. YEAST PRODUCTS OF INDUSTRIAL IMPORTANCE

With industrialization of the brewing industry toward the end of the eighteenth century, brewers had to dispose of excess yeast by separate means because it contributed a substantial proportion of the waste load. Thus these materials often were sprayed on agricultural land as fertilizer.

Toward the end of the nineteenth century, several studies indicated that brewer's yeast contained high protein levels satisfying both human and animal nutritional requirements. More recent studies have demonstrated that brewer's yeast is an excellent source of vitamin B complex. Many large brewers who were impressed by these findings began extensive investigations to produce value-added products for commercial use. The success of these efforts was first apparent in England and Germany with the first commercial introduction of nutritional and flavoring agents in the food market. Further developments during the last few decades have prompted introduction of other commercial products of yeast origin, for example, products of therapeutic value, of flavor-potentiating ability, with growth-promoting characteristics for use in microbiological media and biochemical in nature. These advances have led to numerous industries that will be discussed in some detail.

A. Products that Impart Flavor

1. Production of Yeast Extracts

Yeast extract is a concentrate of soluble material derived from yeast following hydrolysis of cell material, in particular the proteins, by using its own enzymes, or by other methods that release the cell contents in a degraded form. It has become popular during the last few decades in part due to its usefulness as a natural flavoring agent and also its ability to enhance growth of microorganisms when present in microbiological media. A major consideration in its use as a flavoring agent is cost-effectiveness related to other flavoring agents on the basis of equivalent flavor intensity. Total output of yeast extract worldwide is approximately 35,000 tons, which amounts to a $130 MM industry.

In many countries, much of the surplus yeast in the brewing industry is used to produce a variety of extracts of commercial value. Such extracts also can be made from other species of yeast like *K. fragilis* grown on whey or *Candida utilis* grown on ethanol or other carbohydrate substrates.

The choice of the starting material for extract production is based on the cost of the yeast and its availability. In general, primary grown yeast is more costly than brewer's yeast for obvious reasons. However, brewer's yeast as it comes from the brewery is likely to contain undesirable flavor characteristics, mainly because of carryover of bitter hop resins from the fermentation. These bitter components remain firmly adsorbed to the outer surface of yeast cells, and any further processing for extract production requires a debittering step which often is costly. However, brewer's yeasts can produce higher yields and flavor characteristics different from those obtained from primary grown yeast.

Several factors influence flavor and acceptability of yeast extracts. These include type and condition of the yeast, presence of extraneous matter, extraction and processing conditions employed, and level of contamination during processing. When extraction of cell material is by autolysis, involving auto-digestion of cell components by endogenous enzymes, high viability of the processing yeast is a prerequisite for acceptable autolytic activity. Presence of extraneous matter can either enhance or adversely affect flavor characteristics of the final product. Accordingly, well-washed baker's yeast gives an extract with a flavor more bland than that from inadequately washed molasses-grown yeast. Likewise, extracts from *K. fragilis* grown on whey permeate is generally bland. Such products are made meaty by addition of the required level of molasses residues before initiation of the extraction procedure. Although there is a low contamination level in a regular autolysis, precautionary measures must be taken at every step to have it under control so that the end product of these contaminants is not sufficiently high to alter the flavor profile of the yeast extract.

Extract manufacturers generally attempt to keep highly confidential all critical information on extract manufacture. Nevertheless, for descriptive purposes, the process may be divided into the following basic operations: (1) cleaning yeast, (2) solubilization, (3) removal of particulate matter, (4) clarification, (5) concentration, and (6) drying.

Brewer's yeast intended for use in extract production must be highly viable and contain only a minimum level of microbial contaminants, hop bitters, trub, and other undesirable particulate matter. Most of the particulate matter may be removed by passing the yeast cream through a screen of 150-200 mesh size. Primary compounds responsible for bitter notes are isohumulones, isomerized products, of humulones and hop acids derived from hops added to a brewery fermentation. These bitter components remain adsorbed to the surface of the yeast cell and generally are released into extract medium during autolysis, if not removed before initiation of solubilization. A common method used to remove hop bitters before autolysis is based on solubility of isohumulones at higher pH values. Accordingly, clear yeast cream is treated with a mild alkali solution to pH 9 to solubilize the hop bitters. The yeast is then washed and finally adjusted to pH 5 before the start of the autolysis.

Brewer's yeasts from different breweries vary in composition, specifically in nitrogen content. Hence, selection of raw materials on the basis of nitrogen content could be critical for certain extracts which have to meet definite protein requirements. Most extract producers use blends of different yeasts to improve the consistency of the final product.

Yeast extracts are then produced by any of three alternative process routes, namely: (a) autolysis, (b) plasmolysis, or (c) hydrolysis.

Autolysis: Most cell constituents in a viable cell are too large to pass through the cell membrane. Hence, these macromolecules have to be solubilized before extraction. Autolysis, which is one of the methods for solubilization can be used to meet this requirement. This is based on application of carefully controlled heat sufficient to kill cells without adversely affecting activity of degrading enzymes within the cell. Accordingly, yeast slurries at 12-15% solids can be made to release their cell components in a degraded form by subjecting them to temperatures of 45-55°C at pH 5-6 for 40-48 hours, sometimes in the presence of autolyzing aids. When yeast cells are killed without inactivating endogenous enzymes, enzyme systems responsible for carrying out metabolism are no longer governed by the delicate control mechanisms that generally exist in normal healthy cells. Death increases entropy and causes disorderliness within the cell. The degrading enzymes then begin to indiscriminately attack their specific substrates causing breakdown of macromolecules like proteins and nucleic acids to their subunits which are soluble in water. At the same time, the highly organized semi-permeable membrane begins to lose its integrity primarily due to degradation of membrane proteins, causing the soluble substances to leak out while retaining macromolecules within the cell. The cell wall, which is made up of glucan and mannan, is highly porous and does not hinder passage of soluble components from within the cell. The internal proteolytic enzymes degrade the protein into polypeptides, peptides, and amino acids. Nucleic acid-degrading enzymes degrade nucleic acid, primarily RNA, to its corresponding nucleotides and nucleosides. All these degraded components diffuse from the cell and become the major components of the yeast extract (Fig. 4).

The extract yield is largely governed by temperature, pH, duration, and type of autolysis aids used during autolysis. Although some parameters influence the flavor profile, optimum conditions to obtain a given flavor profile may not exactly coincide with those previously arrived at for achieving optimal yield.

The type of self-digestion already described may often be augmented by addition of proteases, for example, papain, at about 0.05% (w/v) of the yeast cream. However, such treatments increase extract yields only in longer autolysis processes as practiced in the United States, United Kingdom, and Australia. Processes often used in France and Japan have shorter autolysis time (20 hours) and hence do not rely on added proteases for improvement of solubilization. Presently, most of the low sodium extracts are made by the use of the autolytic procedure. Although these products are bland, they can produce a variety of flavor characteristics when in combination with other food ingredients. In addition, low sodium extracts are presently being used extensively in special applications such as food for convalescents or infant food formulations.

Plasmolysis: This method is often used by extract manufacturers for rapid initiation of the cell degradation process. The most accepted chemical used to achieve this effect is salt. It is also customary to use certain organic solvents, such as ethyl acetate or isopropanol, to facilitate the

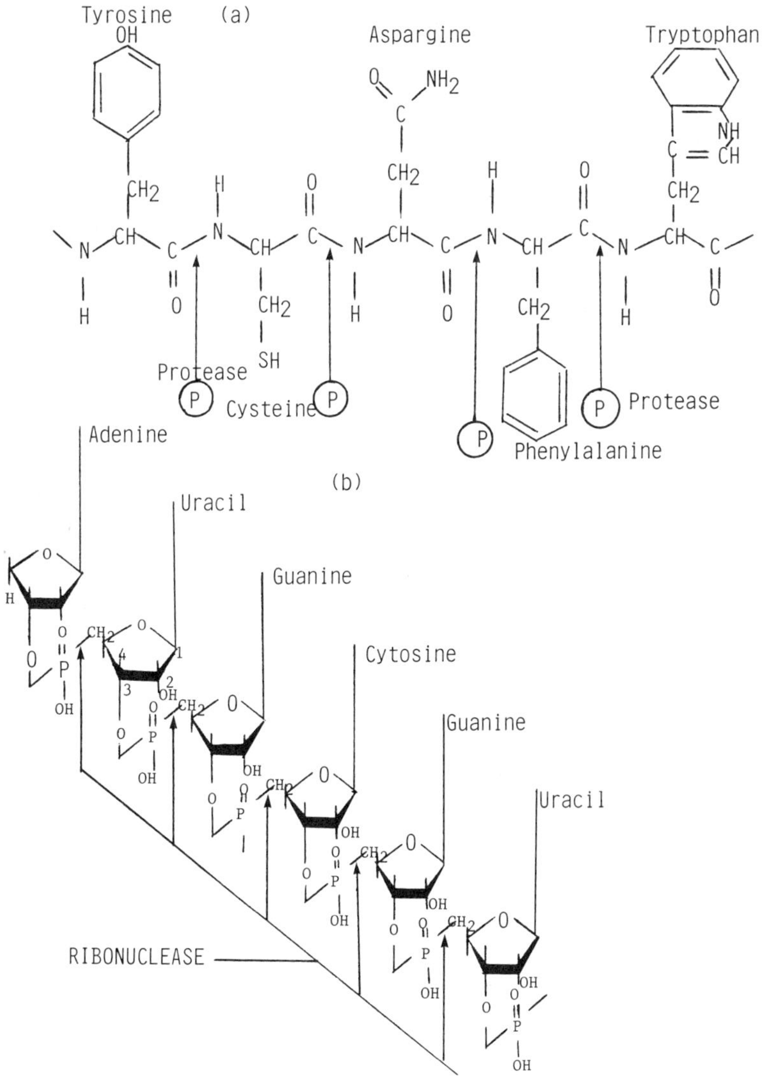

FIGURE 4. Enzymatic degradation pattern of (a) protein and (b) nucleic acids during an autolysis. P = protease.

autolysis. When these solutes are present in the medium, the yeast cell begins to lose water in an attempt to balance the osmotic pressure differential. Under extreme conditions, the cytoplasm tends to separate from the cell wall, a condition known as plasmolysis. When this condition persists long enough, yeast cells rapidly die marking the beginning of the degradative process. In practice, salt is often added to pressed yeast at the time of liquefaction to achieve best results.

Addition of salt as a plasmolyzing agent has the advantage of faster initiation of autolysis, resulting in faster solubilization, and also has a bacteriostatic or bacteriocidal effect reducing growth of any contaminating microorganisms which may otherwise cause clarification problems, or sometimes affect flavor characteristics of the final product. However, because of the high salt content, these products have limited use in the food industry. Presently, there is a great demand for low sodium extracts, and such products are generally made by autolysis rather than plasmolysis.

Hydrolysis: Although this is the most efficient method to solubilize yeast, it is the least practiced in commercial extract production. It makes use of the action of hydrochloric acid on raw yeast at specific temperature and pressure. Typical hydrolysis begins with dried yeast slurried with water to a solids concentration of 65-80%, followed by acidification with concentrated hydrochloric acid. Hydrolysis is carried out at 100°C in a wiped film evaporator equipped with a reflux condenser until the required level of amino nitrogen is achieved. This generally takes place in 6-12 hours, converting 50-60% of the total nitrogen to amino nitrogen [32]. Neutralization of the hydrolysate to pH 5-6, usually with NaOH, is followed by filtration and concentration into a syrup of 45% solids, or to a paste of about 80% solids, or spray-dried to give a product of about 5% moisture.

Although process yields are higher than those for autolysis or plasmolysis, several disadvantages make the process difficult and cost intensive. The corrosive nature of the acids makes it essential that vessels and equipment used for handling the highly acidic media be lined with glass or another unreactive material. Pressure vessels used for hydrolysis should be specially constructed to achieve the required safety standards. Hence, the high capital cost necessary for factory construction makes the process unattractive for extract production. Furthermore, the high salt content, poor flavor characteristics, high carbohydrate level in the final product, and destruction of certain amino acids and vitamins have made the process less appealing than the other two procedures described previously.

2. Production of Autolysates

Solubilized yeast after autolysis is termed an "autolysate" and has, in addition to soluble extract material, a substantial proportion of particulate matter which, in essence, is the cell wall fraction. This cell debris is generally separated from the extract during production of yeast extracts. In production of autolysates, the entire product of autolysis is concentrated and dried for commercial use. Although the product has an inferior flavor profile compared to extracts, it is extensively used in flavor formulations, primarily to take advantage of the water-binding characteristics of the product mainly because of the presence of cell wall material.

3. Yeast Proteins with Reduced Nucleic Acid

Although yeast has 40-55% protein, its high nucleic acid content makes it unattractive as a protein source in the human diet. Ingestion of large amounts of nucleic acid causes certain physiological problems. Humans do not have the uricase enzyme to degrade uric acid formed as a result of ingestion of excessive levels of nucleic acid-containing yeast. Uric acid causes arthritic or gouty conditions or, in some instances, renal stones in the urinary tract of humans.

In yeasts, nucleic acid consists mainly of RNA with a limited amount of DNA. Certain strains of *Candida* (e.g., *C. utilis*) contain 10-15% RNA [33]. Likewise, *Saccharomyces* strains have RNA in the 6-11% range.

The safe limit for nucleic acid intake in the human diet has been established to be 2 g/day [31] corresponding to about 6-7 mg of uric acid in 100 mL of blood plasma, which is generally considered the upper limit of the range for the U.S. male population [34]. Protein-rich products isolated from yeast also contain 12-15 g nucleic acid/100 g crude protein [35,36]. To make these yeast-derived protein-rich products useful as functional food additives, their nucleic acid content must be reduced substantially to avoid health hazards. For production of low RNA yeast proteins, the process involves rupturing cells and separating insoluble cell debris from soluble substances, which are made up predominantly of nucleic acid and proteins. The RNA which is the major nucleic acid in yeast is intimately associated with proteins, and reduction of RNA can be achieved by three techniques described in the literature.

The preferred method for nucleic acid degradation involves use of endogenous nucleases under conditions most conducive for nuclease activity. These conditions, which are generally not optimal for endogenous proteases, allow extraction of proteins unharmed with the minimum level of nucleic acid. The protein is precipitated at a low pH level, leaving behind the degraded RNA in the soluble form. The soluble fraction containing most of the RNA fragments becomes the extract fraction that has commercial value for its flavor-potentiating characteristics. The second method of reducing RNA involves treatment of RNA-containing soluble substances with alkali to reduce the protein/nucleic acid interactions followed by acidification to precipitate only the proteins. Still another method of producing low RNA proteins is to treat soluble materials with malt sprouts as a source of exogenous nuclease. This treatment causes RNA to hydrolyze, and protein may be precipitated by acidification. Use of malt sprout extracts for production of flavor enhancers will be described in the next section.

4. 5'-Nucleotide-Rich Extracts

In 1960, manufacture of two flavor enhancers, 5'-IMP and 5'-GMP, by enzymatic hydrolysis of RNA, was established in Japan. The enzymes used for this hydrolysis were of microbial origin. Yeast was used as the source of RNA because its nucleic acid consisted mainly of RNA (2.5-15%) with a relatively small level (0.03-0.52%) of DNA [37,38]. Its economic production as compared to other microorganisms has made yeast the organism of choice for production and supply of RNA for the 5'-nucleotide industry.

Candida utilis with an RNA content of 10-15% on a dry solids basis is used commercially as the RNA source. Baker's yeast, *S. cerevisiae*, which contains a somewhat lower (8-11%) RNA concentration also can be used for 5'-mononucleotide production. It is highly desirable to harvest the yeast in the logarithmic growth phase, at which time the RNA content is maximum due to high protein biosynthesis [39,40]. Endogenous nucleases inherently present in yeast can hydrolyze the RNA to 3'-nucleotides or nucleosides, which do not impart any characteristic flavors. However, under certain controlled conditions, RNA can be hydrolyzed in the presence of certain exogenous enzymes to 5'-mononucleotides, some of which are flavor potentiators important for the food industry (Fig. 5).

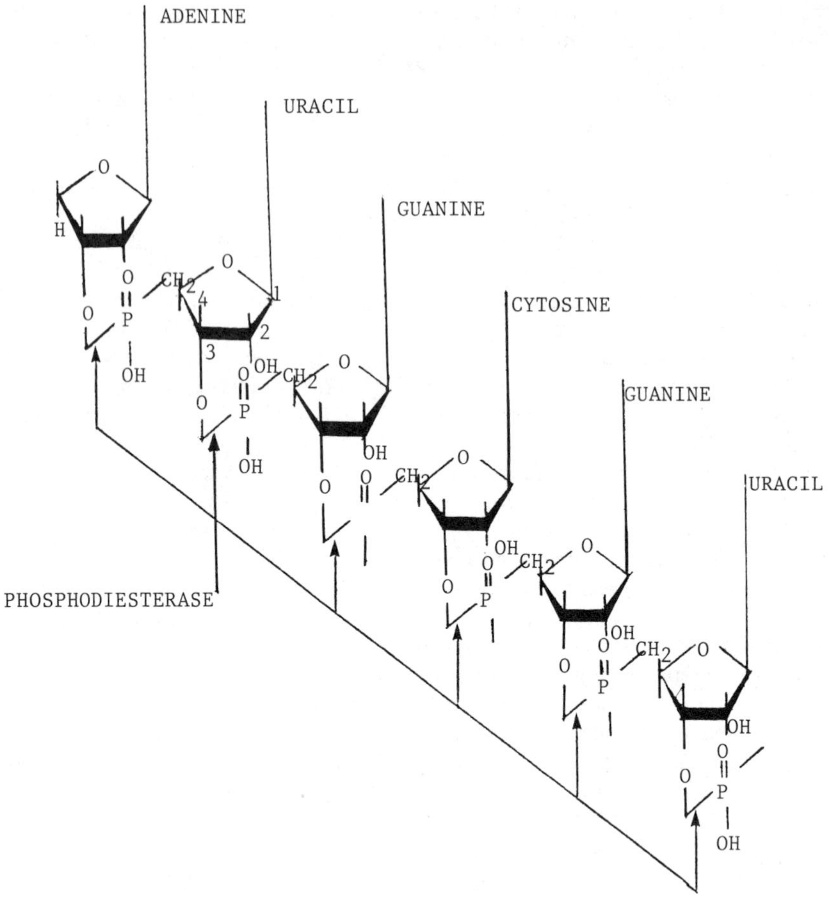

FIGURE 5. Degradation pattern of nucleic acid in the presence of the enzyme phosphodiesterase.

According to the conventional method, RNA from dried yeast cells is extracted with hot alkaline (5-15%) sodium chloride solution. Yeast cells are then separated, and RNA in the extract is precipitated by addition of acid or ethanol [41]. The precipitate is then neutralized and the crude extract is made into a powder by drying.

Kuninaka et al. [42] noted a serious problem associated with this method of RNA extraction. Each polynucleotide chain of the extracted RNA contained several percent of 2',5'-phosphodiester linkages in addition to the predominant 3',5'-phosphodiester linkages. Kuninaka et al. [43] also have observed that such isomerization reactions occur when extraction of RNA is done using the hot NaCl treatment. The disadvantage is that these 2',5'-phosphodiester linkages are resistant to 5'-phosphodiesterase activity. In subsequent studies, Kuninaka et al. [44] were successful in isolating 2',5'-linkage-free RNA by treating yeast packed in a column containing celite with dilute alkali solution containing salt and operating below 40°C. Isomerization was prevented by neutralization of RNA that occurs at the point of release from the yeast.

Another process to prepare crude RNA includes selective precipitation of RNA by heating the yeast slurry followed by treatment with acids. A common disadvantage in such a process is that the heat and acid treatments could cause DNA and a considerable proportion of RNA to decompose causing reduced recovery of RNA.

Under regular autolyzing conditions with the pH of media controlled at 6-6.6, 50-80% of intracellular RNA exists in its native state [45]. Intracellular RNA can easily be extracted from autolyzed yeast cells by heating the suspension at 90-100°C for 1-3 hours.

Enzymatic hydrolysis of crude RNA results in production of four nucleotides, only two of which are important in the food industry as flavor enhancers. These are the 5'-mononucleotides GMP (guanosine 5'-monophosphate) and IMP (inosine 5'-monophosphate). The latter is formed through deamination of the 5'-adenylic acid. The two 5'-mononucleotides, CMP and UMP, are employed as starting material to produce pharmaceutically valuable compounds.

Early studies [46] have shown that the phosphodiesterase of snake venom can hydrolyze RNA into the respective 5'-mononucleotides. However, this enzyme source obviously was unsuitable for industrial application. Fungi, such as *Penicillium citrinum* and actinomycetes like *Streptomyces aureus* have appreciable 5'-phosphodiesterase activity, and with genetic improvement, they have become the principal sources for preparation of 5'-phosphodiesterase enzymes. *Penicillium* enzyme, nuclease P1, has been used on an industrial scale in an immobilized form to produce 5'-mononucleotides. However, the conventional method of batch hydrolysis using this enzyme is not economically feasible.

Schuster [47] has observed that germs of some plants, particularly grasses, contain enzymes which can degrade the ribonucleic acids to 5'-mononucleotides. This finding led to patents issued in France, the United States, Germany, and Japan for production of 5'-nucleotide-rich extracts from yeast using malt sprout extracts as the source of nucleases or 5'-phosphodiesterase. According to these findings, aqueous extracts of malt rootlets produced under various conditions and with different pretreatments are used to hydrolyze nucleic acid to generate the 5'-mononucleotides. Malt rootlet extracts rich in phosphodiesterase serve as a cheap enzyme source and can be used either to hydrolyze crude RNA to produce 5'-mononucleotides at higher concentrations or to hydrolyze RNA in dilute extracts to produce 5'-nucleotide-rich extracts. The aqueous extract of malt sprouts contains the thermostable 5'-phosphodiesterase enzyme and thermolabile RNAs that degrade RNA into compounds that do not have the ability to enhance flavors. Another contaminating enzyme is phosphatase, which can dephosphorylate 5'-mononucleotides into nucleosides. Fortunately, these competing enzymes can be inactivated at 60°C without seriously affecting the overall activity of 5'-phosphodiesterase enzymes. Accordingly, when aqueous solutions containing RNA are incubated with crude, heat-treated malt rootlet extracts, predominantly 5'-mononucleotides are formed. The four 5'-mononucleotides (5'-AMP, 5'-GMP, 5'-UMP, and 5'-CMP) can be separated by use of anion exchange chromatography. Although 5'-AMP is not a flavor-enhancing nucleotide, it is generally converted to IMP, which enhances flavor although its flavor intensity is lower than that of 5'-GMP. The AMP to IMP conversion is achieved by treating with adenylic deaminase often extracted from a mold like *Aspergillus*:

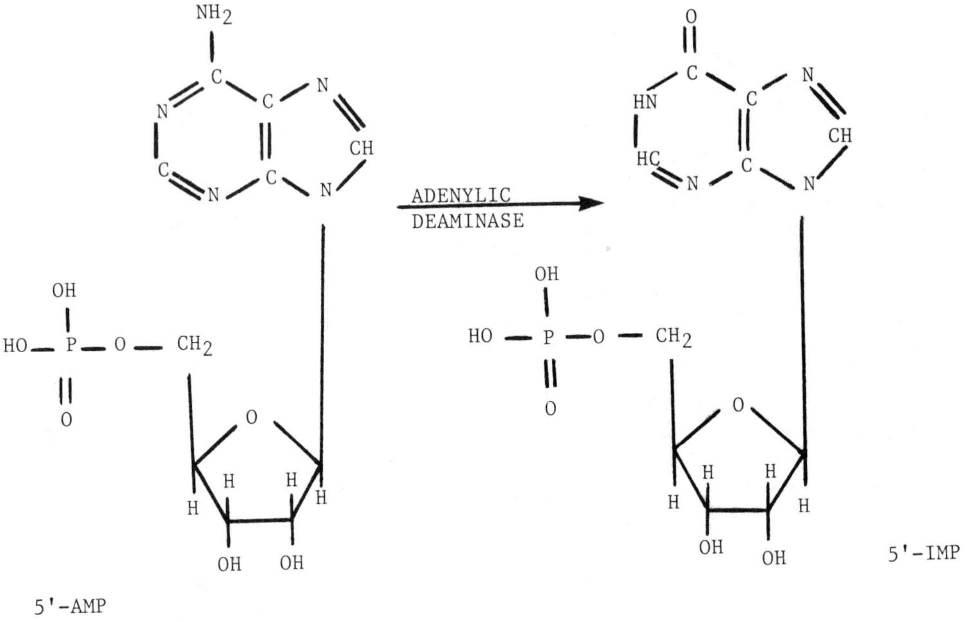

B. Colorants Derived from Yeast

1. *Phaffia rhodozyma*

There is considerable interest worldwide in developing food colorants from natural sources. Although there are only a few pigmented yeasts in nature, these can serve as good sources of natural colors. Yeasts also offer considerable advantages, because they can be produced in any quantity and are not subject to the vagaries of nature. The best known genera among the red yeasts are *Cryptococcus*, *Rhodotorula*, *Rhodosporidium*, *Sporidiobolus*, and *Sporobolomyces* with their characteristic pigmentations due to presence of pigments like β-carotene, α-carotene, torulene, torularhodin, plectaniaxanthin, 2-hydroxy-plectaniaxantin, etc. A recently identified yeast [48] known as *Phaffia rhodozyma* has received much attention. It is strikingly different from other pigmented yeasts because of its ability to produce a carotenoid pigment known as astaxanthin.

Interest in astaxanthin is primarily due to its presence in the animal kingdom. It is conspicuously displayed in the plumage of birds like flamingos, in the exoskeleton of marine invertebrates like lobsters, crabs, and shrimps, or among fishes such as trout and salmon where the astaxanthin is responsible for their red flesh color. When these fish are raised

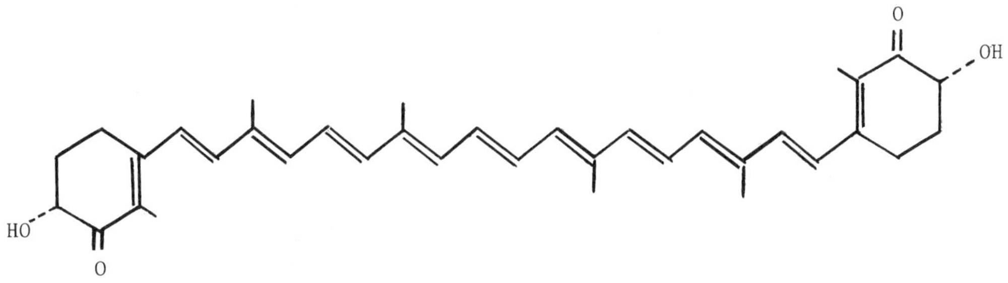

ASTAXANTHIN (3R, 3R' ISOMER)

in pens, they often lack the desirable red flesh color. Johnson et al. [49] found that a preparation of *P. rhodozyma* is a potentially important source of astaxanthin to restore red color in flesh of pen-reared salmonids. Although yeast astaxanthin has the opposite chirality at 3R and 3R' (i.e., the hydroxyl group on the cyclohexane end groups) to that of lobster astaxanthin, this does not affect its acceptability as a colorant in feed formulations for lobsters [50].

Johnson and Lewis [51] studied the effect of culture conditions on astaxanthin production by *P. rhodozyma* with a view to optimizing pigment production. According to their observations, astaxanthin production is directly related to growth of the yeast. Optimum growth and pigment formation occurred at 20-22°C, at pH 5 and with adequate aeration. The best carbon source, according to these studies, was D-cellobiose. Yield of astaxanthin from *P. rhodozyma* was approximately 0.5 mg/g dry yeast [52]. Slow growth and low astaxanthin yield from the yeast make commercial production highly uneconomical.

C. Enzymes from Yeast

1. Invertase

Invertase (β-fructofuranosidase, E.C. 32126) is located in the periplasmic space or bound to the outer surface of the cytoplasmic membrane of most yeasts. It can hydrolyze sucrose to glucose and fructose or the raffinose molecule to fructose and melibiose. This enzyme, which has wide application in production of high-test molasses or in the soft-center candy industry, is commercially produced from baker's yeast.

The enzyme is not excreted by the yeast cell into the environment and is not affected during a short autolysis period. Hence, an economical way of extracting the enzyme is by autolysis of the yeast cell. The autolysate is filtered and the enzyme is concentrated by precipitation with ethanol or isopropyl alcohol. Invertase is commercially available in the dry form as an invertase-rich dry yeast or as a liquid containing 50% (v/v) glycerol as stabilizer.

Invertase enzyme is encoded by six genes of the SUC polymeric system (SUC 1 to SUC 5 and SUC 7). Most strains do not carry all six SUC genes, but any one of the SUC genes confers sucrose-fermenting ability to the yeast cell [53]. Hence, a given industrial strain may contain one or more

SUC genes with the number of copies dictated by the ploidy of the cell. Grossman and Zimmermann [54] have demonstrated the effect of gene dosage on invertase production.

Each SUC gene encodes an internal, nonglycosylated invertase enzyme and an external invertase enzyme which becomes glycosylated during its passage to the cell surface. More recent studies have shown that the two types of invertase enzymes are encoded at two different regions of the SUC gene [55]. Thus, mRNA for external invertase additionally encodes a signal sequence and sites for glycosylation. The external invertase enzyme is largely under glucose regulation. Baker's yeast strains selected for invertase production are generally hyper invertase producers as a result of gene dosage.

Enzyme activity for commercially produced products is expressed as the velocity constant K calculated according to the AOAC method 31024 [56]. Thus the calculation is made based on data determined by polarographic measurements of the rate of sucrose hydrolysis by the extract under a given set of conditions. Based on this assay, invertase-rich dry yeasts yield K values in the 3-5 range.

2. Lactase

Lactase (β-D-galactosidase) is an enzyme found in yeasts such as *K. fragilis*, *C. pseudotropicalis*, *S. lactis*, and certain species of *C. utilis* [57]. It can catalyze hydrolysis of lactose to glucose and galactose. The food-approved yeast, *K. fragilis*, is a very good source of the enzyme and so is suitable for commercial production.

The low sweetness and insolubility of lactose limits its use as a food ingredient in products such as ice cream, confectionery, and animal feed. Lactose also crystallizes in milk-based products during frozen storage. The worst problem is that there is widespread lactose intolerance especially among Asians, Africans, and South American Indians because of the deficiency of lactase in their digestive system. In such individuals, ingested lactose reaches the intestine without being affected and then is subjected to microbial fermentation in the lower intestine causing abdominal bloating, rumbling, diarrhea, and other gastrointestinal disturbances. To improve suitability of milk-based products for these populations, the lactose content must be reduced significantly. The high lactose content of cheese whey also makes its disposal difficult. These problems have received widespread attention from both nutritional and commercial viewpoints and are now being handled through use of the lactase enzyme.

Production of the lactase enzyme from *K. fragilis* is similar to the procedure described earlier for production of invertase from baker's yeast. A short autolysis of the yeast is followed by a filtration and precipitation of the enzyme from the supernatant liquid. The product is commercially available most often as a liquid concentrate.

D. Nutritional Yeast

Over the centuries, extensive use of yeast in various forms of food in the human diet has clearly proven its value as a natural food ingredient. Although excessive intake of yeast has caused human health problems, it is generally recognized as an excellent source of amino acids and vitamins as well as many other nutritional requirements when consumed within safe limits. The problems associated with high nucleic acid consumption have been discussed elsewhere in this chapter.

1. Vitamins

The term "vitamin" is employed to denote a vital trace substance required for normal cell function like maintenance, growth, and reproduction which some species are unable to synthesize and must obtain from exogenous sources. Yeast contains predominantly vitamins of the B-complex group and can serve as an excellent source of those vitamins for human and animal nutrition. The term "vitamin B-complex" refers to a group of water-soluble substances which are derivatives of pyridine, purines, and pyrimidines (B_1, B_2, B_6, niacin, folic acid) or complexes like porphyrin-nucleotide (B_{12}), and amino acids-carboxylic acids (biotin, pantothenic acid). They function as enzyme activators and coenzymes (B_1, B_6, B_{12}, niacin, biotin, folic acid, and pantothenic acid), as redox agents in enzyme reactions (B_2, B_{12}, niacin, and folic acid), in nucleic acid synthesis (B_{12}, biotin, and folic acid), and as mitochondrial agents (B_2 and niacin). Although B-complex vitamins are required in minute quantities to support metabolism, inadequate intake can cause serious health problems.

Use of excessive amounts of yeast in the human diet is restricted by levels of nucleic acid present in the yeast. A safe upper limit for most normal adults, as described previously, has been identified as 2 g of yeast nucleic acid per day. This is equivalent to about 20 g of dry yeast based on the assumption that the product has 10% nucleic acid. Consumption of more yeast by humans requires either isolation of yeast proteins, reduction of nucleic acid content of the whole cell, or use of mutant strains for SCP production which inherently contain lower levels of nucleic acid.

TABLE 1. Percent of RDA for Vitamins Satisfied by Daily Intake of 20 Grams of Nutritional Yeast

Vitamins	Dry yeast[a] (μg/g)	Dry yeast (mg/20 g)	U.S. RDA (mg)	% RDA in 20 g dry yeast
Thiamine (B_1)	120	2.40	1.5	160
Riboflavin (B_2)	40	0.80	1.7	47
Niacin	300	6.00	20.0	30
Pyridoxine (B_6)	28	9.56	2.0	28
Pantothenic acid	70	1.40	10.0	14
Biotin	1.3	0.026	0.3	9
Folic acid	13	0.260	0.4	65
Vitamin B_{12}	0.001	0.000002	0.006	0.3

[a]NFX (regular) primary dried nutritional yeast, Universal Foods Corporation.

Dry yeast is commonly used as a vitamin supplement rather than a protein supplement in human nutrition. Some dry yeast products are even fortified with vitamins B_1, B_2, and niacin to meet certain requirements of tablet manufacturers. Table 1 lists concentrations of different vitamins in selected yeasts, recommended daily allowance (RDA) values for certain vitamins, and the percent satisfied in the diet by 20 g of dry yeast based on RDA. These data suggest that different components of the vitamin B-complex contributed by 20 g of dry yeast is substantial in meeting the complex vitamin requirements. Since yeast is deficient in certain other vitamins, dry yeast has to be used in combination with other food ingredients as is the usual for human diets. (Yeast does not contribute vitamin C, vitamin B_{12}, or fat-soluble vitamins like vitamins A, E, K, and D.)

Some of the sterols, like ergosterol, produced by yeast can be transformed through irradiation with ultraviolet light into vitamin D_2 (calciferol).

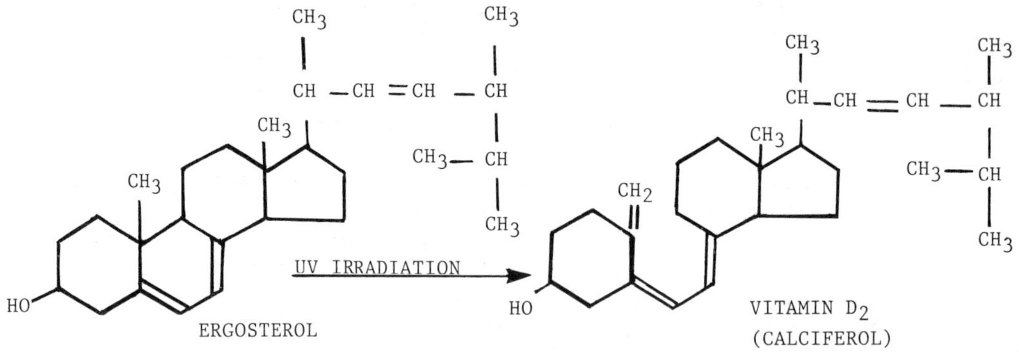

Selected strains of *Saccharomyces cerevisiae* can produce 7-10% ergosterol on a dry solids basis, and such yeast strains can be used with ultraviolet light for producing vitamin D_1-rich products.

Certain strains of the yeastlike organisms *Fremothecium ashbyii* and *Ashbya gossyopii* can overproduce riboflavin. At present these organisms are being used as crude sources of riboflavin for use in animal feed formulations. Although some strains of yeast belonging to the genera *Cryptococcus* and *Rhodotorula* can synthesize significant quantities of β-carotene, the concentration is too low to be economically important. Production of β-carotene by fungi belonging to zygomycetes (*Blakeslea trispora*) is more promising than producing it with yeast.

2. Yeast Proteins

Yeast contains 7-9% nitrogen on a dry solids basis, and the protein value of 40-50% given in the literature and corresponding to N × 6.25 represents crude protein because the nitrogen value also includes the nucleic acid fraction, which usually is 12-15% by weight of the crude protein. Yeast proteins are consistently high in lysine content and low in sulfur-containing amino acids. The protein efficiency ratio (PER) of baker's yeast is 2.02. Addition of 0.16 and 0.5% by weight of DL-methionine increases PER values to 2.27 and 2.77, respectively. These data are based on a PER of 2.5 for casein. Hence, nutritional yeast has 81% of the PER of casein.

TABLE 2. Amino Acid Content of Selected Yeasts (% Protein)

Amino acid	*Candida utilis*[a] (sulfite water liquor)[a]	*Candida utilis*[b]	*Kluyveromyces marxianus*[c]	*Saccharomyces cerevisiae*[d]
Alanine	5.8	5.5	—	—
Arginine	5.4	5.4	—	5.0
Aspartic acid	9.7	8.8	—	—
Cystine	—	0.4	—	1.6
Glutamic acid	15.6	14.6	—	—
Glycine	3.6	4.5	—	—
Histidine	1.2	2.1	2.1	4.0
Isoleucine[e]	3.8	4.5	4.0	5.5
Leucine[e]	7.6	7.1	6.1	7.9
Lysine[e]	4.8	6.6	6.9	8.2
Methionine[e]	1.1	1.4	1.9	2.5
Phenylalanine[e]	8.6	4.1	2.8	4.5
Proline	6.0	3.4	—	—
Serine	5.0	4.7	—	—
Threonine[e]	5.4	5.5	5.8	4.8
Tryptophan[e]	2.4	1.2	1.4	1.2
Tyrosine	6.2	3.3	2.4	5.0
Valine[e]	3.8	5.7	5.4	5.5

[a]From Ref. 58.
[b]From Ref. 59.
[c]From Ref. 60.
[d]From Ref. 61.
[e]Amino acid essential for human nutrition.

This suggests that yeast proteins without supplementation are inferior to animal proteins. This difference can be minimized in a mixed diet.

During digestion, proteins are hydrolyzed to amino acids, which in turn are assimilated by the body and are used for growth and repair processes continually taking place throughout life. Table 2 gives the amino acid composition, expressed as a percentage of protein, of four yeasts of industrial importance. Although there are minute variations in levels of corresponding amino acids among the various species of yeasts, from a nutritional standpoint, such differences can be regarded as insignificant. Nevertheless, all yeasts have a high content of lysine and are deficient in methionine as compared to other rich proteins, making it the limiting amino acid in yeast proteins.

TABLE 3. Mineral Content of Dried Yeast

Micronutrients	Yeast (mg/g)	Trace elements	ppm
Sodium	0.12	Copper	8.0
Calcium	0.75	Selenium	0.1
Iron	0.02	Manganese	8.0
Magnesium	1.65	Chromium	2.2
Potassium	21.00	Nickel	3.0
Phosphorus	13.50	Vanadium	0.04
Sulfur	3.90	Molybdenum	0.04
Zinc	0.17	Tin	3.0
Silicon	0.03	Lithium	0.17

The 20 g of dry yeast per person per day recommended as safe would account for approximately one-sixth of the daily allowance of 65 g of protein/day for a 70-kg adult male. Use of higher levels of yeast protein in the human diet requires reduction of the nucleic acid level to a minimum. Currently, the processes available for nucleic acid reduction are so costly and inefficient that it is almost impossible for yeast proteins to compete with casein or vegetable proteins in the world market.

3. Mineral Yeast

The mineral or ash content of dried yeast amounts to approximately 8% on a dry solids basis. The analytical data for the ash components are shown in Table 3. Although the predominant minerals in yeast are potassium and phosphorus, with lesser amounts of Ca, Mg, and S, the mineral contribution made by 20 g of dry yeast in a daily serving may seem minor considering the proportion of inorganic minerals contributed by the remaining part of an average human diet. Yet, in recent studies, the importance of trace elements like selenium, chromium, molybdenum, and zinc in yeast as essential for both animal and human nutrition has been recognized.

Chromium: Brewer's yeast has long been recognized as a rich source of essential nutrients able to make a notable contribution to human and animal nutrition. Three decades ago, Mertz and Schwartz [62] observed that rats fed certain diets had impaired tolerance for glucose. This condition of intolerance to glucose was reversed by feeding brewer's yeast.

Insulin, the hormone secreted by the pancreas, is responsible for promoting entry of energy-rich glucose into cells of the body. The pancreas of juvenile diabetic patients does not secrete insulin normally, and this disorder can only be corrected by administering insulin. The other disorder associated with glucose intolerance strikes its victims in midlife. This condition differs from the previously described disorder in that insulin is secreted into the bloodstream normally. Nevertheless, the subject cannot reduce the glucose level in the bloodstream. This problem, however, has been corrected by supplementing the diet with brewer's yeast. The active component in brewer's yeast is a trivalent chromium complex called the "glucose tolerance factor" (GTF). There is a general tendency for the chromium content in the human body to decrease with age. This could result in a decrease in the GTF causing an increase in vulnerability to late-onset diabetes.

Importance of chromium as a trace mineral in the human diet was demonstrated by Jeejeebhoy et al. [63] and Freund et al. [64]. A number of chemical trials have demonstrated the efficacy of supplementation of brewer's yeast in improving abnormal glucose tolerance in humans [65].

Analysis of GTF-rich brewer's yeast concentrates revealed the presence of nicotinic acid, trivalent chromium, and the amino acids glycine, cysteine, and glutamic acid [66]. The precise mechanism by which GTF controls glucose metabolism is not clear. However, studies to date postulate that GTF functions as a cofactor for insulin by enhancing insulin binding to receptor sites on the membrane of insulin-sensitive tissues [66]. Furthermore, these authors predict formation of a bridge between -SH groups of insulin receptors on the cell and -S-S-groups on the A-chain of circulating insulin.

The observation that GTF potentiates insulin-mediated uptake and use of glucose by adipose tissue and muscle in vitro suggests a possible mechanism for the observed effect of GTF on lowering the blood glucose level in mice in vivo [67]. The complex between GTF and insulin is not stable, and it is not known what keeps it stable in brewer's yeast.

In addition to the observed effect on plasma glucose, administration of GTF also lowers the nonfasting plasma triglyceride and cholesterol levels in diabetic mice. Similar results have been observed among humans on feeding brewer's yeast containing the GTF [65]. Its effect on lipid and sterol metabolism is not well understood.

The medical community does not yet suggest use of brewer's yeast or trivalent chromium as a possible treatment for sugar or lipid disorders. There is also insufficient evidence to believe that normalization of blood lipid chemistry through brewer's yeast supplementation might alleviate atherosclerosis. Yet research conducted to date has demonstrated the beneficial effect of brewer's yeast rich in chromium on human metabolism. Currently, the concept is beginning to be accepted by the health conscious community.

Selenium: During the past few years, there has been an increasing awareness of the biological importance of selenium in the human diet. Apparently it is closely, but not completely, associated with vitamin E. It is only in the past few years that selenium has been added to the list of essential trace elements in human as well as animal nutrition. Research studies on experimental animals have shown that continuous supplementation of diets with selenium has prevented formation of certain types of tumors. Accordingly, selenium has become recognized as an anticarcinogen.

Epidemiological studies where cancer death rates and selenium levels in food were correlated have been reported by Dickson and Tomlinson [68]. and Shamberger and Fronst [69]. These studies have demonstrated an inverse relationship between the selenium level in blood and human cancer death rates. These data suggest that selenium can inhibit certain types of cancer, and that incidence of such cancers could thus be diminished by adequate dietary supplementation with selenium, especially in countries where selenium intake through the normal diet is inadequate.

Elemental selenium as selenite or selenate is generally not effective due to poor absorption and slow incorporation into tissues of internal organs. However, it has been reported that an organically bound form, selenomethionine, is four times as effective as selenite to control certain diseases [70]. Organically bound selenium, regardless of the source from which it was obtained, is effective for use as a dietary supplement in human nutrition.

An important biochemical function of selenium in animals results from the fact that it is an essential constituent of glutathione peroxidase. This enzyme, present in the cytosol and mitochondrial matrix, plays an important role in destroying peroxides before they attack cellular membranes. Glutathione peroxidase utilizes reducing equivalents from GSH in the reduction of hydrogen peroxide, lipid hydroperoxides, and sterol hydroperoxides according to the following general reaction [71]:

$$\mathrm{ROOH + 2GSH \xrightarrow{GSHpx} ROH + H_2O + GSSG}$$

Vitamin E exerts a similar effect within membranes by preventing autooxidation of membrane lipids. Thus, both selenium and vitamin E protect biological membranes from oxidative degradation.

Selenium compounds may be used as additives to the substrate used for growing certain food yeasts to produce a food-grade product that contains intracellular selenium compounds in an organically bound ingestible form. The exact composition of the selenium-containing product is not yet known. However, the organically bound intracellular selenium is readily assimilable and an efficient dietary source in mammalian diets without exhibiting the usual toxic consequences associated with diets supplemented with inorganic forms of selenium.

Production of selenium-rich yeast by conventional type batch propagations is restricted by several factors. For example, a high concentration of selenium salts in the substrate used to grow the yeast has an inhibitory effect on its growth, thereby adversely affecting yield. Selenium levels present in the organically bound form in yeast propagated by batch processes also are relatively low. Nevertheless, a procedure to propagate edible food yeast having a high intracellular selenium content has been developed by Nagodawithana and Gutmanis [72]. Yeast made by this process contains the organically bound selenium essentially free of elemental selenium, which, if present, can be toxic to humans. This product is currently being used as a human nutritional supplement.

The process to produce selenium-rich yeast comprises a continuous and incremental feeding of a soluble selenium salt to a growth medium containing yeast. Adequate aeration accompanies incremental addition of a carbon source such as molasses to support growth of the yeast. The level of selenium salt added to the medium is determined by the concentration of organically bound selenium expected in the final product. The growth medium is further characterized by supplementation with a low level of inorganic sulfur nutrients to facilitate uptake of selenium by the yeast as a reactionary measure to sulfur deficiency. Like sulfur, selenium is in group

VIB of the periodic table. Because of its structural similarity to sulfur and the ability to form similar salts, selenium may be expected to replace sulfur in yeast metabolism. The extent to which selenium can replace sulfur in yeast-derived compounds is, however, not yet clear. Yeast products with an intracellular selenium content of 1000 ppm or more have been achieved by use of the aforementioned procedure, even though the composition of these seleno compounds is not known.

Amino acids are synthesized by a complex series of enzyme-catalyzed reactions, and it is assumed that selenium can replace sulfur to some extent in these biosynthetic reactions. Accordingly, besides making methionine, cysteine, homocysteine, and cystathionine, in the presence of inorganic selenium, yeast cells also make selenomethionine, selenocysteine, selenohymocysteine, and selenocystathionine, respectively. Korhola et al. [73] studied incorporation and distribution of selenium in yeast using radioactive selenium (^{75}Se). Analysis of the protein fraction of selenium-rich yeast showed that selenium was present in all the major soluble proteins. Selenomethionine was identified as the major selenium-containing compound in the protein fraction as well as in whole yeast cells.

E. Products of Pharmaceutical and Cosmetic Value

1. Skin Respiratory Factor (SRF)

A yeast-derived product has been used in hemorrhoidal preparations for approximately 40 years because of its probable effectiveness as a wound-healing agent. The active ingredient of such extracts is often referred to as "skin respiratory factor" (SRF), although some call it "live yeast cell derivative" (LYCD). SRF has certain biological activity associated with oxygen consumption of the skin, which could, in turn, have a significant influence on wound healing. More recent reports have shown that the product can stimulate wound epithelialization and early angiogenesis. As a result of these findings, so important for skin care, the product is beginning to receive a great deal of attention from cosmetic and pharmaceutical industries.

Production of a crude preparation of SRF was first described by Sperti [74]. It was first prepared as an alcoholic extract of live baker's yeast, *Saccharomyces cerevisiae*, by refluxing yeast cells with 95% ethyl alcohol for 4 hours at 60-70°C. After filtration of the extract, the residue was again refluxed with 50% ethyl alcohol under similar conditions, and the product was filtered to recover maximum SRF activity. The two filtrates were then combined and concentrated under reduced pressure at a temperature not exceeding 60°C. The concentrated extract can be evaporated to dryness at 70°C to obtain a powder. The product is commercially available as a paste, powder, or in granular form.

Early studies have shown that the SRF is unaffected by excessive heat provided the pH is maintained between 5.6 and 7.3. A unit of SRF activity is calculated as the amount of SRF required to increase the oxygen uptake of 1 mg (dry weight) of rat abdominal skin by 1% at the end of one hour in a Warburg chamber. Generally, a gram of SRF has 8000-12000 units of activity, and an effective level of 2000 units per ounce in a variety of preparations is claimed to accelerate the healing of wounds. In 1972, the

Food and Drug Administration (FDA) reviewed the efficacy data of SRF, and results were published in the *Federal Register*, Vol. 43, no. 1511. The panel found that, although SRF was safe for topical applications, insufficient data were available for the FDA to permit final classification of its effectiveness as a skin protectant for over-the-counter sales as a wound-healing aid for humans.

The first systematic study to show that SRF has the general properties expected of a wound-healing aid was done by Goodson et al. [75]. Their evaluation of SRF included use of a number of established methods to study its efficacy in wound healing. These comprised a series of tests done on skin from three patients at excisional surgery, on human fibroblasts, on human polymorphonuclear leukocytes, and on skins of white rats and rabbits. The effect of SRF activity on increased collagen production in excised human skin fragments in vitro was demonstrated by use of ^{14}C-labeled proline. Under these conditions, any radioactivity detected in hydroxyproline was taken as a direct measure of oxidation of proline to hydroxyproline and hence incorporation into collagen. Results indicated a 70% increase in uptake of ^{14}C proline by SRF-treated human skin samples obtained at excisional surgery. In the same study, the oxygen uptake rate of cultured human fibroblasts was increased by 137% with addition of SRF to the test medium. Addition of SRF to leukocytes also caused a similar increase in oxygen consumption. Additional studies conducted with skin of rats and rabbits demonstrated the beneficial effect of SRF in aiding wound healing. However, corroboration of these beneficial effects of SRF for wound healing by human subjects, thus far, is not available.

The first clinical trials to determine the effect of SRF on donor wound sites after skin grafting on humans with burns was done by Trunkey [76]. A conclusion from this investigation was that even though both SRF-containing ointment and the control base ointment tended to reduce wound-healing time, the rate of wound healing on skin graft donor sites where the SRF treatment was used was significantly faster than for the control.

In a study by Kaplan [77], human skin graft donor sites in nine patients were compared in a double-blind, randomized, single-center, inpatient study. Their donor sites were used as a model for superficial wound healing. Statistically significant earlier proliferation of new blood vessels and epithelialization occurred on donor sites treated with SRF as compared with those on the same patient treated simultaneously with the control base ointment. Each patient received both the active and control treatment simultaneously on two separate areas of tissue, thus making the experimental design a paired comparison with time as the measured response variable. The study demonstrated the ability to accelerate wound healing in humans beyond its normal rate.

Although crude yeast extract preparations rich in SRF have been used for decades in a number of OTC formulas, there is no information available in the published literature on isolation or characterization of the active principle found in these crude preparations. Limited knowledge thus far acquired is insufficient to characterize the active principle of the SRF. A recent European patent described a dialysis step for final extraction of an active component from yeast capable of wound healing. This factor may or may not be identical with SRF. Nevertheless, the factor isolated must have a low molecular weight because it passed through the dialyzing membrane. Addition of KCN to SRF caused a decreased oxygen uptake rate suggesting a relation to the cytochrome system. The product also has been shown to be stable at high temperature.

Lack of adequate research for further identification and characterization of SRF may partly be due to the complexity and high costs associated with assaying the efficacy of SRF. However, there is a need for identification of the active principle of SRF for development of more active products. Such concentrated products would undoubtedly have a great potential in the cosmetic and pharmaceutical industries.

2. Glycan

Glycan is the name given to comminuted, washed, pasteurized, and dried cell walls of brewer's or baker's yeast. It is generally produced from intact live yeast cells, or from the insoluble substances remaining after manufacture of autolyzed yeast extract [78]. Yeast cell walls can bind water to thicken aqueous food systems.

Glycan is made up principally of glucan and mannan polymers, which are nutritionally inactive in the human digestive system. One important desirable character of glycan is its ability to make aqueous food systems viscous when added in moderate levels. This also provides a fatlike mouthfeel. These properties have enabled food manufacturers to develop novel low-fat, low-calorie food analogues which retain their desirable creamy or fatlike mouthfeel. Such products include low-fat, low-calorie salad dressings, frozen desserts, cheese analogues, ice creams, etc. [79].

The major byproduct of the extract industry is the insoluble cellular debris which can be recovered commercially by centrifugation. The product at this stage does not have the desired ability to thicken food systems, and most often has an undesirable flavor. Repeated washings of this cellular debris can bring about a more bland flavor, but does not improve the thickening characteristics. However, if the insoluble cellular debris is comminuted, washed with water under alkaline conditions, and recovered, the insoluble cell wall fraction will have the desired ability to thicken aqueous food systems while maintaining a bland flavor. This is a highly desirable characteristic, especially when the product is intended for formulation of low-fat, low-calorie food analogues. Comminution is commonly done by repeated passages of the cell wall fraction through a homogenizer. Several passes are necessary to achieve maximum release of the unwanted flavor constituents after the alkali treatment. A pressure of approximately 10,000 psig is maintained during homogenization. The pH, solids concentration, temperature, and period of homogenization are all regarded as critical for achieving improved viscosity in the final product. The proximate analysis of glycan on a dry solids basis is as follows: carbohydrate, 80%; crude protein, 14%; nucleic acid, 2%; lipids, 1%; and ash, 3%.

Although initial studies have demonstrated the effectiveness of glycan as a low-fat, low-calorie food thickener, more recent research has begun to show unique properties that are likely to permit its use in a variety of biological functions. These specific functional properties have resulted from their physicochemical properties, which are primarily controlled by the molecular structure and arrangement of monomeric components and the linkages coupling them.

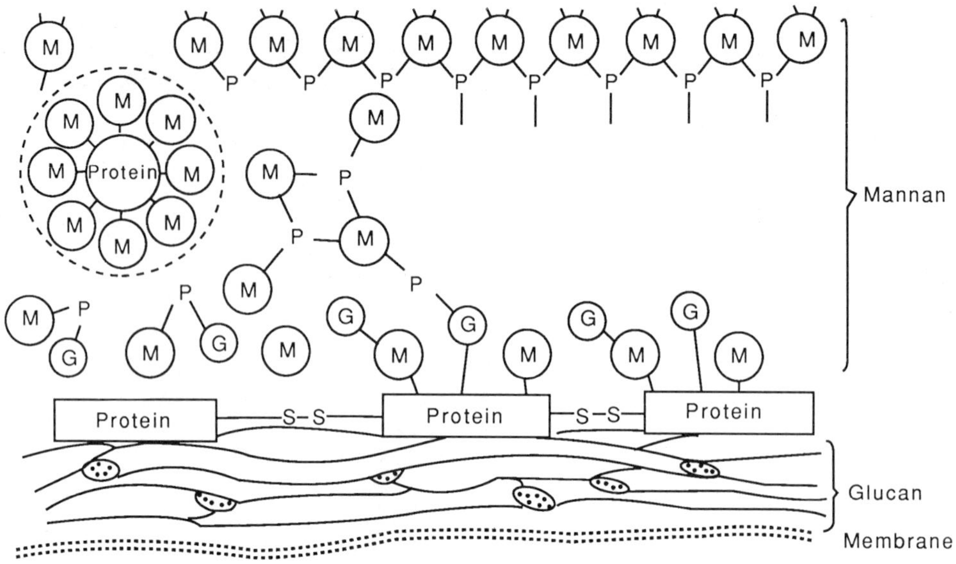

FIGURE 6. Schematic structure of the yeast cell wall.

Glucan, mannan, and chitin account for more than 80% of the dry weight of most yeast glycans [80]. The β-glucan which is made up of glucose polymers is the predominant polysaccharide. This structural component is divided into two major fractions on the basis of solubility in alkali solutions. The alkali-soluble fraction is a minor component probably occurring at a nonfibrillar glucomannan providing the function of a filler in the structure of the cell wall. The major glucan component is an alkali-insoluble fraction responsible for structural integrity of the cell wall. These are organized to form a microfibrillar network providing the specific shape for the cell (Fig. 6). Zymolyase is a commercially available lytic enzyme preparation derived from *Arthrobacter luteus* which can hydrolyze glucans present in the core of the yeast cell wall. This is a crude β-(1→3) glucanase which is specific for β-(1→3) linkages in the glucan polymer.

Yeast glucan contains branched and unbranched chains of glucose units linked by (1→3), (1→4), and (1→6) glucosidic bonds that may be of either α or β type. X-ray defraction studies have demonstrated that particulate glucans exist in the triple helical configurations [81-83]. In the triplex, the polysaccharide chains have a parallel orientation and are wound together in phase and strongly stabilized by extensive interstrand hydrogen bonds (Fig. 7a and b). In the crystalline packing, the structure of (1→3) β-D glucan is hexagonal (Fig. 7c). Approximately 20% of the water present in the helical structure is mobile and noncrystalline surrounding and separating the triplexes. Because of the continually changing positions of water molecules, hydroxyls of polysaccharide chains located on the periphery of the triplexes are rotationally disordered. Hence, there may not be fixed hydrogen bonds between the triplexes [83]. The triple helical structure of (1→3) β-D-glucan could account for special gel properties it generally imparts to foods and, perhaps, in some way explain its reported antitumor activity.

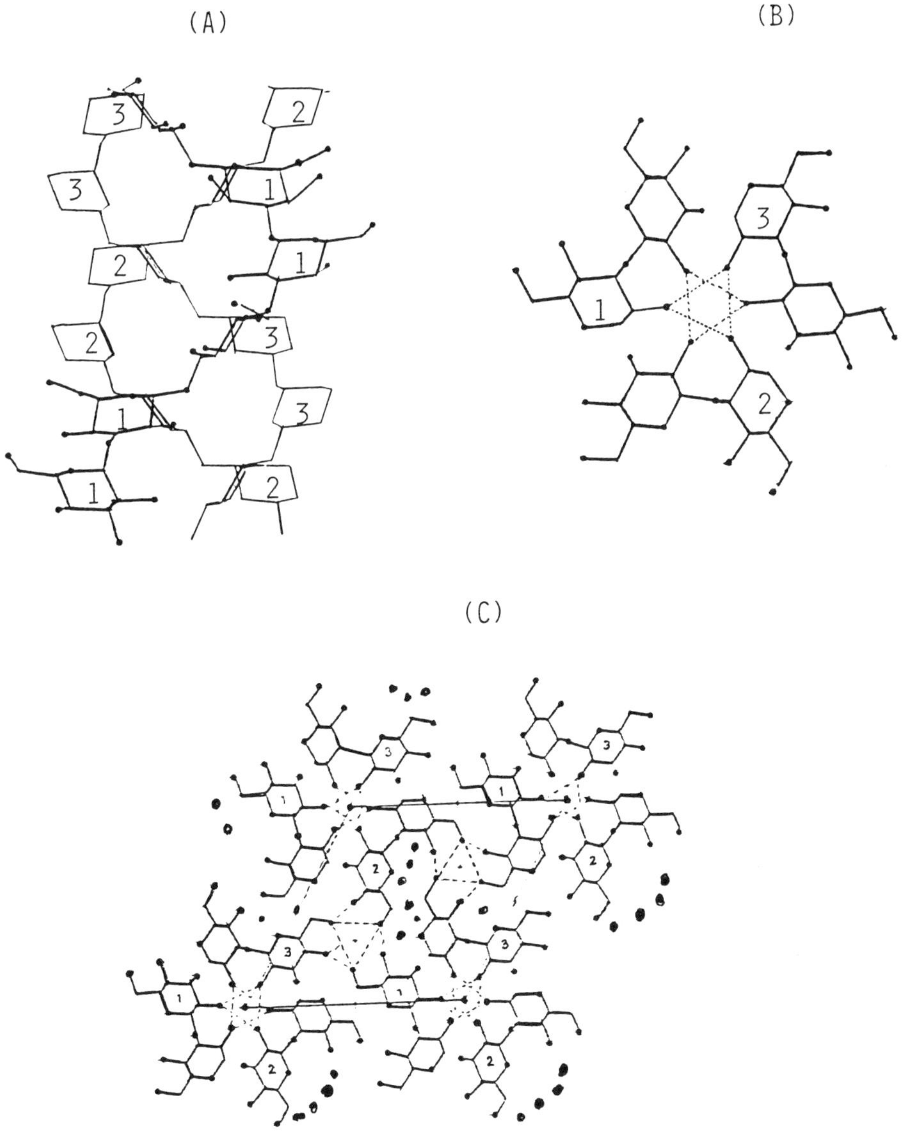

FIGURE 7. Triple-helical structure of (1→3)-β-D glucan. (A) Projection of the triple helix in the XZ plane. (B) Projection of the triple helix in the XY plane. (C) Probable arrangement of triple helices in the crystalline matrix. Black dots represent the water molecules. (From Ref. 81.)

In recent years, attempts have been made to alter the course of neoplasia by augmenting host defense mechanisms. The reticuloendothelial system (RES), which plays an important role in host resistance, influences various conditions such as infections, shock, and development of tumors.

When RES is blocked by particles such as tumor cells, expended erythrocytes, endotoxins, or is depressed by cortisones or other immunosuppressive agents, the outcome for the disease-causing organism will differ significantly from that of individuals where reticuloendothelial functioning is normal. It can be assumed that if RES functions are enhanced, an improved restoration from the pathological state would result. Numerous RES-depressing substances are known, but only a few with the opposite action are available. One substance known to activate RES is glucan derived from baker's yeast. It has been demonstrated experimentally that glucan's nonspecificity could enhance host resistance, potentiating both humoral and cellular immunity to malignant tumor growth as well as to various bacterial and viral infections.

All these therapeutic effects of glycan appear to be due to stimulation of the reticuloendothelial system to produce increased amounts of macrophages, which play a key role in the body's natural immune system. These macrophages execute an immune response by absorbing and destroying foreign particles through phagocytosis.

When particulate glucan was administered to animals, some side effects became apparent. Particulate glucan also produced a high degree of acute toxicity, even causing high mortality among experimental animals. In view of these disadvantages, extensive studies were undertaken to develop a soluble β-(1→3) polyglucose which might be nontoxic, induce no pathological effect, and yet retain significant immunological properties. A lifetime of research by DiLuzio [84] has resulted in development of a low molecular weight phosphorylated soluble glucan preparation able to exert pronounced immunological responses when administered to animals and humans. These new soluble phosphorylated glucans, which are devoid of the triple helical structure common to particulate glucan, have been shown to immunostimulate macrophage activity with resulting activation of the other immunoactive cells in the RES system. Because of these unique properties, the author claims that the soluble phosphorylated glucans can be particularly useful for prophylactic and therapeutic application against a variety of diseases caused by bacteria, viruses, fungi, and parasitic organisms as well as a number of neoplastic conditions.

It is clear that one of the major efforts for the future will be to define in comparative terms the immunopharmacological activity of various glucan preparations so that ultimate judgment can be made as to which properties of glucan are essential for their unique pharmacological actions. Requirements such as interglucosidic linkages, molecular weight and molecular structure of the glucan, and obtaining high yields at low cost to improve process economics are all essential in development of glucans for therapeutic use.

The hypocholesterolemic effect of yeast glycan has also been demonstrated [78]. The exact nature of the mechanism by which glycan imparts this unique character is not well established. Nevertheless, it is generally believed that the hypocholesterolemic effect of many dietary fibers is due to their ability to interact with cholic acid and that the decrease in the fiber content of the human diet in this century may relate to the increase in coronary heart disease.

3. Coenzyme A—Synthesizing Protein Complex (CoA-SPC)

Methods that provide early detection of cancer before symptoms become apparent are helpful in successful control of the disease. One such method provides early diagnosis by detecting the level of B-protein in the blood serum, which is an early indication of cancer [85].

A biologically active preparation referred to as coenzyme A-synthesizing protein complex (CoA-SPC) prepared from baker's yeast has found application in preparation of reagents for the above test. CoA-SPC is characterized by its ability to interact with L-cysteine, D-pantothenic acid, and ATP to produce the active binding protein which can complex with B-protein present in blood serum drawn from cancer victims. Use of labeled L-cysteine, D-pantothenic acid, or ATP could provide the capability to identify the B-protein complex needed for early detection of cancer.

CoA-SPC is an insoluble complex which has a molecular weight of approximately 200,000 and has been identified in the particulate fraction after lysing baker's yeast cells. This particulate fraction also contains other insoluble proteinaceous materials and other unwanted cell debris which make the CoA-SPC purification difficult. It is, however, necessary to devise techniques to selectively solubilize the undesirable components found in the particulate fraction without solubilizing CoA-SPC. This has been accomplished by simple agitation in the presence of salts such as chlorides, nitrates, and acetates of K, Na, Mg, Ca, and Mn. These treatments result in a particulate fraction rich in CoA-SPC.

A low molecular weight component extracted from baker's yeast has been found suitable for solubilizing CoA-SPC from the CoA-SPC-rich particulate fraction which has already been treated to remove other insoluble proteinaceous materials. This fraction, referred to as the "t-factor," is found in baker's yeast and is generally released as a soluble, low molecular weight (400-1000) component when yeast cells are originally lysed by any of the standard methods used in the industry. Purity of the CoA-SPC extracted is determined by purity of the "t-factor," and the degree to which the CoA-SPC-containing particulate fraction was cleaned to eliminate proteinaceous and other impurities. The procedure for preparation of this product is covered by the patent granted to Bucovaz et al. [86].

4. Genetically Engineered Products from Yeast

Production of useful proteins of therapeutic value by exploitation of rDNA technology has previously been associated with *Escherichia coli*. However, in addition to being a fecal organism, this bacterium also produces pyrogenic factors that must be eliminated from any potentially useful pharmaceutical product before it can be used on humans. These undesirable characteristics have caused problems in gaining wide acceptance for *E. coli* as a suitable recombinant organism for producing genetically engineered products for human use. Such considerations have led to increased interest in identifying alternative host/vector systems. Among eukaryotic organisms suitable for exploitation, perhaps the easiest to manage is yeast. This organism offers the researcher considerable advantages over bacteria. It is inexpensive to grow on an industrial scale, and recent studies have yielded a great deal of information on its genetics at a molecular level. Additionally, much creative effort has gone into development of ways to manipulate and modify the yeast genome. Considering these unique features, yeast has now begun to emerge in the forefront of rDNA technology with the possibility of becoming the principal vehicle to produce value-added products of commercial importance.

A few yeast-derived products like rennin (chymosin), insulin, and interferon are already in the marketplace, and these products are likely to show a profit potential with high sales volumes in the future. The industry is likely to gain credibility as more and more products are approved and introduced into the marketplace. Although most capital is concentrated in relatively few but highly visible genetic engineering companies, new investments are being made in less significant companies whose novel products and profits are yet to be realized. Total product sales reached almost $500 million in 1986. Although diagnostic devices received only about 10% of the overall R&D expenditure, this field currently accounts for 55% of all sales. Despite heavy R&D support for development of novel products for human therapy, these pharmaceuticals represent a much smaller segment of sales, primarily due to the more complicated procedures for their testing and approval.

Novo Industries (Denmark), one of the world's largest manufacturers of insulin, has been selling porcine insulin with a single amino acid substituted to make it identical to the human form of insulin. The company plans to replace this form of insulin with a genetically engineered form produced by a baking strain of yeast. This hormone is presently made by Eli Lilly through the use of a genetically engineered strain of *E. coli*. The single chain of insulin made by yeast is folded correctly, and an enzymatic process provides the required double chain for the formation of the active form of the human insulin molecule.

A milk-clotting enzyme found in the lining of the fourth stomach of calves slaughtered for veal is traditionally used in cheese manufacturing. This enzyme, which is also termed chymosin, is in short supply. Although fungal enzymes which can coagulate milk are also used in the cheese industry, these do not give the same flavor to cheese as calf rennin. By use of rDNA techniques, it has been possible to produce rennin identical to that from calves. Collaborative Research's patented process uses yeast cells as hosts which secrete rennin in a fully active, nonglycosylated form. When the protein is produced recombinantly in bacteria, the enzyme is not secreted, insoluble, and nonoxidized, and the yields are low. Genencor (California) has used a filamentous *Aspergillus* as a recombinant host for production of rennin.

Another important recombinant protein of significant therapeutic value is α-1-antitrypsin (AAT). This product is being tested on patients suffering from emphysema caused by deficiency of AAT in the body. AAT's function is to destroy excess elastase and other proteolytic enzymes in the lungs. When elastase is optimally produced, it helps clear lungs of microorganisms and particulate matter from air pollutants and tobacco smoke. If sufficient AAT is not available to maintain the optimal esterase level, the latter will attack lung tissue causing emphysema. AAT is now made with recombinant yeast genetically engineered by some companies like Zymo Genetics (Seattle) and Synergen Inc. (Colorado).

The AAT synthesized in recombinant yeast is not glycosylated and remains within the cell without being secreted into the environment. Hence, the extraction procedure is costly. Unlike bacteria, recombinant yeast can produce glycosylated proteins, which may then be excreted into the environment. In most instances, their sugar attachment patterns are different from glycosylated proteins in human cells. It is quite likely that yeast-produced AAT has a glycosylated pattern significantly different from that of human AAT, which could result in an immune reaction often harmful to the patient. Cooper Biomedical (California), who had ZymoGenetics (Washington) clone and express the AAT in yeast under a research contract, has demonstrated that AAT does not produce any immune response in baboons. The effect of the glycosylated yeast AAT in animals and humans remains to be established.

Interferon is a virus inhibitor produced by mammals or cultured cells as a defense mechanism when infected with viruses. This protein is part of the human immune system and interferes with spread of virus. The reticuloendothelial system provides the bulk of interferon during most viral infections. Extraction of interferon from mammalian tissue for commercial production is nearly impossible.

Although *E. coli* has been used as the host organism in pioneering work on production of α-interferon, geneticists now consider yeast as better suited for commercial production of the product. Interferon Sciences Inc., a subsidiary of National Patent Development Corporation, is seeking approval from the FDA to market a recombinant α-interferon product made from recombinant yeast for use in genital herpes. This relieves pain and shortens the duration of viral shedding from recurrent herpes lesions. Two clinical studies on recombinant α-interferon have confirmed it is effective in reducing chances of catching cold, but its side effects can be dramatically reduced by using it in a nasal spray. The side effects are reduced by shortening the period necessary to produce the desired effect through using the spray.

A recent product which has received approval from the FDA for sale in the United States and certain other countries is the genetically engineered hepatitis-B vaccine. Hepatitis-B is a serious viral infection that has caused nearly 200 million cases of chronic viral infection worldwide. In some cases, the disease is fatal. The plasma-derived hepatitis-B vaccine has been marketed since 1982, but its supply is limited. The supply of plasma-derived vaccine depends on availability of hepatitis-B carrier blood, a raw material in limited supply, for extracting the necessary surface antigen needed to prepare the vaccine. It also requires 65 weeks to be released as an acceptable product. In contrast, the genetically engineered yeast-derived vaccine can be processed in about 10 weeks, and production can be scaled up to meet any demand. Merck Sharp and Dohm has received approval to market its genetically engineered yeast-derived hepatitis-B vaccine in the United States.

Recombinant yeast is presently being tested at the pilot plant level to produce superoxide dismutase (SOD) in kilogram quantities. This enzyme prevents tissue damage after blood flow is restored to the coronary arteries of heart attack patients for treating edema in the brain. There are some research groups investigating the possibility of cloning bovine interleukin-2 in yeast. This product is expected to be useful in restoring the immune system in cattle against bovine leukosis virus. Researchers at Phillips Petroleum have recently produced streptokinase in *Pichia pastoris*, a methanol-utilizing yeast used in single-cell protein production. Streptokinase is currently used for dissolving blood clots associated with heart attacks.

V. CONCLUSIONS

During the early part of this decade, the world witnessed the beginning of a technological revolution which should eventually bring about substantial progress in the world economy. Advances which have already been made in microbial genetics are being exploited rapidly by many industrialists to produce a few low-volume value-added proteins through use of yeast, the most widely studied eukaryotic organism available for commercial purposes.

Progress made in application of newer technologies by many yeast-related industries, particularly the brewing industry, has lagged considerably behind that of many academic institutions. Much of the application of fundamental principles in yeast genetics has been done by universities more as exercises to satisfy curiosity than with the intent of supporting an industry. This has been due to the conservatism which has existed among brewers and which has consequently delayed introduction of novel methods and improved strains of yeast to the industry. This trend has now begun to change in the brewing and other yeast-related industries with the testing of new strains and more cost-effective and improved processes. For example, the possibility of incorporating the starch-utilizing DEX genes into brewing yeast strains, without introducing other undesirable genes, to produce an acceptable low calorie beer is currently being investigated in several brewing research laboratories. Although several specially constructed starch-utilizing *S. cerevisiae* strains tested have produced undesirable flavor characteristics in beer, the brewing industry is hopeful that a successful brewing strain with starch-utilizing ability will eventually emerge for brewing light beer.

Genetic engineering techniques used so successfully to transform yeasts have so far been used only to a very limited extent to improve performance of yeast in making bread, distillery alcohol, and wine. This has been due to an inadequate understanding of the genetics responsible for several yeast characteristics important to these three industries. Although, knowledge of the precise genetic mechanisms responsible for most of these phenomena is fragmentary, in such complex situations a random or "shotgun" approach like protoplast fusion may prove successful. However, to achieve the desired objectives in a completely predictable manner rather than relying on chance, knowledge of the precise genes and their locations will be invaluable. These studies are being pursued actively in several research laboratories throughout the world, and their findings should facilitate strain improvement through the use of more predictable techniques such as cloning.

With the excess demand for cultivated carbohydrates coupled with their high cost of production, it is likely that the price of carbohydrates will remain high as will proteins. Accordingly, single-cell protein (SCP) production from plant crops such as corn, wheat, potatoes, etc. will not be economically attractive. The prospect for profitable SCP production seems unlikely to improve under the present circumstances, and the technique will thrive only when special considerations make it worthwhile. The alternative will be use of waste materials from other industries to minimize the cost of the SCP product. This would result in an overall reduction of effluent treatment cost and additional revenue from the sale of SCP products. It is in this area that one can envisage most future research dealing with SCP.

Extensive research is being done on selection of yeast strains which have broader carbohydrate assimilation spectra or higher yield coefficients than currently available strains. A serious drawback for food yeast is its low sulfur amino acid content (principally methionine) compared to other proteins like casein. With genetic manipulation it has been possible to develop mutants of *Candida tropicalis* with a methionine content increased by 40% over that of the parent strain [87].

A reduction in wall material of yeast cells should increase the protein content relative to its total biomass. This should not only improve the nutritional value of the whole yeast cell but also make extraction of cell material simple and economical. Such mutants with a low carbohydrate level and higher crude protein content than the parent cells have been reported [88]. There also have been several other reports of new types of regulatory and overproducing yeast mutants having several characteristics of industrial importance. Yet, there seem to be serious obstacles which keep these products from the marketplace. Despite these problems, there is an upsurge of research in this field since the profit potential for any successful product is highly attractive.

Our understanding of flavor chemistry as related to yeast extracts is in its infancy, and it is reasonable to state that no serious study—academic or commercial—has been undertaken to scientifically characterize any of the yeast-based flavors currently available. A single flavor can result from the presence of numerous organic compounds, in a blended form, in low concentrations, which can generally be separated by gas chromatography. Reconstitution of these separate components according to their ratio should provide the original flavor of the food product. Although this seems straightforward for certain less complex defined flavors, it becomes more difficult to define or formulate complex flavors as found in yeast extracts. One way to approach this problem is to be consistent in the production procedure. Extensive studies have already been done on autolysis with the intent of achieving consistent flavor profiles of the product for a given application. This approach does not, however, provide a complete understanding of the flavor or the mechanism by which different components interact to give the final flavor. Any attempt to characterize the flavor and interpret its chemical composition should provide the capability of providing a more consistent flavor but also should permit development of novel flavors through adequately controlled reactions or by symbiosis.

Similarly, there is also a need to study solubilization of yeast cell mass with the objective of obtaining higher extract yields in shorter process times. Any modification of the process or use of newer more potent enzymes should make the process more economical. Nevertheless, such commercially available enzymes are at present very expensive, and their use may sometimes be economically unfeasible.

With growing interest in convenience and prepared foods, the food industry may wish to exploit the newly recognized flavor effects of yeast extracts to improve palatability and acceptability of a particular food product. This includes sparing of salt without decreasing flavor quality or intensification of seasoning flavors, such as mustard and horseradish, so that less of such seasonings is required to achieve the same flavor profile. Some flavor chemists describe the salt-sparing effect of yeast extracts as due to depression of the acid peak, which delays the perception of acid notes so that the underlying salt flavor is detected immediately without interference. Presently, yeast-derived flavors include cheese, beef, and baked bread flavors, and a chicken flavor is being developed. More recently developed 5'-nucleotide-rich extracts should broaden our scope in development of new flavors for the food industry.

With the progress of flavor chemistry and the rapid rise in our understanding of changes that occur during autolysis, much attention is directed toward investigating flavor precursors so biochemical reactions can be used to impart fresh flavors to different food products. It is hoped that once flavor precursors and the corresponding enzyme systems are identified, their addition to dried or concentrated products will improve the flavor and acceptability even in the presence of off-flavors.

Recent studies have also shown the therapeutic value of certain yeast-derived products when used in crude form. Yeast cell wall material has functional properties beneficial for certain applications. Some studies have demonstrated the effectiveness of diets rich in yeast cell wall material in reducing the cholesterol content in blood, which could in turn reduce the risk of heart diseases. Current interest, however, centers on use of cell wall material (glycan) as an antitumor agent to combat certain forms of cancer. Several reports suggest that soluble phosphorylated yeast glucans are effective against neoplastic, bacterial, viral, and fungal diseases. Clinical trials for these products are presently underway.

The spectacular advances that have been made during the last decade with rDNA technology have led to many ambitious research programs to produce value-added proteins predominantly for use in the pharmaceutical industry. Although production of nonyeast materials by yeast is in its earliest infancy, a few recombinant products like insulin, α-interferon, and rennin have already reached the marketplace, while a large number of newly developed recombinant products are awaiting FDA approval. The application of rDNA technology to yeast will continue to increase, so there may be large-scale production of specific proteins of significant commercial value. The impact is likely to be broadly felt in human and animal health care, food and nutrition, agriculture, and the chemical industry.

Although yeast-derived products are used in hemorrhoidal and skin care creams, the active component, commonly referred to as "skin respiratory factor"(SRF), has not yet been identified and isolated. Likewise, certain components in brewer's yeast collectively referred to as "glucose tolerance factor" (SRF), has not yet been identified and isolated. Likewise, cer-of diabetic patients. Substantial research effort is presently being made to identify and possibly purify these active components. These concentrated products will be valued higher on the basis of activity, but also could show a higher sales potential within the cosmetic and pharmaceutical industries.

It is thus clear that yeast is continuing to act as mankind's oldest and safest ally in the microbial world. Traditionally, yeast has been extensively used in the food and the beverage industries. With recent developments in biotechnology and genetic engineering, a second generation of value-added yeast-derived products is beginning to emerge. These new techniques also are adding unprecedented scope and precision to efforts to modify yeasts to increase their usefulness. It is hoped that these developments will result in major advances in such areas as food and nutrition, health care, and industrial chemistry in the not-too-distant future. Yet how many of these expectations will eventually become reality remains to be seen.

REFERENCES

1. Mackenzie, D. A. The exquisite artistry of ancient Egypt. In *Wonders of the Past*, Vol. 1 (J. A. Hammerton, ed.), Fleetway House, London, 1925, pp. 253-268.
2. Mackenzie, D. A. The soul's journey to paradise. In *Wonders of the Past*, Vol. 1 (J. A. Hammerton, ed.), Fleetway House, London, 1925, pp. 239-251.
3. Kreger-Van Rij, N. J. W. General classification of the yeasts. In *The Yeasts, a Taxonomic Study* (N. J. W. Kreger-Van Rij, ed.), Elsevier Science Publ., Amsterdam, 1984, pp. 1-44.
4. Nagodawithana, T. W. Yeasts: Their role in modified cereal fermentations. In *Advances in Cereal Science and Technology*, Vol. VIII (Y. Pomeranz, ed.), Am. Assoc. Cereal Chemists, St. Paul, NM, 1986, pp. 25-204.
5. Trivedi, N. B., Jacobson, G. K., and Tesch, W. Baker's yeast. In *CRC Critical Reviews in Biotechnology*, Vol. 24, Issue 1 (G. G. Stewart and I. Russell, eds.), CRC Press, Boca Raton, FL, 1986, pp. 75-109.
6. Reed, G., and Peppler, H. J. *Yeast Technology*, AVI Publ. Co., Inc., Westport, CT, 1973.
7. Sanderson, G. W. Yeast products for baking industry of today and tomorrow. Paper presented at the International Symposium on Advances in Baking Science and Technology, Kansas City, KS, 1984.
8. Stewart, G. G. The place of yeast in biotechnology. *Indian J. Microbiol.*, *21*:171-210, 1981.
9. Stewart, G. G., and Russell, I. One hundred years of yeast research and development in the brewing industry. *J. Inst. Brew.*, *92*:537-558, 1986.
10. Bevan, A., and Markower, M. The physiological basis of the killer character in yeast. In *Genetics Today* (S. J. Goerts, ed.), Pergamon Press, Oxford, 1963, pp. 202-203.
11. Nagodawithana, T. W., and Steinkraus, K. H. Influence of the rate of ethanol production and accumulation on the viability of *Saccharomyces cerevisiae* in rapid fermentation. *Appl. Environ. Microbiol.*, *31*:158-162, 1976.
12. Benda, I. Wine and brandy. In *Prescott and Dunn's Industrial Microbiology*, 4th ed. (G. Reed, ed.), AVI Publishing Co., Westport, CT, 1982.

13. Winkler, A. J. *General Viticulture*, University of California Press, Berkeley and Los Angeles, 1962.
14. Castelli, T. The organisms responsible for wine fermentation. *Arch. Mikrobiologie, 20*:218-223, 1954.
15. Brechot, P., Chauvet, J., and Girard, H. Identification des levures dun mout de Beaujolais au cours de sa fermentation. *Ann. Tech. Agric., 11*:235-244, 1962.
16. Minarik, E., Laho, L., and Navara, A. The yeast flora of grapes, must and wines. *Mitt. Rebe Wein, Ser. A.* (Klosternevberg), *10*: 218-223, 1960.
17. Minarik, E. Ecology of yeasts and yeast like microorganisms on secondary habitats in Czechoslovakia. Antonie Van Leeuwenhoek, J. Microbiol. Serial 35 (Suppl. Yeast Symp.), D7, 1969.
18. Domerq, S. Etude et classification des levures de vin de la Gironde. *Ann. Technol. Agric., 6*:5-58, 139-183, 1957.
19. Ulbrich, M., and Saller, W. Investigations of the practicality of aeration of pure cultures of commercial wine yeasts. *Mitt. Rebe Wein, Ser. A* (Klosterneuburg), *1*:94-104, 1951.
20. Adams, A. M. A simple continuous propagation for yeast. Rept. Ontario Hort. Expt. Sta. Product Lab., 1953, pp. 102-103.
21. Thoukis, G., Reed, G., and Bouthilet, J. R. Production and use of winery fermentations. *Am. J. Enol. Viticul., 16*:1-8, 1963.
22. Caster, J. G. Experimental development of compressed yeast fermentation. *Starter and Wines and Vines, 34*(8):27; (9):33, 1953.
23. Radler, F., Schutz, M., and Doelle, H. W. Die beim Abbau von L-apfelsäure durch Milchsäurebakterien entstehenden Isomeren der Milchsäure. *Naturwissenschaften, 12*:672, 1970.
24. Kunkee, R. E., and Snow, S. R. Method for reducing fusel oil in alcoholic beverages and yeast strains useful in that method. U.S. Patent 4,374,859 (1983).
25. Ingraham, J. L., and Gaymon, J. F. The formation of higher diphetic alcohols by mutant strains of *S. cerevisiae. Archives of Biochem. and Biophys., 88*:157-166, 1960.
26. Scrimshaw. In *Single Cell Proteins* (R. I. Mateles and R. Tannenbaum, eds.), MIT Press, Cambridge, MA, 1968, pp. 3-7.
27. Ridgway, J. A., Lappin, T. A., Benjamin, B. M., Corns, J. B., and Akin, C. Single cell protein material from ethanol. U.S. Patent #3,865,691 (1975).
28. Litchfield, J. H. Production of single cell proteins for use in food or feed. In *Microbial Technology* (H. J. Peppler and D. Perlman, eds.), Academic Press, New York, 1979, pp. 93-156.
29. Bernstein, S., and Plantz, P. E. Ferments whey into yeast. *Food Eng., 49*(11):74-75, 1977.
30. Bernstein, S., and Everson, T. C. Protein production from acid whey via fermentation. Environmental Protection Technology series. U.S. Environmental Protection Agency, EPA 660/2-74-025, 1974.
31. Anon. Single cell proteins, Protein Advisory Group Guidelines No. 4, United Nations, New York, 1970.
32. Ziemba, J. V. Tailored hydrolysates—How made, how used. *Food Eng., 19*(1):82-85, 1967.

33. Akiyama, S., Doi, M., Arai, Y., Nakao, Y., and Fukuda, H. Production of yeast biomass. U.S. Patent 3,909,532 (1975).
34. Waslien, C. I., Calloway, D. H., Margen, S., and Costa, F. Uric acid levels in men fed algae and yeast as protein source. *J. Food Sci.*, *35*:294-298, 1972.
35. Robbins, E. A. Manufacture of yeast protein isolate having a reduced nucleic acid content by a thermal process. U.S. Patent 3,991,215 (1976).
36. Robbins, E. A., and Seeley, R. D. Process for the manufacture of yeast glycan. U.S. Patent 4,122,196 (1978).
37. Nakao, Y. Microbial production of nucleosides and nucleotides. In *Microbial Technology*, Vol. 1 (H. J. Peppler and D. Perlman, eds.), Academic Press, New York, 1979, pp. 312-348.
38. Kuninaka, A. Nucleic acids, nucleotides and related compounds. In *Biotechnology*, Vol. 4 (H. J. Rehm and G. Reed, eds.), Verlag Chemie, FL, 1986, pp. 72-86.
39. Katchman, B. J., and Fetty, W. O. Phosphorus metabolism in growing cultures of *Saccharomyces cerevisiae*. *J. Bacteriol.*, *69*:607-615, 1955.
40. Chayen, R., Chayen, S., and Robert, E. A. Observations on nucleic acid and polyphosphates in *Torulopsis utilis*. *Biochim. Biophys. Acta*, *16*:117-126, 1955.
42. Kuninaka, A., Fujimoto, M., and Yashino, H. Adenylyl (5-2) adenosine 5' phosphate (2-5 p-A'A) in crude crystals of 5' adenylic acid isolated from nuclease P_1 penicillium nuclease digest of technical grade yeast RNA. *Agric. Biol. Chem.*, *39*:597-601, 1975.
43. Kuninaka, A., Fujimoto, M., and Yoshino, H. Formation of 2-5 phosphodiester linkages in polyribonucleotides. *Agric. Biol. Chem.*, *41*: 679-684, 1977.
44. Kuninaka, A., Fujimoto, M., Uchida, K., and Yashino, H. Extraction of RNA from yeast packed into column without isomerization. *Agric. Biol. Chem.*, *44*(8):1821-1827, 1980.
45. Tanekawa, T., Takashima, H., and Hachiya, T. Production of yeast extracts containing flavoring. U.S. Patent 4,303,680 (1981).
46. Cohn, W. E., and Volkin, E. On the structure of ribonucleic acids. *J. Biol. Chem.*, *203*:319-332, 1953.
47. Schuster, L. Rye grass nucleases. *J. Biol. Chem.*, *229*:289-303, 1957.
48. Miller, M. W., Yoneyama, M., and Soneda, M. *Phaffia*, a new yeast genus in the Deuteromycotina (Blastomycetes). *International J. of Syst. Bacteriol.*, *26*:286-291, 1976.
49. Johnson, E. A., Conklin, D. E., and Lewis, M. J. The yeast *Phaffia rhodozyma* as a dietary pigment source for salmonids and crustaceans. *J. Fisheries Res. Board of Can.*, *34*:2417-2421, 1977.
50. Goodwin, T. G. *The Biochemistry of the Carotenoids*, Vol. 1, Plants, Chapman and Hall, London-New York, 1982.
51. Johnson, E. A., and Lewis, M. J. Astaxanthin formation by the yeast *Phaffia rhodozyma*. *J. Gen. Microbiol.*, *115*:173-183, 1979.

52. Johnson, E. A., Villa, T. G., Lewis, M. J., and Phaff, H. J. Simple method for the isolation of astaxanthin from the basidiomycetous yeast *Phaffia rhodozyma*. *Appl. Environ. Microbiol.*, *35*:1155-1159, 1978.
53. Carlson, M., Osmond, B. C., and Botstein, D. SUC genes of yeast: A dispersed gene family. *Cold Spring Harbor Symp. Quart. Biol.*, *45*:799-812, 1981.
54. Grossman, M. K., and Zimmermann, F. K. The structural genes of internal invertase in *Saccharomyces cerevisiae*. *Mol. Gen. Genet.*, *175*:223-229, 1979.
55. Taussig, R., and Carlson, M. Nucleotide sequence of yeast SUC 2 gene for invertase. *Nucleic Acids Res.*, *11*:1943, 1983.
56. AOAC. *Activity of Invertase Solutions*, #31024, 11th ed., 1970, p. 529.
57. Prescott, S. C., and Dunn, C. G. *Industrial Microbiology*, McGraw-Hill, New York, 1959.
58. Peppler, H. J. Amino acid composition of yeast grown on different spent sulfite liquors. *J. Agr. Food Chem.*, *13*:24-36, 1965.
59. Amoco Food Co., Torutein Product Bulletin, Chicago, Illinois, p. 1974.
60. Bernstein, S., and Plantz, P. E. Ferments whey into yeast. *Food Eng.*, *49*(11):74-75, 1977.
61. Reed, G., and Nagodawithana, T. W. *Yeast Technology*, 2nd ed. Van Nostrand Reinhold, N.Y., 1990, p. 387.
62. Meitz, W., and Schwartz, K. Impaired glucose tolerance as an early sign of dietary necrotic overdegradation. *Arch. Biochim. Biophys.*, *58*:504-506, 1955.
63. Jeejeebhoy, K. N., Chu, R. C., Marliss, G., Greenberg, R., and Bruce-Robertson, A. Chromium deficiency, glucose intolerance and neuropathy reversed by chromium supplementation in a patient receiving long term total parental nutrition. *Am. J. Clin. Nutr.*, *30*:531-538, 1977.
64. Freund, H., Atamian, S., and Fischer, J. E. Chromium deficiency during total parental nutrition. *Am. J. Med. Assoc.*, *241*:496-498, 1979.
65. Doisy, R. J., Streeter, D. H. P., Freiber, J. M., and Schneider, A. J. *Chromium Metabolism in Man and Biochemical Effects in Trace Elements in Human Health and Disease*, Vol. 2 (A. S. Prasad, ed.), Academic Press, New York, 1976, pp. 79-104.
66. Mertz, W., Toepfer, E. W., Roginski, E. E., and Polansky, M. M. Present knowledge of the role of chromium. *Fed. Proc.*, *33*:2275-2280, 1974.
67. Nagodawithana, T. W., and Gitmanis, F. Method for the production of selenium yeast. U.S. Patent 4,530,846 (1985).
68. Dickson, R. C., and Tomlinson, R. H. Selenium in blood and human tissue. *Clin. Chim. Acta*, *16*:311-321, 1976.
69. Shamberger, R. J., and Fronst, D. V. Possible protective effect of selenium against human cancer. *Can. Med. Assoc. J.*, *104*:682, 1969.
70. Cantor, A. H., Langevin, M. L., Noguchi, T., and Scott, M. L. Efficacy of selenium in selenium compounds and feed shells for prevention of pancreatic fibrosis in chicks. *J. Nutr.*, *105*:106-111, 1975.
71. Combs, G. F., and Combs, S. B. The nutritional biochemistry of selenium. *Ann. Rev. Nutr.*, *4*:257-280, 1984.

72. Nagodawithana, T. W., and Gutmanis, F. Method for the production of selenium yeast. U.S. Patent 4,530,846 (1985).
73. Korhola, M., Vainio, A., and Edelmann, K. Selenium yeast. *Annals of Clinical Res.*, *18*:65-68, 1986.
74. Sperti, G. Toilet preparation. U.S. Patent 2,320,478 (1943).
75. Goodson, W., Hohn, D., Hunt, T. K., and Leung, Y. K. Augmentation of some aspects of wound healing by a "skin respiratory factor." *J. Surgical Res.*, *21*:125-129, 1976.
76. Trunkey, S. Anorectal drug products for over-the-counter human use. Establishment of a monograph. *U.S. Food and Drug Administration*, *45*:35652, 1980.
77. Kaplan, J. Z. Acceleration of wound healing by a live yeast cell derivative. *Archives of Surgery*, *119*:1005-1008, 1984.
78. Robbins, E. A., and Seeley, R. D. Process for the prevention and reduction of elevated blood cholesterol and triglyceride levels. U.S. Patent 4,251,519 (1981).
79. Sidoti, D. R., Landgraf, G. M., and Khalifa, R. A. Functional properties of baker's yeast glycan. Presented at the 33rd Annual Meeting of the Institute of Food Technologists, Miami, FL, June 6-14, 1973.
80. Phaff, H. J. Cell wall of yeasts. *Ann. Rev. Microbiol.*, *17*:15-30, 1963.
81. Deslandes, Y., Marchessault, R. H., and Sarko, A. Triple helical structure of (1→3)-β-D-glucan. *Macromolecules*, *13*:1466-1471, 1980.
82. Chuah, C. T., Darko, A., Deslandes, Y., and Marchessault, R. H. Triple-helical crystalline structure of curdlan and paramylon hydrates. *Macromolecules*, *16*:1375-1382, 1983.
83. Sarko, A., Wu, H. C., and Chuah, C. T. Multiple helical glucans. *Biochem. Soc. Trans.*, *11*:139-142, 1983.
84. DiLuzio, N. R. Soluble phosphorylated glucan. International Publication WO 87/01037, 1987.
85. Bucovaz, E. T., and Morrison, J. C. Application of protein-protein interaction as an assay for the detection of cancer. U.S. Patent 4,160,817 (1979).
86. Bucovaz, E. T., Morrison, J. C., Whybrew, W. D., and Tarnowski, S. J. Process for the preparation of COA-SPC from baker's yeast. U.S. Patent 4,284,552 (1981).
87. Okinshi, M., and Gregory, K. F. Isolation of mutants of *Candida tropicalis* with increased methionine content. *Can. J. Microbiol.*, *16*: 1139-1143, 1970.
88. Jenkins, P. G., and Raboin, D. Glycogen-deficient mutants of the yeast *Saccharomyces lipolytica*. *J. Appl. Bact.*, *44*:279-284, 1978.

INDEX